P9-DUH-606

Atlas of Infrared Spectroscopy of Clay Minerals and their Admixtures

H.W. van der Marel

Soil Mechanics Laboratory, Delft,
Soil Survey Institute, Wageningen,
The Netherlands

AND

H. Beutelspacher

Institute of Soil Biochemistry,
Agricultural Research Centre,
Braunschweig-Völkenrode, Germany

(in collaboration with E. Rietz, Braunschweig,
Germany, and P. Krohmer, Überlingen, Germany)

Atlas of Infrared Spectroscopy of Clay Minerals and their Admixtures

ELSEVIER SCIENTIFIC PUBLISHING COMPANY
Amsterdam – Oxford – New York 1976

ELSEVIER SCIENTIFIC PUBLISHING COMPANY
335 Jan van Galenstraat
P.O. Box 211, Amsterdam, The Netherlands

AMERICAN ELSEVIER PUBLISHING COMPANY, INC.
52 Vanderbilt Avenue
New York, New York 10017

Library of Congress Card Number: 73-85228

ISBN: 0-444-41187-9

Printed in The Netherlands

ACKNOWLEDGEMENTS AND NOTATIONS

The authors are greatly indebted to Professor Dr. W. Flaig, Director of the Institute for Biochemistry, Braunschweig, Germany, Ir. W.C. van Mierlo, Director of the Laboratory for Soil Mechanics, Delft, The Netherlands, the late Professor Dr. F.A. van Baren, Institute for Soil Science, Utrecht, The Netherlands, Dr. F.W.G. Pijls, Director of the Institute of Soil Survey, Wageningen, The Netherlands, and Professor Ir. E.C.W.A. Geuze, Rensselaar Polytechnical Institute, Troy, Mich., U.S.A., for their great encouragement in the enterprise of this second Atlas on clay minerals and their admixtures, and to Dipl. Phys. E. Rietz, Braunschweig, Germany, and Dr. P. Krohmer, Uberlingen, Germany, for critical reading the manuscript.

The authors are also indebted to Professor Dr. Ir. C. Boelhouwer and Mr. J.F. Goosen, Laboratory of Chemical Technology, University of Amsterdam, for the facilities they allowed.

Grants to sustain this research were kindly received from: the Deuthsche Forschungsgemeinschaft, Bad Godesberg, Germany, the Ministry of Agriculture, The Hague, The Netherlands, the Centrale Organisatie Toegepast Wetenschappelijk Onderzoek, The Hague, The Netherlands, and the Laboratorium voor Grondmechanica, Delft, The Netherlands.

Various colleagus have kindly provided precious samples, among whom:

A. Alietti, Italy
S. Aomine, Japan
J.M. Axelrod, U.S.A.
J.A. Bain, Great Britain
A. Behar, Bulgaria
K.S. Birell, New Zealand
W.F. Brindley, U.S.A.
J.M.M. van den Broek, Netherlands
Mrs. A.M. Byström-Asklund, Sweden
I. Cerný, Czechoslovakia
J.K. Dixon, New Zealand (†)
D. Eisma, Netherlands
J. Erdélyi, Hungaria
W. von Engelhardt, Germany
J.F. de Ferrière, France
U. Hofmann, Germany
P.S. Keeling, England
W.D. Keller, U.S.A.
H. Kodama, Japan
J. Konta, Czechoslovakia
R.A. Kühnel, Netherlands
A.A. Levison, U.S.A.
F. Lippman, Germany
H.G. Midgley, England
K. Oinuma, Japan
B.A.O. Randall, Great Britain
R.H.S. Robertson, Great Britain
S. Shimoda, Japan
T. Sudo, Japan
Smithsonian Institute, U.S. Natl. Museum
J.G. Ubaghs, Netherlands (†)
F. Veniale, Italy
K. Wada, Japan
T.L. Webb, S. Africa
A. Weiss, Germany

We mention gratefully the great care and accuracy of the technical assistants: J. Ariesse, J. Boers, K. Bubeck (W. Germany), B. Bunschoten, M. Claes, H. Engberts, S.J. van der Gaast, J. Jas, Mrs. V. Kwee, B.F. Laamers, H. Labrie, R.J. Nadorp, T.H. Nater and M. Willemsen, for making the infrared analyses and the designs, as well as those of Mr. A. Schoondorp for his assistance in gathering together the literature, and Mrs. W.N. Polling-Vest for correcting the translation of the Dutch manuscript.

Notations

The separates $< 2\ \mu m$ were obtained by first shaking the sample in a flask to which some NaOH was added up to a slight alkaline reaction, and then turning over in a shaking machine for 2 hours. The suspension stayed over in Atterberg cylinders for 24 hours at a level of 30 cm in the cylinder. It then was siphoned off, the exess of NaOH was removed by suction and washing out with water. Afterwards the separate was dried at ca 80°C.

The *amount of the sample* used for the manufacture of the pellets with a surface area of 1 cm^2 is given on 300 mg kaliumbromide.

Impurities of the mineral investigated are mentioned. They are based on infrared-, X-ray and electron microscope observations.

An *asterisk* in the spectrum means that the band only occurs incidentally.

Fig. E.M. This indicates the reference to the Figure number of the electron micrograph as published in: H. Beutelspacher and H.W. van der Marel, 1968. *Atlas of Electron Microscopy of Clay Minerals*, Elsevier, Amsterdam, 333 pp.

For reasons of brevity the titles of the references of the chapters 1 to 5 are omitted.

Formulae mentioned in this Atlas are mainly after H. Strunz, 1970. *Mineralogische Tabellen*, Geest and Portig, Leipzig, 621 pp.

CONTENTS

INTRODUCTION

In 1800, the existence of infrared rays (IR) was discovered by William Herschel. The apparatus used by him (Herschel, 1800a,b,c) consisted of some thermometers and a prism producing a spectrum by exposing it to sun rays. He found that red rays not only gave a heat effect, but that this phenomenon persisted also beyond that region and that its intensity there even increased. This invisible heat radiation of 1—100 μm wavelength, which followed the same laws of reflection and refraction as visible light, was called infrared radiation.

In 1881, Abney and Festing published absorption spectra of 52 compounds. Certain radicals were found to have distinctive absorption bands. Julius (1892) found that the kind of atoms and the manner of their grouping in the molecule determine its character of absorption.

The first collection of IR spectra of mainly organic molecules was made by Coblentz (1905—1910) and published by the Carnegie Institution of Washington. Nowadays about 800 publications appear yearly on infrared analysis.

Barnes' collection was published in 1944 (Barnes et al., 1944) and contains 363 NaCl spectra and 2701 references. But it was not until 1950 that the first large collections appeared — published by the American Petroleum Institute, Pittsburgh (Pa). Others, among which monographs, followed: see Bibliography. The large collections with thousands of spectra giving mainly data of organic molecules are regularly supplemented with new data.

There is a great need for infrared spectra in organic chemistry. Besides their use in the research on the structure of molecules, they are also needed in the analytical field: biochemistry, cosmetics, dairy products, foods, petroleum industry, pharmaceutics, rubber, textiles etc.

The need of infrared spectra of inorganic compounds is smaller and so is the demand in industry for them. Moreover, the bands of characteristic absorptions are rather limited and broad because the molecules cannot vibrate freely in a crystal.

The sample should be "mulled" with an inert liquid but then extra bands are added. Not until the use of the KBr disc technique of Stimson (1952) and Schiedt (1952) in which the inorganic sample is compressed to transparent pellets, the investigations have increased.

But because the other two restrictions remained, there has not been made so much progress as for organic compounds. As compared with organics the amount of investigations on minerals and especially on clay minerals, i.e. the youngest science of IR application, the infrared collections of inorganic substances and especially those of clay minerals are of restricted size. But in recent years there has been an increased use of clays and related minerals for various purposes: ceramics, civil engineering, catalysts, drilling muds, masonry, filler in paper and rubber industry, decolorizers, insulators, geology, agriculture, clarifications, foundries, refractories, etc; see Bateman (1951), Robertson (1960), and Grim (1962). Infrared analyses can distinguish phases which are quite similar, e.g., *polymorphs* of $CaCO_3$ (calcite, vaterite, aragonite), SiO_2 (quartz, tridymite, cristobalite, coesite, stishovite), TiO_2 (rutile, anatase, brookite), *isomorphs* of phosphates ($AlPO_4$, $GaPO_4$, $FePO_4$, $BaPO_4$). Mullite ($3Al_2O_3 \cdot 2SiO_2$) may easily be distinguished from sillimanite ($Al_2O_3 \cdot SiO_2$), which is difficult to determine by X-ray analysis. Very small amounts of kaolinite not detectable by X-ray analysis can be found by the infrared method.

The O-H adsorption band of H_2O is of high intensity and therefore it is used for the quantitative determination of small amounts of H_2O.

Absorption of IR rays depends on the vibration of atoms situated in a certain environment. Small differences in the individual atom ordering may be easily found e.g., varieties of kaolinite etc.

X-ray analysis is more sensitive to the periodic arrangement of the atoms in a crystal and thermal analysis to the strength of decomposition of the bands between the various atoms and atom groups in a certain crystal (heat of decomposition).

Only some mg of sample are required for an infrared analysis as against some hundreds for an X-ray or thermal analysis. Because of the many merits of the IR method, X-ray analyses of minerals are mostly supplemented with IR analysis.

All the above reasons have resulted in an increased use of the infrared method in recent years for the identification and quantitative estimation of clay minerals and their admixtures.

The examples given in this Atlas are of clay minerals and their admixtures. They were selected from about 4000 samples totally investigated. Most are from well-known localities as described in literature. It was almost impossible to give spectra of only monominerals, because many minerals are found in nature only with others.

References

Abney, R.E. and Festing, R.E., 1881. On the influence of the atomic grouping in the molecules of organic bodies on their absorption in the infrared region of the spectrum. Phil. Trans., 172:887—918.

Barnes, R.B., Gore, R.C., Liddel, U. and Williams, V.Z., 1944. Infrared Spectroscopy. Industrial Applications and Bibliography. Rheinhold, New York, N.Y., 236 pp.

Bateman, A.M., 1951. Economic Deposits. Wiley, New York, N.Y., 916 pp.

Coblentz, W.W., 1905. Investigation of infrared spectra. Publ. Carnegie Inst., 35.

Coblentz, W.W., 1906. Investigation of infrared spectra. Publ. Carnegie Inst., 65.

Coblentz, W.W., 1908. Investigation of infrared spectra. Publ. Carnegie Inst., 97.

Coblentz, W.W., 1910. Selective radiation from various solids. Natl. Bur. Stand., U.S.A., 6:301—319.

Grim, R.E., 1962. Applied Clay Mineralogy. McGraw-Hill, New York, N.Y., 422 pp.

Herschel, W., 1800a. Investigations on the powers of the prismatic colours to heat and illuminate objects. Phil. Trans., 90: 255—283.

Herschel, W., 1800b. Experiments on the refrangibility of the invisible rays of the sun. Phil. Trans., 90:284—292.

Herschel, W., 1800c. Experiments on the solar and the terrestrial rays that occasion heat. Phil. Trans., 90: 293—326; 439—538.

Julius, W.H., 1892. Bolometrisch onderzoek van absorptie-spectra. Versl. K. Akad. Wetensch., 1:1—49.

Robertson, R.H.S., 1960. Mineral Use Guide or Robertson's Spider's Webs. Cleaver-Hume Press, London, 44 pp.

Schiedt, U. and Reinwein, H., 1952. Zur Infrarot-Spektroskopie von Aminosauren. Z. Naturforsch., 7b: 270—277.

Stimson, M.M. and O'Donnell, M.J., 1952. The infrared and ultraviolet spectra of cystine and isocystine in the solid state. J. Am. Chem. Soc., 74:1805—1808.

1. BASIC CONSIDERATIONS ABOUT ELECTROMAGNETIC RADIATION AND MOLECULE VIBRATIONS

Electromagnetic radiation is a wave motion or it may be considered to be a stream of particles called quanta or photons. It extends from cosmic and γ rays with a wavelength below 10^{-2} Å to radio waves of 10^3 m (Fig. 1.1).
The visible region is very small: from 4000 to 8000 Å. The infrared region extends from ca. 1 to ca. 100 μm (10 000 to 100 cm^{-1}).
Because of this great difference various units are introduced: m, cm, mm, μm, Å and kilo (10^3), mega (10^6), giga (10^9) cycles, etc.
In infrared spectroscopy the number of waves per cm, called wavenumber (cm^{-1}) or Kayser unit (K) after H. Kayser, is commonly used: $\tilde{\nu}(cm^{-1}) = 10\,000/\lambda(\mu m)$. Thus, 1 μm = 10 000 cm^{-1}, 10 μm = 1000 cm^{-1}, 100 μm = 100 cm^{-1} (Append. I).
Several names are given for certain categories of light. The visible light comprises only a very small region — from 0.38 to 0.78 μm — but it is the most important because it can be observed by the human eye. Radar, television and radio waves are named after their practical use. Cosmic rays are named after their cosmic origin, etc.
The infrared region is divided into three parts e.g., near-infrared = 0.78—3 μm, middle-infrared = 2—15 μm and far-infrared = 15—300 μm.
When a molecule is irradiated with infrared light it can absorb radiation if the radiation has the same energy as the vibrational or rotational transitions of the molecule. In the far-infrared mainly rotational energy is absorbed. In the middle-infrared region rotational as well as vibrational energy. In the near-infrared region mainly multiples of the vibrational energy (overtones) are absorbed. Rotational transitions have only weak energies = ca. 10^{-2} X those of the vibrational. But their number is large. They can only be observed with highly sensitive grating instruments. Also they occur only at molecules moving freely (gases) because in liquids and especially in crystals the rotations are disturbed and even suppressed.
Most vibrations of minerals are found in the 250 to 4000 cm^{-1} region (40 to 2.5 μm). It requires three coordinates to describe the position of an atom in space. The atom has three degrees of freedom. A system of N free moving atoms has 3 N degrees of freedom. For its rotational and transitional movements, three degrees of freedom of each are needed. Thus 3 N—6 degrees of freedom are left to describe the vibrations of the molecule. These vibrations are called the normal or fundamental vibrations.
But only a non-linear molecule e.g., H_2O, has 3 N—6 vibrations. For a linear molecular, e.g., CO_2, only two rotational coordinates are needed. Thus 3 N—5 vibrations are possible.
Stretching vibrations are those where the changes of the atomic distances occur mainly along the chemical bond connecting the atoms. They give a periodic extension and contraction of the chemical bond.
Vibrations other than stretching are called deformation vibrations. They give a periodic bending of the molecule. Deformations are of lower energy because the stretching forces are larger than the bending forces. There are various kinds of deformation vibrations: bending or scissoring, rocking, wagging, twisting (Fig. 1.2).
According to the selection rules a vibration will be infrared-active if a change in dipole moment of the molecule occurs during the vibration, e.g., a change in the position of the centers of the positive and negative charge. The vibration of symmetric diatomic molecules like N_2, H_2 etc., is infrared-inactive because no change in the dipole moment occurs. The vibration of the NO molecules is infrared-active because of the dipole moment changes. Infrared-inactive vibrations may be Raman-active when the motion produces a change in the polarizability of the molecules. Raman-active vibrations give a shift in the frequency of the scattered rays when the sample is irradiated with monochromatic ultraviolet rays, e.g., the blue line of a Hg lamp (4358 Å).
The frequency of a vibration in a molecule depends on the mass of the atoms, the forces acting between them and the interactions with the surroundings. In the above it is assumed that the vibrational motions are strictly harmonic during the whole motion. But this is not always what happens; especially in the higher-energy levels there may be a non-linear change of dipole moment with atomic displacement. In that case weak additional bands may also appear in the infrared spectrum, e.g., combination bands: $\nu_1 + \nu_2$, $2\nu_1 + \nu_2$ etc., for instance CO_2: $\nu_1 + \nu_3$ = 3716 cm^{-1} and overtones: $2\nu_1$, $3\nu_1$, etc., for instance H_2O: $2\nu_1$ = 7251 cm^{-1} and $3\nu_1$ = 10 631 cm^{-1}.
The larger the anharmonicity term, the stronger are the intensities of the combination and overtone bands. Anharmonicity increases with temperature.
Degenerate vibrations are those where the energy level of two or more vibrations of the same polyatomic molecule are identical: doubly degenerate (CO_2 = 668 cm^{-1}), triply degenerate, etc. Fermi resonance (Fermi, 1931) of polyatomic molecules is due to an interaction between vibrational modes when two levels of vibration lie close together. The two bands are

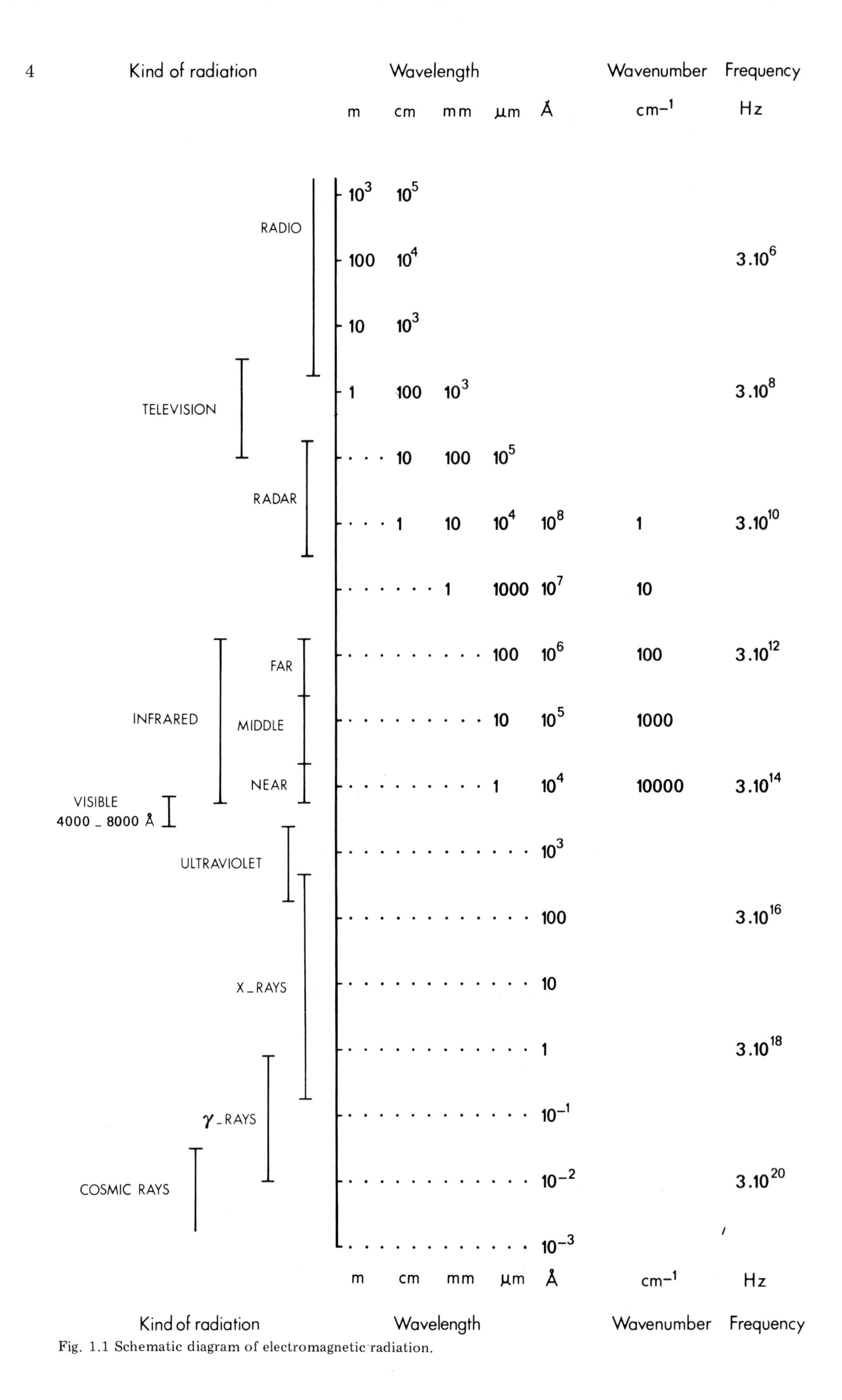

Fig. 1.1 Schematic diagram of electromagnetic radiation.

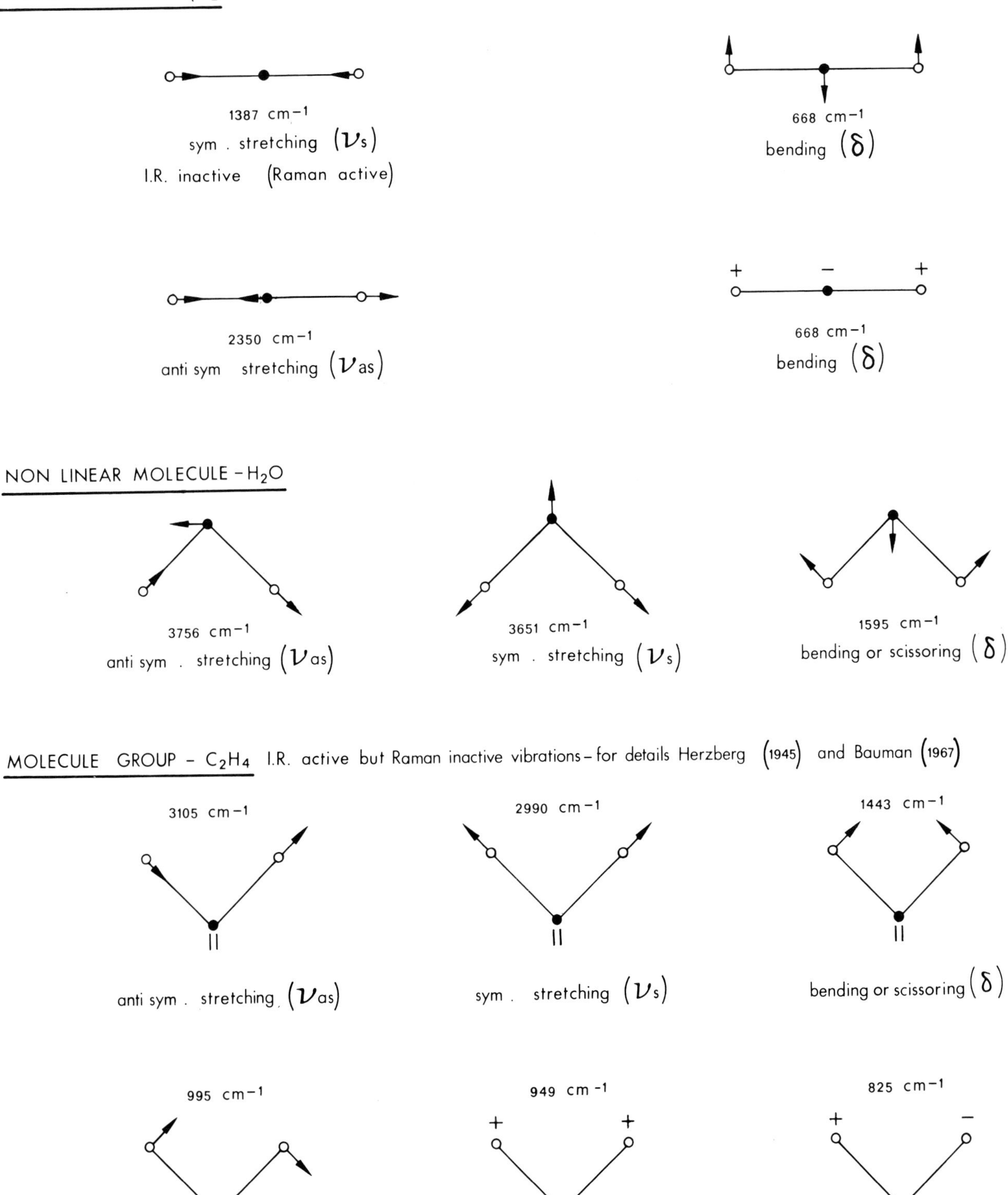

Note : + and − represent motions in a direction perpendicular above and below the plane of the paper respectively

* from interpretation of the Raman band at 1656 cm−1

Herzberg.G. 1945 Infrared and Raman spectra of polyatomic molecules pp 107 and 325−328 D.van Nostrand Comp. Inc New york 632 pp

Bauman R.P. 1967 Absorption spectroscopy p 299 John Wiley & Sons Inc New York 611 pp :

Fig. 1.2. Vibration modes and wavenumbers (cm^{-1}) of: a linear molecule ($3\,N-5$), a non-linear molecule ($3\,N-6$), and a molecule group. (N = number of atoms.)

perturbed in frequency as well as in intensity. As a result they appear separated. Only vibrational levels of the same symmetry type can perturb one another which considerably restricts the Fermi-resonance phenomenon.
According to Einstein the following relation exists:

$E = h\nu$, and because $c = \lambda\nu$ this gives $E = hc/\lambda = hc\tilde{\nu}$

where h = Planck constant = 6.626×10^{-34} joule sec; c = velocity of light = 2.9979×10^{10} cm sec^{-1}; ν = frequency of the radiation (Hz); λ = wavelength of the radiation (cm); $\tilde{\nu}$ = wave number (cm^{-1}).
Thus the energy of 1 $cm^{-1} \sim 1.986 \times 10^{-23}$ joule $\sim 1.24 \times 10^{-4}$ eV.
The energy levels for a molecule are not variable or continuous, but they are quantized. For a harmonic oscillator only transitions between two adjacent levels are allowed:

$E = (u + \frac{1}{2})h\nu$

where u = vibrational quantum number (0,1,2, 3...). When $u = 0$ the oscillator has a finite rest energy of $\frac{1}{2}h\nu$ = the zero point energy.
In the ideal case of a diatomic molecule with masses m_1 and m_2, the following relation exists for the energy of an absorbed photon when the molecule moves from one quantized state to another:

$$E = \frac{h}{2\pi}\sqrt{f_0/\mu_m}$$

where f_0 = bond force constant between two atoms and μ_m = the reduced mass = $(m_1 \times m_2)/(m_1 + m_2)$ where m_1 = mass atom m_1, and m_2 = mass atom m_2.
According to Badger (1934) the following empirical relation exists:

$$f_0 = \frac{1.86 \cdot 10^5}{(r - d_{ij})^3} \text{ dyne/cm}$$

where r = interatomic distance in Å; d_{ij} = a characteristic constant depending on the position of the atoms in the periodic system.
Another relation given for f_0 by Gordy (1946) is:

$$f_0 = 1.67\, N(x_A x_B/r^2)^{3/4} + 0.30 \text{ dyne/cm} \cdot 10^{-5}$$

where x_A and x_B = electronegativities of the atoms A and B; N = the order of the bond between the two atoms, and r their internuclear distance (Å).
Bending vibrations will have a smaller force constant than stretching vibrations and thus occur at lower wave numbers in the spectrum.

For double bonds and increasingly for triple bonds the reverse happens: C—C ca. 1000 cm^{-1}; C=C ca. 1650 cm^{-1}; C≡C ca. 2200 cm^{-1}.

From the E equation it can easily be recognized that the energy of the vibration increases with the force constant and decreases with the masses, for instance C—Cl = ca. 730 cm^{-1}, C—C = ca. 1000 cm^{-1}, C—H = ca. 3000 cm^{-1}.

For a HCl molecule with f_0 = 506 Nm^{-1} and a reduced mass of 0.972 mass units ($1.66 \cdot 10^{-27}$ kg):

$$E = \frac{6.626 \cdot 10^{-34}}{2\pi \times 1.60 \cdot 10^{-19}} \sqrt{\frac{506}{0.972 \times 1.66 \cdot 10^{-27}}}$$

$$= 0.37 \text{ eV molecule}^{-1} = 8.5 \text{ kcal mole}^{-1}$$

A free moving HCl molecule (gas) absorbs radiation at 2886 $cm^{-1} \sim 0.36$ eV/molecule which is nearly the same number as has been calculated.
The calculation for multi-atomic molecules is far more complicated, e.g., for benzene a 4th-degree equation is needed and for o-chlorophenol a 33rd.
Moreover, the frequency of a vibration is not only determined by the mass of the atoms and the forces acting between them but also by its geometric arrangement and the forces acting with the surroundings.
For bending, wagging, rocking, twisting movements, the calculations of the vibration energies are in most cases nearly impossible.
Therefore, the above equations and considerations about the possible vibrations of a molecule give only a simple approximation of what actually happens for liquids or crystals where the vibrations of a certain molecule or molecule group are influenced by that of others.
For instance, by association with other atoms or atom groups, the bands are shifted to lower energy levels e.g., the 3300 cm^{-1} O—H stretching vibration of ethylalcohol moves to 3630 cm^{-1} of its monomer in the vapour state and to 3500 cm^{-1} in dioxane solution.
The O—H stretching band at 3756 cm^{-1} of monomeric H_2O molecules in vapour or in non-polar solvents, is replaced by a very intensive doublet at 3515 and 3575 cm^{-1} of its polymer, when water is dissolved in dioxane (Errera and Sack, 1938). The O—H stretching vibration may even be absent in organics like o-nitro-phenol, salicylaldehyde and 2.6-dinitrophenol as the protons are in this case in a "chelated" position between two oxygen ions (Hilbert et al., 1935).
According to Kienitz (1955) the γCH vibration of benzene (671 cm^{-1}) is shifted by several sol-

vents to higher frequencies in relation to the factor $(\epsilon - 1)/(2\epsilon + 1)$ in which ϵ = dielectric constant of the solvent.

Vibrations of somewhat higher or somewhat lower energy levels may also result from perturbations by neighbouring components, anharmonic coupling interactions, crystal dislocations, strain, stress, vacant holes, Frenkel and Schottky defects, isomorphous substitution of incidental character (impurities), disorder in the arrangement of the atoms, etc. which all modify the electronic structure in their neighbourhood.

A shifting or broadening of the corresponding infrared bands is then possible. The intensity of a band is furthermore higher and sharper when temperature is lower, which is mainly a thermal motion (collision) effect. Moreover, the band can shift to higher wave numbers when temperature is decreased. Thus the maximum intensity of the NH_4^+ band of NH_4Cl is moved from 1762 cm^{-1} at 20° C to 1818 cm^{-1} at −150° C thereby increasing largely in intensity and decreasing in band width (Bovey and Sutherland, 1949).

Crystals have a large number of low-energy (below 200 cm^{-1}) lattice vibrations. They broaden the bands of sharp harmonic vibrations by anharmonic coupling with lattice vibrations or even cause forbidden vibrations to appear in the spectrum by splitting or by combination, e.g., the strong NH_4^+ band at 1762 cm^{-1} of NH_4Cl crystals (Hornig, 1948).

As a result infrared bands of crystals and especially those of clay minerals which have mostly disordered structures are, when compared with those of vapours or liquids, broad, and their number small because of overlapping.

References

Badger, R.M., 1934. A relation between internuclear distances and bond force constants. J. Chem. Phys., 2: 128—131.

Bovey, L.H.F. and Sutherland, G.B.B.M., 1949. Infrared evidence for free rotation in the solid state. J. Chem. Phys., 17: 843.

Errera, J. and Sack, H., 1938. Molecular association studied in the infrared. Trans. Faraday Soc., 34: 728—742.

Fermi, E., 1931. Z. Phys., 71: 25—259.

Gordy, W., 1946. A relation between bond force constants, bond orders, bond lengths and the electronegativities of the bonded atoms. J. Chem. Phys., 14: 305—320.

Hilbert, G.E., Wulf, O.R., Hendricks, S.B. and Liddel, U., 1935. Spectroscopic methods for detecting some forms of chelation. Nature, 135: 147—148.

Hornig, D.F., 1948. The vibrational spectra of molecules and complex ions in crystals. J. Chem. Phys., 16: 1063—1076.

Kienitz, H., 1955. Infrared and Raman spectroscopy as an analytical aid. Microchim. Acta, 1955: 728—733.

2. INSTRUMENTAL

Monochromators

To observe absorption phenomena in gases, liquids and crystals which are caused by resonance motions between the atoms or atom groups and the electromagnetic waves, it is necessary to have monochromatic radiation. Moreover, this radiation should extend over a large region. Resolving power of the radiation produced in the instrument should also be large, constant and of good reproducibility.

Monochromatic light may be produced by prisms, gratings and filters or by a combination of these elements.

Prisms

Fig. 2.1 shows a schematic representation of the path of light in a prism. Because of the different index of refraction for light of different wave-

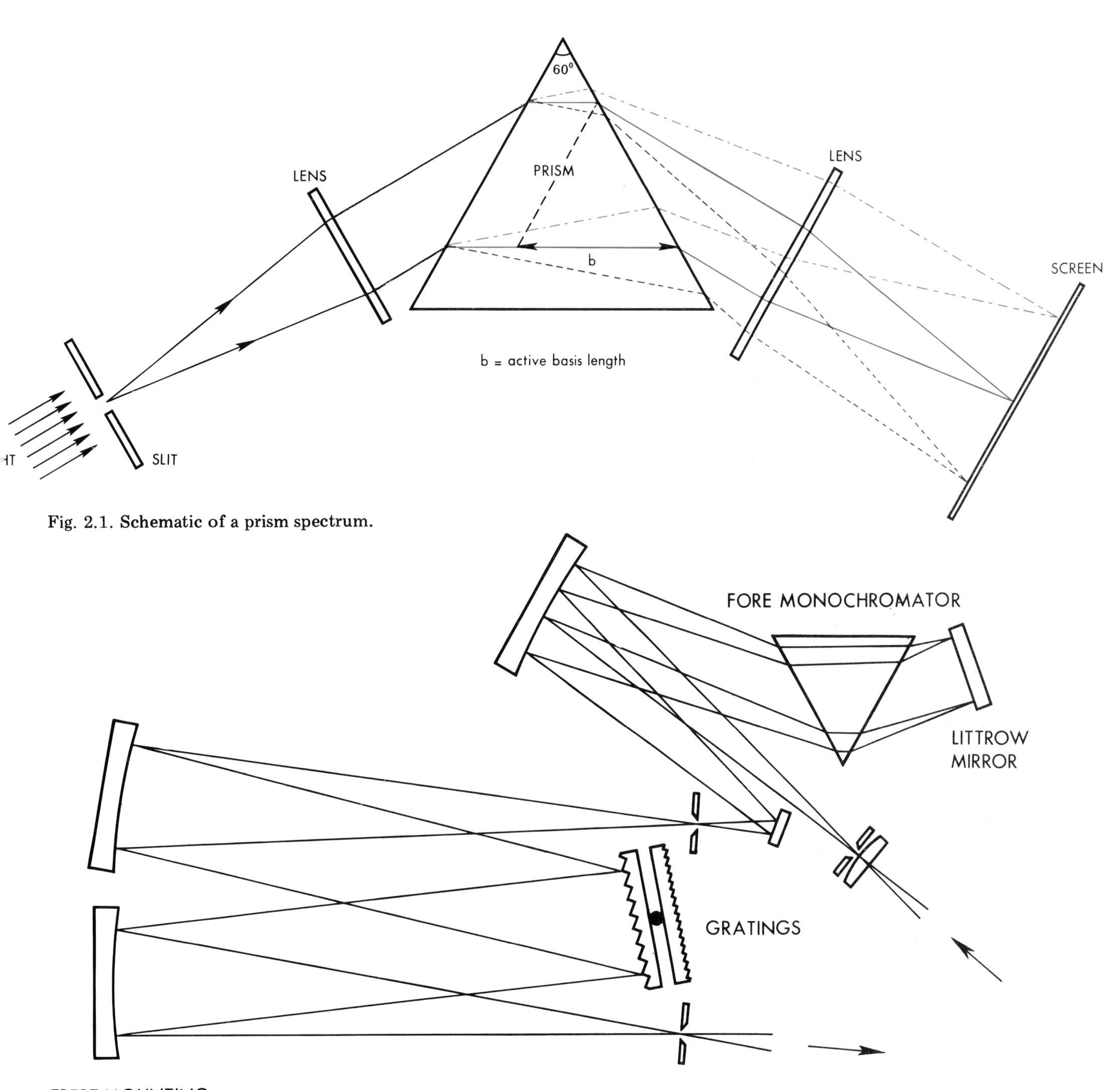

Fig. 2.1. Schematic of a prism spectrum.

Fig. 2.2. Schematic of a grating spectrometer with concave mirror in Ebert mounting and KBr fore monochromator.

lengths the original white light is split into its components, thus forming a spectrum. Red radiation is the least aberrated and violet the most. Lenses are not used in IR spectroscopy in the 3—25 μm range. Mirrors are used instead because lenses are not sufficiently achromatic over the whole wave length region and glass or quartz lenses have no transmission below 4000 cm^{-1}. Besides, mirrors are less sensitive to moisture than lenses. To decrease the still remaining aberration of the mirror, the beam is run twice over it, thus correcting at the second reflection for the curvature in the wave front caused by the first reflection (Ebert, 1889, mounting; Fig. 2.2). In general prisms no wider than 60° are used, which represents a compromise between increased loss by reflection and increased gain by dispersion.

Base lengths of prisms are not more than 10 or 15 cm, which also forms a compromise between gain by increased active base length and cost of preparing the large crystals. NaCl and especial KBr prisms are very sensitive to water and so should be stored in a heated compartment or above dried silica-gel. LiF and CaF_2 prisms are less sensitive to water.

Before the IR instrument is used, the temperature of the newly installed prism and that of the thermostated prism room should be equal, which needs ca. 4—6 hours depending on the thickness of the prism.

The angular dispersion of a prism $d\theta/d\lambda$ is the rate of change of the angle of emergence θ of the beam from the prism with its wavelength λ. The linear dispersion $dl/d\lambda$ = the distance dl of separation in the spectrum of two close lines of similar intensity that can just be resolved and differing in wavelength $d\lambda$.

$dn/d\lambda$, the change of the index of refraction (n) of the incident light beam with wavelength (λ) is commonly called the material dispersion.

The resolving power of a prism $\bar{\lambda}/d\lambda = b(dn/d\lambda) \times 1000$ where $\bar{\lambda}$ = the mean wavelength of two vibrations of similar intensity just distinguishable from each other, $d\lambda$ = the difference of the two vibrations, and b = the effective base length of a prism in mm.

Example: a NaCl prism of 10 cm effective base length and governed by a Littrow mirror (basis-length = 2 × 10 cm), has a theoretical resolving power at 10 μm $(dn/d\lambda) = 0.0066\ \mu^{-1}$ (Kohlrausch, 1955) of 200 × 0.0066 × 1000 = 1320. Or: $\Delta\tilde{\nu}(cm^{-1}) = 1000/1320 = 0.76\ cm^{-1}$.

However, the ultimate resolution which is obtained with an instrument also depends on its slit width, signal to noise ratio, quality of the mirrors, etc.

Each prism has its own maximum working region.

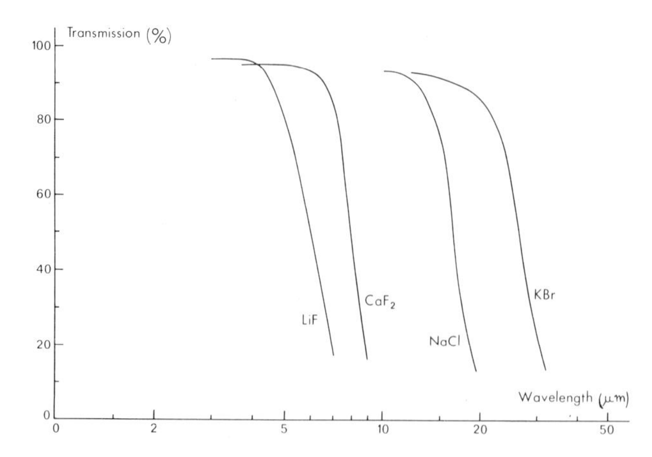

Fig. 2.3. Transmission (%) of various prisms at different wavelengths. (Data of Harshaw Chemical Co., Cleveland — Synthetic Optical Crystals.)

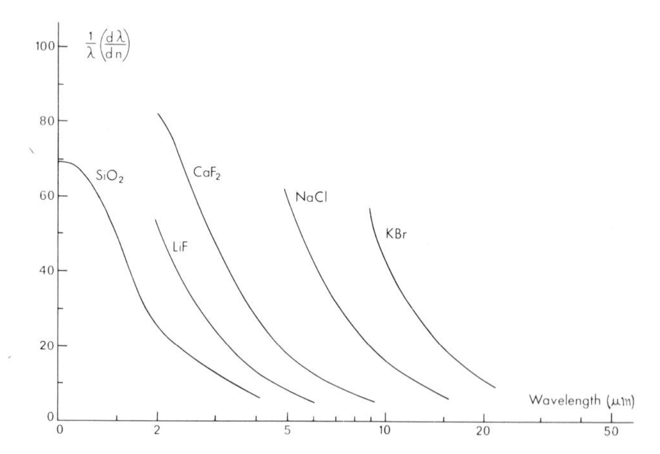

Fig. 2.4. Schematic representation of the dispersive number ($\lambda\ dn/d\lambda$) for various wavelengths. (Data of Harshaw Chemical Co., Cleveland — Synthetic Optical Crystals.)

It is determined by its transmission (Fig 2.3) and relative optical performance (dispersive number = $\lambda(dn/d\lambda)$, Fig. 2.4, or its reciprocal). Thus, for optimum resolution the CaF_2 (and LiF) prism is used in the 2.5—6 μm (4000—1600 cm^{-1}), the NaCl prism in the 6—15 μm (1600—660 cm^{-1}), the KBr prism in the 15—28 μm (660—360 cm^{-1}), region.

Gratings

As early as 1802 Thomas Young (Young, 1802) already experimented with a glass plate in which 20 parallel running grooves per mm (Fig. 2.5) were scratched, and thus obtained a spectrum.

Founder of grating spectroscopy is Fraunhofer (1823), who constructed very precise gratings, made many experiments and gave theoretical interpretations of the results.

Rowland (1882) perfected the grating technique and came even to gratings of 10 cm length with 1720 lines/mm. The deviation of the distances between the grooves was better than 1‰.

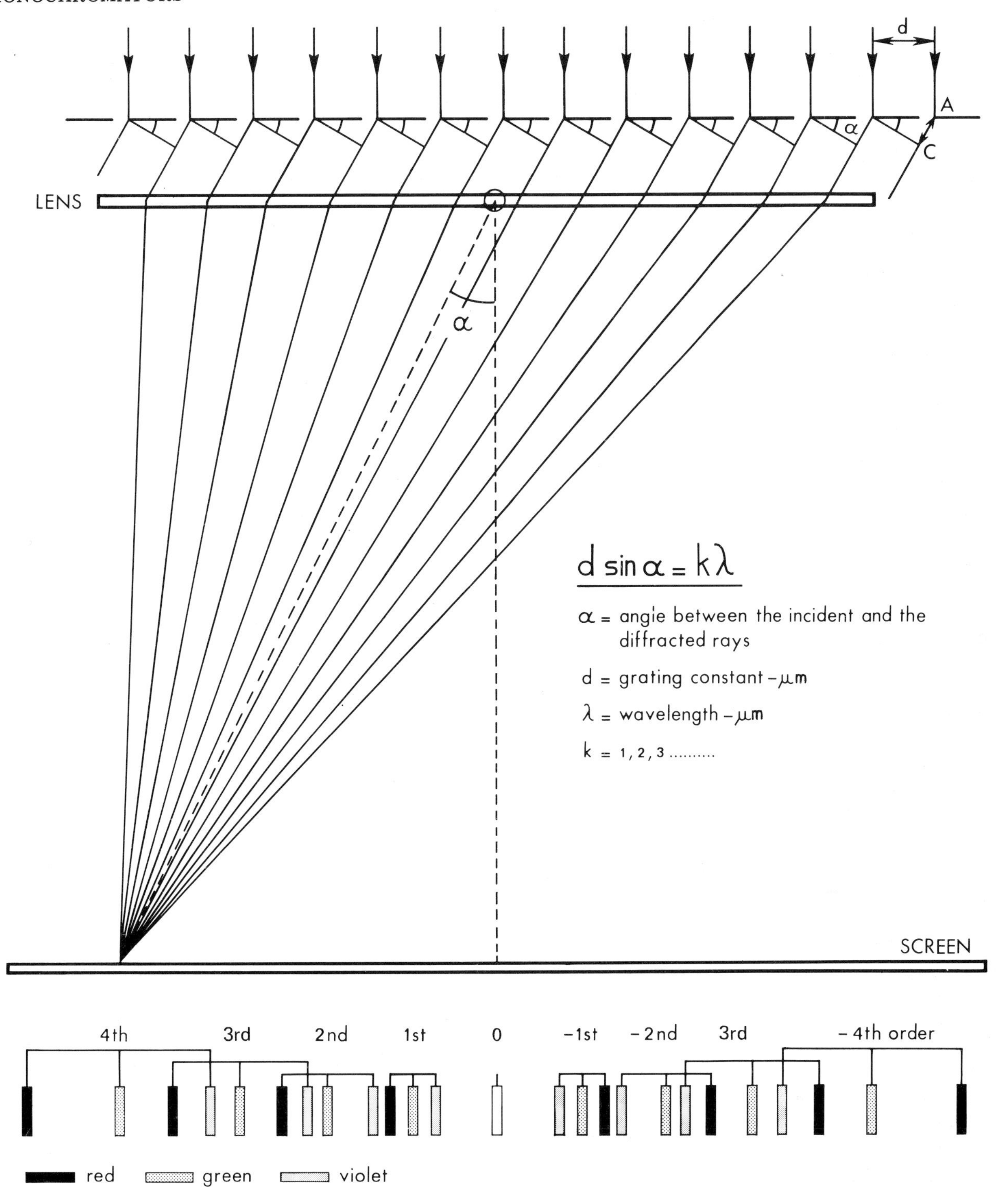

Fig. 2.5. Schematic of a transmission grating spectrum.

In contrast to prisms, gratings are far less influenced by temperature changes. However, they are very sensitive to damage of the grooves when they are cleaned.

When a beam of light falls on a grating with distances between the grooves d (grating constant) about that of some wave lengths, each opening will act as a new source of electromagnetic radiation which is emitted in all directions. If the wavelength is greater than the groove width, the grating acts as a mirror and if the wavelength is much smaller than the groove, the white spot opposite the slit becomes larger and the spectra of the higher orders becomes weaker. The following relation exists:

$AC = \mathrm{d} \sin \alpha = \mathrm{k}\lambda$

where AC = difference in path length between two successive rays; α = the angle between the incident and the diffracted ray; λ = wavelength

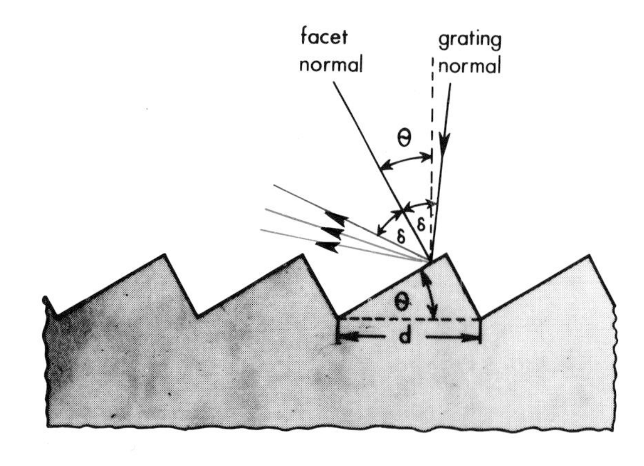

$$k\lambda = 2d \sin\theta \cos\delta$$

λ = wave length – μm
k = order number
θ = blaze angle between facet and grating normal
δ = angle between the incident rays and the facet normal
d = grating constant – μm

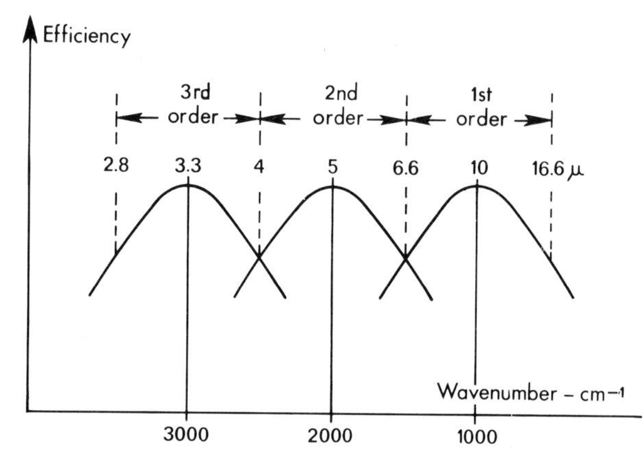

Fig. 2.6. Schematic of the light path for a reflecting echelette grating monochromator with blaze angle θ and maximum intensities at 10 μ, 5 μ, 3.3 μ, etc.

(μm); k = 1,2,3,.,.,.; d = grating constant, that is, the spacing between the grooves (μm).

In contrast to prisms, where the spectrum is caused by dispersion of the prismatic material, a grating spectrum is caused by diffraction and interference resulting from a certain constant arrangement of light permeable and non-permeable grooves.

A further contrast is that a prism only produces one spectrum, but a grating several which overlap each other. Thus the energy of the incident light is spread over several spectra.

Violet light (lower λ) will be less diffracted than red light. The reverse happens for a prism where violet light is the most deviated.

The resolution of a grating is at a slit width approaching zero:

$$\lambda/d\lambda = N.k.b.$$

where N = lines per mm; k = order of the spectrum; b = length of grating in mm.

The effective resolution is 75 to 90% of the theoretical resolution. Thus, e.g., for a 8-cm grating with 100 lines/mm (ca. 10-μm region) the resolution will be for its second order = ca. 13 000, which is ca. 10 × as large as that of a NaCl prism of basis length 10 cm at the same wavelength (10 μm).

Echelette gratings were first constructed by Wood (1910) and Trowbridge (1910) at Johns Hopkins University, U.S.A. They have shallow-shaped grooves; one side wide and the other narrow and are used as reflection gratings (Fig. 2.6). As a result of the above construction instead of ca. 30%, ca. 70—80% of the energy of the incident rays is concentrated into a restricted spectral region.

The blaze wavelength of a grating is the wavelength at which the energy of the diffracted beam is at its maximum in the first order. This will happen at a certain blaze angle = the angle between the facet normal and the grating normal. The relations are (see also Krüger et al., 1960):

$$k\lambda = 2\ d \sin\theta \cos\delta$$

where λ = wave length (μm); k = order number of the spectrum (1,2,3,.,.,); d = grating constant (μm); θ = blaze angle = angle between the facet normal and the grating normal; δ = angle between the incident or the exciting rays and the

facet normal. When supposing $\delta = 0°$, the blaze angle can be calculated ($\sin \theta = \lambda/2d$).
An Echelette grating which has its first-order blaze length at 20 μm has its second order at 10 μm, its third at 6.66, etc.
Grooves are cut by a diamond in a thermostated environment (ca. 0.01°C) on a perfectly flat glass surface coated before with Al. By adjusting the diamond, the blaze angle may be chosen at any angle desired, thus producing gratings with different blaze wavelength maxima.
As a result of modern techniques at the construction of plastic replicas, it is possible to-day to make many reproductions from one original at low cost. Optical properties of the blazed grating depend on the blaze angle and the number of grooves per unit length. Echelette gratings are in general used from 2/3—2 times their blaze wavelength. To cover the whole infrared region, two or more gratings are used which can be changed automatically, e.g., for the 500—4000 cm^{-1} range a grating with 300 lines/mm (4000—2000 cm^{-1}) and a grating with 100 lines/mm (2000—500 cm^{-1}).
A KBr prism fore monochromator or filters are used to prevent overlapping of the orders and to remove short waves of higher orders of the gratings (or short-wave stray light from the KBr fore monochromator). The filters in use are long-wave pan filters, i.e., filters that transmit at wavelengths longer than a sharp cutoff and reject radiation of shorter wavelengths or higher orders.

Sources of light

In absorption spectrometers it is necessary to use continuous sources of radiation. The thermic sources follow the Planck distribution law.
The general distribution of energy emitted by a black body is determined by its temperature.
The following sources of light are the most commonly used.

Nernst filament

A Nernst filament consists of a rod about 2 × 25 mm long of mixed rare-earths oxides: zirconium (mainly), cerium, thorium, erbium and yttrium. The oxides are weakly conducting at room temperature and so should be heated at the start. When heated to just red heat the electrical resistance is decreased and the filament can be maintained glowing at its own electrical current monitored by a phototube. Its working temperature is 1700—1800°K.

Globar

A globar consists of a silicon-carbide rod (5 × 50 mm) which conducts the electrical current when cold. However, it must be cooled because of the large amount of heat released at its working temperature, at ca. 1500°K.

Nickel-chrome wire

Ni-chrome coils are very often used in the smaller instruments. Their temperature is about 1000—1100°K.

Detectors

Non-selective "black" detectors have their receiving elements blackened to ensure that radiation is uniformly absorbed at all relevant wavelengths.
The most commonly used are: thermocouples, bolometers and Golay cells.

Thermocouples

The thermocouples used are tellur-platinum, tellur-constantan or tellur-gold. They are enclosed in a vacuumized tube thus minimizing heat loss caused by convection and conduction. A thermopile consists of a series of thermocouple junctions.

Bolometers

A bolometer consists of a platinated metal strip or a metal strip coated with antimon, tantalum or wismuth in vacuo. The resistance of the electrical circuit in which the strip is enclosed, changes with temperature when IR radiation falls on the voluminous platinum or metal coatings.

Golay cell

This pneumatic detector consists of a chamber filled with a gas of low thermal conductivity f.i. Xenon. In the center of the chamber is a thin plastic film coated with a film of Al to absorb the incident radiation. When the gas expands by the incoming radiation it distorts a mirror membrane on which a beam of light is directed. The change in intensity of the reflected light of this mirror is registered by a photo cell.
Selective detectors are highly sensitive but they can only be used in a small region, for instance: Ge (0.8—1.6 μm), PbS (1—3 μm), PbSe (1—4.5 μm), PbTe (1—5.2 μm).

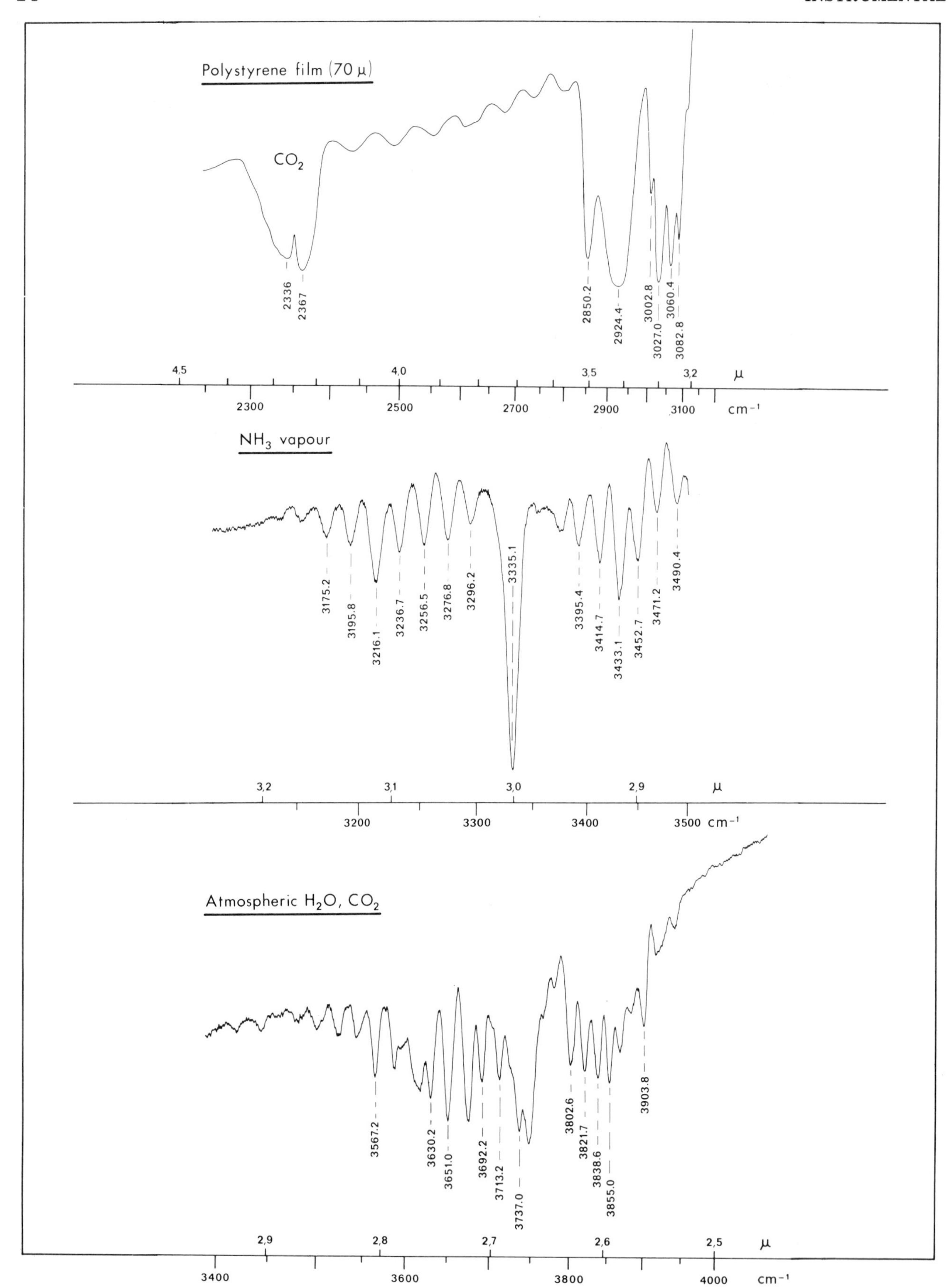

Fig. 2.7. Wavenumbers (air) of: polystyrene, NH_3 vapour, and atmospheric H_2O and CO_2 for calibration of the IR spectrophotometer.

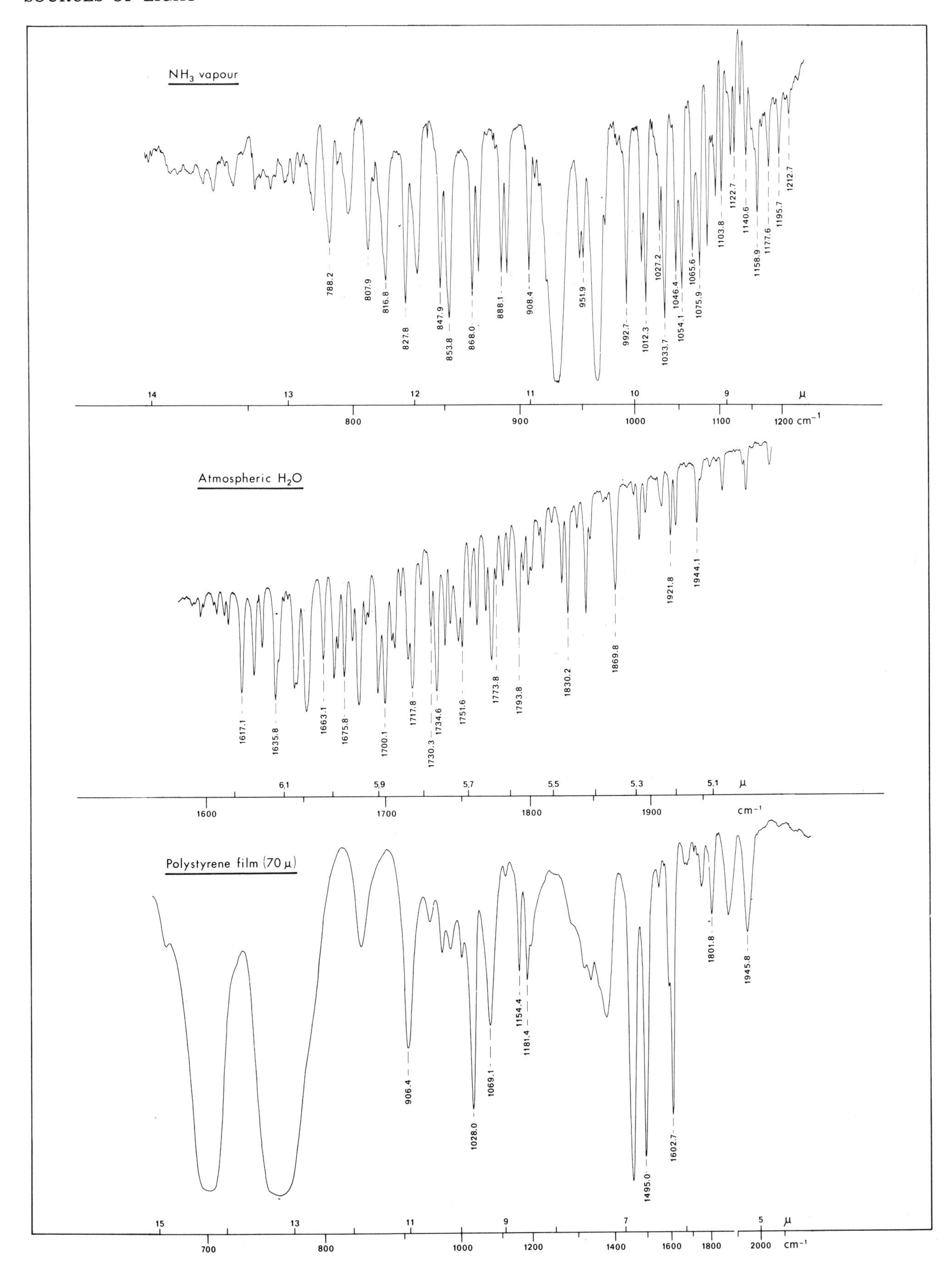

Fig. 2.8. Wavenumbers (air) of: NH_3 vapour, atmospheric H_2O and polystyrene for calibration of the IR spectrophotometer.

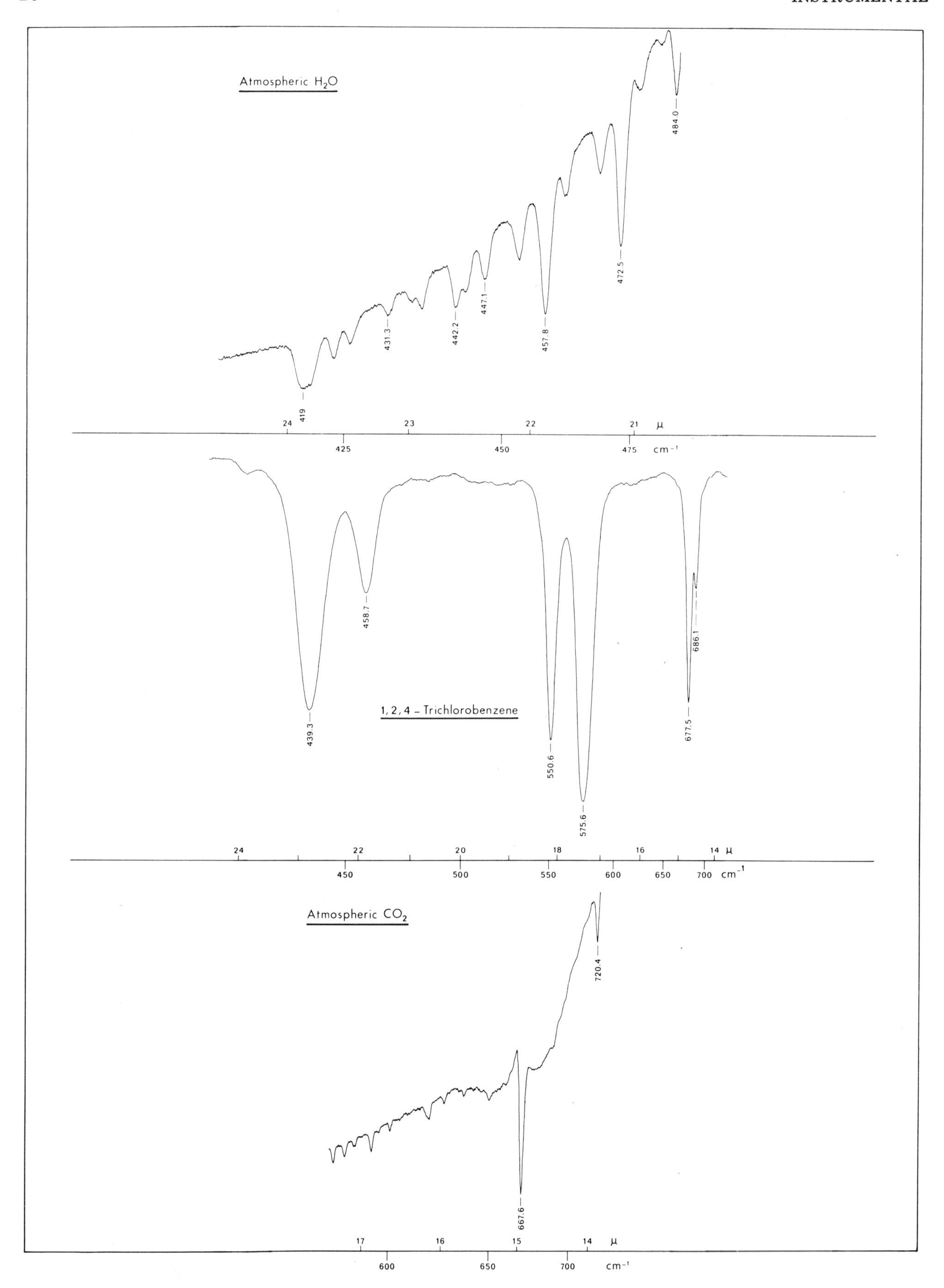

Fig. 2.9. Wavenumbers (air) of atmospheric H_2O, trichlorobenzene, and atmospheric CO_2 for calibration of the IR spectrophotometer.

Calibration

In 1907, the red cadmium line of 6438.4696 Å for standard air (dry air at a normal pressure of 760 mg Hg and at a temperature of 15°C and with 0.03 per cent CO_2 by volume) was adopted by the International Union for Co-operation in Solar Research as the primary standard of wavelength. In 1960, the orange krypton-86 line of 6056.1252 Å in standard air (6057.8021 in vacuum) was chosen by the Paris Conference on Weights and Measures. A 198-Hg lamp constructed by Meggers and Westfall (1950) is easy to operate; it has a long life, ca. 1000 hours, and the green line = 5460.7532 Å (standard air) is intense and isolated. The better IR instruments have at present a resolution of about 0.2 cm^{-1} at ca. 1000 cm^{-1}. But as the spectra of minerals are not very sharp, but broad and banded, the accuracy needed in this case is of the order of 1 cm^{-1}.

Most standards are published with their wave number data in vacuum and their wavelengths in air or standard air. Because the refractive index in air is larger than that in vacuum, the wave numbers in air are larger than in vacuum.

Conversion tables for wavelengths standards in air to wave numbers in vacuum have been calculated by Coleman et al. (1960) from Edlèns (1953) formula for the indices of refraction of standard air at several wave numbers.

The corrections are small, e.g., for 500 cm^{-1} = 0.1 cm^{-1}; for 1200 cm^{-1} = 0.3 cm^{-1}; for 1800 cm^{-1} = 0.5 cm^{-1}; for 2600 cm^{-1} = 0.7 cm^{-1}; for 3200 cm^{-1} = 0.9 cm^{-1}; for 3800 cm^{-1} = 1.1 cm^{-1}.

To convert wavelengths in air into those in standard air a further correction should be applied according to the equation of Penndorf (1957) because conditions of standard air are not fulfilled in the laboratory and the air in the monochromator-room has a temperature of ca. 35°C. But this correction is so small = < 0.1 cm^{-1} for the 400—5000 cm^{-1} range, that it may be neglected.

At the calibration of the IR spectrophotometer first the wavelength drum is fixed on the green Hg line at 5460.75 Å. After that the drum readings are calibrated against bands of spectra from several standards as published in literature. They are also presented in Fig. 2.7, 2.8, 2.9). Also minerals (and liquids) with sharp bands can be used, e.g., talc (3680), pyrophyllite (3678), kaolinite (3694, 3621) benzene (671), cristobalite (1098, 792, 622), quartz (1082), magnesite (886), gypsum (668, 606) cm^{-1}. Through the several points a smoothed curve is drawn representing the relation between drum scale readings and wave numbers (cm^{-1}) in air. Calibrations should be repeated from time to time to ensure the accuracy of the registered band positions. If there are small deviations they are corrected by the Littrow mirror; of course the spectrometer should not be calibrated on one band, but on several bands.

Spectrophotometers

Description of the Leitz prism infrared-spectrophotometer Model III (Fig. 2.10)

The Leitz IR-spectrophotometer Model III is a double-beam instrument with prism monochromator in Littrow arrangement. It works on the optical zero principle.

A glow-cylindre of rare-earth oxides (Nernst glower) provides the light source (*1*, Fig. 2.10C). The radiation emitted by the glower is split by the mirrors *2* and *3* into two beams. The sample to be investigated (*4*) is placed into the "measuring" beam reflected by mirrors *2*, *5* and *6*. The reference beam is guided by mirrors *3*, *7* and *8* and passes through the measuring diaphragm *9*, which behaves as the optical zero. A reference cell (liquid) or any standard (*4a*) may be inserted in the reference beam for the compensation of not-sample-specific effects.

The spherical mirrors *5* and *7* project the Nernst glower, at a 1/1 ratio, into the entrance-slit (*10*) of the monochromator. The section mirror *6* rotates with 12.5 sec^{-1} and reflects during half a revolution the measuring beam and allows passage of the reference beam during the other half revolution. Now the tracks of the two beams are identical. They pass through the monochromator with a modulation of 12.5 sec^{-1} applied by the section mirror and finally arrive at the detector.

A field lens (*11*) collects divergent beams in front of the monochromator entrance-slit by projecting the mirror surfaces *5* and *7*, respectively, into the monochromator mirror surface. The field lens furthermore closes the monochromator chamber against the photometer area.

The beam coming from the entrance-slit *10* (secondary light source) is reflected by mirror *12* onto the collimator (monochromator) mirror *13*. Mirror *13* images the "secondary light source" located inside its focal plane into infinity, i.e., it creates parallel beams. By tilting the monochromator mirror slightly (against the vertical plane) a small angle is created between the main beams of entrance and reflected light in vertical direction. The parallel beam now on "a second platform" can pass without loss of vignetting through the tilted prism (*14*). It falls on the Lit-

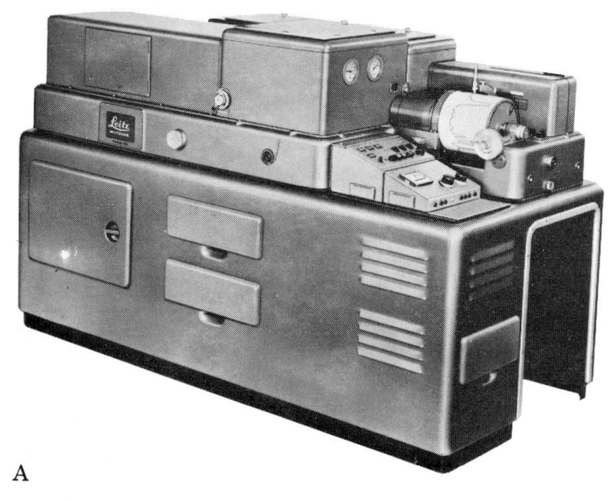

A

Fig. 2.10. The Leitz prism infrared spectrophotometer Model III. A. Photograph of the instrument. B. The internal mechanism. C. Schematic of B.

trow mirror (*15*) and then passes through the prism a second time and is concentrated by collimator mirror (*13*) onto the exit-slit (*16*). The white light (frequency mix) is split into its spectral colors by the prism so that the rays of different wavelengths leave the prism under different angles. The position of the Littrow mirror defines the radiation wavelength guided to the detector system.

A cam adjusted to the dispersion characteristics of the 60° NaCl prism and attached to the axis of the registration drum (*17*) pilots the Littrow mirror in a manner that continuously different wavelengths pass through the exit slit. In most cases this cam is curved to give a linear wavelength, less often to give linear wave number presentation of the spectra.

Light coming out of the exit-slit is reflected by mirror *18* and concentrated by collecting mirror *19* into the thermocouple (*20*). One half of reflector *18* consists of a normal aluminum surface mirror; the other half of a LiF-"Reststrahlen"-plate. At 13.5 μm both halves are automatically exchanged. The LiF plate serves as a stray light filler in the longer wavelength sections.

The thermocouple transforms radiation energies into heat and creates a thermovoltage proportional to the arriving radiation intensity. The resulting electrical signal is processed by a three-step resonance amplifier (*21,22,23*) to allow a coupled motor driving an attenuation diaphragm adjusting the reference beam intensity to the measuring beam intensity. The attenuator is furthermore coupled by a cam with the write-out of the registration device.

By the continuous wavelength advance and the corresponding position of the writer according to sample transmittance, the IR spectrum of the sample is produced.

Description of the Perkin Elmer prism infrared spectro-photometer Model 13 (Fig. 2.11)

Fig. 2.11 shows the layout of the optical system and a block diagram of the complete Perkin Elmer instrument. Energy from a phototube-monitored Nernst glower (*1*) is focused on an axis at 4 × magnification by source mirror *2*. For double-beam operation, a beam-splitting mirror (*3*) in the return beam directs half of the available energy into the sampling area (*4*) where an image of the source is conveniently formed. The remaining energy is passed over (*3*) to (*5*) which directs the beam into the reference area (*6*). Mirror *7* directs the sample beam over mirror *8* to combining mirror *9*. Energy from the reference beam is reflected by *8* to mirror *9*. Mirror *9* re-images the combined beams, and diagonal *10* places the image at the entrance slit. Energy passing through the variable width (automatically controlled) entrance slit *11*, is collimated by the off-axis parabola (*12*) and refracted by the prism (*13*). In normal operation the dispersed energy is scanned automatically by the Littrow mirror (*14*), which is controlled by a wavelength drive mechanism. The Littrow mirror returns the selected energy through the prism to the off-axis parabola. The parabola focusses energy on the variable slit-width exit slit (*15*) by means of a diagonal mirror (*16*). A second diagonal mirror (*17*) directs radiation to a thermocouple ellipsoid mirror (*18*) which focusses the energy on the detector target (*19*).

The detector converts the radiant energy into a composite electrical signal. After amplification, the signal components are separated by a pair of breaker-type synchronous rectifiers, which are 90° in quadrature phase relationship. This mechanical rectifier assembly acts as a phase discriminator of phase sensitive demodulator. The mechanical rectifiers are actuated by rotation of a chopper shaft.

In order to eliminate the effects of ambient and sample temperature, and provide an a.c. signal at the detector, a radiation chopper (13 rps) is mounted in the paths before the sampling area. The chopper acts as a switch, alternately interrupting the radiation to the reference path, and then the radiation to the sample path in the conventional shutter-in shutter-out beam switching principle. Because the beams are chopped 90° out of phase with each other, the a.c. signal can be readily separated into I and I_0 components in the electronic system. The outputs of the rectifiers are filtered and fed to the recorder. In normal ratio-recording operation, the I signal is applied to the recorder amplifier and the I_0 signal to the recorder slide-wire. The pen automatically

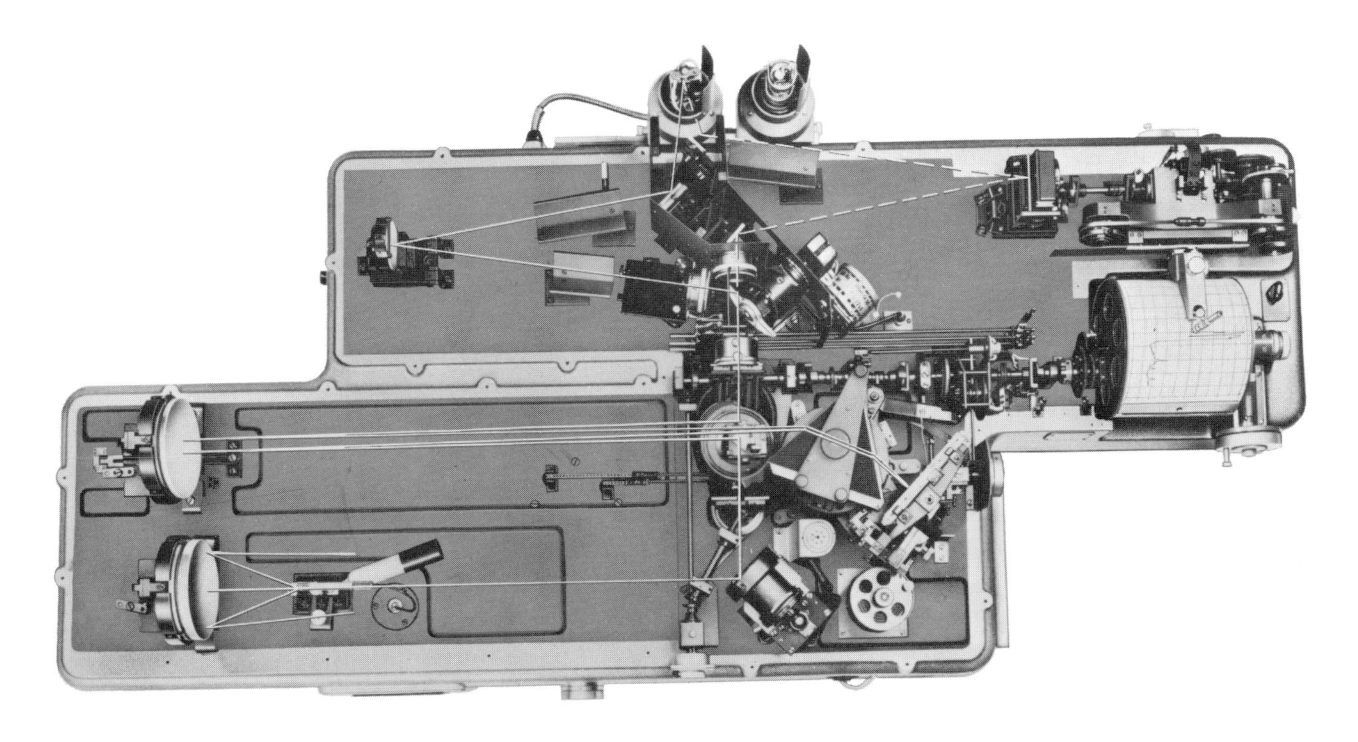

B

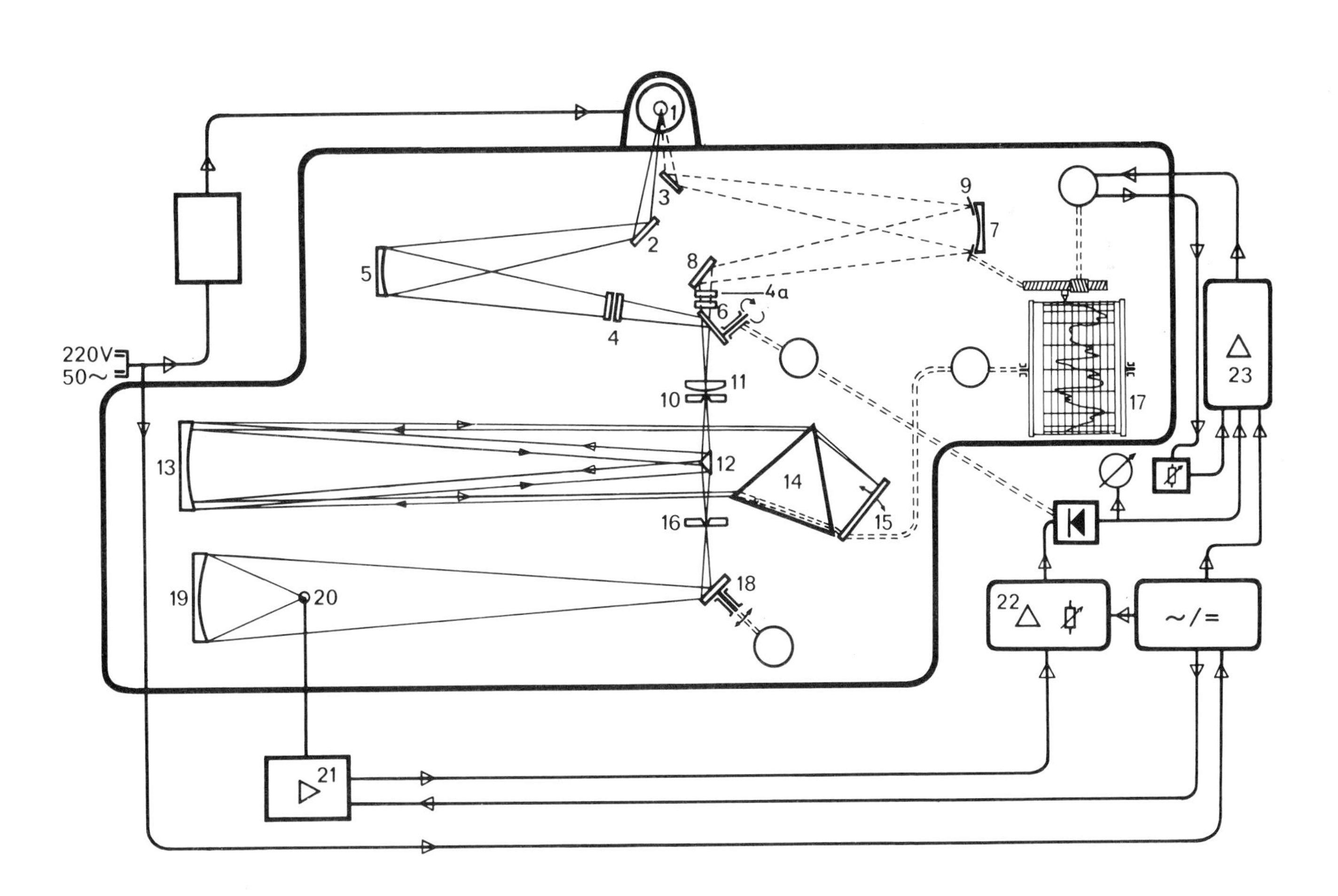

C

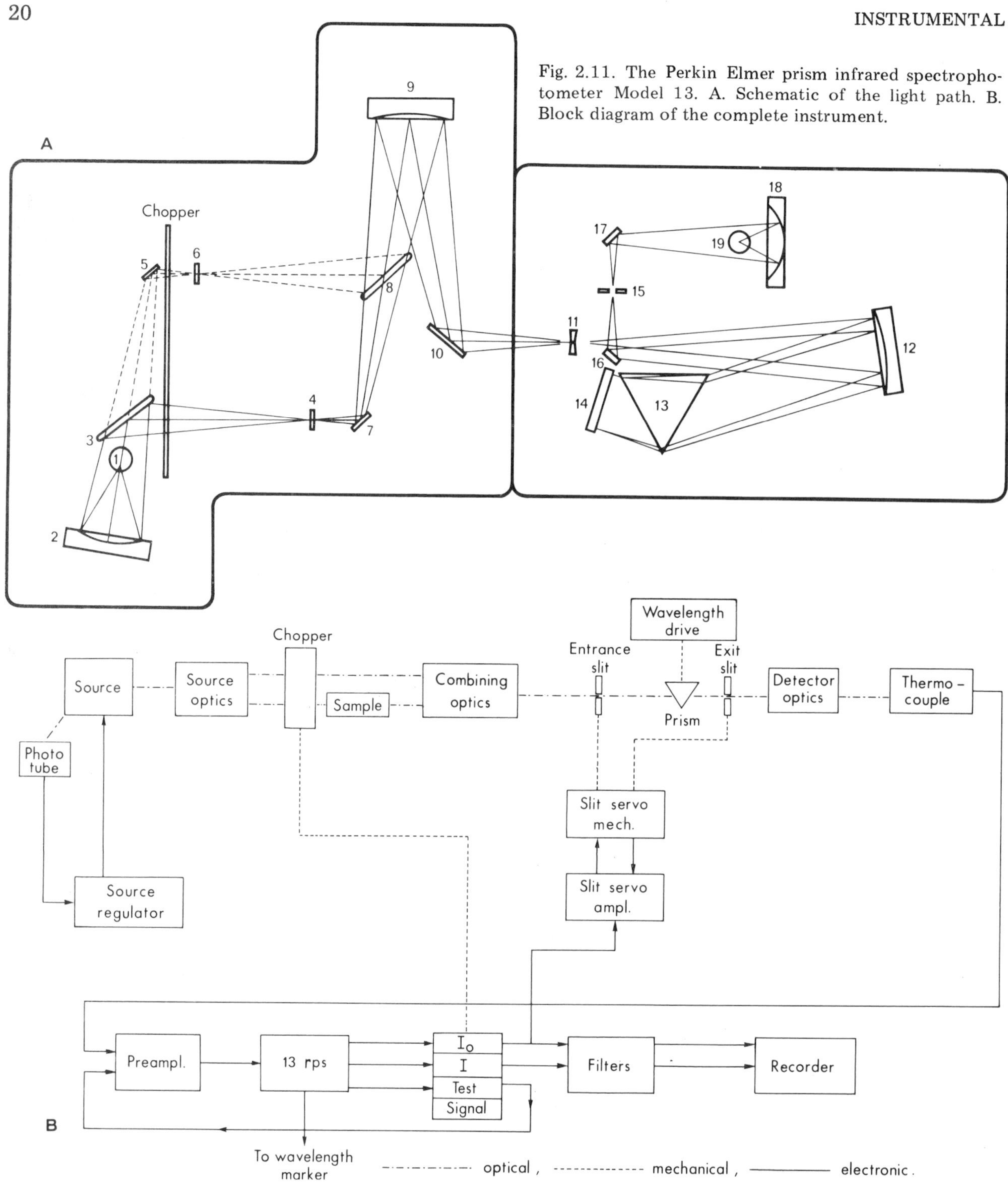

Fig. 2.11. The Perkin Elmer prism infrared spectrophotometer Model 13. A. Schematic of the light path. B. Block diagram of the complete instrument.

plots the ratio of the I signal to the I_o signal to give a continuous record in transmittance. It is possible to obtain the difference between the two signals. Here the I and I_o signals are fed in opposition to the recorder, the slide-wire voltage (10 mV) being supplied by a battery. A Leeds and Northrup or Brown recorder is employed, modified as necessary for the various applications.

When the instrument is employed as a single-beam, its basic operation does not change. However, a special full-aperture mirror may be used in place of (*3*) and (*8*) removed. The path of radiation would then be as shown by the solid lines in the figure.

In single-beam operation the pulsating signal at the detector is amplified, rectified, and filtered as before. The I signal is applied to the recorder amplifier and a battery voltage applied to the slide-wire. The pen plots the ratio of the I signal to a fixed battery reference voltage.

Description of the Perkin Elmer grating spectrophotometer Model 225 (Fig. 2.12)

This instrument is based on the double-beam principle with optical null. To the right there is the source part containing the radiation source (*1*), an air-cooled silicon-carbide rod (trade name: Globar). The radiation is split into sample and reference beam. To the left follows the sample compartment (*2*) and the photometer part, where the two beams are recombined by means of the rotating chopper (*3*). The sample is located near the first source image which has a size of approx. 9 × 12 mm^2.
The optical attenuator (*4*) is located in a pupil, in the image of the first-source mirror.
The fore monochromator in Littrow arrangement (*5*) is equipped with a thermostated 60° KBr prism (*6*). It eliminates undesired grating orders and suppresses stray light to an extremely low level. In the 4th recording range of the instrument (450—200 cm^{-1}), a scatter plate (*7*), monitored by a motor (*8*), is used instead of the prism. In this range, the instrument functions as a filter-grating monochromator. For the grating monochromator (at the left in the illustration), the Ebert-Fastie arrangement was chosen, which guarantees good imaging properties and thus a high resolution. The radiation detector (*9*) is either a large-area thermocouple fitted with a CsI lens or, by request, a Golay detector.
The scan motor (*10*), drives the monochromator and the recorder drum via gears (*11,12,13*) and magnetic clutches. Simultaneously with the monochromator, the wave-number indicator (*14*) is driven, the slit width programmed, and grating change and insertion of the scatter plate (*7*) in the 4th range controlled automatically.
The entrance slit of the fore monochromator (*15*) is coupled mechanically with the two slits *16* and *17* of the grating monochromator. They are driven by motor *18*.
The mechanical slit width can be observed on the slit-width indicator *19*. Besides, the wave number indicator *14* also permits the reading of the prevailing linear dispersion or, more precisely, of the reciprocal value of the linear dispersion. By multiplying both values, the spectral slit width at any given wave number is obtained.
The grating monochromator contains two echelette gratings (*20*) with an area of 84 × 84 mm^2 each. The two gratings with 30 lines/mm and 150 lines/mm, respectively, are mounted back to back. Each of the gratings which can be changed by a motor (*21*) is used in its 1st and 2nd order. This provides four wavelength ranges. For linear wavelength presentation the gratings are controlled by a cam which effects a complete revolution for each range.

The four ranges that can be recorded are: 1st range: 5000 cm^{-1}—2000 cm^{-1}; 2nd range: 2500 cm^{-1}—1000 cm^{-1}; 3rd range: 1000 cm^{-1}—400 cm^{-1}; 4th range: 450 cm^{-1}—200 cm^{-1}.

The instrument permits the recording of a continuous spectrum without any visible interruption over four ranges, i.e., from 4000 to 200 cm^{-1}.
The block diagram shows the servo circuit that is typical for instruments with optical null: the detector (*9*), on which the two beams, due to the rotating chopper (*3*), are projected alternately with a frequency of 13 rps, transforms the optical signal into an electrical one. This signal shows different intensities modulated at 13 rps, whenever the sample and the optical attenuator have different transmittance. In the following amplifier (*22*), the resulting alternating current is further processed. The amplified 13 rps voltage is rectified by a phase-sensitive demodulator. The amount of the resulting voltage is direct proportional to the difference in intensities between sample and reference beam. After further transformation, it drives the pen servo-motor (*23*) which, by means of gear (*24*), acts on the optical attenuator (*4*) and the pen (*25*). Optical attenuator and pen are coupled mechanically. If the transmittance in the reference beam is higher than in the sample beam, the attenuator is automatically closed until the alternating signal disappears, i.e., until the transmittance of the optical attenuator corresponds exactly to that of the sample. In the inverse case, the attenuator is opened accordingly. Since this attenuation process takes place rapidly, the position of the attenuator and of the pen at every moment of the scan corresponds to the sample transmittance.

History

The first commercial infrared spectrophotometer was constructed by Twyman of the Hilger Company (England) in 1913. The apparatus was very simple. More advanced instruments were produced by Kipp and Zn (The Netherlands) and General Electric (U.S.A.) in 1933. In 1944, Grubb and Parsons (England) and after the war Beckman (U.S.A.) and Perkin Elmer (U.S.A.) introduced spectrometers on the market.
Instead of the glass prism of Herschel (ca. 1800) a NaCl prism was made (ca. 1850). Afterwards a KBr, a CaF_2 (or LiF) and, in more recent years, a KRS-5 (mixed crystal of thallium bromide and thallium iodide) followed, which were better suited each for its own wavelength region. Fraunhofer (1823) was the founder of the grating spectroscopy. Rowland fabricated

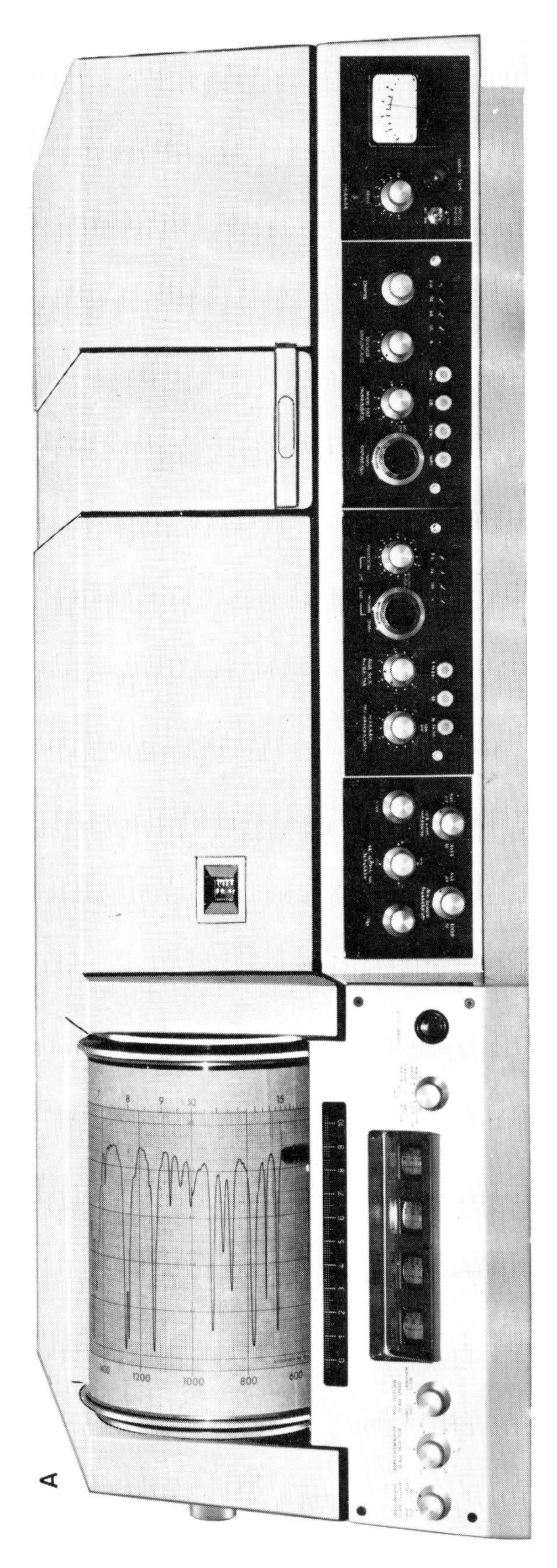

Fig. 2.12. The Perkin Elmer grating spectrophotometer Model 225. A. Photograph of the instrument. B (p. 23). Schematic diagram of the light path.

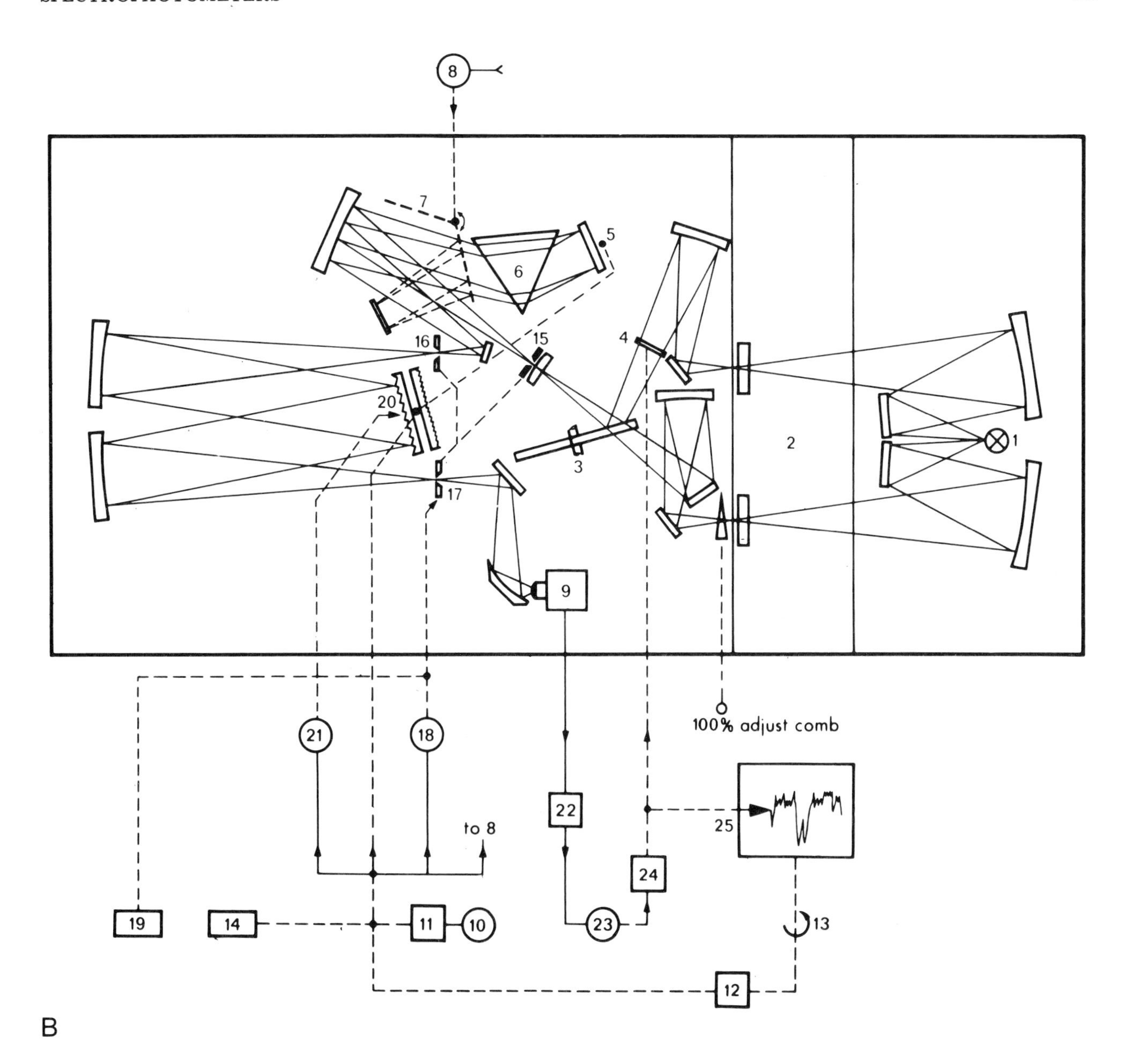

B

gratings with 1720 lines/mm (1882). In 1910 R.W. Wood introduced echelette gratings. But it was not until ca. 1950 that the grating spectrophotometer became a common commercial product. Instead of a thermometer as used by Herschel (1800), a thermocouple was introduced by Seebeck (1823, '26) and a thermopile by Nobili (1830). A bolometer by Svanberg (1849) was afterwards rediscovered and improved by Langley (1881). For details, see Barnes and Bonner (1936).

Another improvement is the construction of a double-beam instead of a single-beam photospectrometer. In this case the spectrum is compensated. It is not disturbed by atmospheric CO_2 and H_2O, the kind of solution in which the sample is dissolved or the thickness of the window cells. The first commercial instrument based on the double-beam principle was constructed by Baird Atomic Corporation (U.S.A.) in 1946.

The recent developments in the perfection of the instruments are largely stimulated by industrial and scientific demands. From a simple laboratory experimental instrument of former years, nowadays an IR photospectrometer is a complicated instrument which can be obtained in various models.

Manufacturers are: Perkin Elmer (U.S.A.), Beckman (U.S.A.), Hilger and Watts (England), Unicam (England), Grubb and Parsons (England), Spectronom (Hungaria), Carl Zeiss, Jena (East Germany), E. Leitz, Wetzlar (West Germany), Hitachi (Japan), Shimadzu (Japan), Japan Spectroscopic Co. (Japan), Models UKC and IKS-22 (Russia).

The most sensitive instruments are with a double beam, Echelette gratings in Ebert mounting, prism fore monochrometer, filters, automatic recording from 200 to 5000 cm^{-1}, and have a resolution of 0.25—0.40 cm^{-1}. The reproducibility is 0.4—0.7 cm^{-1}.

Besides common IR spectra the modern IR spectrograph may also give spectra with polarized or with reflected light. For quantitative analyses they are equipped with a logarithmic integrator. The number of IR spectrometers has increased from some hundreds in 1950 to several ten thousands nowadays.

References

Monochromators

Ebert, H., 1889. Zwei Formen von Spektrographen. Wiedemann's Ann. Phys. Chem., 38: 489—493.

Fraunhofer, J., 1823. Kurzer Bericht von den Resultaten neuerer Versuche über die Gesetze des Lichtes und die Theorie derselben. Gilbert's Ann., 74: 337—378.

Kohlrausch, F., 1955. Praktische Physik, 2: 528—529.

Krüger, H.G., Maennchen, K., Reichel, A. and Volkman, H., 1960. Ein Gitter Zusatz zum Leitz-Infrarot Spektrographen, Leitz Mitt. Wiss. Tech., 1: 100—105.

Rowland, H.A., 1882. Preliminary notice of the results accomplished in the manufacture and theory of gratings for optical purposes. Johns Hopkins Univ., Circ. 16, Phil. Mag., 13: 469—474.

Trowbridge, A. and Wood, R.W., 1910a. Groove-form and energy distribution of diffraction gratings. Phil. Mag., 20: 886—898.

Trowbridge, A. and Wood, R.W., 1910b. Infra-red investigations with the echelette grating. Phil. Mag., 20: 898—901.

Wood, R.W., 1910. The echelette grating for the infrared. Phil. Mag., 20: 770—778.

Young, T., 1802. On the theory of light in colours. Phil. Trans., 2: 12—48.

Calibration

Coleman, B. and Meggers, W.F., 1960. Natl. Bur. Stand. Monogr. 3, 1: 2000—7000 Å; 2: 7000—10000 Å.

Commission of Molecular Structure and Spectroscopy, 1961. Tables of wave numbers for the calibration of infrared spectrometers. (Reprint from Pure Appl. Chem., 1: 537—699).

Edlén, B., 1953. The dispersion of standard air. J. Opt. Soc. Am., 43: 339—344.

Meggers, W.F. and Westfall, F.O., 1950. Lamps and wavelengths of mercuri 198. J. Res. Natl. Bur. Stand., 44: 447—455.

Penndorf, R., 1957. Tables of the refractive index for standard air and the Rayleigh scattering coefficient for the spectral region between 0,2 and 20,0 μm and their application to atmospheric optics. J. Opt. Soc. Am., 47: 176—182.

History

Bowling Barnes, R. and Bonner, L.G., 1936. The early history and the methods of infrared spectroscopy. Am. Phys. Teacher, 181—189.

Fraunhofer, J., 1823. Kurzer Bericht von den Resultaten neuerer Versuche über die Gesetze des Lichtes und die Theorie derselben. Gilberts Ann., 74: 337—378.

Herschel, W., 1800a. Investigations of the powers of the prismatic colours to heat and illuminate objects. Phil. Trans., 2: 255—283.

Herschel, W., 1800b. Experiments on the refrangibility of the invisible rays of the sun. Phil. Trans., 2: 284—292.

Herschel, W., 1800c. Experiments on the solar and terrestrial rays that occasion heat. Phil. Trans., 2: 293—326; 439—538.

Langley, S.P., 1881. The bolometer. Proc. Ann. Acad. (Boston), 16: 342—359.

Rowland, H.A., 1882. Preliminary notice of the results accomplished in the manufacture and theory of gratings for optical purposes. Johns Hopkins Univ., Circ. 16, Phil. Mag., 13: 469—474.

Seebeck, T.J., 1823. Gilbert's Ann. 73: 115, 430.

Seebeck, T.J., 1826. Poggend. Ann., 6: 133, 153.

Wood, R.W., 1910. The echelette grating for the infrared. Phil. Mag., 20: 770—778.

Trowbridge, A. and Wood, R.W., 1911a. Groove form and energy distribution of diffraction gratings. Phil. Mag., 20: 886—898.

Trowbridge, A. and Wood, R.W., 1911b. Notes on infrared investigations with the echelette grating. Phil. Mag., 20: 898—901.

3. PREPARATION OF THE SAMPLES

Scattering

When the particle size is larger than the wavelength of the incident rays in this case 2.5 — 25 μm, they will act as mirrors and thus influence the intensity of the IR bands by reflection. A decrease of the intensity of the transmitted rays may also be caused by Tyndall scattering (1869) of the rays by particles 10—15 × smaller than the wavelength.
Hollow microcavities which have lost their water by heating, as in heated quartz (Bambauer et al., 1969), micro-discontinuities (crystal defects) and small local inhomogeneities of the refraction index in fluids due to fluctuations of density or concentration of the molecules, caused by random thermal motions, are other sources of scattering.
Scattering is a function of particle size, refractive index and wavelength.
According to Pfund (1934) ZnO particles of 0.088 μm diameter in suspension, which was obtained by first mulling with crepe rubber and afterwards dissolving in xylene and diluting with Nujol, scatter light of wavelength from 0.4 to 2 μm according to Rayleigh's (1871, 1899, 1910) equation (see Valasék, 1960):

$$I_s = \frac{I_0(n_1 - n_2)^2}{n_2{}^2} (1 + \cos^2\theta) \frac{\pi^2 NV^2}{r^2 \lambda^4}$$

where: I_s = intensity scattered rays; I_0 = intensity incident rays; n_1 = refractive index of the scattering particles; n_2 = ditto of the embedding medium; V = volume of each scattering particle; N = number of scattering particles; λ = wavelength of the scattered radiation; θ = scattering angle; r = distance from the sample to the observer.

Rayleigh's law of scattering is only valid for particles which are small in comparison to the wavelength.
For a given granular powder of uniform size embedded in a medium with a given refractive index, there will be for a certain wavelength a certain maximum of scattering (opacity) at a certain particle size. At high wavelengths as used here in infrared spectroscopy (3—25 μm), scattering will be negligeable for small clay particles (< 2 μm) because of the inverse 4th-power law for wavelength and the direct 6th-power law for particle size.
The reflection losses in a powder sample may be much higher than the absorption losses. Moreover, the peak can be displaced. Consequently, the interpretation of a reflection spectrum may be very complicated (for details, see Phillippi, 1968, 1969).

Christiansen effect

When an absorption band is approached from the long-wavelength side, the refraction index of the sample increases until a maximum is reached. Thereafter it decreases very sharply.
As the amount of scattered radiation is proportional to the second power of the difference in the refractive index of the particles and the embedding medium, the band will be asymmetric. The transmission is higher at the high-frequency side of the band and the maximum is shifted to longer wavelengths.
The above effect called after Christiansen (1884), its discoverer, decreases when particle size decreases to a diameter very small compared to the wavelength of the light.

Grinding

To decrease loss of radiation by reflection of the infrared rays and to avoid the Christiansen effect, the particles of the samples should be smaller than the wavelength of the rays.
To reduce particle size mechanical vibrators are applied, e.g.: Von Ardenne vibrator, Wig-L-Bug dentist amalgamator (Crescent Dental Manufacturing Co., Chicago), micronizer (Sturtevant Mill Co., Boston, Mass.), vibrating mill (Perkin Elmer), micro-dismembrator (Braun, Melsungen,

Fig. 3.1. Mechanical vibrator to reduce particle size. (B. Braun, Melsungen, German Federal Republic.)

Fig. 3.2. Hydraulic hand press used for the production of mineral KBr pellets. (P. Weber, Stuttgart, German Federal Republic.)

THE PARTS:

1 Press form exterior with outlet pipe
2 Interior parts of press form
3 Lower die
4 Base plate
5 Pressure plate
6 Upper die with plate
7 Hollow rubber spring sleeve
8 Ring for ejecting interior parts
9 Rubber seals

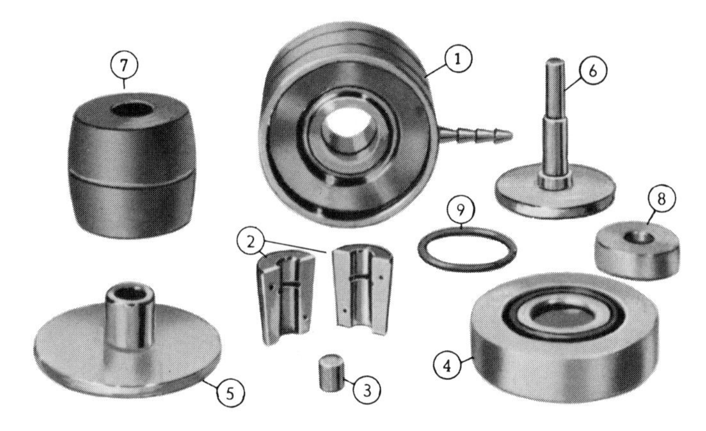

Germany) — Fig. 3.1. The sample is ground in volatile solutions, e.g., aceton, benzene, etc. Intensive grinding and in particular dry grinding can give decreased intensity, bad resolution and shifting of the bands because of deformations in the crystals. Moreover, a very broad band appears at ca. 3430 cm^{-1} because of absorption of H_2O molecules. Hand grinding with a mortar and pestle of agate is also done.

Powder film method

In this method, which was proposed by Hunt and Turner (1953) a paste is made of the ground mineral with isopropanol or amylacetate. The paste is smeared on a NaCl-window and after that it is smoothed out with a polished glass slide. After the liquid has evaporated a thin powder film is left adhering to the window (surface-cast film method). A reference film with only the matrix material is also made.

Mull method

The mull method is one of the oldest methods. In an agate mortar a sample is ground and simultaneously dispersed with a suitable not or only slightly absorbing liquid untill a slurry or mull is

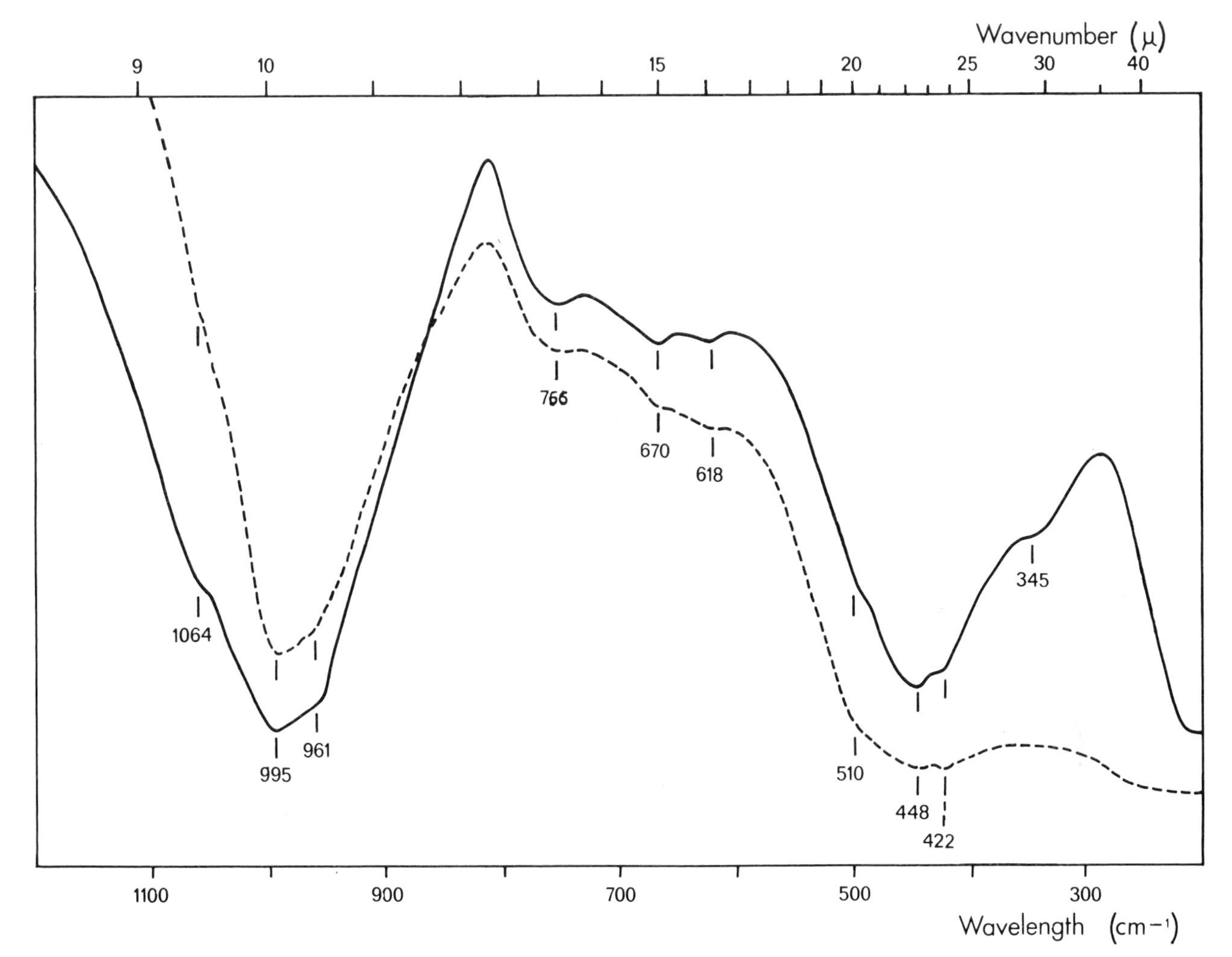

Fig. 3.4. Infrared spectra of lepidomelane after the KBr pellet (———) and the FMIR method with a KRS-5 crystal (- - - - - -). (Courtesy P. Krohmer.)

obtained. After the sample is transferred to a NaCl window the plate is covered by a second and the mull squeezed between the two plates until the film is thin enough to be analyzed.

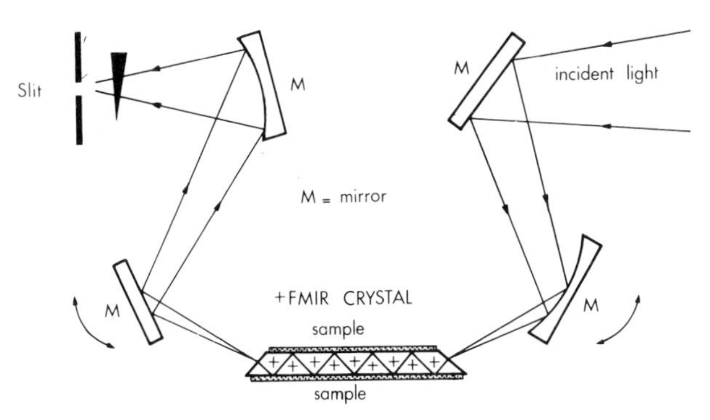

Fig. 3.3. Schematic presentation of the beam path in a Perkin Elmer FMIR accessory.

Mulling liquids are e.g. Nujol[1], perfluorokerosene (fluorolube oil), hexachlorobutadiene. Suspensions of CS_2 + Al-stearate are also used to disperse the solid sample. Unfortunately, these mulls give notwithstanding the use of a reference cell, bands which may seriously interfere when they occur in the region of the bands of the sample to be investigated. Another disadvantage of the mulling method is that fibrous materials, such as chrysotile, attapulgite and sepiolite, give inhomogeneous suspensions; this also happens when the minerals have been pulverized before. In favour of this technique of dispersing a solid in a liquid is that the amount of light scattered may be reduced and band-broadening due to scattering may be prevented.

Pellet method

This method was simultaneously published by Stimson and O'Donnell (1952) and Schiedt and Reinwein (1952). It is now the most frequently method used in IR spectroscopy of solid materials. The sample is mixed in an agate mortar with aethanol and an excess of KBr (1 : 200 to 1 : 700) of analytical quality for IR spectroscopy (Merck, Mallinckrodt, Baker). After drying the

[1] Medical paraffine oil fabricated by Plough Inc.

mixture at 120° C it is pressed in a die for ½ to 1 min at ca. 10 ton/cm^2 under simultaneous suction of ca. 0.1 mm Hg (to remove entrapped air) into a transparent pellet of 13 or 20 mm diameter and ca. 1 mm thickness and again dried for 1 h at 120° C to remove the last traces of absorbed H_2O. A reference pellet with only the matrix is also made. Instead of KBr also other matrices are used, e.g., NaCl, KCl. For bands down to 300 cm^{-1}, CsBr is better suited. Because of vigorous grinding and the high pressure the matrix material may interact with the sample material.
The hydraulic laboratory hand press PW No. 20 of Paul Weber, Stuttgart, was used for this purpose — Fig. 3.2. Other presses and dies are made by (among others) Perkin Elmer (U.S.A.), Research and Industrial Instruments Corporation, R.I.I.C. (Scotland), and Hilger, Watts (London) etc.

Reflection method

Reflectance techniques were already applied seventy years ago by Rubens (1897), Rosenthal (1899) and Aschkinass (1900). The sample is mounted on a plate of polished steel against which light is reflected. Generally the reflected spectra are very weak compared to transmission spectra. The attenuated total reflection method (A.T.R.), however, yields better spectra. Comparable to the transmission method strongly absorbing minerals, pastes, margarines, paints, etc., can also be investigated successfully. The attenuated total reflection (A.T.R.) technique was developed by Fahrenfort (1961). It is based on the following principle. When radiation passing a medium of relatively high index of refraction, e.g., AgCl (n = 2.0[1]), KRS-5 (n = 2.38[1]), germanium (n = 4[1]). strikes the interface of a medium of lower index at a greater angle than that of the critical angle, it is reflected from the interface with very little attenuation, e.g., from a germanium-quartz (n = ca. 1.5[1]) interface ca. 79% is reflected. The thickness of the sample has in general no influence on the intensity of the spectrum because the rays will only penetrate to a depth of a few microns in the sample. The refractive index of the sample changes with the wavelength of the radiation and runs parallel to a corresponding change in the degree of attenuation, i.e., in the intensity of the reflected energy. Because the absorption maxima of the sample investigated corresponds closely to the change in the refractive index, the reflected radiation thus provides the same information about the sample investigated as the transmitted light.

In the FMIR (Frustrated Multiple Internal Reflectance) technique (Fig. 3.3) the beam is reflected 7—25 times at the AgCl, KRS-5 or Germanium crystal. This yields in total a more than 8-fold increase in the sensitivity, which is an advantage in case the sample has a bad contact with the crystal surface (fibres, hard plastics, powders) or when the absorption of the rays by the sample is weak. The KRS-5, AgCl and Ge crystal can be used up to 300 cm^{-1}, 500 cm^{-1} and 870 cm^{-1}, respectively (for details, see Harrick, 1967). In Fig. 3.4 are spectra of lepidomelane which were obtained with transmitted radiation (KBr pellet method) and reflected radiation (FMIR method).[2]

References

Aschkinass, E., 1900. Über anomale Dispersion im ultraroten Spektralgebiete. Ann. Phys., 1900: 42—68.

Bambauer, H.U., Brunner, G.O. and Laves, F., 1969. Light scattering of heat-treated quartz in relation to hydrogen containing defects. Am. Mineralogist, 54: 718—724.

Christiansen, C., 1884. Untersuchungen über die optischen Eigenschaften von fein verteilten Körpern. Wiedemann Ann. Phys., 23: 298—306.

Fahrenfort, J., 1961. Attenuated total reflection. A new principle for the production of useful infrared reflection spectra of organic compounds. Spektrochim. Acta, 17: 698—709.

Harrick, N.J., 1967. Internal Reflection Spectroscopy. Interscience, New York, N.Y., 327 pp.

Hunt, J.M. and Turner, D.S., 1953. Determination of mineral constituents of rocks by infrared spectroscopy. Anal. Chem., 25: 1169—1174.

Lindberg, J.D. and Smith, M.S., 1973. Reflectance spectra of gypsum sand from the white sands. National Monument and basalt from an nearby lava flow. Am. Mineralogist, 58: 1062—1064.

Lord Rayleigh (J.W. Strutt), 1871a. On the light from the sky, its polarization and colour. Phil. Mag., 41: 107—120, 274—279.

Lord Rayleigh (J.W. Strutt), 1871b. On the scattering of light by small particles. Phil. Mag., 41: 447—454.

Lord Rayleigh (J.W. Strutt), 1899. On the transmission of light through an atmosphere containing small particles in suspension and on the origin of the blue of the sky. Phil. Mag., 47: 375—384.

Lord Rayleigh (J.W. Strutt), 1910. The incidence of light upon a transparant sphere of dimensions comparable with the wavelength. Proc. R. Soc., A., 84: 25—46.

Pfund, A.H., 1934. Rayleigh's law of scattering in the infrared. J. Opt. Soc. Am., 24: 143—146.

Phillippi, C.M., 1968. Optical effects underlying the analytical infrared spectra of solid materials. Air Fr. Mater. Lab., Tech. Rep., 67: 437.

Phillippi, C.M. 1969. The absorption and reflection of infrared bands of powdered inorganic materials. Air Fr. Mater. Lab., Tech. Rep., 68: 317.

[1] Refractive index at 5 μm.

[2] Courtesy of P. Krohmer, Perkin Elmer & Co., G.m.b.H., Uberlingen (W. Germany).

Rosenthal, H., 1899. Über die Absorption, Emission und Reflexion von Quartz. Wiedemann Ann. 18: 783—800.
Rubens, H. and Nichols, E.F., 1897. Reststrahlenmethode. Versuche mit Wärmestrahlen von grosser Wellenlänge. Wiedemann Ann., 60: 418—462.
Schiedt, U. and Reinwein, H., 1952. Zur Infrarotspektroskopie von Aminosäuren. Z. Naturforsch., 7b: 270—277.
Stimson, M.M. and O'Donnel, M.J., 1952. The infrared and ultraviolet spectra of cystine and isocystine in the solid state. J. Am. Chem. Soc., 74: 1805—1808.
Tyndall, J., 1869. On the blue colour of the sky and the polarization of light. Proc. R. Soc. Lond., 17: 223—233.
Valasék, J., 1960. Theoretical and Experimental Optics. Wiley, London, 454 pp.

4. ASSIGNMENT OF INFRARED BANDS

Correlations

Several attempts have been made to ascribe a certain IR band to a certain molecular vibration. This is only easy for simple molecules like H_2O, CO_2, SiO_2, etc. but not for more complicated structures. The frequency of a molecular vibration is determined by the masses of the atoms (ions), their distances, symmetry, bond strength, surroundings (geometry). The intensity depends on the change of the dipol momentum during the vibration, particle size, wavelength, and the difference in refractive index between the absorbing substance and the dispersion medium; moreover, on the direction of the vibrational movement against the electric vector of the incident light. Broadening, loss of intensity and decrease of frequency are caused by isomorphous random substitutions and poor crystallinity. The substitution of one element for another in the mineral crystal introduces other variables than those only inherent to the two atoms. As a result, not in all cases absorption bands can be correlated to a type of vibration.

In the following are some correlations between the wave number of a vibration (or its shift) and:

— the O-H---O distance in several organic compounds: Badger (1934, 1940); Errera (1937); Errera and Sack (1938); Fox and Martin (1940); Rundle and Parasol (1952); Lord and Merrifield (1953); Glemser and Hartert (1953, 1955, 1956); Nakamoto et al. (1955); Lippincott and Schroeder (1955); Hofacker and Glemser (1955); Hartert and Glemser (1956); Huggins and Pimentel (1956); Pimentel and Sederholm (1956); Schwarzmann (1962); White and Roy (1964);

— the Pb/SiO_2 and Ba/SiO_2 ratio in lead and barium silicate glasses: Matossi and Bluschke (1938); Matossi and Bronder (1938); Pepperhoff (1954);

— the chemical composition of the plagioclases: Thompson and Wadsworth (1957);

— the radius of alkali halides: Price et al. (1958);

— the mass of the metal of several stearates: Ellis and Pyszora (1956);

— the iron contents of chlorites: Tuddenham and Lyon (1959);

— the Fe, Mg/Fe, Mg, Al ratio of montmorillonite, vermiculite and glauconite: Touillaux et al. (1960);

— the electronegativities of the metal ions in various hydroxides: West and Baney (1960);

— the Al/AlSi quotient of tectosilicates: Milkey (1960);

— the amount of Al-substituting for silicon in muscovite, lepidolite, biotite, phlogopite and chlorite: Lyon and Tuddenham (1960);

— the Si-O band in the 600-750 cm^{-1} region of saponites and the amount of the substitution in tetrahedral coordination: Stubičan and Roy (1961);

— the $Si\text{-}O\text{-}Al^{VI}$ band in the 520-550 cm^{-1} region of muscovites and the amount of Mg ions in octahedral coordination: Stubičan and Roy (1961);

— the Si-O band in the 550-800 cm^{-1} region of chlorites and the amount of Al in octahedral and tetrahedral coordination: Stubičan and Roy (1961);

— the M (metal)-OH_2 distance of several chlorides and sulfates: Gamo (1961);

— the ionization potential of various substituted benzenes absorbed on silica powder: Basila (1961);

— the Fe, Al, and Mg contents of chlorites: Kodama and Oinuma (1963), Oinuma and Hayashi (1966), Hayashi and Oinuma (1965, 1967);

— the ionic radius of the exchangeable cations of zeolites: Kiselev and Lygin (1966);

— the Fe contents in clino amphibioles: Burns and Strens (1966);

— the iron contents of biotites: Liese (1963);

— the Mg/Mg + Fe ratio of biotites: Wilkins (1967);

— the chemical composition of micas: Vedder (1964);

— the mass and the ion radius of different carbonates: Elderfield (1971).

References

Badger, R.M., 1934. A relation between internuclear distances and bond force constants. J. Chem. Phys., 2: 128—131.

Badger, R.M., 1940. The relation between the energy of a hydrogen bond and the frequency of the O-H bands. J. Chem. Phys., 8: 288—289.

Basila, M., 1961. Hydrogen bonding interaction between adsorbate molecules and surface hydroxyl groups on silica. J. Chem. Phys., 35: 1151—1158.

Burns, R.G. and Strens, R.G.J., 1966. Infrared study of the hydroxyl bands in clinoamphiboles. Science, 153: 890—892.

Elderfield, H., 1971. The effect of periodicity on the infrared adsorption frequency ν_4 of anhydrous normal carbonate minerals. Am. Mineralogist, 56: 1600—1606.

Ellis, B. and Pyszora, H., 1956. The infrared spectrum of zinc stearate and the vulcanization of natural rubber. J. Polymer Sci., 22: 348—350.

Errera, J., 1937. Infrared spectrographic examination of intermolecular forces. Physica, 4: 1097—1104.

Errera, J. and Sack, H., 1938. Molecular association study in the infrared. Trans. Faraday Soc., 34: 728—742.

Fox, J.J. and Martin, A.E., 1940. Infrared absorption of the hydroxyl groups in relation to inter- and intramolecular hydrogen bonds. Trans. Faraday Soc., 36: 897—913.
Gamo, I., 1961. Infrared spectra of water of crystallization in some inorganic chlorides and sulfates. Bull. Chem. Soc. Japan, 34: 760—764.
Glemser, O. and Hartert, E., 1953. Knickschwingungen der O-H Gruppe im Gitter von Hydroxyden. Naturwissenschaften, 40: 552—553.
Glemser, O. and Hartert, E., 1955. Untersuchungen über die Bindung des Wassers in kristallisierten Salzhydraten. Naturwissenschaften, 42: 534.
Glemser, O. and Hartert, E., 1956. Untersuchungen über die Wasserstoffbrückenbindung in kristallisierten Hydroxyden. Z. Anorg. Allgem. Chem., 283: 111—122.
Hartert, E. and Glemser, O., 1956. Ultrarotspektroskopische Bestimmung der Metall-Sauerstoff-Abstände in Hydroxyden, basischen Salzen und Salzhydraten. Z. Electrochem., 60: 746—751.
Hayashi, H. and Oinuma, K., 1965. Relationship between infrared absorption spectra in the region 450—900 cm^{-1} and chemical composition of chlorite. Am. Mineralogist, 50: 476—483.
Hayashi, H. and Oinuma, K., 1967. Si-O absorption band near 1000 cm^{-1} and OH absorption bands of chlorite. Am. Mineralogist, 52: 1206—1210.
Hofacker, L. and Glemser, O., 1955. Über den Bindungsmechanismus bei O-H...O Wasserstoffbrücken. Naturwissenschaften, 42: 369.
Huggins, C.M. and Pimentel, G.C., 1956. Systematic of the infrared spectral properties of hydrogen bonding systems: frequency shift, half width and intensity. J. Phys. Chem., 60: 1615—1619.
Kiselev, A.V. and Lygin, V.I., 1966. Infrared spectra and adsorption by zeolites. In: L.H. Little (Editor), Infrared Spectra of Absorbed Species. Academic Press, London-New York, pp. 352—381.
Kodama, H. and Oinuma, K., 1963. Identification of kaolin minerals in the presence of chlorite by X-ray diffraction and infrared adsorption spectra. Clays Clay Minerals. Proc. 11th Natl. Conf. Clays Clay Minerals Natl. Acad. Sci. Natl. Res. Council, 1963: 236—249.
Liese, H.C., 1967. Supplemented data on the correlation of infrared absorption spectra and composition of biotites. Am. Mineralogist, 52: 877—880.
Lippincott, E.R. and Schroeder, R., 1955. One-dimensional model of the hydrogen bond. J. Chem. Phys., 23: 1099—1106.
Lord, R.C. and Merrifield, R.E., 1953. Strong hydrogen bonds in crystals. J. Chem. Phys., 21: 166—167.
Lyon, R.J.P. and Tuddenham, W.M., 1960. Determination of tetrahedral aluminium in mica by infrared absorption analysis. Nature, 185: 374—375.
Matossi, F. and Bluschke, H., 1938. Das ultrarote Reflexionsspektrum von Gläsern. Z. Phys., 108: 195—313.
Matossi, F. and Bronder, O., 1938. Das ultrarote Absorptionsspektrum einiger Silikate. Z. Phys., 111: 1—17.
Milkey, R.G., 1960. Infrared spectra of some tectosilicates. Am. Mineralogist, 45: 990—1007.
Nakamota, K., Margoshes, M. and Rundle, R.E., 1955. Stretching frequencies as a function of distances in hydrogen bonds. J. Am. Chem. Soc., 77: 6480—6486.
Oinuma, K. and Hayashi, H., 1966. Infrared study of clay minerals from Japan. J. Univ. Tokyo, General Educ. Natl. Sci., 1—15.
Pepperhoff, W., 1954. Infrared reflectance spectra of binary lead silicate glasses. Z. Elektrochem., 58: 520—522.
Price, W.C., Sherman, W.F. and Wilkinson, G.R., 1958. Infrared studies of water adsorbed on alkali halides. Proc. R. Soc. Lond., 247: 467—468.
Pimentel, G.C. and Sederholm, Ch.H., 1956. Correlation of infrared stretching frequencies and hydrogen bond distances in crystals. J. Chem. Phys., 24: 639—641.
Rundle, R.E. and Parasol, M., 1952. O-H stretching frequencies in very short and possibly symmetrical hydrogen bonds. J. Chem. Phys., 20: 1487—1488.
Schwarzmann, E., 1962. Zusammenhang zwischen OH-Valenzfrequenzen und OH-OH Abständen in festen Hydroxyden. Naturwissenschaften, 49: 103—104.
Stubičan, V. and Roy, R., 1961. Isomorphous substitution and infrared spectra of the layer-lattice silicates. Am. Mineralogist, 46: 33—50.
Thompson, C.S. and Wadsworth, M.E. 1957. Determination of the composition of placioclase feldspars by means of infrared spectroscopy. Am. Mineralogist, 42: 334—341.
Touillaux, R., Fripiat, J.J. and Toussaint, F., 1960. Etude en spectroscopie infra-rouge des minéraux argileux. Trans. 7th Int. Congr. Soil. Sci., U.S.A., IV (VII-3): 460—467.
Tuddenham, W.M. and Lyon, R.J.P., 1959. Relation of infrared spectra and chemical analysis for some chlorites and related minerals. Anal. Chem., 31: 337—380.
Vedder, W., 1964. Correlations between infrared spectrum and chemical compositions of mica. Am. Mineralogist, 49: 736—768.
West, R. and Baney, R.H., 1960. The relationship between O-H stretching frequency and electronegativity in hydroxydes of various elements. J. Phys. Chem., 64: 822—824.
White, W.B. and Roy, R., 1964. Infrared spectra-crystal structure correlations, Comparison of simple polymorphic minerals. Am. Mineralogist, 49: 1670—1687.
Wilkins, R.W.T., 1967. The hydroxyl stretching region of the biotite mica spectrum. Mineral Mag., 36: 325—333.

Adsorbed water and crystal water (Fig. 4.1)

Minerals which contain loosely adsorbed water molecules will give strong bands at ca. 3435 cm^{-1} (O-H stretching vibration) and ca. 1630 (O-H bending vibration). The amount of adsorbed water, the kind of the mineral and the exchangeable cation to which the water is bonded, all have influence on the wave number of the above vibrations. For monomeric unassociated (vapour) H_2O molecules are found 3756 and 1595 cm^{-1}; for liquid H_2O molecules 3455 and 1645 cm^{-1} and for ice molecules 3256 and 1655 cm^{-1}, respectively.
When water molecules are tightly bonded to the mineral surface as a monolayer a 3200—3250 cm^{-1} band appears (montmorillonite). This band is also found in minerals with crystalwater: vermiculite, hydrous biotite, attapulgite, sepiolite, zeolites (heulandite, chabasite, desmine), gypsum, melanterite, ($FeSO_4 \cdot 7H_2O$).
An exception is mirabilite ($Na_2SO_4 \cdot 10H_2O$) which is caused by its small hydration energy (K = 75, Na = 94.5, Li = 120, Ca = 360, Mg = 440 kcal/g ion). Hisingerite, obsidian, deweylite, opal wood and silicic acid have H_2O molecules in very small pores. The smaller the pores the stronger the H_2O molecules are bonded to the surface of the capillars e.g. Φ 3 μm ~ 1 kg/cm^2, Φ 0.003 μm ~ 1000 kg/cm^2, Φ 0.0003 μm (3 Å) ~ 10.000 kg/cm^2.
In the very tightly bonded water of hisingerite, attapulgite, sepiolite, melanterite and gypsum there is in addition an O-H bending vibration at 1650—1685 cm^{-1} and the O-H stretching vibration is decreased to 3385—3400 cm^{-1}.
There exists a relation between the O-H---O distance of various organic and inorganic substances and the wave numbers of the band (see p. 31).
The 3220 cm^{-1} band corresponds with an O-H---O distance of 2.77 Å which points to an ice-like structure (liquid H_2O = 2.86 Å, ice = 2.76 Å) of the adsorbed water on the surface or in the small pores of the mineral particles. There is only one layer of these ice-like water molecules: Van der Marel (1961).
The O-H stretching vibration may even shift to 1775 cm^{-1} as in nickel dimethyl-glyoxime, if the distance is very small: Rundle and Parasol (1952).
About the arrangement of these disturbed H_2O molecules at the liquid mineral contact several suggestions have been made, e.g., "fit" of the ice-like lattice of the adsorbed H_2O molecules on top of the oxygen net of the basal planes of the clay minerals: Macey (1942), Forslind (1948); adsorbed H_2O molecules in hexagonal nets related to the hexagonal oxygen nets of the clay minerals: Hendricks and Jefferson (1938); adsorbed H_2O molecules forming tetrahedra with the bases of the linked silica tetrahedra of the clay minerals: Barshad (1949). As compared with common (liquid) H_2O molecules the monolayered H_2O molecules have a larger viscosity, smaller diffusiveness, lower dielectric constant (ca. 3 as against 80). It freezes at a lower temperature and evaporates at a higher temperature (150—200°C). Also its dielectric loss (tgδ), nuclear magnetic signal (line width), conductivity and polarizability differ from that of common (liquid) H_2O molecules. A highly controversional point in literature is whether its density is higher or lower than that of common H_2O molecules.

Remarks

The property of a soil to adsorb H_2O has caused many difficulties in agriculture and soil mechanics. Heavily, easily swelling and contracting montmorillonitic soils are the black regurs (India), tir noirs (Morocco), margalites (Java), black turf (S. Africa), badobs (Egypt). Mexico City has its sinking skyscrapers (0.5 m/year) caused by contraction of wet montmorillonitic clay as a result of water removal for living-purposes. South Africa has the problem of its cracking buildings which were accidentally built on montmorillonitic alluvial sediments (De Vereniging district). In the rainy season rainwater from the outside oozed through down to under the building, where the temperature is lower and the relative vapour pressure higher (thermo-osmotic effect).
The energy effect for the adsorption of H_2O molecules by the mineral surface and the hydration of the exchangeable Na-ions of pure Na-saturated montmorillonite and illite = ca. 150 erg/cm^2 (Van der Marel, 1966).
Thus for the above montmorillonite soils with a specific surface of ca. 400 m^2/g this will be per ton of soil = ca. 14000 kcal which equals the heat of ca 2 m^3 natural gas. The pressure exerted by dry montmorillonite is about 30 kg/cm^2 or equal to the weight of a flat building of ca 300 floors.

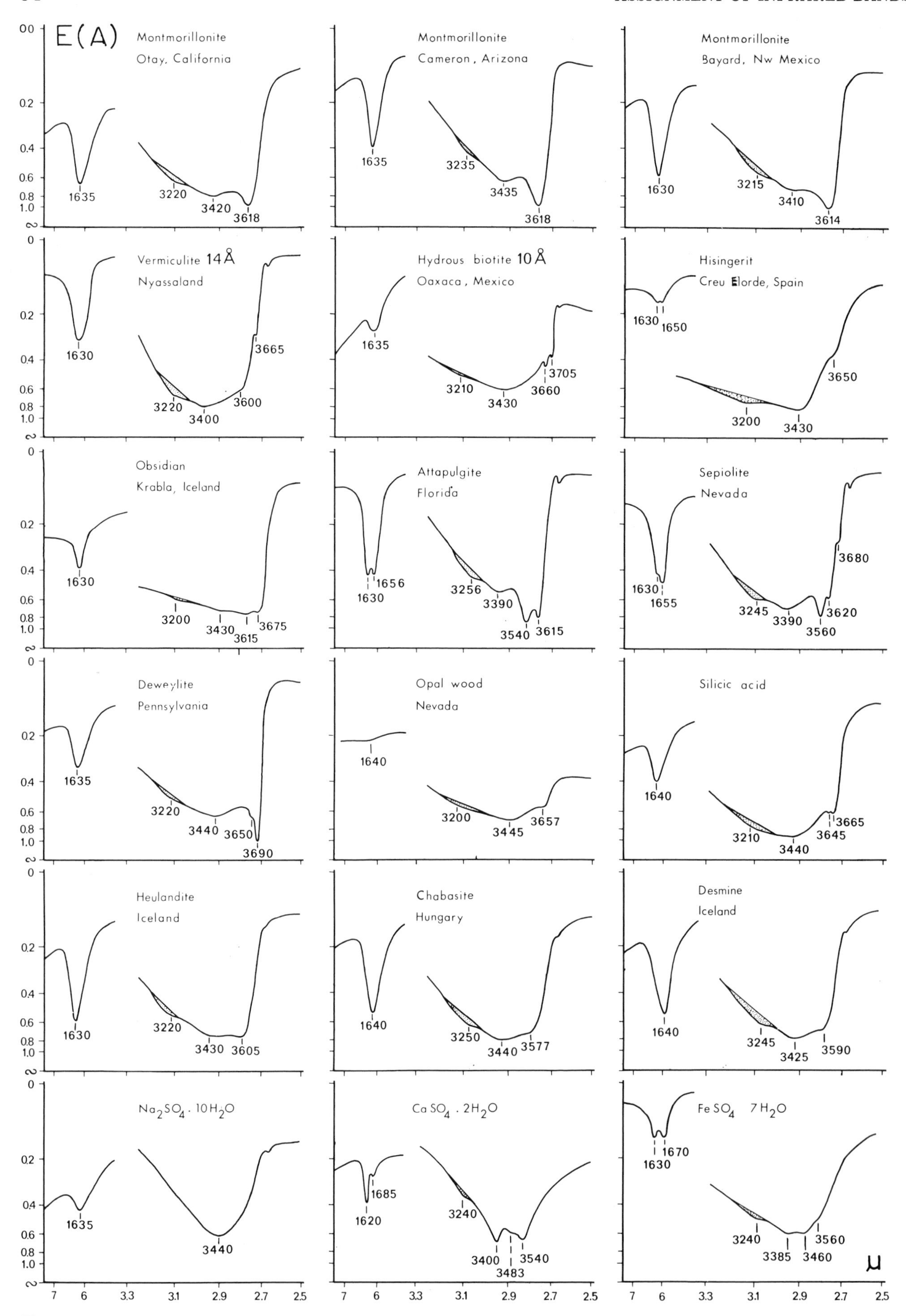

Fig. 4.1. O-H stretching and O-H bending vibrations of surface- and cation-bonded H_2O molecules, of H_2O molecules in pores and canals, and of interlayered H_2O molecules. Pellets (also for Fig. 4.2—4.15): 1.73 mg/300 mg KBr.

References

Barshad, I., 1949. The nature of lattice expansion and its relation to hydration in montmorillonite and vermiculite. Am. Mineralogist, 34: 625—684.

Forslind, E., 1948. The clay/water system and water absorption of clay minerals. Swedish Cement Concrete Res. Inst. Stockholm, Bull., 11: 20 pp.

Hendricks, S.B. and Jefferson, M.E., 1938. Structures of kaolin and talc-pyrophyllite hydrates and their bearing on water sorption of the clays. Am. Mineralogist, 23: 863—875.

Macey, H.H., 1942. Clay/water relationships and the internal mechanism of drying. Trans. Brit. Ceram. Soc., 41: 73—121.

Rundle, R.E. and Parasol, M., 1952. O-H stretching frequencies in very short and possibly symmetrical hydrogen bonds. J. Chem. Phys., 20: 1487—1488.

Van der Marel, H.W., 1961. Quantitative analysis of the clay separate of soils. Acta Univ. Carolinae, Geol. Suppl., 1: 23—82.

Van der Marel, H.W., 1966. Cation exchange, specific surface and absorbed H_2O molecules. Z. Pflanzernähr., Düng. Bodenk., 114: 161—175.

Heating

Heating the sample is sometimes useful for the assignment of an absorption band of a mineral. For instance, the Si-O group will resist better heating than the Si-OH group. The Fe-OH group will decompose more easily as the Al-OH group, etc. Minerals of the feldspar group can be distinguished from each other when heated before at increasing temperatures in the 700—1400°C traject.

Dickite, kaolinite, fire clay mineral, halloysite (Fig. 4.2, 4.3)

The heated minerals show a sharp decrease of their IR intensities at 550°C. Only for dickite and the well-crystallized Zettlitzer kaolinite there is still some IR activity left. For dickite and (to a lesser degree) for the Zettlitzer kaolinite, there is, when heated at 550°C a comparatively greater decrease in the intensity of the outer O-H vibration (ca. 3700 cm^{-1}) than in that of the inner O-H vibrations (3650 and 3620 cm^{-1}). The opposite effect happens for the inner O-H band of halloysite at ca. 3620 cm^{-1}.

Sericite, illite, glauconite (Fig. 4.4)

When these minerals are heated, there is a sharp decrease in the intensity of the bands for sericite at 650°C. For illite and glauconite the intensity already decreases at 550°C. When heated, the 3620 cm^{-1} vibration of glauconite of inner Al...O-H gradually increases in intensity compared to the inner Fe...O-H vibration at 3540 cm^{-1}.

Montmorillonite, hectorite, nontronite (Fig. 4.5)

The large H_2O band of adsorbed water at ca. 3440 and 3416 cm^{-1} of these minerals, masks their O-H stretching bands inherent to Al...O-H, Mg...O-H and Fe...O-H vibrations, respectively. This holds in particular for hectorite.
By heating at 350°C the surface-adsorbed H_2O of the above three minerals and also the more tightly bonded (crystal) water (ca. 3235 cm^{-1}) of montmorillonite and hectorite is lost.
Nontronite (Fe) is largely decomposed at 550°C, montmorillonite (Al) at ca. 650°C whereas hectorite (Mg, Li) with all its octahedral holes fully occupied remains quite stable at this temperature.

Sepiolite, vermiculite (Fig. 4.6)

The band at ca. 3400 cm^{-1} of adsorbed water in both minerals, largely masks the O-H bands inherent to the crystal at 3690, 3630, 3556 cm^{-1} and 3708, 3663, 3605 cm^{-1}, respectively. When heated at 350°C the water is lost inclusive the strongly adsorbed water at the surface of these minerals at ca. 3235 cm^{-1}. Thus a better understanding is obtained of the O-H vibrations in the crystal.
The relative intensities of the sepiolite bands remain nearly constant when heated to 350°C. But there is a decrease in their intensities. After 350°C there is a large decrease in the intensity of the 3515 cm^{-1} band. The relative intensities of the vermiculite bands remain nearly constant when heated to 250°C. Also their intensities remain constant. After that there is a sharp decrease in the intensities of the bands, especially of that of the 3610 cm^{-1} band.
The E(ca. 3400)/(ca. 1630) and E(ca. 3240)/(ca. 1655) quotients of sepiolite show a strong decrease when the sample is heated. Also for the vermiculite sample there is a strong decrease in the E(ca. 3410)/(ca. 1630) quotient.

Deuteration

This method is very useful for the identification of an OH group in a mineral and for the assignment of an O-H band in a mineral.
By substitution of an atom by its isotope the mass changes, from which results a shifting of the infrared band. The largest effect will be obtained with atoms of low masses, e.g., ^1H and ^2H and the smallest with high masses, e.g. ^{35}Cl and ^{37}Cl. For our purpose of the recognition of an O-H or a metal—OH vibration in the IR spectrum of minerals, only the OH/OD and metal-OH/metal-OD substitutions are of importance. To get deuterated minerals ca. 2 g samples are treated with ca. 4 cc D_2O (purity = 99.7%) in an autoclave for 10 days at ca. 200°C and ca. 12 atmospheres. The excess D_2O is removed by washing them with water and the samples are dried [1].
The wavenumber $\tilde{\nu}$ (cm^{-1}) of the vibration of a theoretical free moving diatomic molecule is proportional to $\sqrt{1/\mu_m}$, in which μ_m is the reduced mass = $m_1 m_2/(m_1 + m_2)$ in which m_1 = mass of atom m_1 and m_2 = mass of atom m_2 (see also p. 6).
The following relation exists between the wave numbers and the reduced masses:

$$\tilde{\nu}_1/\tilde{\nu}_2 = \sqrt{(\mu_1/\mu_2)}.$$

[1] The authors are greatly indebted to Professor Dr. Ir. C. Boelhouwer and Mr. J.F. Goosen, Laboratory of Chemical Technology, University of Amsterdam, for facilities to use the autoclaves.

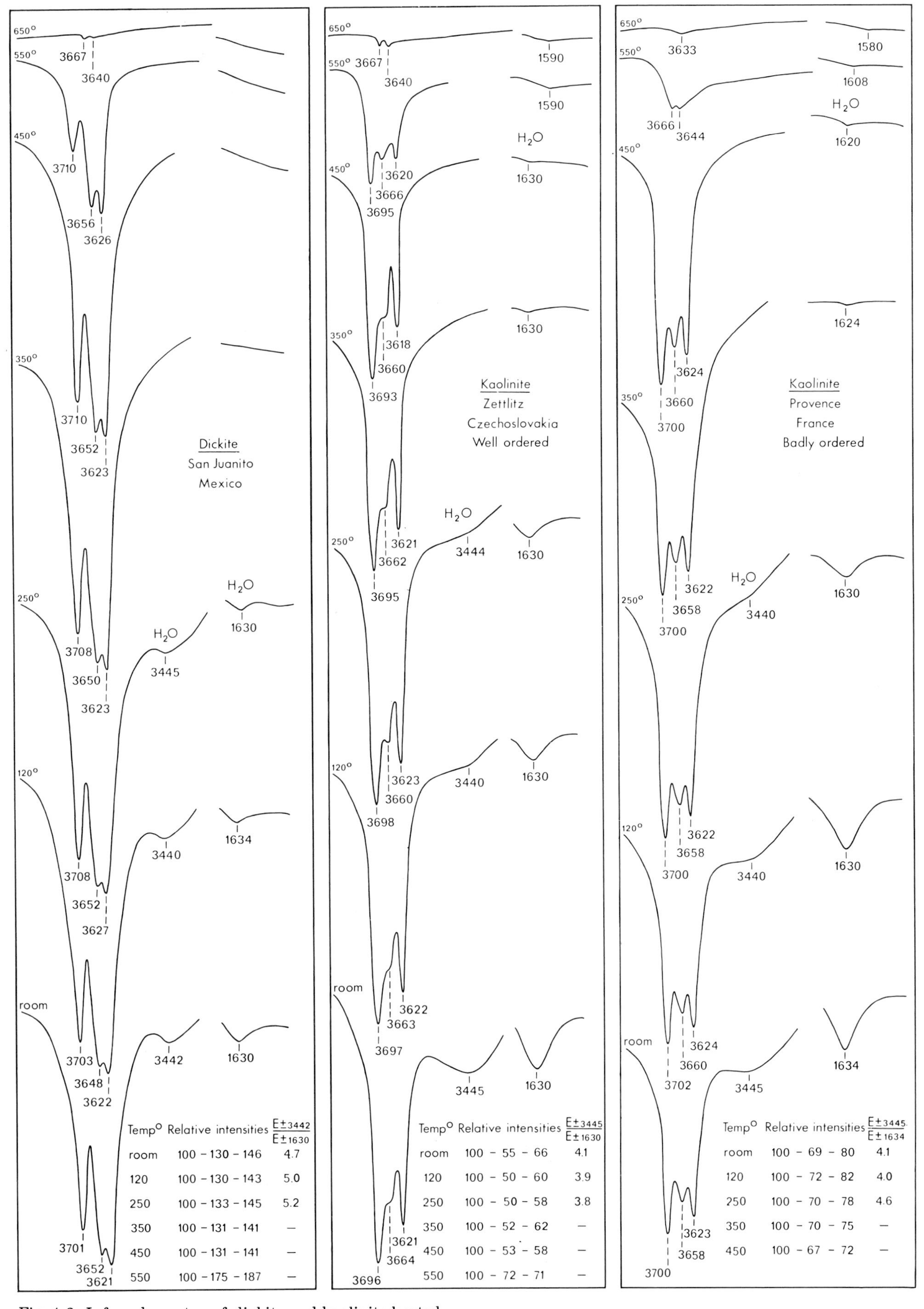

Temp°	Relative intensities	E±3442/E±1630
room	100 – 130 – 146	4.7
120	100 – 130 – 143	5.0
250	100 – 133 – 145	5.2
350	100 – 131 – 141	–
450	100 – 131 – 141	–
550	100 – 175 – 187	–

Temp°	Relative intensities	E±3445/E±1630
room	100 – 55 – 66	4.1
120	100 – 50 – 60	3.9
250	100 – 50 – 58	3.8
350	100 – 52 – 62	–
450	100 – 53 – 58	–
550	100 – 72 – 71	–

Temp°	Relative intensities	E±3445/E±1634
room	100 – 69 – 80	4.1
120	100 – 72 – 82	4.0
250	100 – 70 – 78	4.6
350	100 – 70 – 75	–
450	100 – 67 – 72	–

Fig. 4.2. Infrared spectra of dickite and kaolinite heated at various temperatures for 1 hour. Relative intensities of the bands from high to low wave numbers and E 3445 cm^{-1}/E 1630 cm^{-1} quotients.

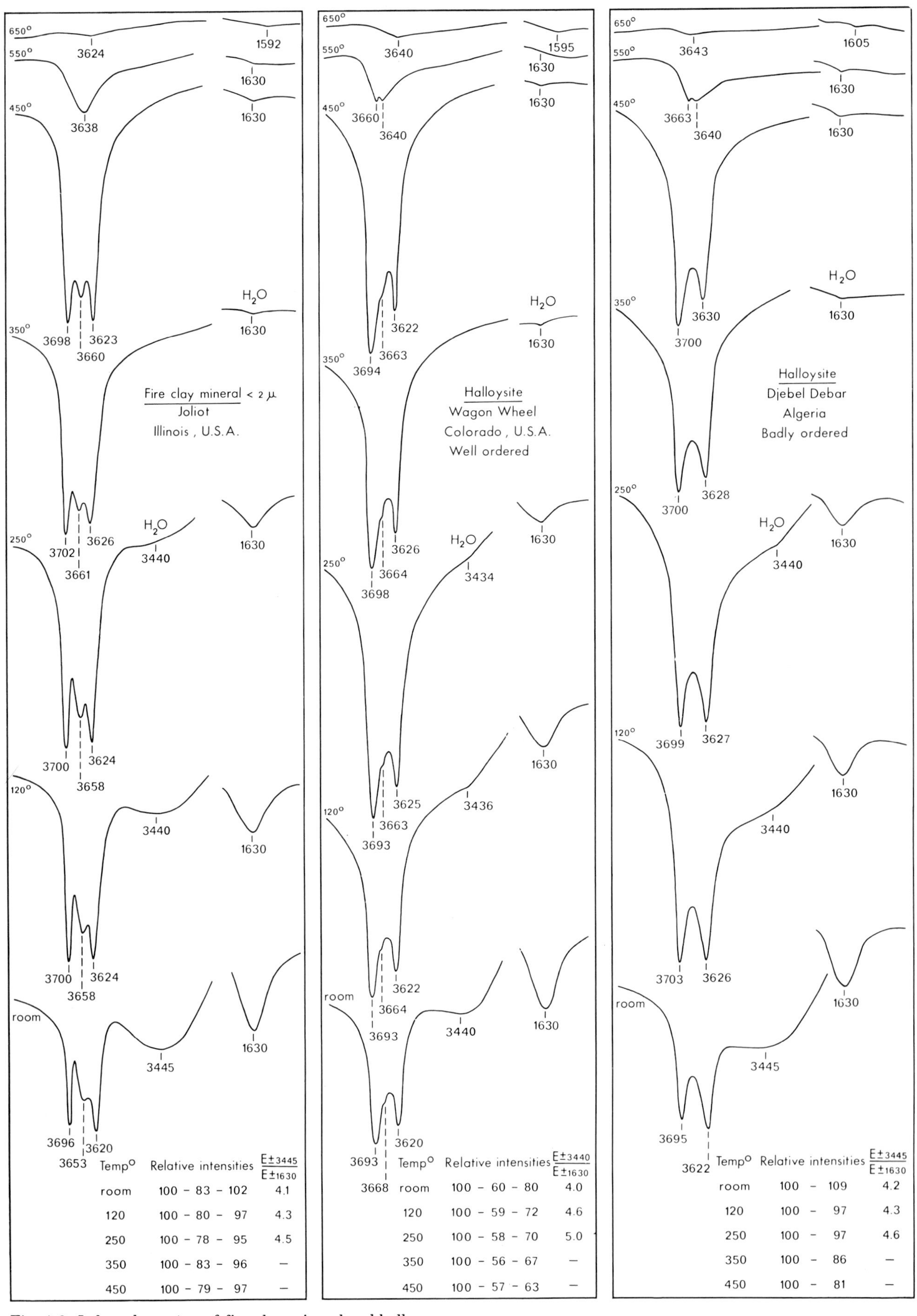

Temp°	Relative intensities	E±3445/E±1630
room	100 - 83 - 102	4.1
120	100 - 80 - 97	4.3
250	100 - 78 - 95	4.5
350	100 - 83 - 96	—
450	100 - 79 - 97	—

Temp°	Relative intensities	E±3440/E±1630
room	100 - 60 - 80	4.0
120	100 - 59 - 72	4.6
250	100 - 58 - 70	5.0
350	100 - 56 - 67	—
450	100 - 57 - 63	—

Temp°	Relative intensities	E±3445/E±1630
room	100 - 109	4.2
120	100 - 97	4.3
250	100 - 97	4.6
350	100 - 86	—
450	100 - 81	—

Fig. 4.3. Infrared spectra of fire clay mineral and halloysite heated at various temperatures for 1 hour. Relative intensities of the bands from high to low wave numbers and E 3445 cm^{-1}/E 1630 cm^{-1} quotients.

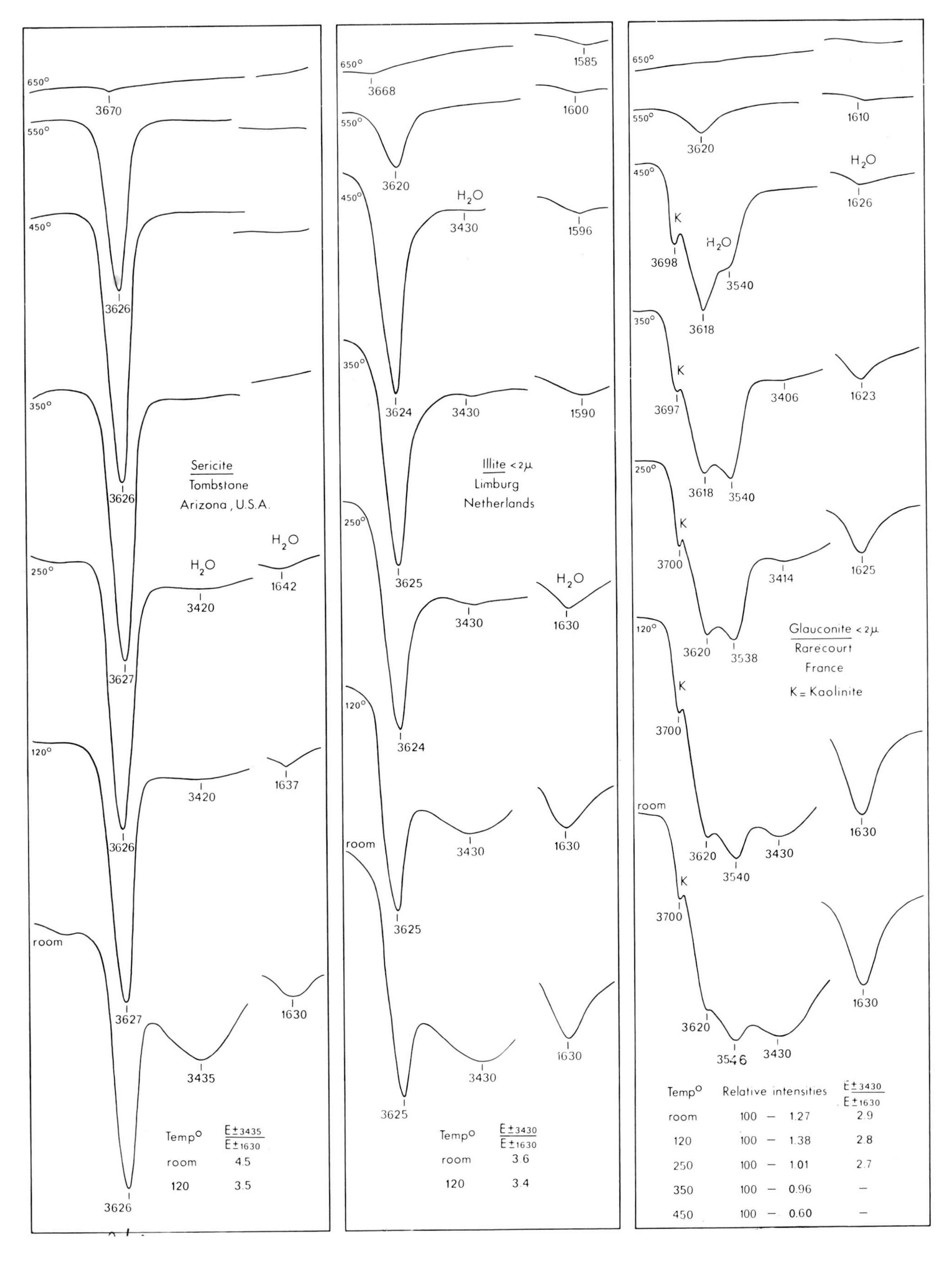

Fig. 4.4. Infrared spectra of sericite, illite and glauconite heated at various temperatures for 1 hour. Relative intensities of the glauconite bands from high to low wave numbers and E 3430 cm^{-1}/E 1630 cm^{-1} quotients.

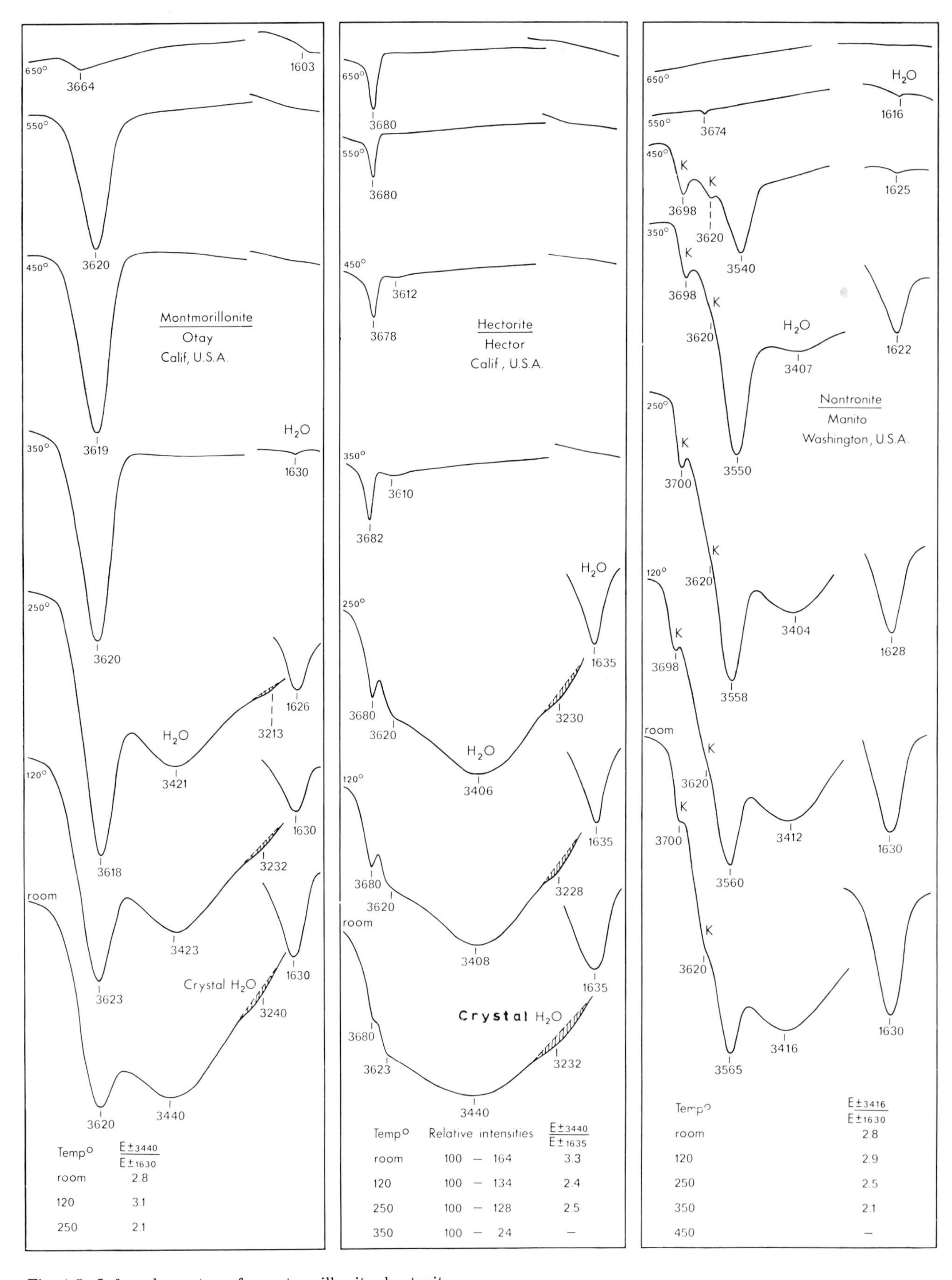

Fig. 4.5. Infrared spectra of montmorillonite, hectorite and nontrinite heated at various temperatures for 1 hour. Relative intensities of the bands from high to low wave numbers and E 3440 (3416) cm^{-1}/E 1630 cm^{-1} qoutients. (K = kaolinite).

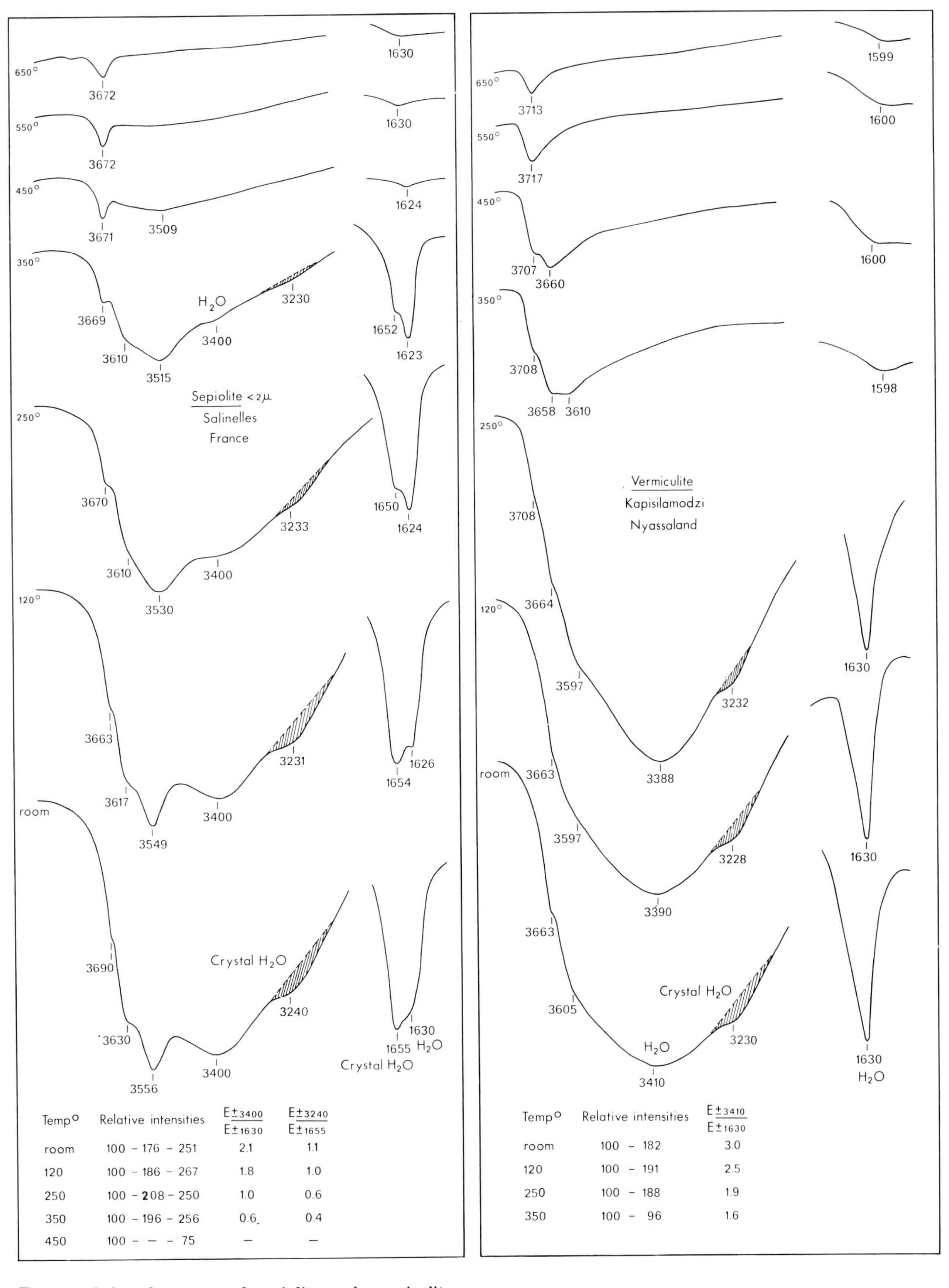

Temp°	Relative intensities	$E\pm3400/E\pm1630$	$E\pm3240/E\pm1655$
room	100 - 176 - 251	2.1	1.1
120	100 - 186 - 267	1.8	1.0
250	100 - 208 - 250	1.0	0.6
350	100 - 196 - 256	0.6	0.4
450	100 - — - 75	—	—

Temp°	Relative intensities	$E\pm3410/E\pm1630$
room	100 - 182	3.0
120	100 - 191	2.5
250	100 - 188	1.9
350	100 - 96	1.6

Fig. 4.6. Infrared spectra of sepiolite and vermiculite heated at various temperatures for 1 hour. Relative intensities of the bands from high to low wave numbers and E 3400 cm^{-1}/E 1630 cm^{-1}, E 3240 cm^{-1}/E 1655 cm^{-1} and E 3410 cm^{-1}/E 1630 quotients.

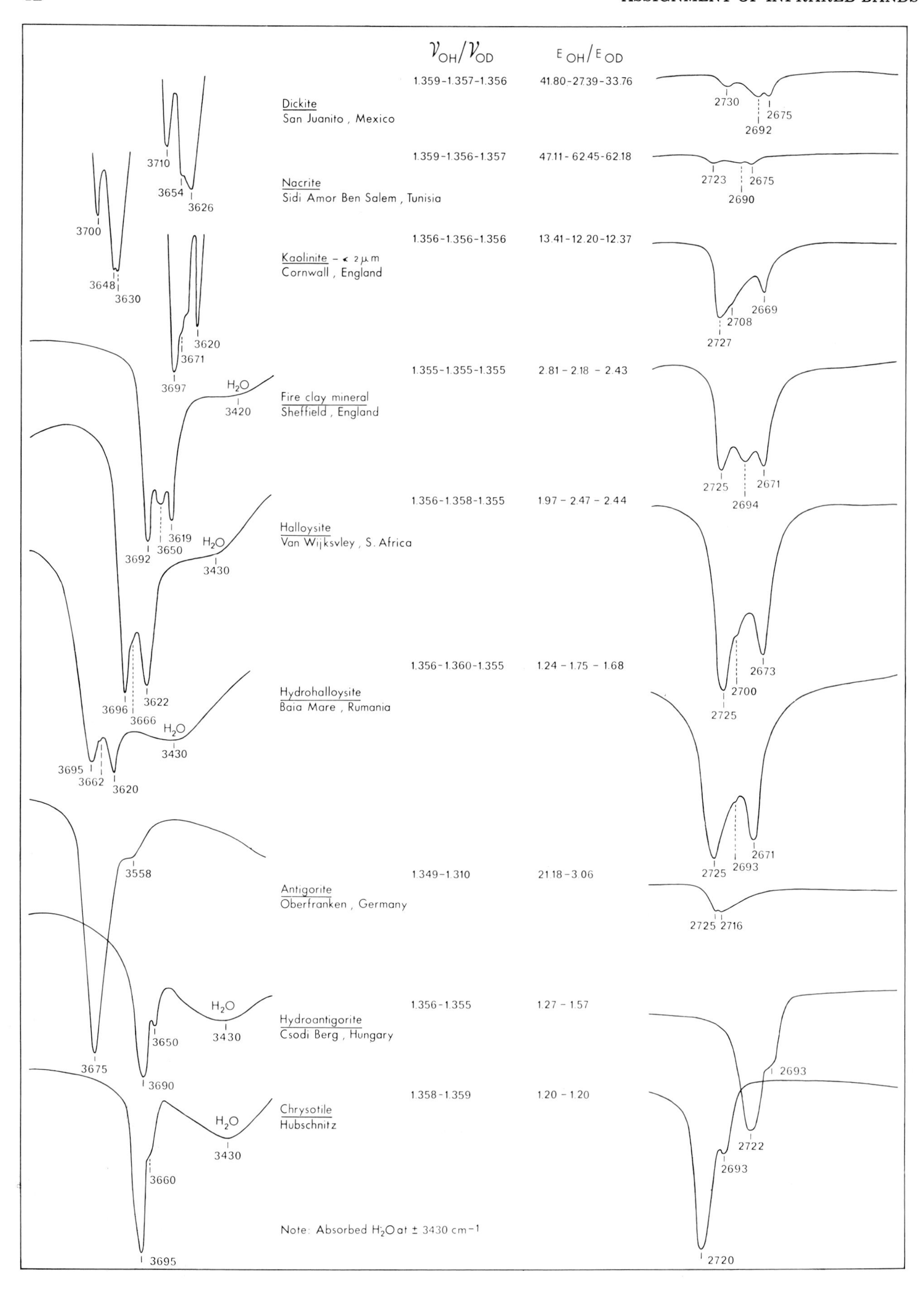

Fig. 4.7. Infrared spectra and ratios of wave numbers $\tilde{\nu}_{OH}/\tilde{\nu}_{OD}$ and intensities of corresponding E_{OH} and E_{OD} vibrations of deuterated kaolin minerals.

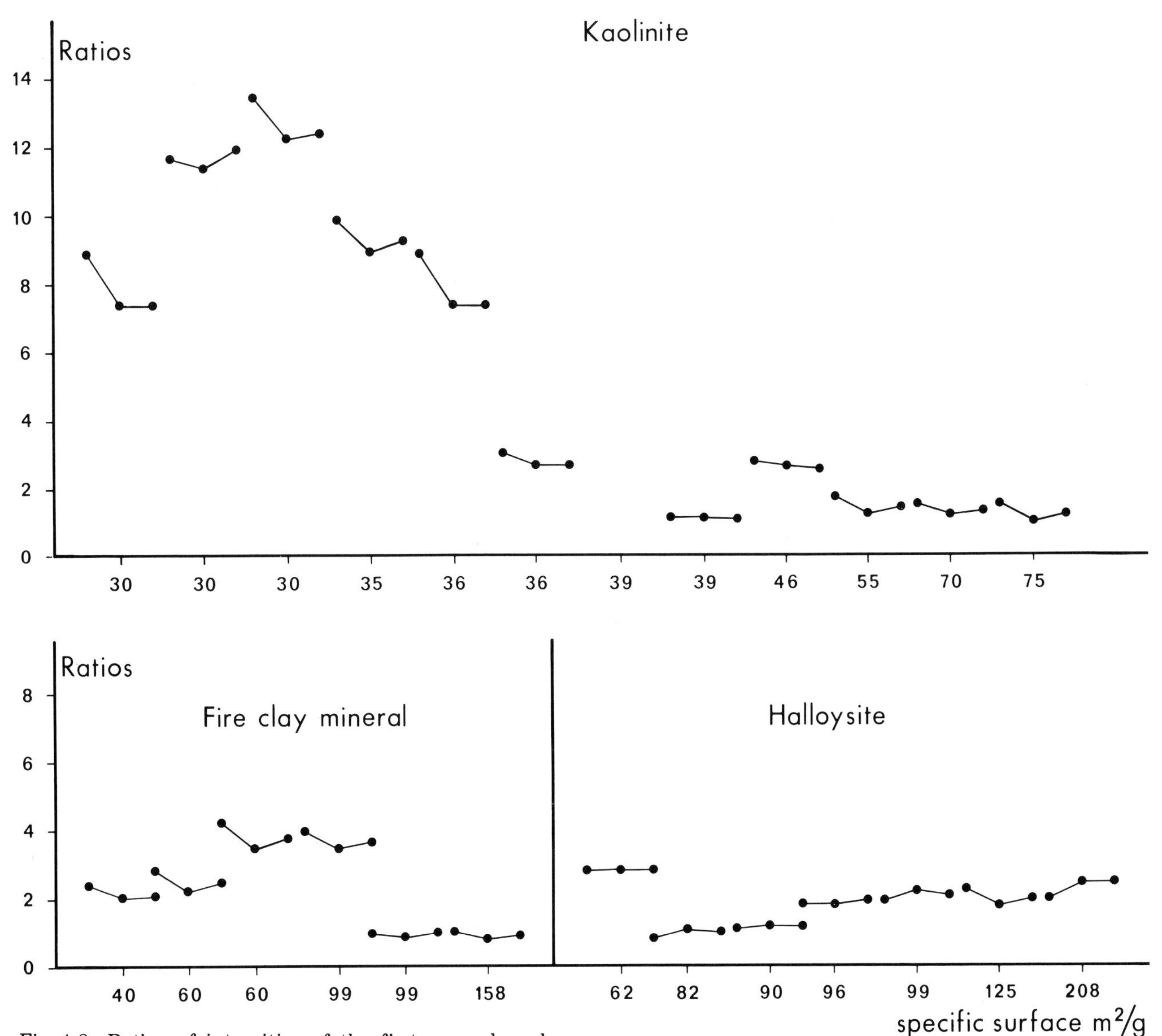

Fig. 4.8. Ratios of intensities of the first, second, and third O-H (E 3695, 3660, 3620 cm^{-1}) and O-D (E 2725, 2698,2672 cm^{-1}) stretching bands of deuterated kaolinite, fire clay mineral and halloysite from various origin and their specific surface (m^2/g). (Aethylene glycol method.)

In the case of free O-H stretching vibrations this will be $\tilde{\nu}_{OH}/\tilde{\nu}_{OD} = \sqrt{1.89} = 1.37$. For O-OH/O-OD vibrations it is $\sqrt{1.028} = 1.1014$ and for Al-OH/Al-OD it is $\sqrt{1.035} = 1.017$.

However, mostly 1.355—1.356 is found for clay minerals, but for antigorite chlorites and limonite and boehmite ca. 1.34.
Dickite and nacrite, coarse (even grinded) and well-ordered minerals, are bad to deuterate. The spectra of the deuterated kaolinite, fire-clay mineral and halloysite, clearly demonstrate that the surface OH groups (ca. 3695 cm^{-1}) of kaolinite, are the least deuterated and the inner OH between the kaolinite plates (ca. 3660 m^{-1}) the most. The octahedral O-H pointing to the empty inner holes (ca. 3620 cm^{-1}) are in between (Fig. 4.7). For the finer kaolinite samples, the fire-clay mineral and the halloysite mineral, the differences are smaller. The ratios of the intensities E_{OH}/E_{OD} of the three O-H stretching bands are the highest for the coarser kaolinite samples. They are the smallest for the finer fire-clay minerals and the halloysites (Fig. 4.8).
The above phenomena may be ascribed to a decrease in the grade of ordering of the atoms in the crystal and thus levelling their activities against OD exchange when passing from well-crystallized coarse kaolinite to less-ordered fire-clay mineral and to badly-ordered halloysite. Antigorite is badly deuterated, hydroantigorite and chrysotile much better (Fig. 4.7). Coarse, well-crystallized sericite, biotite, phlogopite, margarite, paragonite and lepidolite, although finely

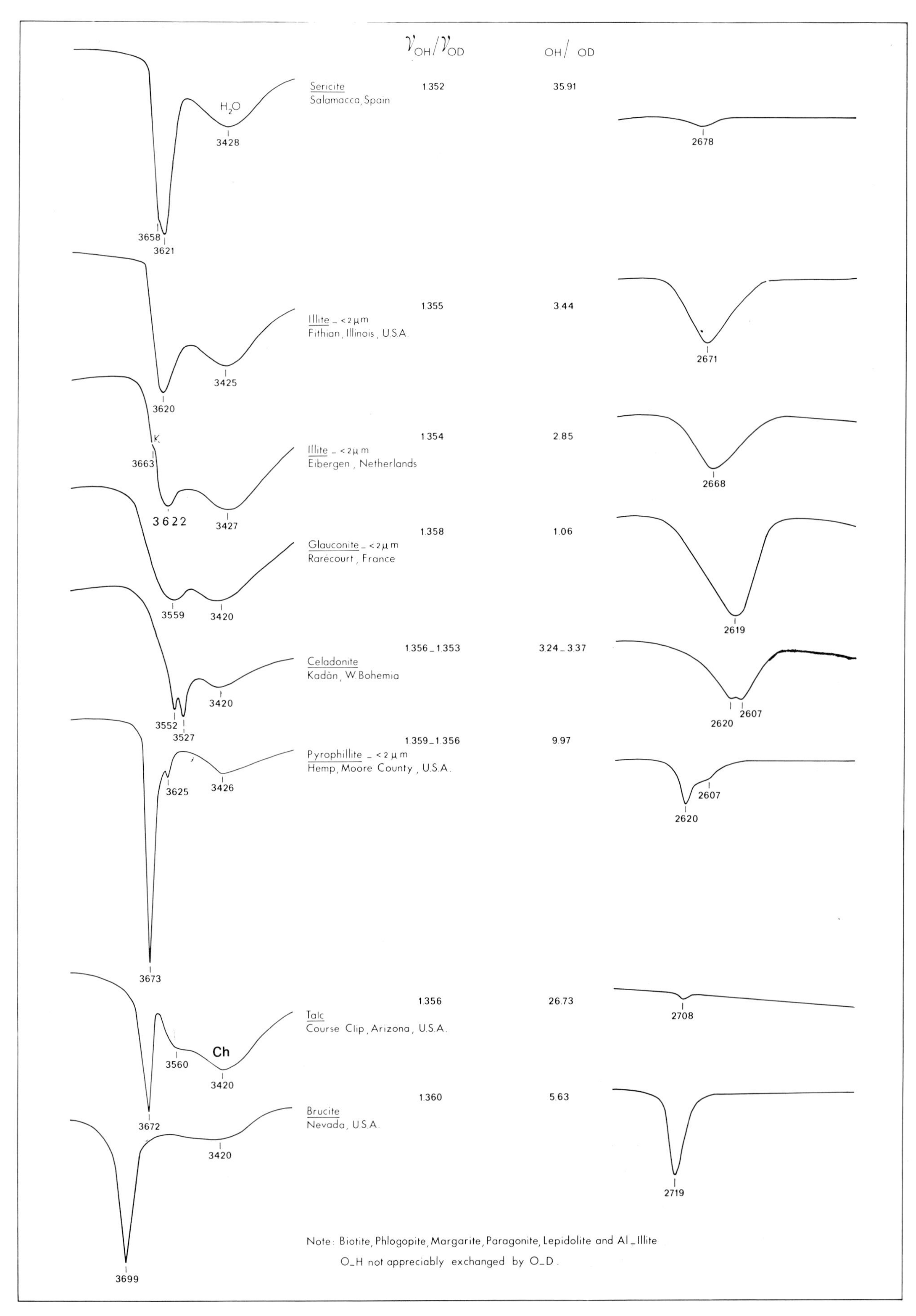

Fig. 4.9. Infrared spectra and ratios of wave numbers $\tilde{\nu}_{OH}/\tilde{\nu}_{OD}$ and intensities of corresponding OH and OD vibrations of deuterated mica minerals, pyrophyllite, talc and brucite. K = kaolinite; Ch = chlorite.

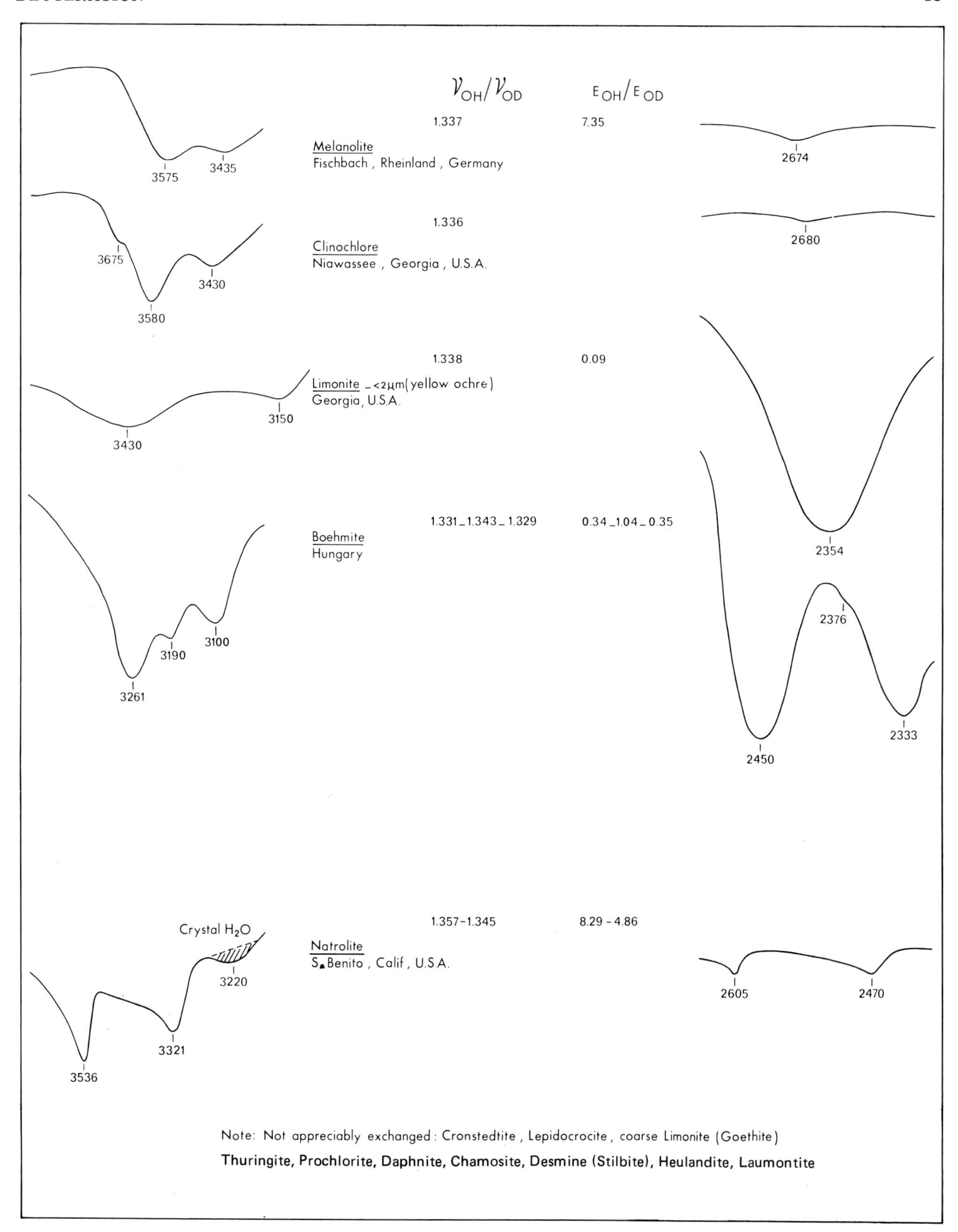

Fig. 4.10. Infrared spectra and ratios of wave numbers $\tilde{\nu}_{OH}/\tilde{\nu}_{OD}$ and intensities of corresponding OH and OD vibrations of some deuterated chlorite minerals, limonite, boehmite, and some zeolites.

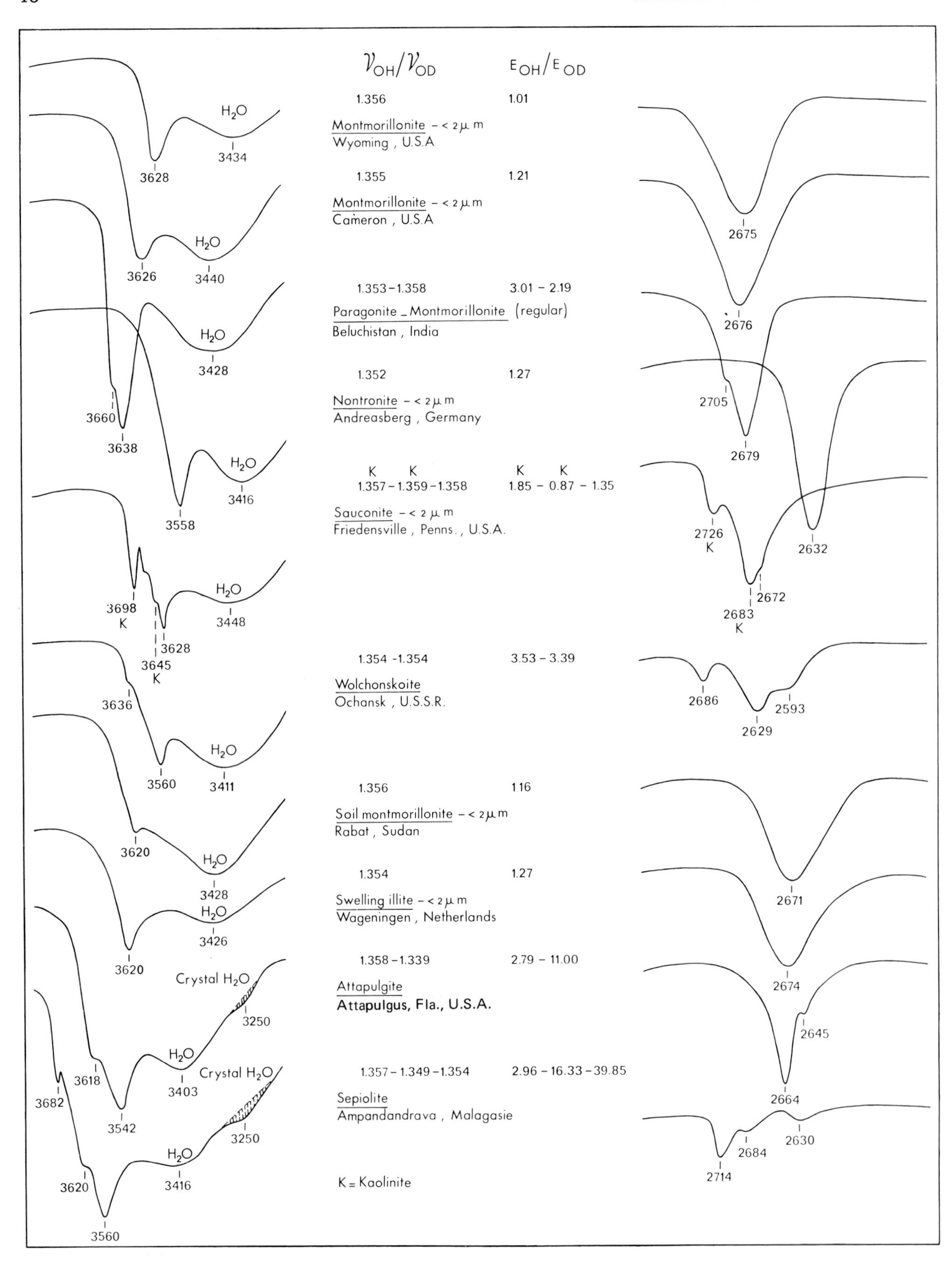

Fig. 4.11. Infrared spectra and ratios of wave numbers $\tilde{\nu}_{OH}/\tilde{\nu}_{OD}$ as well as of intensities E_{OH}/E_{OD} of corresponding OH and OD vibrations of several deuterated swelling minerals, attapulgite and sepiolite.

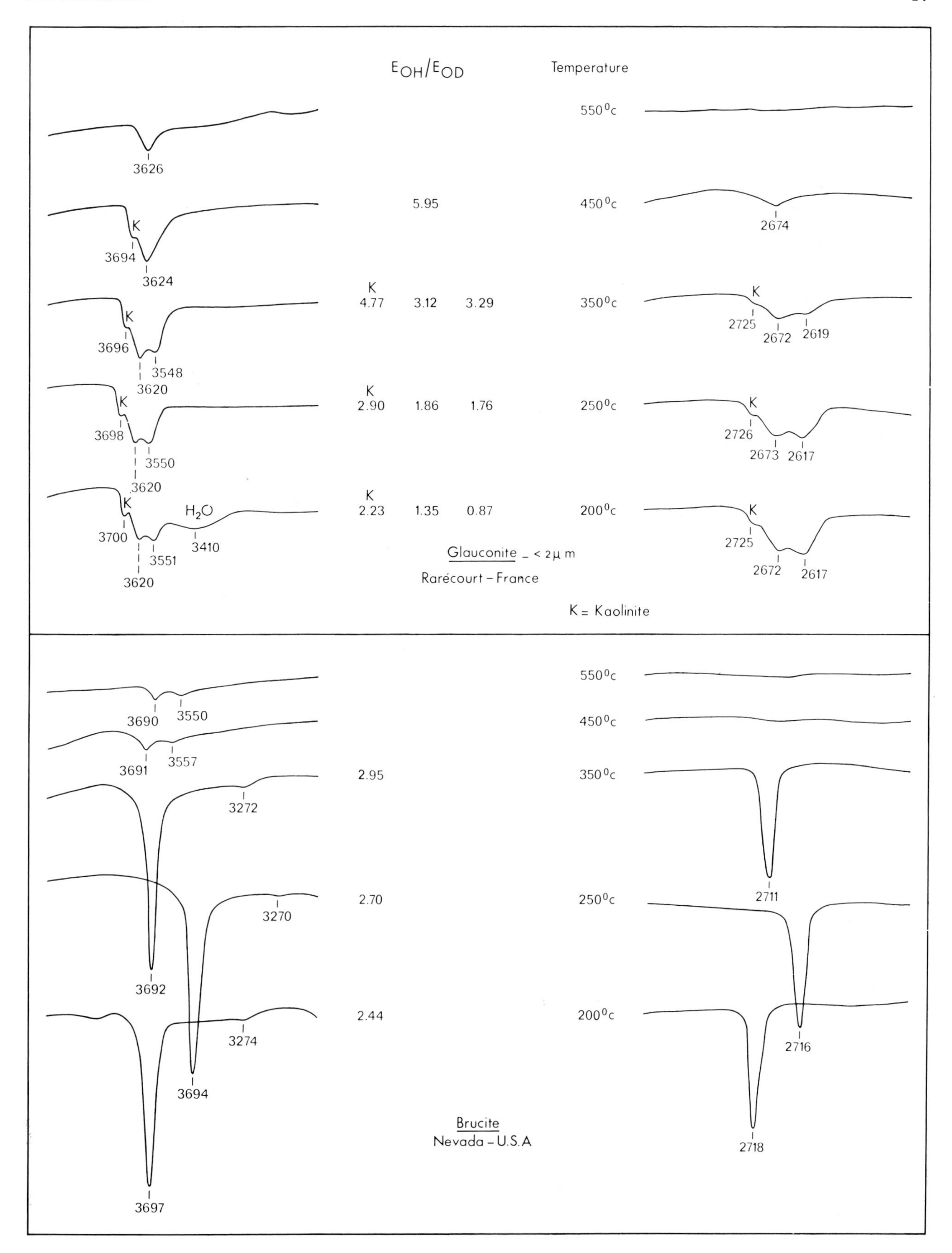

Fig. 4.12. Infrared spectra and ratios of intensities of corresponding OH and OD vibrations of deuterated glauconite and brucite heated at different temperatures for 1 hour.

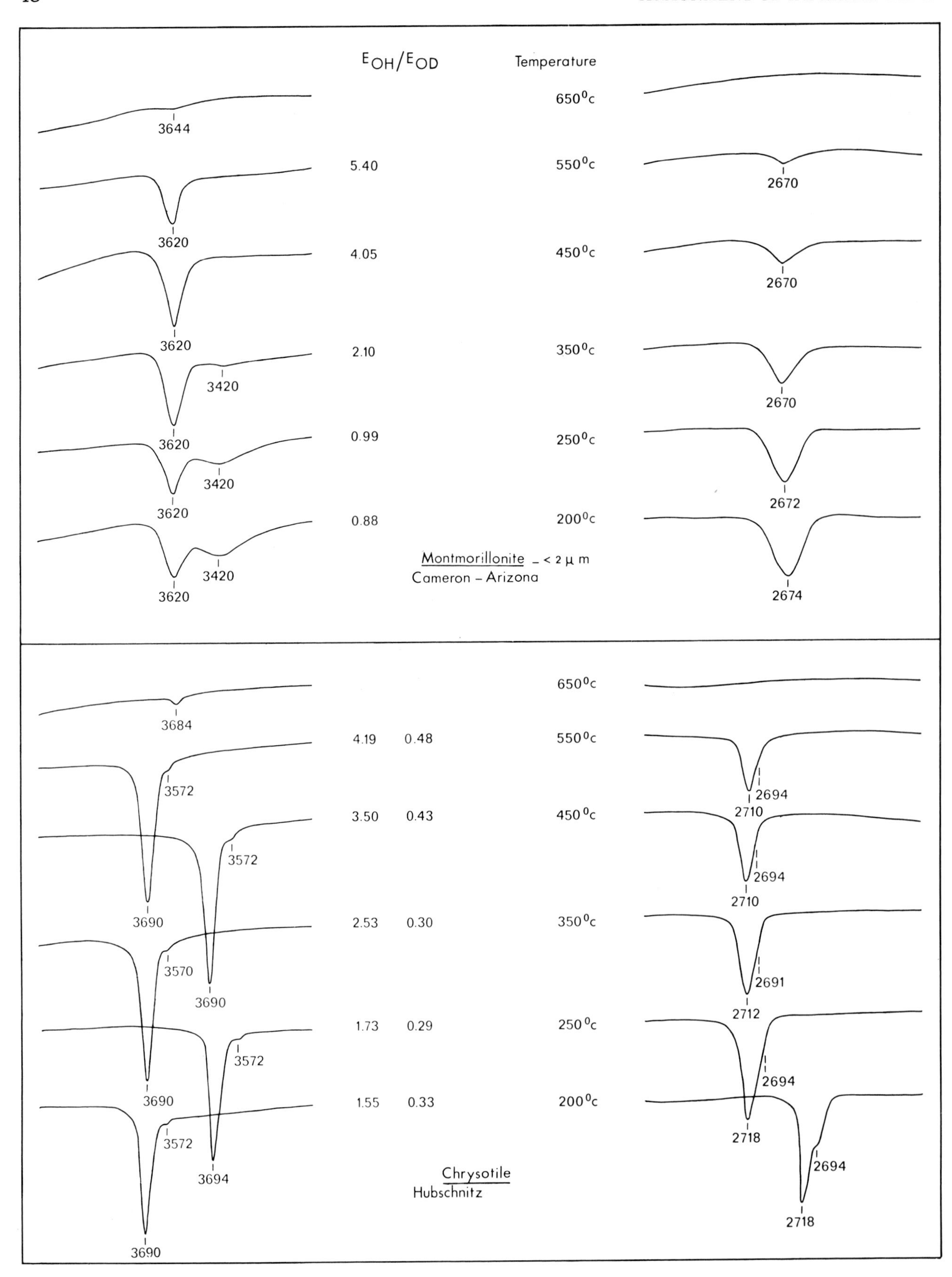

Fig. 4.13. Infrared spectra and ratios of intensities of corresponding OH and OD vibrations of deuterated montmorillonite and chrysotile heated at different temperatures for 1 hour.

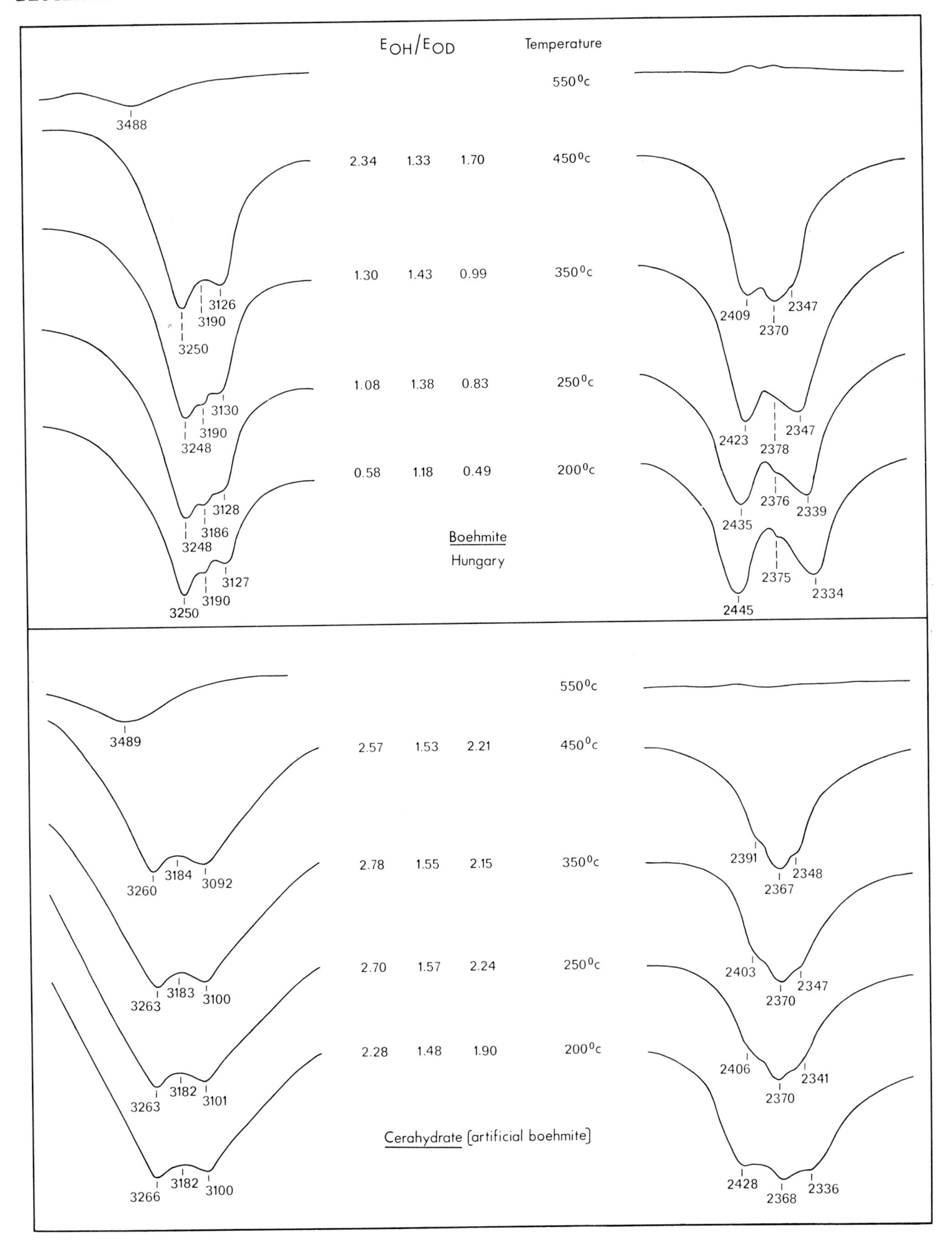

Fig. 4.14. Infrared spectra and ratios of intensities of corresponding OH and OD vibrations of deuterated boehmite and cerahydrate (artificial boehmite) heated at different temperatures for 1 hour.

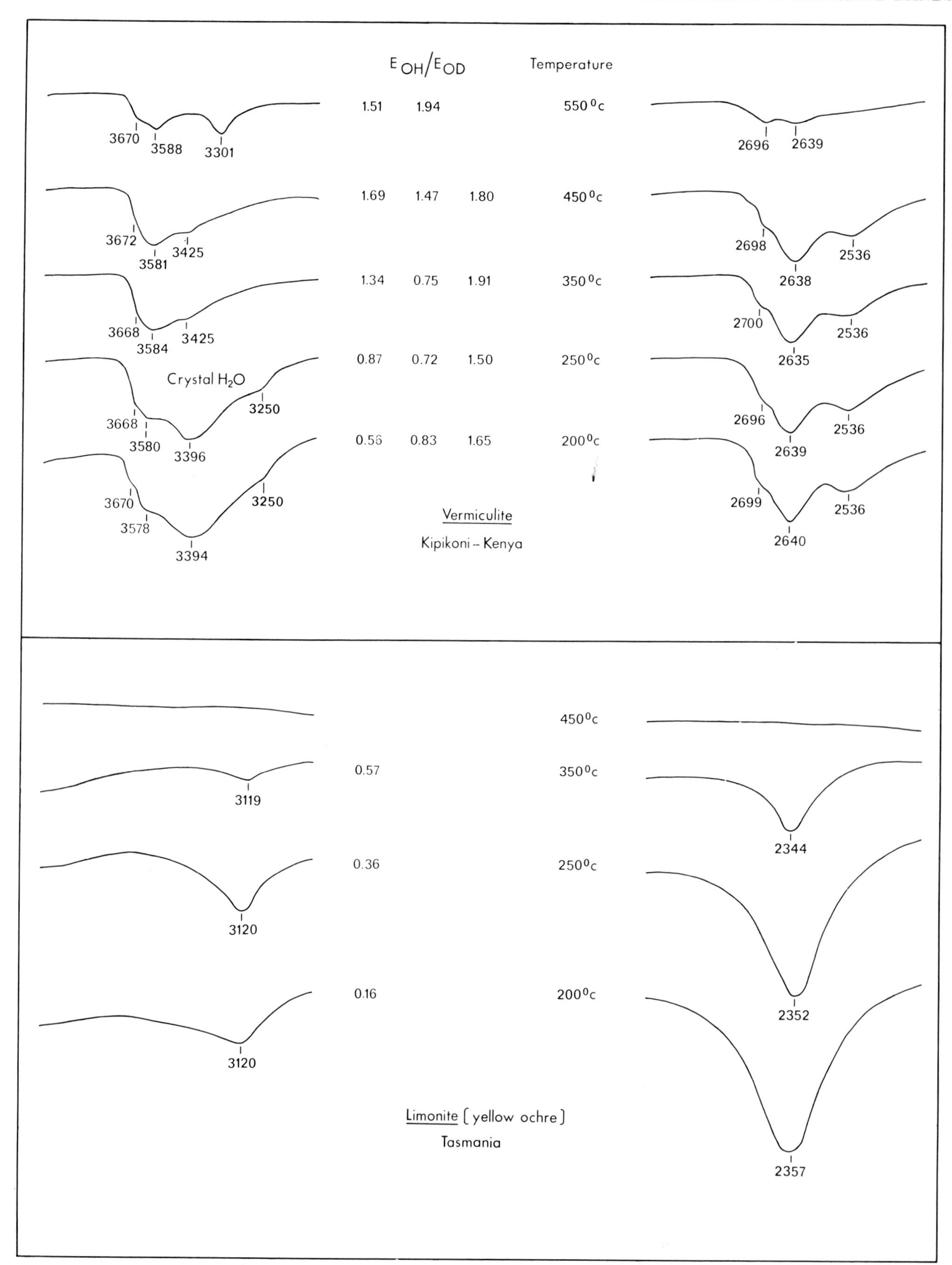

Fig. 4.15. Infrared spectra and ratios of intensities of corresponding OH and OD vibrations of deuterated vermiculite and limonite heated at different temperatures for 1 hour.

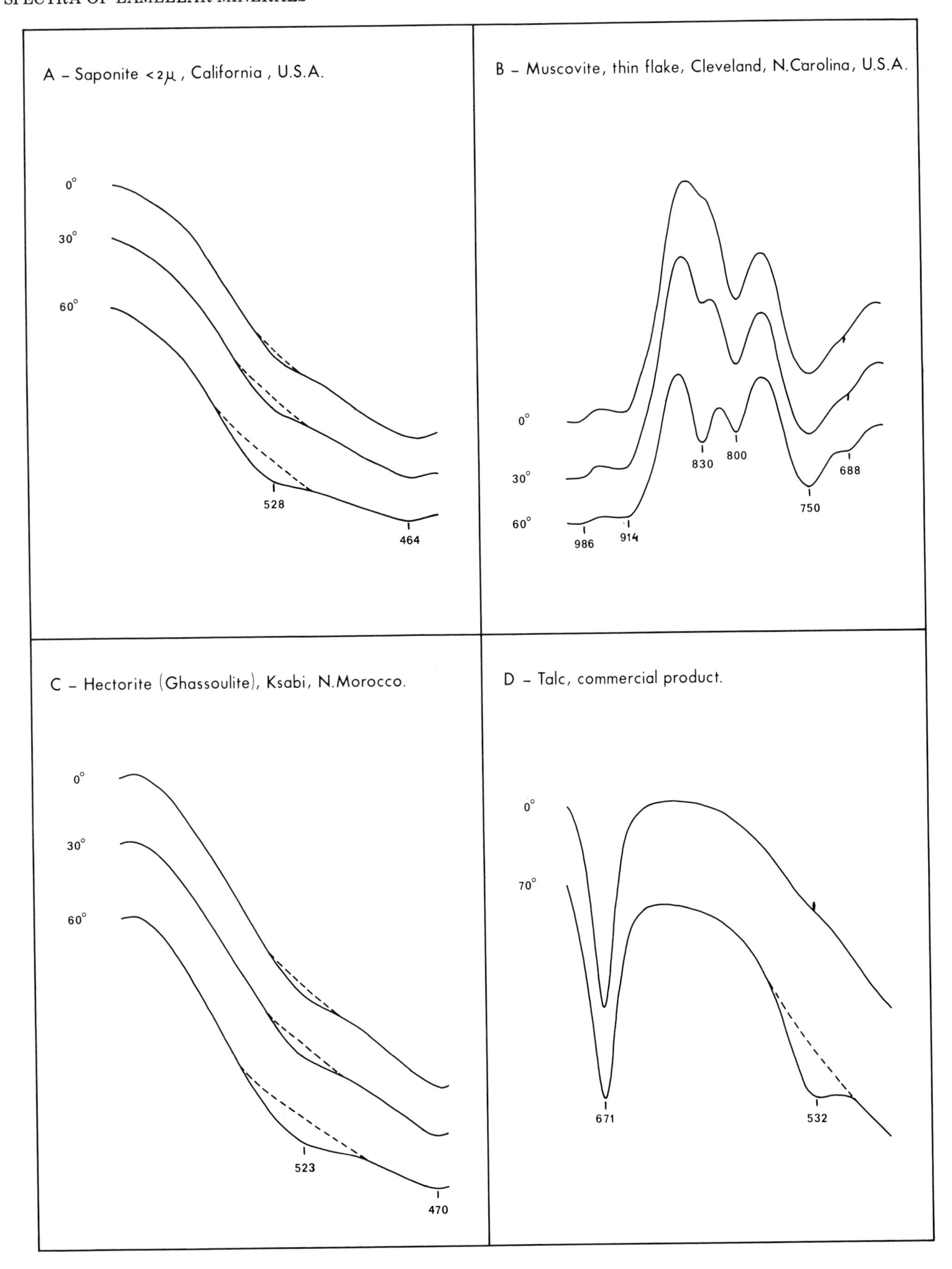

Fig. 4.16. Infrared spectra of some lamellar minerals at various inclinations of the flake. Lamellar orientation obtained by deposition of a suspension of the mineral in ethyl alcohol on a KBr disk.

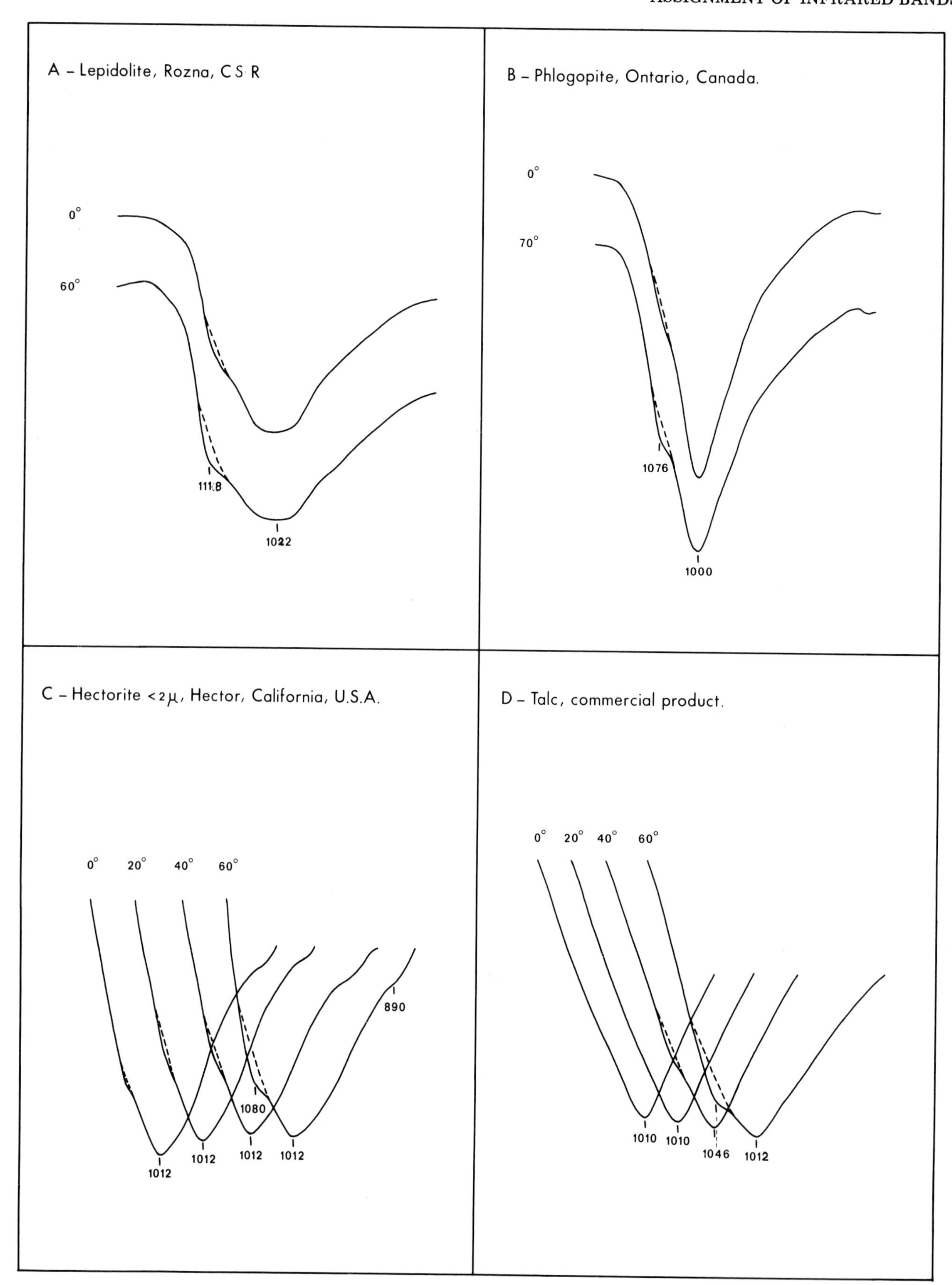

Fig. 4.17. Infrared spectra of some lamellar minerals at various inclinations of the flake. Lamellar orientation obtained by deposition of a suspension of the mineral in ethyl alcohol on a KBr disk.

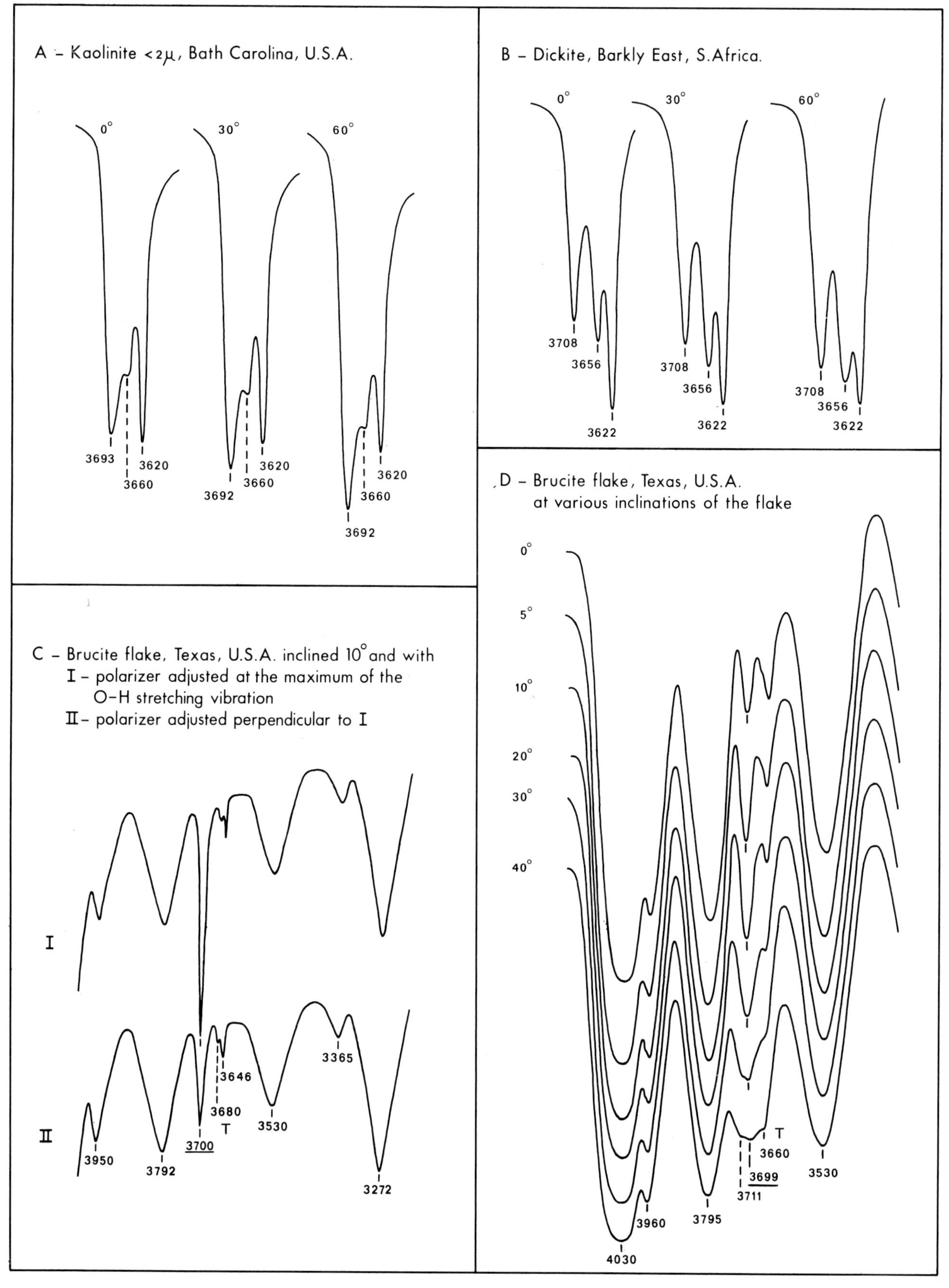

Fig. 4.18. Infrared spectra of some lamellar minerals at various inclinations of the flake. A.B. Lamellar orientation obtained by deposition of a suspension of the mineral in ethyl alcohol on a KBr disk. C. Infrared spectra of a brucite flake with polarized light. (Courtesy P. Krohmer.) D. Infrared spectra of a brucite flake at various inclinations of the flake. Broad bands are caused by interference effects. T = talc.

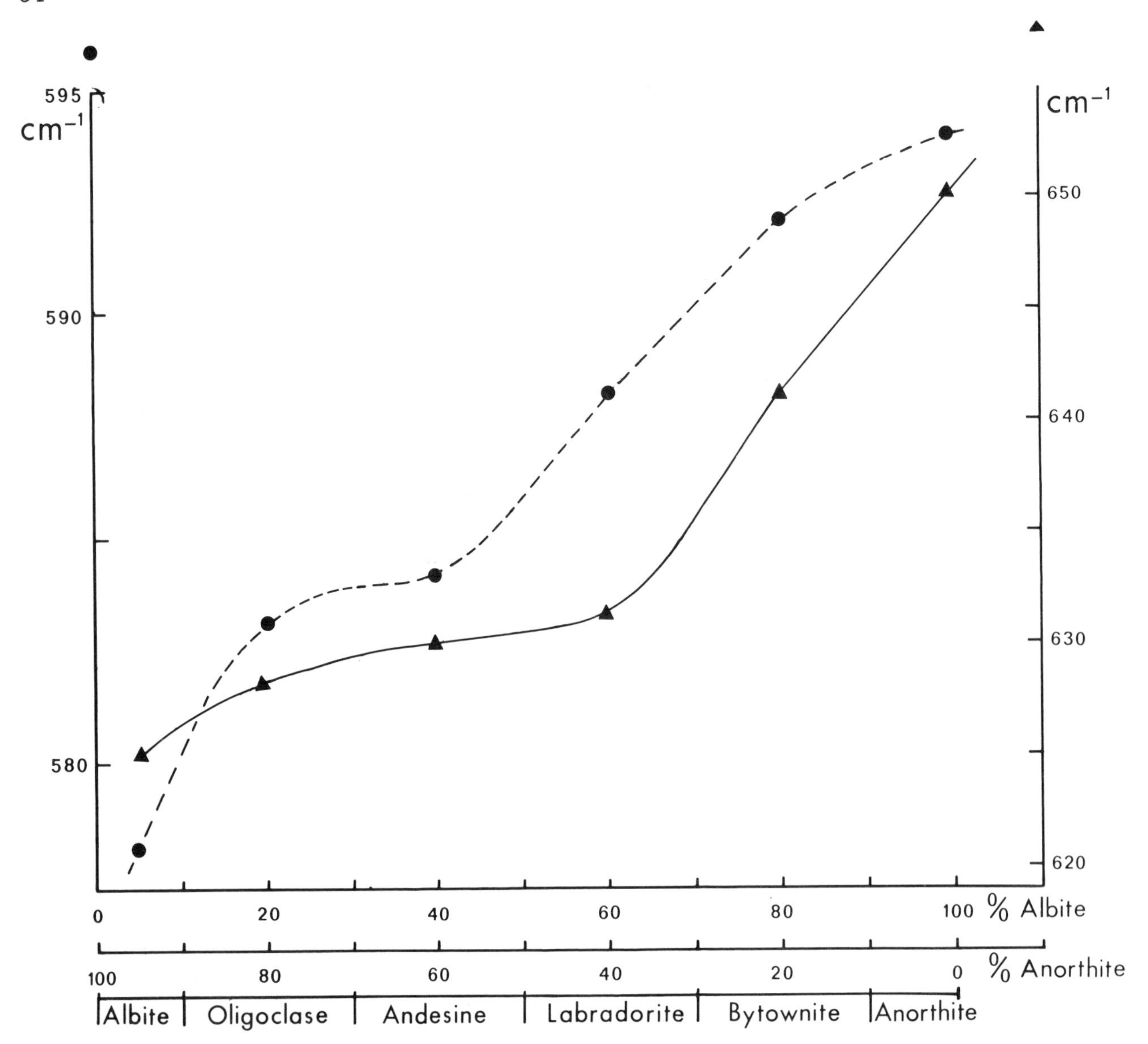

Fig. 4.19. Infrared bands of feldspar series in cm^{-1} in relation to albite/anorthite contents for 578-594 cm^{-1} (●) and 625-650 cm^{-1} (▲) regions.

powdered before, are very badly deuterated (Fig. 4.9). Illite is better deuterated but the coarse Al-illite from Sarospatak very badly. Glauconite $< 2\ \mu m$ is well-deuterated but celadonite less. Pyrophyllite, brucite and especially talc are very badly deuterated. The same holds for the chlorite minerals. Fine crystallized limonite $< 2\ \mu m$ is very easily deuterated but not coarse crystallized goethite, cronstedtite and lepidocrocite. Boehmite is easily deuterated but natrolite and other zeolites moderately to badly (Fig. 4.10). Minerals of the smectite group (montmorillonite, nontronite, sauconite) are all easily deuterated. But wolchonskoite and the regular paragonite-montmorillonite are somewhat less affected (Fig. 4.11). Also soil montmorillonite and swelling illite are easily deuterated. Attapulgite and sepiolite are moderately deuterated, but their inner O-H groups badly (Fig. 4.11).

Reviewing the results of all the analyses it may be concluded that in particular resistant minerals are badly deuterated and minerals with a loose, easily accessible disordered structure, the easiest. For boehmite, glauconite, limonite and montmorillonite there is, when heated, a large decrease in the OD contents of the mineral as compared with the O-H — see the increase in the E_{OH}/E_{OD} quotients when the temperature increases (Fig. 4.12 through 4.15). For chrysotile and vermiculite it is smaller and for artificial boehmite and brucite it is the least.

Natural and artificial boehmite, when deuterated, show an appreciable shift in their vibrations as compared to those of the non-deuterated samples (between brackets = non-deuterated): boehmite = 3250, 3190, 3127 (3280, absent, 3090); boehmite artificial = 3266, 3182, 3100 (3275, absent, 3080).

Electric vector of the incident rays and the transition moment of the vibration

This method is used to get information about the orientation of atoms or molecules in crystals or in long-chain organics. Thus it may reduce appreciably the number of possible structures for a certain sample. This holds in particular for the study of the structure of high polymers where many configurations may be possible.
When the electric vector (E) of the radiation makes an angle θ with the direction of the transition moment (M) of the vibration, the absorption is proportional to $\cos^2 \theta$. Thus maximum absorption occurs when the transition moment of a certain vibration and the electric vector are parallel and the frequency of the vibration corresponds with the resonance frequency. Zero absorption will occur when the two vectors are perpendicular to each other. In this way the intensity of the O-H vibration of the inclined OH groups in the dioctahedral muscovite crystal is only slightly increased by change of the electric vector. Such in contrast with the perpendicular to the cleavage flake-oriented octahedral OH groups in trioctahedral biotite, which show a large increase in intensity when the direction of the lamellar-oriented particles is changed (dichroism).
Polarized radiation can be obtained by setting a pile of very thin transparent parallel oriented films so that the incident beam strikes the surface at the Brewster angle $\tan \alpha = n$ (n = refractive index). Polarizers have been made using selenium (α = 68°), AgCl (α = 63.5°), KRS-5 (α = 67°), Tellur (α = 78°), Germanium (α = 76°) and polyethylen (α = 55.5°).
Polarized radiation can also be obtained by the use of a AgBr wire-grid polarizer. This new type of polarizer has some advantages over the Brewster type.
When the sample is first oriented and then turned vertical against the path of the beam, it can be investigated on diochroism. Orientation can be obtained by using a lamellar flake of the layer mineral to be investigated or by first deposition of a suspension of the fine mineral powder in aethylalcohol on a KBr disk, which afterwards is evaporated leaving a thin film on the disk. Because of polarization effects of the light by spectrographic optics, in particular by grating monochromators, measurements have to be made carefully. Another objection is that the angle of crystal inclination can not be increased to more than 40—50° without loss of light.

In Fig. 4.16, 4.17 and 4.18 are some examples. They clearly show the electric vector effect on the transition moment of the mineral vibration. From this can be obtained the orientation of a certain atom or a certain atomgroup in the crystal.
The mineral flake of brucite shows besides the electric vector effect of its Mg...O-H vibration at 3700 cm^{-1} also several other bands. They result from interferences because of reflections at the external surface of this coarse material.

Feldspars

The plagioclase minerals form an isomorphous series from albite (Ab) = $NaAlSi_3O_8$ to anorthite (An) = $CaAl_2Si_2O_8$ by substitution of Na and Si by Ca and Al. Disorder increases in the same direction because of the replacement of the smaller Si (r = 0.38 Å) by the larger Al (r = 0.45 Å).
In Fig. 4.19 are the wave numbers for the 594 and 650 cm^{-1} band of albite moving to 578 and 625 cm^{-1} respectively for anorthite. There is a discontinuity at ca. 30—50% albite (70—50% anorthite). (See also Thompson, C.S. and Wadsworth, M.E., 1957: Determination of the composition of plagioclase feldspars by means of infrared spectroscopy. Am. Mineralogist, 42: 334—341.)

Carbonates

The carbonates may structurally be divided into three groups:

calcite: calcite (Ca), magnesite (Mg), smithsonite (Zn), siderite (Fe), rhodochrosite (Mn), octavite (Cd), cobaltite (Co);
dolomite: dolomite (Ca,Mg), ankerite (Ca,Fe), huntite ($CaMg_3$);
aragonite: aragonite (Ca), strontianite (Sr), witherite (Ba), cerussite (Pb),

which is also the order of decreasing symmetry, e.g., ditrigonal skalenoedric, trigonal rhomboedric and ortho-rhombic dipyramidal.
The third group has the heavier cations and the two former groups the lighter ones. Calcium (radius = 0.99 Å) stands between them and occurs in two modifications: calcite and aragonite. The trigonal double-carbonate minerals (dolomite group) consist of ditrigonal-like layers.
Radius, mass and electro negativity of the single cations are factors which all influence the wavelength (frequency) of the CO_3^{2-} ion vibration in the carbonate minerals. Another factor is the structural arrangement (coordination) of the other components in the crystal against the

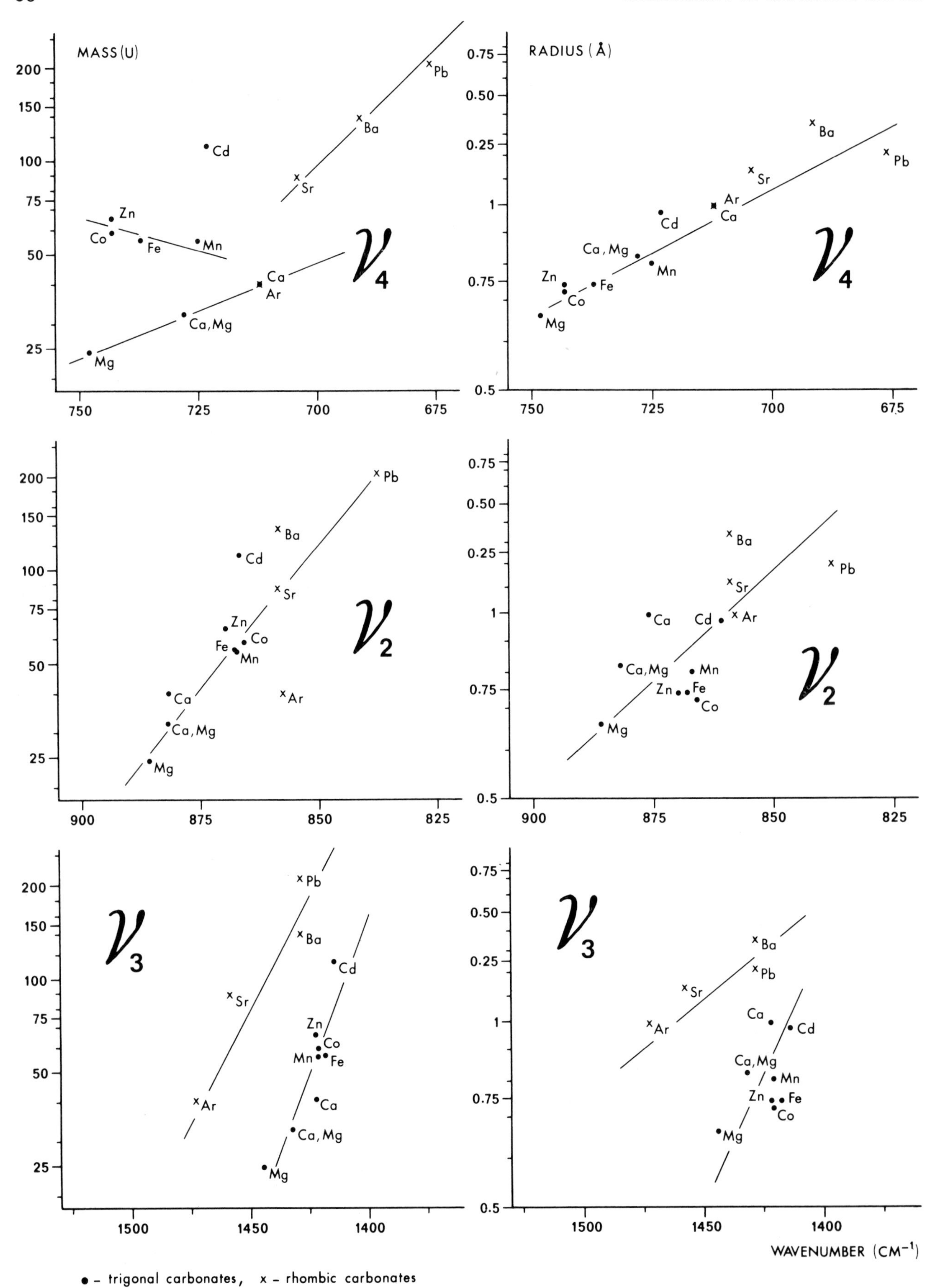

Fig. 4.20. Wavenumbers of ν_3, ν_2 and ν_4 vibrations of CO_3^{2-} ions of various carbonates, their mass and radius.

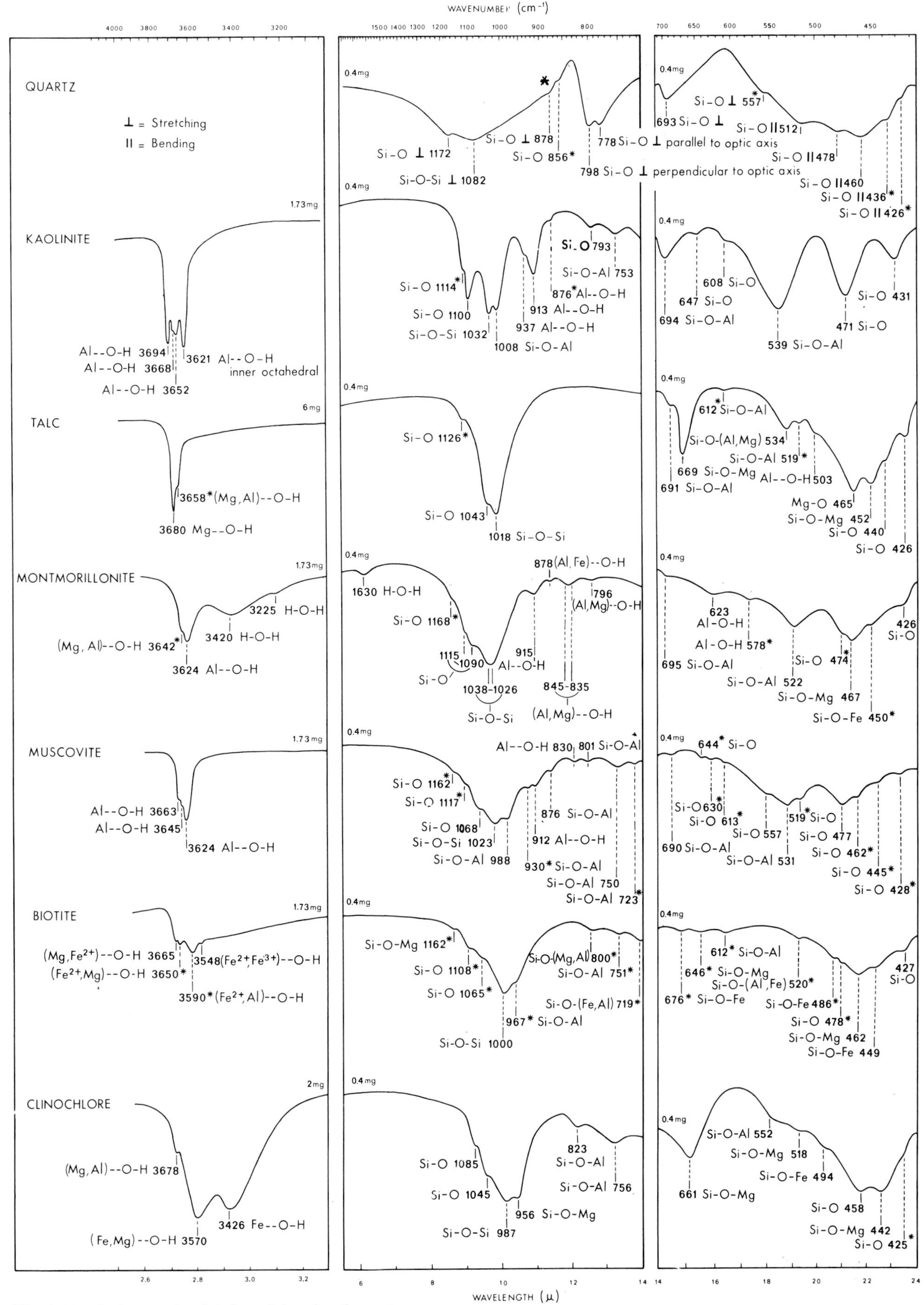

Fig. 4.21. Assignment of infrared bands of most common clay minerals (according to literature and own observations).

CO_3^{2-} ion. Consequently, the environment of a certain mode of vibration may be quite different for the various carbonate minerals.
Fig. 4.20 and Table 7.I (see p. 242, Carbonates) contain the results of the ν_2 (non-degenerate out of plane bending), ν_3 (doubly degenerate asymmetric stretching) and ν_4 (doubly degenerate planar bending) vibration of various carbonate minerals in relation to the mass or the radius of their cations. There exists a correlation between the two but this is not valid for all carbonate groups. The radius effect is stronger than the mass effect. (See also Elderfield, H., 1971: The effect of periodicity on the infrared absorption frequency ν_4 of anhydrous normal carbonate minerals. Am. Mineralogist, 56: 1600—1606.)

Clay minerals

The indentification of the infrared bands in clay minerals remains very difficult because of their complicated and non-constant composition. The weak intensities which are commonly found in this category of minerals and which are caused by substitutions and badly ordering, form another obstacle. In general: stretching vibrations need a higher energy than bending vibrations. Consequently the former will be found in the higher wave number part of the infrared spectrum and the latter in the lower part. The presence of a foreign ion in some of the tetrahedra will distort both the structure of neighbouring tetrahedra and thus the form and frequency of the original band. In Fig. 4.21 are the IR spectra of some of the most investigated minerals compiled from literature and according to own observations. By different structural arrangements (quartz, cristobalite, tridymite) or isomorphous substitution (montmorillonite (Al), saponite (Mg), nontronite (Fe) etc.), the spectra are significantly changed. Differences in the distance of the ions and in their mass, have an opposite effect on the position of the bands. When the direction of the vibration stays perpendicular to the direction of the incident light, the band is of lower intensity, etc.
Much experimental work on OH-OD exchange, dichroism, heating and chemical relations is still needed to get better information about the assignment of the bands also for the other minerals.

5. QUANTITATIVE ANALYSIS

The intensity of an absorption band is proportional to the square of the change in the electronic configuration of the molecule, e.g., in the total dipole moment (transition moment). The latter is the sum of an ionic, a polar and an inductive component.

Symmetric vibrations will be inactive. Broad bands are caused by coupling of the motions with those of neighbouring molecules. The intensity also depends on the direction of the electric vector of the incident radiation against the transition moment. Moreover, the intensity may change with particle size and the differences in refractive index between the sample and the embedding medium because of scattering effects. The following relation exists when monochromatic radiation passes through an inert solution in which a substance is dissolved:

$$I = I_0 \exp(-acl)$$

(law of Bouguer, 1729; Lambert, 1760 and Beer, 1852; for details Perkin, 1948).

I = intensity of transmitted light, I_0 = intensity of incident light; c = concentration mole/l; l = length of sample cell (cm); a = absorption coefficient which gives when going from Naperian to Briggsian logarithms:

Absorbance $(A) = \log I_0/I = 0.4343\ acl = \epsilon cl$.

where ϵ = molar extinction coefficient ($cm^2\ mole^{-1}$) = $1/cl \log I_0/I$.

For a system in which solid particles of the substances are dispersed in a non-absorbing fluid medium or pressed with KBr to a transparent pellet, the following relation exists (Duyckaerts, 1955, 1959; Otvos et al., 1957; Bonhomme, 1955; and Bonhomme and Duyckaerts, 1955):

$$E \text{ (extinction)} = A \text{ (absorbance)} = \log(I_0/I)$$

$$= -\frac{m}{KS\rho d} \log[(1-K) + K\theta]$$

$$= \frac{m}{KS\rho d} \times \frac{Kkd}{2.303...} = \frac{m}{S\rho} \times \frac{k}{2.303}$$

which relation is the same as that of the Bouguer-Lambert-Beer law when for $m/\rho = cl$ and for $k/S \times 2.303... = \epsilon$ is written.

m = mass of the sample dispersed in the KBr disc; ρ = density of the particles; d = cubic length of the cubic particles; S = surface area of the disc normal to the beam; K = the geometrical fraction of the surface normal to the incident rays that is covered by the particles; θ = transmittance of the particles, for cubes = e^{-kd}; for spheres = $2(1-(kd+1)e^{-kd})/(k^2d^2)$; k = absorption coefficient:

$$\log[(1-K) + K\theta] = -\frac{Kkd}{2.303}$$

Suppose:

$$K(1-\theta) = x\,, \qquad (1)$$

then:

$$\log[(1-K) + K\theta] = \log(1-x) \qquad (2)$$

When expanding $\ln(1-x)$ in a series:

$$\ln(1-x) = -x - \frac{x^2}{2} - \frac{x^3}{3} \ldots$$

and neglecting higher terms than the first, results:

$$\ln(1-x) = -x \qquad (3)$$

For cubes $\theta = \exp(-kd)$ which introduced in (1) yields:

$$-x = -K[1-\exp(-kd)] \qquad (4)$$

Expanding $\exp(-kd)$ in a series yields:

$$\exp(-kd) = 1 - kd + \frac{(kd)^2}{2} - \frac{(kd)^3}{3} \ldots . \qquad (5)$$

When third and higher terms are neglected (4) becomes:

$$-x = -K(1-1+kd) = -Kkd \qquad (6)$$

So equation (2) can be written:

$$\ln[(1-K) + K\theta] = \ln(1-x) = -Kkd \qquad (7)$$

or when Briggsian logarithms are used instead of Naperian:

$$\log[(1-K) + K\theta] = -\frac{Kkd}{2.303}$$

The Bouguer-Lambert-Beer law is restricted to monochromatic light, non-interacting absorbers, very dilute solutions (no appreciable change in the refractive index), absence of a Christiansen effect, particle size smaller than the used wavelength, absence of scattering, linearity of the detector and the assumption that the slits are infinitely small.

Absorption bands suited for quantitative analyses should be of medium to high intensity and not overlapped by neighbouring bands of other minerals.

The concentration at which the error of E is a minimum in respect to the concentration is found as follows (Willard et al., 1958, pp. 47–48):

$$\frac{I}{I_0} = \exp(-2.303\epsilon cl) \qquad (1)$$

$$dI = I_0 \exp(-2.303\epsilon cl)\,(-2.303)\epsilon l dc \qquad (2)$$

or when A (absorbance) = $\log I_0/I = \epsilon cl$ is introduced:

$$dI = -2.303 \exp(-2.303\epsilon cl)\, I_0 A \frac{dc}{c} \qquad (3)$$

Introduction of R = the ratio of the relative concentration error to the relative photometric error in (3) yields:

$$R = \frac{dc/c}{dI/I_0} = \frac{-0.4343 \exp(2.303\epsilon cl)}{A} \qquad (4)$$

The concentration giving a minimum error is found by setting the derivative of R with respect to c equal to zero:

$$\frac{dR}{dc} = \frac{d}{dc}\,\frac{-0.4343 \exp(2.303\epsilon cl)}{\epsilon cl}$$

$$= \frac{\exp(2.303\epsilon cl)}{c}\left(\frac{1}{2.303\epsilon cl} - 1\right)$$

$$= \frac{\exp(2.303\epsilon cl)}{c}\left(\frac{0.4343}{A} - 1\right) \qquad (5)$$

So $dR/dc = 0$ when $0.4343/A - 1 = 0$ or when $A = 0.4343$ which is equal to a transmittance (transmission) of 36.8%.

Correction of maximum extinction for background absorption (base line) is made according to the Wright (1941) and Heigl et al. (1947) procedure (Fig. 5.1). Fig. 5.2 shows examples representing the straight correlation which exists between absorbance and concentration for some minerals, following the laws of Bouguer, Lambert and Beer.

In Fig. 5.3 and 5.4 absorbances of various kaolin minerals are plotted against their specific surface. Although the particle sizes are in this case smaller than the wavelength of the used radiation, there is a decrease of the intensity when particle size decreases. It is caused by increased disorder of the finer particles and increased amorphous (IR-inactive) Beilby layer on the surface of the particles. The particle-size effect is the smallest for the inner Al--O-H vibrations = at 3622 cm^{-1}. The other bands 3694, 793, 753, 694 cm(;1 are caused by outer Al--O-H, Si-O, Si-O-Al and Si-O-Al vibrations, respectively.

The integrated adsorption intensity A of a band between the two frequencies (ν_2 and ν_1) is:

$$A = \int_{\nu_1}^{\nu_2} \epsilon_\nu d\nu = \frac{1}{cl}\int_{\nu_1}^{\nu_2} \log (I_0/I)\, d\nu$$

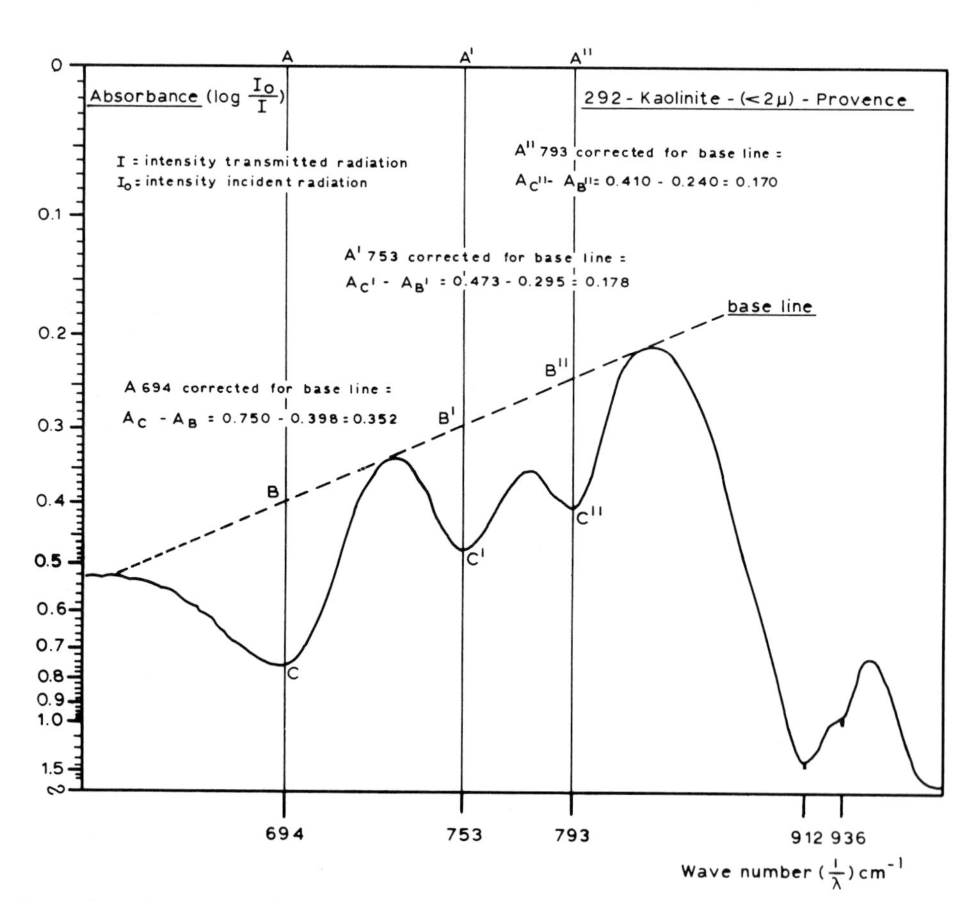

Fig. 5.1. Correction of maximum absorbance for background absorption (base line). (After Wright, 1941, and Heigl et al., 1947.)

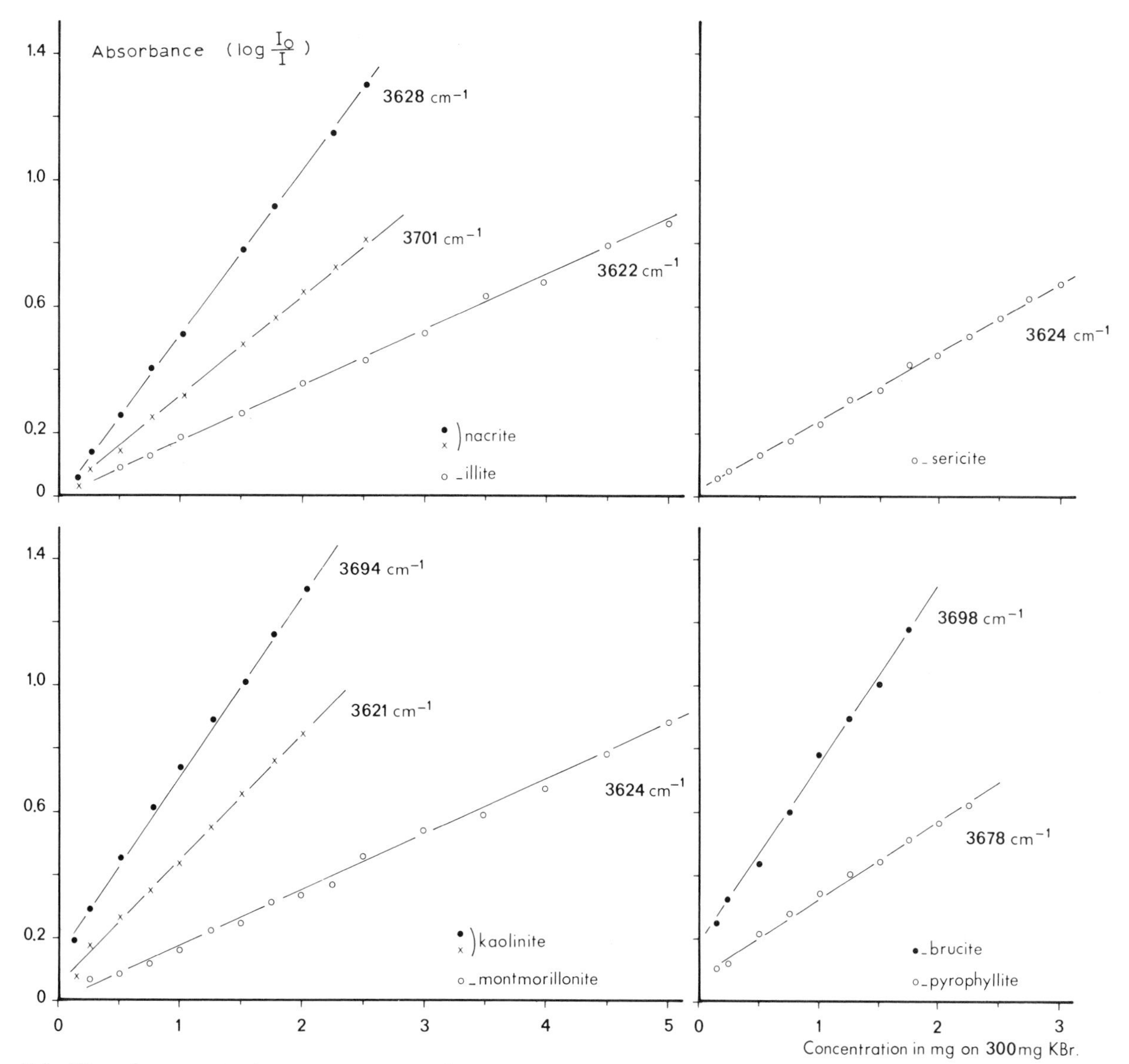

Fig. 5.2. Mineral concentration in mg on 300 mg KBr and absorbance corrected for base line: Lambert-Beer law.

The measurement of the integrated intensity of an absorption band provides a closer approach to the real intensity of the vibrational movement than peak intensities, such in particular for broad, asymmetric bands. It furthermore greatly reduces the error due to finite slit width. Calculation by graphical way is very tedious, but modern instruments can be provided with logarithmic integrators with a high degree of accuracy. A useful approximation can be obtained by multiplying peak height × width at half height. If a double-beam instrument is used with always the same slit width, pellet surface and excess of matrix, the addition of a standard mineral is not needed unless the absorption law is not followed. An advantage of the standard additive method is that the effect of material losses during the preparation of the sample is annihilated. Standards giving sharp bands are needed in the mull method. The most frequently used are $CaCO_3$, polystyrene, polyethylene, Na-azide, dl-alanine, K-cyanide, Pb-thiocyanate, K-thiocyanate. The bands of a mineral are not always isolated but may overlap each other. By the derivative method it is possible to get a better differentiation in the spectrum, and the differential (compensation) method enables to compensate the bands of other minerals in the mixture. For that purpose the disturbing mineral is put as much in the reference beam that the disturbing band is compensated totally.

This method can also be used to determine quantitatively the concentration of a component in a mixture when increasing amounts of that component are put into the reference beam. The concentration of the mineral in the sample is the same as in the reference beam when the band is compensated totally.

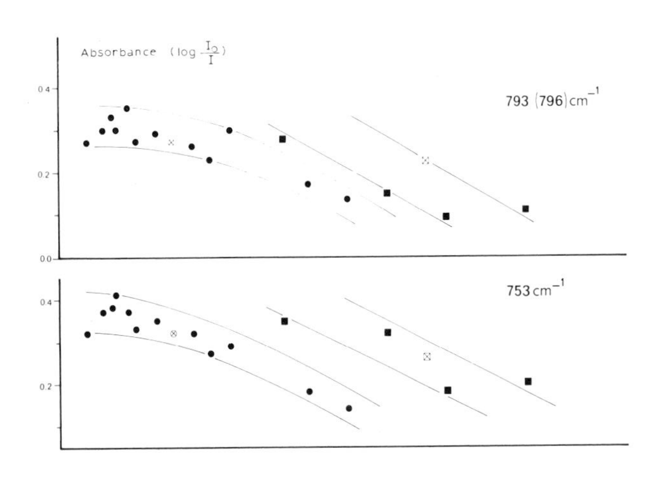

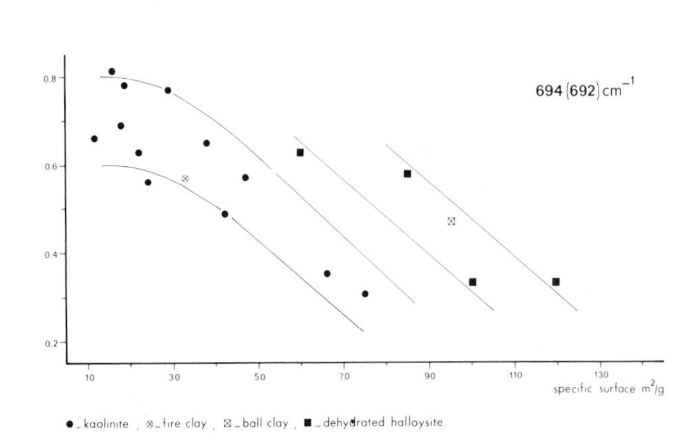

Fig. 5.3. Absorbance corrected for base line, and specific surface for kaolinite (793, 753, 694), fire clay (793, 753, 694) and dehydrated halloysite (796, 753, 692).

Absorbance (log $\frac{I_0}{I}$)

3694, 3695 cm^{-1}

● _ kaolinite deposits, ■ _ kaolinite soils — 3694; △ _ fire clay, ball clay, pottery clay — 3695

3621 cm^{-1}

● _ kaollinite deposits, ■ _ kaollinite soils, △ _ fire clay, ball clay, pottery clay — 3621

3695, 3708, 3701 cm^{-1}

◐ _ halloysite deposits, ◭ _ halloysite soils — 3695; x _ dickite — 3708; + _ nacrite — 3701

3623, 3622, 3628 cm^{-1}

◐ _ halloysite deposits, ◭ _ halloysite soils — 3623; x _ dickite — 3622; + _ nacrite — 3628

specific surface m^2/g

Fig. 5.4. Absorbance corrected for base line of various kaolin minerals and their specific surface.

References

Beer, A., 1852. Bestimmung der Absorption des rothen Lichts in farbigen Flüssigkeiten. Poggendorff's Ann. Phys. Chem., 86: 78—88.

Bonhomme, J., 1955. Contribution à l'analyse quantitative par le spectrus d'absorption infrarouge de poudres. Spectrochim. Acta, 7: 32—44.

Bonhomme, J. and Duyckaerts, G., 1955. The importance of grain size in the infrared spectroscopy of powders. Ind. Chem. Belge, 20: 145—150.

Bouguer, P., 1729. Essai d'Optique sur la Gradation de la Lumière. Paris, 164 pp.

Duyckaerts, G., 1955. Contribution à l'analyse quantitative sur les spectres d'absorption infrarouge des poudres. Spectrochim Acta, 7: 25—31.

Duyckaerts, G., 1959. The infra-red analysis of solid substances. Analist, 84: 201—214.

Heighl, J.J., Bell, M.F. and White, J.U., 1947. Application of infrared spectroscopy to the analysis of liquid hydrocarbons. Anal. Chem., 19: 293—298.

Lambert, J.H., 1760. Photometria, sive de Mensura et Gradibus Luminis, Colorum et Umbrae.

Otvos, J.W., Stone, H. and Haro, W.R., 1957. Theory of radiant-energy absorption by randomly dispersed discrete particles. Spectrochim. Acta, 9: 148—156.

Perkin, F.H., 1948. Whose adsorption law. J. Opt. Soc. Am., 38: 72—74.

Willard, H.H., Merritt, L.L. and Dean, J.A., 1958. Instrumental Methods of Analysis. Van Nostrand, 626 pp.

Wright, N., 1941. Application of infrared spectroscopy to industrial research. Ind. Eng. Chem. Anal. Ed., 13: 1—8.

6. CLAY- AND RELATED MINERALS

Koaline minerals (kandites)

The minerals of this group consist of unit layers formed by the linkage of one Si(Al,Fe)-O tetrahedral sheet with one Al(Mg,Fe)-OH octahedral sheet in regular succession. This kind of minerals are also indicated as 1:1 minerals (Plate I, pp. 190—191).

Dickite and nacrite; formula: $Al_2(OH)_4Si_2O_5$

The crystal structure (dickite = monoclinic, nacrite = approaching orthorhombic) and dimensions resemble those of kaolinite. But the structural elements of the unit layer are better ordered and the interlayer forces are stronger. This effects, especially in the case of dickite, a better growth in the direction of the c-axis perpendicular to the basis plane {001} and consequently, thicker crystals.

Occurrence. Both kaolin minerals are rarely found and may originate under hydrothermal or pneumatolytic (nacrite) conditions. As a result of these favourable genetic conditions the crystals are well-developed and the atoms well-ordered.

Infrared spectra (Fig. 6.1, 6.2, 6.11). The spectra of both minerals are nearly alike. Bands (in cm^{-1}) of large intensity which they have in common are (nacrite in brackets): 3627 (3628), 3622 (3620), 1117 (1118), 1002 (1001), 913 (913), 796 (798), 754 (754), 693 (694), 608 (608), 469 (470), 429 (428). Bands of large intensity to distinguish dickite from nacrite (nacrite between brackets) are: 3708 (3701), 3656 (3649), 1098 (1102), 1035 (1038), 966 (956), 935 (932), 538 (535). The geode from Keokuk, Missouri[1] (Fig. 6.4) has as well specific kaolinite (3695, 3653, 1032, 1008, 937 cm^{-1}), as nacrite (1102, 535 cm^{-1}) and dickite (468 cm^{-1}) characteristics.

Remarks. Dickite is named after A.B. Dick, while the mineral name nacrite is derived from its nacreous appearance (Greek nakros).

Kaolinite; formula: $Al_2(OH)_4Si_2O_5$

Occurrence. Kaolin may be formed by the weathering of K- and Na- feldspars from magmatic and metamorphic rocks or by hydrothermal action of carbonic and sulfuric acid solutions on feldspars and micas. All these minerals occur in many kinds of rocks (granite, gneiss, porphyry, etc.) so that the resulting kaolinite is widespread. On the other hand, kaolinite may be formed by silicification of hydrargillite by silicic acid solutions.

Kaolinite deposits which do not yield plastic masses even after prolonged submersion in water but which break in conchoidal fractures, are called flint clays. Kaolinite deposits consisting of poorly-ordered kaolinite polluted with varying amounts of carbon and coal-like products, quartz and mica are called fire clay (pottery clay), and the corresponding monoclinic kaolin mineral which it contains, fire-clay mineral. Other names are: *b*/3 disordered kaolinite, livesite[2]. mellorite[3] and pM kaolinite.

Samples of poorly-ordered kaolinite in which the carbon and mica components are absent or nearly absent, are called ball clay. Kaolinite and fire-clay mineral are of frequent occurrence in soils and the thin, platy particles may be very small (to 0.05 μm) and are sometimes curled.

Infrared spectra (Fig. 6.3—6.7, 6.12, 6.13). Bands of large intensity are for kaolinite and fire clay (ball clay) at 3694, 3621, 1100, 1032, 1008, 913, 694, 539, 471, 431 cm^{-1}. The spectra show wide variations in their intensities; especially at the higher wave numbers (3695, 3620 cm^{-1} region). The weakest are those of the fireclays (ball clay) and of most of the heavily weathered latosols and red earths, which are also the least ordered according to X-ray analysis. (Fig. 12,13.) However, the Provence kaolinite which according to X-ray and infrared analysis is badly ordered, consists of very small well-developed hexagonal plates. (Fig. E.M.24.)

Many latosols and red earths contain besides kaolinite, also some hydrargillite, limonite, quartz and mica. Hydrargillite and limonite are easily found by the I.R. method in amounts $>$ 3—5%. For quartz more is needed to observe the 798—780 cm^{-1} doublet. A band at ca 794 cm^{-1} may be from cristobalite, tridymite, silicic acid, very small amounts of quartz still detectable by X-ray analysis, and a weak vibration inherent to kaolinite, montmorillonite and mica minerals.

Remarks. The name kaolinite is derived from the Kaoling hills near Jauchau-Fu, China. High-graded kaolinite is an important filler or a coating pigment in paper industry (good paper coaters are of good crystallinity) and rubber industry. Kaolinite (and fire clay) are used in refractories, pottery, porcelains and foundry molds.

---

[1] Gift from W.D. Keller, Missouri (U.S.A.)

[2] After Prof. Livesey, formerly at the University of Leeds; see Carr et al. (1952) and Roberts (1958).

[3] After Mellor (1911).

Halloysite; formula: $Al_2(OH)_4Si_2O_5 2H_2O$

The lattice structure of halloysite is more disordered than that of the fire-clay mineral. It, furthermore, contains weakly bound interlayered H_2O molecules. Two species exist, namely hydrated halloysite (also called endellite) and dehydrated halloysite (also called metahalloysite).

Occurence. Pure halloysite in large amounts is only found in some deposits, e.g. Eureka, Wagon Wheel and Tintic district (U.S.A.), Djebel Debar (Algeria), Capalbio (Italy), Martinsberg (Hungary). In these cases it is the result of sulfuric or carbonic acid solutions on kaolinite or in general on alumosilicates.

Infrared spectra (Fig. 6.8, 6.9, 6.10, 6.14). The spectra of halloysite show wide variations in their intensities, especially in the higher wave numbers. The halloysites from sedimental origin, which are the less-ordered according to X-ray analysis, have also the weakest bands.
Bands of large intensity are: 3695, 3623, 1094, 1033, 1012, 913, 692, 540, 471, 432 cm^{-1}.
Halloysite may be distinguished from kaolinite and fire clay mineral by the following bands (kaolinite and fire clay mineral in brackets): 1094 (1100), 1012 (1008), 941* (937), 692 (694), 650* (647), 562* (absent).
The Les Eyzies and Tasmania samples (Fig. E.M.40) which consist of plates and tubes have according to infrared analysis kaolinite as well as halloysite characteristics (Fig. 6.10). Les Eyzies: kaolinite = 1100, 938, 647 cm^{-1}; halloysite = 692, 561 cm^{-1}. Tasmania: kaolinite = 1100, 647 cm^{-1}; halloysite = 942, 692, 562 cm^{-1}. So they should not consist of kaolinite plus uprolled kaolinite plates as suggested in literature, but of kaolinite plus halloysite. Most halloysite sediments investigated here contain small amounts of kaolinite which cannot be detected by the infrared method, nor by X-analysis. The only way is to use electron microscopy.

Remarks. This kaolin mineral is named after its discoverer, the Belgian mineralogist M. Omalius d'Halloy. The name endellite is after K. Endell. Pure halloysite is of rare occurrence and of little commercial significance. Some finer sorts are used accidentally in the paper industry or as a coating material.

Hydrohalloysite; formula: $Al_2(OH)_4Si_2(O,OH)_5$

The above mineral, as described by Erdélyi (1962) which is supposed to contain an excess of hydroxyl, is said to be the first crystallization product of allophane transformation into halloysite.

Occurrence. Up till now hydrohalloysite has been known to occur only at Baia Mare (Rumania).

Infrared spectra (Fig. 6.9, 6.14). Hydrohalloysite cannot be distinguished from the badly ordered halloysites. Only the 1012, 562* cm^{-1} bands of the latter fail in the hydrohalloysite spectrum. But the excess of O-H, only some percent, is too small to be detected by the infrared method.

Remarks. The precise structure of this mineral needs further investigation.

Serpentine minerals

The various serpentine minerals form a series, ranging from monoclinic (clino chrysotile), to monoclinic with super lattice (antigorite), to orthorhombic (orthochrysotile, parachrysotile), to pseudo-orthohexagonal one-layer (lizardite) and to orthohexagonal six-layer (Unst-type mineral). Moreover, most of the samples are polluted with other minerals.

Occurrence

Serpentine minerals are formed by hydrothermal action on ultrabasic Mg-rich rocks with high olivine contents (picrite, peridotite, dolerite). They weather easily and, apart from magnesite, may yield also sepiolite, palygorskite, saponite or deweylite.

Serpentine; formula: $(Mg,Al,Fe)_3(OH)_4Si_2O_5$

Serpentine is a rock, either of fibrous or platy composition. Mixtures are also found.

Occurrence. Serpentine rocks are not widespread.

Infrared spectra (Fig. 6.15). The spectrum of the Cardiff rock is different from that of the other, the former consists of antigorite and the latter are chrysotile or lizardite — see further.

Remarks. Serpentine rocks may contain Ni ores of commercial value. Polished massive serpentinite is used for ornamental and interior decoration purposes.

Chrysotile; formula: $(Mg)_3(OH)_4Si_2O_5$

This mineral is the well pronounced fibrous variety of the serpentine group. The material filling the hollow fibres and that filling the voids between the fibres should be amorphous.

Occurrence. The mineral is not widespread.

Infrared spectra (Fig. 6.16,6.17). Bands of large intensiy are at 3690, 1080, 958, 610, 438 cm^{-1}. Tremolite- and amphibole asbestos which both are morphological like chrysotile, may be easily distinguished from chrysotile.

Remarks. Fibrous chrysotile (asbestos) is used in the manufacture of non-combustible textiles and paints, filters, brake-linings for cars and several products of the asbestos-rubber industry.

Hydro-chrysotile; formula: $Mg_3(OH)_4Si_2(O,OH)_5$

The above mineral is supposed to contain an excess of hydroxyl.

Occurrence. Up till now hydro-chrysotile has been found only at Perkupa, (Hungary).

Infrared spectra (Fig. 6.17). The spectrum is like that of chrysotile. But also in this case the excess of OH is only some %, which is too small to be detected by the infrared method.

Remarks. This mineral is supposed to be a new variety of the serpentine group. (Erdélyi and Veniale, 1970.)

Antigorite; formula: $(Mg,Al,Fe)_3(OH)_4Si_2O_5$

Antigorite is the platy variety of the serpentine minerals.

Occurrence. Antigorite is a rare mineral.

Infrared spectra (Fig. 6.18, 6.19). Bands of large intensity are at: 3673, 1076, 988, 960, 647, 622, 567, 462, 448, 436 cm^{-1}. Antigorite may easily be distinguished from chrysotile by the following bands (chrysotile between brackets): 3673 (3690), 3650 weak (3650 but somewhat stronger), 1200 (absent), 1127 (absent), 1076 (1080), absent (1017), 988 (absent), 778 (absent), 647 (absent), 622 (610), 567 (absent), absent (552), 508 (absent), absent (486), 448 (absent).

Remarks. The mineral is named after Antigorio, a small village in Italy. Iddingite (ferroan antigorite) is said to be an iron-rich variety and garnierite, noumeite and nepouite (schuchardtite) — Glasser, 1907 — Ni-rich varieties.

Hydroantigorite; formula: $(Mg,FeAl)_3(OH)_4Si_2(O,OH)_5$

This mineral, which is waxy and rose-coloured, is Fe- and Al-poor: ($FeO + Fe_2O_3 = 0.58\%$, $Al_2O_3 = 0.31\%$). It is supposed to contain an excess of OH, according to Erdélyi et al. (1959).

Occurrence. The mineral is known to occur only at the type locality Dunabogdány (Hungary).

Infrared spectra (Fig. 6.20). Hydroantigorite has typical chrysotile as well as antigorite characteristics. Chrysotile: between brackets: 3687 (3690), 1080 (1080), 610 (610), 583 (585*), 556 (552), 485 (486). Antigorite: between brackets: 647 (647), 462 (462). The 999 cm^{-1} reflection belongs to neither.

Remarks. The precise structure of this nearly Unst-type mineral needs further investigation.

Lizardite; formula: $(Mg,Al,Fe)_3(OH)_4Si_2O_5$

The white-coloured glittering sample from Kennack Cove (England) is poor in Al and Fe ($FeO + Fe_2O_3 = 0.48\%$, $Al_2O_3 = 2.26\%$). The FeAl-rich sample from another locality at The Lizard ($Fe_2O_3 = 9.03\%$, $Al_2O_3 = 3.46\%$) has a red-brown colour. The sample from St. Margherita (Italy) ($FeO + Fe_2O_3 = 3.45\%$, $Al_2O_3 = 3.98\%$) is dark green and the resinous Fe, Al-poor sample from Snarum (Norway) ($FeO + Fe_2O_3 = 3.26\%$, $Al_2O_3 = 0.36\%$) is yellow-green.

Occurrence. This mineral is rare (see above).

Infrared spectra (Fig. 6.20). Characteristic bands are at: 3690, 3658, 461, 443 and 427 cm^{-1}. There is a large spreading in the 1092-1072 and 992-940 cm^{-1} range which is caused by the large variation in composition of the lizardites. The bands of lizardite come the most near to those of chrysotile.

Remarks. The mineral is named after the type locality The Lizard, England.

Unst-type mineral (6-layer orthohexagonal); formula: $(Mg,Fe)_3(OH)_4Si_2O_5$

The sample from Unst ($FeO + Fe_2O_3$) = 3.04 to 3.42%, Al_2O_3 (= 0.10%) has a greenish colour.

Occurrence. This mineral is very rare.

Infrared spectra (Fig. 6.20). The bands are nearly alike although not identical to those of chrysotile. X-ray analysis gives a better identification: see the many super-lattice reflections.

Remarks. The mineral is named after the type locality Unst (Shetland Isles).

Amesite; formula: $(Mg,Al,Fe)_3(OH)_4(Si,Al)_2O_5$

Amesite is said to be a trioctahedral mineral in which about half of the Al^{3+} in the octahedral sheets would be replaced by Mg^{2+} and some Fe^{2+}. In the tetrahedral sheets half of the tetravalent Si is said to be replaced by trivalent Al and the vacant holes are filled up by Mg^{2+} and Fe^{2+} so compensating the charge.

Occurrence. Amesite is a very rare mineral. Only a few localities are known in the world.

Infrared spectra (Fig. 6.21). The spectra show wide differences between the various samples. According to X-ray analysis the Ural sample (no. 103312 Natl. Museum, Washington DC) and the sample from the Emery Mine in Chester, Mass. (80715 Natl. Museum Washington DC) consist mainly of 14 Å chlorite. The sample from the well-known locality Lydenburg, Transvaal consists mainly of a mineral of the amphibole group.

Remarks. This mineral, but then obtained from better localities, needs further investigation.

Hydro-amesite; formula: $(Mg,Al,Fe)_3(OH)_4(Si,Al)_2(O,OH)_5$

The above mineral, as described by Erdélyi et al. (1959), is supposed to contain an excess of hydroxyl.

Occurrence. Up till now hydro-amesite has been found only at Halap-Berg, Hungary.

Infrared spectra (Fig. 6.21). The spectrum is different from antigorite, chrysotile, hydro-chrysotile, lizardite and Unst mineral. However, according to X-ray analysis, the sample belongs to the above types.

Remarks. The precise nature of this mineral needs further investigation.

Greenalite; formula: $(Fe,Mg,Al)_3(OH)_4Si_2O_5$

In this mineral nearly all of the Al in the octahedral kaolinite sheets are said to be replaced by Fe^{2+} and the vacant holes filled up with Mg and some Fe^{3+} to produce a charge-free crystal (FeO = ca. 34%, Fe_2O_3 = ca. 2%, MgO = ca. 6%).

Occurrence. Greenalite is a very rare mineral.

Infrared spectra (Fig. 6.21). The samples from the well-known localities at Port Arthur (Canada) and Gilbert, Minnesota, are very impure. Consequently, only a few bands could be identified e.g. 3685, 3645*, 3626, 1036, 998, 736*, 650, 612*, 560*, 438*.

Remarks. The precise nature of greenalite needs further investigation. Better samples should be used for this purpose.

Cronstedtite; formula: $(Fe^{2+},Fe^{3+})_2(OH)_4(Si,Fe^{3+})_2O_5$

In this mineral all Al in the octahedral layers are replaced by Fe^{2+} and Fe^{3+} and about half of the Si in the tetrahedral sheets is replaced by Fe^{3+}. Moreover, the vacant holes in the octahedral sheets are partly filled with Fe^{2+} and Fe^{3+} to produce a charge-free structure (Fe_2O_3 = ca. 30%, FeO = ca. 42%).

Occurrence. This mineral is very rare.

Infrared spectra (Fig. 6.22). The mineral may easily be distinguished from the other kaolin minerals by the bands at 3510 and 3245 cm^{-1}. Other important bands are at: 1022, 935, 677, 565, 508, 478, 426 cm^{-1}.

Remarks. The above mineral is named after A. Cronstedt.

Deweylite (gymnite); formula: $(Mg,Fe,Al)_3(OH)_4(Si_2O_5)nH_2O$

This name is given to white to brownish-coloured, waxy to resinous Mg-rich kaolin minerals of various composition and disordered lattice structure. They may contain amorphous matter and small cavities filled with water.

Occurrence. The mineral is occasionally found in serpentinites and Mg-rich carbonaceous rocks.

Infrared spectra (Fig. 6.23). The bands of this mineral are broad and their number is restricted which points to high disorder in the crystals.

The non-uniform occurrence and relative intensities of the bands in the 955—1085 and 426- 478 cm^{-1} trajects are caused by the non-uniform composition. Those in the 3653—3692 traject are more uniform. Bonds of importance are at 3692, 3653, 1084, 1003*, 983*, 638, 610, 565, 460 and 436 cm^{-1}.
The 3440 cm^{-1} band is from adsorbed H_2O in pores. Some mica and intermediate which after X-ray analysis should also occur, fail in the I.R. spectra.

Remarks. Minerals indicated as serpophite, deweylite, gymnite are essentially identical. The first name was given by Lodočhnikov (1936), who was apparently unaware of the existence of the other names. They all three indicate a certain group of crypto-crystalline serpentine minerals of varying crystal ordering and composition. However, because the name deweylite is older it should have priority. The name gymnite is derived from the locality Bare Hills, Maryland (Greek: gymnos = naked).

Monothermite

Structure. According to Beliankin (1932) this mineral should resemble the English ball clays. In a later publication (1938) it is suggested that the mineral should be a mixture of kaolinite, quartz and hydromica. Petrov (1958) believes that it should also contain montmorillonite and that there are varying degrees of "monothermiteness"; some samples being close to fire clay or livesite.

Occurrence. The type locality is the large Donetz basin in the U.S.S.R.

Infrared spectra (Fig. 6.23). The spectrum shows mainly well-ordered kaolinite bands and also some quartz. Characteristic mica bands (1023, 988 cm^{-1}) which according to X-ray analyses should also occur at an amount up to ca. 20%, are absent with the exception of that at 832 cm^{-1}.

Remarks. The name monothermite should not be used because it does not represent a certain well-defined mineral, but a mixture of minerals.

Anauxite

Structure. According to Smirnoff (1907), the mineral should resemble kaolinite. Mellor (1911) assumes that anauxite consists of a mixture of colloidal silica gels and clay minerals. Anauxite is further said to contain a higher SiO_2: Al_2O_3 ratio than the 2:1 ratio for kaolinite.

Occurrence. This rare Al-silicate mineral was found in fissures of eruptive pyroxene rocks at Bilin in Bohemia (Czechoslovakia).

Infrared spectrum (Fig. 6.23). The spectrum from the well-known type locality at Bilin shows mainly kaolinite. Characteristic bands of montmorillonite (522 cm^{-1}) or cristobalite (622 cm^{-1}), which minerals should also occur according to X-ray analysis, are absent. Silicic acid, which should also occur (Langston and Pask, 1969) and which should account together with montmorillonite for the excess of SiO_2 over Al_2O_3, could neither be identified by the infrared method in this mixture: this was only possible by X-ray and EM-analysis (E.M. fig. 69). SiO_2 dissolved by NaOH: spec. wt. 1.035 for 1 h at ca. 100° C = 19.4%.

Remarks. The mineral from the above locality in Bohemia was named anauxite by Breithaupt (1838).

References

Beliankin, D.S., 1932. Bull. Ceram. Inst., 1: 10—15.
Beliankin, D.S., 1938. On the characteristics of the mineral monothermite. C. R. Acad. Sci. U.S.S.R., 18: 673—676.
Breithaupt, A., 1838. Bestimmung neuer Mineralien. J. Prakt. Chem., 15: 325—338.
Carr, K., Grimshaw, R.W. and Roberts, A.L., 1952. Yorkshire fire clays. Trans. Br. Ceram. Soc., 51: 334—344.
Erdélyi, J., Koblencz, V. and Varga, N.S., 1959a. Hydroantigorit ein neues Serpentinmineral und metakolloidaler Brucite vom Csodi-Berg bei Dunabogdány (Ungarn). Acta Geol. Hungariacea, 6: 65—93.
Erdélyi, J., Koblencz, V. and Varga, N.S., 1959b. Hydroamesite, ein neues Mineral aus den Hohlräumen des Basaltes vom dem Haláp Berge am Plattensee Gebiet. Acta Geol. Hungariacae, 6: 95—106.
Erdélyi, J., 1962. Hydrohalloysit (Hydroendellit) ein neues Mineral der Halloysitgruppe aus dem Matra-Gebrige (Ungarn) und von Baia Mare (Nagybánya) in Rumänien. Chem. Erde, 21: 321—343.
Erdélyi, J. and Veniale, F., 1970. Idro-crisotile: un nuovo minerale del gruppo del serpentino. Rend. Soc. Ital. Mineral. Petrol., 26: 403—404.
Glasser, E., 1907. Sur un nouvelle espèce, la nepouite. Bull. Soc. Fr. Mineral., 30: 17—40.
Langston, R.B. and Pask, J.A., 1969. The nature of anauxite. Clays Clay Minerals, 16: 425—436.
Lodočnikov, W.N., 1936. Serpentine und Serpentinite der Iltschirlagerstätte und im allgemeinen und damit verbundene petrologische Probleme (Russia). Centr. Geol. Prosp. Inst. Trans., 38: 817 pp.
Mellor, J.W., 1911. The chemical constitution of the kaolinite molecule. Trans. Ceram. Soc., 10: 94—120.
Petrov, V.P., 1958. The problem of monothermite; the history of the investigation of monothermite and its nature. The Investigation and Utilization of Clays. Publ. House Lvov Univers., pp. 84—91.
Petrov, V.P., 1958. Genetic types of white clays in the USSR and laws governing their distribution. Clay Minerals Bull., 3: 287—296.
Roberts, A.L., 1958. Mineralogie des argiles refractaires; la livesite. Bull. Soc. Fr. Ceram., 41: 29—32.
Smirnoff, W.P., 1907. Über ein kristallinisches Verwitterungsprodukt des Augits. Z. Krist. Mineral., 43: 338—346.

Dickite
Oury, Colo., U.S.A.
pure

Barkly East, Cape Province, S.A.
pure
Fig. E.M. 33

Pholerite
Neurode, Silesia,
Poland
pure

San Juanito, Mexico
pure

Vogtland, Saxony, Germany
pure

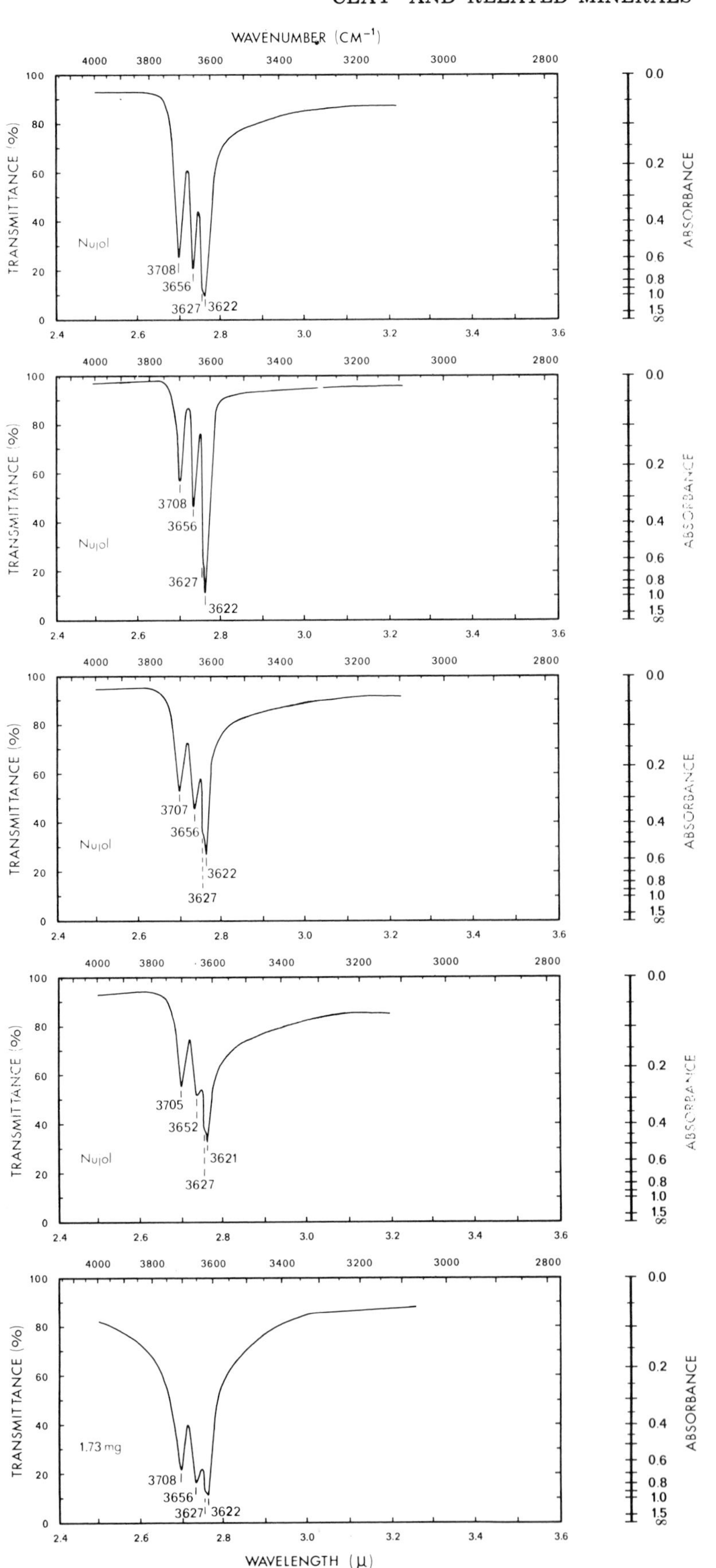

Fig. 6.1. Infrared spectra of dickite from different origin.

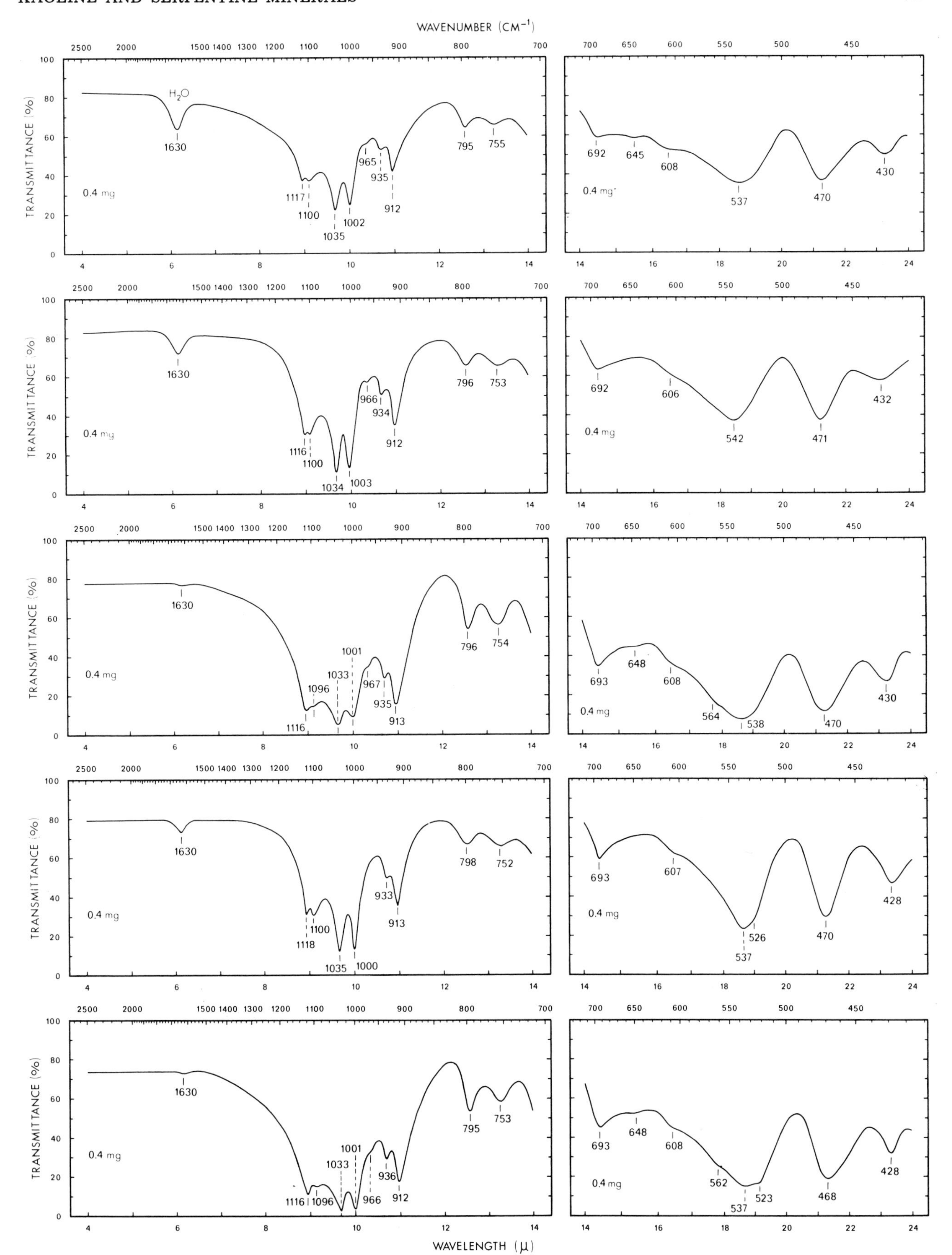
WAVENUMBER (CM⁻¹)
TRANSMITTANCE (%)
WAVELENGTH (μ)
0.4 mg
H₂O
1630
1117
1100
1035
1002
965
935
912
795
755
692
645
608
537
470
430

Dickite
Milikov, Poland
quartz

St. George, Utah, U.S.A.
pure

Nacrite
Cerain, Spain
pure

Pikes Peak, Colo., U.S.A.
quartz

Sidi Amer Ben Salem, Tunesia
some quartz

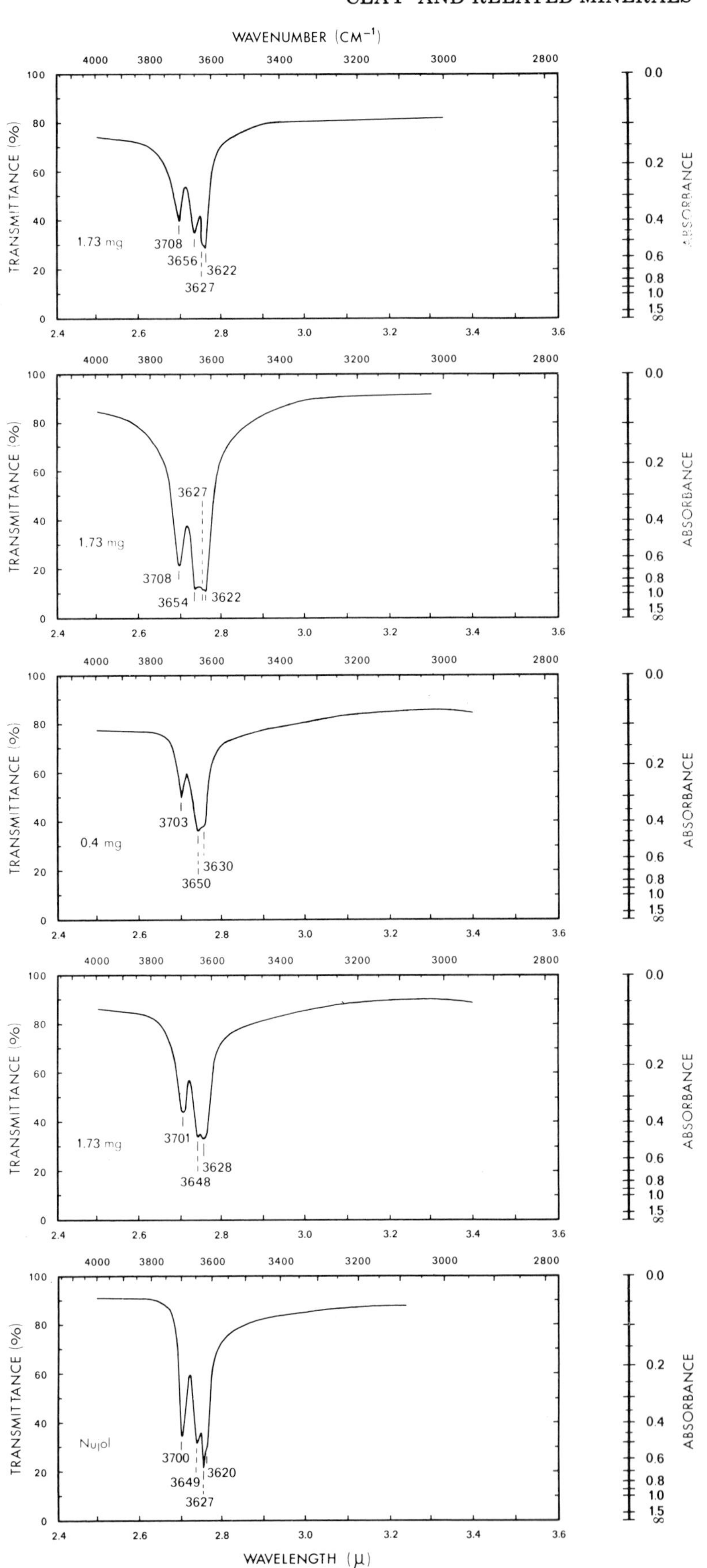

Fig. 6.2. Infrared spectra of dickite and nacrite from different origin.

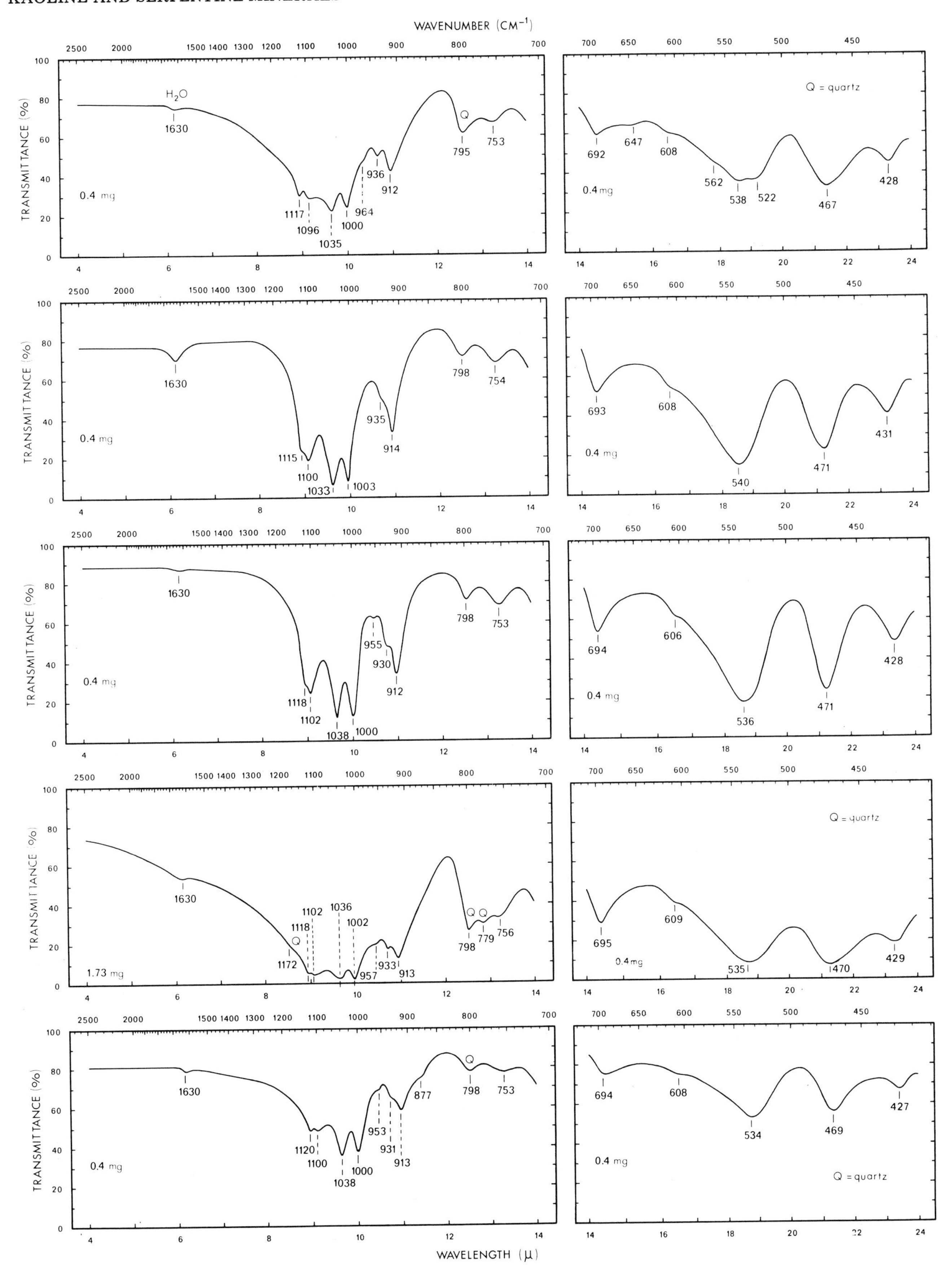
WAVENUMBER (CM⁻¹)
TRANSMITTANCE (%)
WAVELENGTH (μ)
H₂O
1630
0.4 mg
Q = quartz
1.73 mg
1117
1096
1035
964
1000
936
912
795
753
692
647
608
562
538
522
467
428
1115
1100
1033
1003
935
914
798
754
693
540
471
431
1118
1102
1038
955
930
606
694
536
1172
1036
1002
957
933
913
779
756
695
609
535
470
429
1120
953
931
877
534
469
427

Kaolinite
deposit, Cornwall, Great Britain
some mica (overlapped) and quartz
Fig. E.M. 23

deposit, Mesa Alta, S.C., U.S.A.
some quartz

deposit, Bath, S.C., U.S.A.
pure
Fig. E.M. 22

deposit, Zettlitz, Czechoslovakia
some mica (overlapped) and quartz
Fig. E.M. 21

deposit, Bangka, Indonesia
some mica (overlapped) and quartz

Lateritic soil
Saramacca region, Surinam
some quartz

deposit, Groszalmeroden, Germany
quartz

deposit, Provence, France
pure
Fig. E.M. 24

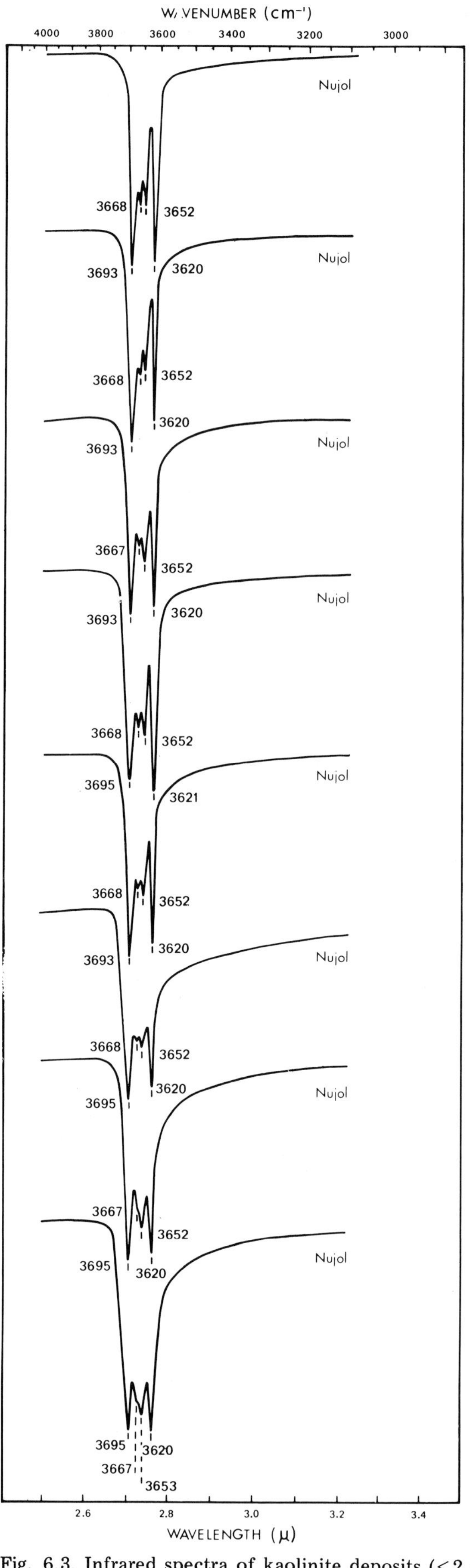

Fig. 6.3. Infrared spectra of kaolinite deposits (<2 μm) from different origin.

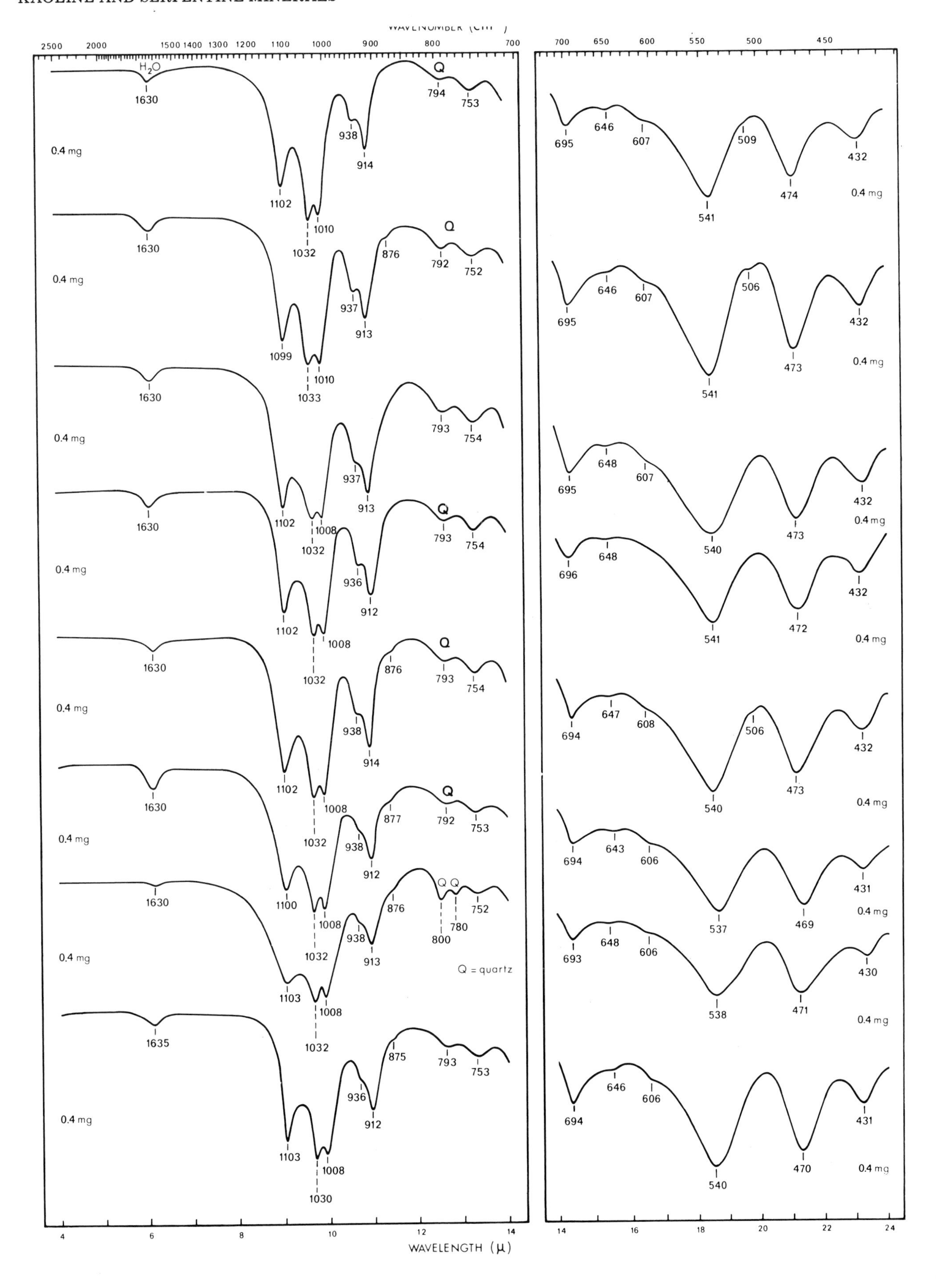

WAVENUMBER (cm⁻¹)
2500 2000 1500 1400 1300 1200 1100 1000 900 800 700
700 650 600 550 500 450
H2O
Q
Q = quartz
0.4 mg
WAVELENGTH (μ)

Kaolinite
geode, Keokuk, Miss., U.S.A.
quartz

deposit, Dry Branch, Ga., U.S.A.
pure

deposit, Mitchell County, S.C., U.S.A.
some mica intermediate and quartz (overlapped)

deposit, Macon, Ga., U.S.A.
some mica intermediate (overlapped)

deposit, Thailand
quartz
Fig. E.M. 26

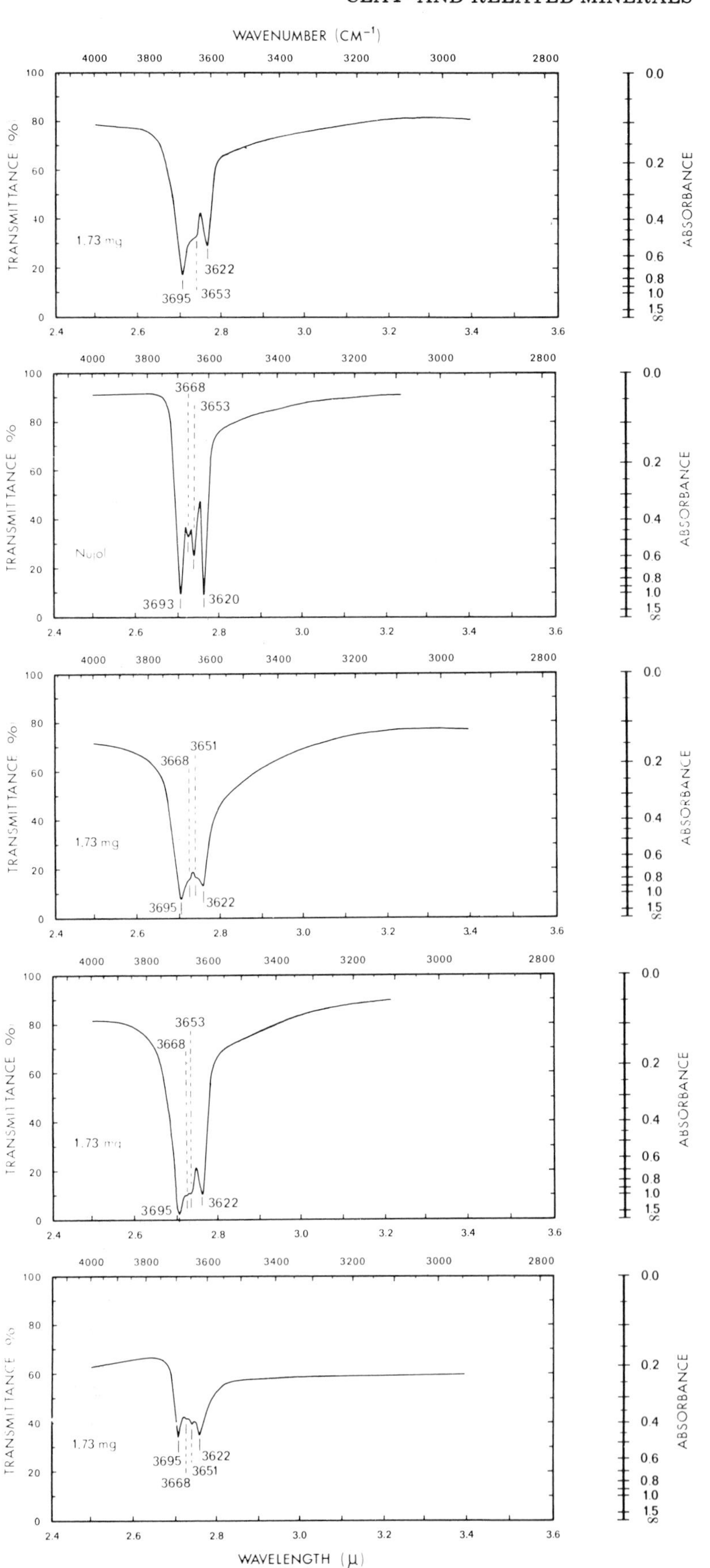

Fig. 6.4. Infrared spectra of kaolinite deposits (< 2 μm) from different origin.

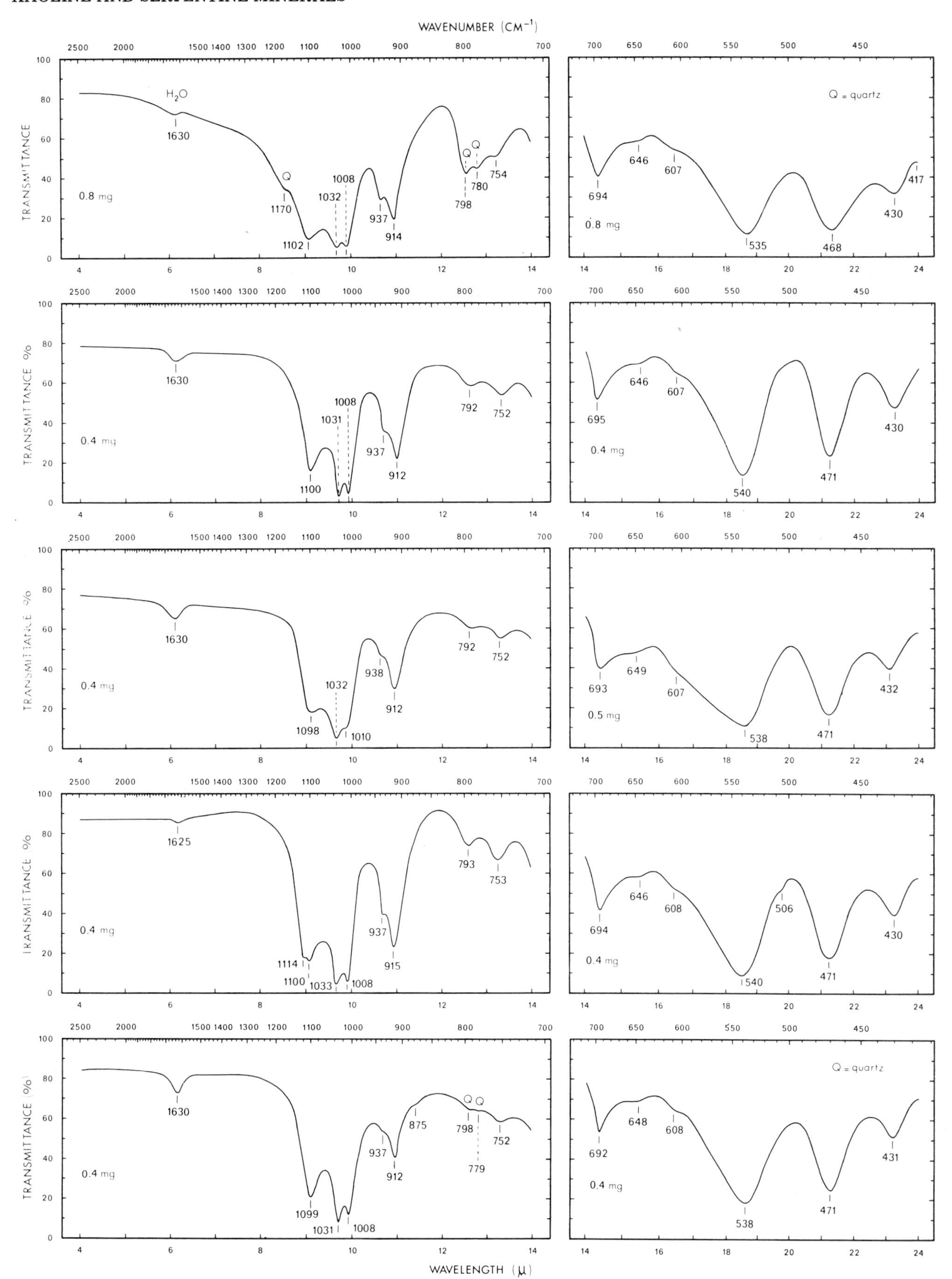
WAVENUMBER (CM⁻¹)
TRANSMITTANCE %
H₂O
1630
Q
1170
1102
1032
1008
937
914
Q
Q
798
780
754
0.8 mg
646
607
694
535
468
430
417
Q = quartz
1100
1031
1008
937
912
792
752
0.4 mg
695
646
607
540
471
430
1630
1098
1032
1010
938
912
792
752
693
649
607
538
471
432
0.5 mg
1625
1114
1100
1033
1008
937
915
793
753
694
646
608
540
506
471
430
1099
1031
1008
937
912
875
798
779
752
692
648
608
538
471
431
WAVELENGTH (μ)

Flint clay
Missouri, U.S.A.
some quartz

Pottery clay
Akeron, Ohio, U.S.A.
quartz and mica (overlapped)

Fire clay
Illinois, U.S.A.
quartz and mica (overlapped)

Sayreville, N.J., U.S.A.
pure

Colorado, U.S.A.
quartz and mica intermediate (overlapped)
Fig. E.M. 28

Vonsov, Czechoslovakia
some quartz and mica (overlapped)

Sheffield, Great Britain
some quartz
Fig. E.M. 27

Ball clay
Dorsetshire, Great Britain
some quartz and mica intermediate (overlapped)
Fig. E.M. 29

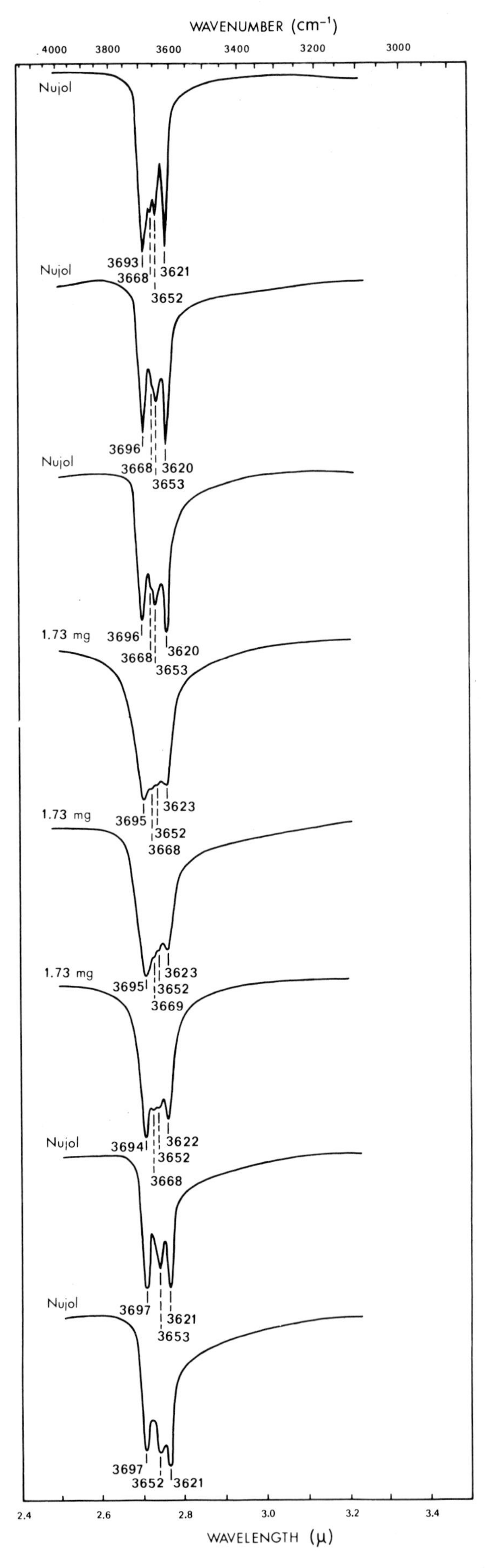

Fig. 6.5. Infrared spectra of flint clay, pottery clay, fire clay and ball clay ($< 2\ \mu m$) from different origin.

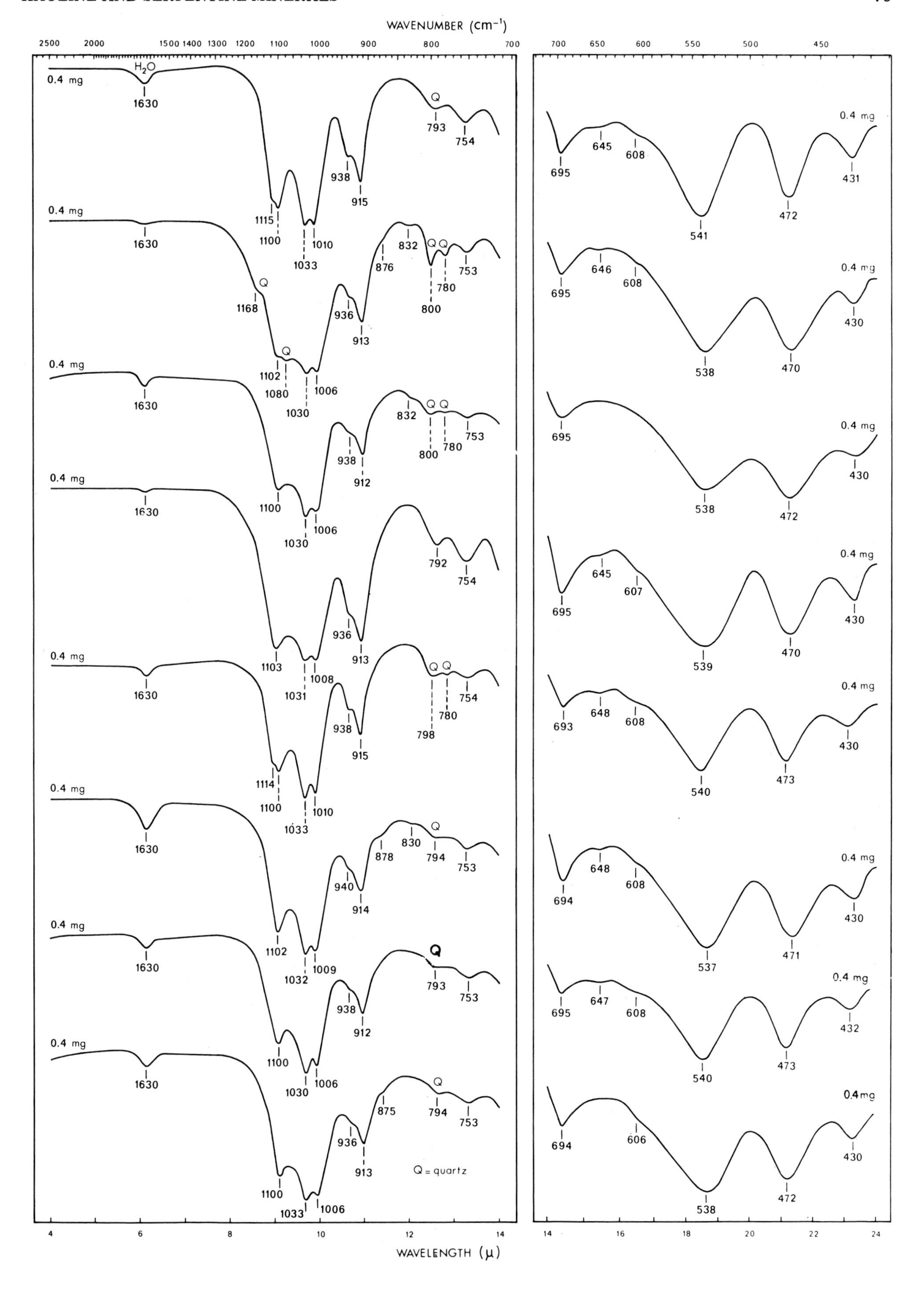

WAVENUMBER (cm⁻¹)
2500 2000 1500 1400 1300 1200 1100 1000 900 800 700
700 650 600 550 500 450
0.4 mg
H2O
1630
Q
793
754
1115
1100
1033
1010
938
915
695
645
608
541
472
431
1168
1102
1080
1030
1006
936
913
876
832
800
780
753
646
538
470
430
1630
1100
1030
1006
938
912
832
800
780
753
695
538
472
430
1103
1031
1008
936
913
792
754
645
607
695
539
470
430
1114
1100
1033
1010
938
915
798
780
754
693
648
608
540
473
430
1102
1032
1009
940
914
878
830
794
753
694
648
608
537
471
430
1100
1030
1006
938
912
793
753
695
647
608
540
473
432
1100
1033
1006
936
913
875
794
753
Q = quartz
694
606
538
472
430
4 6 8 10 12 14
14 16 18 20 22 24
WAVELENGTH (μ)

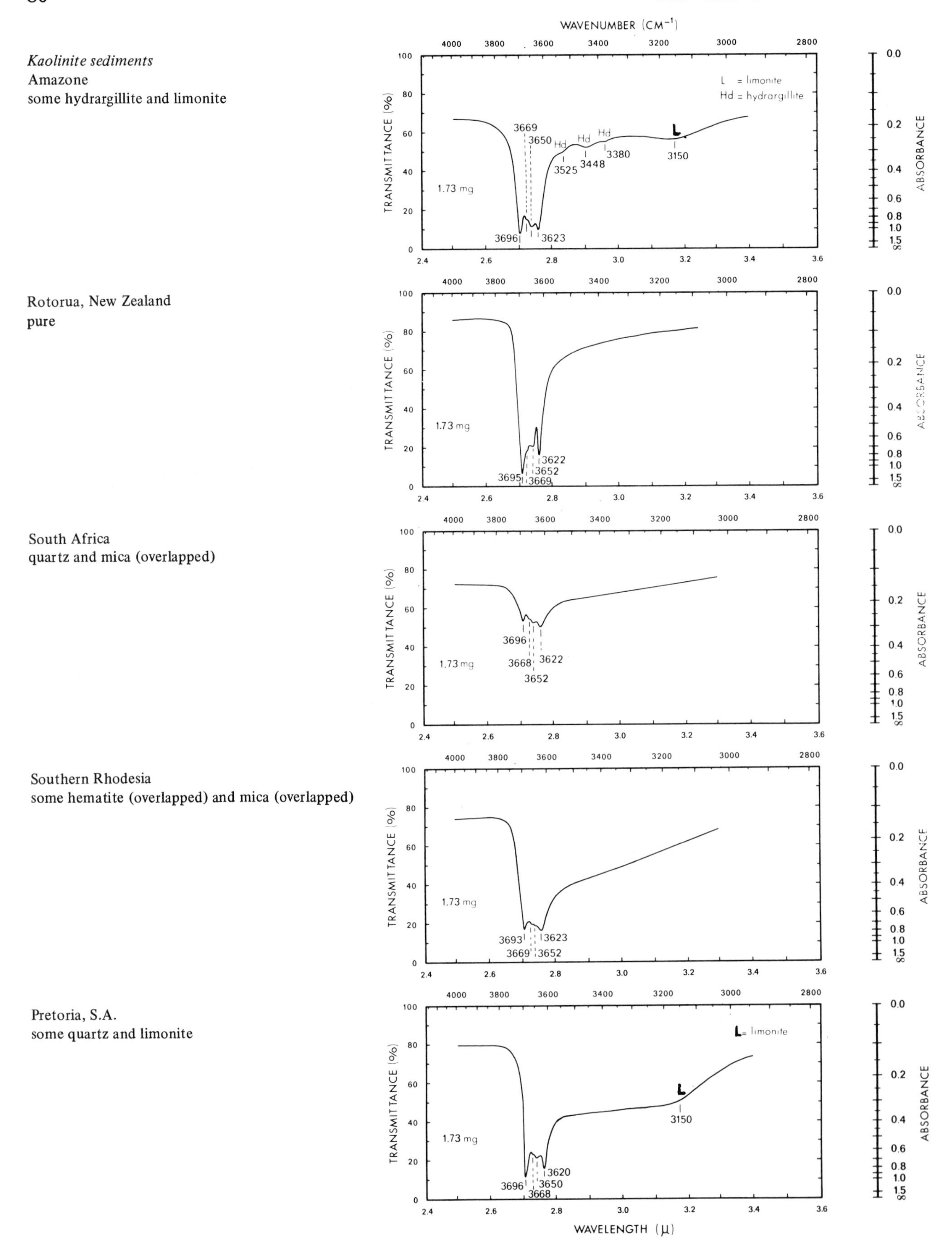

Fig. 6.6. Infrared spectra of kaolinite sediments (< 2 μm) from different origin.

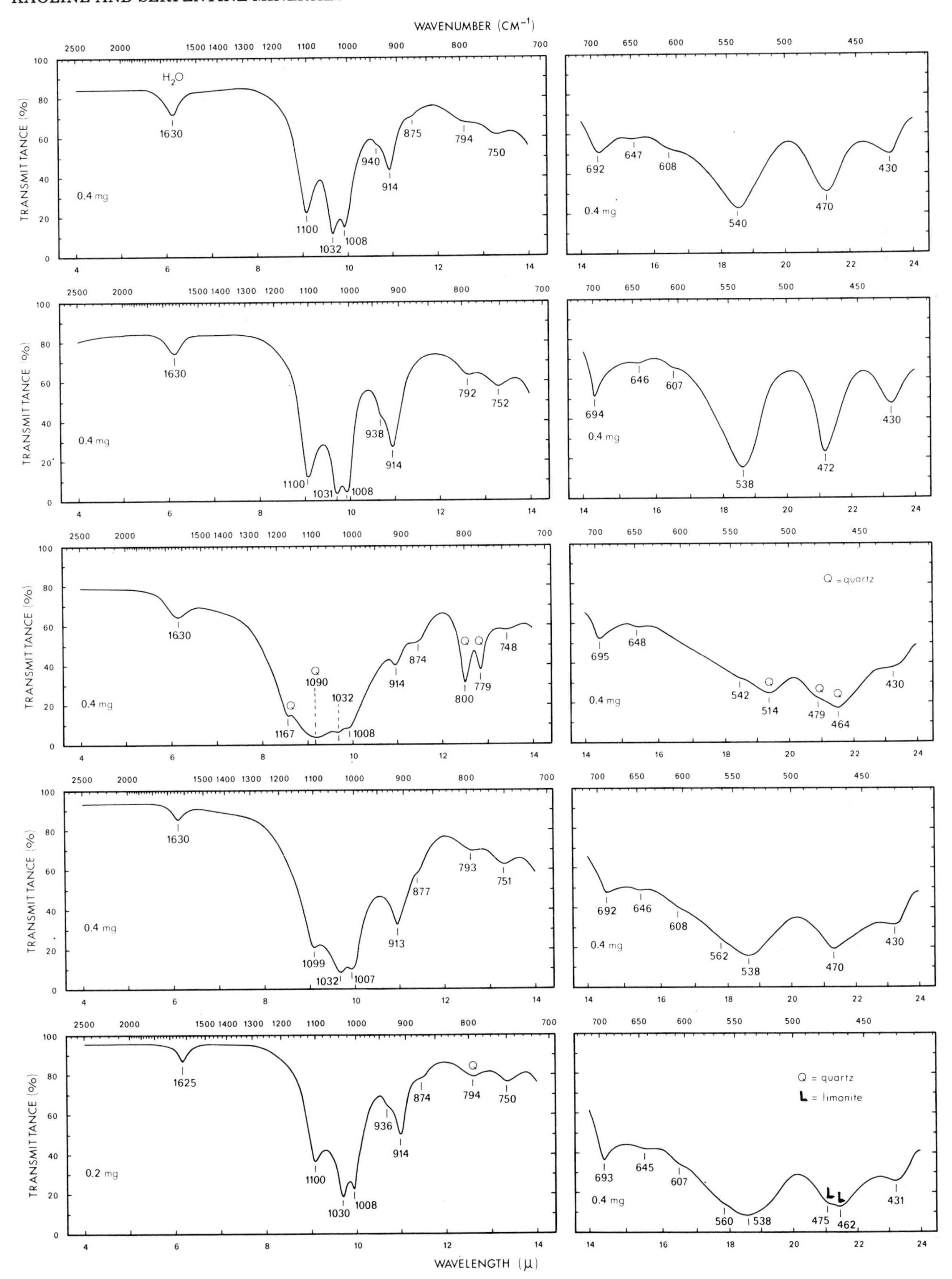

WAVENUMBER (CM⁻¹)
TRANSMITTANCE (%)
H₂O
1630
1100
1032
1008
940
914
875
794
750
0.4 mg
692
647
608
540
470
430
0.4 mg
1630
1100
1031
1008
938
914
792
752
0.4 mg
694
646
607
538
472
430
0.4 mg
1630
1167
1090
1032
1008
914
874
800
779
748
0.4 mg
Q = quartz
695
648
542
514
479
464
430
0.4 mg
1630
1099
1032
1007
913
877
793
751
0.4 mg
692
646
608
562
538
470
430
0.4 mg
1625
1100
1030
1008
936
914
874
794
750
0.2 mg
Q = quartz
L = limonite
693
645
607
560
538
475
462
431
0.4 mg
WAVELENGTH (μ)

Kaolinite sediments
Morocco
some quartz, mica (overlapped), hydrargillite and hematite

Terra rossa, Lisbon, Portugal
some hematite (overlapped), limonite, illite (overlapped) and quartz
Fig. E.M. 30

Terra rossa, Algeria
some mica, quartz and limonite

Terra rossa, Brazil
some hydrargillite, hematite (overlapped) and quartz
Fig. E.M. 31

Cooks Islands, Rarotongga
some hydrargillite, limonite

Borneo, Indonesia
some hydrargillite, quartz and limonite

Ecuador
some hematite (overlapped), hydrargillite, quartz and limonite
Fig. E.M. 32

Cecil loam, U.S.A.
hydrargillite, quartz and limonite

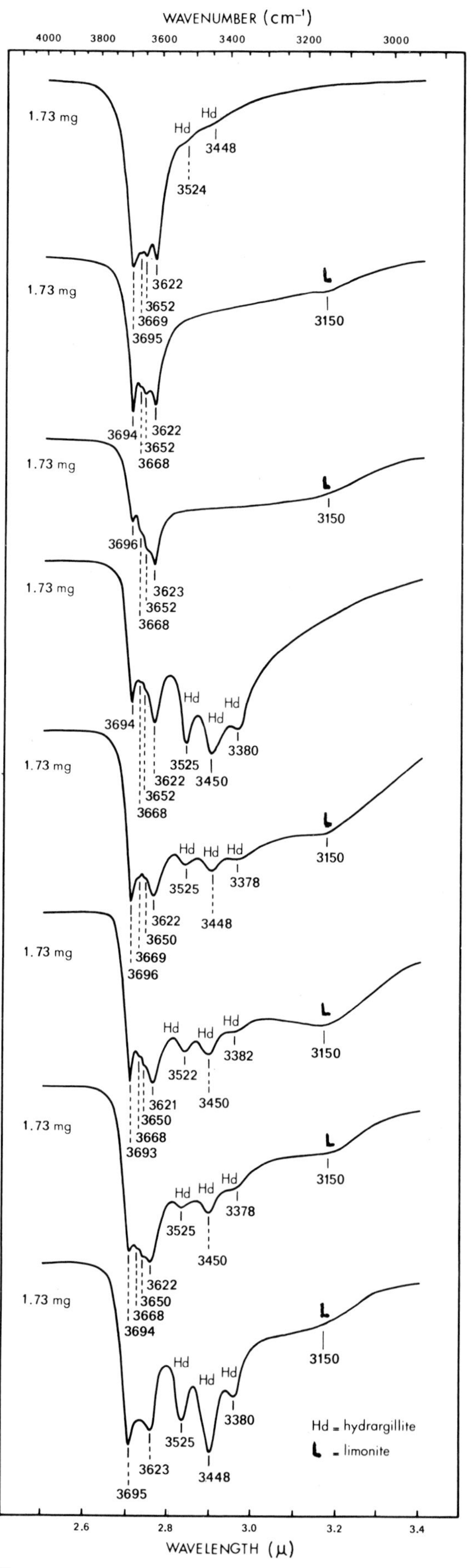

Fig. 6.7. Infrared spectra of kaolinite (< 2 μm) from different sediments.

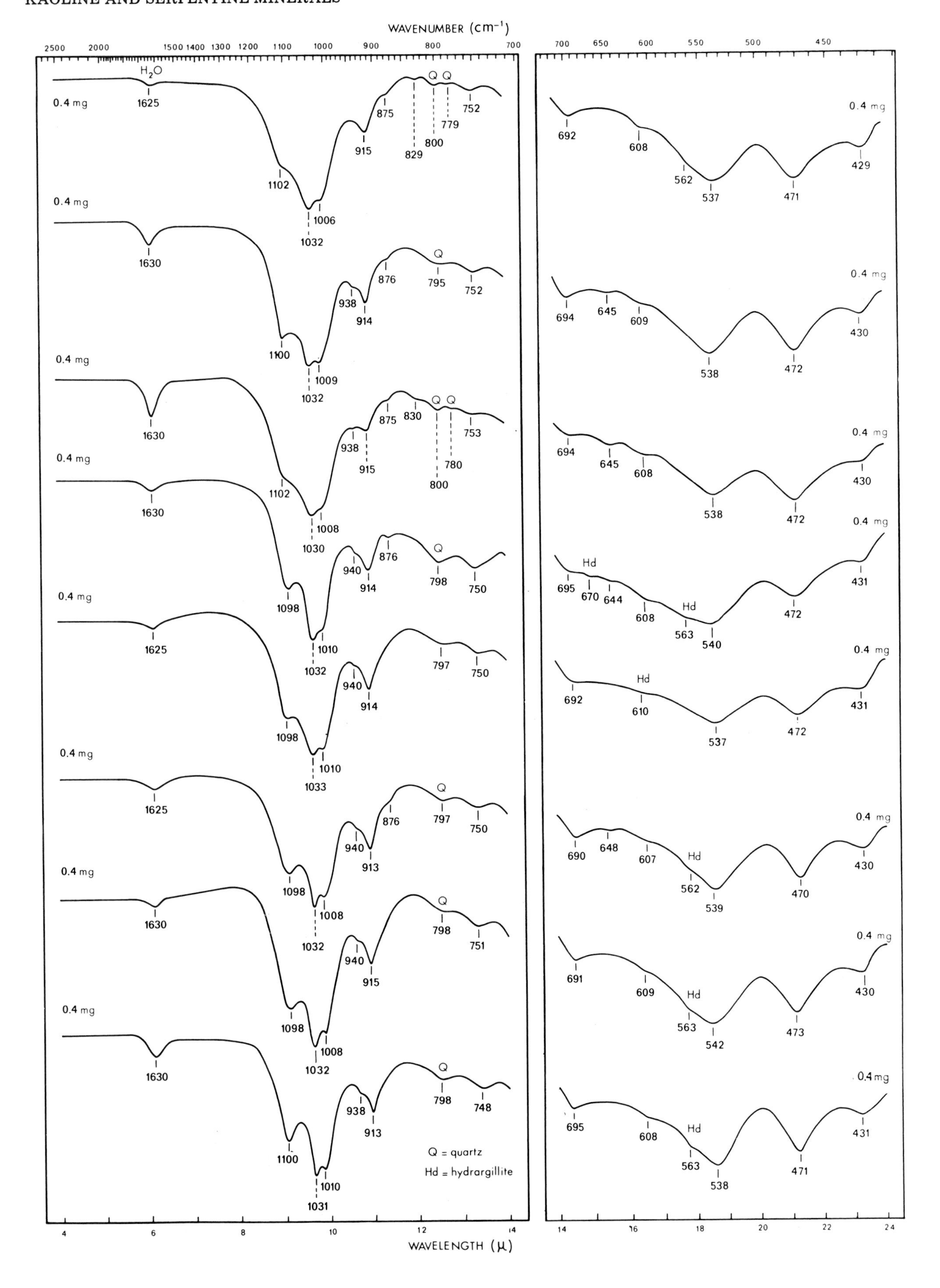
WAVENUMBER (cm^{-1})
WAVELENGTH (μ)
H_2O
Q = quartz
Hd = hydrargillite
0.4 mg

Halloysite
deposit, Tintic, Utah, U.S.A.
pure

Louveigne, Belgium
pure

deposit, Bayonne, France
pure

deposit, Djebel Debar, Algeria
pure
Fig. E.M. 37

deposit, Eureka, N.J., U.S.A.
some hydrargillite
Fig. E.M. 38

Brokopondo, Surinam
some hydrargillite and kaolinite (overlapped)

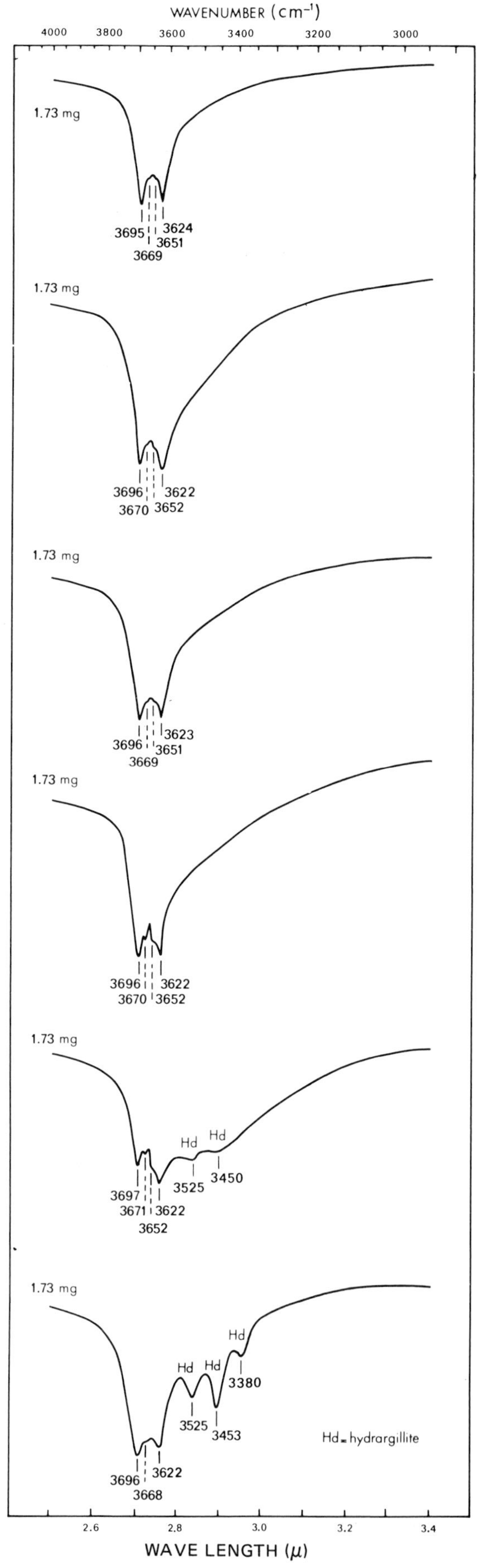

Fig. 6.8. Infrared spectra of halloysite from different origin.

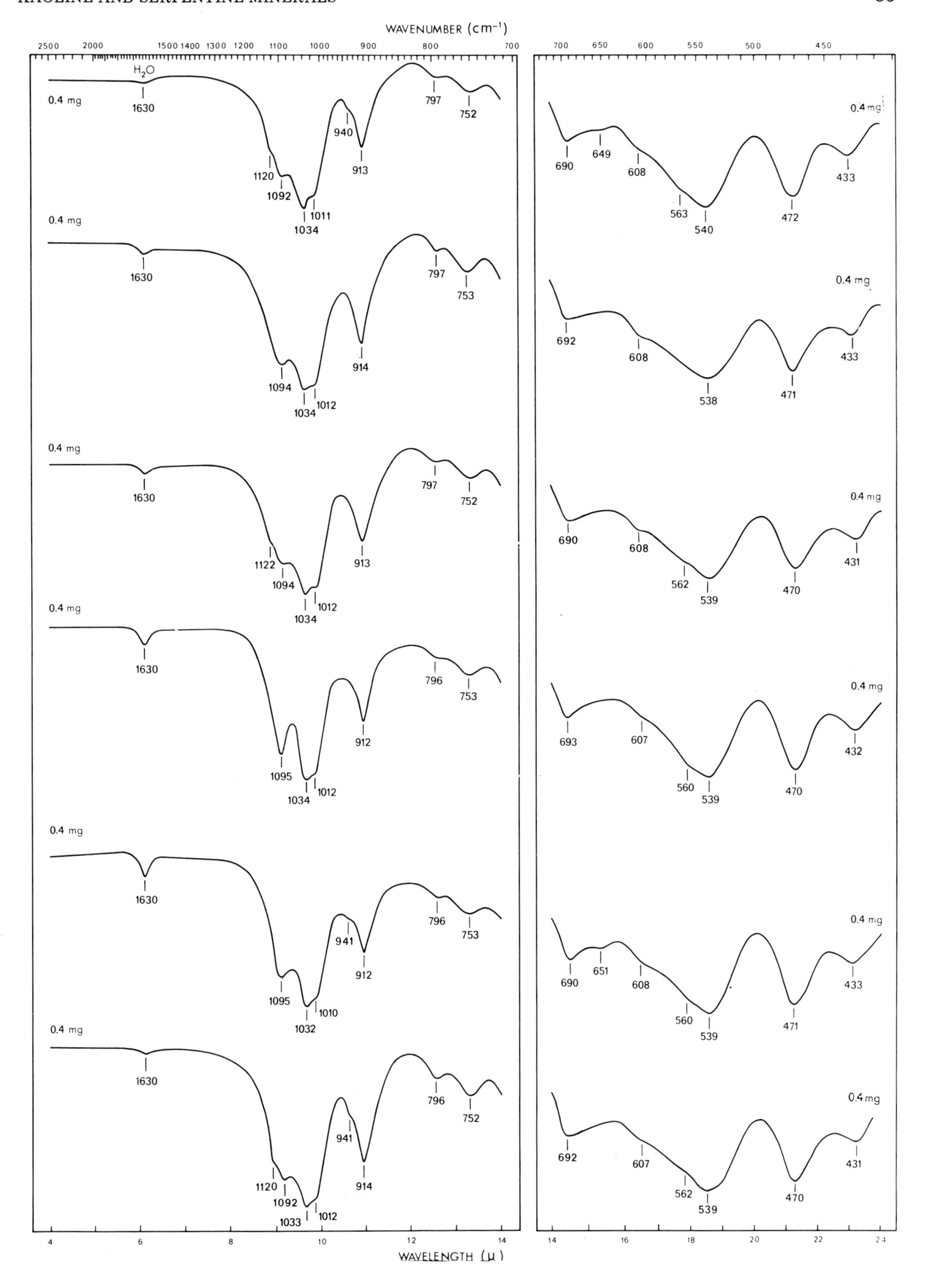
WAVENUMBER (cm^{-1})
WAVELENGTH (μ)
H_2O
0.4 mg
1630
1120
1092
1034
1011
940
913
797
752
690
649
608
563
540
472
433
1094
1012
914
753
692
538
471
1122
562
539
470
431
1095
912
796
693
607
560
432
1032
1010
941
651
1033
562

Halloysite
deposit, Vanwijksvlei, Cape Province, S.A.
some quartz
Fig. E.M. 35

deposit, Capalbio, Italy
some quartz
Fig. E.M. 36

deposit, Martinsberg, Hungary
some quartz
Fig. E.M. 39

deposit, Wagon Wheel, Colo., U.S.A.
some cristobalite (overlapped)

Hydro-halloysite
Baia Mare, Rumania
pure
Fig. E.M. 44

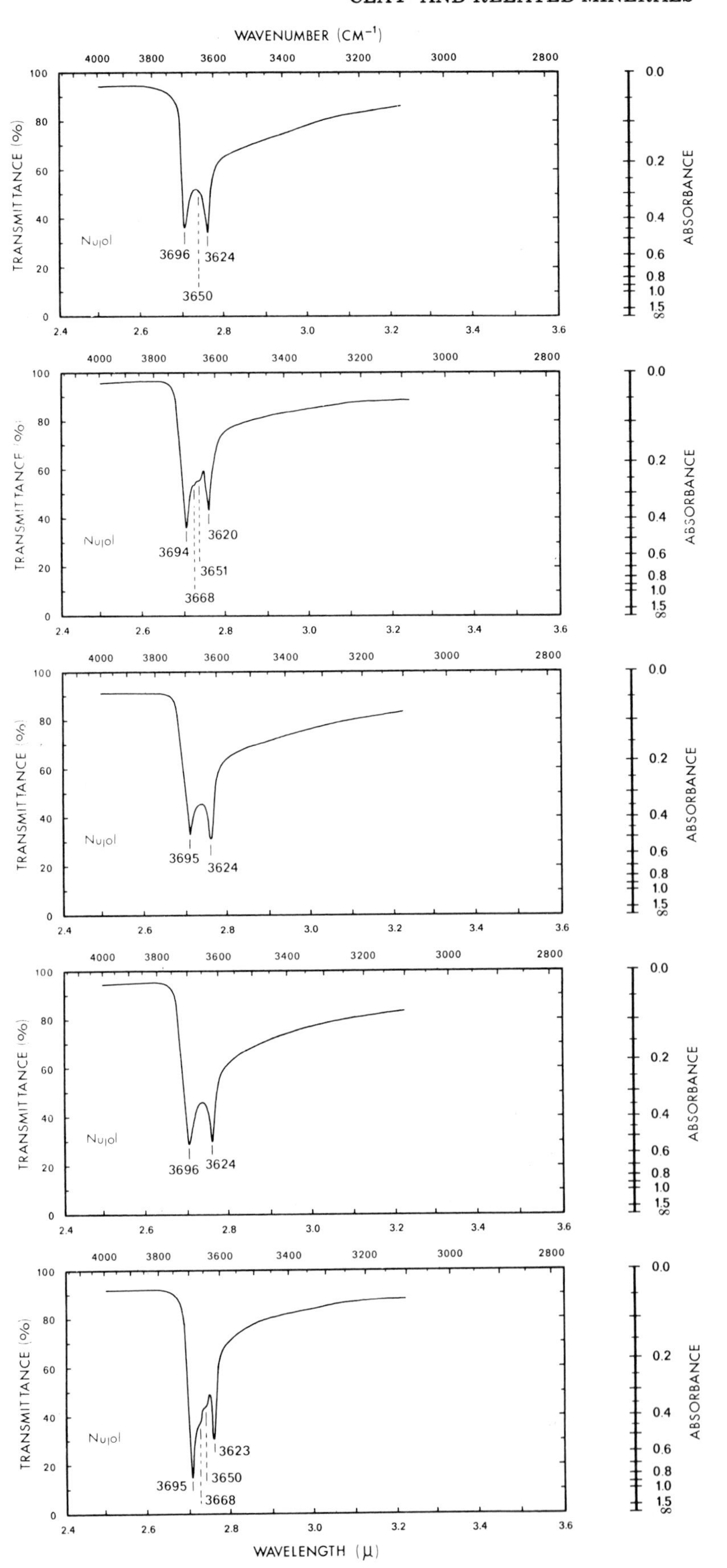

Fig. 6.9. Infrared spectra of halloysite and hydro-halloysite from different origin.

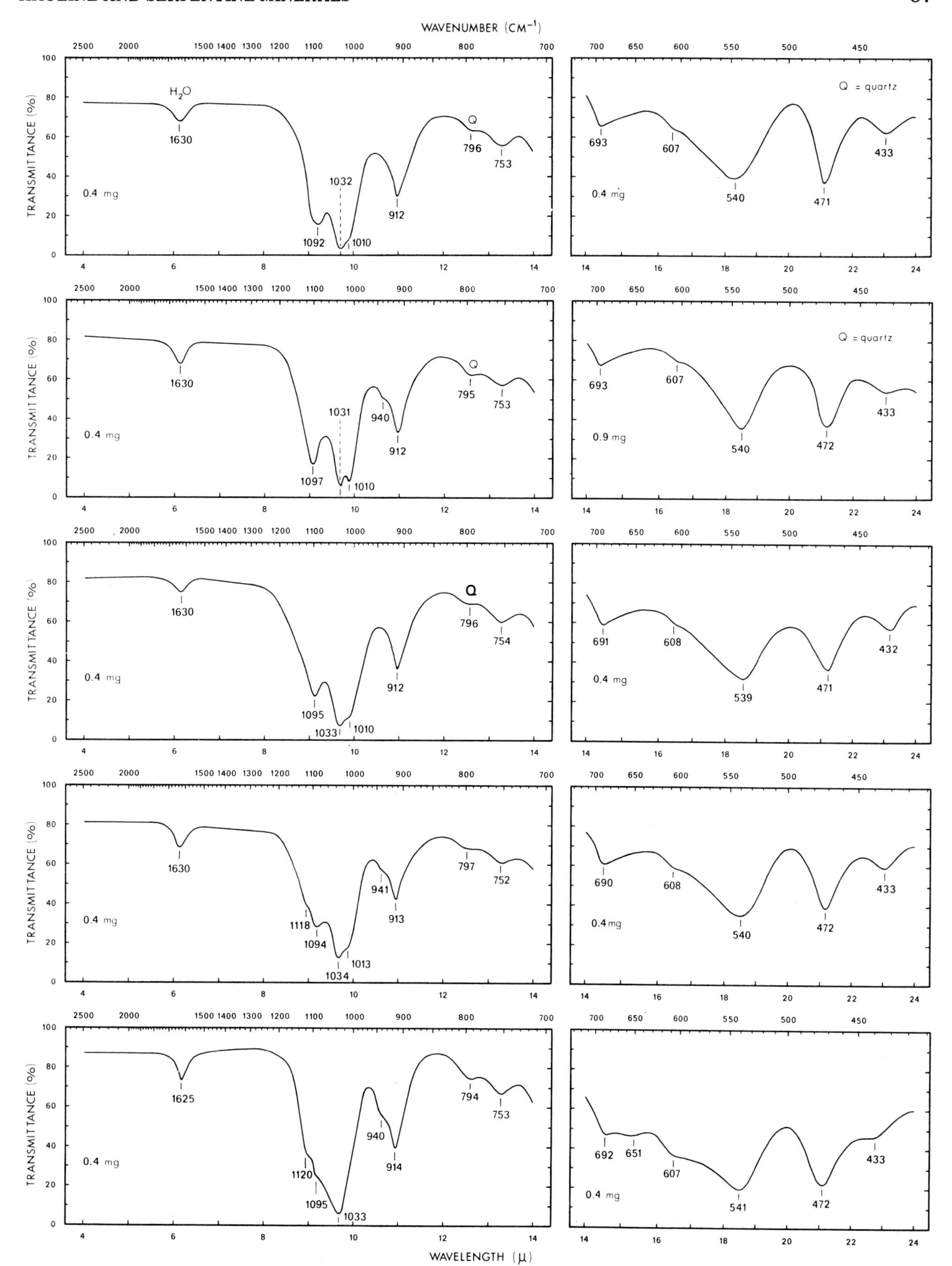
WAVENUMBER (CM⁻¹)
TRANSMITTANCE (%)
WAVELENGTH (μ)
Q = quartz
H2O
0.4 mg
0.9 mg
1630
1625
1092
1032
1010
912
796
753
693
607
540
471
433
1097
1031
940
795
472
1095
1033
754
691
608
539
432
1118
1094
1034
1013
941
913
797
752
690
1120
914
794
692
651
541

Halloysite sediments
Sng. Dadap, N.W. coast of Sumatra, Indonesia
some hydrargillite and quartz

Djokjakarta, Indonesia
some cristobalite (overlapped), kaolinite and allophane (overlapped)
Fig. E.M. 43

Tasmania
some kaolinite (overlapped), hydrargillite, hematite (overlapped) and limonite
Fig. E.M. 40

Les Eyzies, France
some quartz and kaolinite (overlapped)

Tindjowan, Sumatra, Indonesia
some quartz, mica and kaolinite (overlapped)

Mt. Hagen, Papua, Australia
some quartz, limonite, hydrargillite and kaolinite (overlapped)
Fig. E.M. 42

Northern Sumatra, Indonesia
some quartz, hydrargillite and kaolinite (overlapped)

Celebes, Indonesia
some hydrargillite

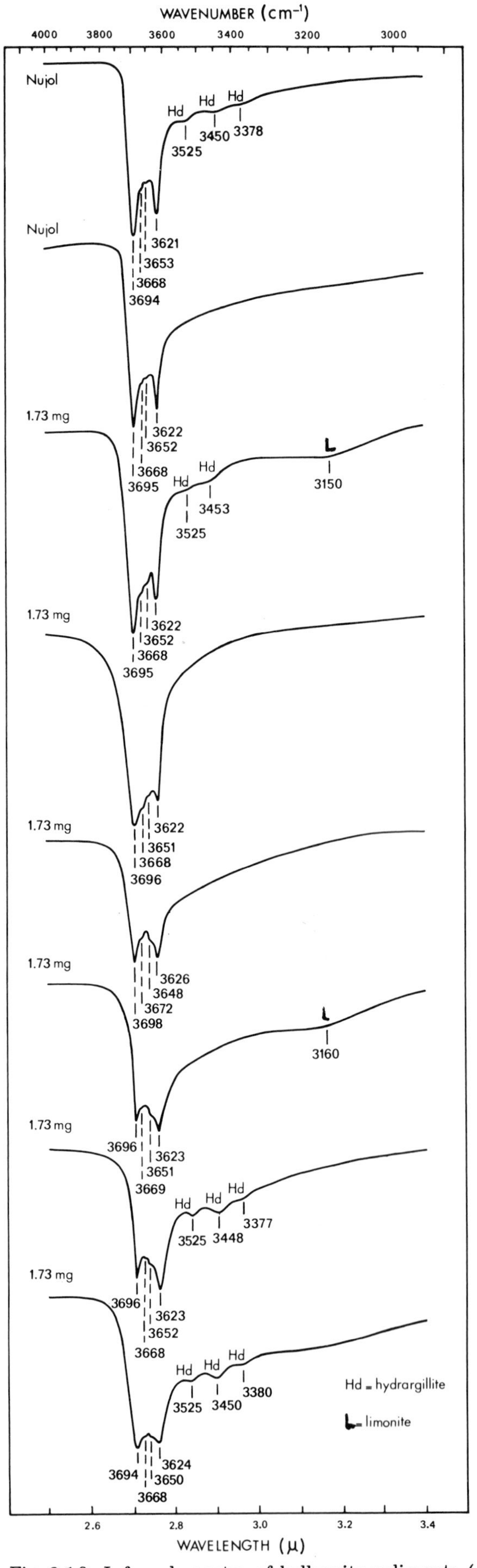

Fig. 6.10. Infrared spectra of halloysite sediments (< 2 μm) from different origin.

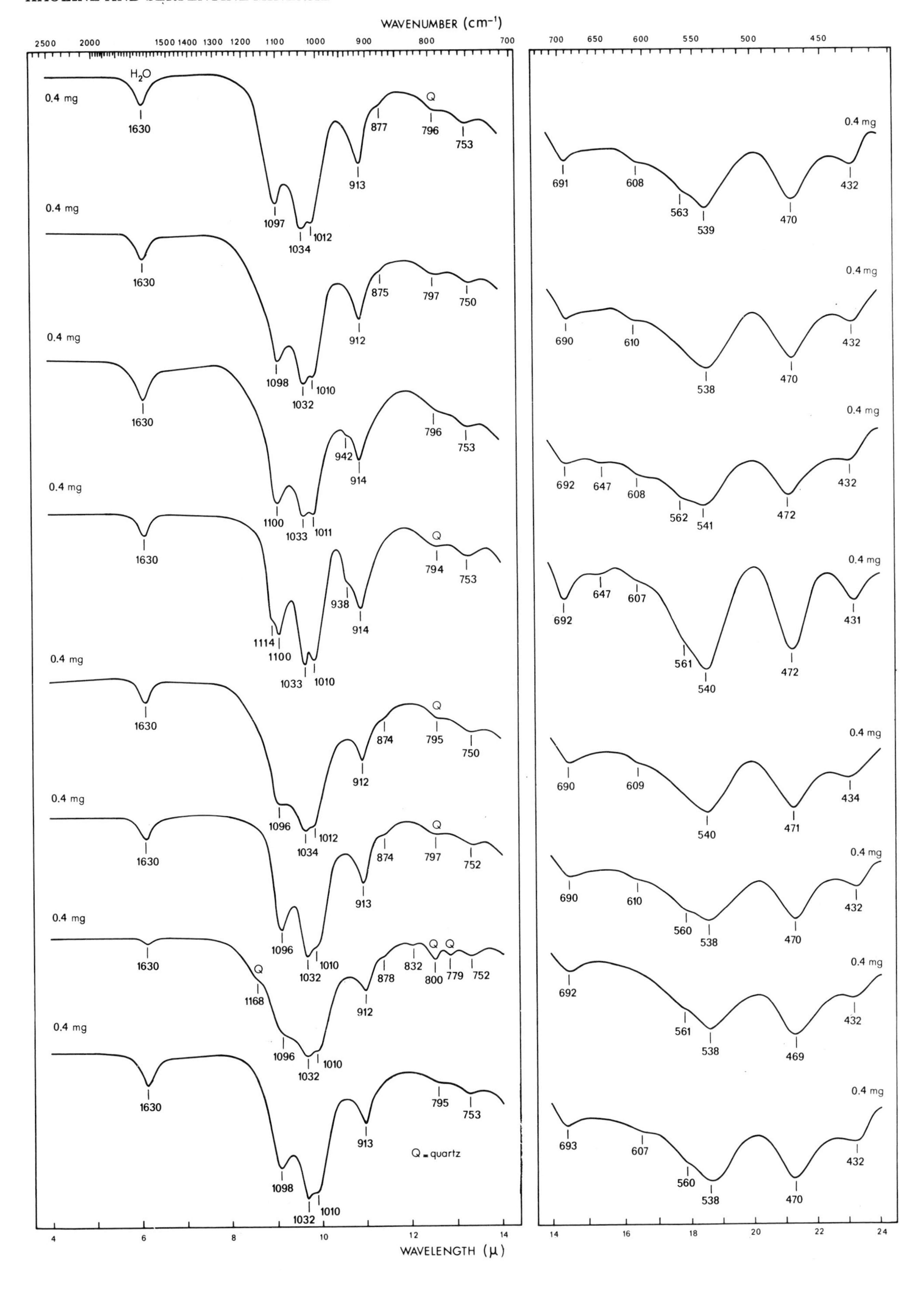
WAVENUMBER (cm⁻¹)
WAVELENGTH (μ)
H₂O
0.4 mg
Q = quartz
1630
1097
1034
1012
913
877
796
753
691
608
563
539
470
432
1098
1032
1010
912
875
797
750
690
610
538
1100
1033
1011
942
914
692
647
608
562
541
472
1114
938
794
607
561
540
431
1096
874
795
609
471
434
752
560
1168
878
832
800
779
469
1098
693

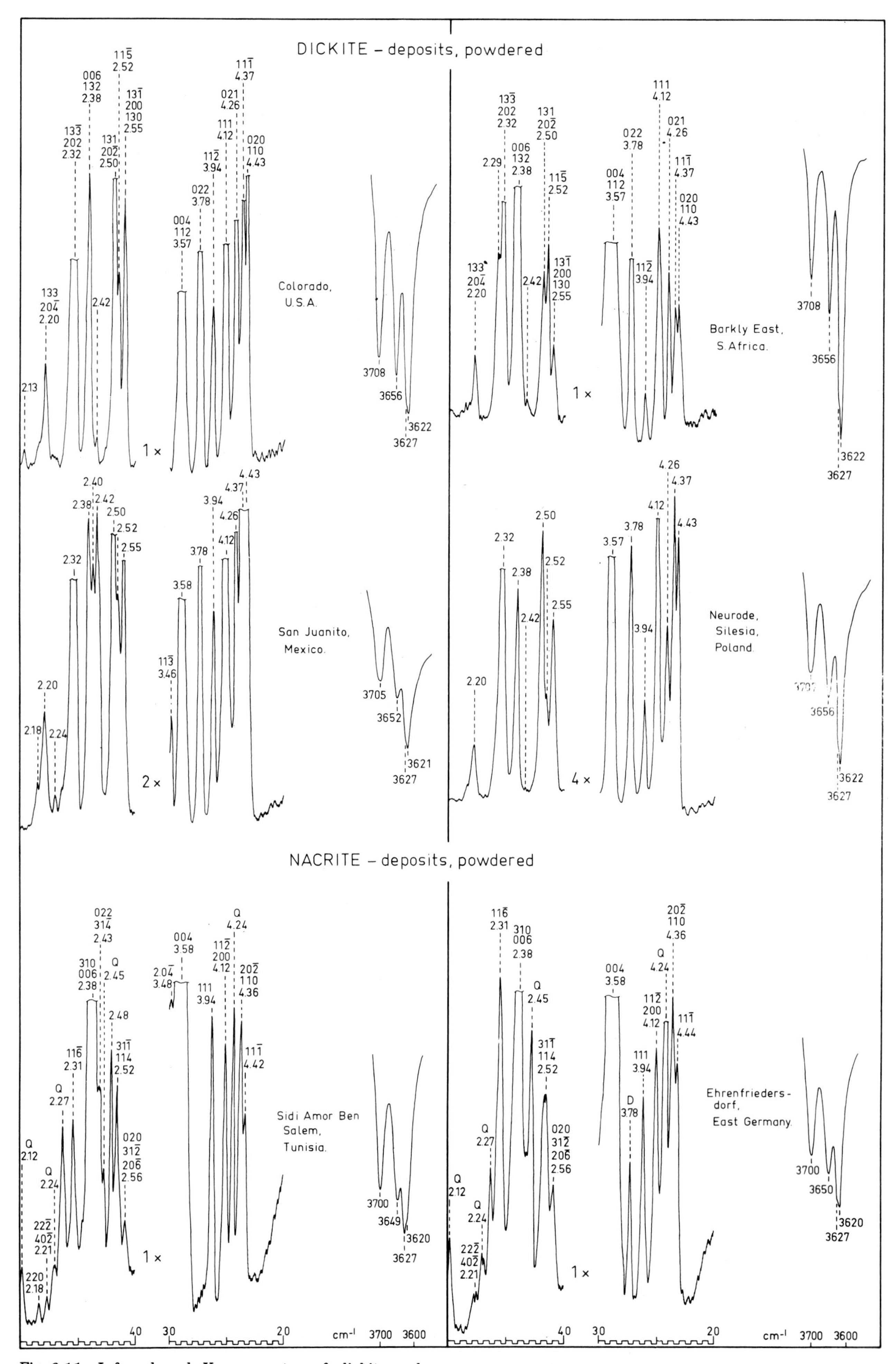

Fig. 6.11. Infrared and X-ray spectra of dickite and nacrite from different deposits. (d spacings in Å.)

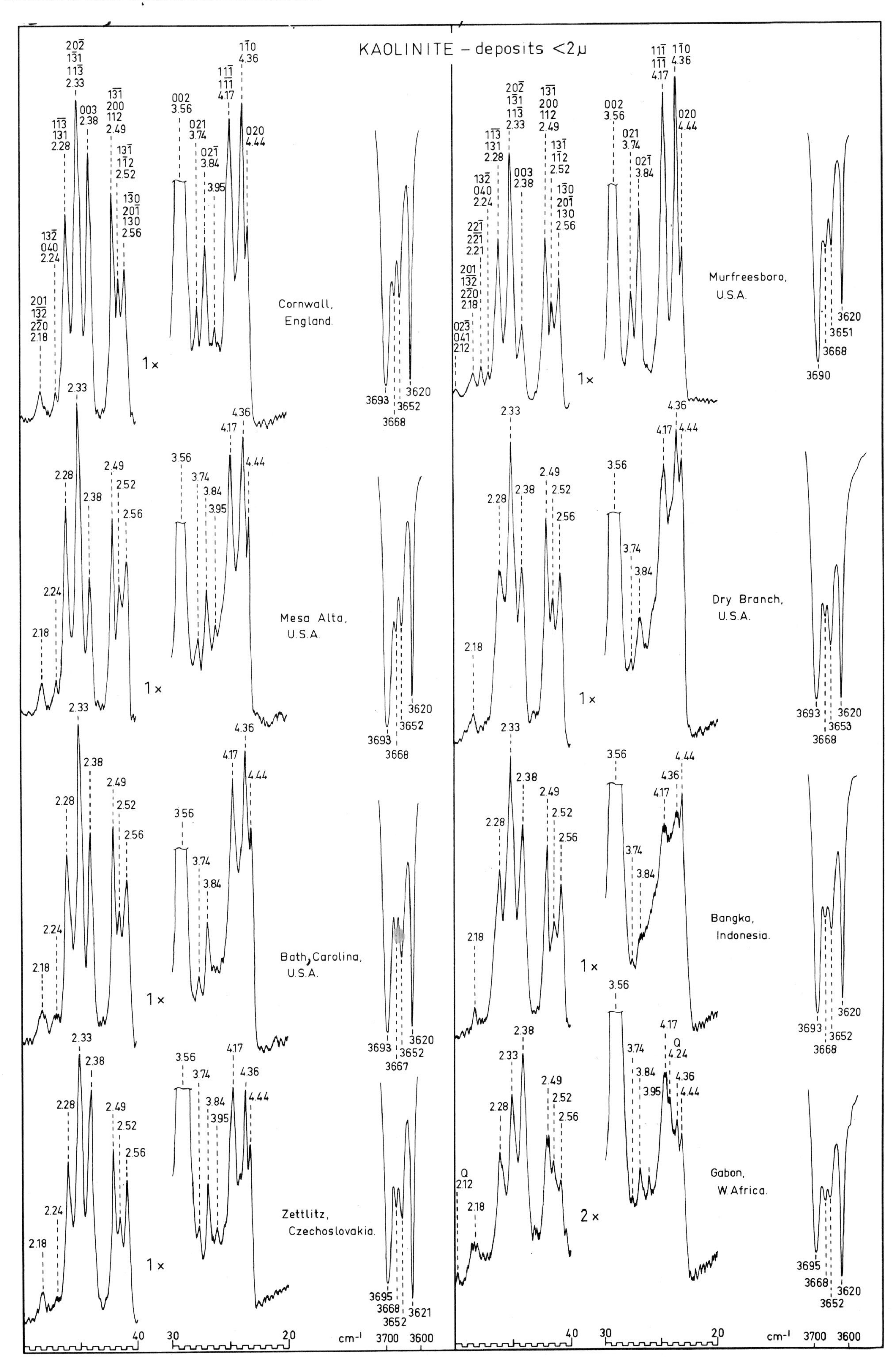

Fig. 6.12. Infrared and X-ray spectra of kaolinite (< 2 μm) from different deposits. (d spacings in Å.)

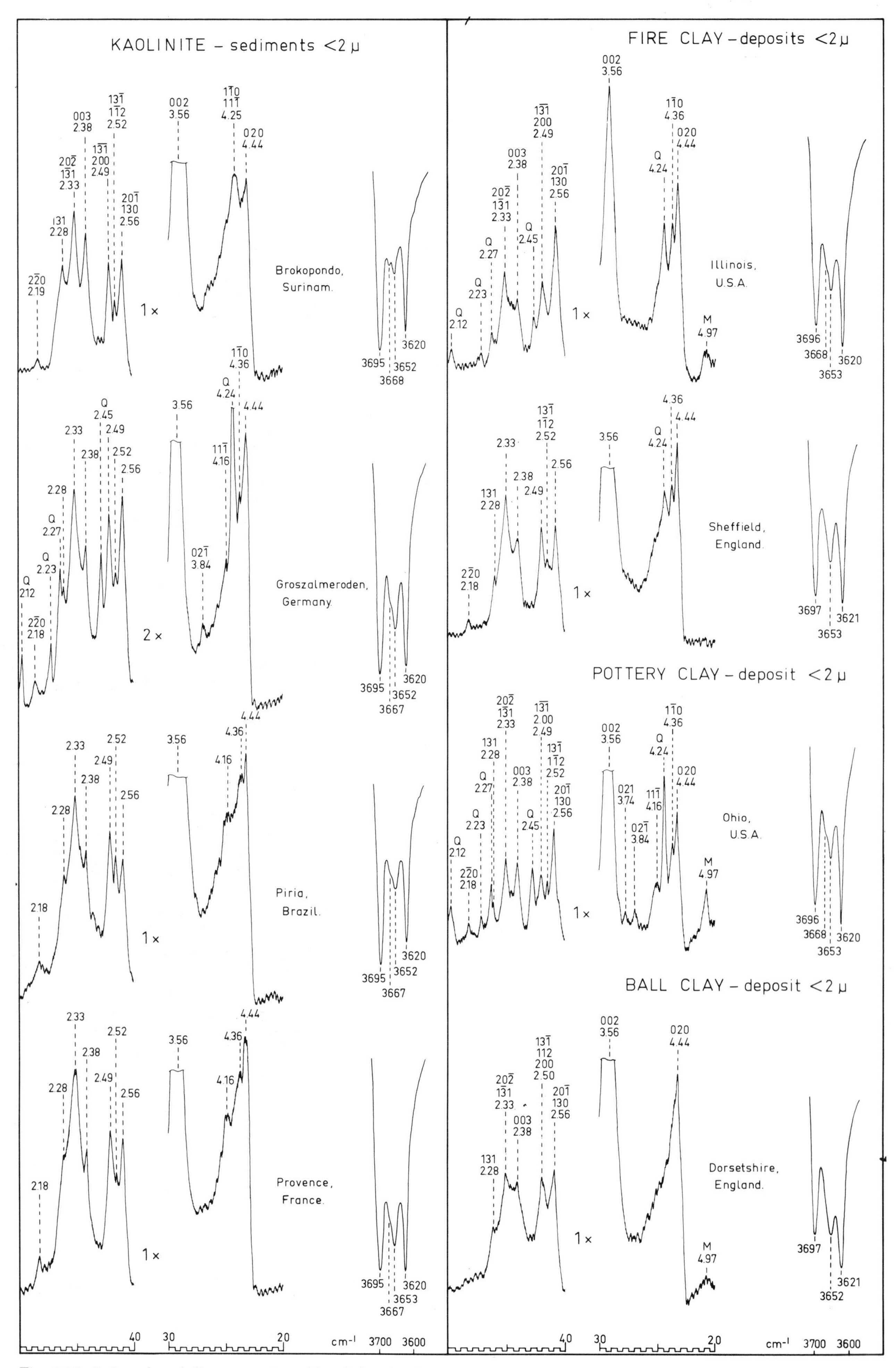

Fig. 6.13. Infrared and X-ray spectra of kaolinite (< 2 μm) from different sediments and of fire clay, pottery clay and ball clay (d spacings in Å.)

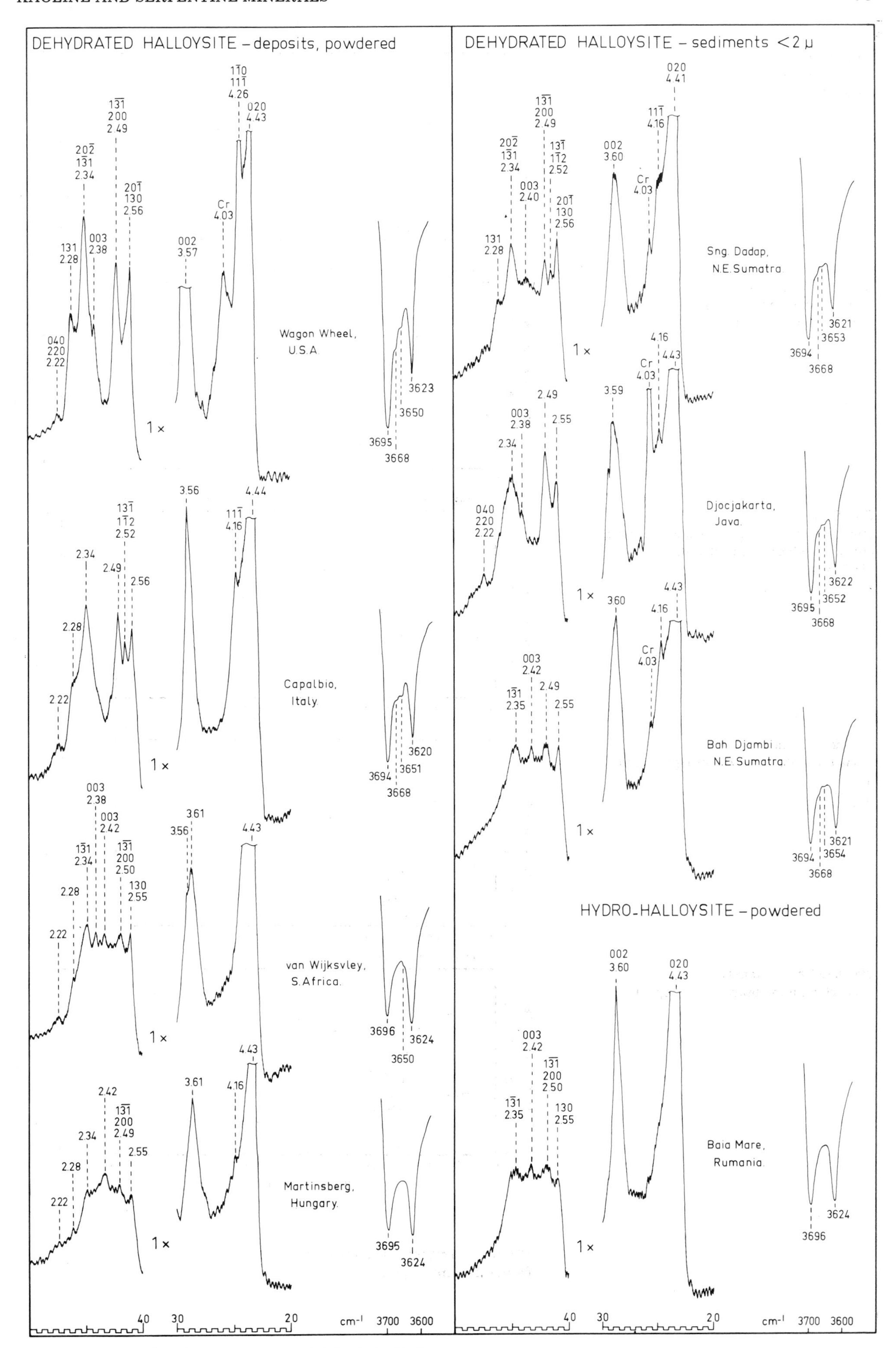

Fig. 6.14. Infrared and X-ray spectra of dehydrated halloysite and hydro-halloysite from different deposits. (d spacings in Å). Cr = cristobalite

Serpentine
Reichenstein, Silesia
some calcite

Kraubath, Styria, Austria
some brucite (overlapped), chlorite (overlapped) and talc (overlapped)

Cardiff, Great Britain
some talc (overlapped), dolomite, magnesite, brucite (overlapped) and chlorite
Fig. E.M. 46

Puy de Vol, France
some dolomite, brucite (overlapped) and chlorite
Fig. E.M. 45

Malili, Celebes, Indonesia
some chlorite and brucite (overlapped)

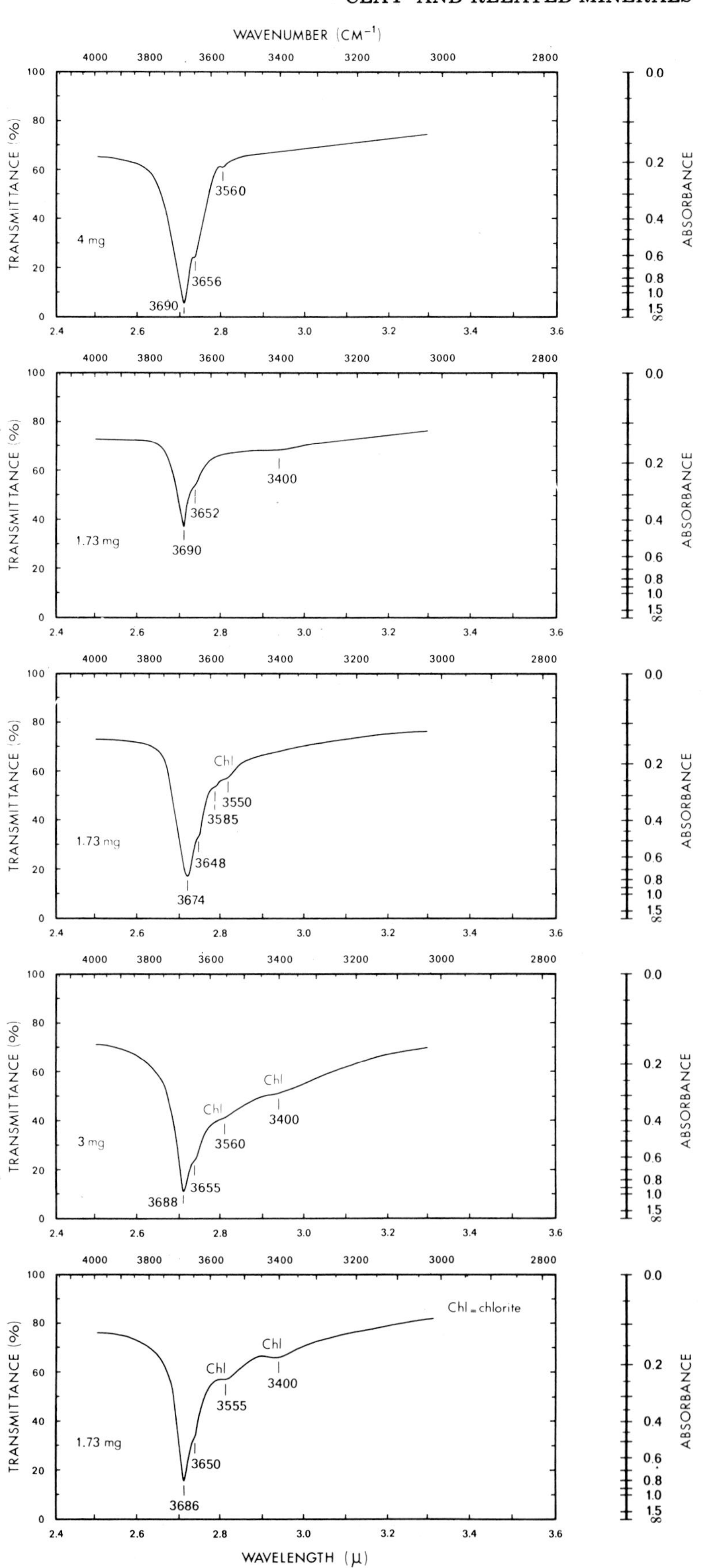

Fig. 6.15. Infrared spectra of serpentine from different origin.

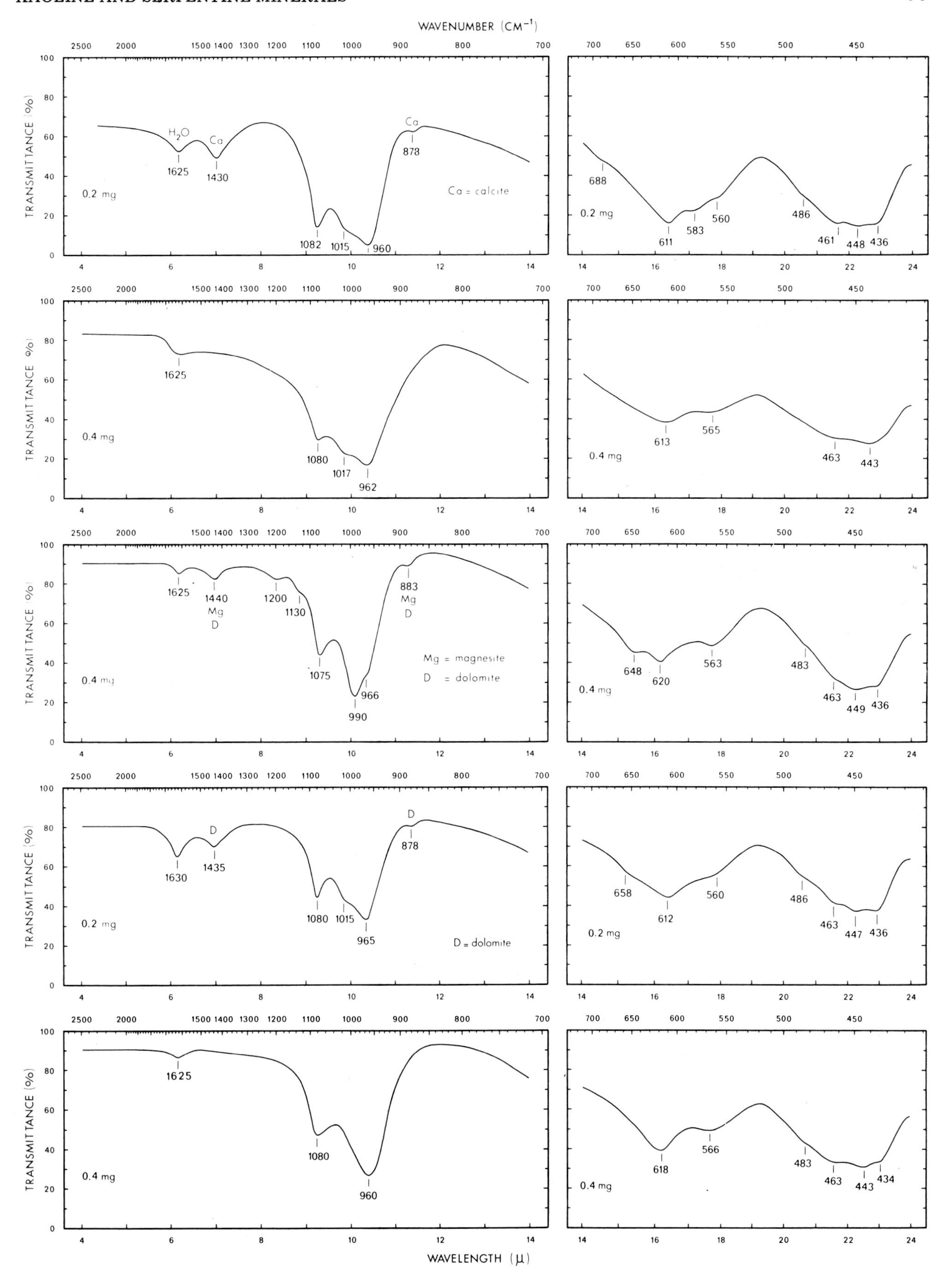

WAVENUMBER (CM⁻¹)
TRANSMITTANCE (%)
WAVELENGTH (μ)
H₂O
Ca
Ca = calcite
1625
1430
878
1082
1015
960
0.2 mg
688
611
583
560
486
461
448
436
1625
1080
1017
962
0.4 mg
613
565
463
443
1440
Mg
D
1200
1130
883
1075
966
990
Mg = magnesite
D = dolomite
648
620
563
483
449
1630
1435
1080
1015
965
D = dolomite
658
612
560
486
447
1080
960
618
566
483
443
434

Chrysotile
Gila, Ariz., U.S.A.
some calcite
Fig. E.M. 47

Quebec, Canada
some calcite
Fig. E.M. 48

Regal Mine, U.S.A.
some calcite

Pinal, Ariz., U.S.A.
some calcite

Davos, Switzerland
pure

Sweden
pure

Zöblitz, Saxony, Germany
chlorite

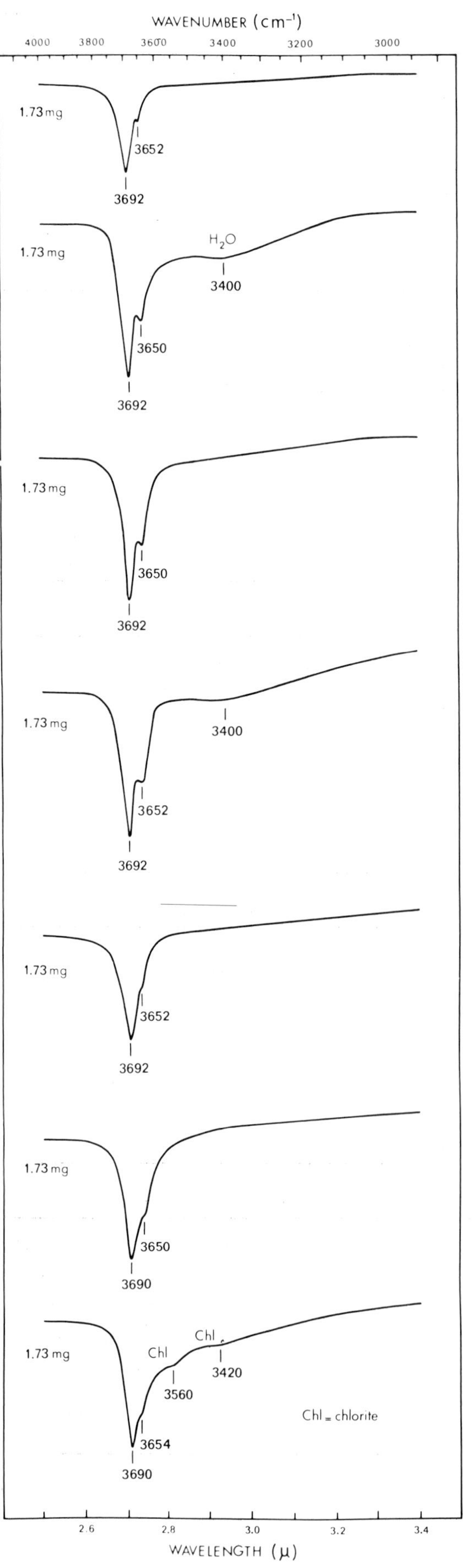

Fig. 6.16. Infrared spectra of chrysotile from different origin.

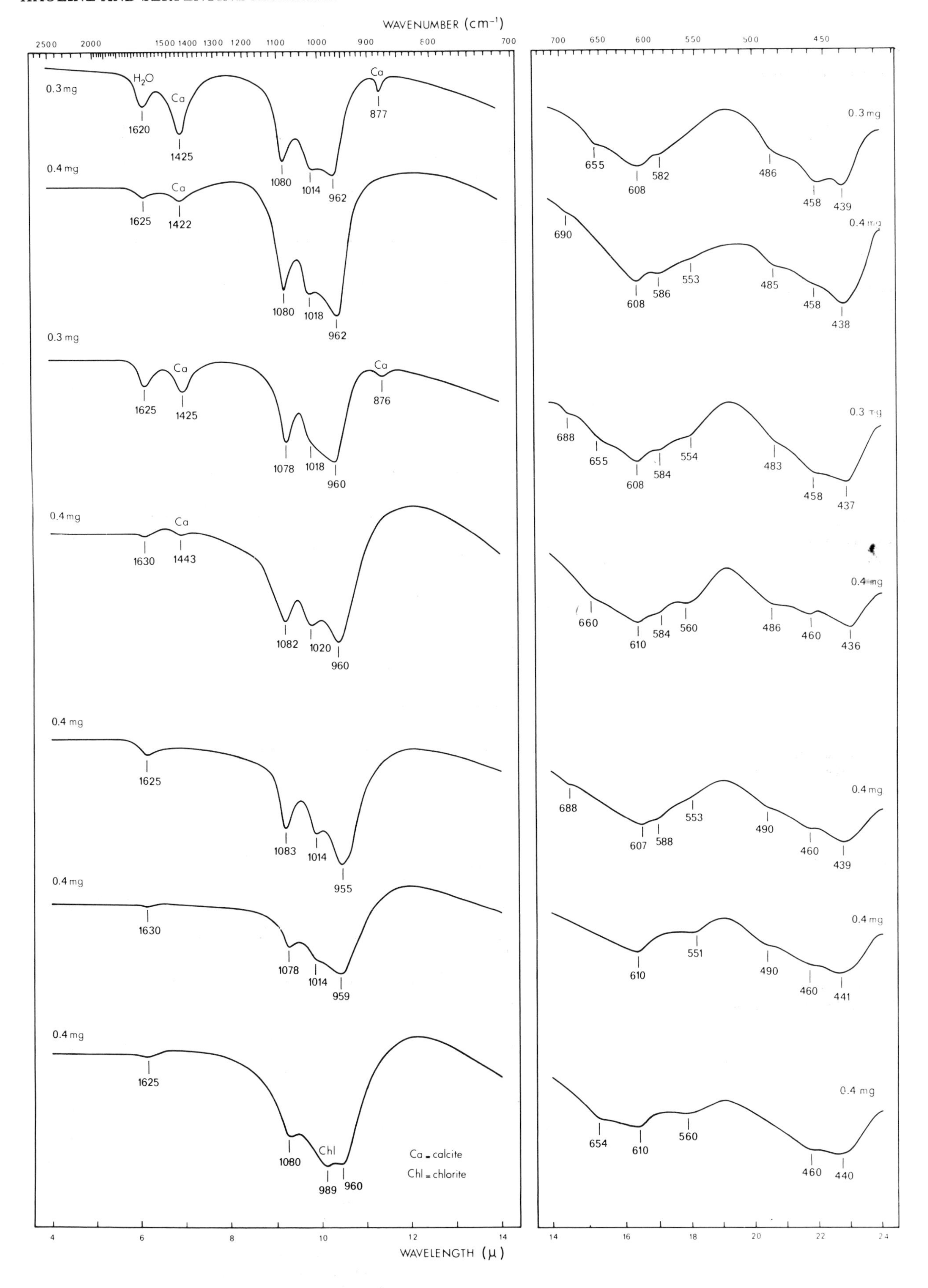
WAVENUMBER (cm⁻¹)
WAVELENGTH (μ)
0.3 mg
H₂O
Ca
1620
1425
1080
1014
962
877
0.4 mg
1625
1422
1018
0.3 mg
1625
1425
1078
1018
960
876
0.4 mg
1630
1443
1082
1020
960
0.4 mg
1625
1083
1014
955
0.4 mg
1630
1078
1014
959
0.4 mg
1625
1080
Chl
989
960
Ca = calcite
Chl = chlorite
655
608
582
486
458
439
690
608
586
553
485
458
438
688
655
608
584
554
483
458
437
660
610
584
560
486
460
436
688
607
588
553
490
460
439
610
551
490
460
441
654
610
560
460
440

Chrysotile
Moravia, Czechoslovakia
pure
Fig. E.M. 49

Zermatt, Switzerland
pure

Hydro-chrysotile
Perkupa, Hungary
pure

Tremolite-asbestos
eastern Pennsylvania, U.S.A.
pure

Amphibole-asbestos
Bozeman, Montana, U.S.A.
pure

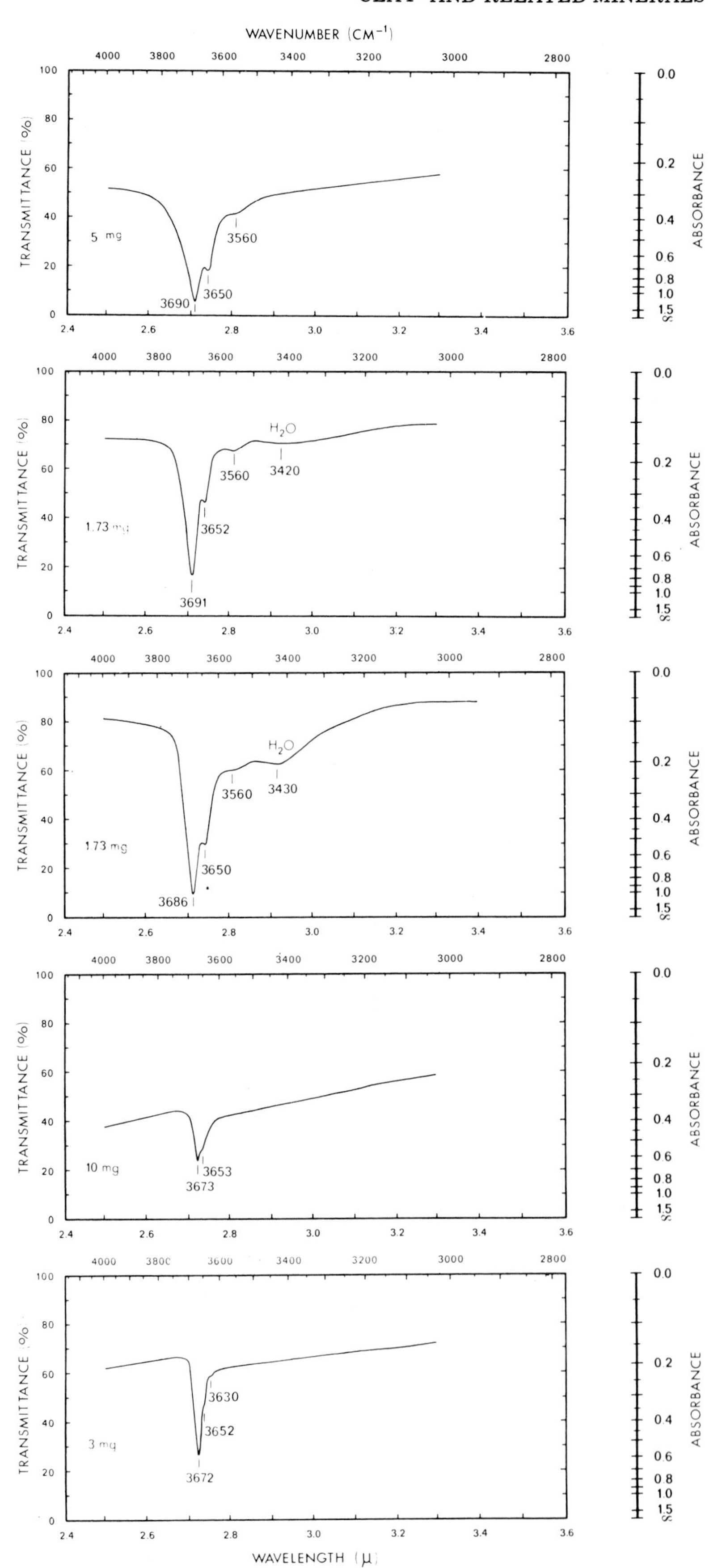

Fig. 6.17. Infrared spectra of chrysotile, hydro-chrysotile, tremolite-asbestos and amphibole-asbestos from different origin.

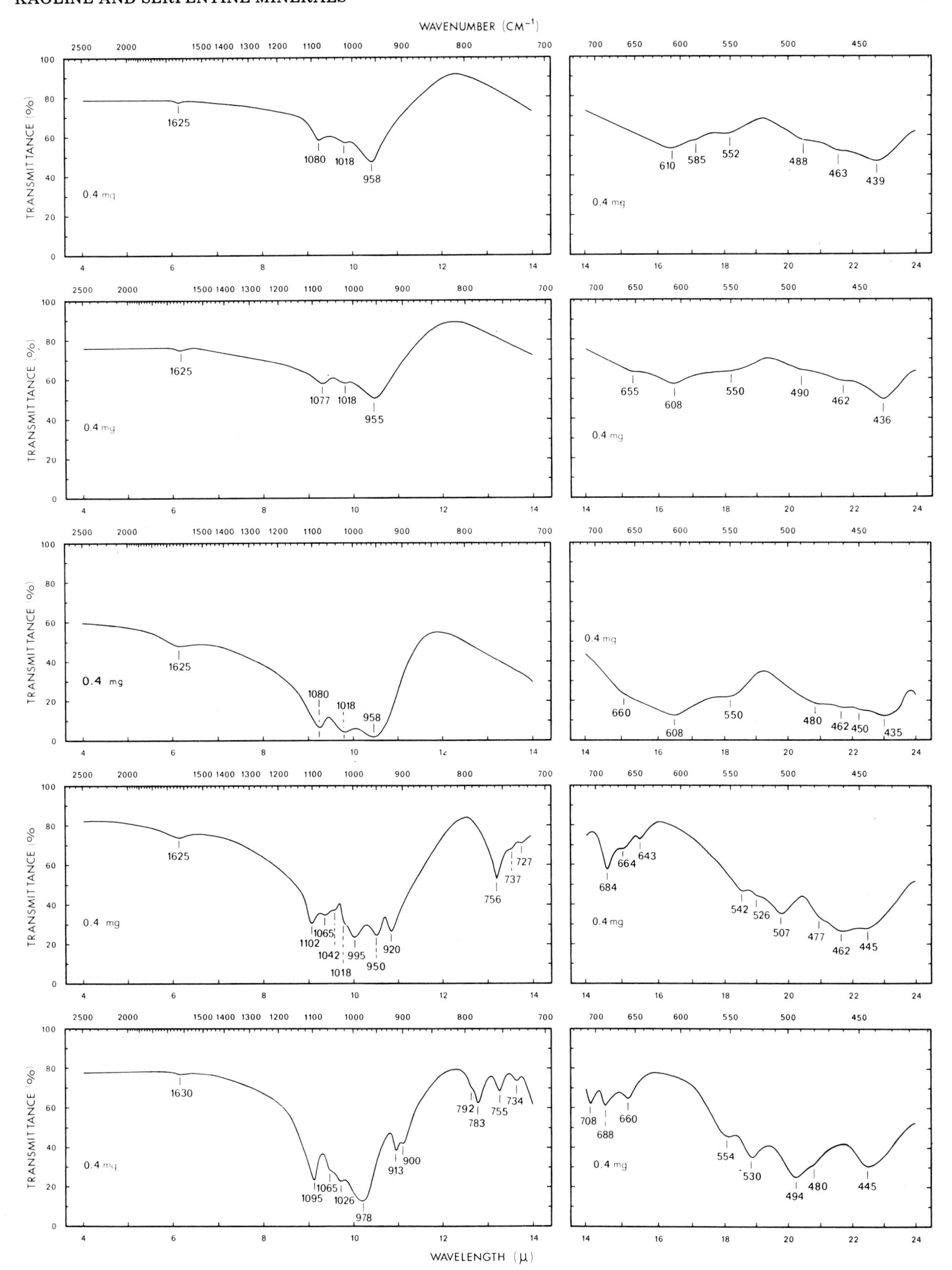
WAVENUMBER (CM⁻¹)
TRANSMITTANCE (%)
WAVELENGTH (μ)
0.4 mg
1625
1080
1018
958
610
585
552
488
463
439
1077
955
655
608
550
490
462
436
660
480
450
435
1102
1065
1042
995
950
920
756
737
727
684
664
643
542
526
507
477
445
1630
1095
1026
978
913
900
792
783
755
734
708
688
554
530
494

Antigorite
Antigorio, Italy
some chlorite and chrysotile
Fig. E.M. 50

Friesach, Austria
some dolomite

Lancaster, Pa., U.S.A.
some chlorite
Fig. E.M. 51

Macedonia, Yugoslavia
pure

Tirol, Austria
some chlorite

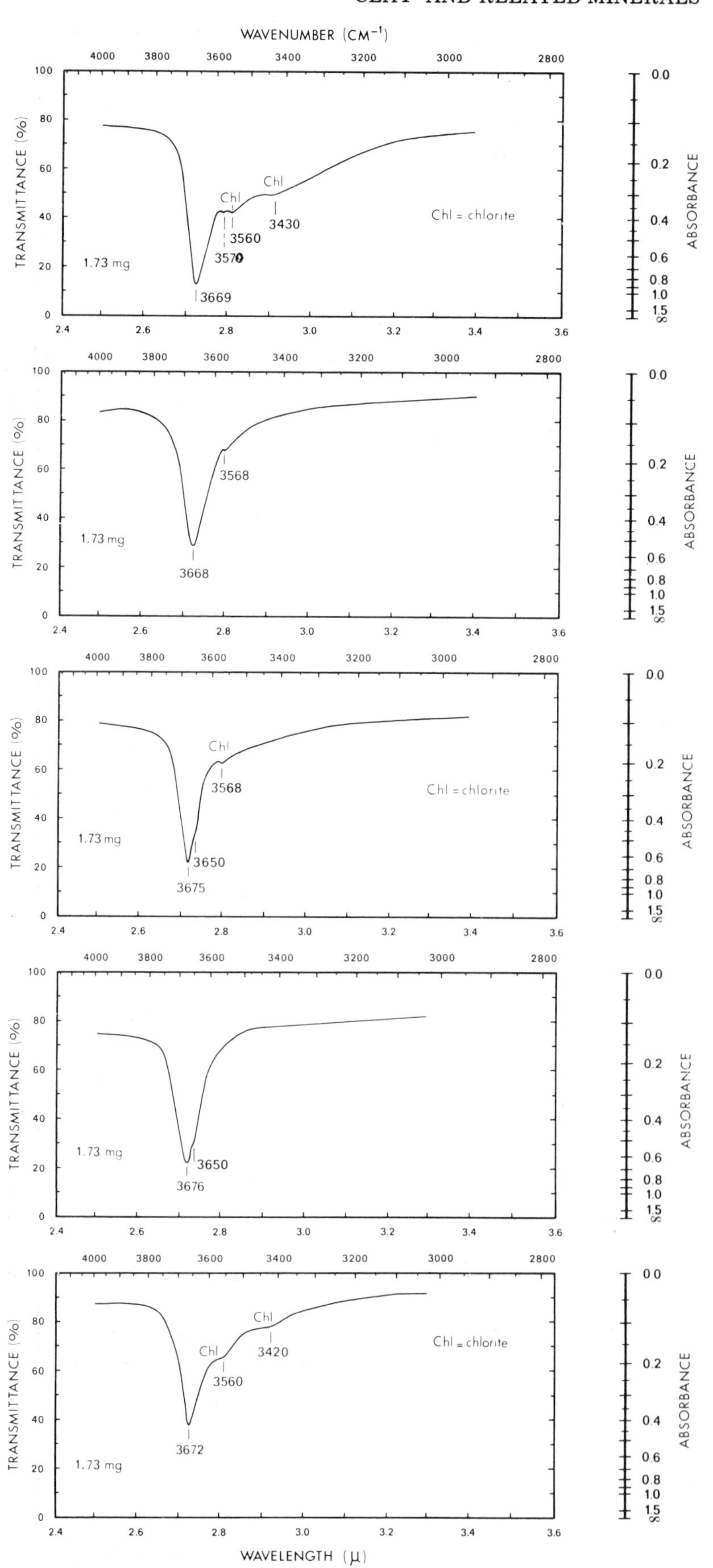

Fig. 6.18. Infrared spectra of antigorite from different origin.

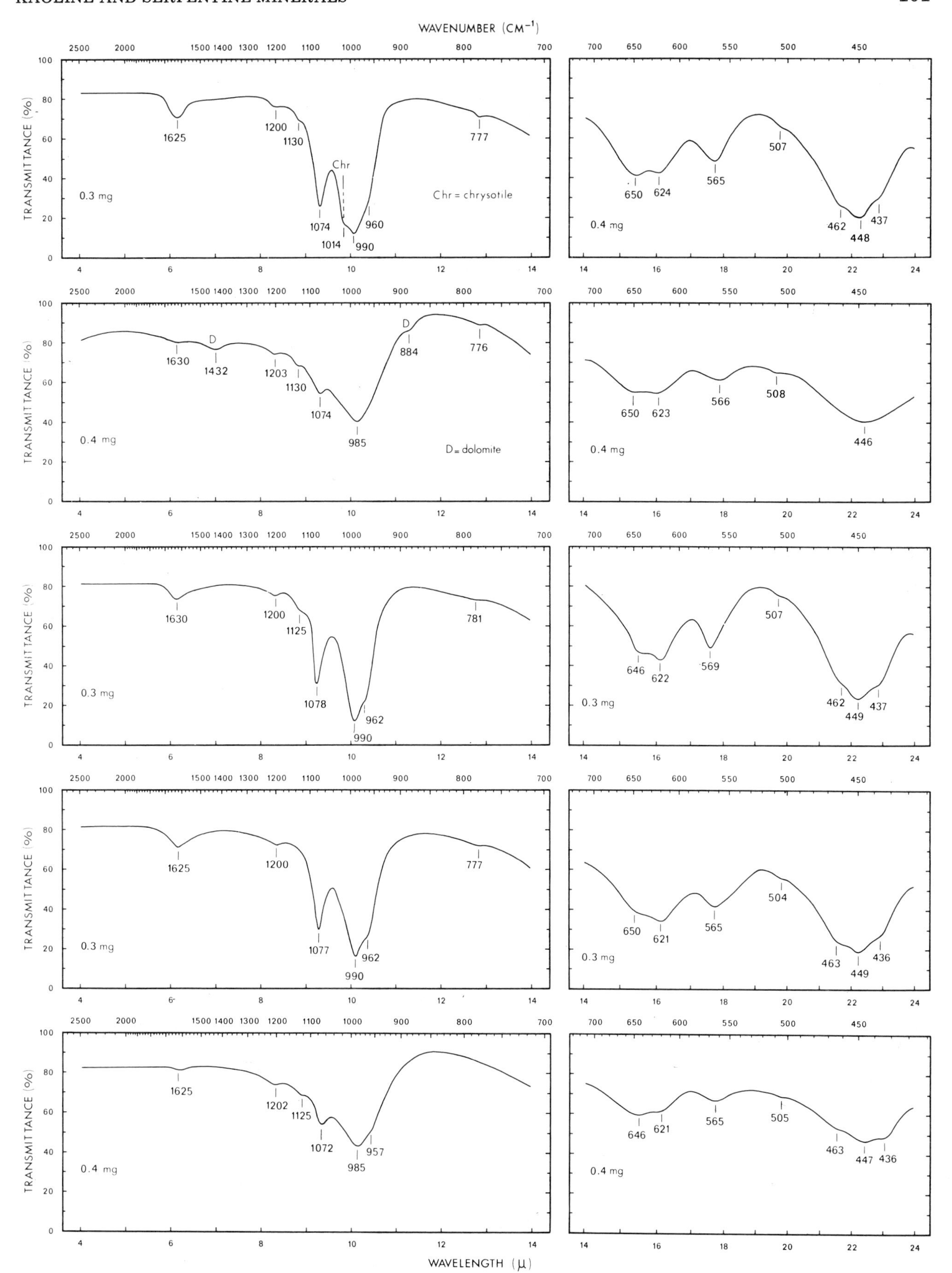

WAVENUMBER (CM⁻¹)
TRANSMITTANCE (%)
WAVELENGTH (μ)
0.3 mg
Chr = chrysotile
Chr
1625
1200
1130
1074
1014
990
960
777
650
624
565
507
462
448
437
0.4 mg
D = dolomite
D
1630
1432
1203
1130
1074
985
884
776
650
623
566
508
446
1630
1200
1125
1078
990
962
781
646
622
569
507
462
449
437
1625
1200
1077
990
962
777
650
621
565
504
463
449
436
1625
1202
1125
1072
985
957
646
621
565
505
463
447
436

Antigorite
Tirol, Austria
pure

Oberfranken, Germany
pure
Fig. E.M. 52

Valtourmanche, Aosta, Italy
pure

Picrolite
Zöblitz, Germany
some talc

Shippman Range, Great Britain
pure
Fig. E.M. 53

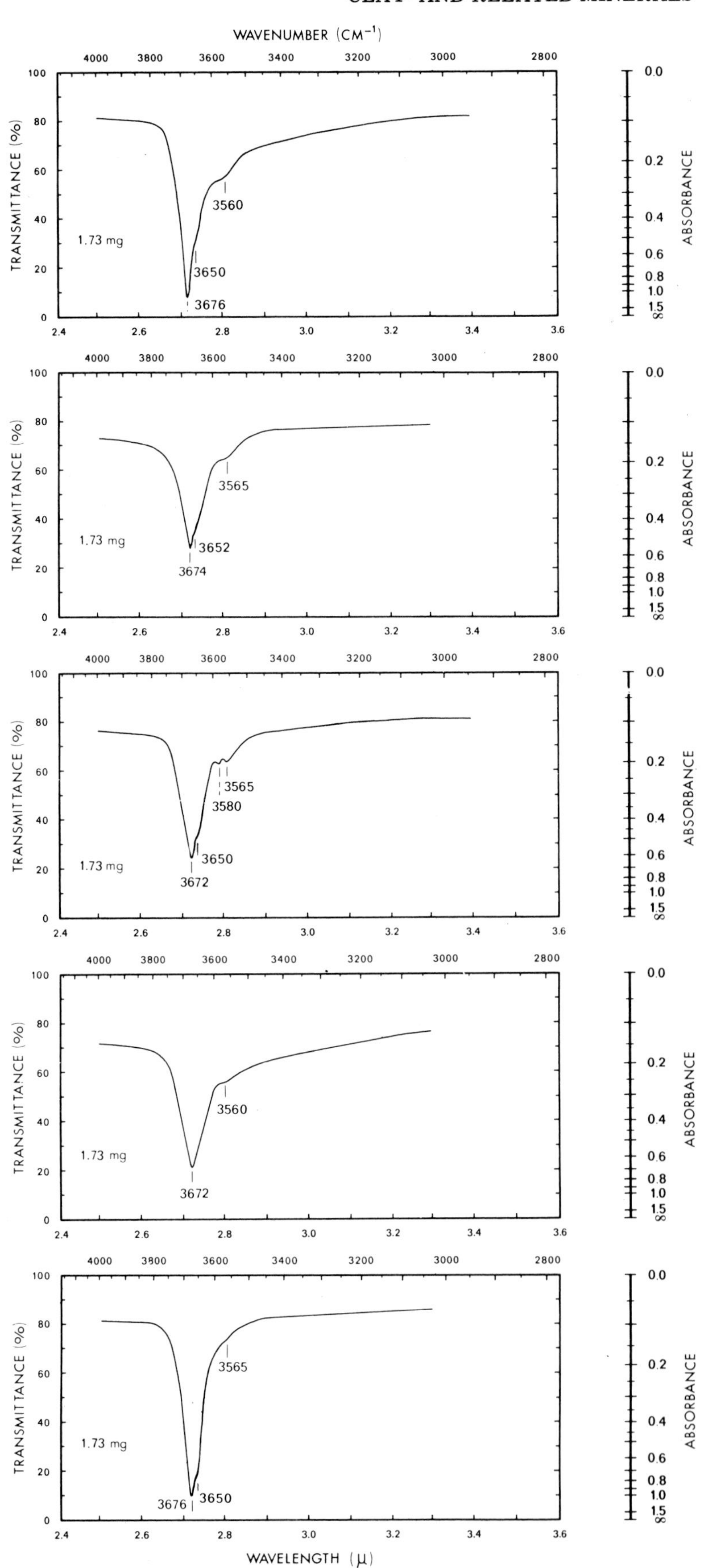

Fig. 6.19. Infrared spectra of antigorite and picrolite from different origin.

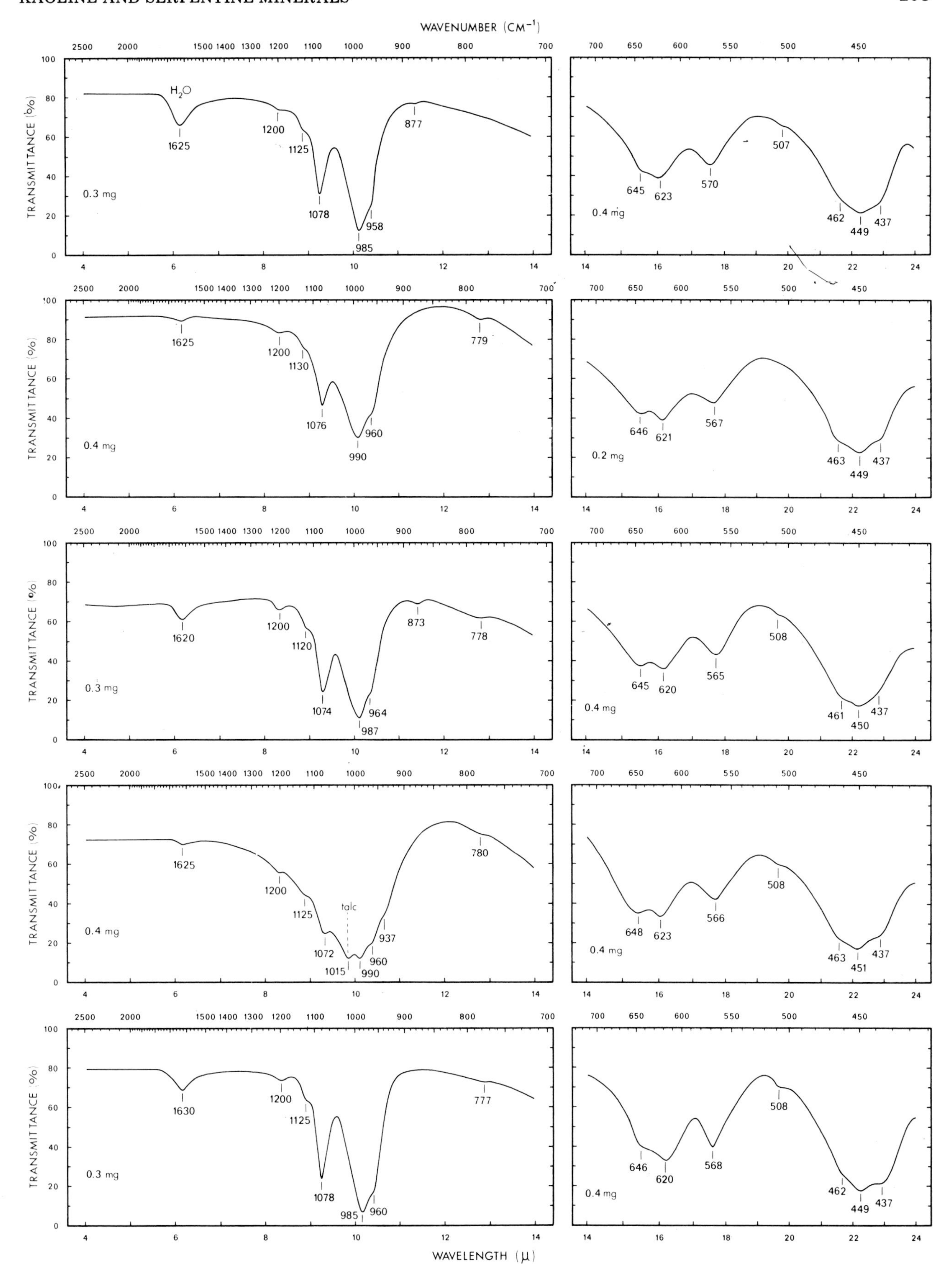
WAVENUMBER (CM⁻¹)
TRANSMITTANCE (%)
H₂O
1625
1200
1125
1078
985
958
877
0.3 mg
645
623
570
507
462
449
437
0.4 mg
1625
1200
1130
1076
990
960
779
0.4 mg
646
621
567
463
449
437
0.2 mg
1620
1200
1120
1074
987
964
873
778
0.3 mg
645
620
565
508
461
450
437
0.4 mg
1625
1200
1125
talc
1072
1015
990
960
937
780
0.4 mg
648
623
566
508
463
451
437
0.4 mg
1630
1200
1125
1078
985
960
777
0.3 mg
646
620
568
508
462
449
437
0.4 mg
WAVELENGTH (μ)

Hydro-antigorite
Dunabogdány, Hungary
some calcite
Fig. E.M. 59

Lizardite
St. Margherita, Pavia, Italy
some chlorite, quartz
Fig. E.M. 56

Snarum, Norway
some dolomite, chlorite and brucite (overlapped)
Fig. E.M. 57

The Lizard, Cornwall, Great Britain
Fig. E.M. 55

Kennack Cove, Great Britain
some talc and chlorite

The Lizard, Kennack Sands, Great Britain
some talc and chlorite
Fig. E.M. 54

Dunabogdany, Hungary
some calcite

Unst-type mineral
Unst, Shetland Isles
some chlorite, talc, and brucite (overlapped)
Fig. E.M. 58

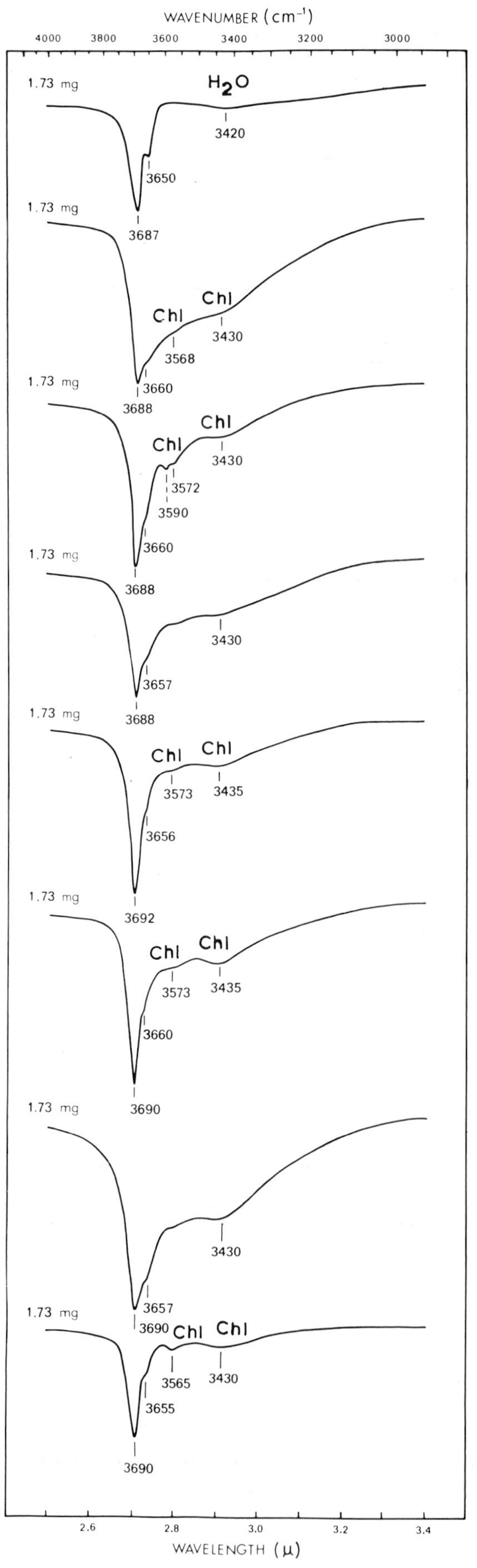

Fig. 6.20. Infrared spectra of hydro-antigorite, lizardite and Unst-type mineral from different origin.

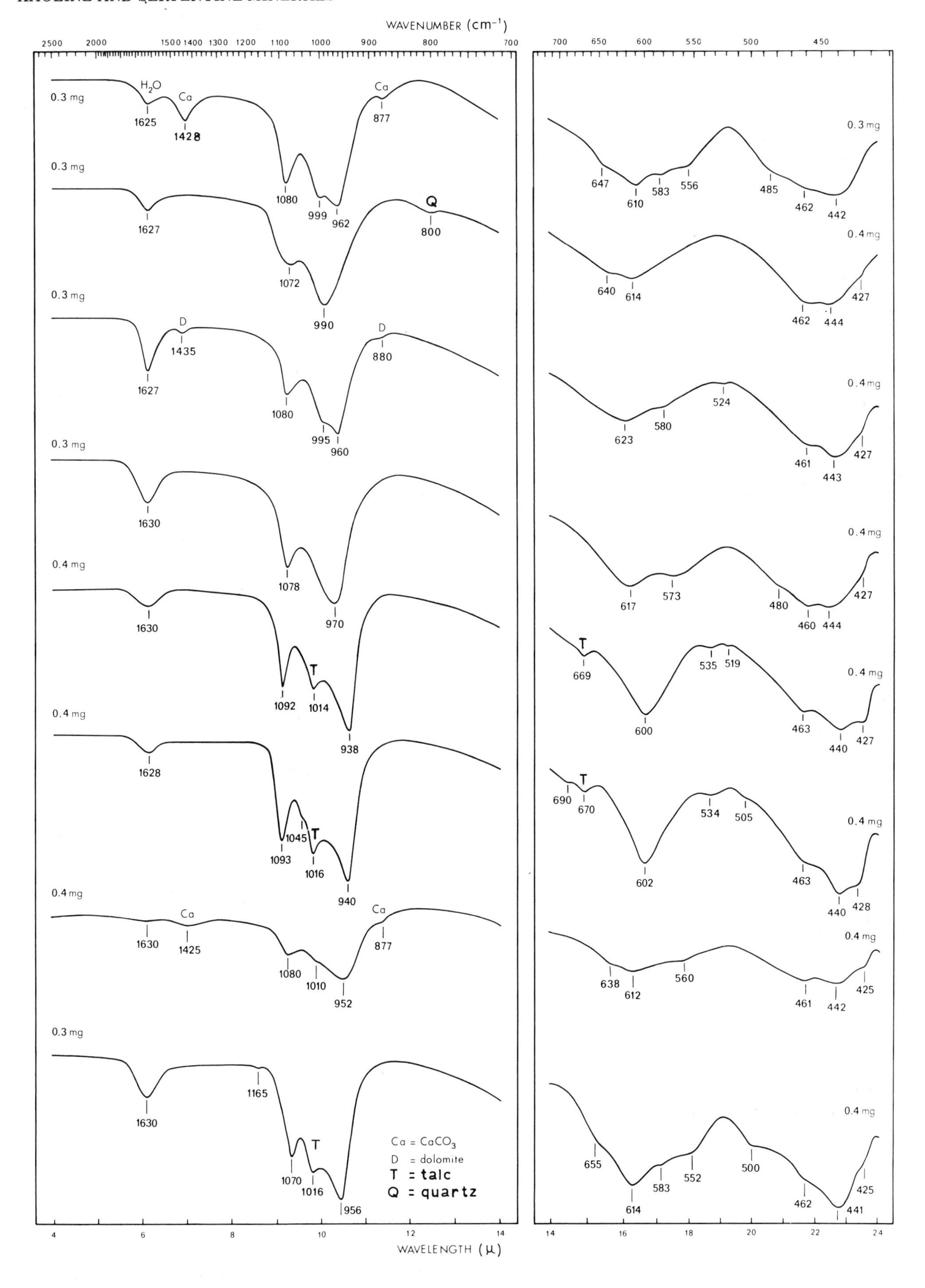
WAVENUMBER (cm⁻¹)
WAVELENGTH (μ)
Ca = CaCO₃
D = dolomite
T = talc
Q = quartz

Amesite
Lijdenburg, Transvaal, S.A.
mainly mineral of amphibole group
Fig. E.M. 65

Ural, U.S.S.R.
mainly chlorite
Museum 1033 12

Chester, Mass., U.S.A.
mainly chlorite
Museum 80715

Hydro-amesite
Halap Berg, Hungary
some calcite
Fig. E.M. 66

Greenalite
Port Arthur, Manchuria
siderite, calcite and quartz
Fig. E.M. 67

Gilbert, Ma., U.S.A.
quartz and some siderite ($FeCO_3$)

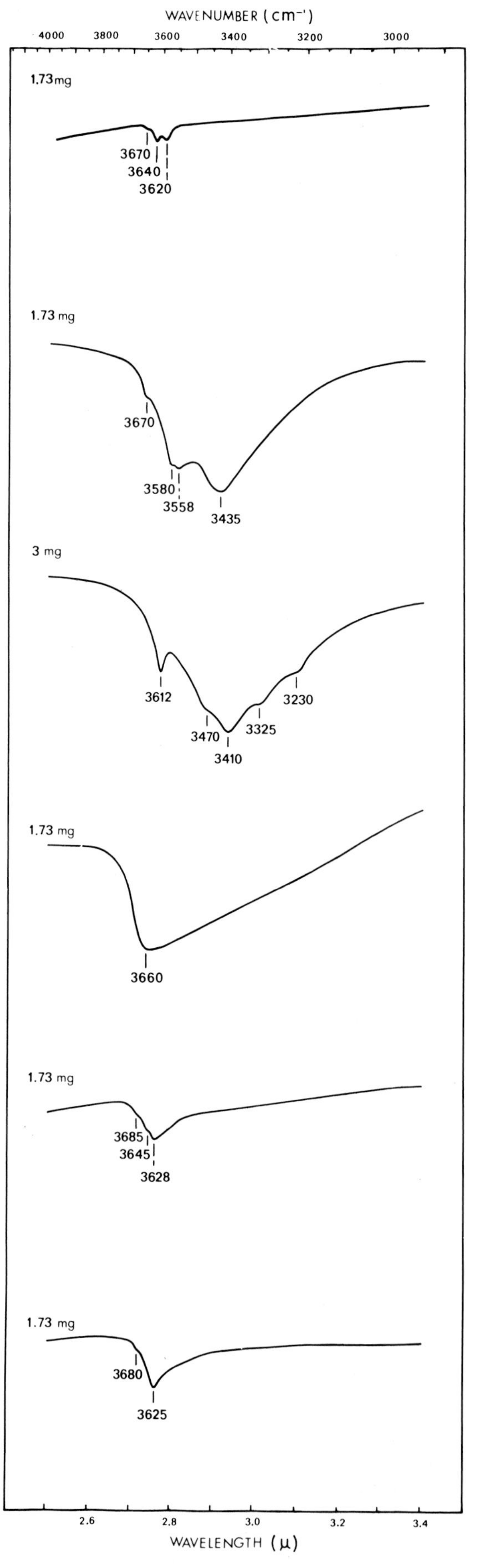

Fig. 6.21. Infrared spectra of amesite, hydro-amesite and greenalite from different origin.

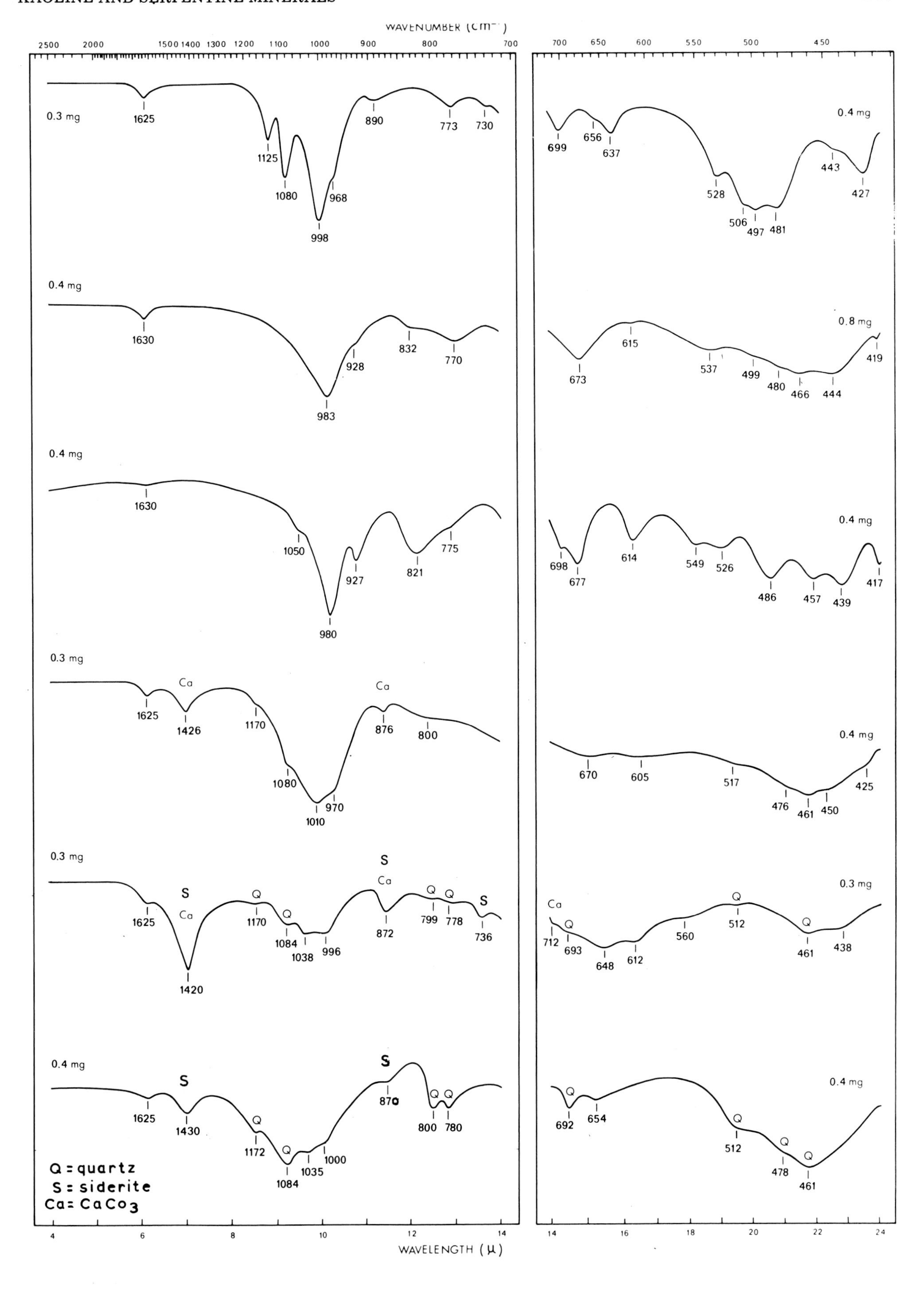
WAVENUMBER (cm⁻¹)
WAVELENGTH (μ)
0.3 mg
0.4 mg
0.8 mg
Q = quartz
S = siderite
Ca = CaCo3

Cronstedtite
Litosvia, Bohemia, Czechoslovakia
pure

Cornwall, Great Britain
pure

Kuttenberg, Bohemia, Czechoslovakia
some siderite
Fig. E.M. 68

Pribram, Bohemia, Czechoslovakia
pure

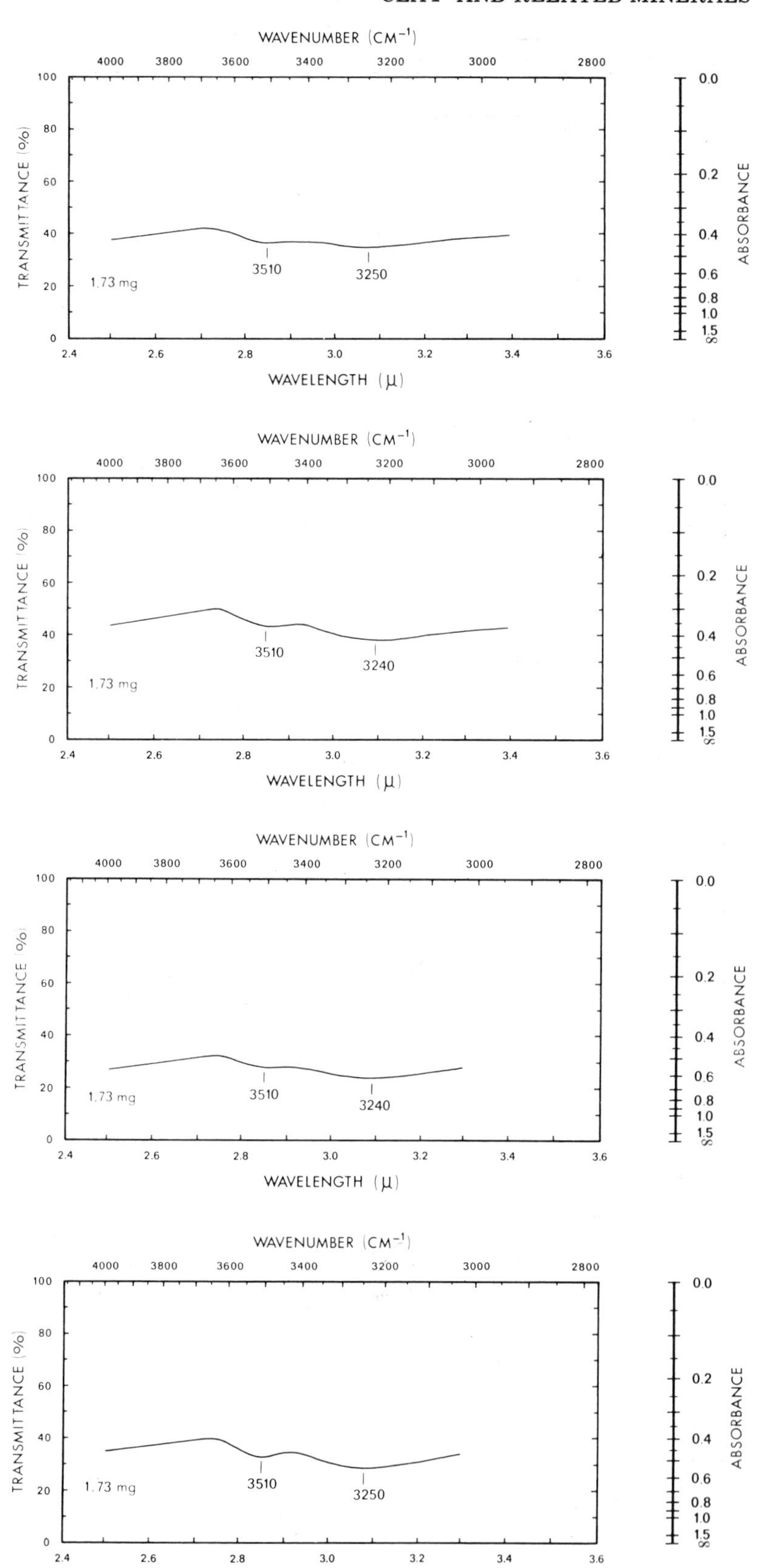

Fig. 6.22. Infrared spectra of cronstedtite from different origin.

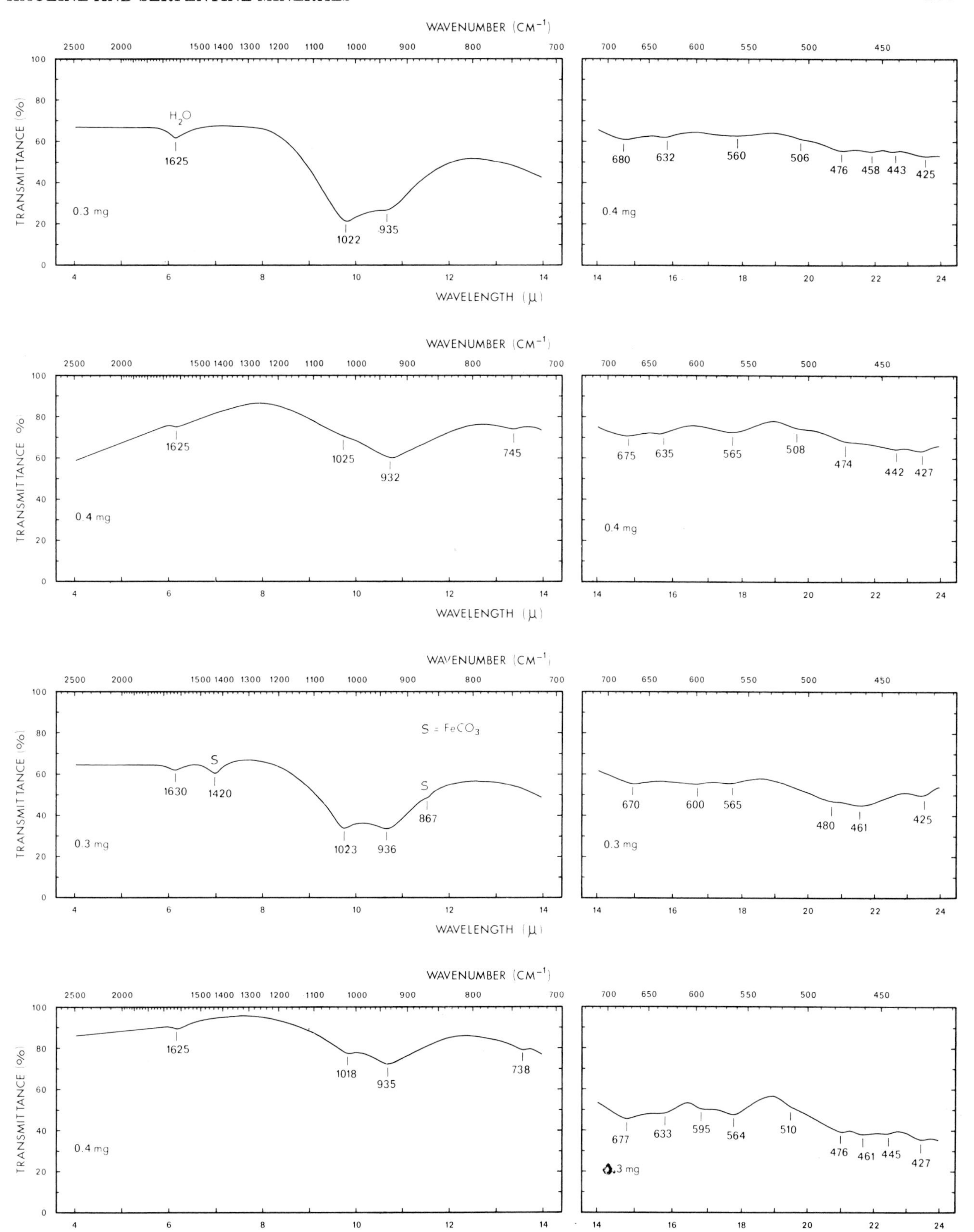
WAVENUMBER (CM⁻¹)
TRANSMITTANCE (%)
WAVELENGTH (μ)
H2O
1625
1022
935
0.3 mg
680 632 560 506 476 458 443 425
0.4 mg
1625
1025
932
745
0.4 mg
675 635 565 508 474 442 427
0.4 mg
S = FeCO3
S
1630 1420
1023 936
867
0.3 mg
670 600 565 480 461 425
0.3 mg
1625
1018
935
738
0.4 mg
677 633 595 564 510 476 461 445 427
0.3 mg

Deweylite
Lancaster, Pa., U.S.A.
some intermediate (overlapped)

Bare Hills, Md., U.S.A.
some quartz, dolomite and intermediate (overlapped)
Fig. E.M. 63, 64

Mladotice, Czechoslovakia
pure
Fig. E.M. 60

Mladotice, Czechoslovakia
pure
Fig. E.M. 61

Monothermite
Donetz basin, U.S.S.R.
kaolinite, mica and some quartz

Anauxite
Bilin, Bohemia, Czechoslovakia
mainly kaolinite and some montmorillonite (overlapped), silicic acid (overlapped), cristobalite (overlapped) and calcite
Fig. E.M. 69

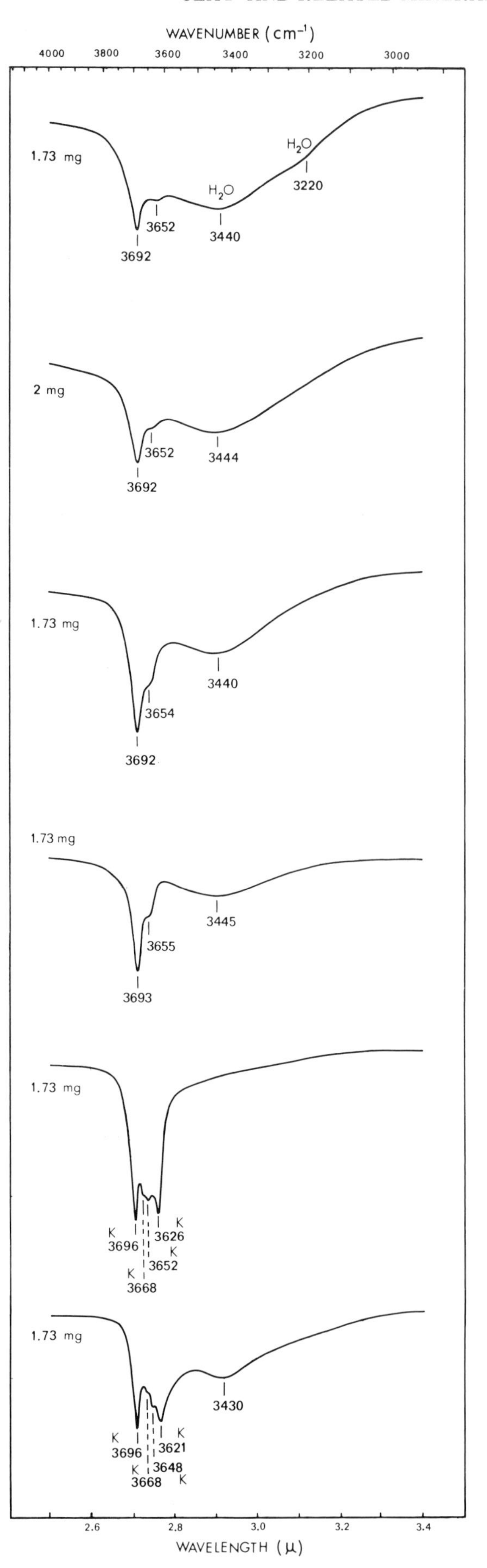

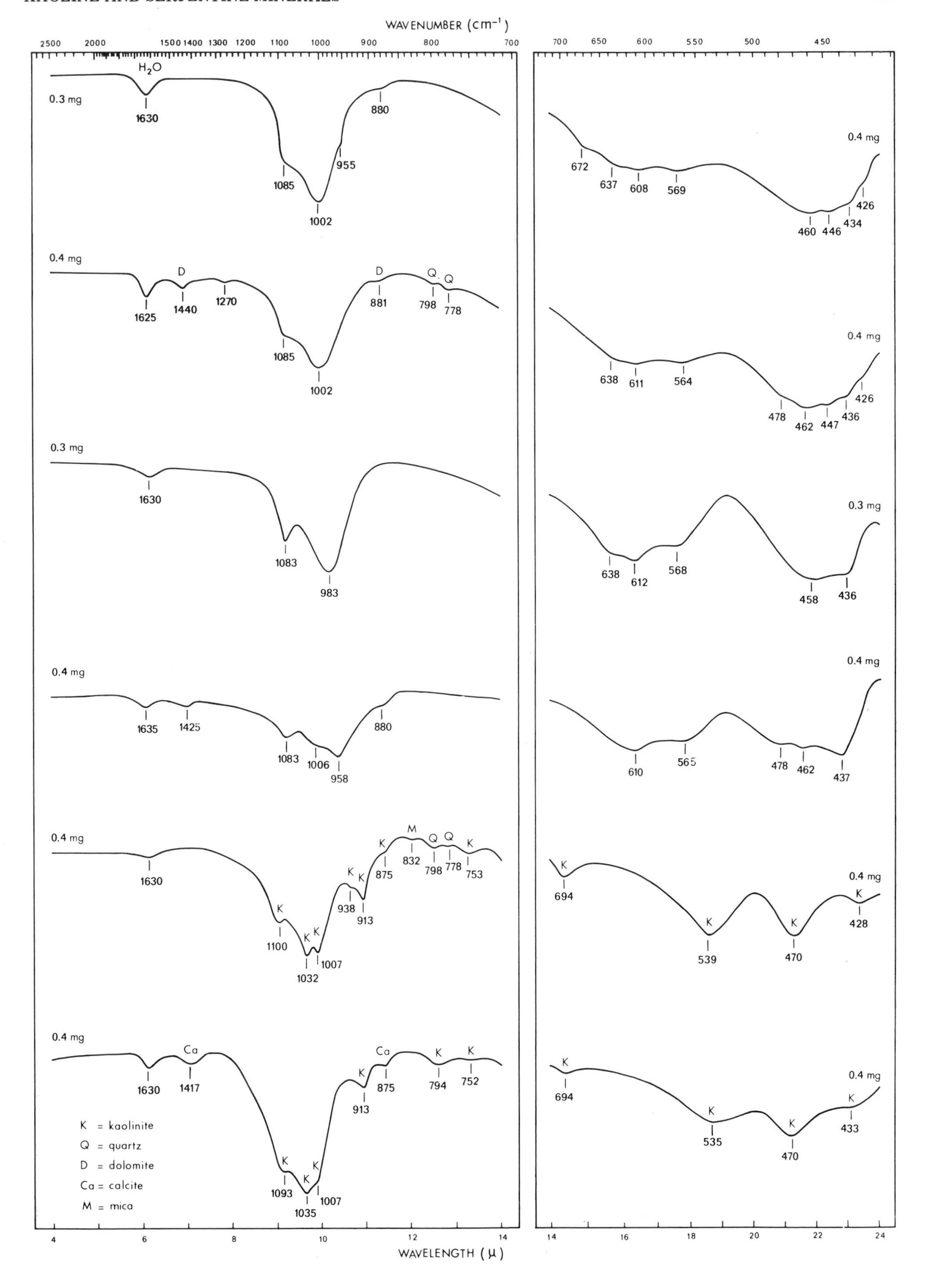
WAVENUMBER (cm^{-1})
WAVELENGTH (μ)
H_2O
0.3 mg
1630
880
955
1085
1002
0.4 mg
672
637 608 569
460 446 434 426
D
1625 1440 1270
881
798 778
1085
1002
638 611 564
478 462 447 436 426
0.3 mg
1630
1083
983
638 612 568
458 436
0.4 mg
1635 1425
880
1083 1006 958
610 565
478 462 437
M
832
1630
875
798 778 753
938 913
1100
1032 1007
694
539 470 428
Ca
1630 1417
875 794 752
913
1093 1035 1007
694
535 470 433
K = kaolinite
Q = quartz
D = dolomite
Ca = calcite
M = mica

Pyrophyllite, talc, brucite

Pyrophyllite and talc

Both minerals have the same basic structural units as the mica and smectite minerals, e.g., two Si-O tetrahedral sheets enclosing one Al- or Mg-OH octahedral sheet. However, as in this case non-saturated charges in the layers are absent, the successive particles are not linked by cations but by Van der Waal's forces (Plate I, pp. 190—191). Neither will the layers swell when treated with H_2O, glycerol or other liquids of high polarizability.

Pyrophyllite; formula: $Al_2(OH)_2Si_4O_{10}$

Occurrence. This mineral is found in fissures and veins of Al-rich rocks (granodiorite, sericite schists, etc.) as a hydrothermal or metamorphic alteration product. It is found also in sediments derived from the above rocks by weathering.

Infrared spectra (Fig. 6.24). Characteristic bands of high intensity are 3678, 1120, 1068, 1049, 948, 539 and 482 cm^{-1}. The small amounts of kaolinite, mica and chlorite which after X-ray analysis should also occur, are absent in the I.R. spectra.

Remarks. Pyrophyllite is commercially used for paper coating, as a filler in the manufacture of ceramic and rubber ware, as a carrier for insecticide dusts and sprays, in refractories and electrical insulators.

Talc; formula: $Mg_3(OH)_2Si_4O_{10}$

Occurrence. This mineral is found in fissures or veins of Mg-rich rocks (dunite, serpentine, chlorite schists, etc), as a hydrothermal or metamorphic alteration product of ultramafic rocks. Moreover, it is found in sediments derived from the above rocks by weathering. Steatite is a massive compact and agalite a fibrous variety of talc. Soapstone is a soft rock which consists of talc but also contains other Mg-components (chlorite, magnesite, enstatite, etc.).

Infrared spectra. (Fig. 6.25, 6.26). Characteristic bands of high intensity are at 3680, 1043, 1018, 669, 465 and 452 cm^{-1}. The mineral may easily be distinguished from pyrophyllite (between brackets): 1043 (1049), 1018 (absent), absent (948), 669 (very weak), absent (578), absent (482). The bands of the degraded talc which is a talc-saponite intergraded mineral (see for details p. 184) is alike to talc but very broad because of great disorder. Commercial talc is highly defiled with chlorite. It has also some quartz and dolomite.

Remarks. Ground, iron-free talc is used in refractories, electrical insulators, cosmetics, in the manufacture of toilet powders, soap, non-combustible and light-resistant paints. Its high absorption properties make this material suitable for the bleaching and degreasing of crude cotton. It is also applied as a filler in paper, linoleum, rubber, as a dressing for skins and leather and as a carrier of insecticide dusts and sprays.

Brucite; formula: $Mg(OH)_2$

This mineral has a layered structure and thus may foliate like mica. As far as it does not contain some Fe it is fibrous (nemalite).
The hydroxyl groups are in hexagonal close packing and each Mg atom is the centre of an octahedron. The layers are so arranged one on another that each hydroxyl group in the upper sheet of the lower layer is in contact with three OH groups in the lower sheet of the layer above.

Occurrence. The mineral is common in serpentine and is further found in metamorphic limestones and chlorite schists.

Infrared spectra (Fig. 6.27). Because of the lack of strong Si-O vibrations, the I.R. bands in the ca. 1000 and 500 cm^{-1} regions are very weak and scarce. The 3275 band is caused by strongly bonded H_2O molecules to Mg. Characteristic bands for brucite are at 3698, 3275, 1282, 1225*. Only the 3698 band is strong. Brucite is usually defiled with talc, degraded and non degraded talc, calcite, dolomite and chrysotile which hinders its identification.

Remarks. Brucite is used in the manufacture of refractory materials.

Smectites (montmorillonite minerals)

The minerals belonging to this group consist of a series of successive layers, each formed by one Al(Mg,Fe,Cu,Cr,Zn,Li,Mn)-OH octahedral sheet enclosed between two (Si,Al)-O tetrahedral sheets (Plate I, pp. 190—191). These kinds of minerals are indicated as 2:1 minerals.
Because of the weak interlayer charge, the large surface (400—800 m^2/g), the large amount of exchangeable cations (60—120 m equiv./100 g) and the poor ordering of the several layers, the smectites will swell when treated with H_2O or other liquids of high polarizability. Swelling oc-

curs most strongly in samples with the smallest particle size and the most hydrated cations (Na^+, Li^+).

Montmorillonite (Al); formula:

$$\begin{bmatrix}(Al,Mg)_2(OH)_2Si_4O_{10}\\(Ca^{2+},Mg^{2+},Na^+)\end{bmatrix} nH_2O$$

Occurrence. Montmorillonite results from the alteration of volcanic ashes and tuffs with an acidic to basic character. Therefore it is found only in some localities in the world called "bentonite beds".

Infrared spectra (Fig. 6.28, 6.29, 6.30, 6.31). The montmorillonites may be defiled with calcite (1422, 877 cm^{-1}), quartz (798—778 cm^{-1}) and kaolinite (3698, 3668, 1008, 752 cm^{-1}).
Small amounts of kaolinite and quartz which are absent in the X-ray spectra, may be found by the IR method. For very small amounts the quartz doublet is decreased to a single band at 796 to 800 cm^{-1}. If by the X-ray method quartz is absent the band is further decreased to ca. 790—796 cm^{-1} (montmorillonite). However, the IR method is less sensible to cristobalite, mica, hematite and feldspar than the X-ray method. Characteristic bands for montmorillonite are at 3642*, 3624, 1115—1090, 1038—1026, 915, 878, 845—835, 796—790, 623, 522, 467 cm^{-1}. The beidellite samples from Nashville and Burns which should contain a larger amount of Al as common montmorillonite, however, are after the spectra alike. The same holds for the "iron-rich" montmorillonite from Atzcapozalco (Mexico) (Fe_2O_3 = 5.38%, FeO = traces) which is polluted with small hematite particles on its surface.

Remarks. This clay mineral occurring at the type locality Montmorillon (France) was first described by Damour and Salvétat (1847). The French indication of this clay is "terre a foulon" after "fouler" (trample) laundry with the foot in a clay-water slurry to degrease them. The English name = Fuller's earth.
The name bentonite was originally given to highly colloidal plastic clays found near Fort Benton in the Cretaceous of Wyoming, U.S.A. The Na-bentonite from Wyoming (volclay) swells far better than the Ca-bentonite from Panther Creek.
The mineral beidellite is rare. It is named by Larsen and Wherry (1925) after the type locality Beidell (Colorado, U.S.A.) where it occurs in rock gangues.
Montmorillonite is an important raw material for various industries, papers, rubber. It is also used to restrict seepage in dams, as a catalyst for cracking petroleum and oils, as a decolorant for oils, as a clarifier for sugars and wines or as an essential component of foundry molds, oil drilling muds, cosmetics, as a carrier for catalysts and insecticides, etc. Its activity can be largely increased by treatment with soda, sulfuric- or hydrochloric acid (activated bentonite). Trade marks: Fulbent, Fulmont, Huttenite, Supergel, Tonsil, Petrisil, Tixoton, Filtrol, etc.
In spite of its high plasticity montmorillonite is of little use for ceramic industries, because its well-pronounced swelling properties strongly damage ceramic forms through strong shrinkage during the drying process.

Soil montmorillonite, soil beidellite (Al,Fe,Mg); formula:

$$\begin{bmatrix}(Al,Fe,Mg)_2(OH)_2(Si,Al)_4O_{10}\\(Ca^{2+},Mg^{2+})\end{bmatrix} nH_2O$$

Occurrence. This mineral is formed as a result of the weathering of basic rocks and ashes under poor drainage conditions. Illite and expanded illite may be intermediate stages in this process. According to specific-surface analyses their dimensions are larger than those of montmorillonite (montmorillonite = 400—600 m^2/g, soil montmorillonite = 250—350 m^2/g). Fe_2O_3 content is up to 12%.

Infrared spectra (Fig. 6.32, 6.33). Soil montmorillonites are mostly defiled with calcite (1422, 877), quartz (798, 778) and kaolinite (3698, 3668, 1008, 752). As compared with common montmorillonite, the soil montmorillonite spectra give practically the same picture. They are only broadened because of larger disorder. As a result several bands of montmorillonite have vanished altogether (796—790, 623, 578*, 474 cm^{-1}).
Other differences are the 1115—1090, the 1038—1026, and the 845—835 cm^{-1} bands of montmorillonite which become a single band at 1106, 1028 and 828 cm^{-1} in soil montmorillonite.
The iron-rich soil montmorillonites from Cuba, Sudan, Ethiopia and India (Fe_2O_3 + FeO = 10—12%) have in addition a 3576 cm^{-1} band due to Fe--O-H.

Remarks. Soil montmorillonite is a common clay mineral in large areas of the world with heavy, easily swelling soils. This typical behaviour which has given many troubles in soil mechanics and plant breeding has been indicated by certain names, e.g. black regur (India), tir noir (Morocco), margalite (Java), black turf (South Africa), badob (Sudan).

Saponite (Mg,Al,Fe); formula:

$$\begin{bmatrix}(Mg,Al,Fe)_3(OH)_2(Si,Al)_4O_{10}\\(Na^+,Ca^{2+},Mg^{2+})\end{bmatrix} nH_2O$$

Saponite is the Mg-rich variety (MgO = 8—25%) of the smectite group minerals. Bowlingite and diabantite are iron-rich members.

Occurrence. Saponite is not widespread. It results from weathering of Mg-rich volcanic rocks (serpentine, amphibolite).

Infrared spectra (Fig. 6.34). By the substitution of Al by Mg and Fe in varying amounts and filling up the holes with Mg to level down the surplus of negative charges, such as happens in the saponite mineral, there are changes in the spectra of montmorillonite (between brackets montmorillonite): 3678 (3624), 1010 (1038—1026), absent (915), 652 (absent), 615 (absent). There is also a large broadening and decrease of intensity in the high-frequency region. Because of the non-constant composition, the spectra are not uniform. The Fe-rich member (Italy) has in addition a 3576 band (Fe--O-H).

Remarks. The name saponite is derived from the Greek sapoun (soap). Minerals of the saponite group were first described by Svanberg (1840). Bowlingite is named after the type locality Bowling (on the Clyde): Hanny (1877). Griffithite is a hematite rich saponitic material which occurs as fillings in the amygdules of a basalt in Griffith Park, Los Angelos (Calif.) and named by Larsen and Steiger (1917, 1928). According to Larsen: MgO = 15.8%, Fe_2O_3 = 7.32%, FeO = 7.83%, Al_2O_3 = 9.05%).

Hectorite (Mg,Li,F); formula:

$$\begin{bmatrix}(Mg,Li)_3(OH,F)_2 \mid Si_4O_{10}\\(Na^+,Ca^{2+},Mg^{2+})\end{bmatrix} nH_2O$$

This mineral is the Mg,Li,F-rich variety of the smectite group minerals.

Occurrence. The mineral is not widespread. It results from weathering of highly Mg-rich volcanic ashes and tuffs.

Infrared spectra (Fig. 6.34). By the introduction of the small Li (r = 0.68 Å) instead of the much larger Mg (r = 0.99 Å) such as in hectorite and ghassoulite, there is a considerable broadening of the spectra. Moreover, several weak bands of saponite have completely vanished, (1160*, 832*, 748*, 615, 565*, 480*) or have become one band: 1100—1090 cm^{-1} = 1078 cm^{-1}.

Remarks: The mineral is named after the type locality Hector (California, U.S.A.). The sample from Djebel Ghassoul in E. Morocco, called ghassoulite, is locally used for laundering. Macaloid is the trade name for a hectorite from which its contents of carbonates has been removed.

Nontronite (morencite, pinguite, faratsihite, chloropal (Fe); formula:

$$\begin{bmatrix}(Fe,Mg)_2(OH)_2 \mid (Si,Al)_4O_{10}\\(Ca^{2+},Mg^{2+})\end{bmatrix} nH_2O$$

This mineral is the iron-rich variety (20—30% Fe_2O_3) of the montmorillonite minerals.

Occurrence. Nontronite is found as a green weathering product where iron-rich rocks alter under poor drainage conditions.

Infrared spectra (Fig. 6.35). By the substitution of Al by Fe^{3+} (and some Fe^{2+}) in montmorillonite there is a considerable decrease in the high wavenumber for nontronite e.g. from 3624 to 3560 cm^{-1}. New bands are at 815, 675, 600 and 492 cm^{-1}. The 778 cm^{-1} band may be masked by the ca. 780 to 800 cm^{-1} band of quartz.

Remarks. The name nontronite is after the type locality in the district of Nontron (France). That of faratsihite is after the type locality Faratsiho (Melawi).

Wolchonskoite (volkonskoite (Cr); formula:

$$\begin{bmatrix}(Al,Fe,Cr,Mg)_2(OH)_2 \mid (Si,Al)_4O_{10}\\(Ca^{2+},Mg^{2+})\end{bmatrix} nH_2O$$

In the above mineral, chromium (2—7% Cr_2O_3) and iron (4—12% Fe_2O_3) partly replace aluminium in the octahedral sheets.

Occurrence. Kämmerer (1831) first found the above mineral in the Ural (U.S.S.R.) as a weathering product of serpentines. Afterwards it was also found in the U.S.A. and Germany, but it is very rare.

Infrared spectra (Fig. 6.36). In this Cr montmorillonite mineral from the type locality Ochansk (Ural), the 3624 cm^{-1} band of mont-

morillonite has decreased to 3560 cm^{-1} like in Fe-montmorillonite (nontronite).
Also in this case of substitutions of unlike atoms, several weak bands of montmorillonite have vanished (1168*, 1115—1090, 845—835, 578*, 474*, 450* cm^{-1}). Or there is only one single broad band (1038—1026 cm^{-1} to 1017 cm^{-1}). New bands are at 901, 853, 780, 685 and 543 cm^{-1}.

Remarks. The mineral is named after the Russian prince A. Wolchonskoy.

Mn-montmorillonite; formula:

$$\begin{bmatrix}(Mg,Mn)_3(OH)_2 \mid (Si,Al)_4O_{10} \\ Mg^{2+}\end{bmatrix} nH_2O$$

Occurrence. This rosa coloured mineral is very rare. The mineral from Turkey as investigated here contains only 0,54% Mn.

Infrared spectrum (Fig. 6.36). The spectrum of this mineral resembles that of montmorillonite, but the Mn content is only very small.

Remarks. none.

Nickel saponite (Mg, Ni); formula:

$$\begin{bmatrix}(Mg,Ni)_3(OH)_2 \mid (Si,Al)_4O_{10} \\ Mg^{2+},Ca^{2+}\end{bmatrix} nH_2O$$

This rosa cloured mineral should contain large amounts of MgO and NiO.

Occurrence. This very rare mineral with ± 15% MgO and ± 10% NiO is found in Baro Alto (Brazil) in serpentine.[1]

Infrared spectra (Fig. 6.36). In the Ni-saponite mineral the 3624 cm^{-1} band of montmorillonite has appreciably decreased to 3540 cm^{-1} like that in the other montmorillonites with heavy metals. Moreover, some bands of montmorillonite have vanished (1168*, 1115—1090, 915, 845—835, 796—790, 578*, 474* cm^{-1}). The 1038—1026 cm^{-1} bands have become 1018 cm^{-1} and the 845—835 bands = 819 cm^{-1}.

Remarks: none.

Sauconite (tallow clay) (Zn); formula:

$$\begin{bmatrix}(Zn,Al,Fe)_3(OH)_2 \mid (Si,Al)_4O_{10} \\ (Ca^{2+},Na^{+})\end{bmatrix} nH_2O$$

This mineral is the Zn-rich variety (ZnO = 30%) of the smectite group minerals.

Occurrence. The above mineral is rare. It is occasionally found in sphalerite (ZnS) ore deposits.

Infrared spectra (Fig. 6.36). The high-graded sample from U.S.A. (ca. 20% ZnO) is somewhat different from the low-graded (ca. 8% ZnO) from Spain. For instance (low-graded between brackets) 1085 (nihil), 1028 (1040), nihil (882, 847), 665 (700, 630), 518 (nihil) cm^{-1}. By the substitution of Al for Zn, the high frequency OH vibration of montmorillonite at 3624 cm^{-1} is not appreciably changed = 3628 and 3630 cm^{-1}. Such although Zn is heavier as iron or chromium = 65.3, 55.8 and 52.0 respectively.
The 1115—1090, 1038—1026 cm^{-1} bands have become 1113 (low-graded), 1085 (high-graded) and 1040 (low-graded), 1028 cm^{-1} (high-graded) respectively. Other montmorillonite bands especially in the high-graded sample, have vanished at all: 878, 845—835, 623, 578*, 474* cm^{-1}.

Remarks. The mineral was first found by Risse (1865) and then named moresnetite after the village of Moresnet near Aachen (Germany). Afterwards it is named sauconite after the Saucos Valley (Pennsylvania), where it was discovered and analyzed by Roepper and described by Genth (1875).

Medmontite; formula:

$$\begin{bmatrix}(Cu,Al)_3(OH)_2 \mid (Si,Al)_4O_{10} \\ Ca^{2+}\end{bmatrix} nH_2O$$

This montmorillonite mineral is said to contain up to 20% CuO.

Occurrence. Medmontite is found in copper ore deposite in the U.S.S.R. and U.S.A.

Infrared spectrum (Fig. 6.36). The sample from Džeskazgan (U.S.S.R.), the type locality, contains large amounts of chrysocolla $CuSiO_3 \cdot nH_2O$ = 3618, 1200, 1032, 676, 467, 429 cm^{-1} (see also the needles on figure E.M.102. It further contains some kaolinite, a montmorillonite mineral and quartz. Some feldspars and mica which after X-ray analysis, should also occur, are not visible by infrared analysis.

Remarks. The mineral is derived from the Russian word "med" (copper) and an abbreviation of the word montmorillonite: Chukhrov and Anosov (1950).

---

[1] Gift from R. Kühnel, Delft (The Netherlands).

References

Chukhrov, F.V. and Anosov, F.Y., 1950. Medmontite, a Cu-bearing mineral of the montmorillonite group. Zap. Vses. Mineral. Obs., 79: 23—27.

Damour, A. and Salvétat, J., 1847. Et analyses sur un hydrosilicate d'alumine trouvé à Montmorillon (Vienne). Ann. Chem. Phys., 21: 376—383.

Genth, F.A., 1875. Mineralogy of Pennsylvania. Second Geol. Surv., Pennsylvania, 120-B.

Hannay, J.B., 1877. On bowlingite a new Scottish mineral. Mineral. Mag., 1: 154—157.

Kämmerer, A., 1831. Auszüge aus Briefen. Neues Jahrb. Mineral. Geol. Paläontol., 2: 420.

Larsen, E.S. and Steiger, G., 1917. Mineralogical notes. J. Wash. Acad. Sci., 7: 6—12.

Larsen, E.S. and Steiger, G., 1928. Dehydration and optical studies of alunogen, nontronite and griffithite. Am. J. Sci. 5th Ser., 25: 1—19.

Larsen, E.S. and Wherry, E.T., 1925. Beidellite, a new mineral name. J. Wash. Acad. Sci., 15: 465—466.

Risse, H., 1865. Korrespondenzbrief. Naturhist. Ver. Rheinland Westfalen, 22: 98—99.

Svanberg, L.F., 1840. Vet. Akad., Stockholm, Handl. p. 153.

Mica minerals

The minerals of the mica group consist of layers, each with two (Si, Al)-O tetrahedral sheets enclosing one Al(Mg,Fe)-OH octahedral sheet. The layers are linked by K-, Mg-, Ca- or Na-cations, which compensate the negative charge surplus resulting from isomorphous replacement of trivalent Al by bivalent Mg or Fe in the octahedra, or of the tetravalent Si by trivalent Al or Fe in the tetrahedra. The succession of the layers is quite regular. This kind is also indicated as 2:1 minerals (Plate I, pp. 190—191).
The mica minerals may be devided in two groups: dioctahedral and trioctahedral minerals. In the first group 1/3 or nearly 1/3 of the octahedral holes are not occupied. In the latter all or nearly all octahedral holes are occupied by cations.
In the octahedral micas the O-H is considerably inclined from an axis perpendicular to the {001} whereas in trioctahedral micas it is nearly perpendicular. As a result only the trioctahedral micas show a dichroic effect, but the dioctahedral micas do not. And as for the former the interlayer cations are more nearer to the H, the trioctahedral micas will weather more easily than the dioctahedral.
Mica minerals are rock-forming components, but in sediments they are also widespread. In addition to the common mica minerals of magmatic origin (muscovite, biotite), a large amount of mica and mica-like minerals with structures derived from them, may be found in sediments too. Mica and mica-like minerals of hydrothermal and metamorphic origin may also be found in them.

Dioctahedral minerals

Muscovite (Al); formula:

$$\begin{bmatrix} Al_2(OH)_2 \mid (Si_3Al)O_{10} \\ K \end{bmatrix}$$

Occurrence. This mica mineral is rich in aluminium and is a common component of many rocks (granite, pegmatite, rhyolite, dacite, etc.).. Mica schists especially are very rich in muscovite. By hydrothermal action on K-feldspars fine-grained muscovite (sericite) is formed. Mica crystals up to 7 tons are found e.g. at Eau Claire, Ontario. Hydrous mica also called hydromica, hydromuscovite, hydroglimmer, Ballater illite, etc., is a K^+ poor but OH or $(H_3O)^+$-rich muscovite mineral. It is found in the clay < 2/μm or silt (2—16 μm) fraction of sediments, where it resulted from weathering of common muscovite.

Infrared spectra (Fig. 6.37, 6.38). Muscovite bands of large intensity are at 3624, 1023, 531, 477 cm^{-1}. Sericite, a fine-grained variety of muscovite can be distinguished from the latter by the 932 cm^{-1} band which is mostly absent for muscovite.
Small amounts of plagioclase can not be found by IR analysis; only by the X-ray method. The 932 band is also lacking in hydrous muscovite and in the rare 3T (3-layer triclinic)polymorph from the Sultan Basin (Smith and Yoder, 1954; Levinson, 1955) which was earlier referred to as being 3 M by Axelrod and Grimaldi (1949). Common muscovite (and sericite) are usually 2 M = 2-layer monoclinic.

Remarks. Sheet mica is mainly used in electrical industry on account of its excellent insulation properties. Scrap mica, which is ground from mica wastes is used in the manufacture of roofing papers, fancy paints, lubricants, filler in rubber industry, etc.

Illite (Al,Fe,Mg); formula:

$$\begin{bmatrix} (Al,Fe,Mg)_2(OH)_2 \mid (Si,Al)_4(O,OH)_{10} \\ (K(K^+,H_3O^+) \end{bmatrix}$$

This mineral is supposed to be poorer in K and richer in Fe,Mg, OH and $(H_3O)^+$ than hydromuscovite. By X-ray analysis illite can be distinguished from muscovite by a broadened basal (001) reflection at 10 Å and a strong endothermal effect at ca. 580°C.
The mineral illite mostly occurs as 1 M (one-layer monoclinic) but when poorly ordered (degraded) it has a 1 Md (one-layer monoclinic, disordered) structure. Other names for the mineral are: hydromica, hydro-muscovite, hydrous mica, hydroglimmer, potassiumbearing clay mineral resembling sericite, sericite-like mineral, Glimmerton, clay muscovite, detrital muscovite, Roxburghshire illite, etc.

Occurrence. Illite is a widespread mineral in sediments. It may be formed by authogenic processes from the weathering products of other minerals. Illite is less resistant to weathering than muscovite or hydrous muscovite.

Infrared spectra (Fig. 6.39). Bands of large intensity are like those of muscovite and sericite (the latter between brackets) at 3622 (3624, 3624), 1022 (1023, 1023), 475 (477, 477) cm^{-1}. The 534* cm^{-1} band of illite comparable to the 531 (533) cm^{-1} band of muscovite (sericite) does not always occur. Instead the 523

cm^{-1} band is more common than the 519* (and 520*) band of muscovite (sericite). Most illites have a 1080 cm^{-1} band instead of the 1068 cm^{-1} band of muscovite (sericite).
The 1117* (1116*) cm^{-1} bands of muscovite (sericite) have moved to 1107* and the 801 (802) cm^{-1} band is absent.
Weak and only accidentally occurring bands of muscovite (sericite) at 723* (722*), 462* (463*), 445* (445*), cm^{-1} are also absent.

Remarks. The term illite was proposed by Grim et al. (1937) as a general name for Al, Mg, Fe-rich mica or sericite-like minerals found in the clay (< 2 μm) fraction of weathered slates in Illinois (U.S.A.).

Swelling illite (Al,Fe,Mg), *ammersooite*; formula:

$$\begin{bmatrix} (Al,Fe,Mg)_2(OH)_2 \mid (Si,Al)_4(O,OH)_{10} \\ (Ca^{2+},Mg^{2+},K^+,H_3O^+) \end{bmatrix} nH_2O$$

This mineral is derived from illite by a reduction of interlayer charge due to hydroxylation (tetrahedral O replaced by OH) or by silification (tetrahedral Al^{3+} replaced by Si^{4+}) because of larger energy constant of Si-O (3,123 kcal/mole) than Al-O (1,793 kcal/mole); Keller (1954). Thereby is a particular replacement of highly polarizable K^+ by less polarizable cations like $(H_3O)^+$, Ca^{2+}, Mg^{2+}.
The resulting end member 14 Å mineral will easily swell when treated with an excess of glycerol or other liquids of high hydrogen bonding capacity, but when treated with an excess of strongly polarizable cations (KCl or NH_4Cl) it contracts its layers to 10 Å.
Degree of swelling with water or glycerol and contraction with K^+ or other cations depend on particle size, interlayer charge (Al-Si substitution) and polarizability of the cations. So no well defined mineral of constant composition should exist but a series of minerals with variable K-contracting and glycerol-expanding properties and which run opposite.
Swelling illite is very often mistaken for common- or for soil montmorillonite because of its similar physical characteristics. But the origin of both groups of minerals is different.

Occurrence. Swelling illite is a common component in poorly drained alluvial sediments, derived from illitic shales where it resulted from silification of the original illitic mineral.

Infra-red spectra (Fig. 6.40). The spectra are like those of illite. But still further bands of weak intensity and only occurring accidentally are absent: 988*, 934*, 644*, 632*, 614* cm^{-1}. Apparently in this mineral the grade of ordering of the atoms has further decreased.

Remarks. The illitic character of this mica soil clay mineral was described by Van der Marel (1952, 1954) and afterwards verified by Rivière et al. (1961). It was first found at a potash experimental field at Ammerzoden (Netherlands) and then named "ammersooite". In contrast to vermiculite the mineral is far better resistent to acids like common illite thus proving its illitic (muscovitic) origin. Other names for the above clay mineral as mentioned in literature are: a mineral very nearly an end member of the montmorillonite series but approaching illite, hydromuscovite gonflante, clay vermiculite with expandable layers, expanded illite mistaken for montmorillonite, quellender illite, illite gonflante, montmorillonite, expanding mineral of the montmorillonite type, highly charged dioctahedral montmorillonite, etc.
An intermediate stage between illite and swelling illite is a mineral named "expanded illite" which highly contracts its layers when treated with an excess of KCl but does not swell appreciably. Other names for this illitic mineral are: vermiculite clay, soil vermiculite, dioctahedral soil vermiculite, hydrated mica (vermiculite), illite ouverte, broadened illite, degraded illite, hydrated illite, dioctahedral vermiculite, vermiculite-like mineral, etc.

Leverrierite (Al); formula:

$$\begin{bmatrix} Al_2(OH)_2 \mid (Si,Al)_4(O,OH)_{10} \\ K(K^+,H_3O^+) \end{bmatrix}$$

Occurrence. This large, K-poor, mica-like mineral is found as wormlike aggregates or as large well-developed crystals: Schüller and Grassmann (1949); Schüller (1951); Hoehne (1957); Saalfeld (1960). They are found in oil shales where they were formed from the weathering products of organic matter in the ancient marshes.

Infrared spectra (Fig. 6.41). The structure of this coarse (some mm) well-crystallized mineral is most like that of sericite. However, weak bands of the latter only occurring accidentally (1164*, 1116*, 722*, 644*, 613*, 520*, 463*, 445*) are absent.
The spectrum of the Hudig biogene illite (see Remarks) is very poorly developed. It is most like that of illite and swelling illite, but with an additional loss of bands. Also according to electron microscopy and X-ray analysis this mineral

is very badly ordered because of its genesis in an impure organic environment.

Remarks. The name leverrierite was given by Termier (1889), who found first this coarse mineral in coal layers at St. Etienne (France) where it results from the debris of plant minerals yielding K, Si, Al, Fe, Mg, e.g., the basic components of mica minerals.
The fine ($< 2\ \mu m$) soil clay mineral of the marshes is named "Hudig biogene illite" after J. Hudig who now 30 years ago was the first to point out its existance (for details, see Hudig, 1964).

Iron illite; formula

$$\begin{bmatrix}(Al,Fe,Mg)_2(OH)_2 \mid (Si,Al)_4(O,OH)_{10} \\ (K,K^+,H_3O^+)\end{bmatrix}$$

Occurrence. This illite mineral was found by Stremme (1951) enclosed in Fe concretions in the B horizon of a redbrown coloured forest soil of diluvial origin in the Rhine plain near Heidelberg (Germany).

Infrared spectrum (Fig. 6.41). The iron-illite mineral is most like illite and swelling illite; see the characteristic band at 1080 cm^{-1}.
Also in this case several bands of illite and swelling illite (between brackets) are absent 3643 (3643), 1166* (1165*), 1107* (1100*), 752 (752) probably because of the disorder caused by the introduction of iron in the crystal. Only large amounts of hematite can be found by the IR method, such in contrast to X-ray analysis.

Sarospatakite (Al); formula:

$$\begin{bmatrix}Al_2(OH)_2 \mid (Si_3Al)(O,OH)_{10} \\ K(K^+,H_3O^+)\end{bmatrix}$$

The above mineral is poor in iron. The E.M. micrograph (figure EM 129) shows perfectly crystallized thin plates with pseudo-hexagonal habit and prominent length growth, thus pointing to excellent growing conditions.

Occurrence. Sarospatakite is found in the Nagybörzsony deposit (Hungary), in the neighbourhood of Sarospatak. It resulted there from hydrothermal action on slates.

Infrared spectra (Fig. 6.41). The spectrum shows broad bands with the characteristics of illite and swelling illite, (the latter between brackets) at: 3644 (3643, 4643), 1076 (1080, 1080*), 1026 (1022, 1026), 915 (912, 913), 565 (561, 562*), 525 (523, 527) cm^{-1}.
Weak illite and swelling illite bands (swelling illite between brackets) at 1166* (1165*), 934* (absent), 875 (875), 644* (absent), 632* (absent), 534* (absent) cm^{-1}, however are lacking. In this case a well-crystallized mineral shows broad bands. But the same contrast is also found for kaolinite from Provence, France, see p. 65.

Remarks. The mineral is named after the above type locality. Some care must be taken with the samples, because of contamination with kaolinite and montmorillonite from the same locality.

Potassium bentonite; formula:

$$\begin{bmatrix}(Al,Fe,Mg)_2(OH)_2 \mid (Si,Al)_4(O,OH)_{10} \\ K(K^+,Ca^{2+},Mg^{2+})\end{bmatrix}$$

The above non-swelling 10 Å mica mineral from Catawba, Virginia, is fine-grained and of illitic lath-like structure. Other names are: metabentonite, highly altered bentonite K-rich, Ordovician (K) bentonite.

Occurrence. The mineral is formed from volcanic ashes and tuffs in sea water. They first weathered to montmorillonite and then have picked up K between the layers.

Infrared spectra (Fig. 6.41). The spectrum contains the characteristic bands of illite and swelling-illite (the latter between brackets) at: 3642 (3643, 3643), 3620 (3622, 3622), 1170 (1166*, 1165*), 1085 (1080, 1080*), 912 (912, 913), 875 (875, 875), 831 (830, 830), 564 (561, 562*), 525 (523, 527), 472 (475, 472) cm^{-1}.
Because of its fineness there is a O-H stretching vibration at 3290 cm^{-1} of tightly bonded water at the surface of the particles.

Remarks. Besides the 10 Å potassium bentonite mineral, also illite-montmorillonite intermediates are formed. The amount of the illite component, increases with Si/Al substitution and with K uptake to balance the charges. The intermediate products are also called K-bentonite, see further p. 184.

Glauconite, celadonite; formula:

$$\begin{bmatrix}(Fe^{3+},Fe^{2+},Al,Mg)_2(OH)_2 \mid (Si_3Al)O_{10} \\ K(K^+,Mg^{2+},Ca^{2+},Na^+)\end{bmatrix}$$

The above minerals, nearly related to each other, are dioctahedral despite the presence of Mg and Fe to the extent of 1/4 to 1/3 of all the holes in octahedral coordination. They contain 4—9%

K_2O between the interlayers by which the charge caused by substitution of trivalent Al by divalent Mg and Fe is compensated. A typical characteristic of glauconite and celadonite is their high amount of ferrous iron = 2—5% as against 15—20% for ferric iron. This feature points to reducing conditions prevailing at their growth. Celadonite has less Al but more Mg than glauconite i.e., 2—6% Al_2O_3, 5—9% MgO as against 10—16% Al_2O_3, 2—4% MgO, respectively.

Glauconite and celadonite may also like illite expand their layers by weathering action and after passing through several intermediate stages the endform is a dioctahedric iron-magnesium-rich montmorillonite mineral.

Occurrence. Glauconite is a common mineral in many sediments. It is formed under anaerobic conditions in marine muds. Its main constituents (Si,Al,Fe) are derived from these muds and the K and Mg from the seawater. Sulfate-reducing bacteria should create the necessary reducing potential. However, glauconite may also result from weathering of biotite.

Celadonite is found in fissures of vesicular olivine basalt where it resulted from weathering of olivine.

Infrared spectra (Fig. 6.42, 6.43). Glauconite can easily be distinguished from celadonite (the latter between brackets) by the bands at: absent (3672), absent (3650), absent (970, 956), 872 (absent), 813 (absent), absent (799), absent (743), 566 (absent). The difference between the above minerals by X-ray and thermal analysis is small.

For glauconite there are appreciable differences in the spectra in the 1015—995 cm^{-1} region which are caused by differences in chemical composition.

The bands of celadonite are much sharper and stronger, especially in the lower frequency range, than those of glauconite, pointing to a better ordering of the atoms in the celadonite crystal, probably because of better growth conditions — see before.

Remarks. The name of the bluish-green glauconite mineral is derived from the Greek word glaukos = blue green. The name celadonite was given by Glocker (1847), and is derived from its greenish colour (French: celadon = sea green).

Greensand, an impure occurrence of glauconite (ca. 6% K_2O), is applied as a local fertilizer. In former years greensand was also used as a base exchanger for the purification of water. But nowadays it is out of the market because the synthetic exchangers have much higher exchange capacities.

Miscellany; formulas:

paragonite	$NaAl_2[(OH,F)_2 \mid (Si_3Al)O_{10}]$
margarite	$CaAl_2[(OH)_2 \mid (Si_2Al_2)O_{10}]$
fuchsite	$(K,Cr)Al_2[(OH,F)_2 \mid (Si_3Al)O_{10}]$
alurgite	$K(Fe,Mg,Mn^{2+},Mn^{3+})Al[(OH,F)_2 \mid (Si_3Al)O_{10}]$

Occurrence. The above minerals are rare; one more than the other.

Infrared spectra (Fig. 6.44). The above minerals can easily be distinguished from each other by their bands at: 3628, 3632, 3620, 3608 cm^{-1}. Paragonite and margarite have many weak and broad bands in the 700 to 425 cm^{-1} section. Fuchsite and alurgite have only some broad bands in this region. But also in the 1100—720 cm^{-1} region there are characteristic differences between the above four minerals.

Remarks. Fuchsite (up to 5% Cr_2O_3), named after the chemist J.H. von Fuchs, is used in alloys.

Trioctahedral minerals

Biotite; formula:

$$\begin{bmatrix} (Mg,Fe,Al)_3(OH,F)_2 \mid (Si_3Al)_4O_{10} \\ K,Mg \end{bmatrix}$$

Occurrence. This coarse Fe, Mg-rich mica mineral is found in many kinds of rocks in varying amounts. By erosion it may also occur in the clay fraction of fluviatile or glacial sediments.
Varities are lepidomelane (Fe-rich) and manganophyllite (Mn-rich).

Infrared spectra (Fig. 6.45, 6.46). The spectra of lepidomelane and Mn phyllite are somewhat different from biotite, e.g. 992, 1018 and 1000 cm^{-1}. The bands in the highest wave numbers of the above minerals are badly developed. Such in contrast to those of the dioctahedral mica minerals. It is caused by the transition moments of the vibration being perpendicular to the electric vector of the incident beam, and nearly all of the octahedral holes occupied by cations.
The poorest are those for the Mn phyllite which is thereby caused by the uptake of the large amount of Mn in the crystal (India = 18%, Japan = ca. 35% MnO).

Remarks. In contrast to muscovite, biotite is of little commercial use. Neither are the Fe and Mn-rich varieties, because of their rareness. By weathering action ferro iron of biotite is oxidized to ferri, or Al of trioctahedral illite (ledikite) is silicified and then the layers may expand like expanded muscovite or illite. The expanded layers can be contracted by treatment with an excess of KCl. In consequence of their identical behaviour finer particles are also called vermiculite, expanded trioctahedral illite, expanded clay biotite, expanded ledikite (name derived from Ledikin, Scotland), biotite-like mineral resembling vermiculite, etc. Designations for the swelling trioctahedral analogue of swelling illite are: cardenite (after Carden Wood, Scotland), expanded swelling ledikite, and trioctahedral soil montmorillonite.
It is not possible to distinguish with the infrared method the expanded from the non-expanded forms as is also found for the dioctahedral mica minerals.

Phlogopite; formula:

$$\begin{bmatrix} Mg_3(OH,F)_2 \mid (Si_3Al)O_{10} \\ Mg,K \end{bmatrix}$$

Occurrence. This coarse Mg-rich mineral is found in many biotite-rich rocks, but it is scarcer than biotite.

Infrared spectra (Fig. 6.46). The mineral can easily be distinguished from biotite (lepidomelane and manganophyllite) by its 3708 cm^{-1} band. Bands of high intensity are at: 1000, 963, 463, 448 cm^{-1}. Also in this case especially the bands of the highest wave numbers are poorly developed.

Remarks. Phlogopite is of little commercial purpose. Because of the interlayered Mg-cations which have a smaller polarizability than K-ions, the mineral can degrade more easily: e.g., expanding its layers by weathering action than biotite. Even leaching of fine phlogopite particles with a normal solution of $CaCl_2$ is sufficient to replace K by Ca and then to move the layers apart. They can again be closed by treating the particle with an excess of KCl or NH_4Cl solution; not a solution of $MgCl_2$, LiCl or NaCl (polarizability in $Å^3$: K = 0,97, NH^+_4 = 1,60, Mg = ca. 0,2, Li = 0,02, Na = 0,21).

Hydrated biotite; formula:

$$\begin{bmatrix} (Mg,Fe,Al)_3(OH)_2 \mid (Si_3Al)O_{10} \\ Mg,K(H_2O) \end{bmatrix}$$

This 10 Å mineral, which contains up to 2% of interlayered H_2O molecules, expands when heated at 800–900°C.

Occurrence. This rare mineral is found in biotite-rich rocks.

Infrared spectra (Fig. 6.47). The bands of this trioctahedral mineral too, which has all its octahedral holes filled with Mg and Fe, are very broad.
Moreover, because of its H_2O contents, the bands of the minerals are masked by the H-O-H vibration at ca. 3430 cm^{-1}. More tightly bonded H_2O gives the weak band at ca. 3220 cm^{-1}.
Bands of high intensity are at 998 and 463 cm^{-1}.

Remarks. The mineral is of slight importance.

Vermiculite; formula:

$$\begin{bmatrix} (Mg,Al,Fe)_3(OH)_2 \mid (Si_3Al)O_{10} \\ Mg^{2+}(H_2O)_4 \end{bmatrix}$$

This coarse Mg-rich mica mineral is expanded to 14 Å because of strong hydration of the interlayered Mg-cations. The mineral contracts to 10 Å when heated at 550° C or when treated with a concentrated solution of highly polarizable cations (K^+,NH_4^+).

Occurrence. This coarse mineral is not widespread. It is found in Mg-rich serpentines and it may result from hydrothermal action on biotite-rich rocks. Owing to erosion the mineral is also found, like biotite, in the clay fraction of fluviatile and glacial sediments.

Infrared spectra (Fig. 6.47). The bands of the vermiculite mineral, like those of the foregoing trioctahedral minerals, are very broad. Its high contents of H_2O (ca. 8%) strongly masks in the highest wave numbers the bands inherent to the mineral itself. The strongly bonded H_2O molecules between the layers have a band at ca. 3230 cm^{-1}. Bands of high intensity are at 1000, 463 and 449 cm^{-1}.

Remarks. The mineral is named by Webb (1824) after its property to expand harmonically to wormlike (vermis = worm) aggregates when heated at 800—900° C. It then was supposed to be a particular kind of talc. The exfolation is caused by the H_2O vapour from the interlayered H_2O molecules. Its volume is then increased up to 16 times. Consequently the heated material is very porous. Vermiculite is commercially used as a soil conditioner and an insulator of sound and heat.

Lepidolite, amblygonite - Li(F); formula:

$$\begin{bmatrix}(Li,Al)_2(F,OH)_2 \mid (Si_3Al)_4O_{10} \\ K\end{bmatrix}$$

Occurrence. This Li-, F-rich mineral (Li_2O = up to 5% and F = up to 6%) is not widespread. It may be found in granite, pegmatite. Most lepidolites have 2M structure, but the lepidolite from Topsham Mine as described by Levinson (1953) is 1M (one-layer monoclinic).

Infrered spectra (Fig. 6.48). Characteristic bands are at 3665, 3628, 1024, 998, 965, 800, 754, 532 and 478 cm^{-1}. The bands of the 1M lepidolite are alike to those of the common 2M lepidolite, but only sharper. It points to a better ordering in the crystal.

Remarks. Lepidolite is used in ceramic bodies. F and Li decrease expansion and increase their strength.

Zinnwaldite - Li,Fe; formula:

$$\begin{bmatrix}(Li,Fe,Al)_2(F,OH)_2 \mid (Si_3Al)_4O_{10} \\ K\end{bmatrix}$$

Occurrence. This Li,Fe,F mineral is also rare. It may also occur in acid magmatic rocks.

Infrared spectra (Fig. 6.48). The bands are very broad; in particular for Sc zinnwaldite. Apparently the introduction of several different atoms (Li,F,OH) and thereby ca. 1% Sc in the Sc zinnwaldite, has caused a large disorder in the crystal. Zinnwaldite can be distinguished from lepidolite by the bands (Zinnwaldite between brackets) at: absent (3640), absent (1072), 965 (absent), 754 (748).

Remarks. The mineral is named after the type locality Zinnwald, Germany.

References

Axelrod, J.M. and Grimaldi, F.S., 1949. Muscovite with small optical axial angles. Am. Mineralogist, 34: 559—572.

Glocker, E.F., 1847. Generum et Specierum Mineralium Secundum Ordines Naturales Digestorum Synopsis. Halle, 348 pp.

Grim, R.E., Bray, R.H. and Bradley, W.F., 1937. The mica in argillaceous sediments. Am. Mineralogist, 22: 813—829.

Hoehne, K., 1957. Leverrierit in Kohlenflözen und seine Erkennung im Mikroanschliffbild. Geologie, 6: 190—202.

Hudig, J., 1964. Vorkommen, Entstehung und Eigenschaften biogener Tone in den Niederungsmooren der Niederlände. Z. Pflanzernähr. Düng., Bodenk., 105: 212—239.

Keller, W.D., 1954. The bonding energies of some silicate minerals. Am. Mineralogist, 39: 783—793.

Levinson, A.A., 1953. Relationship between polymorphism and composition in the muscovite-lepidolite series. Am. Mineralogist, 38: 88—107.

Levinson, A.A., 1955. Polymorphism among illites and hydrous mica. Am. Mineralogist, 40: 41—49.

Rivière, A., Vernhet, S. and Van der Marel, H.W., 1961. Sur les "illites gonflantes". Identification, analogies et nomenclature. C.R. Séances Acad. Sci., 252: 159—161.

Saalfeld, H., 1960. Röntgenuntersuchungen zum Leverrierit Problem. Heidelb. Beitr. Mineral. Petrogr., 7: 63—71.

Schüller, A., 1951. Die Tonsteine aus den Steinkohlflözen von Dobrilugk und ihre Entstehung. Heidelb. Beitr. Mineral. Petrogr., 2: 413—427.

Schüller, A. and Grassmann, H., 1949. Über den Nachweis von echtem Leverrierit in Tonsteinen aus unterkarbonischen Steinkohlflözen von Dobrilugk. Heidelb. Beitr. Mineral. Petrogr., 2: 269—278.

Smith, J.V. and Yoder, H.S., 1954. Theoretical and X-ray study of the mica polymorphs. Am. Mineralogist, 39: 343—344.

Stremme, H.E., 1951. Quantitative Untersuchungen über Zersetzung und Bildung von Mineralien im braunen Waldboden. Z. Plfanzernähr., Düng., Bodenk., 53: 193—203.

Termier, P., 1889. Sur une phyllite nouvelle, la leverrierite et sur les Bacillarites du terrain houiller. C. R., 108: 1071—1073.

Van der Marel, H.W., 1952. In: J. Temme and H.W. van der Marel, Potassium fixation in two long continued experimental fields. Versl. Landbouwk. Onderz., 58(6): 11—53.

Van der Marel, H.W., 1954. Potassium fixation in Dutch soils. Mineralogical analysis. Soil Sci., 78: 163—179.

Webb, Th.H., 1824. New localities of tourmalines and talc. Am. J. Sci., 7: 55.

Pyrophyllite
Vestana, Sweden
some mica (overlapped)

Visp, Wallis, Switzerland
some mica (overlapped)

Africa
some mica (overlapped), kaolinite (overlapped) and chlorite (overlapped)

Ardennen, Belgium
some mica (overlapped), kaolinite (overlapped) and chlorite (overlapped)

Mariposa County, Calif., U.S.A;
some kaolinite (overlapped)

Robbins, N.C., U.S.A.
some mica (overlapped), kaolinite (overlapped) and quartz
Fig. E.M. 133

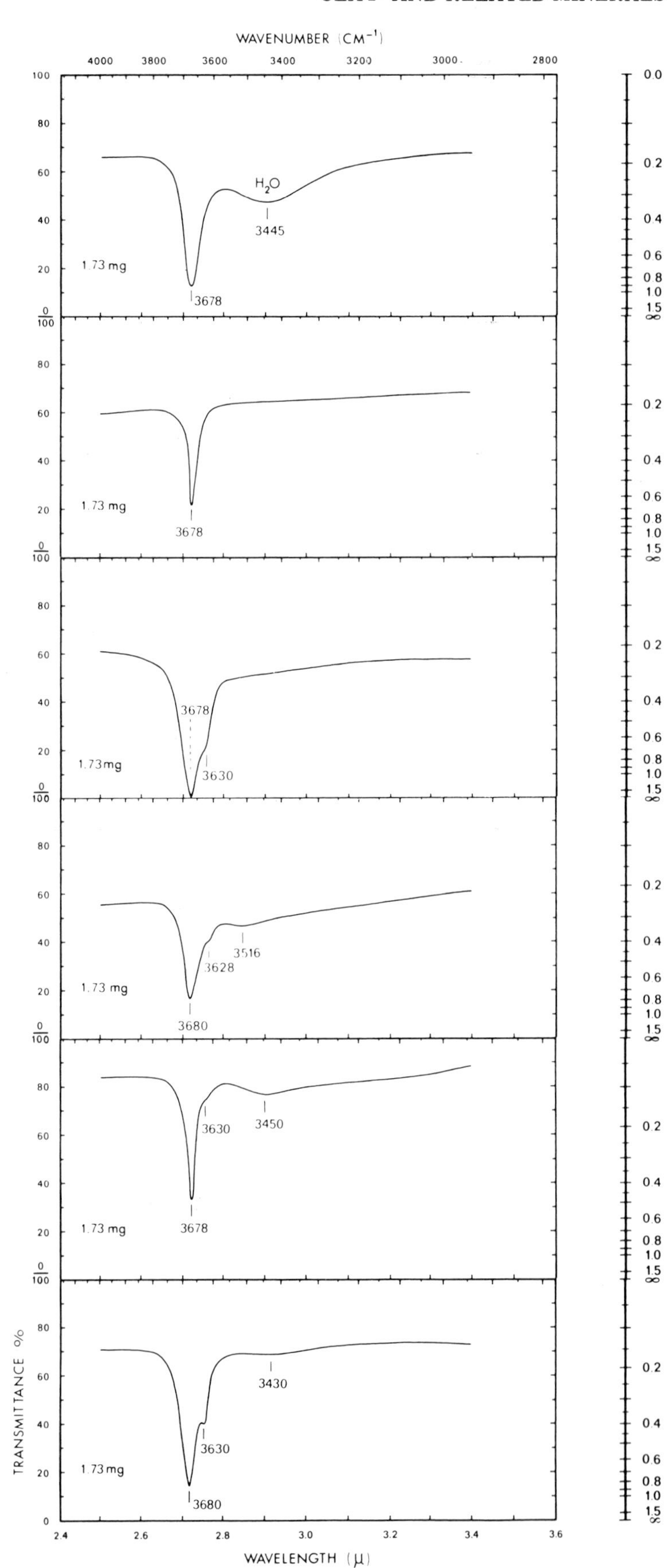

Fig. 6.24. Infrared spectra of pyrophyllite from different origin.

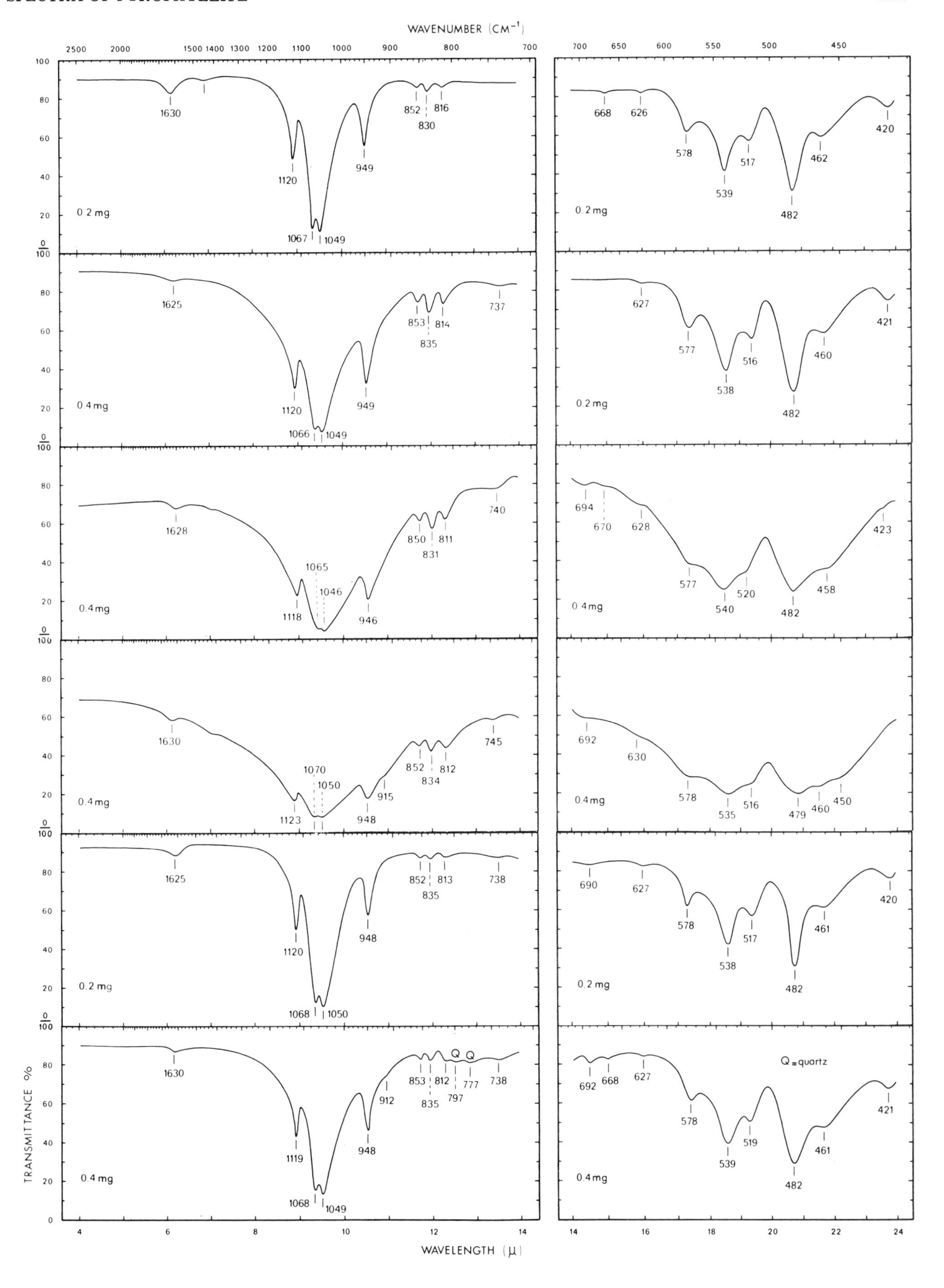
WAVENUMBER (CM⁻¹)
WAVELENGTH (μ)
TRANSMITTANCE %
0.2 mg
0.4 mg
Q = quartz

Talc
St. Lawrence County, U.S.A.
pure

Transvaal, South Africa
pure
Fig. E.M. 135

Moravia, Czechoslovakia
chlorite

Cornwall, Great Britain
some magnesite

Swerdlowsk, Ural, U.S.S.R.
pure

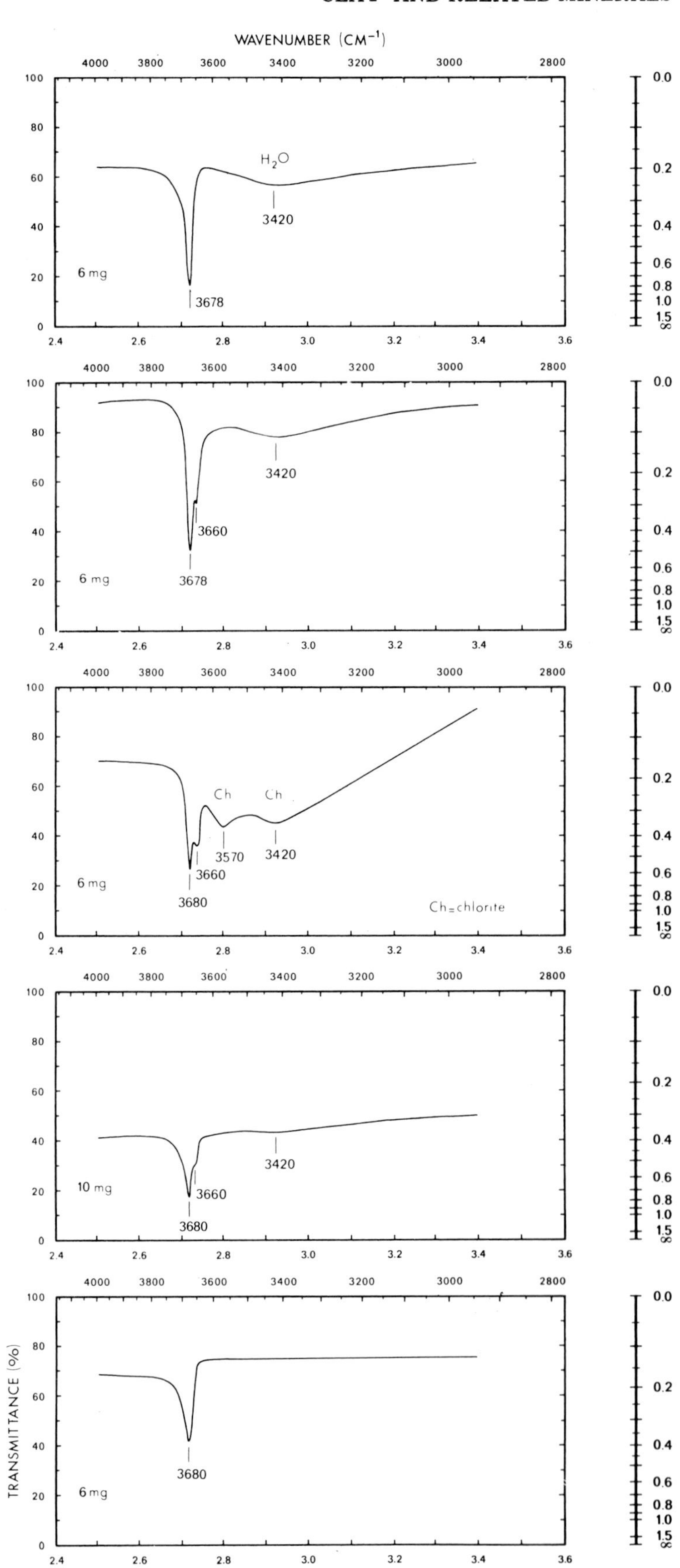

Fig. 6.25. Infrared spectra of talc from different origin.

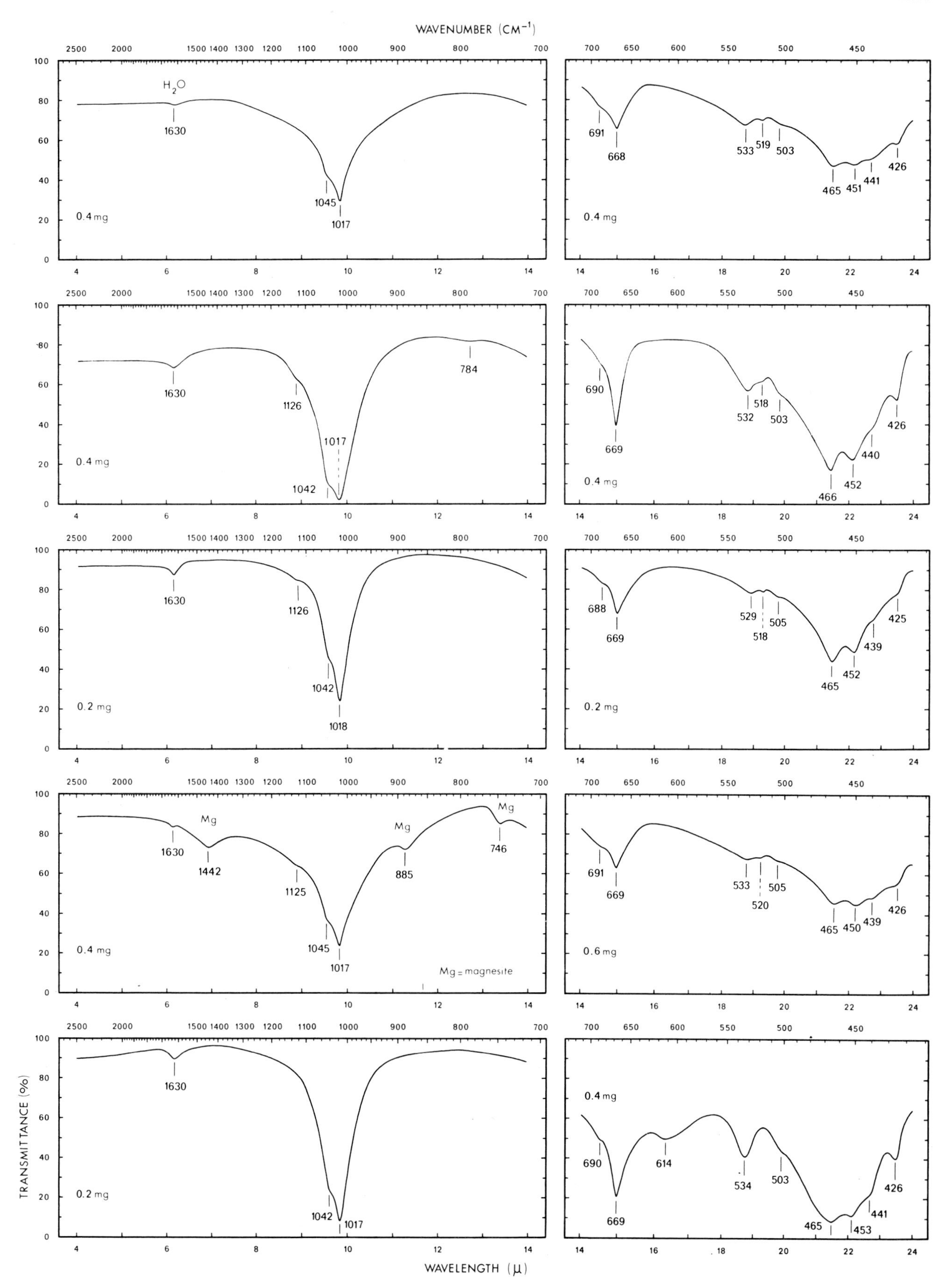
WAVENUMBER (CM⁻¹)
H_2O
1630
1045
1017
0.4 mg
691
668
533
519
503
465
451
441
426
0.4 mg
1630
1126
1017
1042
784
0.4 mg
690
669
532
518
503
466
452
440
426
0.4 mg
1630
1126
1042
1018
0.2 mg
688
669
529
518
505
465
452
439
425
0.2 mg
Mg
1630
1442
1125
1045
1017
Mg
885
Mg
746
Mg = magnesite
0.4 mg
691
669
533
520
505
465
450
439
426
0.6 mg
TRANSMITTANCE (%)
1630
1042
1017
0.2 mg
0.4 mg
690
669
614
534
503
465
453
441
426
WAVELENGTH (μ)

Steatite
S. Andreas Mountain, N. Mex., U.S.A.
some calcite

Soapstone
Arizona, U.S.A.
some chlorite

Georgia, U.S.A.
pure

Commercial talc
chlorite and some dolomite

chlorite and some quartz

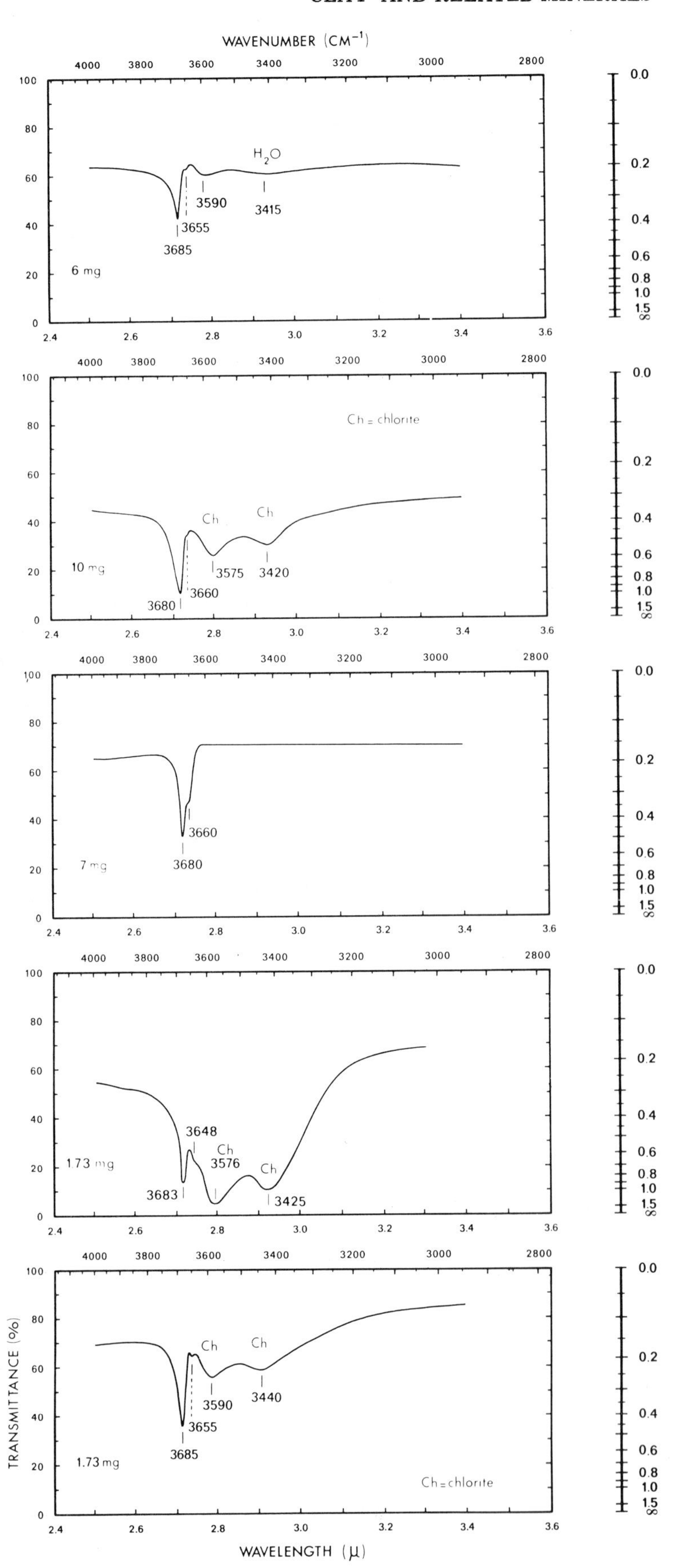

Fig. 6.26. Infrared spectra of steatite, soapstone and commercial talc from different origin.

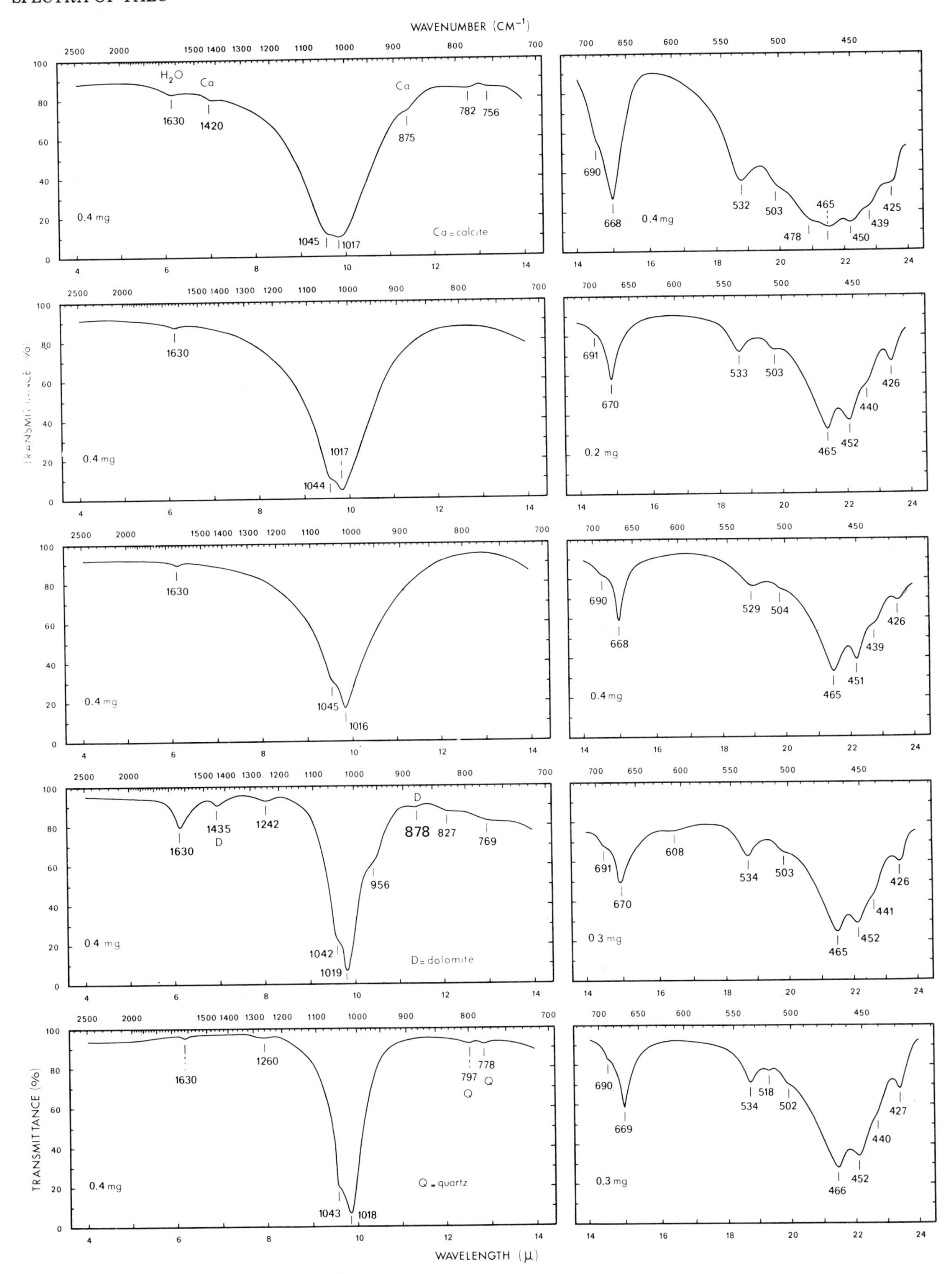
WAVENUMBER (CM⁻¹)
H₂O
Ca
1630
1420
875
782
756
1045
1017
0.4 mg
Ca = calcite
690
668
532
503
465
478
450
439
425
1630
1044
0.2 mg
691
670
533
426
440
452
1016
529
504
451
1435
D
1242
878
827
769
956
1042
1019
D = dolomite
608
534
441
0.3 mg
1260
797
778
Q
1043
1018
Q = quartz
518
502
466
427
TRANSMITTANCE (%)
WAVELENGTH (μ)

Brucite
Wakefield, Que., Canada
degraded and non-degraded talc, some calcite, dolomite and chrysotile

Chevalah, Wash., U.S.A.
degraded talc, some chrysotile and dolomite

Texas, U.S.A.
degraded and non-degraded talc, chrysotile and some calcite, dolomite

Brewster, N.Y., U.S.A.
degraded and non-degraded talc and some chrysotile

Nevada, U.S.A.
degraded talc and pyrophyllite, talc, chrysotile and dolomite

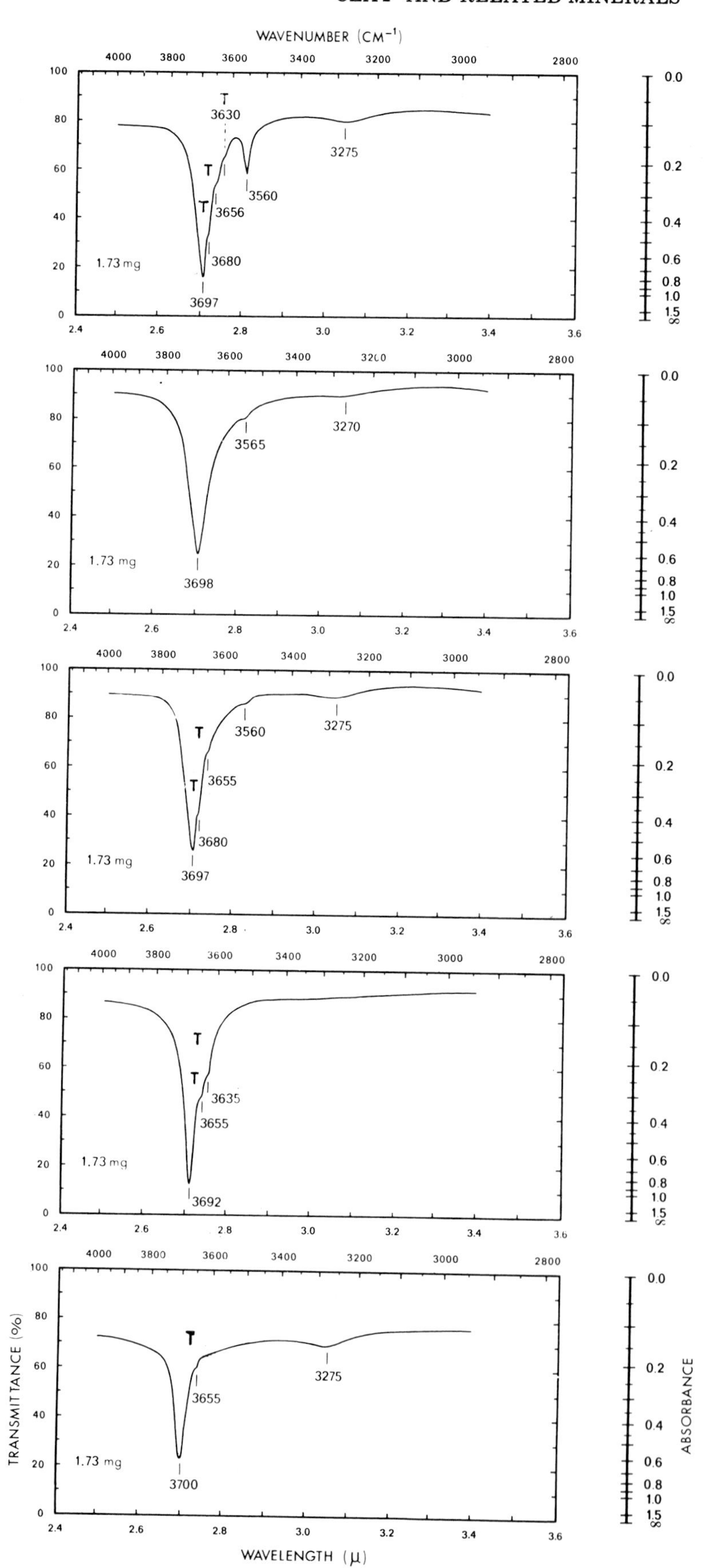

Fig. 6.27. Infrared spectra of brucite from different origin.

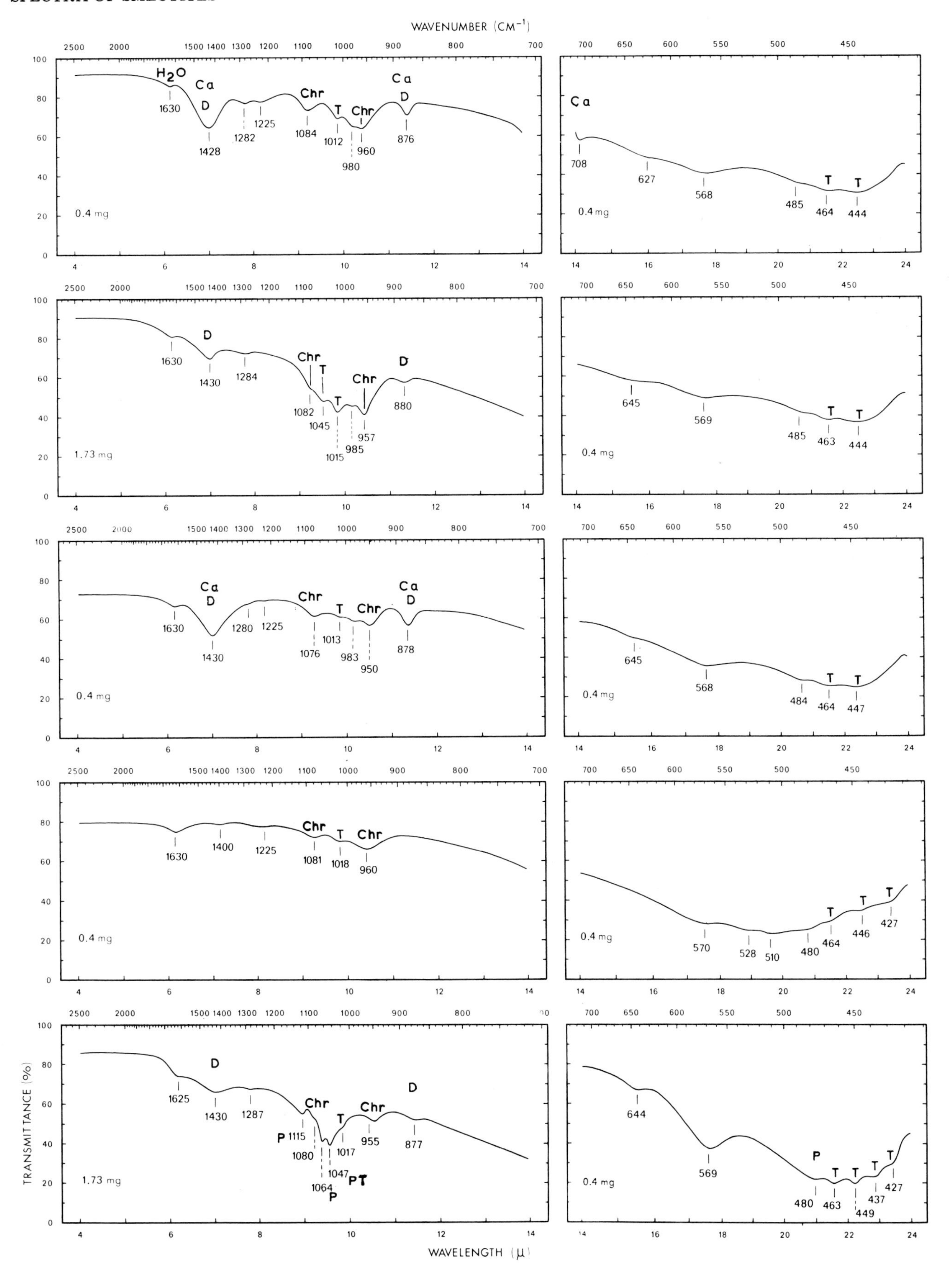

Montmorillonite
Cornwall, Great Britain
pure
Fig. E.M. 72

Atzcapozalco, Mexico
some cristobalite and hematite (both overlapped)
Fig. E.M. 80

Hütenite
Düsseldorf, Germany
some quartz

Moosburg, Germany
pure
Fig. E.M. 79

Cadouin, France
some kaolinite and quartz
Fig. E.M. 97

Bantam, Java, Indonesia
some kaolinite and quartz
Fig. E.M. 76

Tarragona, Spain
some kaolinite and quartz

Pontevedro, Spain
kaolinite and some quartz
Fig. E.M. 73

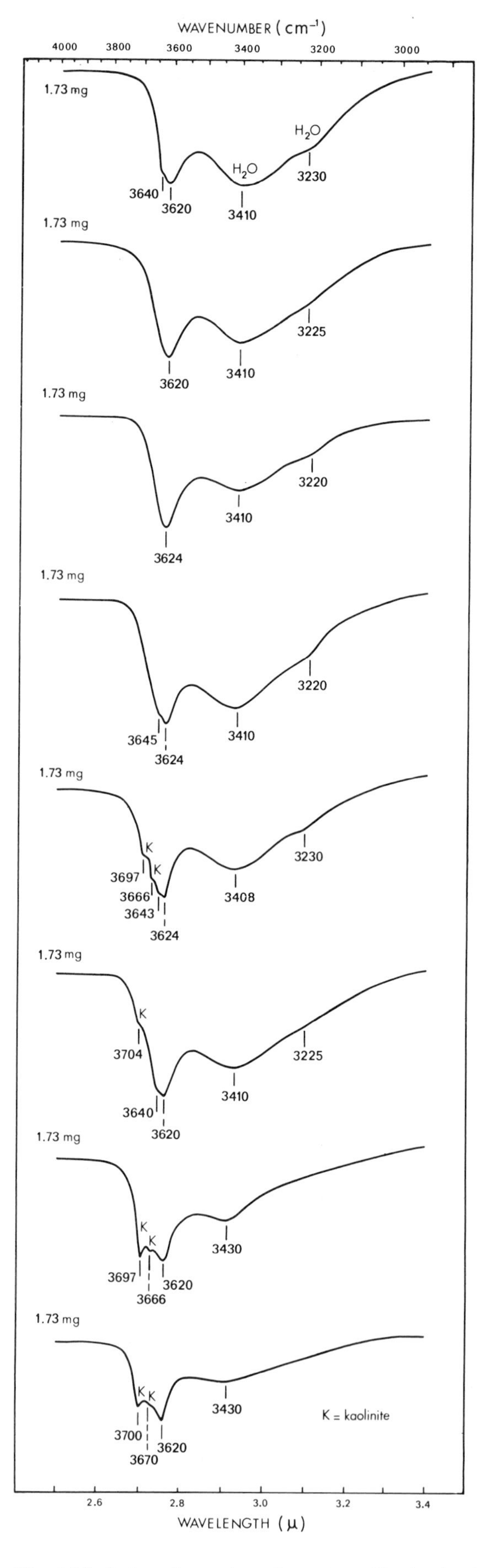

Fig. 6.28. Infrared spectra of montmorillonite (< 2 μm) from different origin.

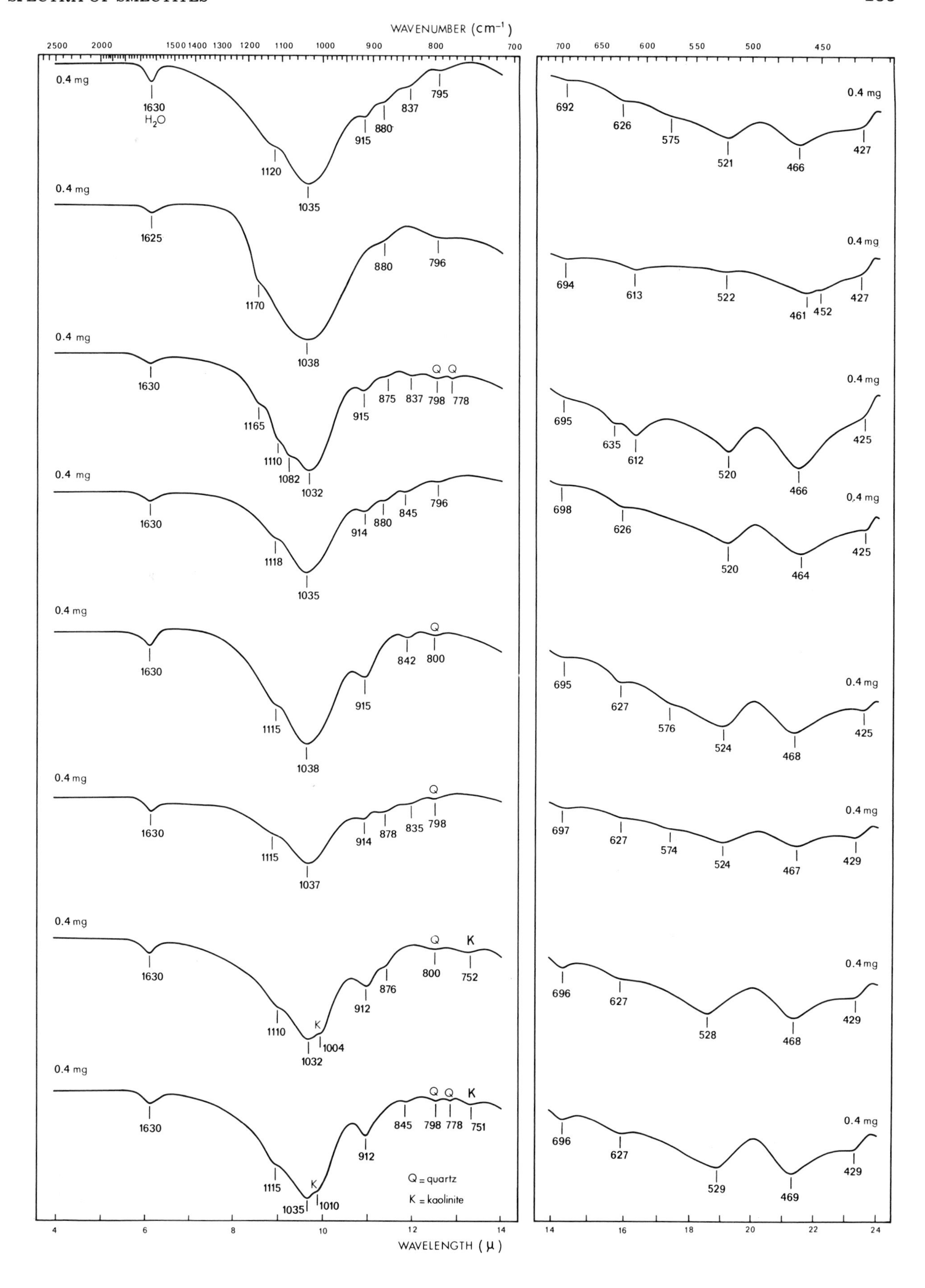
WAVENUMBER (cm⁻¹)
2500 2000 1500 1400 1300 1200 1100 1000 900 800 700
700 650 600 550 500 450
0.4 mg
1630
H2O
1120
1035
915
880
837
795
692
626
575
521
466
427
1625
1170
1038
880
796
694
613
522
461
452
1165
1110
1082
1032
875
837
798
778
Q Q
695
635
612
520
466
425
1118
914
845
796
698
626
464
1115
842
800
Q
576
524
468
1037
878
835
798
697
627
574
467
429
K
1004
912
876
752
696
528
1010
751
529
469
Q = quartz
K = kaolinite
4 6 8 10 12 14
14 16 18 20 22 24
WAVELENGTH (μ)

Montmorillonite
Morocco
some calcite and cristobalite (overlapped)

Camp Bertaux, Morocco
some mica (overlapped), calcite, feldspar (overlapped) and kaolinite
Fig. E.M. 74

Besanje, Yugoslavia
some quartz and kaolintie
Fig. E.M. 71

Bulgaria
pure

Lodborany, Bohemia, Czechoslovakia
pure

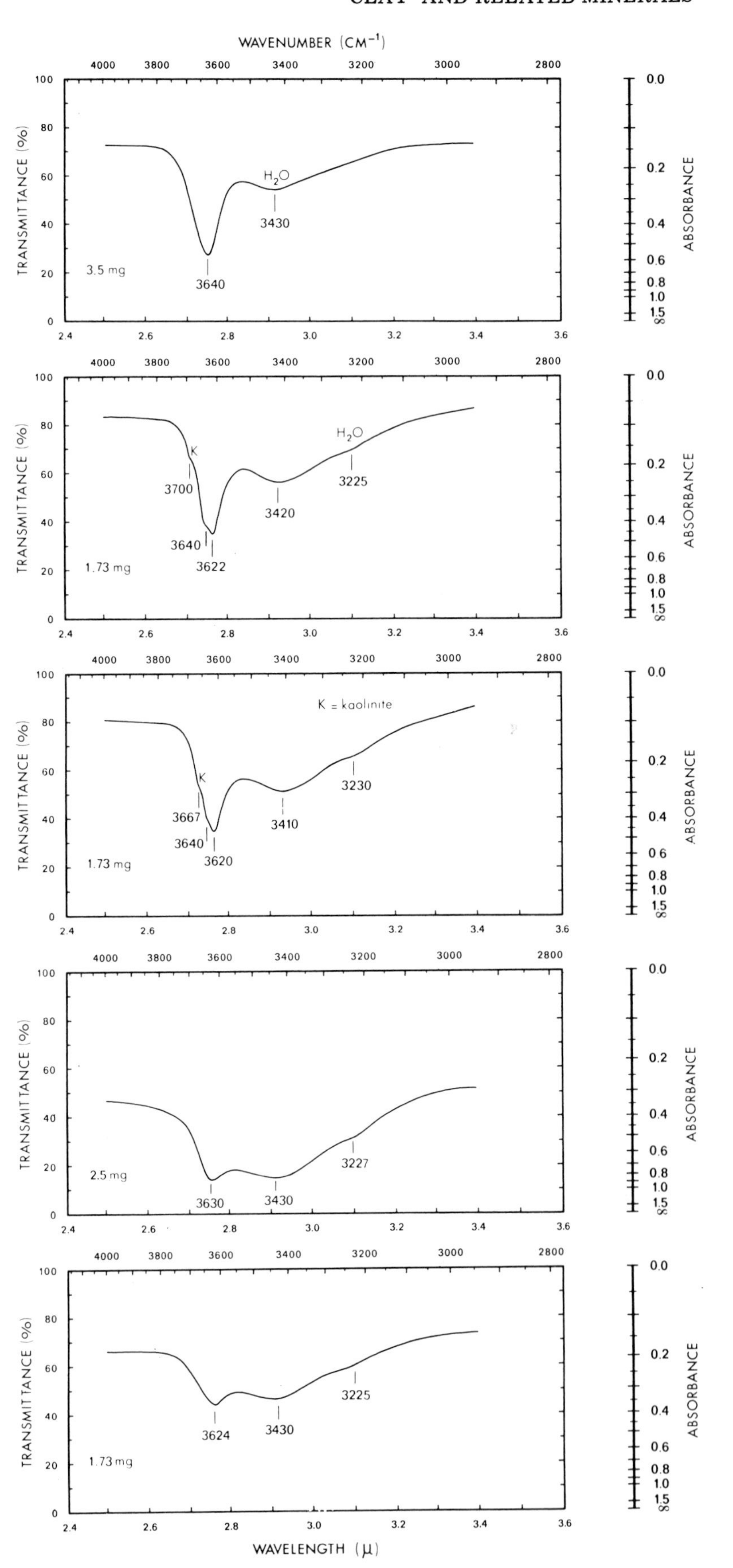

Fig. 6.29. Infrared spectra of montmorillonite (< 2 μm) from different origin.

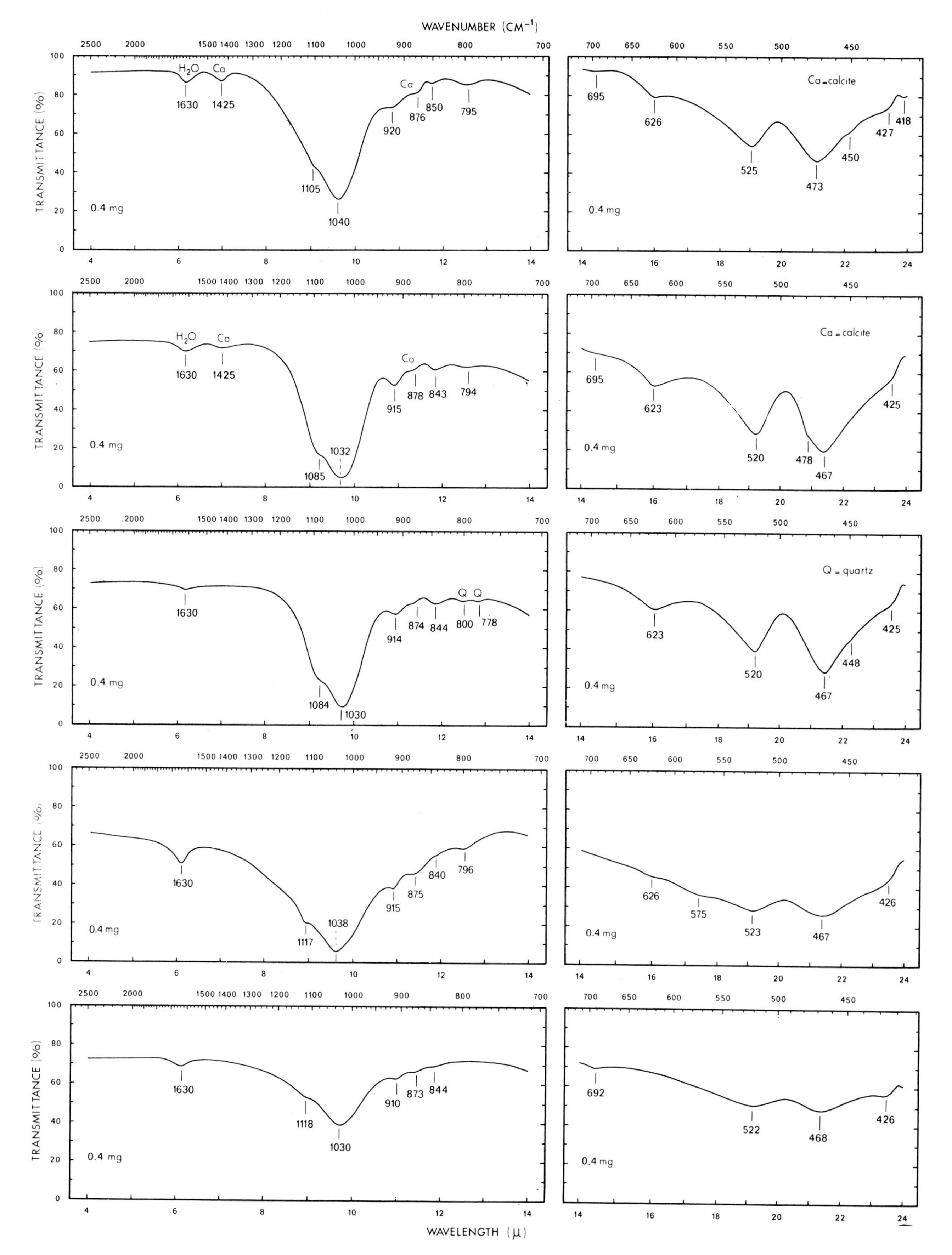
WAVENUMBER (CM⁻¹)
TRANSMITTANCE (%)
H₂O
Ca
1630
1425
Ca
920
876
850
795
1105
1040
0.4 mg
Ca = calcite
695
626
525
473
450
427
418
1085
1032
915
878
843
794
623
520
478
467
425
Q = quartz
1084
1030
914
874
844
800
778
448
1117
1038
875
840
796
575
523
426
1118
910
873
692
522
468
WAVELENGTH (μ)

Montmorillonite
Upton, Wyo., U.S.A.
some quartz

Osage, Wyo., U.S.A,
some quartz,
Fig. E.M. 70

Amory, Mississippi, U.S.A.
some kaolinite

Polkville, Mississippi, U.S.A.
some kaolinite

Bayard, N.Mex., U.S.A.
some cristobalite

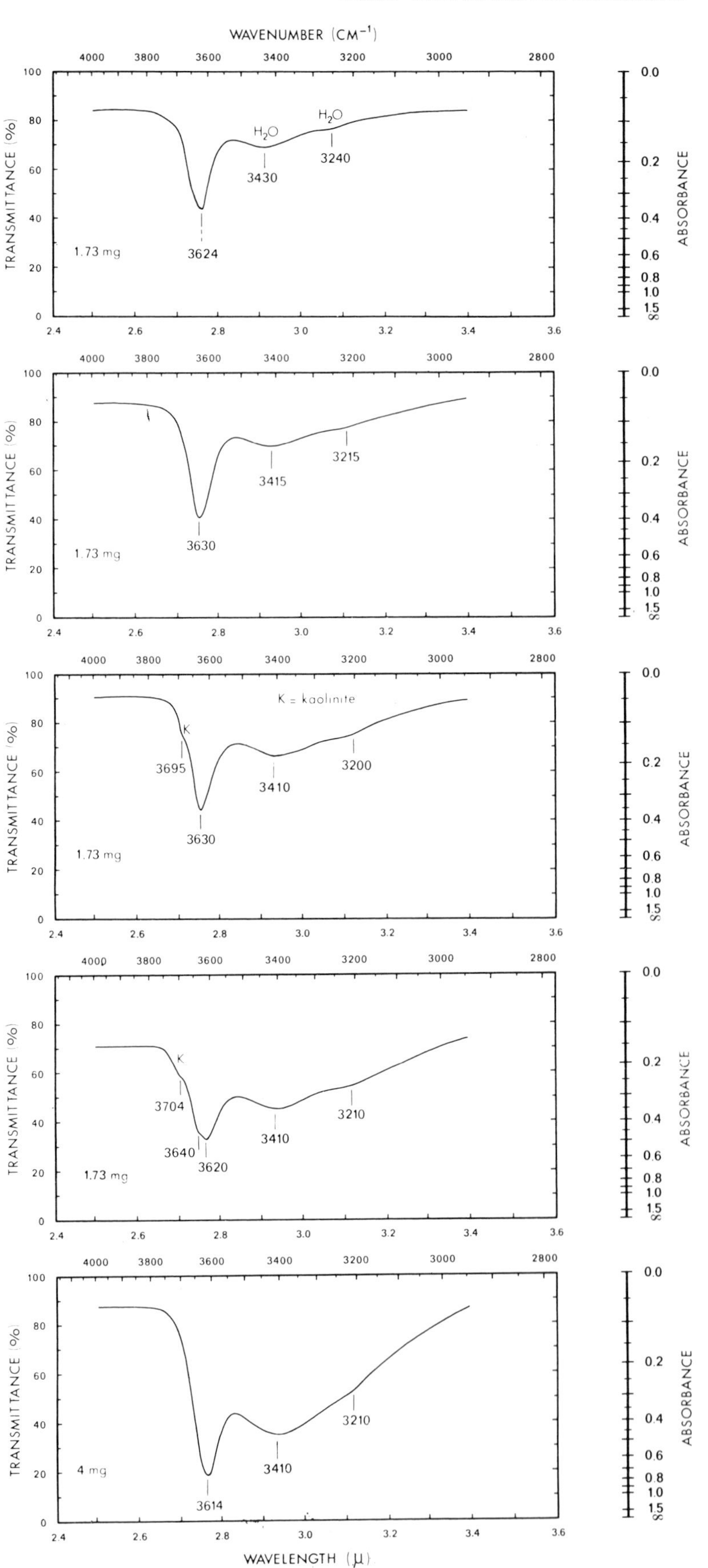

Fig. 6.30. Infrared spectra of montmorillonite (< 2 μm) from different origin.

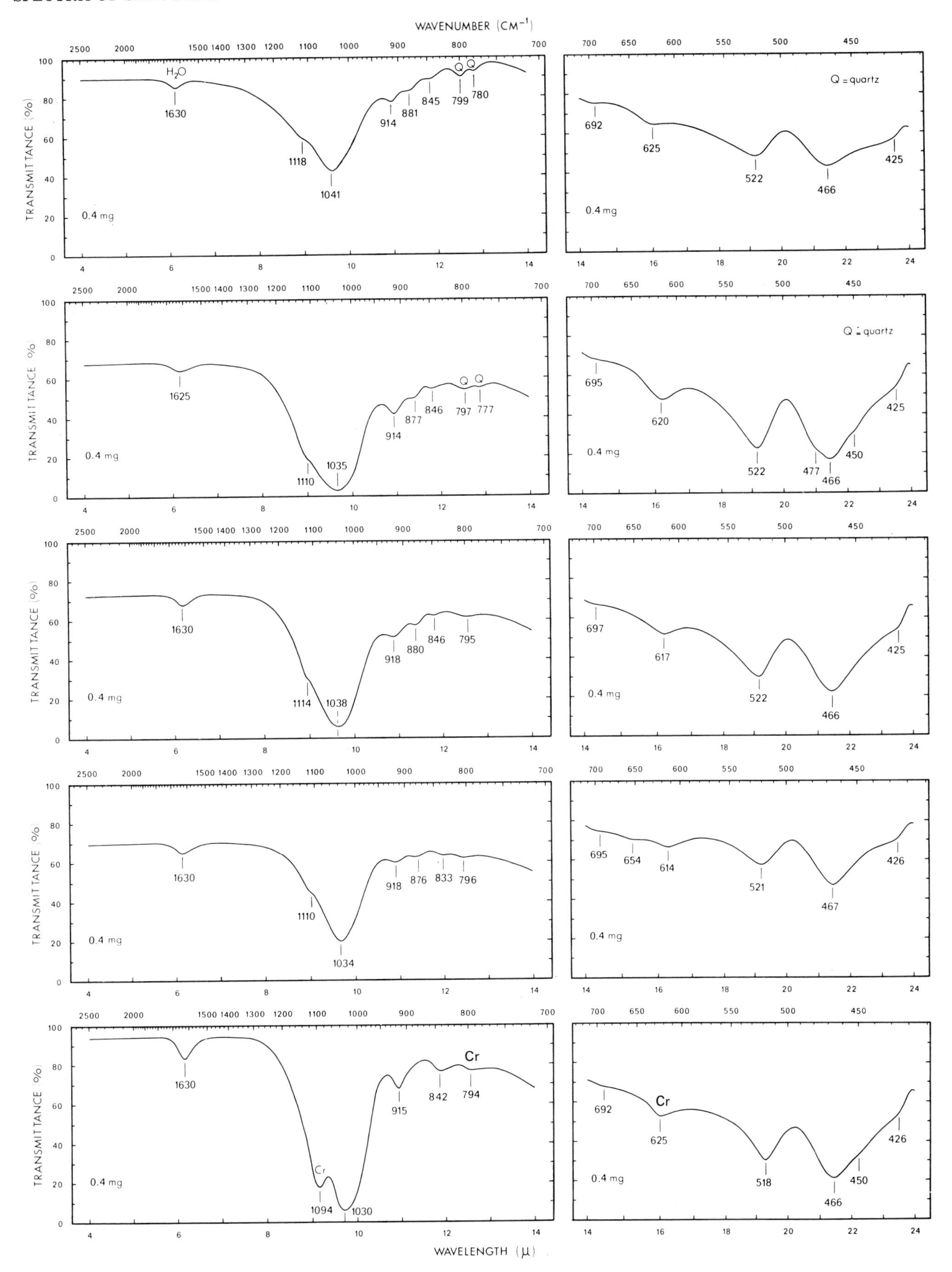
WAVENUMBER (CM⁻¹)
TRANSMITTANCE (%)
WAVELENGTH (μ)
Q = quartz
0.4 mg
H2O
1630
1118
1041
914
881
845
799
780
692
625
522
466
425
1625
1110
1035
877
846
797
777
695
620
477
450
1114
1038
918
880
795
697
617
876
833
796
654
614
521
467
426
Cr
1094
1030
915
842
794
518

Montmorillonite
Javapai, U.S.A.
pure
Fig. E.M. 78

Otay, Calif, U.S.A.
pure
Fig. E.M. 236

Cheto, Ariz., U.S.A.
some calcite

Chambers, Ariz., U.S.A.
pure
Fig. E.M. 77

Cameron, Ariz., U.S.A.
some quartz and kaolinite

Long Bell, Ark., U.S.A.
some kaolinite

Beidellite

Nashville, Ark., U.S.A.
some kaolinite and quartz
Fig. E.M. 92

Burns, Miss., U.S.A.
some quartz

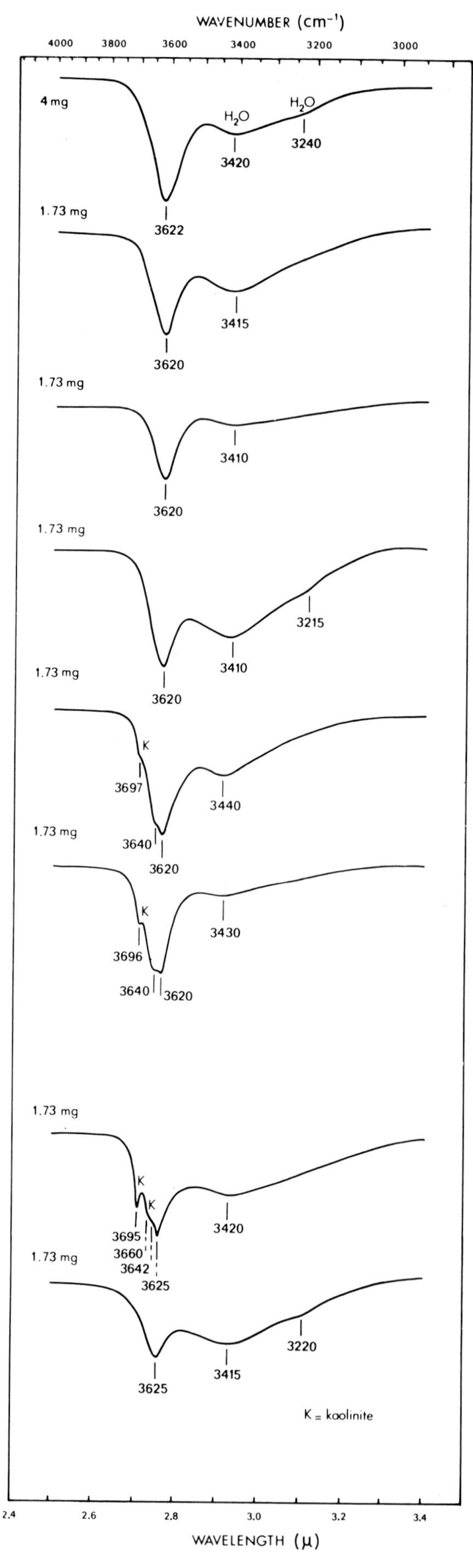

Fig. 6.31. Infrared spectra of montmorillonite and beidellite (< 2 μm) from different origin.

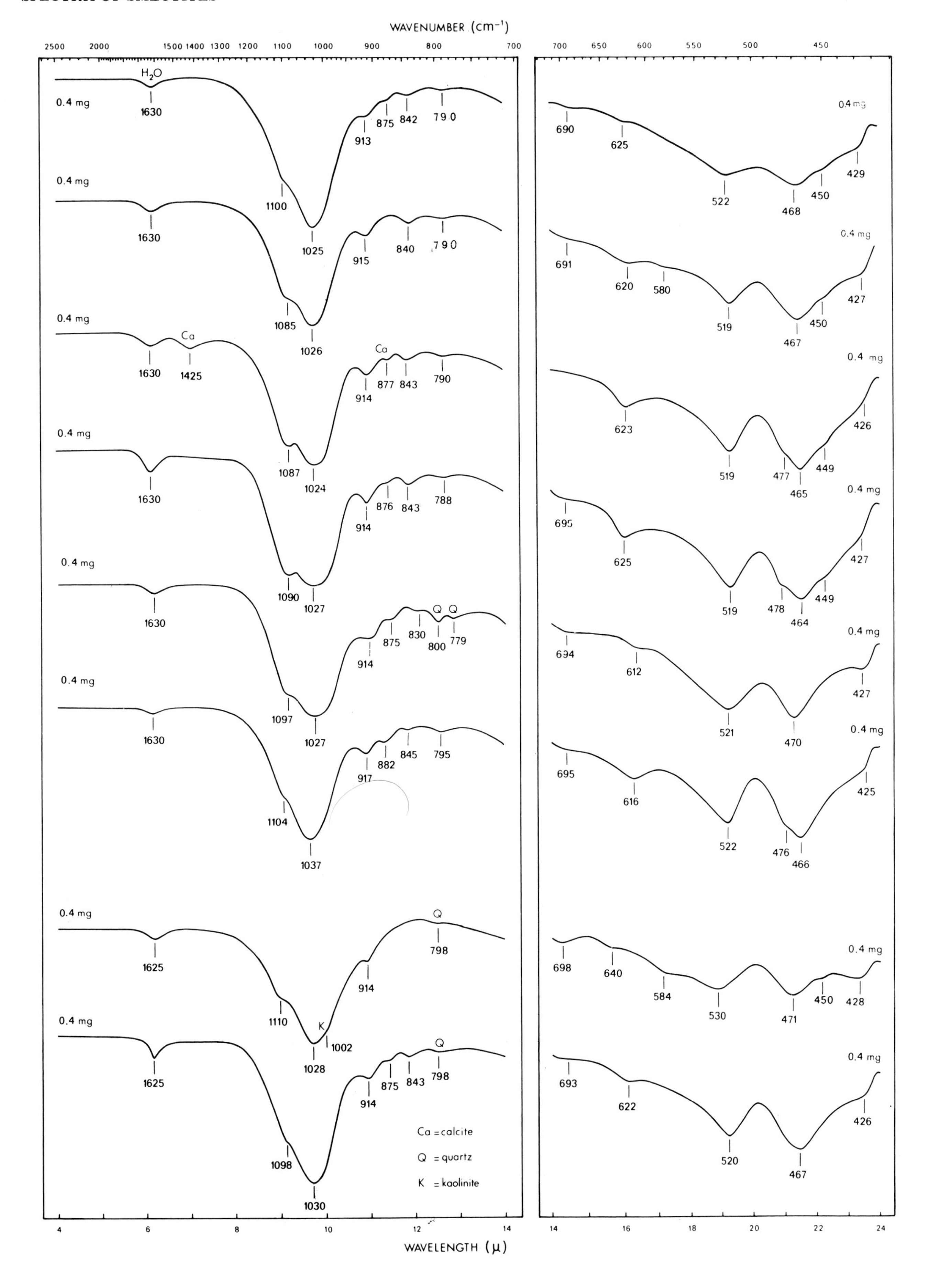
WAVENUMBER (cm⁻¹)
WAVELENGTH (μ)
Ca = calcite
Q = quartz
K = kaolinite

Soil montmorillonite
Pro Santa Clare, Cuba
some quartz, kaolinite and calcite

Madioen, Java, Indonesia, margalite
some cristobalite (overlapped) and kaolinite

Souk el Arba, Morocco, tir noir
some kaolinite, quartz and calcite
Fig. E.M. 84

Berkel, The Netherlands
some quartz, kaolinite and feldspar (overlapped)

Wad Medani, Sudan, badob soil
some kaolinite, quartz and calcite

Addis Abeba, Ethiopia, black soil
some mica (overlapped), kaolinite, quartz and feldspar (overlapped)
Fig. E.M. 81

Poppoh, Java, Indonesia, margalite
some quartz, cristobalite (overlapped) and kaolinite

Texas, U.S.A., black clay
calcite and some quartz and kaolinite
Fig. E.M. 87, 223

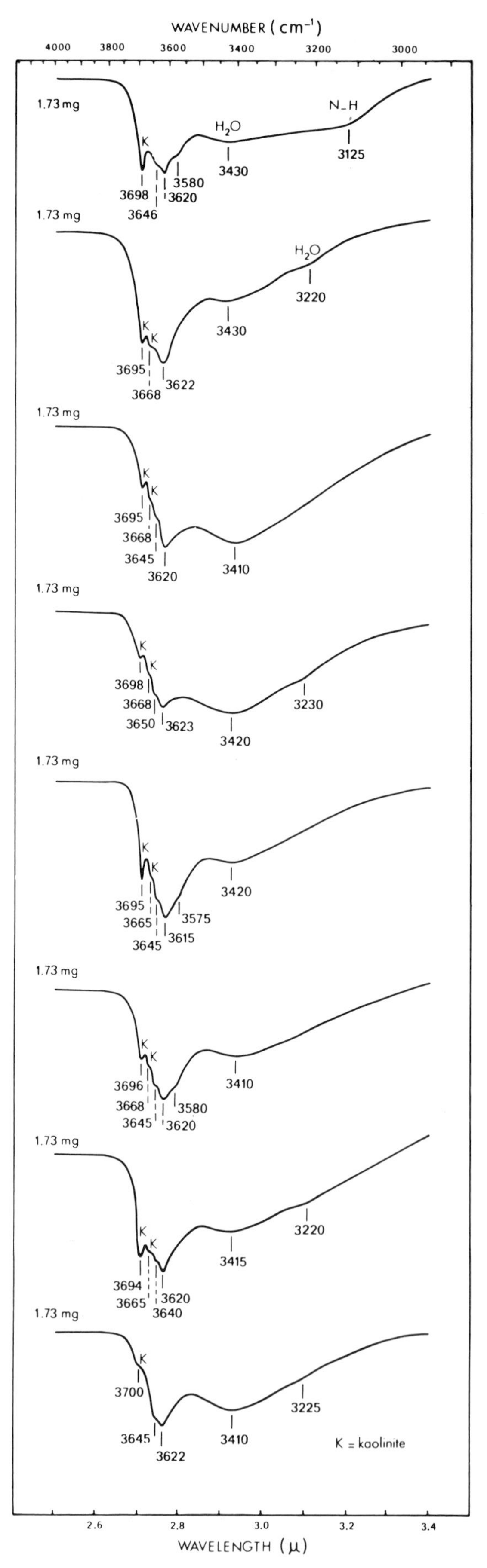

Fig. 6.32. Infrared spectra of soil montmorillonite (< 2 μm) from different origin.

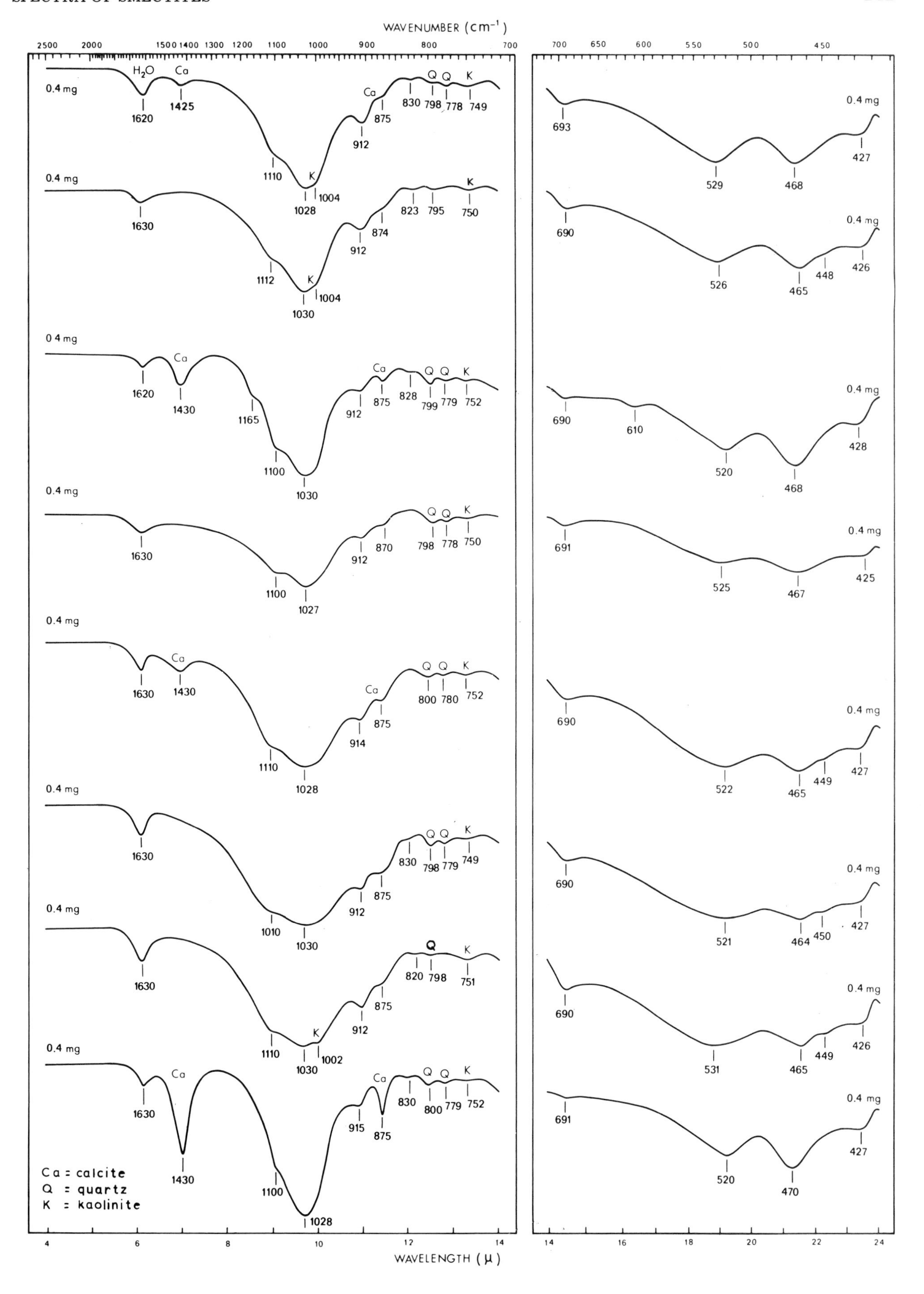
WAVENUMBER (cm^{-1})
WAVELENGTH (μ)
0.4 mg
H_2O
Ca
Q
K
Ca = calcite
Q = quartz
K = kaolinite

Soil montmorillonite
Bombay, India
some quartz and kaolinite

Nile delta, Egypt
some quartz, kaolinite, mica (overlapped) and calcite
Fig. E.M. 90

Accra Plains, Nigeria
some calcite, quartz and kaolinite

Edmonton, Canada, black soil
some mica (overlapped), quartz and kaolinite

Rhodesia, black turf
some kaolinite and quartz
Fig. E.M. 82

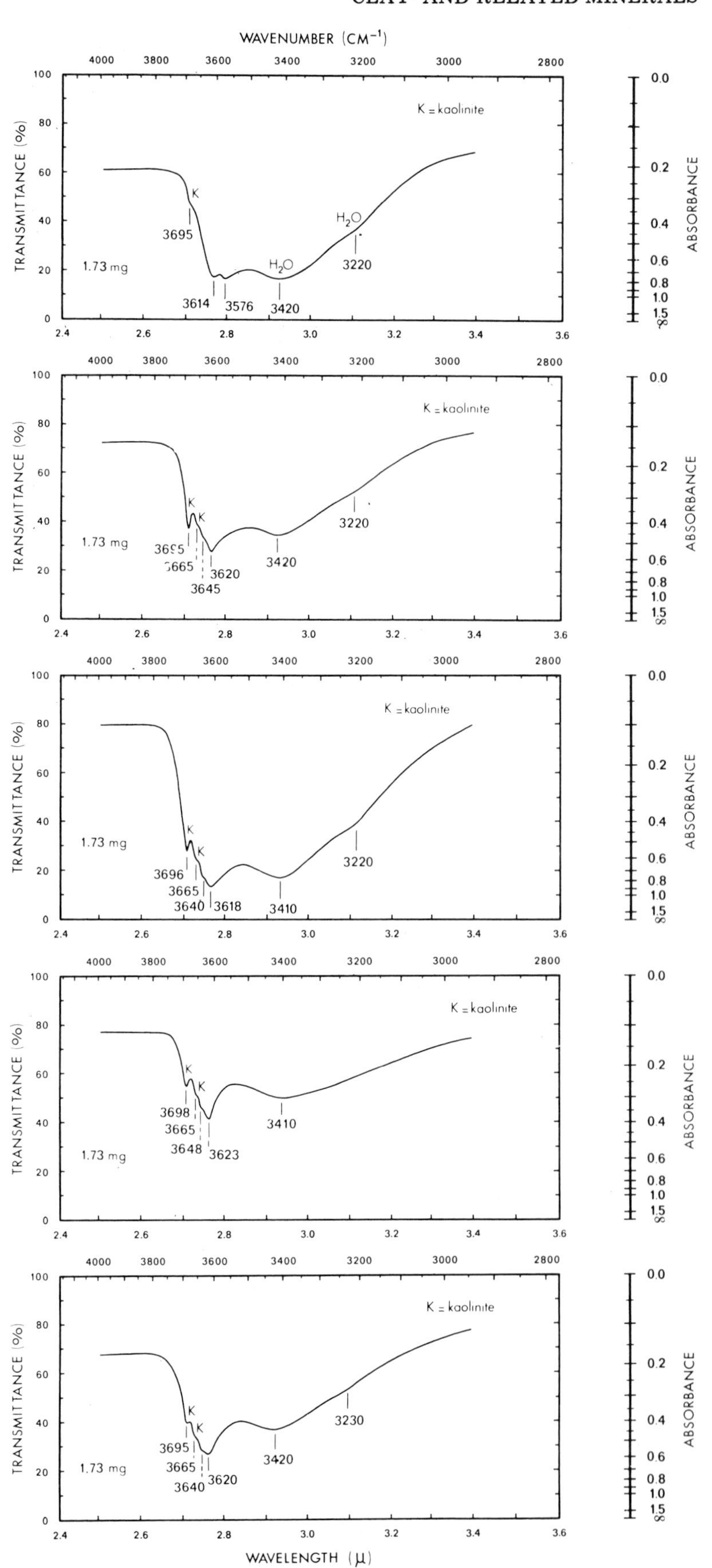

Fig. 6.33. Infrared spectra of soil montmorillonite (< 2 μm) from different origin.

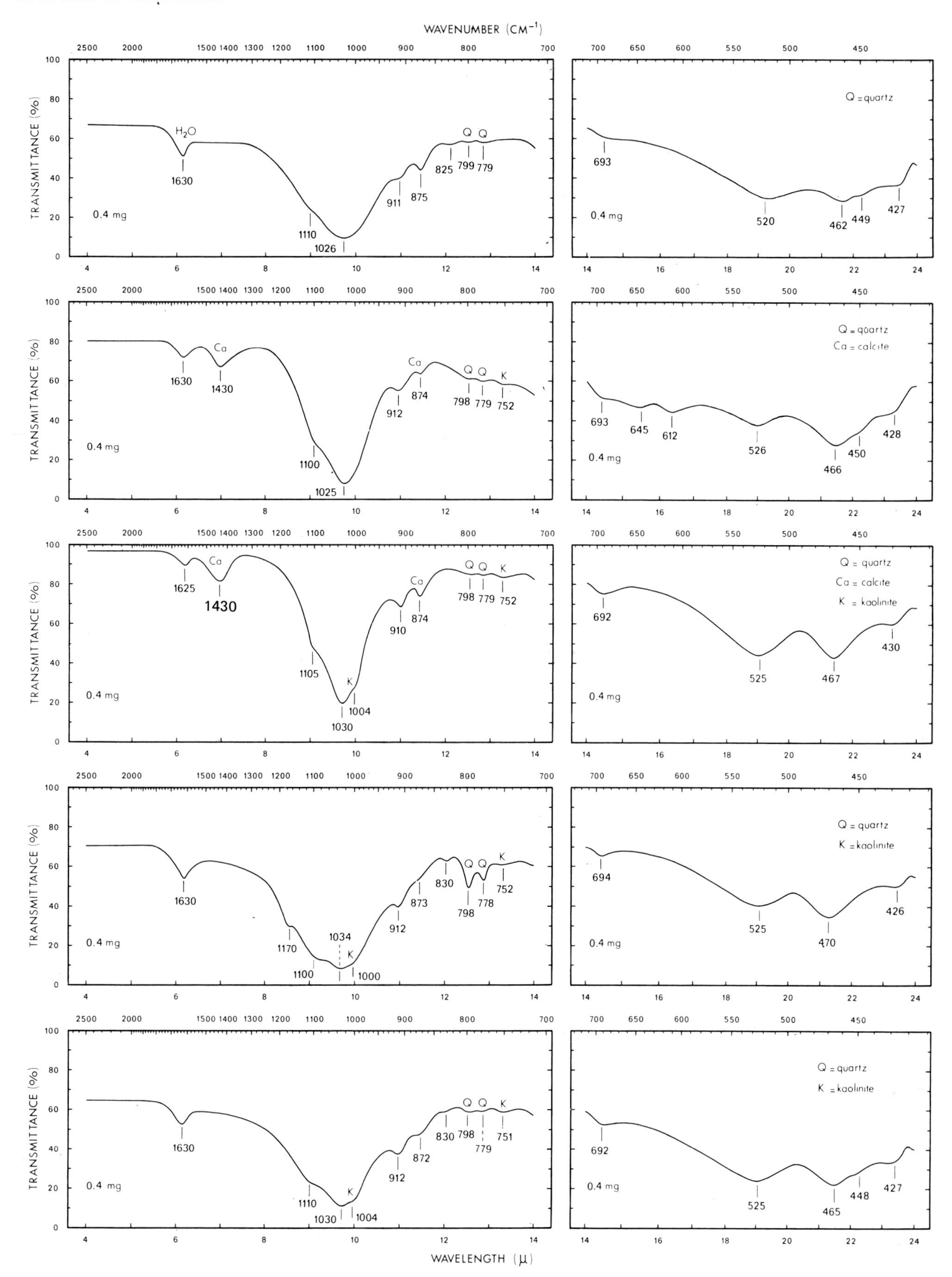
WAVENUMBER (CM⁻¹)
TRANSMITTANCE (%)
WAVELENGTH (μ)
Q = quartz
Ca = calcite
K = kaolinite
0.4 mg
H2O
1630
1110
1026
911
875
825 799 779
693
520
462 449
427
1430
1100
1025
912
874
798 779 752
645
612
526
466
450
428
1625
1105
1030
1004
910
692
525
467
430
1170
1034
1000
873
830
798
778
694
470
426
872
751
448
465

Saponite
Krugersdorp, Transvaal
some quartz
Fig. E.M. 93

California, U.S.A.
some quartz, calcite and dolomite
Fig. E.M. 94

Apennines, Italy
some quartz and hematite (overlapped)

St. Margherita, Pavese, Italy
some quartz, chrysotile and calcite

Reggione, Italy
talc, some calcite and quartz

Griffithite
Griffith Park, Calif., U.S.A.
hematite (overlapped) and feldspar (overlapped)

Hectorite
Hector, Calif., U.S.A.
some calcite
Fig. E.M. 96

Ghassoulite
Ghassoul, Morocco
some quartz, dolomite, and calcite
Fig. E.M. 98

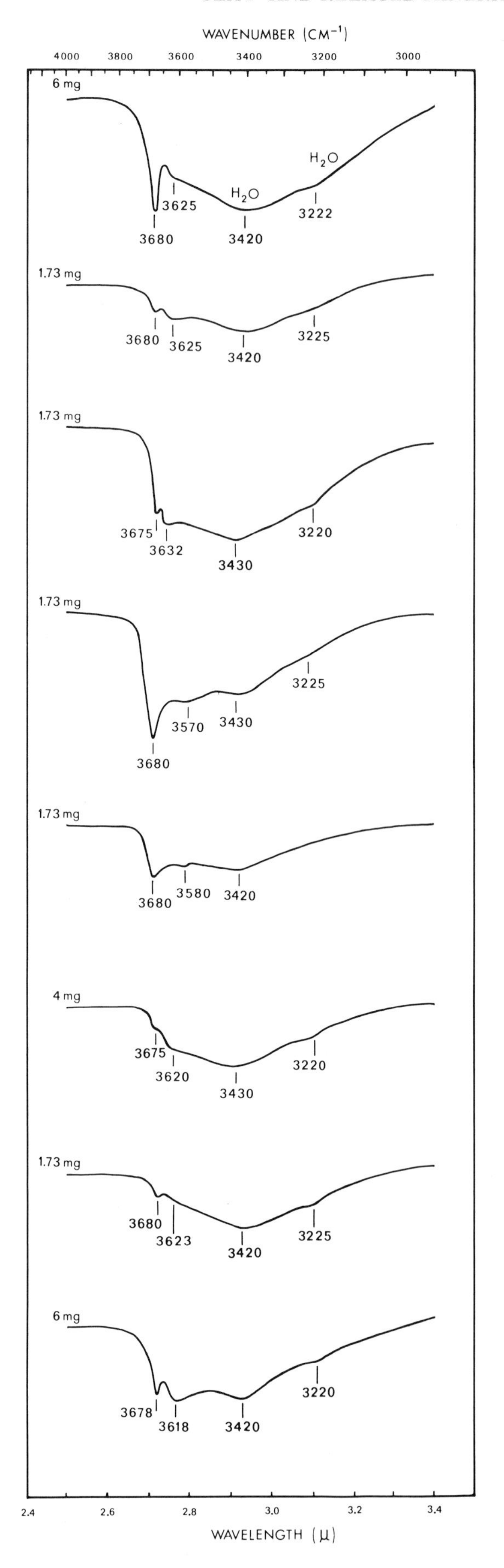

Fig. 6.34. Infrared spectra of saponite (< 2 μm) from different origin, griffithite, hectorite and ghassoulite (< 2 μm).

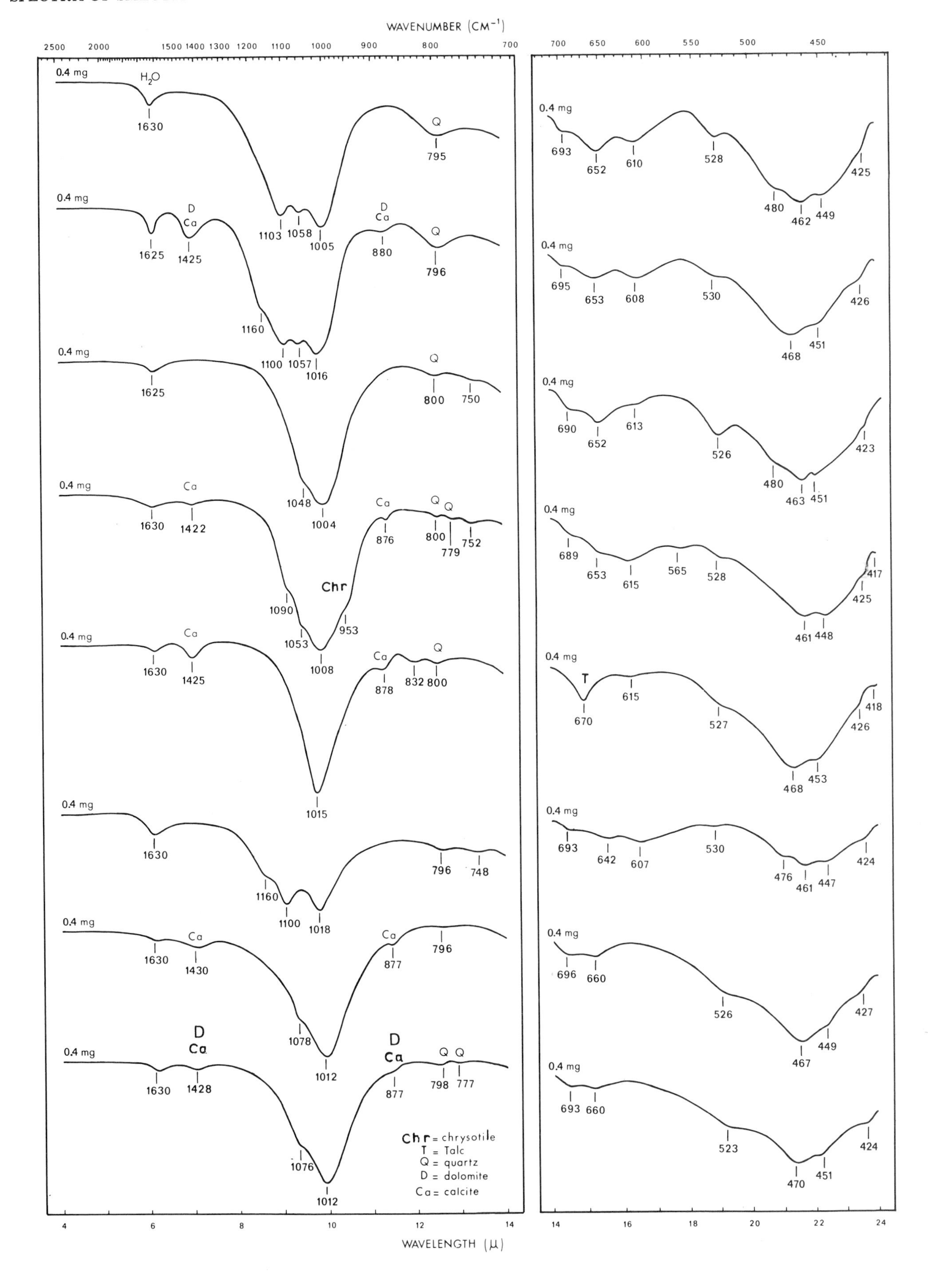
WAVENUMBER (CM⁻¹)
WAVELENGTH (μ)
Chr = chrysotile
T = Talc
Q = quartz
D = dolomite
Ca = calcite

Nontronite
Andreasberg
some quartz

Hohenhagen, Germany
some quartz
Fig. E.M. 99

Garfield, U.S.A.
pure
Fig. E.M. 100

Allentown, Pa., U.S.A.
some kaolinite and quartz

Manito, Wash., U.S.A.
some kaolinite and quartz
Fig. E.M. 101

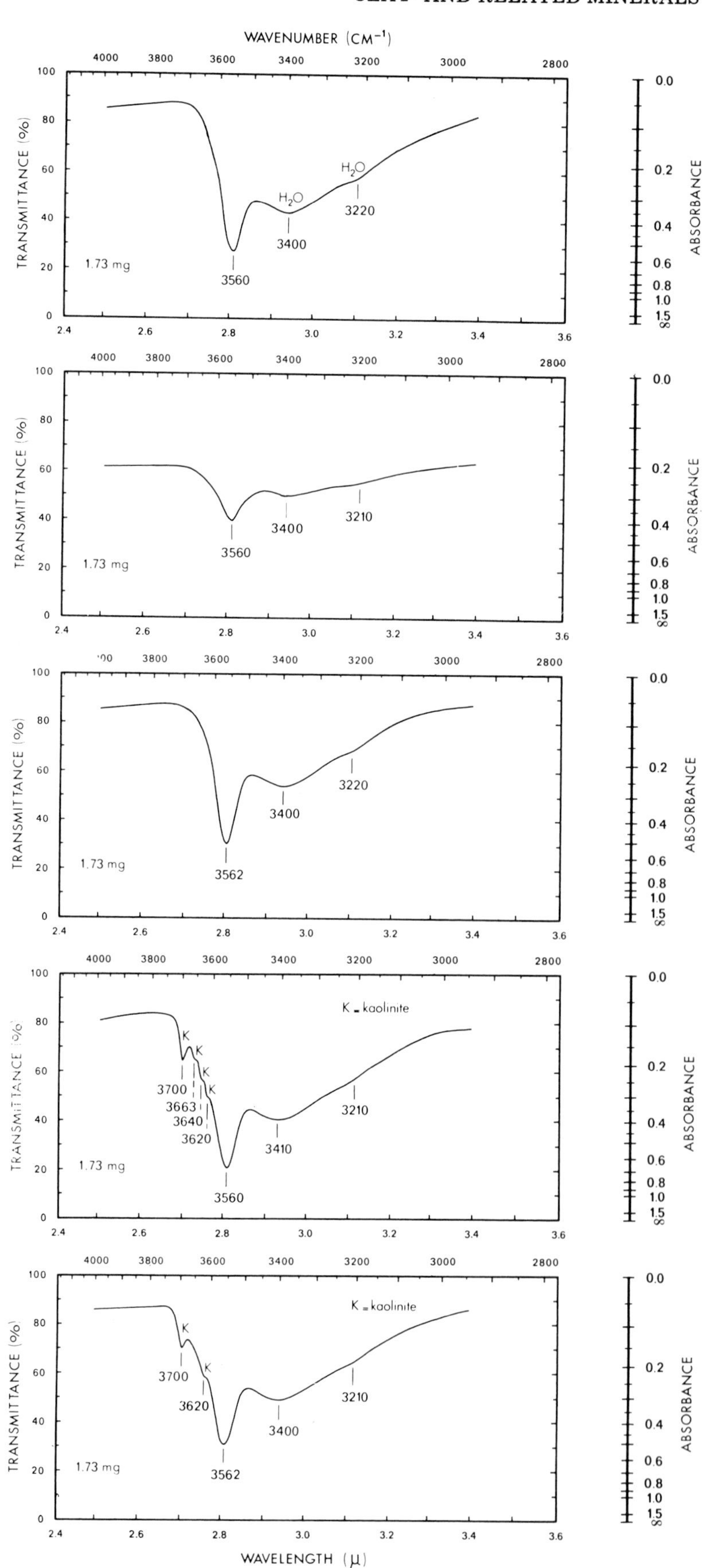

Fig. 6.35. Infrared spectra of nontronite ($< 2\ \mu m$) from different origin.

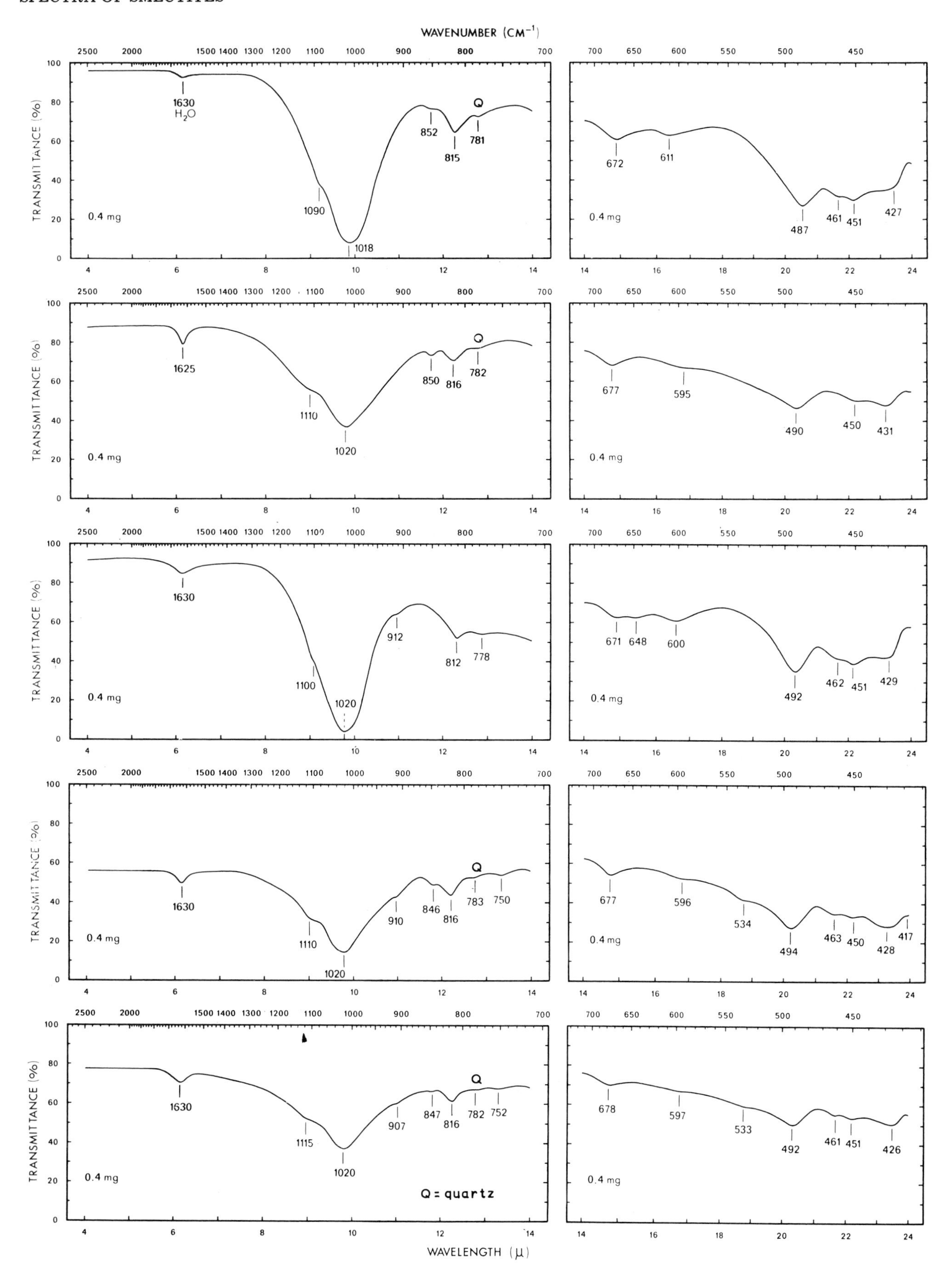
WAVENUMBER (CM^{-1})
TRANSMITTANCE (%)
1630
H_2O
1090
1018
852
815
781
Q
672
611
487
461
451
427
0.4 mg
1625
1110
1020
850
816
782
677
595
490
450
431
1630
1100
1020
912
812
778
671
648
600
492
462
451
429
1630
1110
1020
910
846
816
783
750
677
596
534
494
463
450
428
417
1630
1115
1020
907
847
816
782
752
678
597
533
492
461
451
426
Q = quartz
WAVELENGTH (μ)

Wolchonskoite
Ochansk, Ural, U.S.S.R.
pure
Fig. E.M. 103

Mn montmorillonite
Tirebolic, Turkey
pure

Ni saponite
Barro Alto, Brazil

Sauconite
Spain
some kaolinite

Friedenville, Pa., U.S.A.
some quartz and kaolinite
Fig. E.M. 95

Montmorillonite

Saponite

Nontronite

Medmontite
Dzeskazgan, U.S.S.R.
chrysocolla, kaolinite, montmorillonite, quartz, feldspar, mica, the last two only detectable by X-ray analysis
Fig. E.M. 102

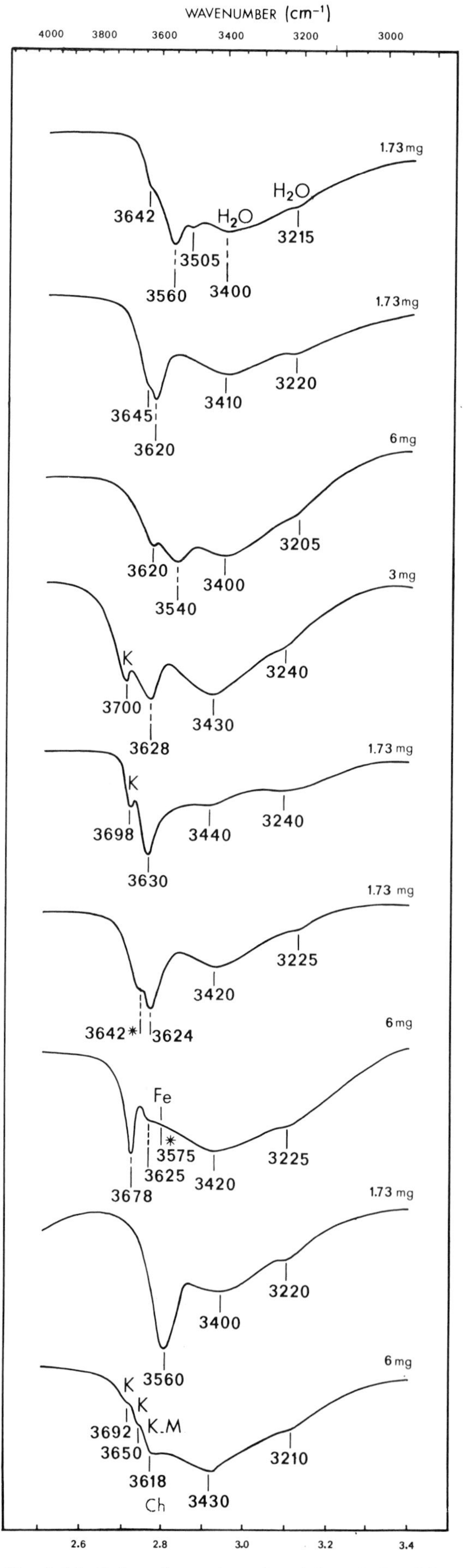

Fig. 6.36. Infrared spectra of wolchonskoite, Mn montmorillonite, Ni saponite, sauconite and medmontite.

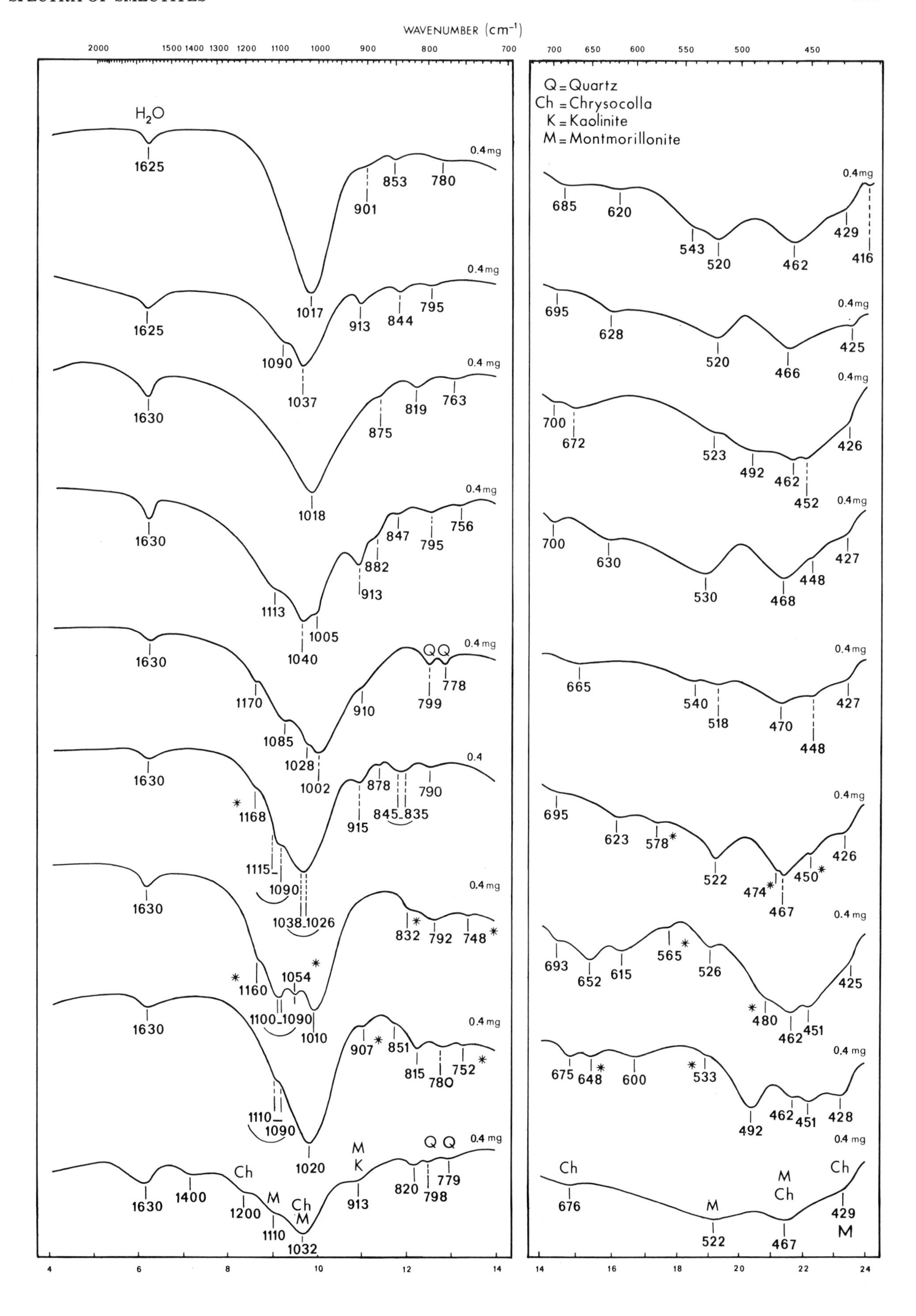
WAVENUMBER (cm⁻¹)
2000
1500 1400 1300 1200 1100 1000 900 800 700
700 650 600 550 500 450
Q = Quartz
Ch = Chrysocolla
K = Kaolinite
M = Montmorillonite
H2O
1625
0.4 mg
901 853 780
1017
685 620 543 520 462 429 416
1625 1090 1037 913 844 795
695 628 520 466 425
1630 1018 875 819 763
700 672 523 492 462 452 426
1630 1113 1040 1005 913 882 847 795 756
700 630 530 468 448 427
1630 1170 1085 1028 1002 910 799 778
QQ
665 540 518 470 448 427
1630 1168 1115 1090 1038 1026 915 878 845 835 790
0.4
695 623 578 522 474 467 450 426
1630 1160 1100 1090 1054 1010 832 792 748
693 652 615 565 526 480 462 451 425
1630 1110 1090 1020 907 851 815 780 752
675 648 600 533 492 462 451 428
1630 1400 Ch 1200 M 1110 Ch M 1032 M K 913 Q Q 820 798 779
Ch 676 M 522 M Ch 467 Ch 429 M
4 6 8 10 12 14
14 16 18 20 22 24

Common muscovite
Salamanca, Spain
some feldspar (overlapped)

Krägero, Norway
some feldspar (overlapped)

Bagnères de Luchon, Spain
some quartz and feldspar (overlapped)

Effingham, Ont., Canada
some feldspar (overlapped)

Georgia, U.S.A.
some feldspar (overlapped)

3T-muscovite
Sultan Basin, U.S.A.
pure

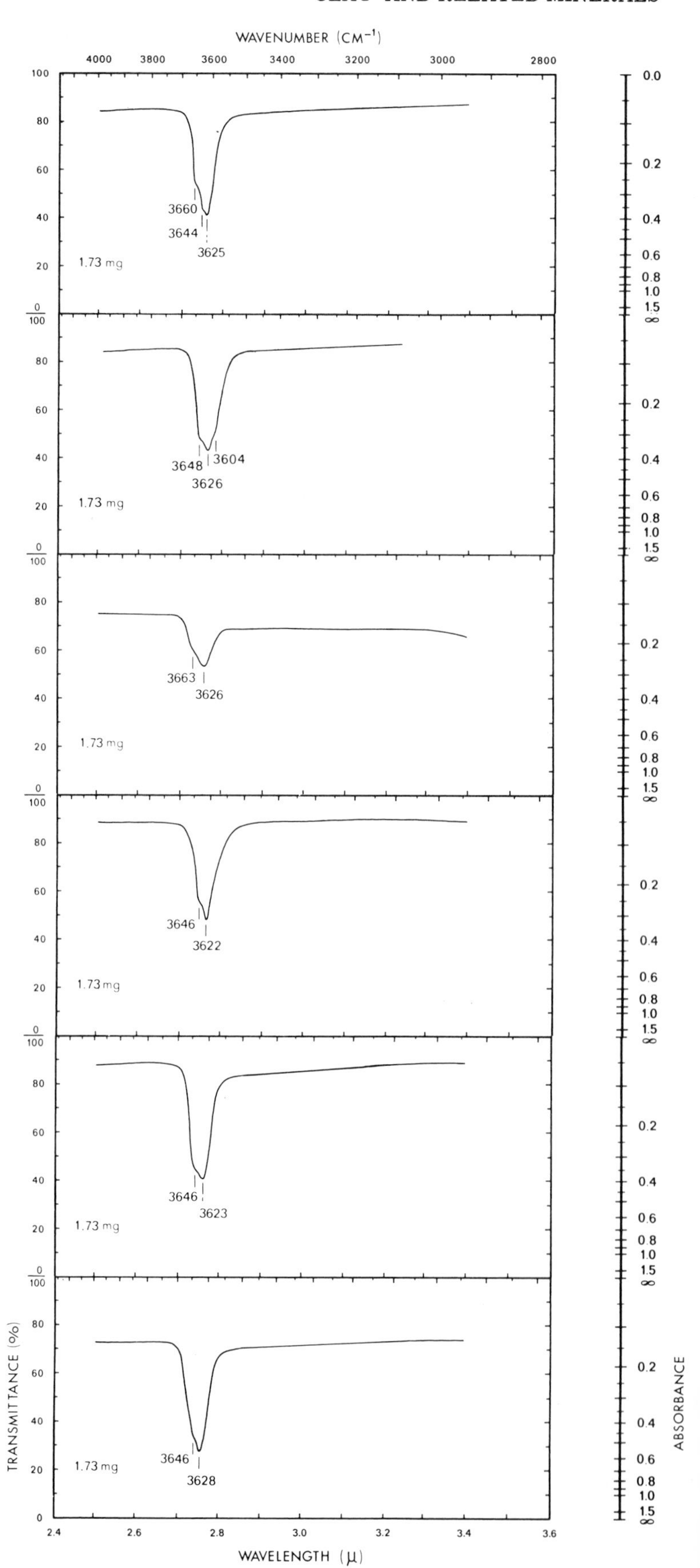

Fig. 6.37. Infrared spectra of common (2 M) muscovite from different origin and of 3T muscovite.

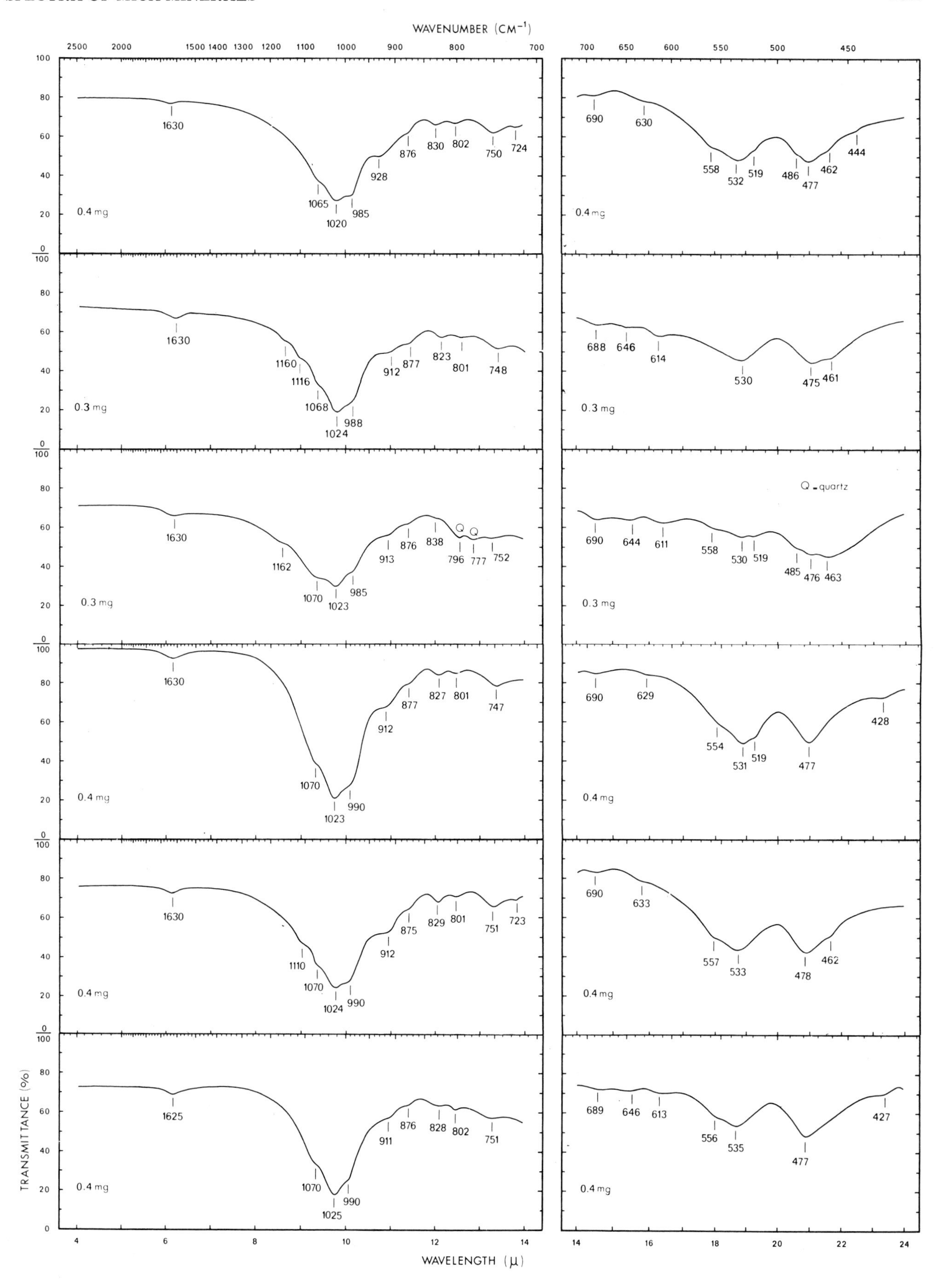
WAVENUMBER (CM⁻¹)
TRANSMITTANCE (%)
WAVELENGTH (μ)
0.4 mg
0.3 mg
Q = quartz

Sericite
Mosambique, Africa
some feldspar (overlapped)

Alamo Crossing, Ariz., U.S.A.
some feldspar (overlapped)

Villar de Puerco, Spain
some feldspar (overlapped)

Tombstone, Ariz., U.S.A.
some feldspar (overlapped)

Hydrous muscovite

Ordley, Scotland
some kaolinite, quartz and hematite (overlapped)
Fig. E.M. 111

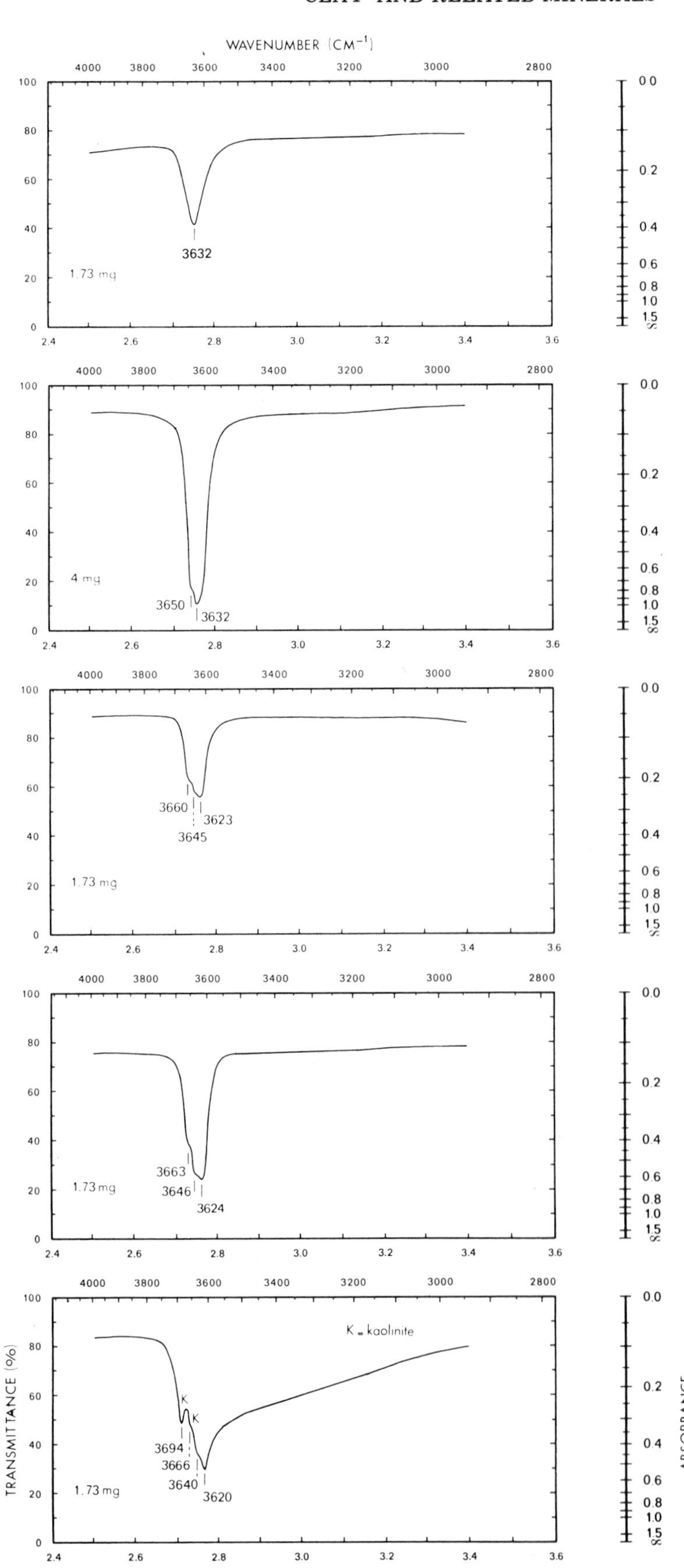

Fig. 6.38. Infrared spectra of sericite from different origin and of hydrous muscovite.

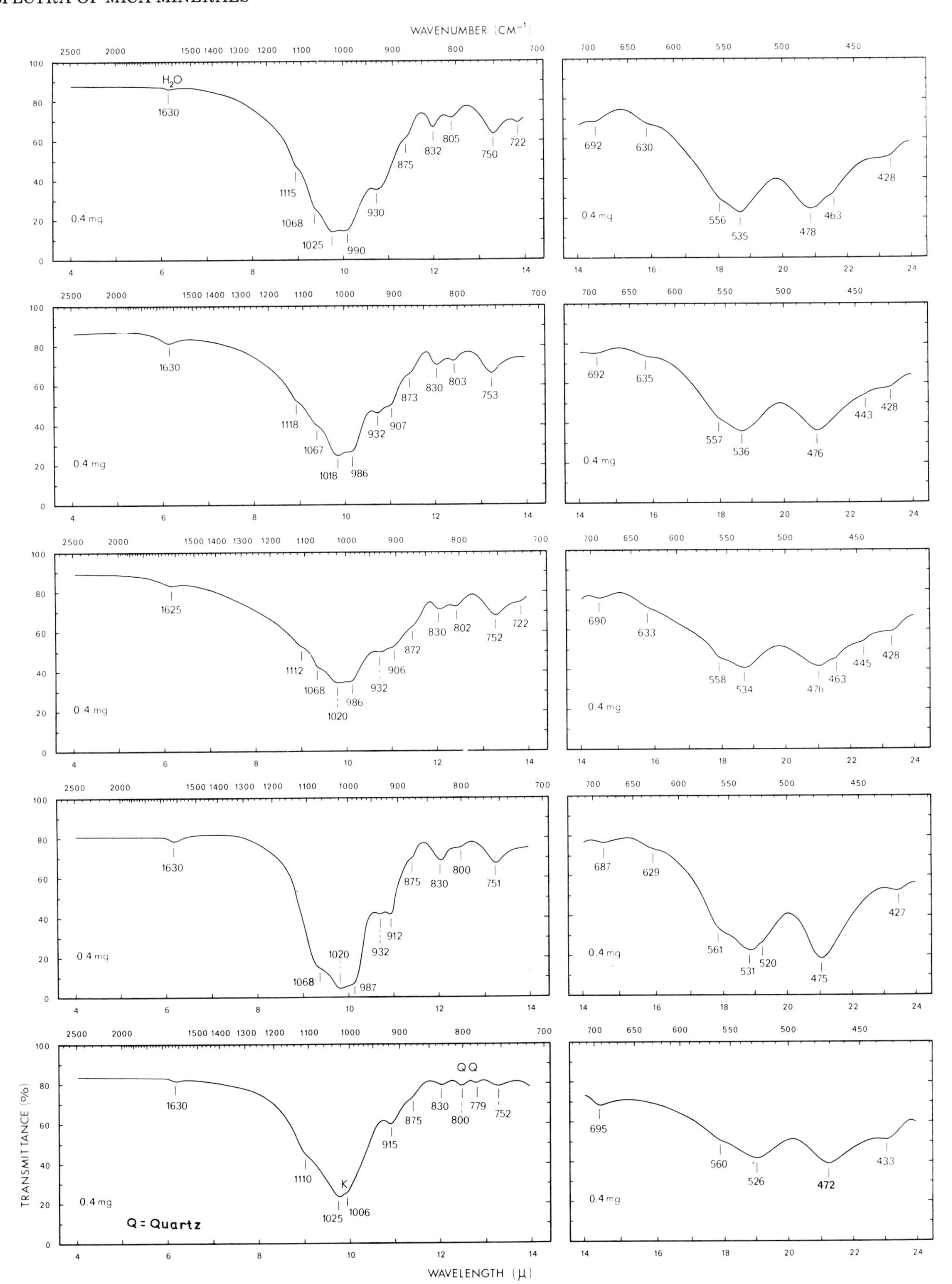
WAVENUMBER (CM⁻¹)
TRANSMITTANCE (%)
WAVELENGTH (μ)
0.4 mg
Q = Quartz

Illite
Ordovician Shale, Prague, Czechoslovakia
some kaolinite, quartz, organic matter and feldspar (overlapped)

Pierre shale, S. Dak., U.S.A.
some calcite, quartz and kaolinite

Alluvial, Pretoria, South Africa
some quartz, kaolinite and hematite (overlapped)
Fig. E.M. 114

Shale, Basbellain, Luxembourg
some quartz, kaolinite and feldspar (overlapped)
Fig. E.M. 115

Shale, Livingstone Co., N.Y., U.S.A.
some quartz, feldspar (overlapped) and organic matter

Fithian Shale, Ill., U.S.A.
some quartz

Shale Heimansgroeve, Epen, The Netherlands
some quartz, calcite, kaolinite and organic matter

Shale Eibergen, The Netherlands
some kaolinite, quartz and calcite

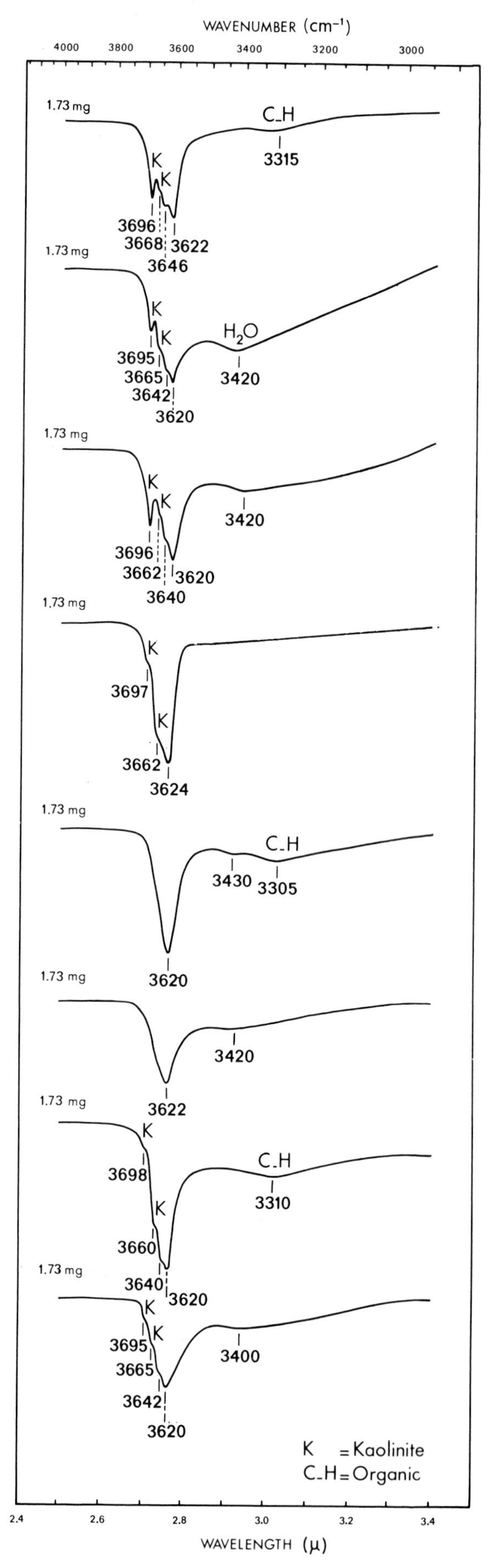

Fig. 6.39. Infrared spectra of Illite shale (< 2 μm) from different origin.

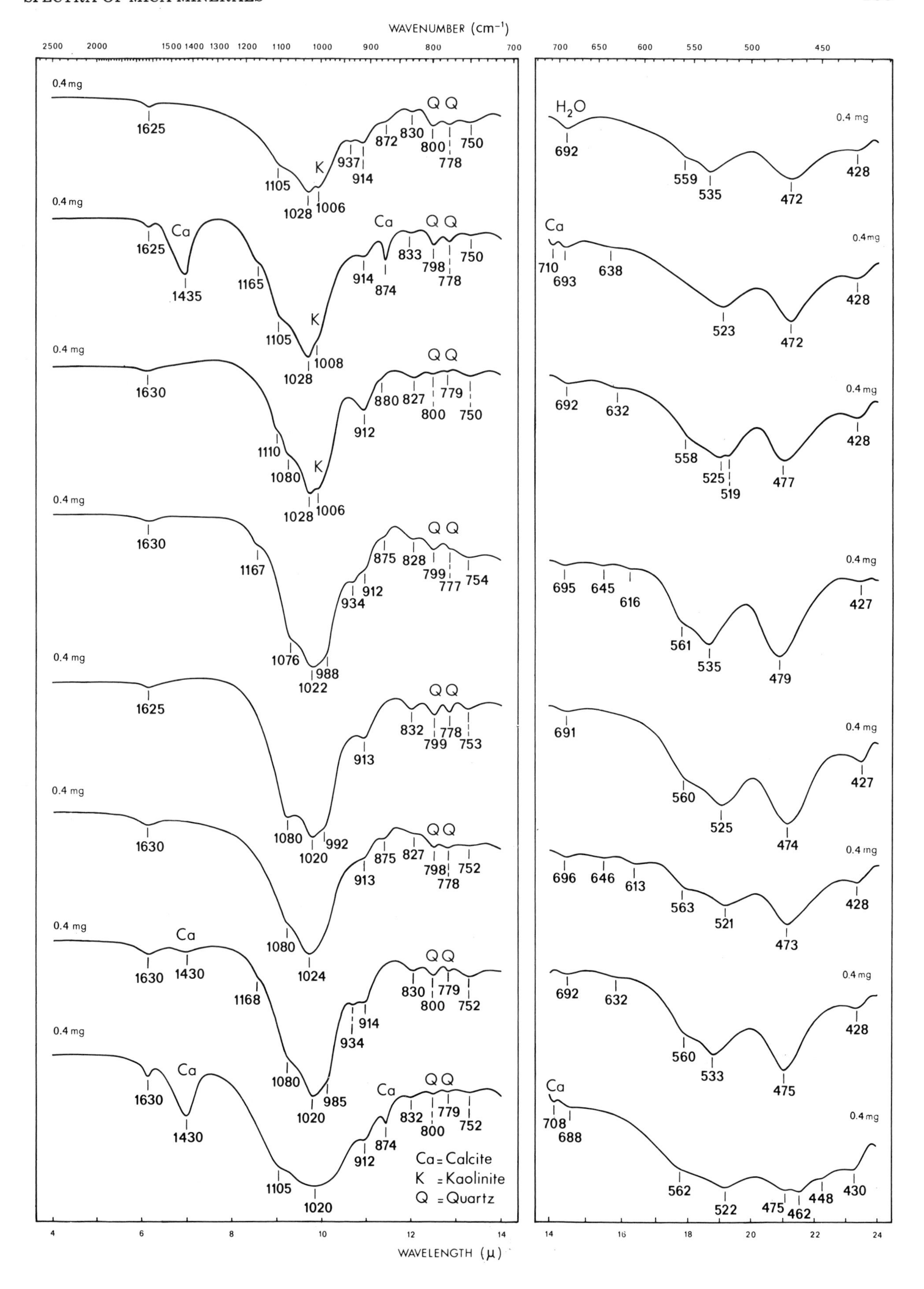
WAVENUMBER (cm⁻¹)
WAVELENGTH (μ)
0.4 mg
H₂O
Ca
K
Q Q
Ca = Calcite
K = Kaolinite
Q = Quartz

Weathered illite
Morris – 10.6 Å
some quartz

Goose lake, U.S.A. (grundite) – 10.2 Å
some kaolinite and quartz

Willalooka, Australia – 11.2 Å
some kaolinite and quartz

Swelling illite

Wageningen, The Netherlands – 14 Å
some kaolinite, quartz, mica (overlapped)

Amerongen, The Netherlands – 14 Å
some kaolinite, quartz, mica (overlapped)

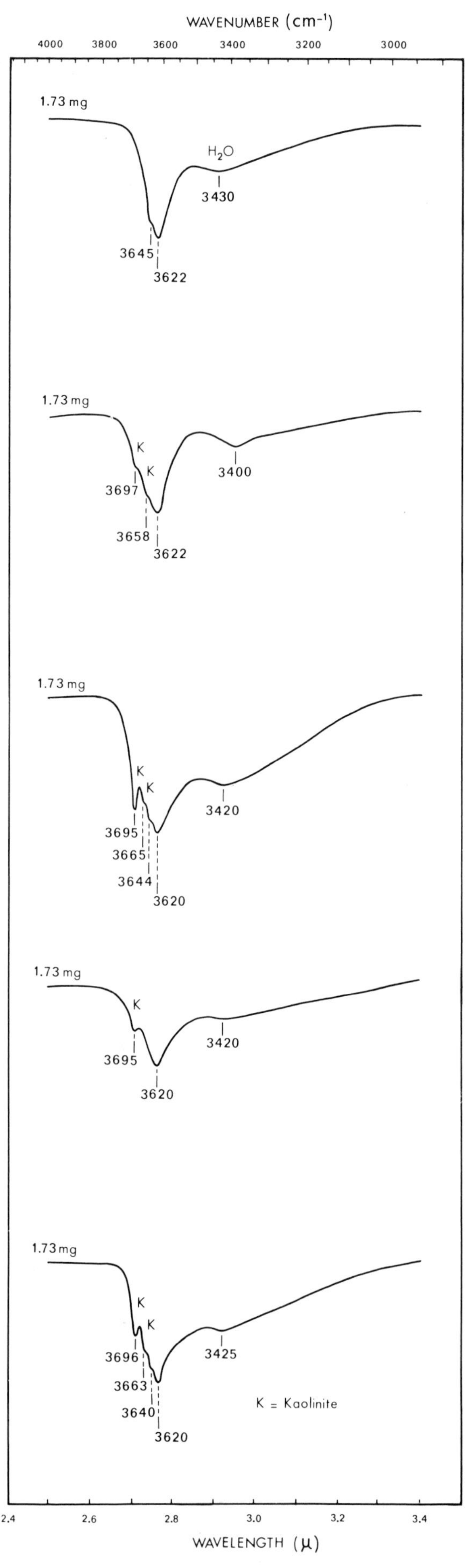

Fig. 6.40. Infrared spectra of weathered illite and swelling illite (< 2 μm) from different origin.

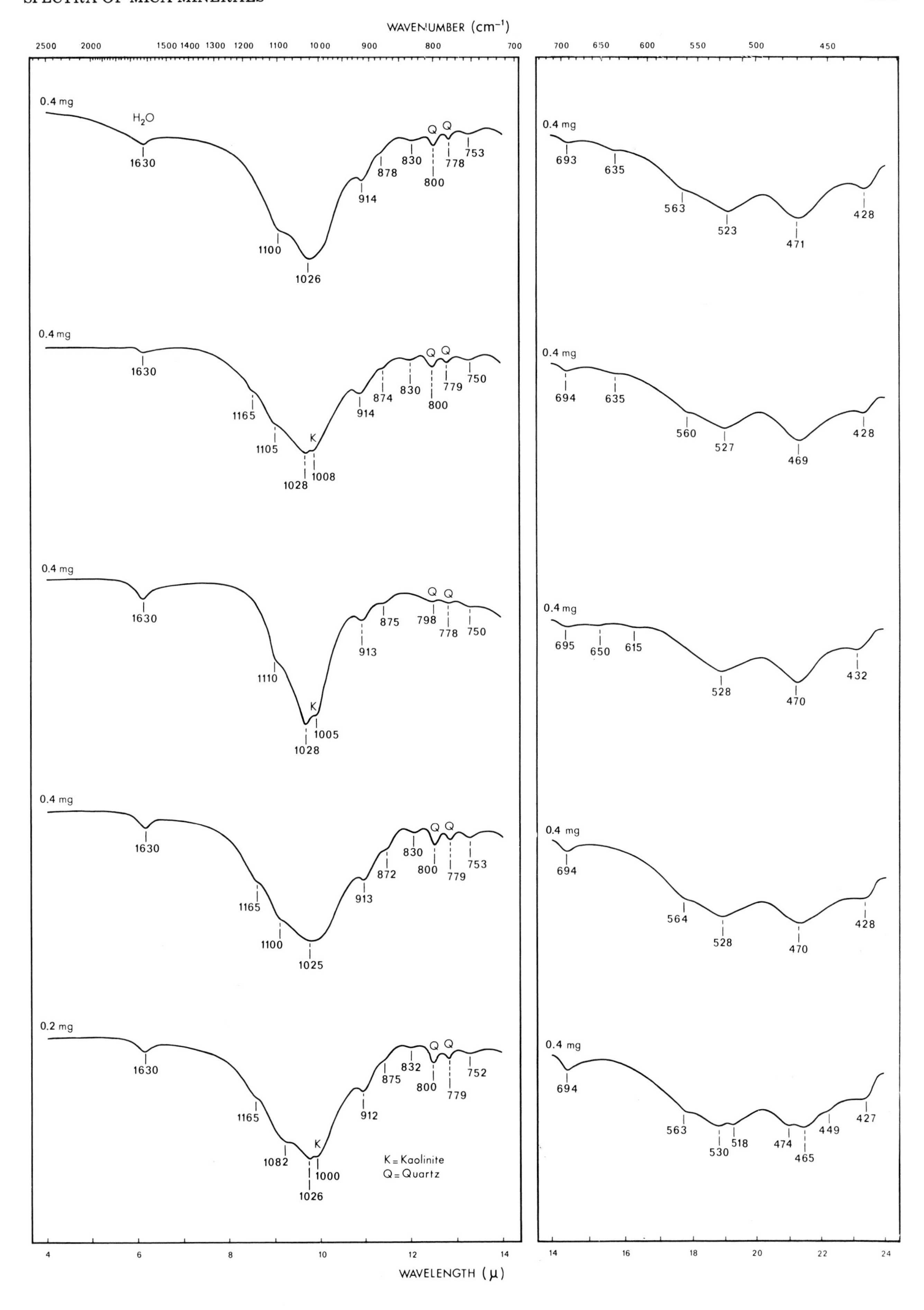
WAVENUMBER (cm⁻¹)
0.4 mg
H₂O
1630
1100
1026
914
878
830
800
778
753
Q Q
693
635
563
523
471
428
1165
1105
1028
1008
K
874
779
750
694
560
527
469
1110
1005
913
875
798
695
650
615
528
470
432
1025
872
564
0.2 mg
1082
1000
912
832
752
563
530
518
474
465
449
427
K = Kaolinite
Q = Quartz
WAVELENGTH (μ)

Leverrierite

Dobrilugk, Germany
some chlorite
Fig. E.M. 124

Völklingen, Saarland, Germany
some organic matter

Hudig biogene illite
Bodegraven, The Netherlands
some kaolinite, quartz

Iron illite
Feuerletten, Unterfranken, Germany
some dolomite and hematite (overlapped)
Fig. E.M. 117

Aluminium illite
Sarospatak
some kaolinite
Fig. E.M. 129

Potassium bentonite – 10 Å
Catawba, Va., U.S.A.
some kaolinite, quartz

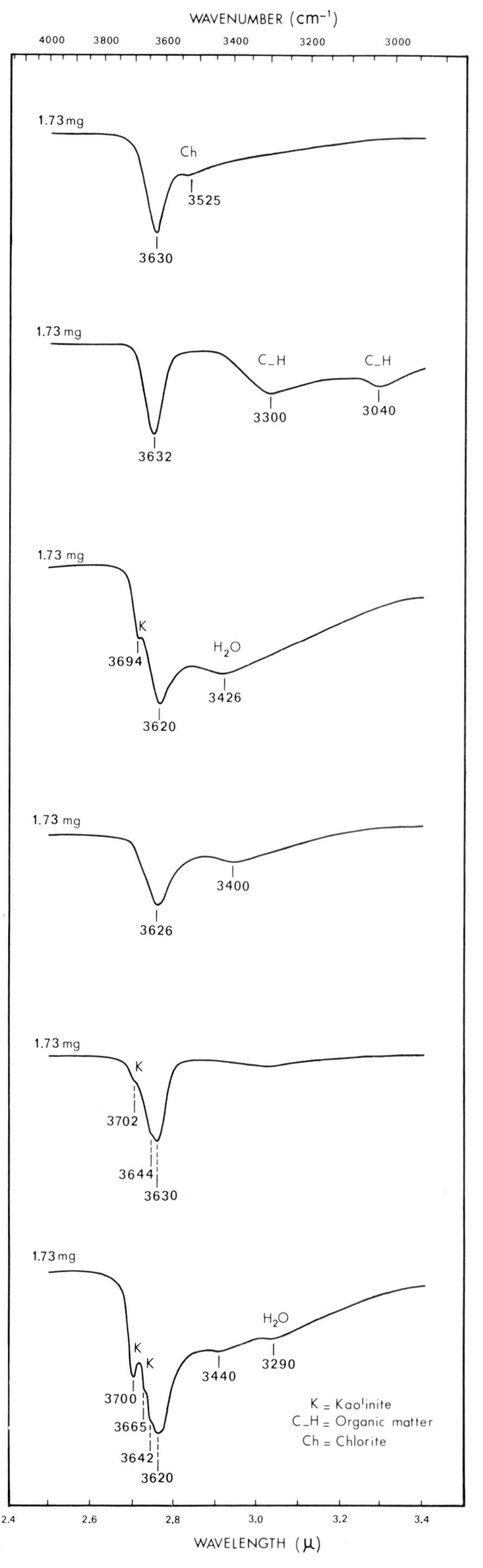

Fig. 6.41. Infrared spectra of leverrierite and of Hudig biogen illite, iron illite, aluminum illite, and potassium bentonite (< 2 μm).

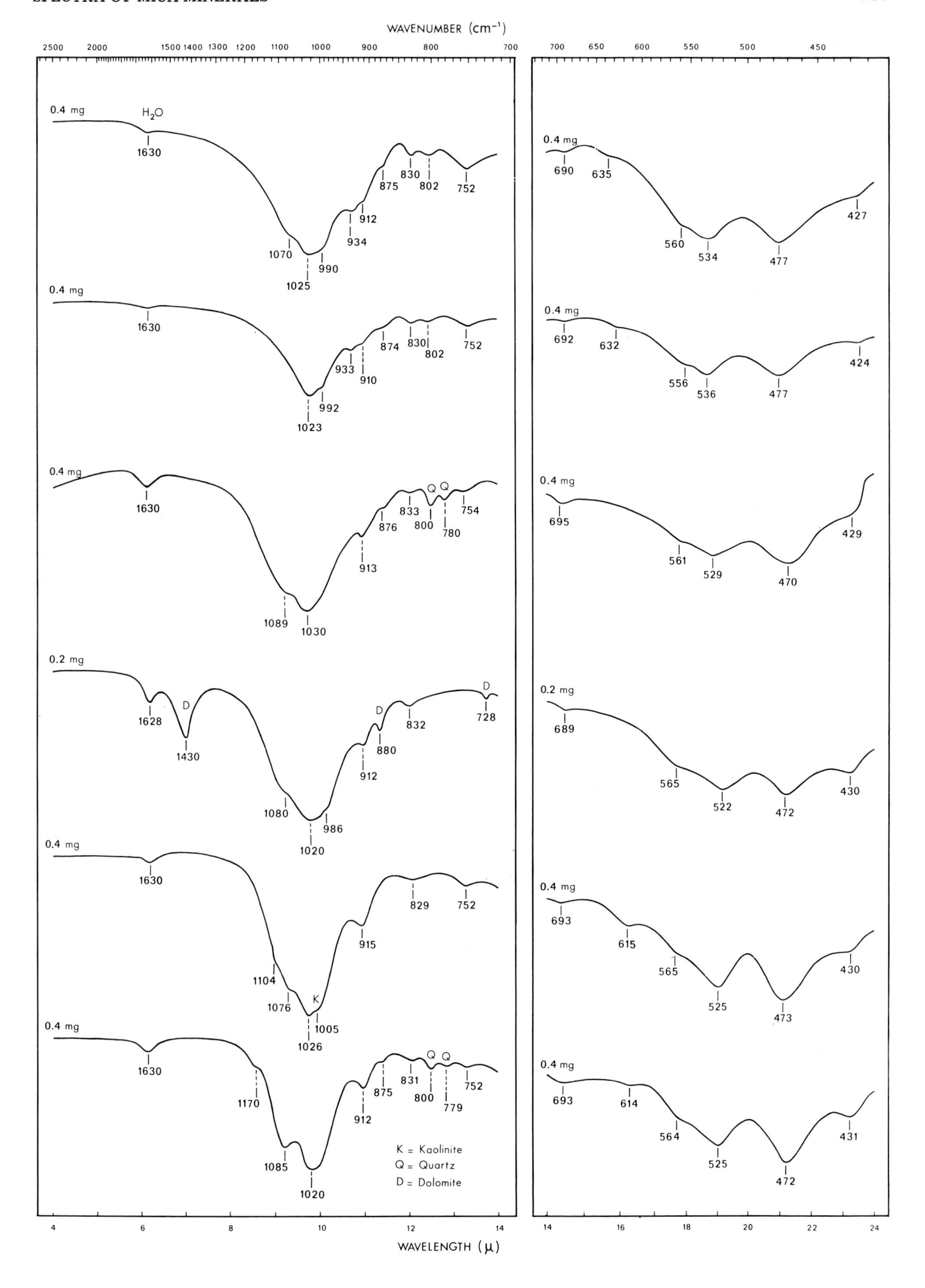
WAVENUMBER (cm⁻¹)
WAVELENGTH (μ)
0.4 mg
0.2 mg
H₂O
K = Kaolinite
Q = Quartz
D = Dolomite

Glauconite
Rarécourt, France
some kaolinite and quartz
Fig. E.M. 125

Hazlet, N.J., U.S.A.
pure

Birmingham, N.J., U.S.A.
some kaolinite, quartz, calcite and
feldspar (overlapped)

Slenaken, The Netherlands
some calcite
Fig. E.M. 126

Commercial glauconite
pure

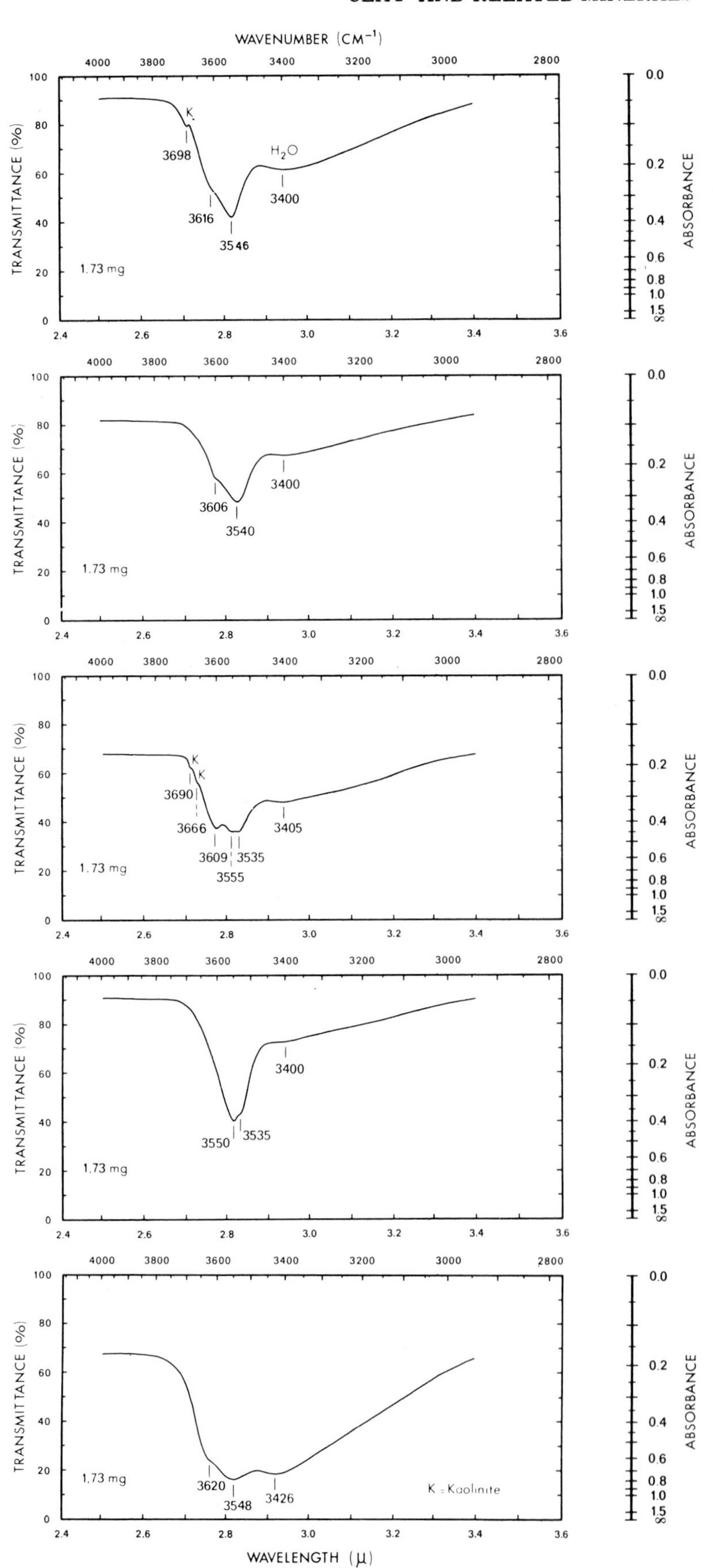

Fig. 6.42. Infrared spectra of glauconite (< 2 μm) from different origin, and of commercial glauconite.

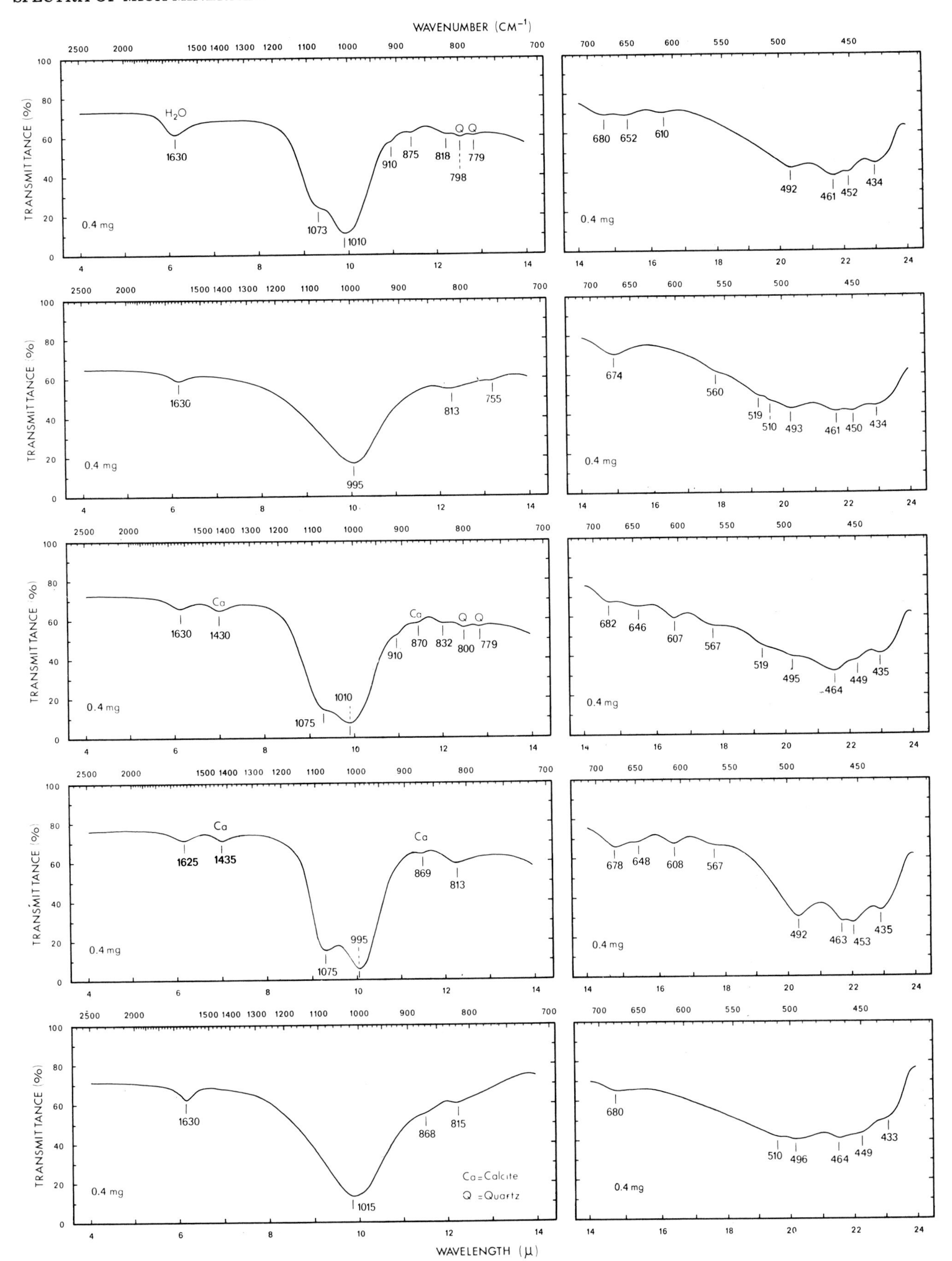
WAVENUMBER (CM^{-1})
TRANSMITTANCE (%)
H_2O
1630
1073
1010
910
875
818
798
779
Q Q
0.4 mg
680 652 610
492 461 452 434
995
813
755
674
560
519 510 493 461 450 434
Ca
1430
1075
870 832 800 779
682 646 607 567
519 495 464 449 435
1625 1435
869
678 648 608 567
492 463 453 435
1015
868 815
680
510 496 464 449 433
Ca = Calcite
Q = Quartz
WAVELENGTH (μ)

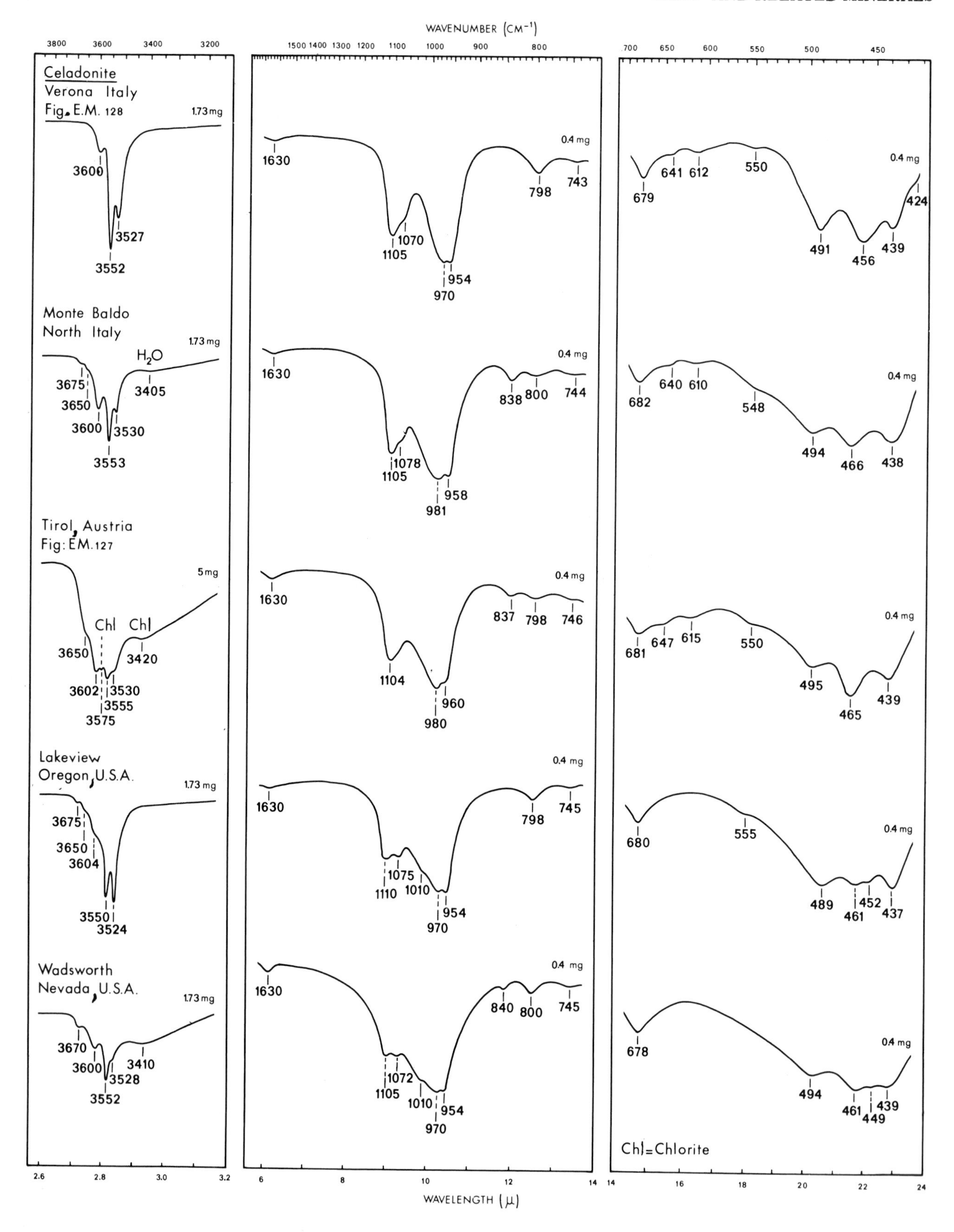

Fig. 6.43. Infrared spectra of celadonite from various origin.

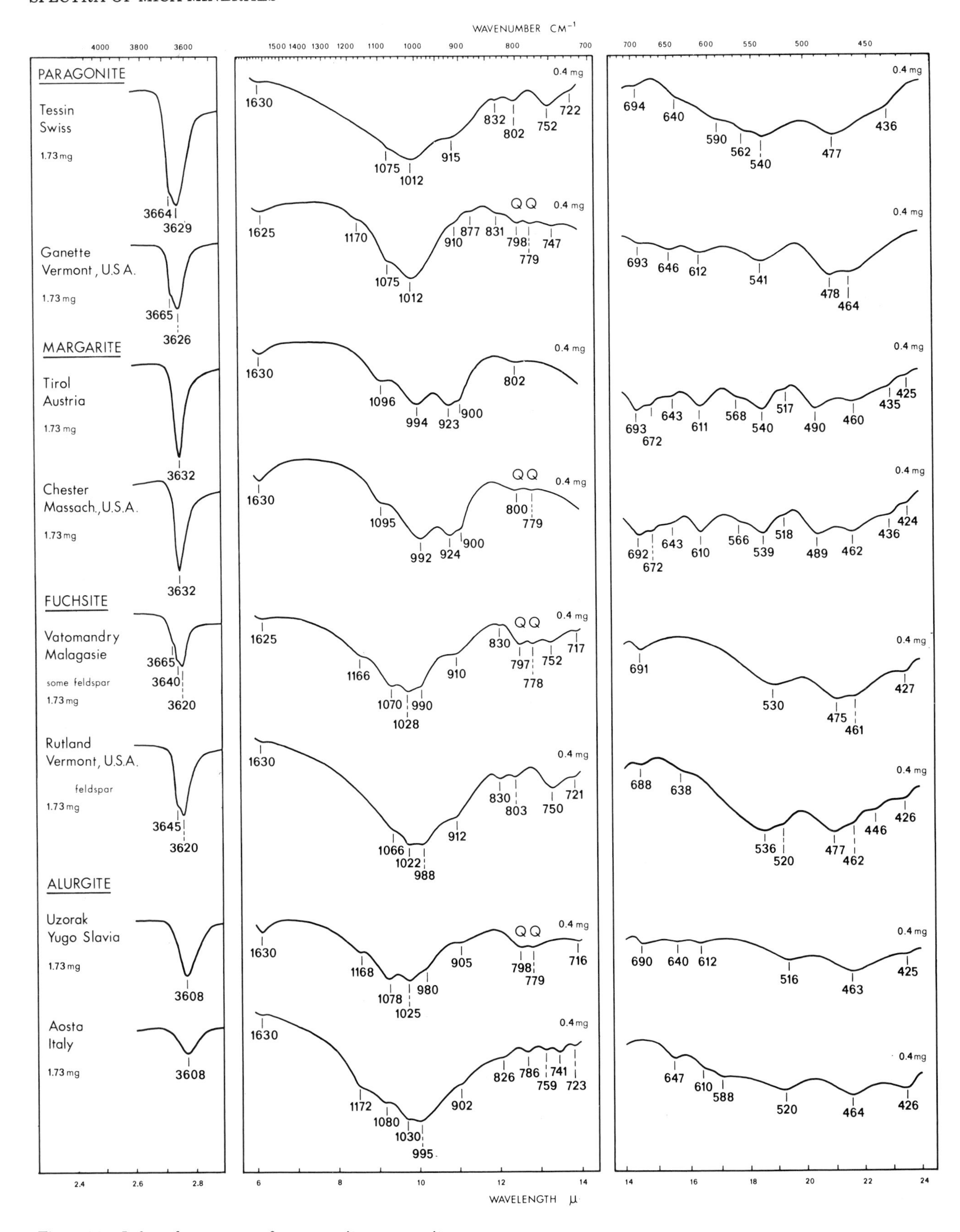

Fig. 6.44. Infrared spectra of paragonite, margarite, fuchsite and alurgite from different origin. (Feldspar impurity overlapped.)

Biotite
Yancey Co., N.C., U.S.A.
some quartz

Bancroft, Ont., U.S.A.
some quartz

Madison, N.C., U.S.A.
some quartz

Miask, Ural, U.S.S.R.
pure

Moss, Norway,
pure

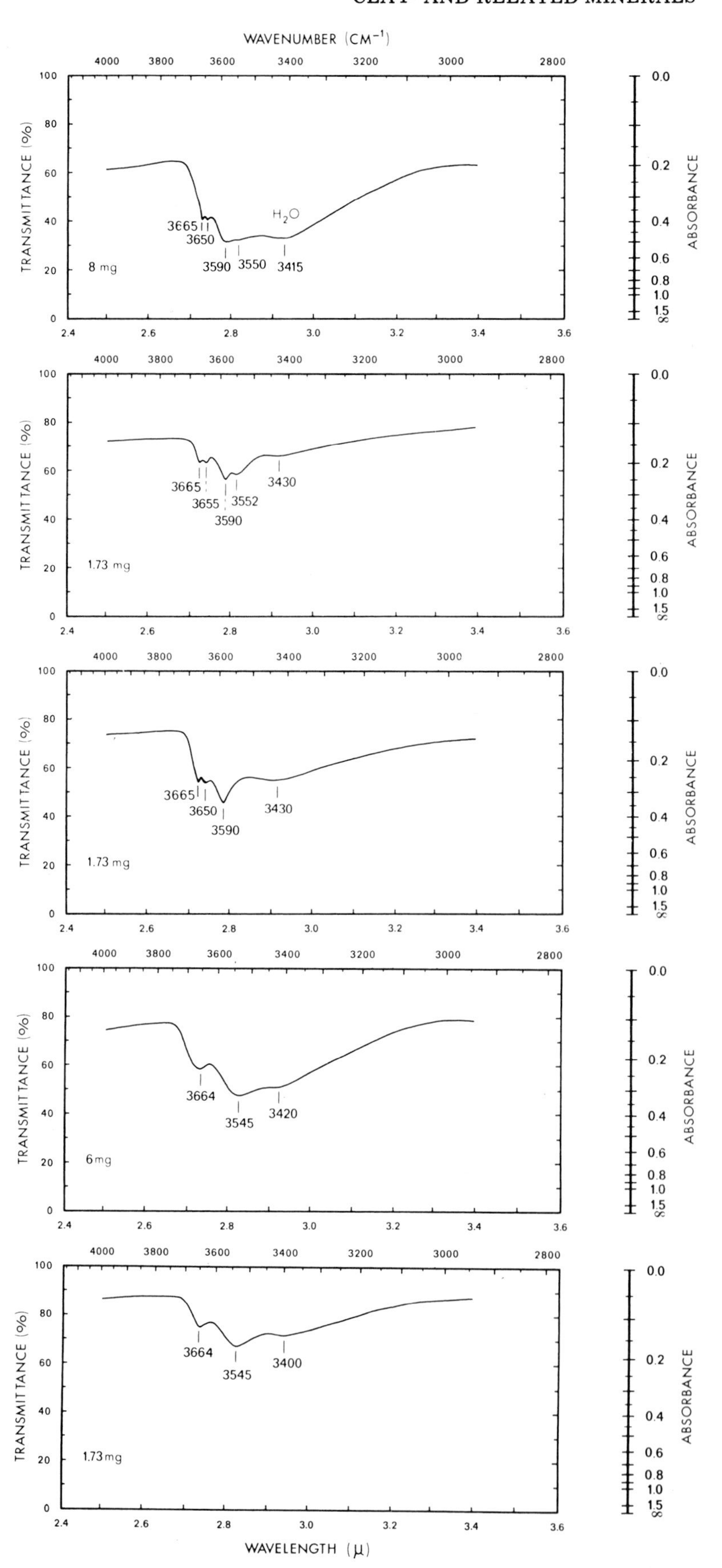

Fig. 6.45. Infrared spectra of biotite from different origin.

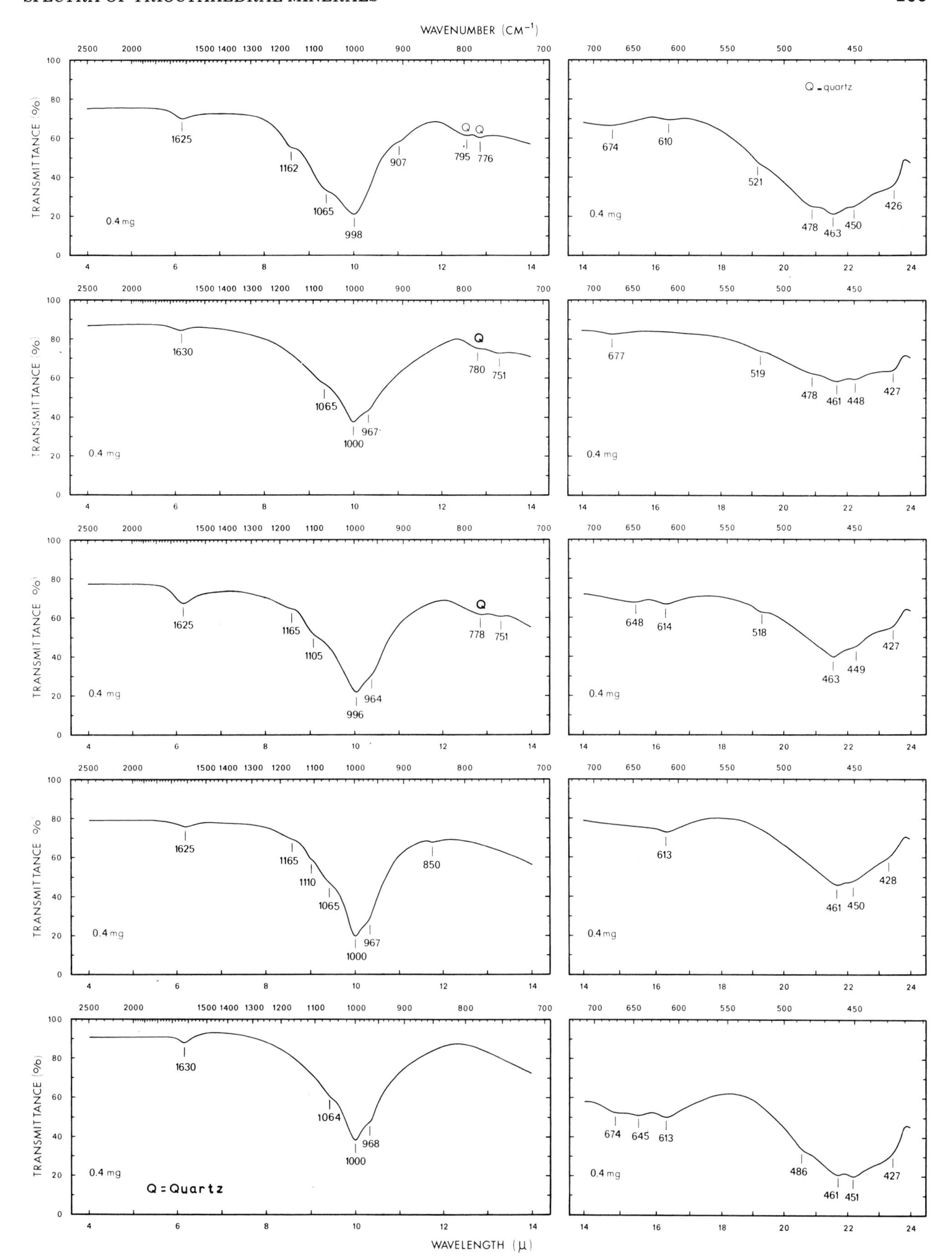
WAVENUMBER (CM⁻¹)
TRANSMITTANCE (%)
WAVELENGTH (μ)
Q = quartz
Q = Quartz
0.4 mg
1625
1162
1065
998
907
795
776
674
610
521
478
463
450
426
1630
1065
1000
967
780
751
677
519
478
461
448
427
1625
1165
1105
996
964
778
751
648
614
518
463
449
427
1625
1165
1110
1065
1000
967
850
613
461
450
428
1630
1064
1000
968
674
645
613
486
461
451
427

Lepidomelane
Ontario, Canada
pure

Barkeviksjor, Norway
pure

Mn-phyllite (biotite)

Taguchi-Mine, Japan
some chlorite (overlapped)

Magpur, India
some chlorite (overlapped)

Phlogopite

Krägero, Norway
pure

Odegaarden, Norway
pure

North Burgess, Ont., Canada
some chlorite (overlapped)

Canada
some chlorite (overlapped)

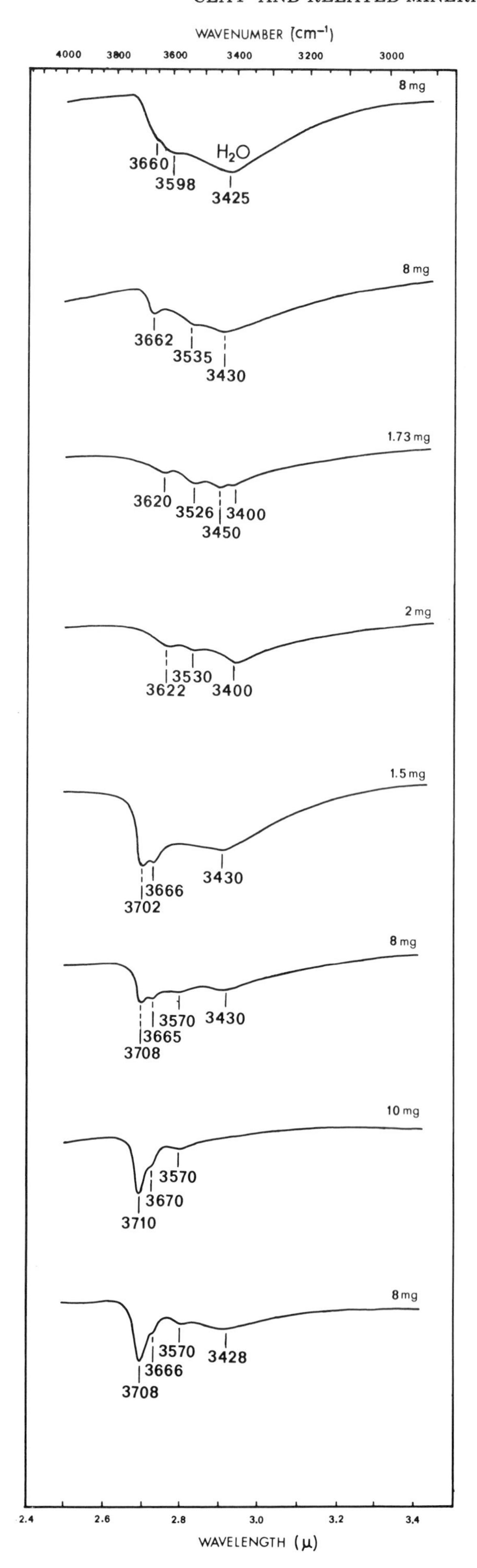

Fig. 6.46. Infrared spectra of lepidomelane, Mn-phyllite and phlogopite.

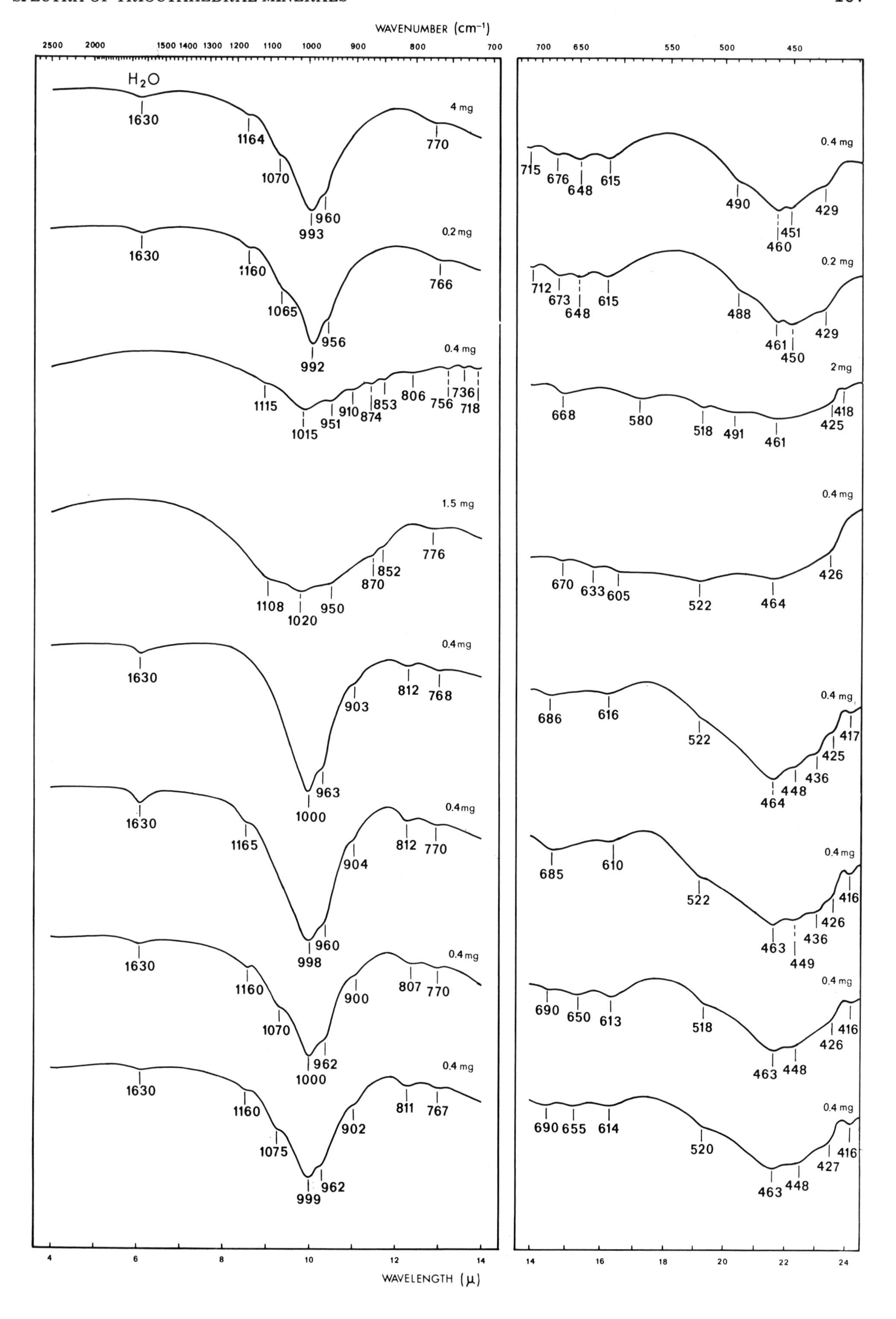
WAVENUMBER (cm⁻¹)
2500 2000 1500 1400 1300 1200 1100 1000 900 800 700
700 650 550 500 450
H₂O
1630 1164 1070 993 960 770 4 mg
715 676 648 615 490 460 451 429 0.4 mg
1630 1160 1065 992 956 766 0.2 mg
712 673 648 615 488 461 450 429 0.2 mg
1115 1015 951 910 874 853 806 756 736 718 0.4 mg
668 580 518 491 461 425 418 2 mg
1108 1020 950 870 852 776 1.5 mg
670 633 605 522 464 426 0.4 mg
1630 1000 963 903 812 768 0.4 mg
686 616 522 464 448 436 425 417 0.4 mg
1630 1165 998 960 904 812 770 0.4 mg
685 610 522 463 449 436 426 416 0.4 mg
1630 1160 1070 1000 962 900 807 770 0.4 mg
690 650 613 518 463 448 426 416 0.4 mg
1630 1160 1075 999 962 902 811 767 0.4 mg
690 655 614 520 463 448 427 416 0.4 mg
4 6 8 10 12 14 16 18 20 22 24
WAVELENGTH (μ)

Hydrated biotite
Oaxacu, Mexico
pure

Libby, Mont., U.S.A.
pure

Vermiculite
W. Chester, Pa., U.S.A.
pure

Tetuan, Morocco
pure

Kenya
pure

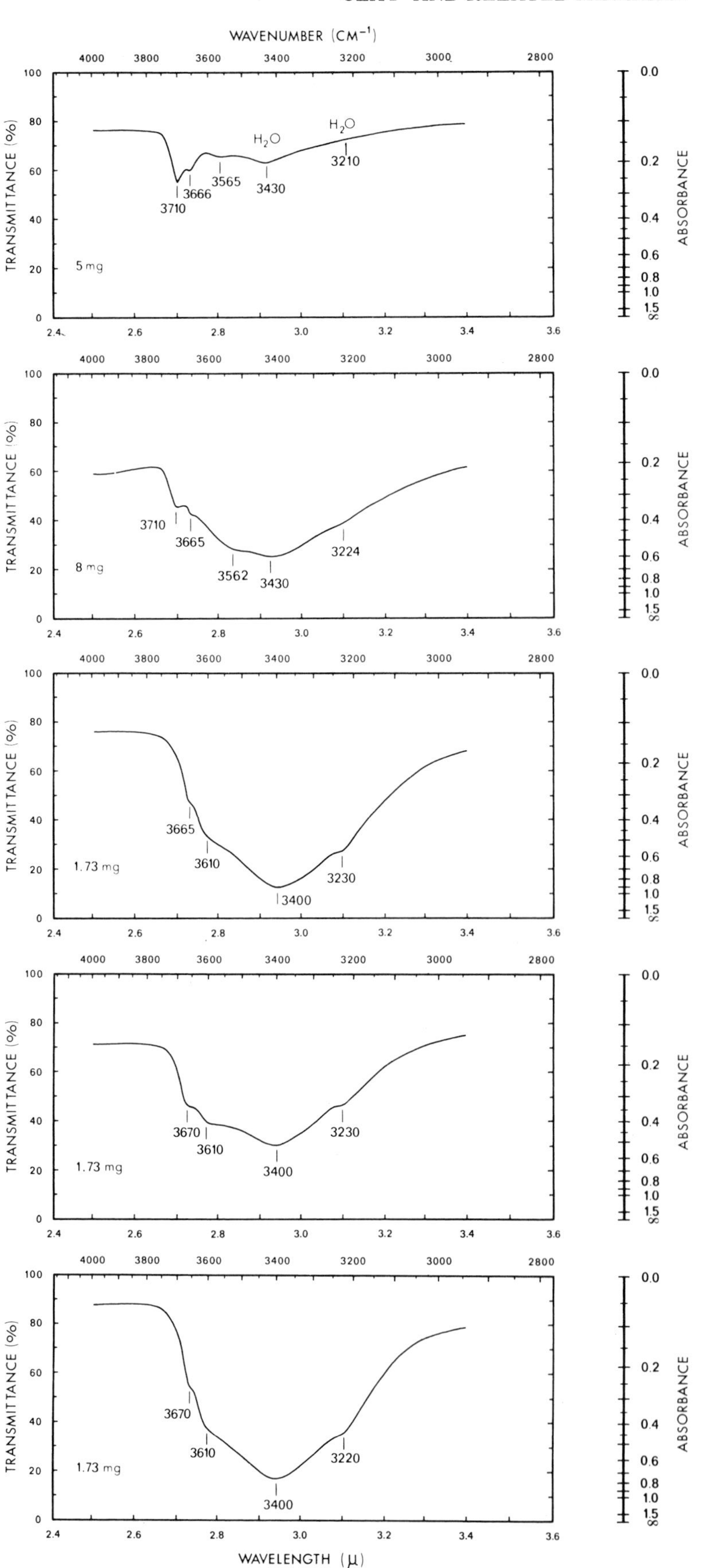

Fig. 6.47. Infrared spectra of hydrated biotite and of vermiculite from different origin.

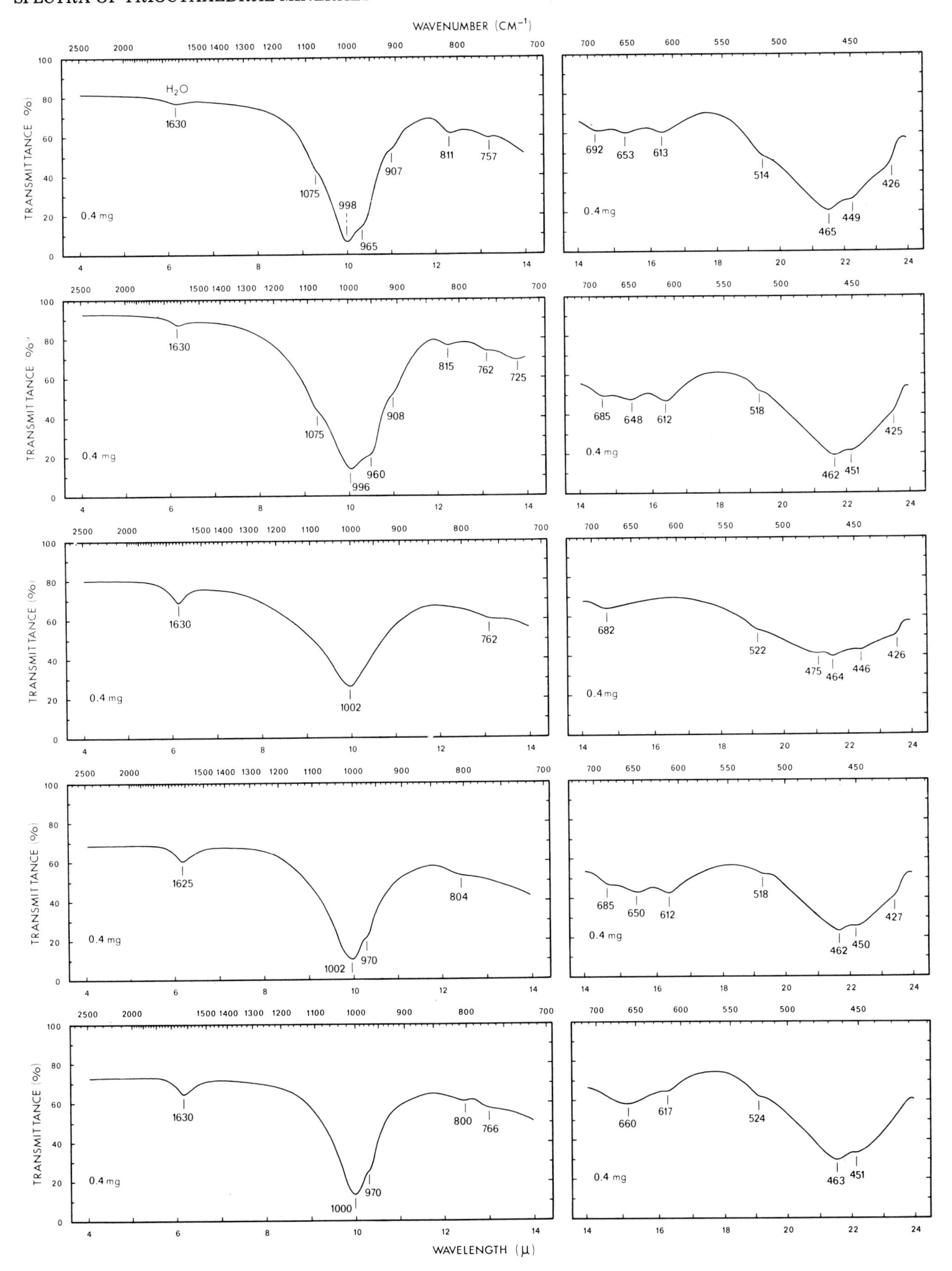
WAVENUMBER (CM⁻¹)
TRANSMITTANCE (%)
WAVELENGTH (μ)
H₂O
1630
1075
998
965
907
811
757
692
653
613
514
465
449
426
0.4 mg
1630
1075
996
960
908
815
762
725
685
648
612
518
462
451
425
0.4 mg
1630
1002
762
682
522
475
464
446
426
0.4 mg
1625
1002
970
804
685
650
612
518
462
450
427
0.4 mg
1630
1000
970
800
766
660
617
524
463
451
0.4 mg

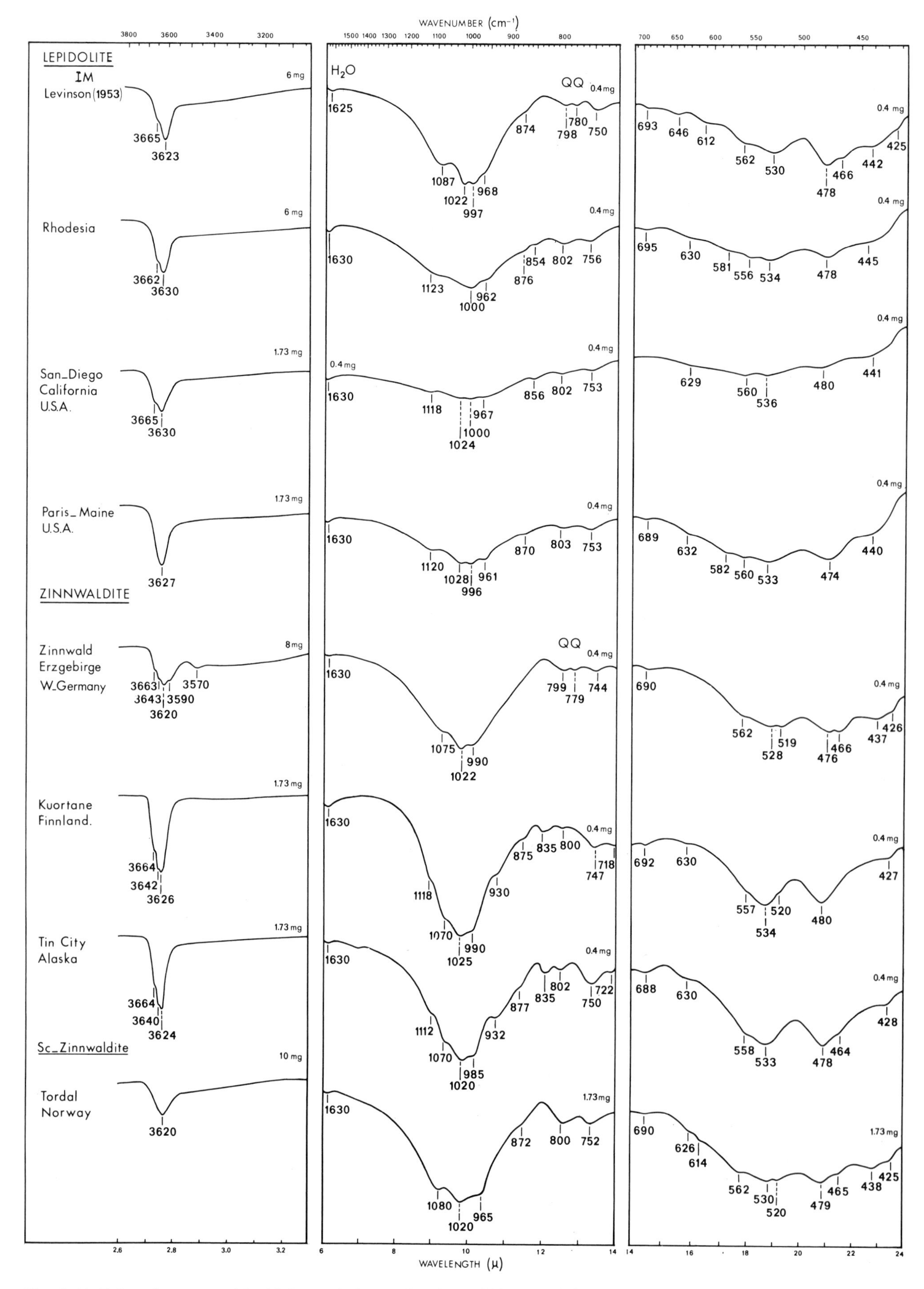

Fig. 6.48. Infrared spectra of lepidolite and zinnwaldite from different origin.

Chlorite minerals

Minerals of the chlorite group (chloros = green) also called 2 : 2 minerals and those related to this group, consist of 2 : 1 layers; two (Si,Al)-O tetrahedral sheets enclosing one (Mg,Al,Fe)-OH octahedral sheet. The layers are not linked by cations, like the mica minerals, but by octahedral sheets of brucite $[Mg_3(OH)_6]$ or brucite-like $[(Mg,Fe,Al)_3(OH)_6]$ composition (Plate I, pp. 190—191). Several chlorite- and chlorite-like minerals are formed by substitutions in the brucite layer, the octahedra and the tetrahedra, which lead to di-di, tri-tri and di-tri and tri-di combinations. The chlorites have a wide range in chemical composition and the names of varieties are given quite arbitrarily. Consequently the classification of the numerous chlorite polytype minerals has given much trouble and controversy. The following is the classification according to Strunz (1970). Nelson and Roy (1953) have suggested to use the term septe-chlorite for 7 Å trioctahedral kaolin minerals (chrysotile, antigorite, amesite, greenalite) and the 7 Å kaolin-chamosite mineral (berthierite), which when heated gives in addition a 14 Å chlorite reflection.

Coarse minerals in rocks and deposits

A. Trioctahedral

a. Talc chlorite Mg/Fe = >1
   Talc-chlorite $Mg_3[(OH)_2 | Si_4O_{10}]Mg_3(OH)_6$
   Penninite (pennine) $(Mg,Al)_3[(OH)_2 | Al_{0.5-0.9}Si_{3.5-3.1}O_{10}]Mg_3(OH)_6$
   Clinochlore (leuchtenbergite) $(Mg,Al)_3[(OH)_2 | AlSi_3O_{10}]Mg_3(OH)_6$
   Sheridanite $(Mg,Al)_3[(OH)_2 | Al_{1.2-1.5}Si_{2.8-2.5}O_{10}]Mg_3(OH)_6$
   Ripidolite (prochlorite) $(Mg,Fe,Al)_3[(OH)_2 | Al_{1.2-1.5}Si_{2.8-2.5}O_{10}]Mg_3(OH)_6$
b. Ferro-chlorite Mg/Fe = <1
   Daphnite (aphrosiderite, bavalite) $(Fe^{3+},Al)_3[(OH)_2 | Al_{1.2-1.5}Si_{2.8-2.5}O_{10}]Fe^{2+}{}_3(OH)_6$
c. Ferro-ferrichlorite (leptochlorite)
   Delessite (melanolite) $(Mg,Fe^{2+},Fe^{3+})[(OH)_2 | Al_{0-0.9}Si_{4-3.1}O_{10}](Mg,Fe^{3+})_3(O,OH)_6$
   Chamosite $(Fe^{2+},Fe^{3+})_3[(OH)_2 | AlSi_3O_{10}](Fe^{2+},Mg)_3(O,OH)_6$
   Thuringite $(Fe^{2+},Fe^{3+},Al)_3[(OH)_2 | Al_{1.2-2}Si_{2.8-2}O_{10}](Mg,Fe^{2+},Fe^{3+})_3(O,OH)_6$
   Pennantite $(Mn,Al,Fe^{3+})_3[(OH)_2 | (Al,Si)Si_3O_{10}]Mn_3(OH)_6$
d. Kämmererite, chrome chlorite
   $(Mg,Cr)_3[(OH)_2 | Cr^{3+}Si_3O_{10}]Mg_3(OH)_6$

B. Dioctahedral

a. Cookeite, lithium chlorite
   $Al_2[(OH)_2 | AlSi_3O_{10}]Li,Al_2(OH)_6$
b. Sudoite, aluminium chlorite, donbassite
   $Al_2[(OH)_2 | AlSi_3O_{10}]Al_2(OH)_6$

Occurrence. Minerals of the chlorite group are mostly found in metamorphic rocks. They are very difficult to get as a mono mineral because they are largely mixed with other minerals like phlogopite, talc, biotite, feldspar, mica, chamosite and especially with siderite.
Chamosite is a common mineral in many oölitic iron ores. Mineralized faecal pellets of a rather Mg-rich poorly-ordered 7 Å chamosite mineral are found in the shallow marine delta of the Niger and the Orinoco and in the Serawak Shelf. Transitions of chlorite to the above chamosite mineral were also found: Porrenga (1967).

Infrared spectra (Fig. 6.49—6.52). There are only small differences in the spectra of the talc chlorites, e.g. chlorite has 3420 cm^{-1}, penninite 3455 cm^{-1}, clinochlore 3426 cm^{-1} and the iron-rich ripidolite member = 3428 cm^{-1} (Fig. 6.49). For daphnite the 3565 to 3570 cm^{-1} band of the foregoing minerals has moved to ca. 3548 cm^{-1} and the number of small bands in the environment of the main bands at ca. 986—1000 cm^{-1} has decreased (Fig. 6.50). For delessite (melanolite) are 3 bands at the highest wave numbers = 3560, 3480, 3400 cm^{-1} (Fig. 6.50). The iron- and (manganese)-rich chamosite mineral has a very variable spectrum which is caused by its non-uniform composition (Fig. 6.51).
That of the sample from the type locality at Chamoson (Swiss) is very poorly developed and has thereby several admixtures. Orthochamosite has a band at 3590 cm^{-1}, which is different from the other chamosite samples. For its other bands there is better agreement. The iron-rich thuringite sample has wave numbers 3680 and 3626 cm^{-1}, by which it can be distinguished from all the foregoing chlorite minerals (Fig. 6.51). A 3680 cm^{-1} band is also found for Mn chlorite (pennantite) and chrom chlorite (kämmererite: Fig. 6.52).
Cookeite (Li) has a strong band at 3525 cm^{-1} by which it can be distinguished from the other chlorite minerals (Fig. 6.52).
Al-chlorite (sudoite, donbassite; Lazarenko, 1940) has a strong band at ca. 1010 cm^{-1}, by which it can be distinguished from all the other chlorite minerals (Fig. 6.52).

Remarks. Also by the X-ray and the thermal method the various chlorite minerals are difficult to be distinguished. Chemical analyses are very useful. Chamosite is explored for its high iron contents = ca. 45—50% FeO + Fe_2O_3. Sudoite is named after its discoverer T. Sudo, Tokio, Japan.

Hydrated chlorite; formula:

$(Mg,Fe,Al)_3[(OH)_2 | (Al,Si)_4O_{10}] (Mg,Fe,Al)_3$-$(OH)_6 - nH_2O$

Occurrence. The mineral is found at Maaninka, Posio, Finland in amphibolite rock.

Infrared spectra (Fig. 6.53). Bands of high intensity are at 3580, 3440, 1080, 1000, 963, 660, 524 and 461 cm^{-1}. The band at 3580, 3440 cm^{-1} is characteristic for a chlorite mineral and not for vermiculite (3670, 3610 cm^{-1}) for which it can be mistaken because of its expanding character on heating.
The small amounts of amphibole and augite can not be found by the I.R. method, only by X-ray analysis.

Remarks. This chlorite mineral expands like vermiculite when heated, but to a lesser degree. Also the 3230 cm^{-1} band of crystal water of vermiculite is absent in the Posio chlorite mineral. Apparently in this case the amount of H_2O molecules, which cause the expansion are only small, as is also the case for hydrated biotite and perlite.

Fine chlorite minerals in sediments; formula:

$(Mg,Al,Fe)_3[(OH)_2 | (Si,Al)_4 O_{10}] (Mg,Al,Fe)_3(OH)_6$

Occurrence. Minerals of the chlorite group may be found as an abrasion product in the clay and silt separates of sediments mixed together with other minerals. The exact identification of the kind of polytype chlorite mineral in this case is very difficult because of interference with other bands and the fact that already an identification of even the pure chlorite minerals has given much trouble — see before.
Fortunately scarce minerals like ripidolite, daphnite, delessite, chamosite, thuringite, pennantite, kämmererite, cookeite and sudoite scarcely occur in appreciable concentrations in sediments.
By weathering action chlorite may degrade to related minerals just as illite, e.g. swelling chlorite, soil chlorite and swelling soil chlorite. A chlorite-like mineral may also be formed by interleaving (intergradation) of a sheet of gibbsite or gibbsite-like composition between layers of vermiculite, expanded illite, swelling illite or soil montmorillonite.

All the above chlorite-like minerals are rare, because they are formed only under certain conditions.

Sedimentary chlorite

This name indicates all kinds of chlorite minerals which, when heated for 2,5 h at 550°C, do not appreciably change their 14 Å reflection. But their intensities are increased. Other names are: Brindley chlorite, well-crystallized chlorite, sedimentary chlorite, etc.

Occurrence. Found in the clay (<2 μm) and coarser fractions of many alluvial and glacial sediments as an abrasion product of various types of coarse chlorite minerals of the talc-chlorite group which is the most common.

Swelling chlorite

In this case the bonds in the chlorite minerals with the interlayered brucite sheets are partly broken (OH^- is converted to H_2O by H^+ action of weathering agents), leaving brucite or brucite-like "islands" and thus giving the mineral more or less swelling properties. The mineral is still heat-resistant, similar to sedimentary chlorite and magmatic chlorite. Other names are: hydrated type of chlorite, quellfähiges chloritisches Tonmineral, weathered chlorite, poorly crystallized expanding chlorite, pseudo-chlorite, etc.

Occurrence. The mineral is rare.

Soil chlorite

For this mineral the gibbsite- or gibbsite-like sheets are not regularly interleaved between the 2 : 1 layers. Consequently the mineral is not resistant to heating at 550°C for a period of 2,5 h. However, soil chlorite will not swell when saturated with an excess of glycerol. Other names for the mineral are: intergradational (intergradient) chlorite-vermiculite, secondary (aluminous) chlorite, chlorite-like mineral, fine grained poorly-crystallized chlorite, blocked expanded illite, sedimentary chlorite poorly-crystallized, chloritized vermiculite or partially Al-interlayered dioctahedral vermiculite, chlorite-like intergrade clay mineral, hydrated mica with partial chloritization, Bradley chlorite, etc.

Occurrence. The mineral is rare.

Swelling soil chlorite

The structure of this mineral resembles that of

soil chlorite and swelling chlorite, because in this case the poorly developed interlayers of various composition are poorly ordered as well as badly linked. The mineral is not only non-resistant to heat but swells more or less with glycerol. Other names for swelling soil chlorite are: mineral with montmorillonite-chlorite interleaved complexes, chloritized montmorillonite, montmorillonite mineral with brucite islands, intergradient chlorite-vermiculite-montmorillonite mineral with brucite islands, vermiculite-montmorillonite mineral, intergradient chlorite expansible 2 : 1 layer silicate, etc.

Occurrence. The mineral is rare.

Infrared spectra (Fig. 6.53). Kaolinite is a common admixture in various kinds of sediments. As moreover, its bands are of high intensity, the identification of small chlorite amounts is difficult. Fortunately, chlorites which are found in the clay separate of soils are mainly of the talc-chlorite type. In this case the strong ca. 3565—3580 and 3420—3455 cm^{-1} bands are very characteristic for their identification.
The X-ray method with a non-treated sample and with samples treated with glycerol and heated at 550°C gives far better results in the identification of the chlorite minerals in sediments. Also small amounts of feldspar, hematite, limonite and mica admixtures can better be identified by X-ray analysis.

Remarks. Because the chlorite-related minerals are very difficult to be distinguished from each other, several names are used for one and the same member of this small group of minerals; 52 out of a total of 80 names are synonymous: Van der Marel (1963).

References

Nelson, B.W. and Roy, R., 1953. New data on the composition and identification of chlorites. Clays Clay Minerals, Publ., 327: 335—348.

Porringa, D.H., 1967. Clay Mineralogy and Geochemistry of Recent Marine Sediments in Tropical Areas. Thesis Univ. Amsterdam, 145 pp.

Strunz, H., 1970. Mineralogische Tabellen. Akademie Verlagsgesellsch., Geest and Portig, Leipzig, 5th edn., 621 pp.

Van der Marel, H.W., 1963. Identification of chlorite and chlorite-related minerals in sediments. Beitr. Mineral. Petrogr., 9: 462—480.

Chlorite

Chester, U.S.A.
pure

Ural, U.S.S.R.
pure

Penninite (pennine)

Langerud, Norway
pure

Zermatt, Switzerland
pure

Clinochlore (Leuchtenbergite)

Murphys, Calif., U.S.A.
pure

Zlatoëst, Ural, U.S.S.R.
some amphibole (overlapped)

Ripidolite (prochlorite)

Chester, Vermont, U.S.A.
pure

Piemont, Italy
pure

Zalida, Colo., U.S.A.
pure

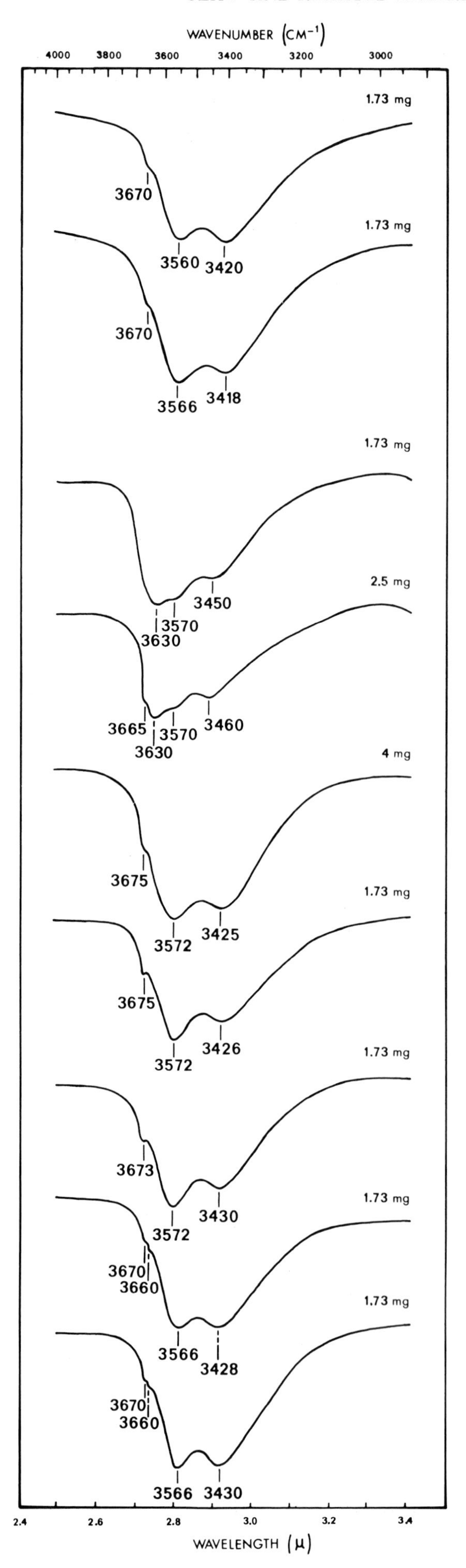

Fig. 6.49. Infrared spectra of chlorite, penninite, clinochlore and ripidolite from different origin.

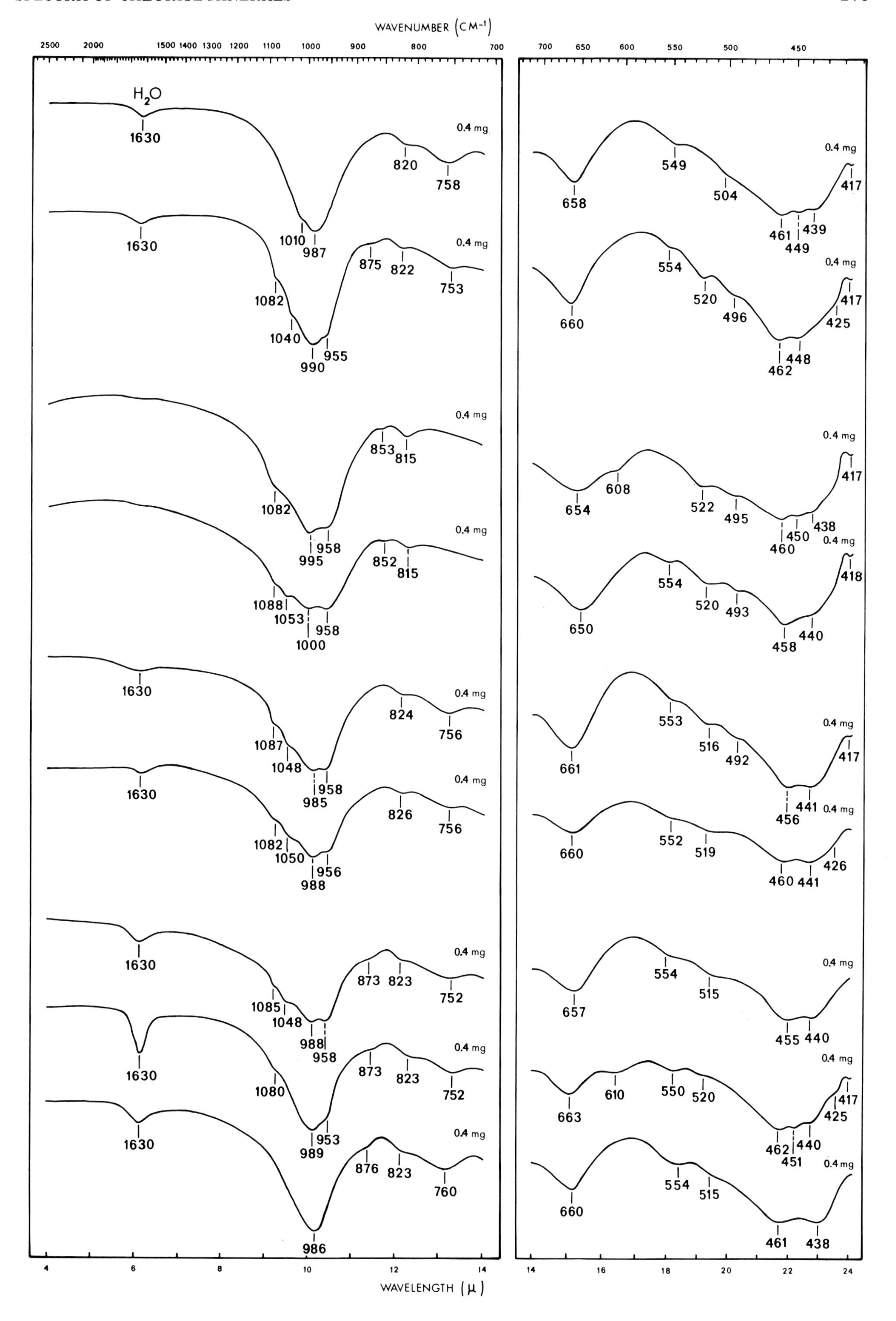
WAVENUMBER (CM-1)
WAVELENGTH (μ)
H2O
0.4 mg
1630
1010
987
820
758
658
549
504
461
449
439
417
1082
1040
990
955
875
822
753
660
554
520
496
462
448
425
1082
995
958
853
815
654
608
522
495
460
450
438
1088
1053
1000
958
852
815
650
554
520
493
458
440
418
1087
1048
985
958
824
756
661
553
516
492
456
441
1082
1050
988
956
826
756
660
552
519
460
441
426
1085
1048
988
958
873
823
752
657
554
515
455
440
1080
989
953
873
823
752
663
610
550
520
462
451
440
425
986
876
823
760
660
554
515
461
438

Daphnite (aphrosiderite, bavalite)

Bas Vallon, France
pure

Hessen, Germany
some calcite

Herborn, Germany
pure

Harz, Germany
pure

Delessite (melanolite)
unknown
some feldspar (overlapped)

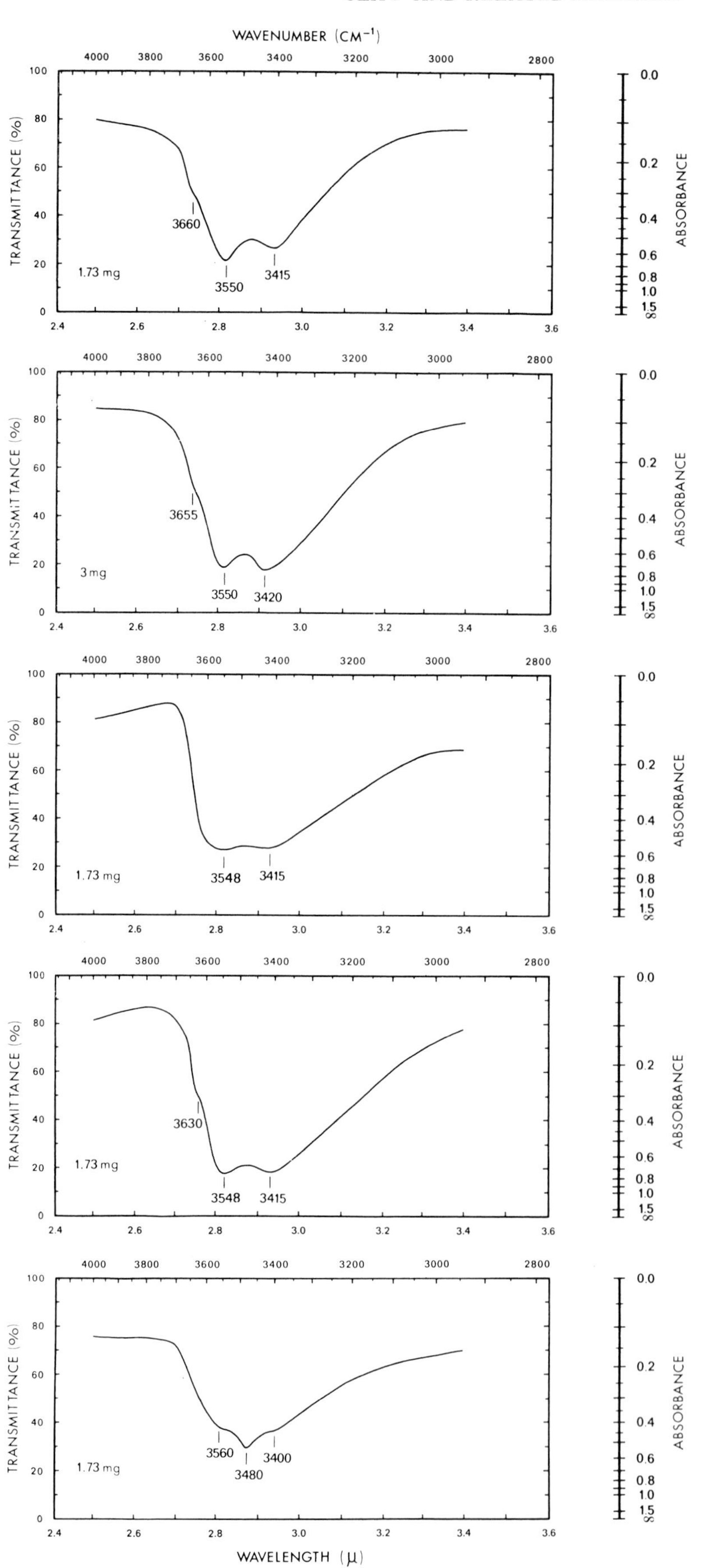

Fig. 6.50. Infrared spectra of daphnite and delessite from different origin.

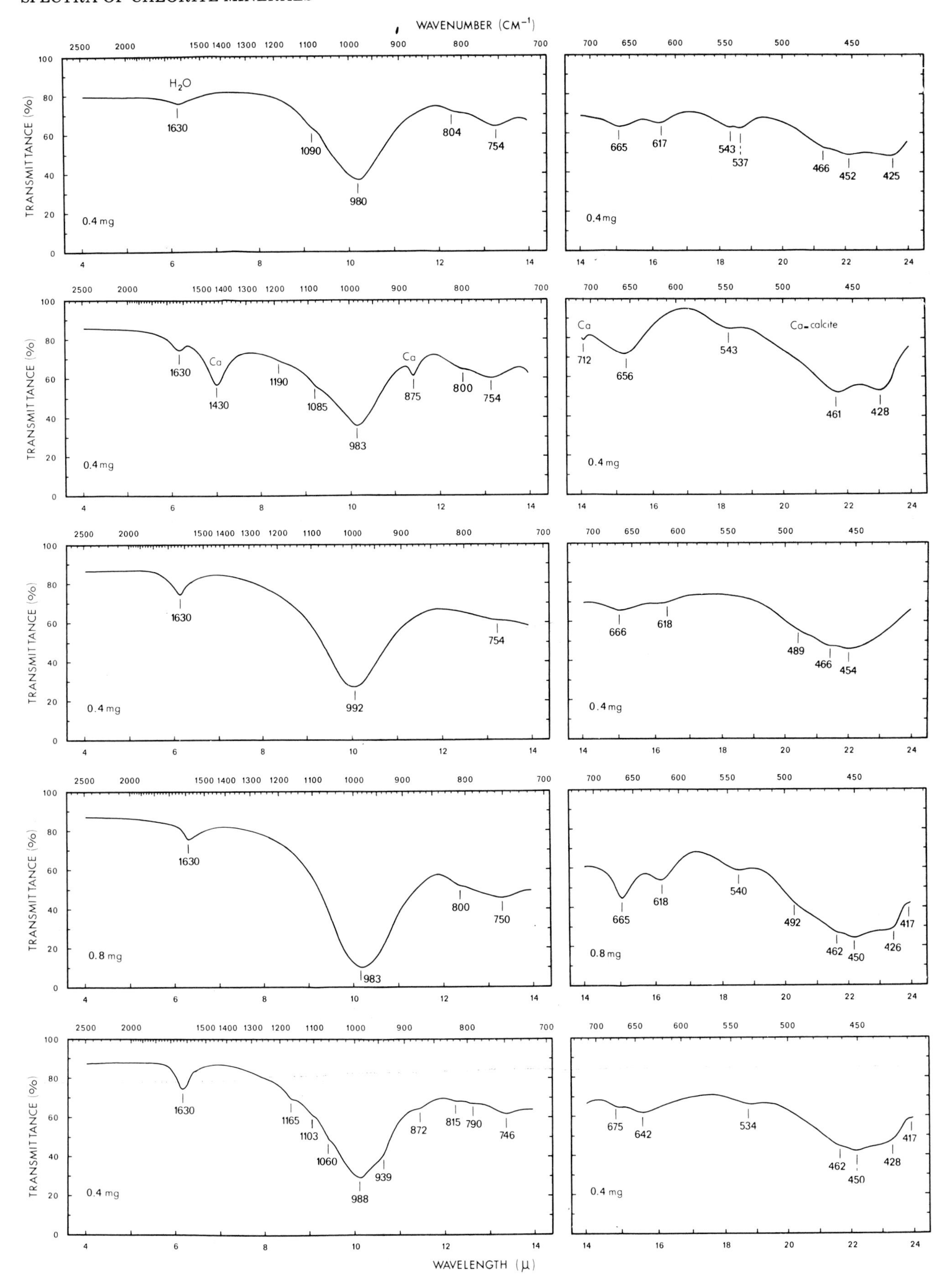

WAVENUMBER (CM^{-1})
TRANSMITTANCE (%)
WAVELENGTH (μ)
H_2O
1630
1090
980
804
754
665
617
543
537
466
452
425
0.4 mg
Ca
1430
1190
1085
983
875
800
712
656
Ca-calcite
461
428
992
666
618
489
454
0.8 mg
750
540
492
462
450
426
417
1165
1103
1060
988
939
872
815
790
746
675
642
534

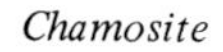

Chamosite

Sahara, Algeria
some mica (overlapped) and quartz

Nucice, Bohemia, Czechoslovakia
siderite

Chamoson, Wallis, Switzerland
some quartz and calcite

Orthochamosite

Kutna Hora, Bohemia, Czechoslovakia
pure

Thuringite
Gobitschau, Bohemia, Czechoslovakia
pure

Dona Ana Co., N. Mex., U.S.A.
pure

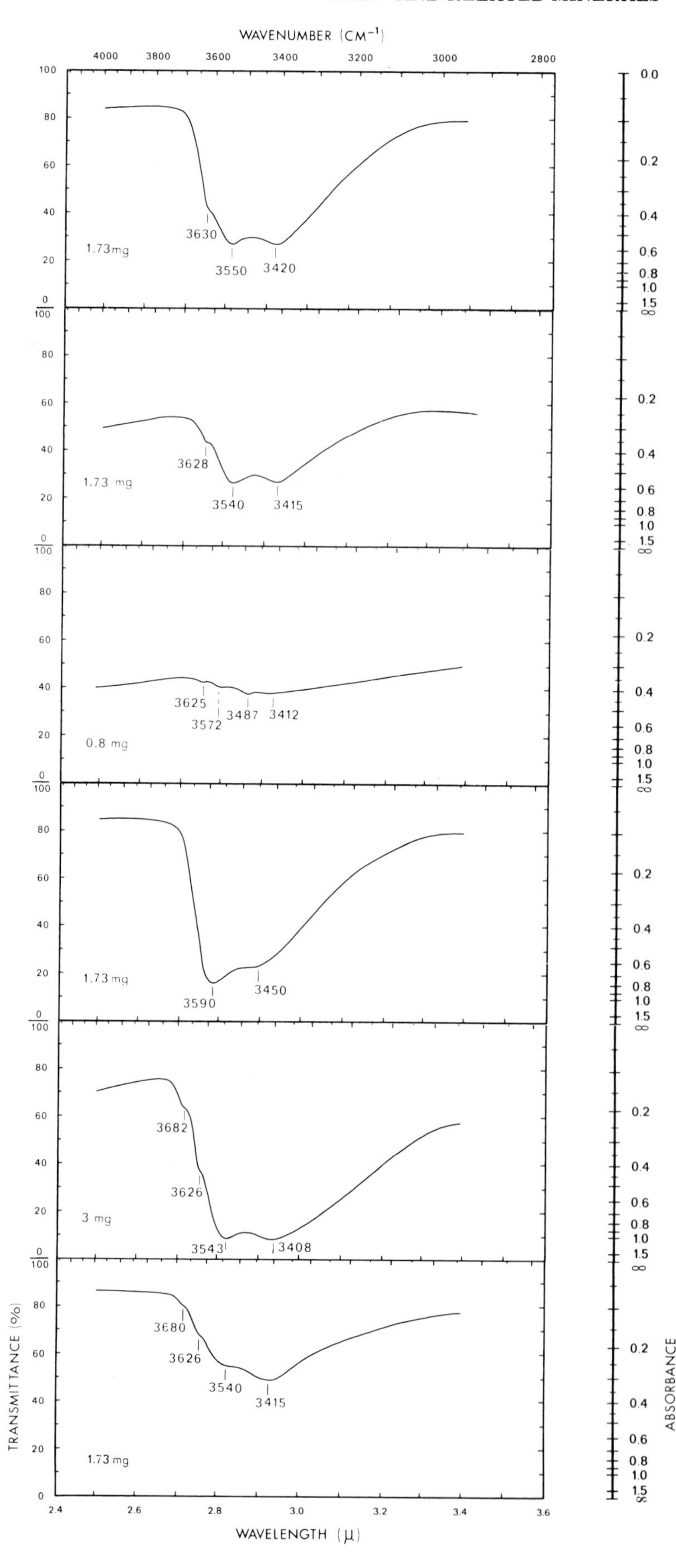

Fig. 6.51. Infrared spectra of chamosite and orthochamosite from different origin.

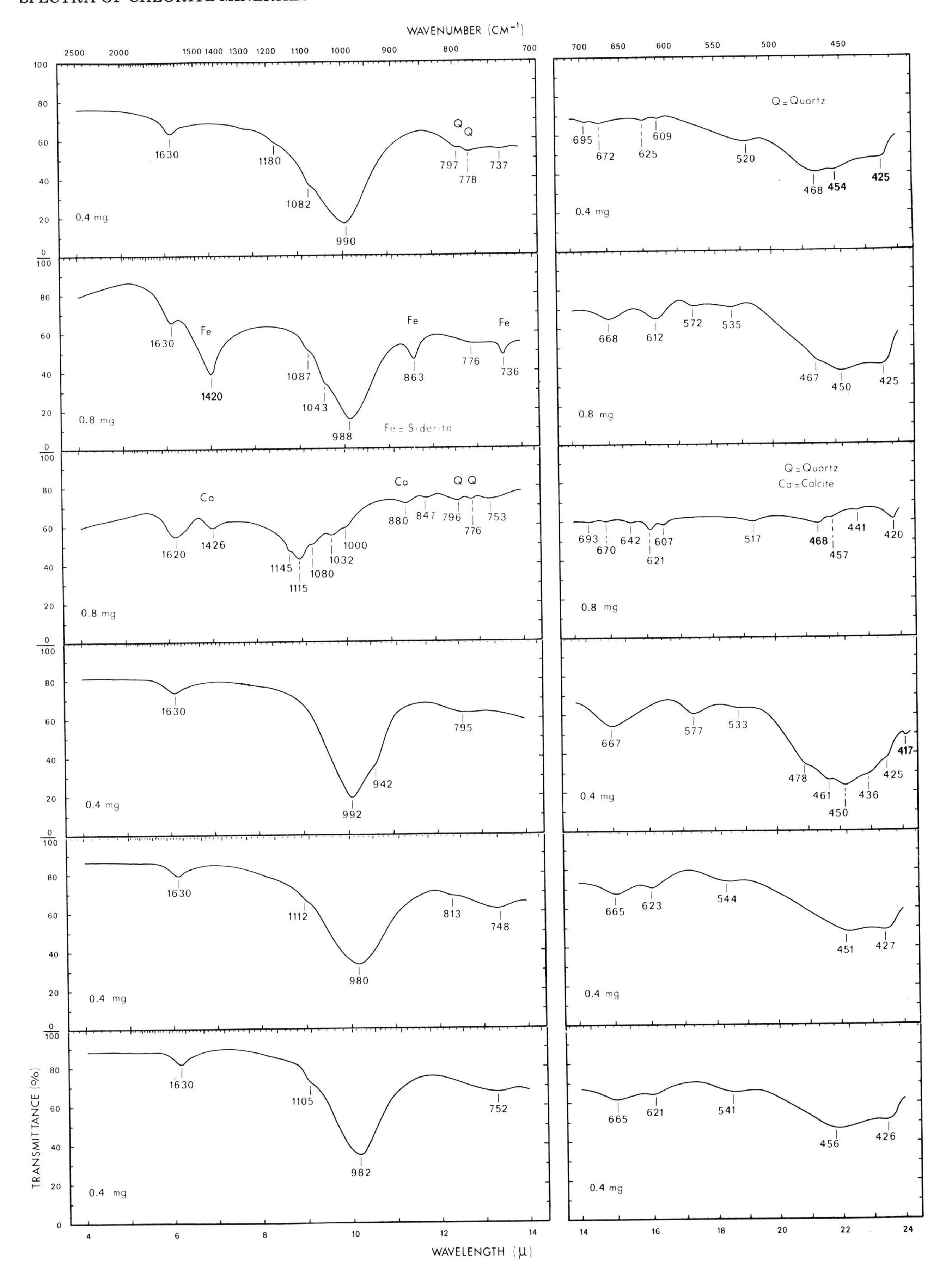
WAVENUMBER (CM⁻¹)
TRANSMITTANCE (%)
WAVELENGTH (μ)
Q = Quartz
Fe = Siderite
Q = Quartz
Ca = Calcite
0.4 mg
0.8 mg
0.8 mg
0.4 mg
0.4 mg
0.4 mg

Pennantite (mangan chlorite)
Pafsberg, Sweden
pure

Kämmererite (chrome chlorite)
Kraubath, Austria
pure

Radusa, Yugoslavia
pure

S. Benito, Calif., U.S.A.
pure

Cookeite (lithium chlorite)
Minas Gerais, Brazil
pure

Al-chlorite (sudoite)
Kamikita Mine, Japan
pure

Kaiserbach, Germany
pure

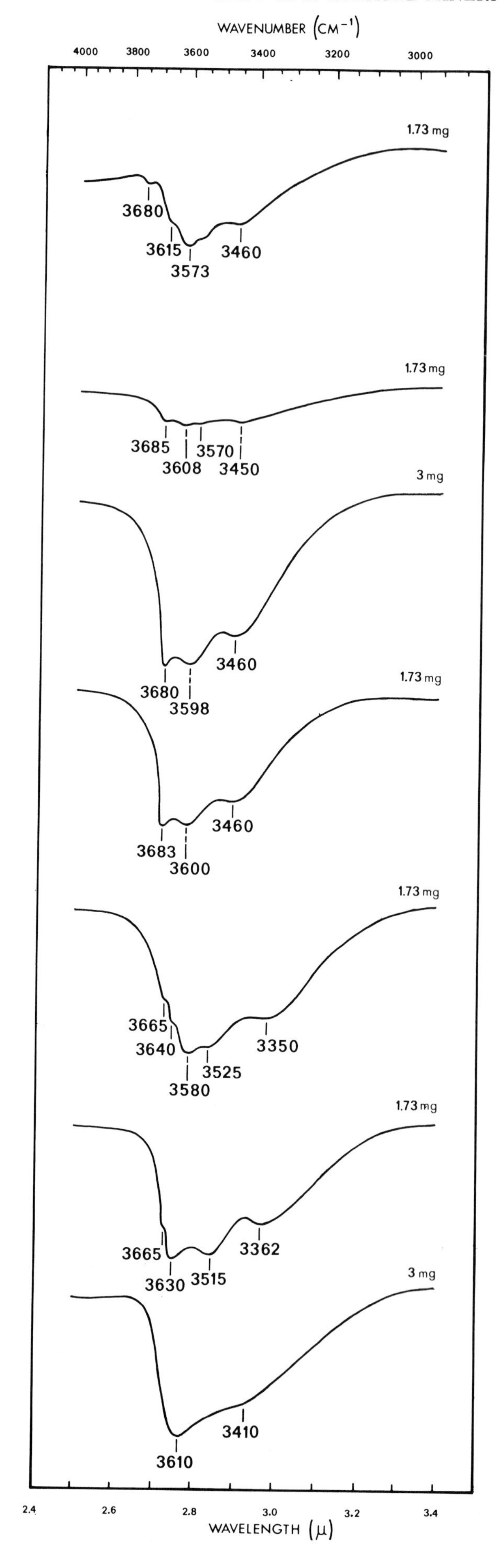

Fig. 6.52. Infrared spectra of pennantite, kämmererite, cookeite and Al-chlorite (sudoite) from different origin.

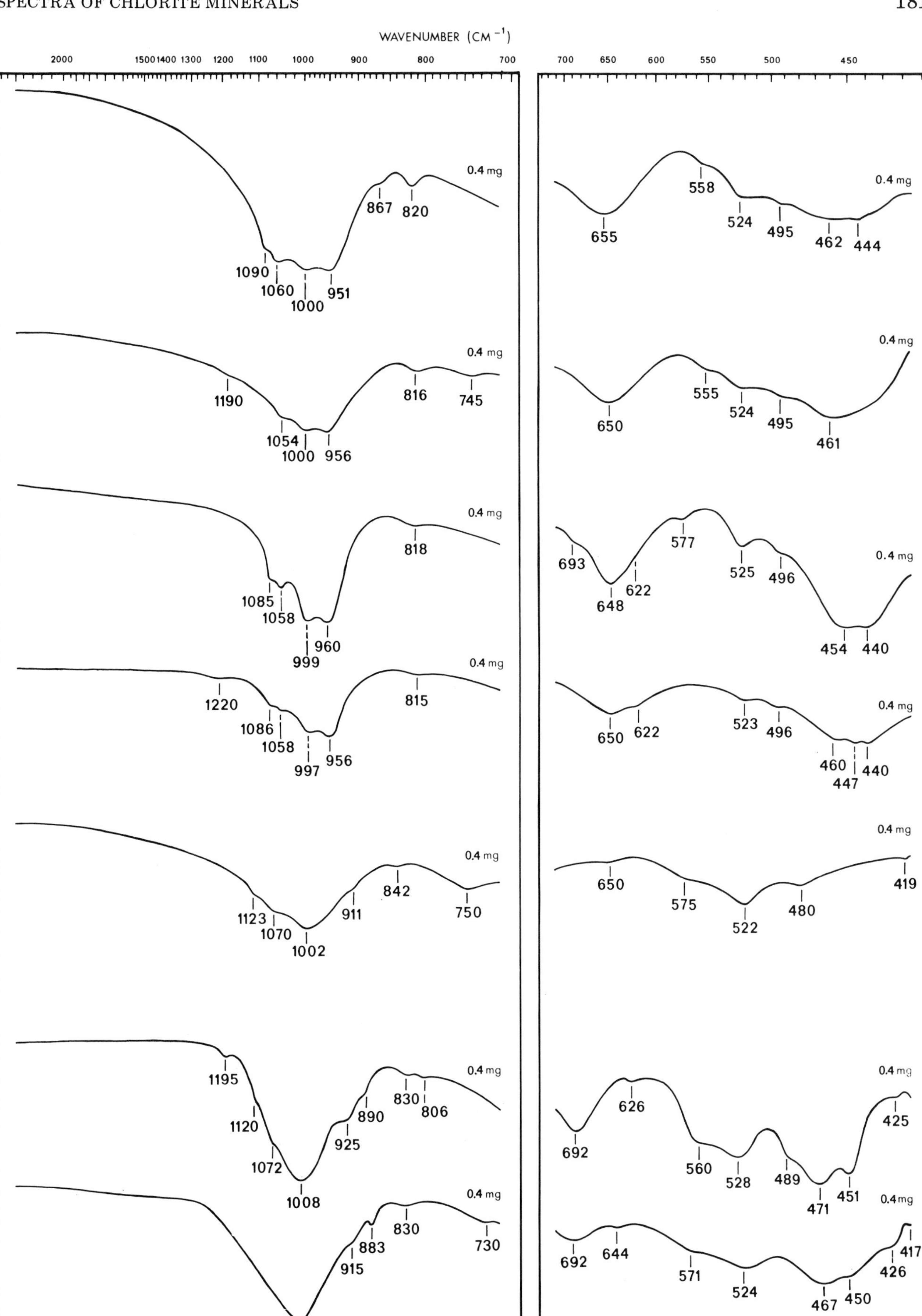
WAVENUMBER (CM⁻¹)
2000 1500 1400 1300 1200 1100 1000 900 800 700
700 650 600 550 500 450
0.4 mg
1090 1060 1000 951 867 820
655 558 524 495 462 444
1190 1054 1000 956 816 745
650 555 524 495 461
1085 1058 999 960 818
693 648 622 577 525 496 454 440
1220 1086 1058 997 956 815
650 622 523 496 460 447 440
1123 1070 1002 911 842 750
650 575 522 480 419
1195 1120 1072 1008 925 890 830 806
692 626 560 528 489 471 451 425
1013 915 883 830 730
692 644 571 524 467 450 426 417
4 6 8 10 12 14
14 16 18 20 22 24
WAVELENGTH (μ)

Hydrated chlorite
Maaninka, Finnland
some quartz, feldspar, amphibole (overlapped) and augite (overlapped)

Sedimentary chlorite

Zagreb, Yugoslavia
separate < 2 μm of weathered chlorite schist
mica and some quartz, feldspar (overlapped) and kaolinite
Fig. E.M. 138

Luxembourg
separate < 2 μm of weathered shale
mica, kaolinite, quartz and some feldspar (overlapped)

Teesta river, India
separate < 2 μm of alluvial sediment,
mica, kaolinite, quartz and some feldspar (overlapped)
Fig. E.M. 137

Swelling chlorite
Maulbronn, Germany
separate < 2 μm of Keuper marl
mica, kaolinite, hematite (overlapped)
Fig. E.M. 139

Soil chlorite

Bathmen, The Netherlands
separate < 2 μm of loam
quartz, kaolinite and some limonite (overlapped)

Paramaribo, Surinam
separate < 2 μm alluvial
sediment
kaolinite, quartz and some mica
Fig. E.M. 142

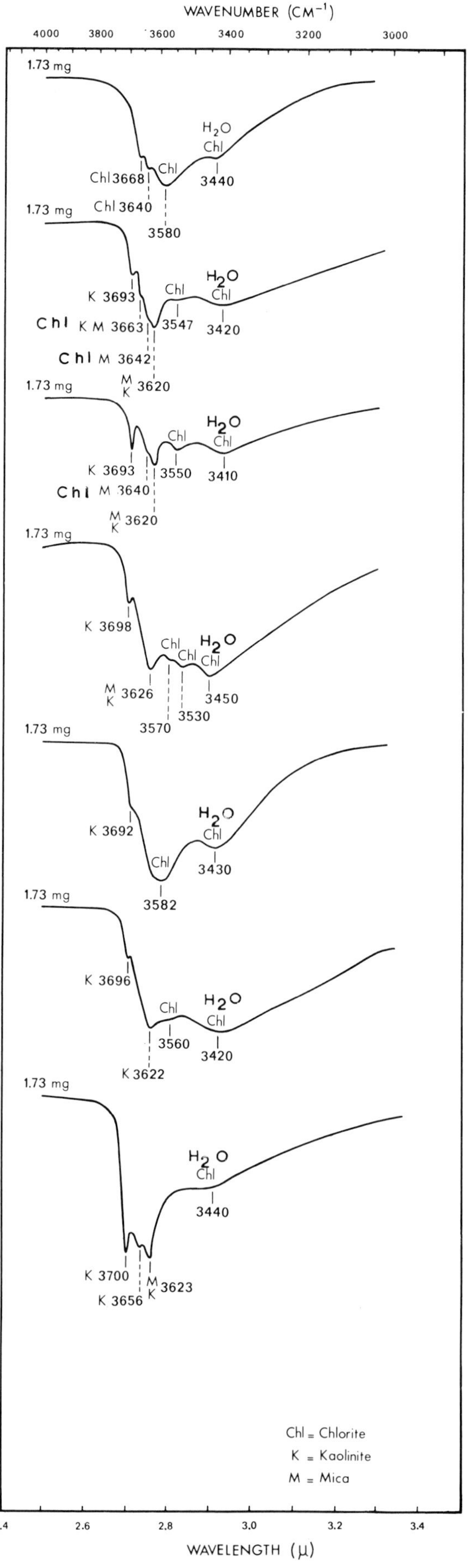

Fig. 6.53. Infrared spectra of various chlorite minerals.

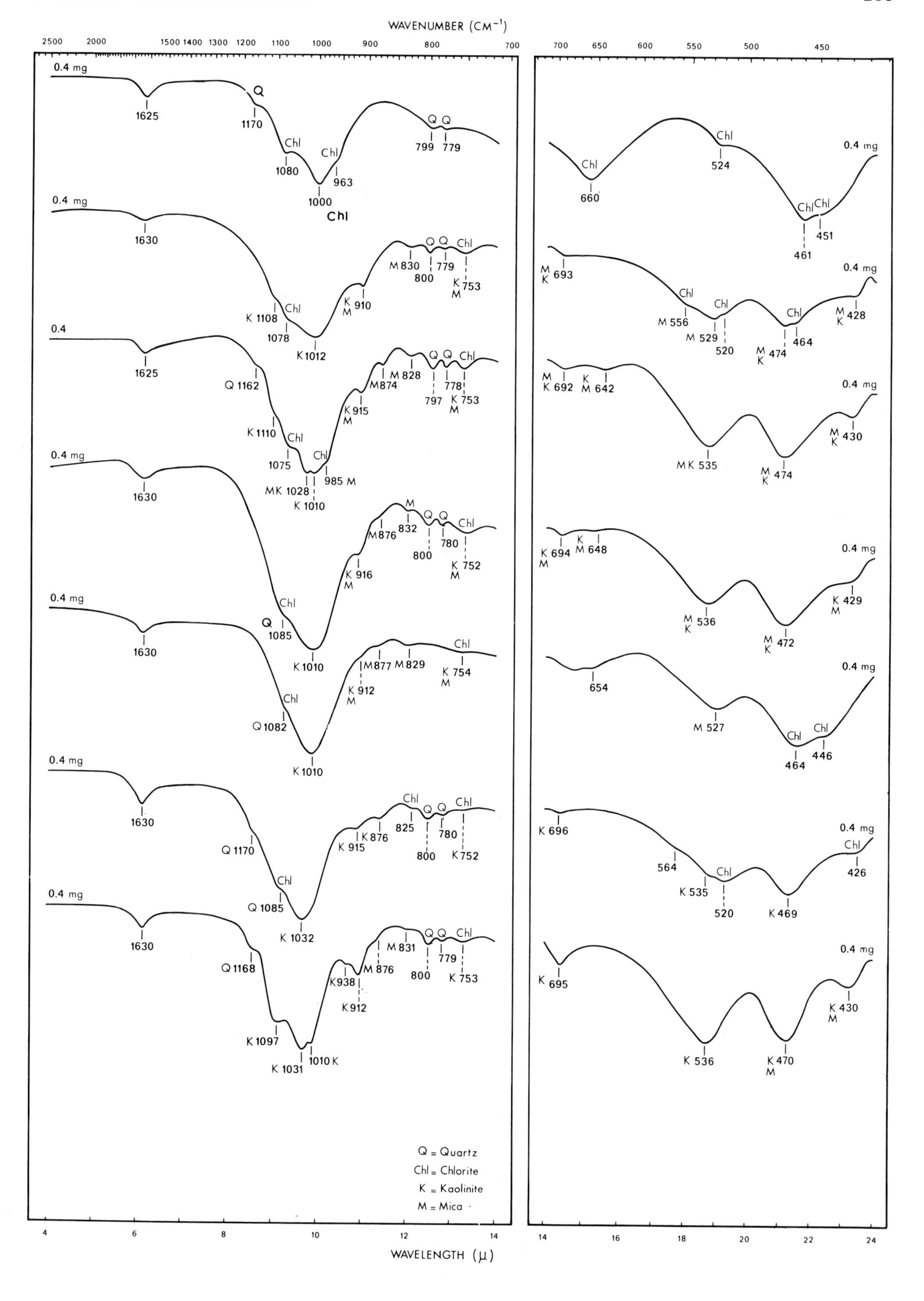
WAVENUMBER (CM⁻¹)
WAVELENGTH (μ)
Q = Quartz
Chl = Chlorite
K = Kaolinite
M = Mica

Interstratified minerals

Under specific conditions clay minerals may be formed which consist of an interstratification of two and sometimes three or four different clay minerals. If the components are stacked regularly the mineral is of the "regular" type; otherwise it is an "irregular", randomly interstratified mineral. The latter is most common in sediments.
Under hydrothermal or metamorphic conditions the formation of regular interstratified minerals is favoured. Usually they are better crystallized and with X-ray analysis may give an integral series of sharp, intensive basal reflections up to even the 19th order.
Illite-morillonite (swelling illite) interstratifications are frequently found, because the basal minerals are widespread.

Non regular illite-montmorillonite (swelling illite); formula:

$$\begin{bmatrix}(Al,Fe,Mg)_2(OH)_2 \mid (Si,Al)_4(O,OH)_{10} \\ (K,K^+,H_3O^+)\end{bmatrix}$$

$$\begin{bmatrix}(Al,Fe,Mg)_2(OH)_2 \mid (Si,Al)_4O_{10} \\ (Ca^{2+},Mg^{2+})\end{bmatrix} nH_2O$$

The grades of interstratification are manifold, from which results a different degree of swelling or contraction when treated with glycerol or potassium chloride respectively.

Occurrence. Bravaisite, Mallard (1878) is found as a rare mineral in coal mines where it resulted from weathering of shales. The Veldriet sample is from an alluvial sediment which is also of illitic origin. The highly interstratified soil montmorillonite (swelling illite)-montmorillonite interlayered with silicic acid and $Al(OH)_3$ polymers, was found in solontchak-like solonetz soils in Bulgaria [1].
Metabentonite resulted from weathering of volcanic ashes and tuffs in seawater into montmorillonite, which then have picked up potassium. Other names for this mineral are: highly altered K-rich bentonite, Ordovician (K)bentonite, etc. (see also p. 119).

Infrared spectra (Fig. 6.54). The spectra of illite and montmorillonite are nearly the same. One difference is that the 915, 524 and 467 cm^{-1} bands of (soil) montmorillonite are for illite weak or lacking, respectively.
Another difference is that muscovite (and sericite), which mineral is mostly found together with illite, have a strong 988 cm^{-1} band, which is lacking in the montmorillonite spectrum. Muscovite, sericite and illite have a strong band at 476 cm^{-1}. According to the above criteria, bravaisite, and the Veldriet sample have illite (muscovite) as well as montmorillonite components. The Bulgarian sample is mainly montmorillonitic.
The metabentonite from High Bridge-Kentucky is also mainly montmorillonitic whereas that from Kinnekulle is more illitic.

Non regular talc-saponite (stevensite); formula:

$$(Mg_3(OH)_2 \mid Si_4O_{10}$$

$$\begin{bmatrix}(Mg,Fe,Al)_3(OH)_2 \mid (Si,Al)_4O_{10} \\ Na^+,Ca^{2+}\end{bmatrix} nH_2O$$

Occurrence. Stevensite is a late stage hydrothermal weathering product of pectolite. It is found in cavities of basalts at North Tyne (Northumberland), Springfield, N.J.

Infrared spectra (Fig. 6.55, 6.56). The degraded talc samples have some saponite components — see the saponite vibration at 3628 cm^{-1} instead of the talc vibration at 3658 cm^{-1} and the low order of the bands for the highest wavenumbers. Reversely the mineral has some talc components — see the talc vibration at 670 cm^{-1} which is lacking for saponite, the 1018 cm^{-1} vibration instead of the saponite vibration at 1010 cm^{-1} and the absence of the saponite vibrations at 1100—1090 cm^{-1}.
The X-ray analysis shows a broadening of the bands and a shift of the 9.36 Å reflection of talc to ca. 9.6 Å (Fig. 6.56). The stevensite sample from North Tyne also is an intermediate between talc and saponite.
The sample from Chauzh-Ural USSR named β kerolite by Ginzburg (1950) and which is supposed to be chrysotile and attapulgite-like, is nearly the same as the stevensite sample from N. Tyne (England). They both are disordered, which picture is also given by their X-ray analysis (Fig. 6.56).

Remarks. The mineral was first discovered at Langban (Sweden) by Berlin (1840) and then named aphrodite (Greek aphros = foam). Afterwards the name was changed into stevensite by Leeds (1873) after the Stevens Institute of Technology at Hoboken.
According to the published data in literature, stevensite is not of constant composition. In one sample the talc component may be larger than the saponite component (the sample described

[1] Gift from A. Behar, Sofia.

by Faust et al., 1959) and the reverse may be the case (the sample described by Brindley, 1955) although it comes from the same locality — Springfield, N.J.
According to Brindley (1955) stevensite is a talc-saponite interlayered mineral with a strong 12 Å and weak 24—25 Å spacing when air dry. Its swelling to 17—18 Å when treated with ethylene glycol (Faust et al., 1959) suggests the mineral to be Mg-montmorillonite with a defective structure arising from a deficiency in the total number of ions in octahedral coordination. The reflection at 24.5—25.5 Å is said to be caused by domains of talc in combination with more abundant domains of stevensite.
Otsu et al. (1963) distinguish two kinds of stevensite: trioctahedral Mg-montmorillonite, and a randomly mixed-layer mineral which does not swell; thus increasing the difficulties about the name stevensite for a certain mineral with certain characteristics.

Regular illite-montmorillonite; formula:

$$\begin{bmatrix}(Al,Fe,Mg)_2(OH)_2 \,|\, (Si,Al)_4(O,OH)_{10} \\ K(K^+,H_3O^+)\end{bmatrix}$$

$$\begin{bmatrix}(Al,Fe,Mg)_2(OH)_2 \,|\, (Si,Al)_4O_{10} \\ (Ca^{2+},Mg^{2+})\end{bmatrix} nH_2O$$

Occurrence. The mineral described here is found over large areas at Burghersdorp, (Cape Provence) in clays associated with intrusions of dolerite into shales. It resulted from illite by hydrothermal action.

Infrared spectra (Fig. 6.57). The mineral is of strong illitic character: see the characteristic illite band at 477 cm^{-1} and the absence of the strong 915 and 467 cm^{-1} bands of montmorillonite. It contains 4.8% K_2O which agrees with the above identification.

Remarks. By X-ray analysis the basal reflection can be followed to the 10th order.

Regular paragonite-like-montmorillonite (allevardite, rectorite); formula:

$$\begin{bmatrix}Al_2(OH,F)_2 \,|\, (Si_3Al)O_{10} \\ (Na)\end{bmatrix}$$

$$\begin{bmatrix}(Al,Mg,Fe)_2(OH)_2 \,|\, (Si,Al)_4O_{10} \\ (Ca^{2+},Mg^{2+})\end{bmatrix} nH_2O$$

Occurrence. The above mineral is found as a rare mineral in rock cavities where it resulted from hydrothermal action on muscovite. Paragonite is formed during the later stages of diagenesis.

Infrared spectra (Fig. 6.57). The above samples have in common a 3640 cm^{-1} vibration whereas paragonite has 3664 and 3628 and montmorillonite = 3624 cm^{-1}, respectively. Apparently in this case the vibrations of the single components have changed. Other important vibrations are at 1022, 544, 481 cm^{-1}, Paragonite has in these regions 1012, 540, 478 cm^{-1}, and montmorillonite 1038—1026, absent, 474*, 467 cm^{-1}.
The sample from Arkansas is very badly ordered pointing to a large disorder in the arrangement of the atoms in the crystal.

Remarks. Rectorite is named after E.W. Rector, formerly connected with Geologic Survey, Arkansas (1891). Allevardite is named after the type locality La Table near Allevard, Savoie, France.
By X-ray analysis the above minerals can be followed to the 16th order of their basal reflection.

Regular talc-saponite (aliettite); formula:

$$[Mg_3(OH)_2 \,|\, Si_4O_{10}]$$

$$\begin{bmatrix}(Mg,Al,Fe^{3+})_3(OH)_2 \,|\, (Si,Al)_4O_{10} \\ (Na^+,Ca^{2+})\end{bmatrix} nH_2O$$

Occurrence. This mineral is very rare. A regular type which resulted from hydrothermal action was found in serpentine rocks at Monte Chiaro (Italy).

Infrared spectra (Fig. 6.58). The mineral has talc characteristics (talc between brackets) at 1040 (1043), 668 (669) cm^{-1}. Saponite characteristics (saponite between brackets) are at 3575 (3575*), 1010 (1010) cm^{-1}, thus proving its dual nature. The spectrum is broad and very badly developed in the higher wave lengths, thus demonstrating the large effect of interaction of the atoms belonging to different groups; even when the mineral is regularly interstratified.

Remarks. The mineral is named after its discoverer Professor A. Alietti, Modena, Italy (1956). The mineral shows an integral series of (001) reflections to the 9th order.

Regular chlorite-swelling chlorite (corrensite); formula:

$$\begin{bmatrix}(Mg,Al,Fe)_3(OH)_2 \,|\, (Si,Al)_4O_{10} \\ (Mg,Al,Fe)_3(OH)_6\end{bmatrix}$$

$$\begin{bmatrix}(Mg,Al,Fe)_3(OH)_2 \,|\, (Si,Al)_4O_{10} \\ (Mg,Al,Fe)_3(OH)_6\end{bmatrix} nH_2O$$

Occurrence. This mineral is very rare. The sample from Monte Chiaro (Italy) resulted from hydrothermal action on serpentine.

Infrared spectrum (Fig. 6.58). This mineral shows the same vibrations as clinochlore with some slight shifts to lower and to higher frequences.
Thus the swelling component can't be identified, which however, is possible by the X-ray method.

Remarks. The mineral is named after Professor Dr. C.W. Correns, Göttingen. An integral series of (001) X-ray reflections up to the 12th order is observed; thus proving its regular structure.

Regular chlorite-saponite and non regular (randomly) saponite-swelling chlorite; formula:

$$\begin{bmatrix}(Mg,Al,Fe)_3(OH)_2 \mid (Si,Al)_4O_{10} \\ (Mg,Al,Fe)_3(OH)_6\end{bmatrix}$$

$$\begin{bmatrix}(Mg,Al,Fe^{3+})_3(OH)_2 \mid (Si,Al)_4O_{10} \\ (Na^+,Ca^{2+})\end{bmatrix} nH_2O$$

Occurrence. The above intermediate minerals are very rare. A regular mineral was found at Monte Chiaro (Italy). It resulted from hydrothermal action on serpentine. A randomly interstratified mineral which resulted from weathering of a Lizardite rock was found at St. Margherita Staffora (Italy).

Infrared spectra (Fig. 6.58). The regular mineral is of more chloritic character than the nonregular — see the strong 3570 cm^{-1} vibration of the regular sample — in common with clinochlore. The strong 3680 cm^{-1} vibration of the nonregular sample is in common with saponite (3678 cm^{-1}). Chlorite bands in common for both are (nonregular between brackets) at: 1084 (1087), 957 (955), 758 (750) cm^{-1} as against the 1085, 956, 756 cm^{-1} vibrations respectively for clinochlore. Saponite bands in common are (nonregular between brackets) at: 1000 (1005), 657 (652), 460 (462), against the 1010, 652, 462 cm^{-1} vibrations respectively for saponite.

Remarks. The regular mineral shows an integral series of (001) reflections to the 14th order.

Regular biotite-vermiculite (hydrobiotite); formula:

$$\begin{bmatrix}(Mg,Fe,Al)_3(OH,F)_2 \mid (Si,Al)_4O_{10} \\ K\end{bmatrix}$$

$$\begin{bmatrix}(Mg,Al,Fe)_3(OH)_2 \mid (Si,Al)_4O_{10} \\ Mg,Mg^{2+}(H_2O)_4\end{bmatrix}$$

This 12 Å mineral contracts its layers to 10 Å when treated with a KCl solution or when heated at 550° C.

Occurrence. The mineral is found in some places in large amounts e.g. Palabora (S. Africa), Libby (Montana). It resulted from thermal action on biotite-phlogopite rocks.

Infrared spectra (Fig. 6.59). The above interstratified mineral is mainly vermiculite-like — see the large band of H_2O at 3410 cm^{-1} with, moreover, the band at 3230 cm^{-1} for structural bonded H_2O.
Other specific vermiculite characteristics (vermiculite between brackets) are at 686 (683*) cm^{-1}. Specific biotite characteristics as e.g. at 3650*, 3590*, 3548 cm^{-1} are absent.

Remarks. An integral series of (001) X-ray reflections up to the 12th order is observed, thus proving its regular structure. The biotite-vermiculite mineral from the deposits at Palabora where huge amounts occur, is handled as "vermiculite" for different purposes — see vermiculite.

Regular Al chlorite-montmorillonite (tosudite); formula:

$$\begin{bmatrix}Al_2(OH)_2 \mid (Si,Al)_4O_{10} \\ Al_2(OH)_6\end{bmatrix}$$

$$\begin{bmatrix}(Al,Mg,Fe)_2(OH)_2 \mid (Si,Al)_4O_{10} \\ (Ca^{2+},Mg^{2+},Na^+)\end{bmatrix} nH_2O$$

The above mineral is said to consist of a regular interstratification of the named components.

Occurrence. This mineral is very rare[1]. It was first discovered in the Kurata fire clay deposit, Japan: Sudo et al. (1954).

Infrared spectrum (Fig. 6.59). The following shows that only a restricted number of bands of the two components appear in the spectrum of the regular interstratified mineral. They are (the latter between brackets): montmorillonite: 3642* (3640), 915 (915), 878 (878), 578* (570), 426 (429) cm^{-1}; Al chlorite: (sudoite): 1120 (1122), 471 (477), 425 (429) cm^{-1}. Other hands among which some of high intensity e.g. at 3624, 1038—1026, 522, 467 cm^{-1} for montmorillonite and at 3630, 3515, 3362, 1008, 692, 528, 451 cm^{-1} for Al chlorite are absent in the spectrum of the intermediate mineral.

[1] Gift from S. Shimoda, Japan.

Remarks. An integral series of (001) X-ray reflections up to the 10th order is observed, thus proving the regular structure of this intermediate. The mineral is named after its discoverer Professor T. Sudo, (Japan).

Regular Al,Mg chlorite-montmorillonite; formula:

$$\begin{bmatrix} Al_2(OH)_2 \mid (Si,Al)_4O_{10} \\ (Al,Mg)_2(OH)_6 \end{bmatrix}$$

$$\begin{bmatrix} (Al,Mg,Fe)_2(OH)_2 \mid (Si,Al)_4O_{10} \\ (Ca^{2+},Mg^{2+}) \end{bmatrix} nH_2O$$

The mineral consists of a regular interstratification of the named components. In this case the hydroxide sheet of the chlorite structure is of the trioctahedral type. For MgO is found 6.44%.

Occurrence. This mineral is very rare[1]. It was first described by Sudo and Kodama (1957) to occur in the Honko ore body of pyrite in Tertiary tuffaceous volcanic rocks in the Kamikita mine, Japan.

Infrared spectra (Fig. 6.59). This sample can easily be distinguished from tosudite by the bands at (tosudite between brackets): 3612 (absent), 3526 (3540), 3350 (3380), 1031 (1043), 704 (absent), 452 (absent).
Moreover, the bands of the former are much weaker, in particular at the highest wavelengths.

References

Alietti, A., 1956. Il minerale a strati misti saponite-talc di Monte Chiaro (Val di Taro, Appennino Emiliano). C.R. Accad. Nazl. Lincei, 21: 201—207.

Brindley, G.W., 1955. Stevensite a montmorillonite type mineral showing mixed layer characteristics. Am. Mineralogist, 40: 239—247.

Faust, G.T., Hathaway, J.C. and Millot, G., 1959. A restudy of stevensite and allied minerals. Am. Mineralogist, 44: 342—370.

Ginzburg, I. and Rukavishnikova, I.A., 1950. β kerolite: $3MgO \cdot 3SiO_2 \cdot 3H_2O$. Zap. Vses. Mineral. Obshch., 1: 33—44.

Leeds, A.R., 1873. Contributions to mineralogy. Am. J. Sci., 6: 22—26.

Mallard, F.E., 1878. Sur la bravaisite, substance minérale nouvelle. Bull. Soc. Fr. Minéral., 1: 5—8.

Otsu, H., Shimazaki, Y. and Ohmachi, H., 1963. On stevensite from Ohori mine. Bull. Geol. Surv., Japan, 14: 27 (591)—35 (599)

Sudo, T., Takahashi, H. and Matsui, H., 1954. Long spacing of 30 Å from fireclay. Nature, 173: 161.

Sudo, T. and Kodama, H., 1957. An aluminium mixed-layer mineral of montmorillonite-chlorite. Z. Kristall., 109: 379—387.

Pseudo layer silicates with chain structure (hormites)

The above minerals are members of a series of Mg-rich minerals with chain structure of which attapulgite is the Al(Fe)-rich and sepiolite the Al-poor member. In contrast to the foregoing minerals with a two-dimensional $(Si_2O_5)^{2-}$ layer structure, they consist of SiO-tetrahedra bound to $(SiO_3)^{2-}$-chains of pyroxene type (Plate I, pp. 190—191).

Attapulgite (palygorskite); formula:

$[(OH_2)_4(Mg,Al,Fe)_5(OH)_2 | Si_8O_{20}]4H_2O$

Attapulgite consists of two double chains of the pyroxene-type $(SiO_3)^{2-}$ like amphibole $(Si_4O_{11})^{6-}$ running parallel to the fibre axis.
However, as opposed to the amphiboles, the attapulgite units are connected to one another by shared oxygen atoms and not by cations (Ca,Mg, etc.). From this results a structure with channels about 3.7 × 6.0 Å wide in cross-section. The interstices between the chains are filled by H_2O molecules arranged partly parallel to the fibre axis (zeolitic water) and partly bound to the Mg cations of the Mg(Al,Fe) brucite ribbon edges (crystal water) bordering the channels, which run along the length of the crystals (Plate I).

Occurrence. Attapulgite is found in Georgia (U.S.A.) in large deposits of Miocene age as an alteration product of volcanic tuffs. The deposit at Mormoiron (France) is alluvial. Smaller quantities of the mineral are found in fissures of rocks. Also in Russia (Ukraine) are large deposits called astrangitegel.
Many sediments in Ethiopia, Egypt, Persia, Iraq, Tunesia and Syria contain attapulgite together with varying amounts of kaolinite, sepiolite, and quartz. In these cases, they resulted from weathering of Mg-rich rocks (dolomite, magnesite, serpentine, etc.).

Infrared spectra (Fig. 6.60, 6.62). Characteristic bands for attapulgite are at: 3620, 3540, 3265, (crystal water), 1037, 987, 648, 513, 482 cm^{-1}. There are great differences in the intensities and broadness of the bands of the various samples. In particular the bands of the highest wavelengths of the Israel samples are poorly developed. In fine sediments, the pseudo layer-silicate mineral found is mainly of the attapulgite type.

Remarks. The name was given by De Lapparent (1935) to a fibrous clay mineral, found in "Fuller's earth" (trademark) of Attapulgus and in "Terre à foulon" of Mormoiron (France). It is, furthermore, suggested that it has the same structure as the fibrous mineral from Palygorsk (Ural, USSR), as described by Tsavtchenkov (1862) and Fersmann (1913).
Attapulgite has excellent purifying and decolorizing properties for oils and greases together with good permeability. It is also used as a catalyst for cracking oils. Furthermore in drilling muds because of its small thixotropic effect which prevents aggregation of the particles in the bore hole by electrolytes.
The name hormite is derived from the Greek hormathos = chain. Attapulgite has been used by "fullers" for "fulling" that is to remove grease from cloths. The name "Fuller's earth" is also used for montmorillonite beds in England. Other trade mark names for attapulgite are floridum, florex, florigel.

Sepiolite; formula:

$[(OH_2)_4Mg_8(OH)_4 | Si_{12}O_{30}]8H_2O$

This mineral is a Mg-rich but Al, Fe-poor chain mineral. Sepiolite consists of three pyroxene-type chains instead of two as in attapulgite, and the oxygens linking the chains together are double-linked, as opposed to the single-linked oxygen atoms in attapulgite. As a result of this different configuration, the channels of sepiolite are wider = 5,6 × 11,0 Å, against 5,4 × 6,5 Å for attapulgite.

Occurrence. Sepiolite is found only in certain deposits. The best known are: Ampandandrava (Malagasy), Salinelles (France), Vallecas (Spain). The mineral occurs further in sediments or in fissures of Mg-rich rocks in the Midlands (Great Britain), Fukuoka Prefecture (Japan), U.S.S.R., etc. Here the mineral resulted from weathering of Mg-rich rocks.

Infrared spectra (Fig. 6.61). As a contrast to attapulgite, the bands of this mineral are less variable. Sepiolite can easily be distinguished from attapulgite (the latter between brackets) at: 3560 (3540), 3245 (3265), 1200 (1190), 1074 (1090), 1025 (1037), absent (940), absent (912), 784 (absent), absent (584), 534 (absent), 503 (513) cm^{-1}.

Remarks. The name was given by Glocker (1847), because of the mineral's resemblance to the light-coloured porous bone of the internal shell of the cuttle fish (Latin: os sepium).
Sepiolite is used for the same purpose mentioned under attapulgite. Its purifying, decolorising, permeable and thixotropic characteristics are even more pronounced than those of attapulgite.

Sea foam (Meerschaum, écume de mer)

Structure. The mineral is supposed to be an earthy, short vibrous variety of sepiolite (figures E.M. 161, 162).

Occurrence. The best known locality is Eskisché-hir, Turkey, where the deposit resulted from weathering of serpentine rock.

Infrared spectra (Fig. 6.63). The spectra show that the 3540 cm^{-1} and 3560 cm^{-1} bands of attapulgite and sepiolite are replaced by a strong 3574 cm^{-1} band. The other bands of meerschaum are most like those of sepiolite. The Meerschaum sample from Arizona has specific chrysotile (3690, 616 cm^{-1}) + antigorite (564, 449 cm^{-1}) bands.

Remarks. The mineral, named para-sepiolite by Fersmann (1908, 1913) floats on water when dry. It is applied in the manufacture of porcelain.

Mountain leather, Mountain cork, Bergkork, Bergholz, Schweizerite

Structure. The above minerals mostly consist of attapulgite. But they may also be a mixture of attapulgite and sepiolite or be sepiolite or even be chrysotile.

Occurrence. They are found in small amounts as coatings, or in fissures of highly Mg-rich rocks.

Infrared spectra (Fig. 6.64). Most of the above named samples consist mainly of attapulgite; Alaska, Washington, New Mexico. But that from Tirol, called Bergkork, has attapulgite, sepiolite and Meerschaum.
The "mountain leather" sample from Texas consists of sepiolite. The Schweizerite samples from Zermatt and Dagnoeska are chrysotile and chrysotile + calcite, respectively.
The I.R. method is far better to distinguish the various fibrous minerals from each other, also in mixtures, than the E.M., the X-ray and the D.T.A. method.

Remarks. The above minerals are named after their morphological appearance. In consequence also other fibrous minerals may be included in this group.

References

De Lapparent, J., 1935. Sur un constituant essentiel des terres à foulon. C. R., 201: 481—483.

Fersmann, A., 1908. Parasepiolite. Bull. Acad. Imp. Sci., St. Petersburg, Ser. 6, 2: 263.

Fersmann, A., 1913a. Über die Palygorskitgruppe. Bull. Acad. Imp. Sci., St. Petersburg, 2: 255.

Fersmann, A., 1913b. Recherches sur les silicates de magnésia (en russe). Mem. Russ. Acad. Sci., St. Petersburg, Chem. Phys. Math.

Glocker, E.F., 1847. Generum et Specierum Mineralium Secundum Ordines Naturales Digestorum Synopsis, Halle, 348 pp.

Tsaftchenkov, T., 1862. Über Palygorskite. Verh. Russ. Kaiserl. Mineral. Ges., St. Petersburg, 102—104.

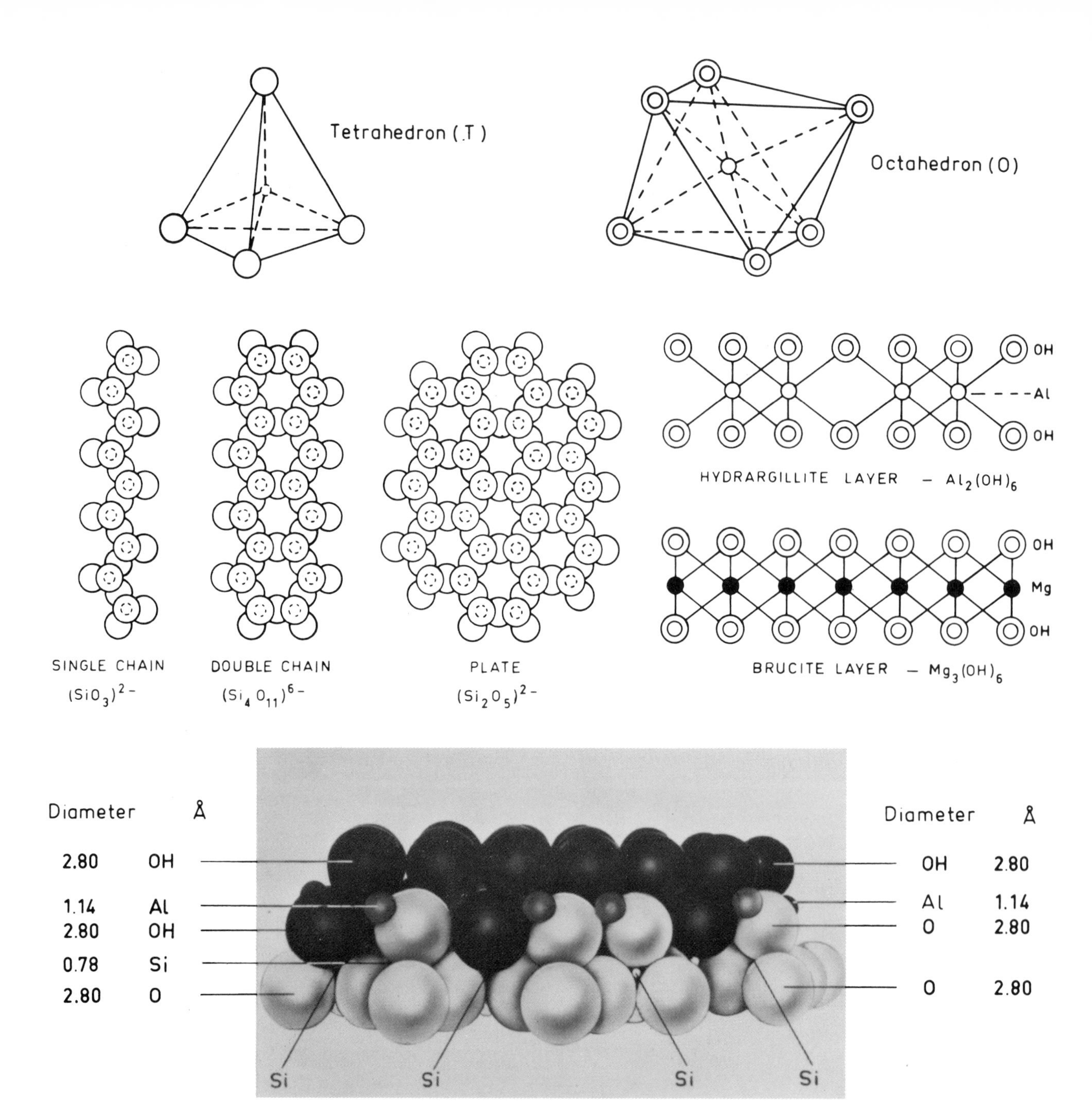

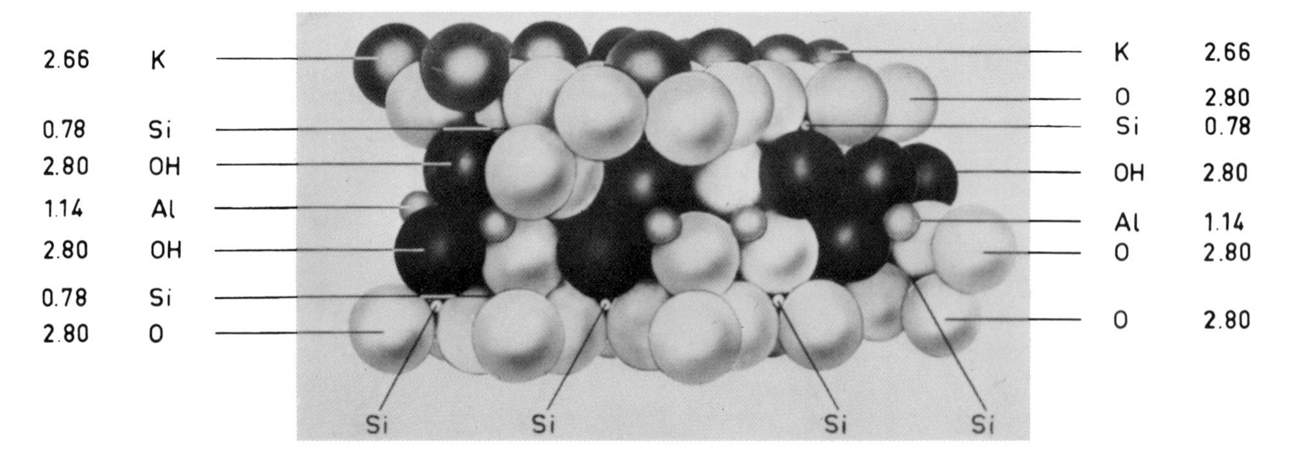

PLATE I. Schematic presentation of the structure of various clay minerals.

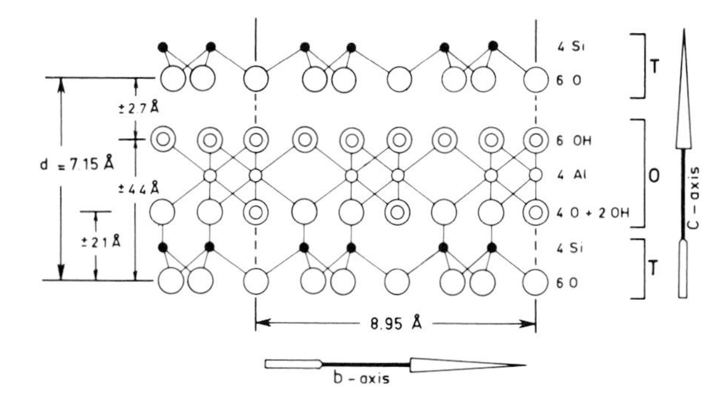

KAOLINITE – G.W. BRINDLEY and K. ROBINSON (1946)

$$2 \left[Al_2 (OH)_4 \mid Si_2 O_5 \right]$$

TRICLINE $a_o = 5.14$ Å, $b_o = 8.93$ Å, $c_o = 7.37$ Å, $\alpha = 91°50'$, $\beta = 104°50'$, $\gamma = 89.9°$

$$d = c_o (1 - \cos^2\alpha - \cos^2\beta)^{1/2} = 7.15 \text{ Å}$$

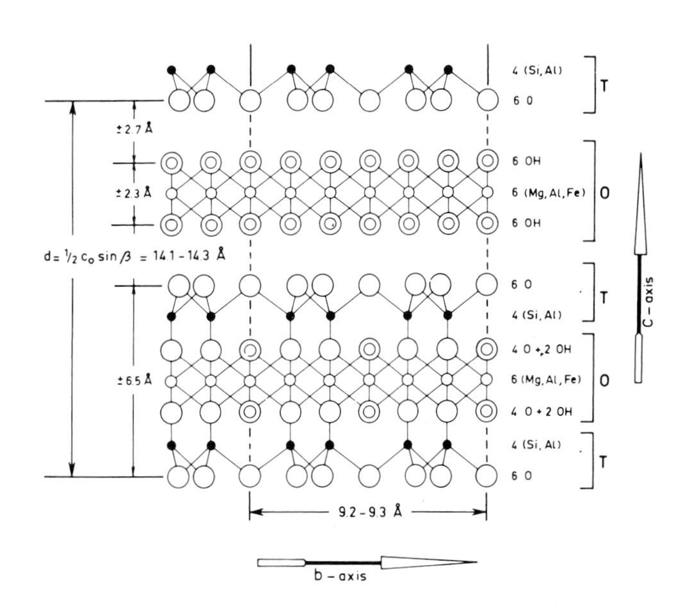

CHLORITE – R.C. McMURCHY (1934)

$$4 \left[\begin{matrix} (Mg,Al,Fe)_3 (OH)_2 \mid (Si,Al)_4 O_{10} \\ (Mg,Al,Fe)_3 (OH)_6 \end{matrix} \right]$$

MONOCLINE $a_o = 5.30 - 5.35$ Å, $b_o = 9.19 - 9.27$ Å, $c_o = 2 \times 14.15 - 14.29$ Å, $\beta = 97°8'40''$
depending on chemical composition

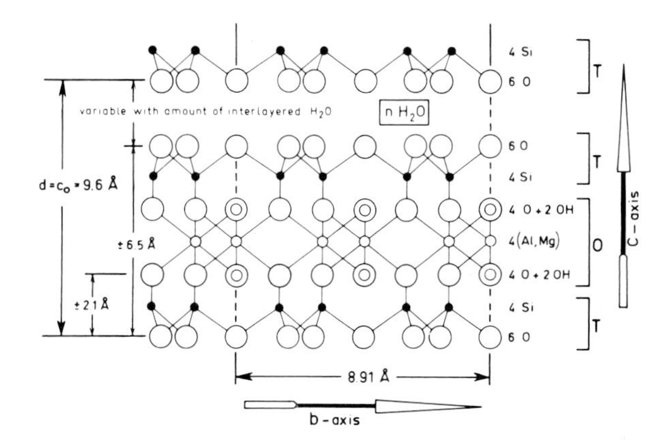

MONTMORILLONITE – U. HOFMANN et al. (1933)

$$2 \left[\begin{matrix} (Al,Mg)_2 (OH)_2 \mid Si_4 O_{10} \\ (Na^+, Ca^{2+}) \end{matrix} \right] n\, H_2O$$

RHOMBIC $a_o = 5.095$ Å, $b_o = 8.83$ Å, $c_o = 9.6$ Å

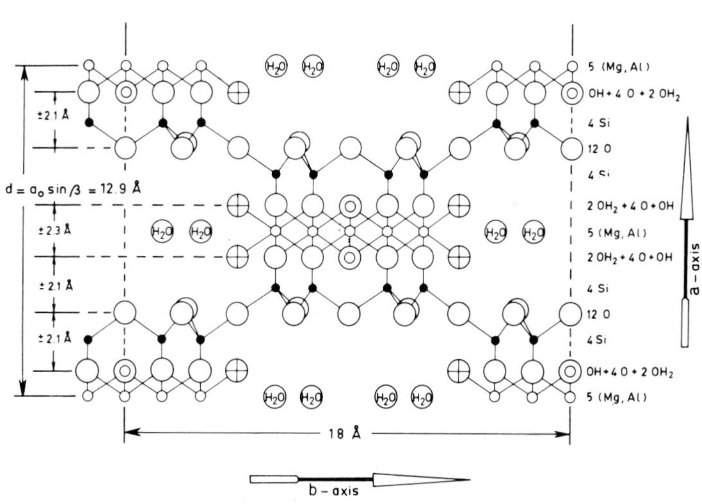

ATTAPULGITE – W.F. BRADLEY (1940)

$$2 \left[(OH_2)_4 (Mg,Al,Fe)_5 (OH)_2 \mid Si_8 O_{20} \right] 4\, H_2O$$

MONOCLINE $a_o \sin\beta = 12.9$ Å, $b_o = 18.0$ Å, $c_o = \sim 5.2$ Å

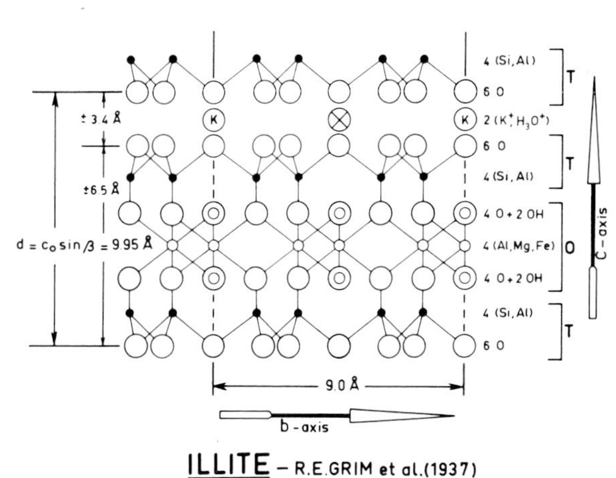

ILLITE – R.E. GRIM et al. (1937)

$$2 \left[(K, K^+, H_3O^+) \mid (Al,Mg,Fe)_2 (OH)_2 \mid (Si,Al)_4 (O,OH)_{10} \right]$$

MONOCLINE $a_o = \sim 5.2$ Å, $b_o = \sim 9.0$ Å, $c_o = \sim 10.0$ Å, $\beta = \sim 95°30'$

Description	•	○				◯	◎	Ⓚ	⊗	⊕	(H_2O)
Kind of ion	Si^{4+}	Mg^{2+}	Al^{3+}	Fe^{2+}	Fe^{3+}	O^{2-}	OH^-	K^+	H_3O^+	OH_2 crystalwater	H_2O freewater
Diameter in Å (Coordination number = 6)	0.78	1.56	1.14	1.66	1.34	2.80	2.80	2.66	2.80	2.80	

Iron minerals

Iron is a most essential element in modern civilization. Iron minerals of various composition are found widespread. Under certain conditions deposits are formed of industrial significance.

Goethite (needle iron ore) and limonite (hydrogoethite, soil goethite); formula: α FeO(OH) and α FeO(OH) · nH_2O.

Goethite is the well crystallized form, whereas limonite contains adsorbed H_2O molecules on its surface and in its pores. It may, moreover, have up to 30 mol % AlOOH substituted in the limonite crystal.

Occurrence. Both minerals are of sedimentary origin. They resulted from weathering of Fe minerals. Organic matter is an important factor in the displacement of iron because of the production of easily soluble ferro-compounds. Furthermore by chelation, by which iron is protected against oxidation and precipitation because of complexing with ring-structural organic compounds.
Bacterial action is another factor. *Bacillacea* and *Pseudomonadacea* reduce ferric compounds. *Spirophyllum ferrugineum*, *Gallionella ferrugineum*, *Leptrothrix ochracea* and *Cladothrix dichotoma* oxidize the ferrous compounds.
Ferrohydroxide (white, green) may be oxidized to α FeOOH or γ Fe OOH depending on the oxygen concentration. A lowering of the redox potential (Eh) increases the activity of Fe^{2+} at the expense of Fe^{3+} and the reverse.

Infrared spectra (Fig. 6.65, 6.66). The spectra of goethite and limonite are quite similar. Those of well-crystallized goethite such as from Pribram, Siegen, Bendorf are poorly developed. Some limonite samples give strong bands which are broad — Conakry, Georgia, Michigan.
Characteristic bands are at 3140—3100, 900—880, 673, 461 and 428 cm^{-1}.
By the D.T.A. method both minerals can be distinguished by their endothermic reactions which are for goethite higher than for limonite = 320 to 425 and 260 to 340°C, respectively.
Most samples contain small amounts of quartz which only can be found by the X-ray method, due to the overlapping of the quartz vibrations by that of the goethite and limonite vibration at 798 cm^{-1}.

Remarks. Goethite and limonite are important basic materials in the manufacturing of numerous kinds of iron products. Deposits with 40% Fe_2O_3 and more, depending on local conditions, are of economic importance.
Goethite is named after Goethe, the German poet and philosopher. Limonite is derived from leimoun = meadow.

Lepidocrocite (Rubin glimmer); formula: γ Fe OOH.

Occurrence. This mineral occurs less frequently than goethite and limonite. Like the foregoing, lepidocrocite is formed by weathering of Fe minerals and is found in sediments. Lepidocrocite is more often found connected with marshes and then associated with limonite. Samples of this type contain usually small amounts of quartz which can only be found after the X-ray method.

Infrared spectra (Fig. 6.67). The spectrum of lepidocrocite from iron rust is similar to that of common lepidocrocite. Both have only weak bands and the spectrum is also nearly like that of goethite and limonite. The most useful bands for identification are those at 1018 and 743 cm^{-1} for lepidocrocite, which are lacking in the goethite (limonite) mineral. By D.T.A. lepidocrocite gives a characteristic endothermal reaction at 300 to 380°C followed by an exothermal reaction at 370 to 450°C, which is absent for goethite (limonite). It is very difficult to detect limonite impurities in lepidocrocite by X-ray analysis. The reverse is easier (lepidocrocite = 6.26 Å).

Remarks. Iron ore from marsh sediments consisting of lepidocrocite and limonite is locally used in gas plants to purify the crude gas. Fe may be substituted by Al in the crystal as in limonite.

Maghemite; formula: γ Fe_2O_3

Occurrence. This mineral is of very restricted occurrence. It may be formed by heating low-temperature (artificial) magnetite or lepidocrocite at ca. 300°C, which by further heating yields hematite at ca. 600°C.
Another way is slowly heating lepidocrocite or goethite (limonite) at ca. 400°C mixed with organic matter (amylum, citric acid, oxalic acid). Maghemite is also found sporadically in marsh soils associated with lepidocrocite, limonite and goethite (limonite) at ca. 400°C mixed with or- the named minerals. The process may be accelerated by fires from reclamations or from ancient habitation.

Infrared spectra (Fig. 6.68). Also the spectrum of this mineral is poorly developed. It resembles that of the foregoing mineral. Maghemite can be distinguished by the 1036, 735, 696, 636, 586, 558 and 444 cm^{-1} bands, which fail in the spectrum of goethite (limonite) and lepidocrocite.
The best way is to use D.T.A. — see the very strong exothermic peak at 750°C.

Remarks. The γ Fe_2O_3 from sediments contains Mn, which increases its resistance to heating. Also Na has the same effect. In this way γ Fe_2O_3 can also be made from Fe-citrate + some NaCl heated even at ca. 600°C.

Hematite; formula: α Fe_2O_3

Occurrence. Hematite is widespread. Most hematite is of alluvial origin, but it may also be found in basic magmas. Hematite formed by rapid dehydration of amorphous ferrihydroxyde in the air at normal temperature, is named ferrihydrite: Chukhrov et al. (1972).
Dehydration of goethite, limonite, lepidocrocite, oxidation of magnetite and heating γ Fe_2O_3 and low-temperature magnetite at ca. 600°C may also yield hematite.

Infrared spectra (Fig. 6.69, 6.70). The spectra show that hematite has several bands in common with those of the foregoing minerals. Moreover, with the exception of the 463 and 425 cm^{-1} band, all the other do not always appear in the spectrum. In this way the identification of hematite is further hindered. But also the X-ray method gives difficulties for mixtures.
The spectra of the heated $Fe(OH_3)$ samples which were obtained by precipitation of $FeCl_3$ by NH_4OH in the heat, filtration and washing out the excess of salt, show a decrease in intensity and broadness when temperature increases. Also the commercial hematite samples show bands of variable broadness and intensity. The strong band at ca. 600 cm^{-1} is probably inherent only to artificial hematites.

Magnetite; formula: Fe_3O_4

Occurrence. Magnetite is mostly found in Fe-rich basic volcanic rocks. But it can also be artificially made. The product, however, is less stable than the natural one against heat. Sea beaches may often contain locally high concentrations of magnetite.

Infrared spectrum (Fig. 6.71). The bands of the natural and the artificial mineral are of low intensity, and they do not always occur. Consequently, the identification of the magnetite mineral together with the other iron minerals is a great problem. But also X-ray analysis gives for mixtures many difficulties in the identification.

Remarks. The name is given to the mineral by Magnes who first observed its magnetic properties. Magnetite may contain up to 25 mole percent of Fe substituted by Ti = titanomagnetite.

Reference

Chukrov, F.C., Zvyagin, B.B., Ermilova, L.P. and Gorshkov, A.I., 1972. New data on iron oxides in the weathering zone. Int. Clay Conf., Madrid, 1: 397–404.

Aluminum minerals

Aluminum hydroxide and oxyhydroxide minerals are not so widespread as those of the corresponding iron components. Bauxite which is named after the deposit at Les Beaux, Rhône (France) is an important raw material for the Al industries.
Bauxite is formed by desilification of laterite. The latter is a common rock type in tropical and subtropical climates with a high rainfall and temperature. Consequently, with a high leaching (weathering) action on the bases of the mother rock leaving the oxides and oxihydroxides of iron and aluminum + kaolinite.
This kind of rock, which is red-coloured (later = red) by some hematite, was first described by Buchanan in India (1807). Bauxite may also be formed by desilification of kaolinite rock, e.g. in the Guyanas or as a precipitate of Al solutions in carbonaceous rocks concentrated in fillings and pockets: Yugoslavia, Hungary, Greece, France.

Gibbsite (hydrargillite); formula: γ $Al(OH)_3$

Occurrence. The above monocline mineral is found in the clay < 2 μm separate of various sediments, especially in areas with a hot rainy climate and a dry period (monsoon). It is also found concentrated in various bauxite deposits differing in purity and extension.

Infrared spectra (Fig. 6.72, 6.73). Bands of high intensity are at 3620, 3521, 3455, 3380, 1020, 668, 560, 450 cm^{-1}. Also the artificial product obtained from $AlCl_3$ and NH_4OH solutions in the heat, gives the same spectrum.

Remarks. Hydrargillite is an important raw material for many aluminum industries.

Nordstrandite; formula: $Al(OH)_3$

Occurrence. This tricline very rare mineral, the third polymorph in the $Al(OH)_3$ series, is found in alkaline environments, such as in Serawak, Guam and Israel, where it occurs in cavities of a weathered limestone just near the contact with the underlying basic volcanic rock.

Infrared spectrum (Fig. 6.72). There is a large difference between the above mineral and the foregoing e.g. (hydrargyllite between brackets) at 3660 (absent), 3558 (absent), absent (3521), 3490 (absent), absent (3455), absent (3380), 3360 (absent), 1060 (absent), 1030 (1020), 823 (absent), 770 (absent), 461 (absent).

Remarks. The mineral is described by Van Nordstrand (1956) and named by Papée et al. (1958).

Bayerite; formula: α $Al(OH)_3$

This monocline mineral is in a meta-stable stage. It is an intermediate product between amorphous $Al(OH)_3nH_2O$ (cliachite, alumogel) and monocline hydrargillite. The mineral is not of constant, well-defined structure.

Occurrence. The mineral is easily prepared by acidification of Na aluminate solutions with carbon dioxide at a pH of ca. 6 or by aging of $Al(OH)_3$ gels. The natural product is very rare. It is found sometimes in veinlets in calcite rocks (Israel).

Infrared spectra (Fig. 6.73). The spectrum of bayerite is practically alike that of hydrargillite. Only the former has a 3660, 1060, 764 and 460 cm^{-1} band, which are absent in the latter, but are also found for nordstrandite (3660, 1060, 770, 461 cm^{-1}).

Remarks. The name bayerite is after the soda Bayer process for the manufacture of Al-hydroxide.

Boehmite; formula: γ $AlO(OH)$

Occurrence. The mineral is found in the clay < 2 μm separate of lateritic or terra rossa soils. Bauxite deposits with mainly boehmite are situated mostly in subtropical regions. They are smaller than those of gibbsite and the boehmite found is usually associated with gibbsite and diaspore.

Infrared spectra (Fig. 6.74, 6.75). Characteristic bands are for boehmite at 3282 (very strong), 3090 (very strong) and 2095 (weak), 1980 (weak). It further has a strong band at 1080 cm^{-1} like diaspore and corundum (hydrargillite and bayerite = 1020 cm^{-1}) and at 747 and 636 cm^{-1}, which are useful for its indentification.
The artificial products have, with the exception of the 3566, 493, 446 cm^{-1} bands, the same bands as the natural mineral.
The spectrum of the bauxite samples from various origin shows that boehmite occurs widespread and it even may be the main mineral together with kaolinite-Dubrovnik (Yugoslavia). But there are others which consist mainly of hydrargillite (Chatanooga, Arkansas and Guyanas). The sample from Nézsa (Hungaria) mainly consists of diaspore and some kaolinite.

Remarks. The mineral is named after Böhm (1925) who first identified its real nature by X-ray analysis of the bauxite deposite at Les Beaux (France). The artificial product is called Cera-hydrate (trade mark Brit. Aluminium Co.).
Large amounts of bauxite are exploited for the aluminum industry. Al_2O_3 contents of the commercial bauxites is about 55—65%; however, when their kaolinite content is larger than ca. 4% they are less valuable in the soda Bayer process to win aluminium hydroxide.

Diaspore; formula: α AlO(OH)

Occurrence. Large amounts of this mineral are found in some bauxite deposits; e.g., Nézsa (Hungary), Rosebud, Mo. (U.S.A.). Thermal action favours its genesis. Diaspore is furthermore found as a minor component in various other bauxite deposits in the world which mainly contain gibbsite or boehmite.

Infrared spectra (Fig. 6.76). Characteristic bands of high intensity are at 2987, 2910, 2115, 2000, 1080, 960, 753, 580*, 567*, 519 cm^{-1}.
Most diaspore samples are defiled with kaolinite mica and chlorite (Rosebud). That of Nevada contains a large amount of pyrophyllite. The infrared spectrum of the Mn diaspore from Cape Provence with ca. 3% Mn is greatly similar to that of common diaspore.

Remarks. Diaspore is an important raw material for many aluminium industries. Bacause of its high Al_2O_3 content = ca. 80% it is also used in the manufacture of refractory bricks.

Corundum; formula: α Al_2O_3

Occurrence. It is found as an accessory mineral in limestone, dolomite, gneiss, slate. Artificial corundum can be obtained by dehydration of Al hydroxide or Al oxihydroxide minerals at 1000—1200°C with the exception of diaspore, for which only 420°C is needed.
Several kinds of polymorphs may be found as intermediate products depending on the heat, the duration applied and the kind of mineral used. They are for:
gibbsite $\rightarrow \chi$ $Al_2O_3 \rightarrow \kappa$ $Al_2O_3 \rightarrow \alpha$ Al_2O_3
boehmite $\rightarrow \gamma$ $Al_2O_3 \rightarrow \delta$ $Al_2O_3 \rightarrow \theta$ $Al_2O_3 \rightarrow \alpha$ Al_2O_3
bayerite $\rightarrow \eta$ $Al_2O_3 \rightarrow \theta$ $Al_2O_3 \rightarrow \alpha$ Al_2O_3
diaspore $\rightarrow \alpha$ Al_2O_3

Infrared spectra (Fig. 6.77, 6.78). Characteristic bands are at 1082*, 1065*, 800, 642, 605, 454 cm^{-1}. The spectra are badly developed which hinders the identification of korund. The spectra of the artificial products are better differentiated. In particular those in the lower regions at 490 and 450 cm^{-1}. They are thereby broad which points to a larger disorder in the arrangements of the atoms.

Remarks. Several corundum species, the colour of which is caused by traces of Fe, Cr, Ti, have found their way in jewelry.
Ruby-red or pigeon's blood, (Cr and Fe), sapphire-blue (divalent iron and Ti), topaz-yellow (trivalent Fe and Ni), emerald-green (di- and trivalent Fe and Ti). They are found in gravels or river beds in Ceylon, Burma, N. Carolina, Georgia, Montana.
Alundum is an artificial abrasive which is made by heating bauxite to 2500—3000°C in electric furnaces. Owing to its great specific surface artificially Al_2O_3 (alumina), is widely used as an isolator, adsorbent and as a catalyst. The catalytic activity is related to acid sites on the surfaces (Lewis and Brönsted-acid type).

References

Böhm, J., 1925. Über Aluminium- und Eisenhydroxyde. Z. Anorg. Allgem. Chem., 149: 203—216.

Buchanan, F., 1807. A Journey from Madras through the countries of Mysore, Kanara and Malabar. London (3 volumes).

Papée, D., Tertian, R. and Biais, R., 1958. Gels and crystalline hydrates of alumina. Bull. Soc. Chim., Fr., 1301—1310.

Van Nordstrand, R.A., Hettinger, W.P. and Keith, C.D., 1956. A new aluminium trihydrate. Nature, 177: 713—714.

Silica minerals

Quartz, cristobalite, tridymite

The structure of the above silicium oxide minerals, including their low (α) and high temperature (β) modifications, is based on a three-dimensional linking of SiO_4 tetrahedra; each silicon being tetrahedrally surrounded by four oxygens and each oxygen shared by two silicons.
In quartz the tetrahedra are closely packed in spirals. The twisting direction determines whether the crystal is left- or right-handed. Quartz (sp.w. = 2.65) has voids of limited size so that only Li, Be, Bo can enter the crystal. In tridymite (sp.w. = 2.27) and cristobalite (sp.w. = 2.33) the tetrahedra have a more open structure. Thus larger atoms (Na, K, Ca) can also be trapped. Cristobalite, which has the widest structure may even contain Rb and Cs.
At the transformation of the α to the β forms, merely the bond angle and bond length are changed, e.g. for cristobalite the bond angle is changed from 150° to 180° and the bond length from 3.067 to 3.080 Å. There is no breaking of bonds and a re-grouping of the silica tetrahedrons. The high temperature modifications are always of higher symmetry than the low ones.

Quartz; formula: SiO_2

Coloured quartz crystals contain minute amounts of foreign atoms or minerals e.g. amethyst = purple or violet (Fe,Mn), rose quartz = pink to rose (Mn, hematite, TiO_2), smoky quartz = grey to darkbrown (Fe), citrine = yellow brown or red brown (colloidal ferric hydroxide), aventurine = glistening (mica, hematite), rutilized quartz (needles of rutile), Jasper = red or brown (ironoxide), tiger's eye = yellow-brownish with pronounced opalescent luster (inclusions of asbest fibres). Cat's eye = greyish with opalescence (inclusions of asbest fibres). Milky quartz has micro discontinuities, by which the rays are scattered. Quartz when heated produces a sharp endothermal reaction (ca. 4.1 cal/g). However, the inversion temperature is not constant. For quartz from igneous rocks it is 573° to 576°C; for authigenic quartz from sediments the inversion temperature may be as low as 563°.
The reaction is not wholly reproducible but shows a hysteresis effect which depends on the way the quartz mineral is formed.
The inversion reaction is caused by a change of the original trigonal structure into a hexagonal modification. The mineral is called chalcedonite or secondary quartz when the above inversion is absent because of stress and strain.

Occurrence. Quartz is the most abundant mineral in the earth's crust. It is especially found in acid magmatic rocks and in metamorphic rocks together with other minerals. Sometimes the weight of well-formed individuals reaches to 10 tons (Brasil).
Because of the strong resistance to chemical weathering, quartz is a common mineral in the clay separate of many sediments.
Quartz can be obtained artificially from amorphous silicic acid by heating at 1000°C with some CaO. The latter acts as an accelerator (mineralizer) to dislocate the crystal structure and thus lowers the transition temperature and the duration of heating. Single quartz crystals of ca. 500 g of high purity and optical characteristics can be made within some months by heating silicic acid with alkaline solutions in an autoclave at ca. 300°C and 500 atmospheres.
In the vicinity of geysers the amorphous precipitated $SiO_2 \cdot nH_2O$ (siliceous sinter) is in the long run transformed into quartz. The reaction may even proceed at 300°C in alkaline solutions. Low temperature quartz (ca. 20°C) may be formed by aging of a highly siliceous matter formed by concentration of dissolved silicic acid by sesquihydroxides or mangan hydroxide.

Infrared spectra (Fig. 6.79, 6.80). Simon and McMahon (1953) found with incident polarized rays parallel to the optic axis only the 780 cm^{-1} band and when the incident polarized rays were perpendicular to the optic-axis only the 800 cm^{-1} band. Other characteristic bands for quartz are at 1172, 1082, 693, 512 and 460 cm^{-1}.
Jasper, amethyst, tiger eye and thunder egg, which minerals are known in jewelry, are all quartz. The bands of the pulverized quartz sample $< 2\ \mu m$ are largely increased.

Remarks. Quartz crystals of high purity and optical character are used in the manufacture of lenses, prisms and for piezo-electric properties. Many coloured species of quartz have found their way as gemstones. Quartzites and quartz-rich sandstones are used for buildings and pavings. Ground quartz or naturally occurring fine sinter quartz "Milowite" (Isle of Milos), "Tripoli" (Missouri) are used in abrasives, as a flux in metallurgical processes and in refractory. Silicosis is a serious fibrous lung disease of man working in mines and stone quarries. It is caused by inhalation of fine ($< 5\ \mu m$) sharp edged quartz particles. As many years will pass before the infection can be recognized, the disease is very treacherous.

Tridymite; formula: SiO_2

This mineral produces two endothermic reactions α/β_1 = 90 to 120°C (0.43 cal/g) and β_1/β_2 = 135°–180° (0.23 cal/g) when heated. The precise temperature and strength depends on the kind and amount of the accelerators added, the duration and temperature of heating and the kind of starting material used to obtain the sample. The reactions are caused by a change of the original monocline and rhombic structure respectively into a hexagonal modification.

Occurrence. The mineral is found, like cristobalite, in fissures of acid effusive rocks and obsidians together with cristobalite. Some chalcedons and opals contain tridymite in small amounts; incidentally together with their high temperature modifications. Tridymite can be obtained in an artificial way e.g. by heating quartz, silicic acid or amorphous SiO_2 glass with small amounts of Na_2WO_4 or with oxides of Li, Na, K at 900 to 1000°C which act as an accelerator.

Infrared spectra (Fig. 6.81). Characteristic bands are at 1100, 790, 480 cm^{-1}. There is a large difference in intensity and broadness for the various artificial samples. Tridymite can easily be distinguished from cristobalite and quartz by its many bands of low intensity which point to disorder in the structure.

Remarks. Tridymite is derived from the Greek tridymos (= threefold), because of its characteristic occurrence in groups of three.

Cristobalite; formula: SiO_2

This mineral produces an endothermic effect (ca. 4 cal/g) between 200° and 280°C. The precise temperature and strength of the reaction depend on the way (kind and amount of added accelerator, temperature and duration of heating, kind of starting material) the cristobalite is obtained. The thermal reaction is caused by a change of the original tetragonal structure into a cubic modification. It is not completely reversible (hysteresis effect) and depends on the way the cristobalite is formed.

Occurrence. The mineral may be found in fissures of acid effusive rocks and obsidians together with tridymite. It is also found in sediments from volcanic rocks where it resisted weathering action.

Cristobalite can be obtained in an artificial way by heating silicic acid or amorphous SiO_2 glass with some oxide of Mg, Ca, Sr, Ba or Mn at ca. 900°C that is ca. 750° below the temperature when no additions are used. Quartz is an intermediate product. These additives act as an accelerator (flux) in the reaction mechanism.

Rock crystal quartz and tridymite are converted to β cristobalite when heated at 1200°C to 1500 °C.

Infrared spectra (Fig. 6.81). Characteristic bands are at 1200, 1166, 1098, 792, 622, 488 cm^{-1}. Cristobalite can easily be distinguished from quartz by its strong 622 cm^{-1} band. Other distinctions are (quartz between brackets) at: 1166 (1172), 1098 (1082), absent (778), 488 (478) cm^{-1}.

Remarks. Because of the large volume change (ca. 3.5%) at a low inversion temperature followed by an insignificant expansion when further heated, artificial cristobalite is widely used in dentistry and in the precision mould-casting process in foundries.

The name cristobalite is after the type locality San Cristobal, Mexico.

Chert, hornrock, chalcedon, agate, flint, opal, geyserite; formula: $SiO_2 \cdot nH_2O$

The above minerals have in common that they are originally precipitated silicic acids, which afterwards have lost partly (opal) or wholly their adsorbed H_2O molecules.

Cherts (hornstein, hornrock) have a dull hornlike structure. They mostly contain quartz. Chalcedon has a microfibrous waxy luster. It contains a large amount of micropores and microcrystallites of quartz. Other species are agate (banded), jasper (red by Fe inclusions), onyx (black), carnelian (yellow to red), heliotrope also called bloodstone (green with red spots), hyalite (transparent like water), fire opal (redbrown), cacholong (enamel white). Flints (grey, brown) have nodular shapes. Geyserite is a silica sinter with micropores up to 500 Å (Fig. E.M. 226).

For opal the myriads of micro cracks filled with water reflect and refract the rays thus giving a great range of colours (iridescense). They contain high, but possibly disordered low cristobalite.

Occurrence. The precious and semi precious minerals are found in alluvial sands and gravels where they were supplied from elsewhere, e.g. cracks and cavities in acid volcanic rocks where the mineral was deposited from hydrothermal water of high silicic acid contents.

Flints (Feuerstein) are found in chalks; mostly

in layers and banks. Here the silicic acid of the chalk was precipitated and afterwards lost its H_2O. Flints are mostly associated with siliceous sinters called pseudokieselguhr.

Infrared spectra (Fig. 6.82, 6.83, 6.84). The spectra show that chert, Hornrock, chalcedon, agate, flint (Feuerstein), wood opal, petrified wood and honeystone mainly consist of quartz. They may also contain tridymite or even mainly tridymite: Feuerstein (Silesia), Holzopal (Rheinland), opalized wood (Nevada), Opal (Rheinland), opal (Nevada, Australia), hyalite (West Carolina, Bohemia). Another common component in the above samples is silicic acid, that, however, can not be detected by X-ray analysis. Several samples which are amorphous to X-rays give an infrared spectrum although somewhat poor: opal (Australia), hyalite (West Carolina), geyserite (Yellowstone Park and Iceland). Tridymite and silicic acid are in these samples the most common found. The sinter opal from Hot Springs has well-developed bands of kaolinite, which hardly occur in the X-ray spectrum of this sample. In this way the infrared method is better suited to detect peculiarities in the genesis of minerals.

Remarks. Many kinds of the above minerals have found their way in jewelry.

Reference

Simon, J. and McMahon, H.O., 1953. A study of the structure of quartz, cristobalite and vitreous silica by reflection in infrared. J. Chem. Phys. 21: 23—30.

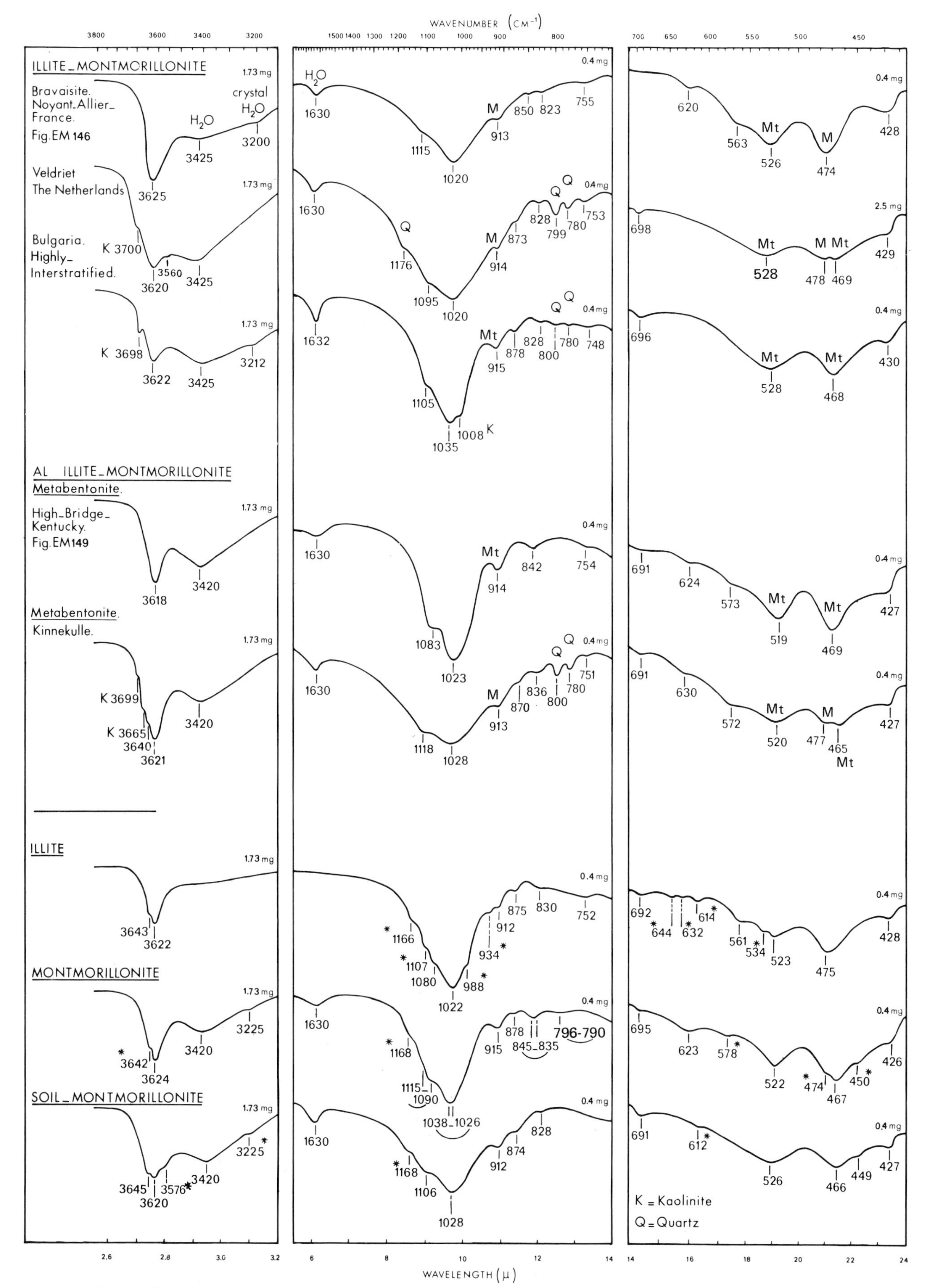

Fig. 6.54. Infrared spectra of non-regular interstratified illite-montmorillonite (swelling illite) and Al-illite-montmorillonite (metabentonite) from different origin. M = mica; Mt = montmorillonite.

Degraded talc

Hrubsice, Czechoslovakia
pure

Vezna
some cristobalite

β-kerolite
Chauzh River, Ural, U.S.S.R.
pure

Stevensite
N. Tyne, Great Britain

Talc

Saponite

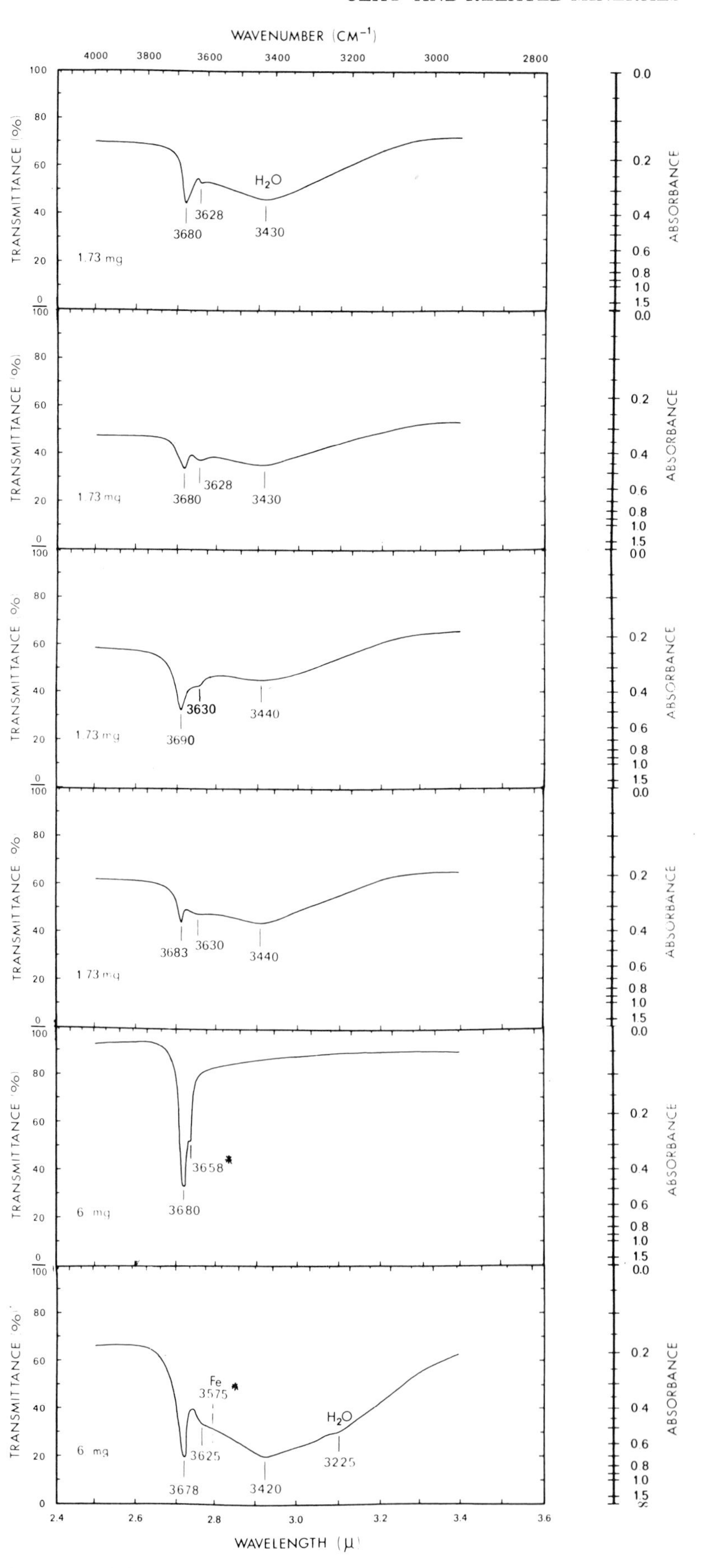

Fig. 6.55. Infrared spectra of degraded talc, β-kerolite and non-regular talc saponite (stevensite) from different origin.

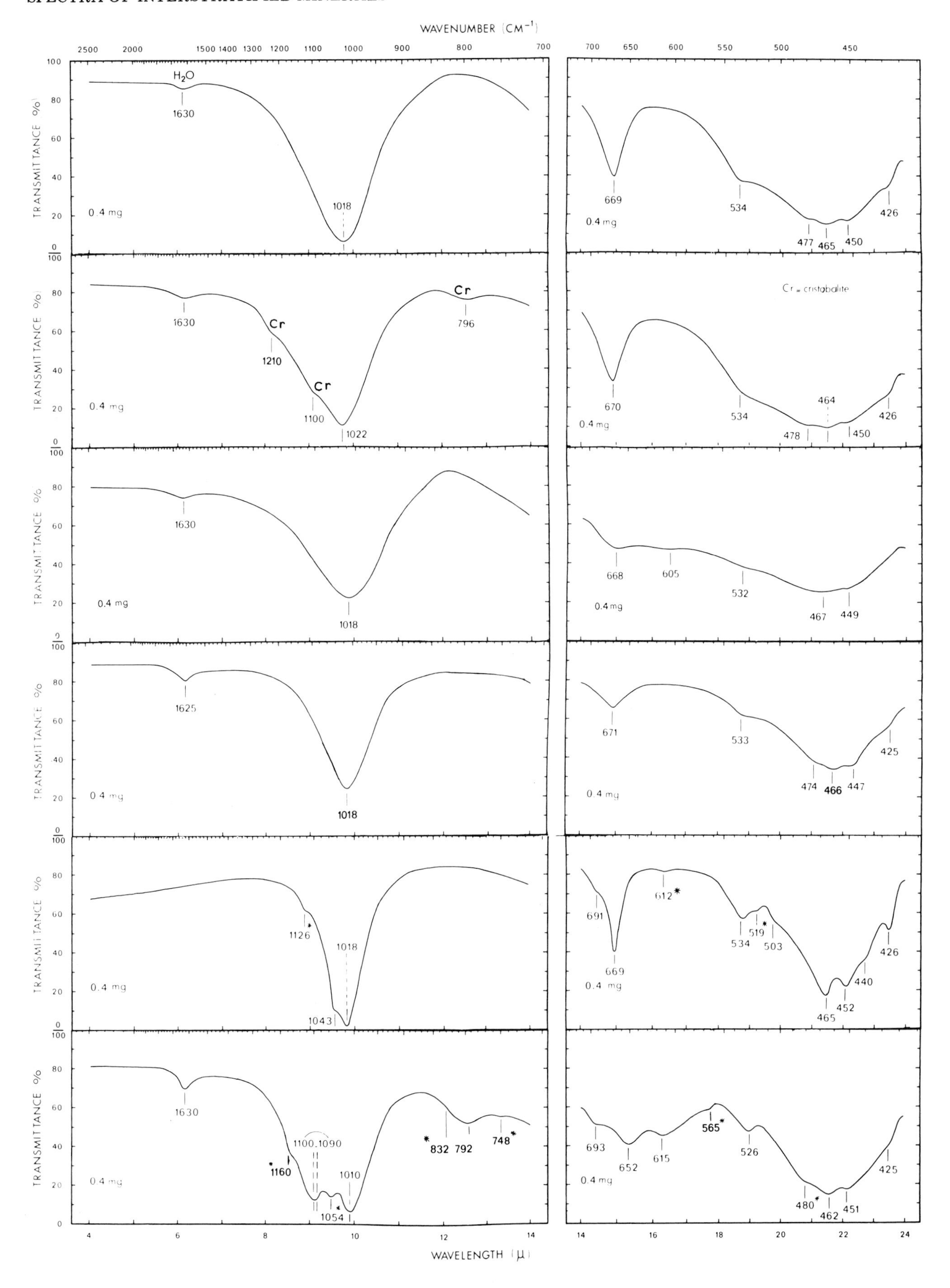
WAVENUMBER (CM⁻¹)
TRANSMITTANCE %
WAVELENGTH (μ)
H2O
1630
1018
0.4 mg
669
534
477
465
450
426
Cr
1210
1100
1022
796
Cr = cristobalite
670
478
464
605
668
532
467
449
1625
671
533
474
466
447
425
1126
1043
691
612
519
503
452
440
1160
1100,1090
1054
1010
832
792
748
693
652
615
565
526
480
462
451

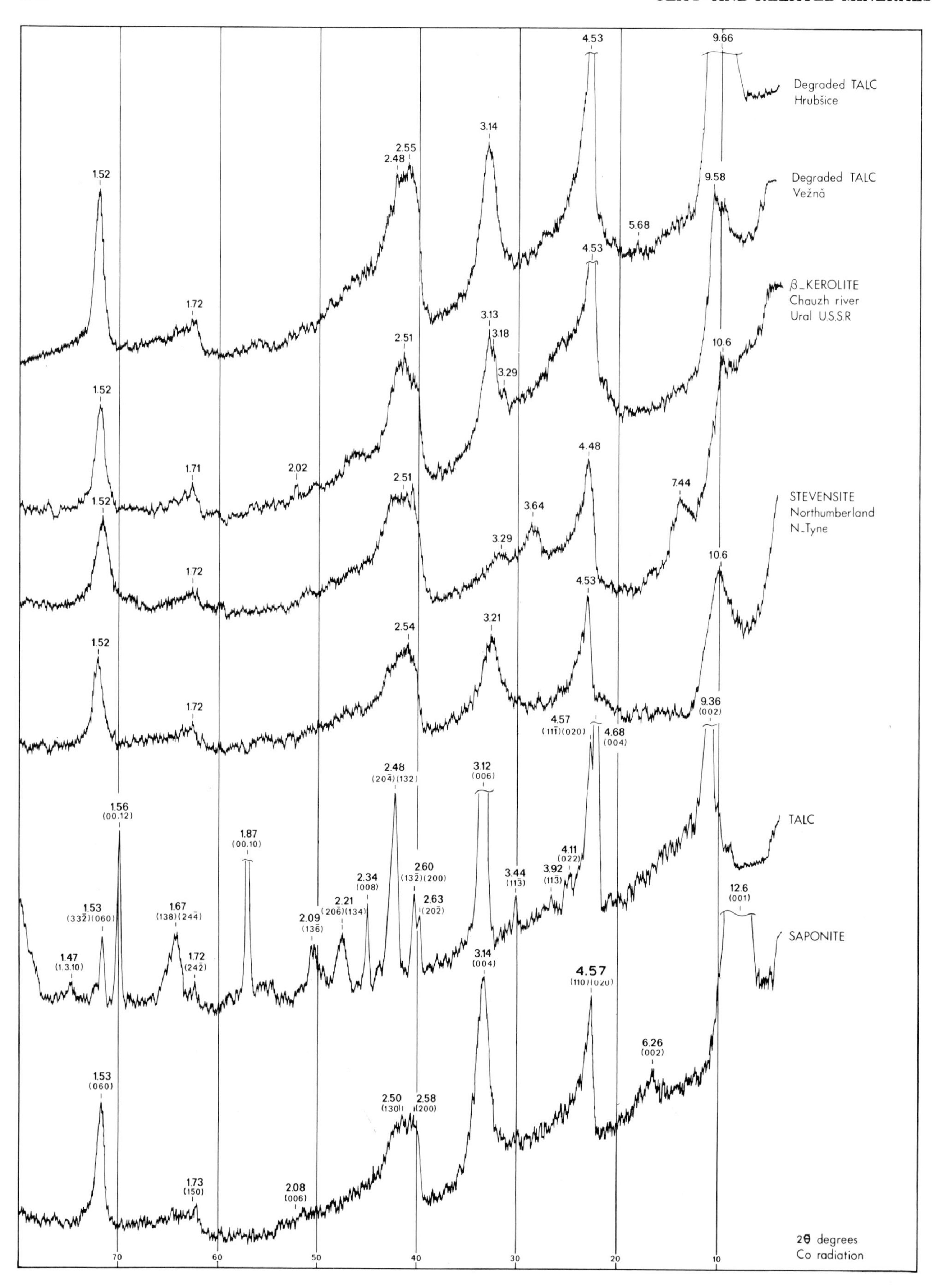

Fig. 6.56. X-ray diffraction spectra of degraded talc, β kerolite, stevensite, talc and saponite; d spacings in Å.

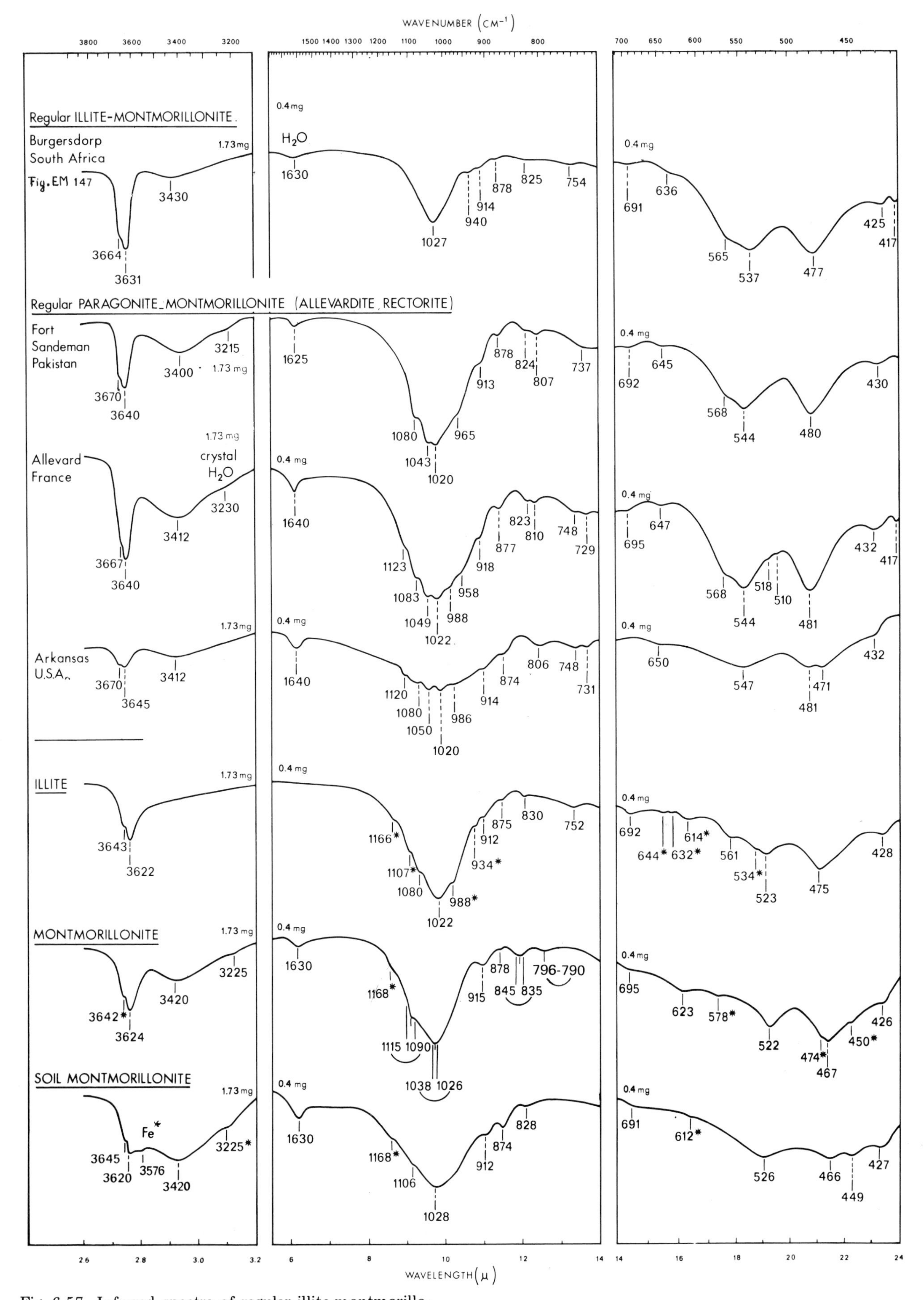

Fig. 6.57. Infrared spectra of regular illite-montmorillonite and regular paragonite-montmorillonite (allevardite, rectorite) from different origin.

Talc-saponite (aliettite) regular
Nure Valley, Italy
pure

Chlorite-swelling chlorite (corrensite) regular
Monte Chiaro, Italy
some calcite and quartz
Fig. E.M. 152

Chlorite-saponite regular
Monte Chiaro, Italy
some quartz
Fig. E.M. 151

Saponite – swelling chlorite (randomly)
St. Margherita, Staffora, Italy
some quartz and calcite

Talc

Clinochlore

Saponite

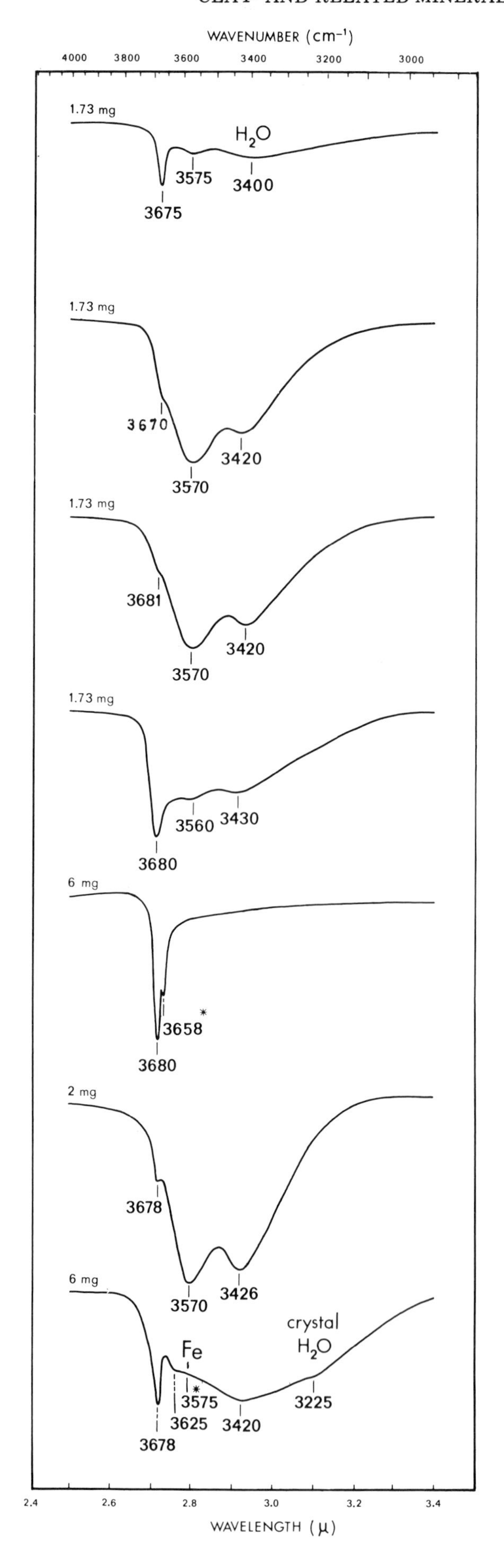

Fig. 6.58. Infrared spectra of regular talc-saponite (aliettite), chlorite-swelling chlorite (corrensite) and of regular chlorite-saponite and non-regular saponite-swelling chlorite.

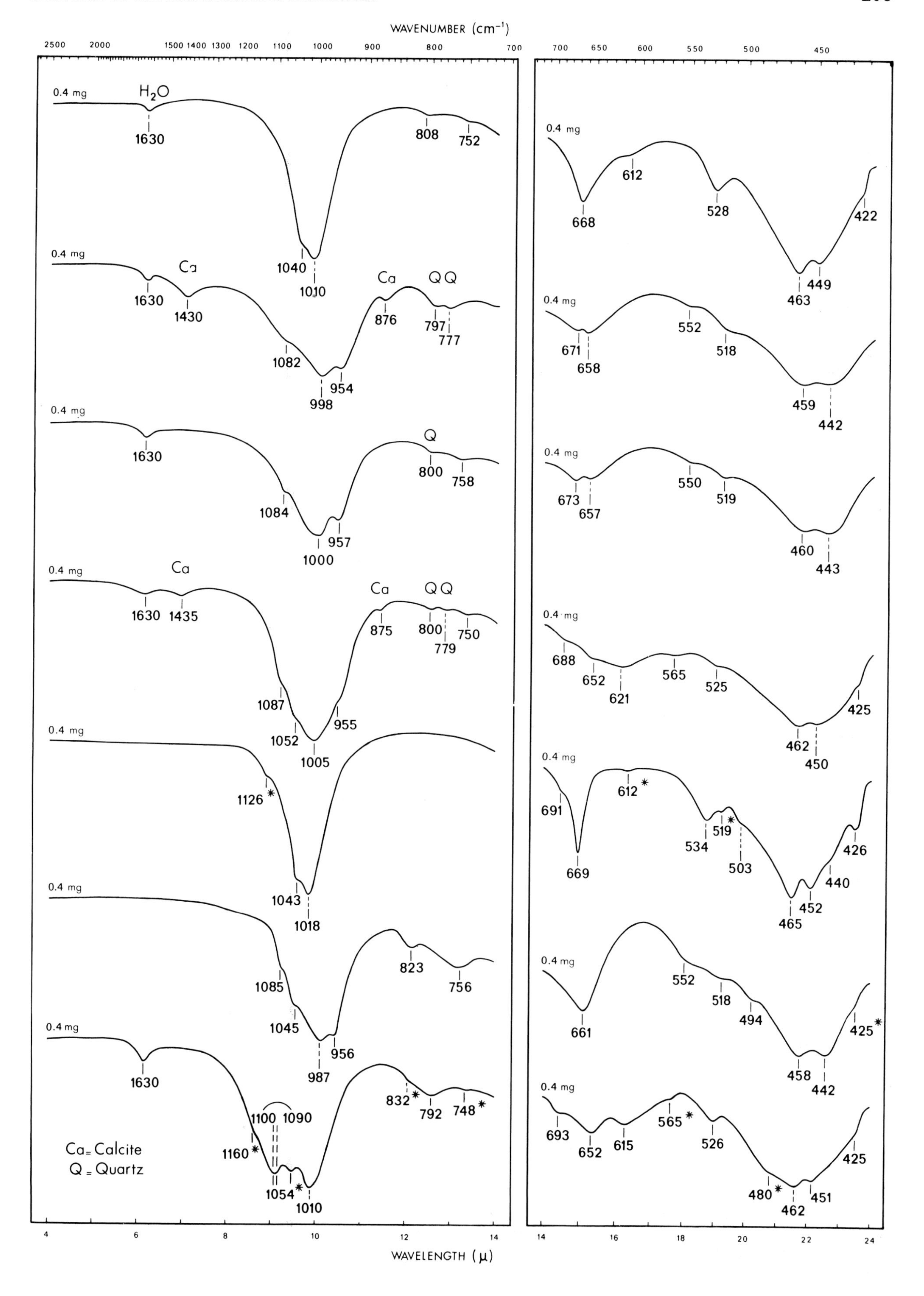
WAVENUMBER (cm⁻¹)
WAVELENGTH (μ)
0.4 mg
H₂O
1630
1040
1010
808
752
Ca
Q Q
1430
1082
998
954
876
797
777
Q
1084
1000
957
800
758
1435
1087
1052
1005
955
875
779
750
1126*
1043
1018
1085
1045
987
956
823
756
1100 1090
1160*
1054*
832*
792
748*
Ca = Calcite
Q = Quartz
668
612
528
463
449
422
671
658
552
518
459
442
673
657
550
519
460
443
688
652
621
565
525
462
450
425
691
669
612*
534
519*
503
465
452
440
426
661
494
458
425*
693
615
565*
526
480*
451

Biotite-vermiculite (hydrobiotite)

Palabora, South Africa
some vermiculite

Libby, Mont., U.S.A.
some vermiculite and biotite (both overlapped)

Al chlorite-montmorillonite (tosudite)
Takatoma mine, Japan

(Al,Mg) chlorite-montmorillonite
Kamikita,mine, Japan

Biotite

Vermiculite

Al chlorite (sudoite)

Montmorillonite

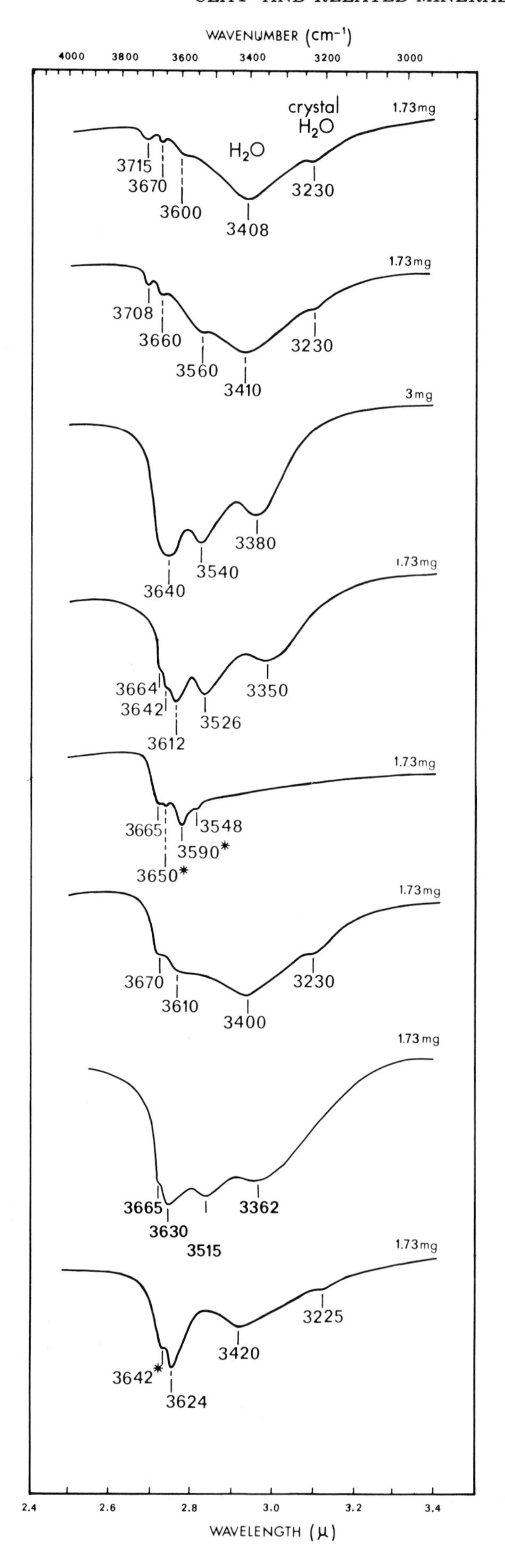

Fig. 6.59. Infrared spectra of regular biotite-vermiculite (hydrobiotite), Al chlorite-montmorillonite (tosudite) and (Al, Mg) chlorite-montmorillonite.

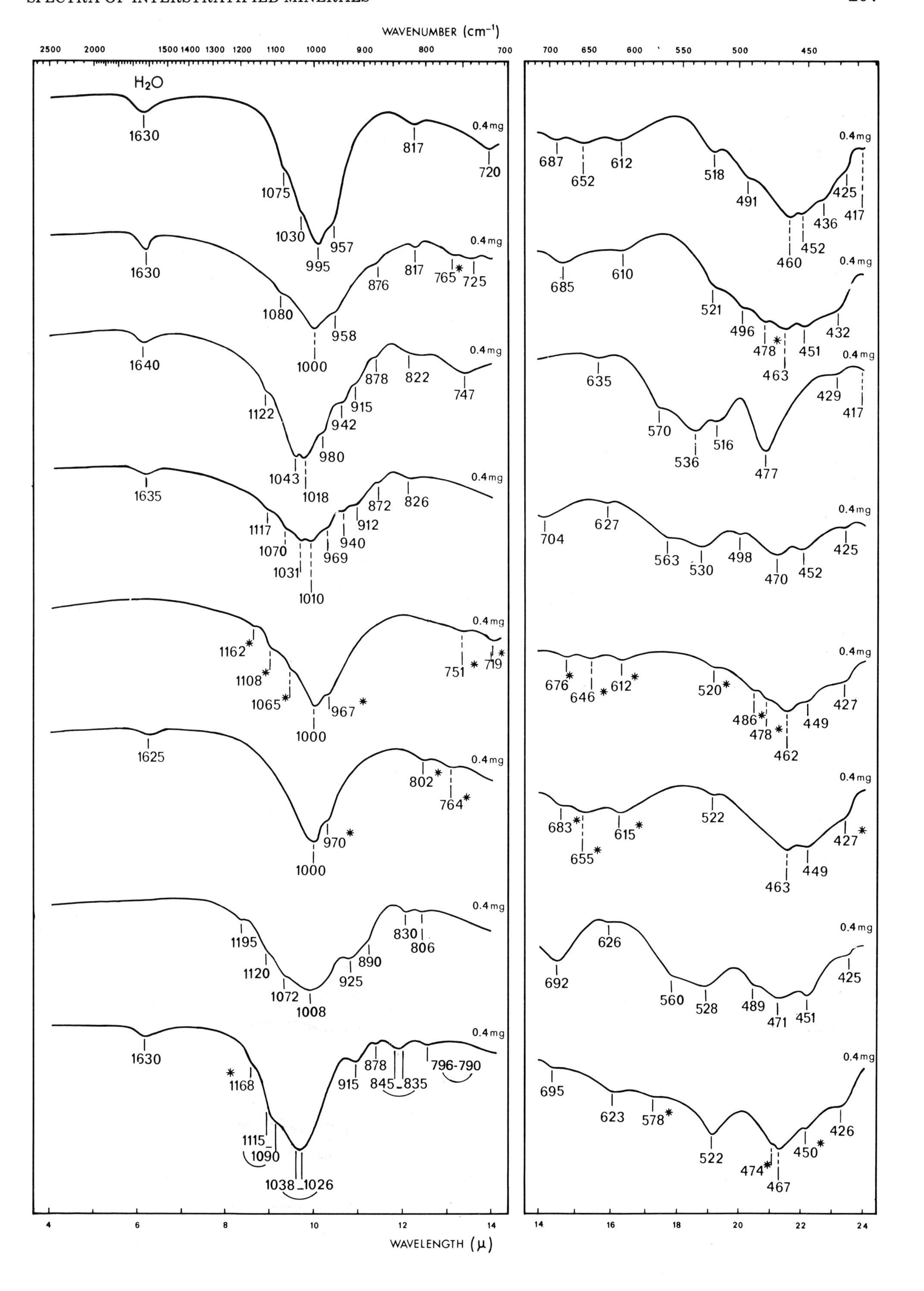
WAVENUMBER (cm⁻¹)
WAVELENGTH (μ)
H2O
0.4mg

Attapulpite (palygorskite)

Hebron, Sinaii, Israel
some quartz and calcite

Anram Massif, Israel
some quartz, calcite and hematite (overlapped)

Gömör, Hungaria
some quartz and calcite

Pinas Altos, New Mex., U.S.A.
some quartz and calcite
Fig. E.M. 156

Attapulgus, Fla., U.S.A.
some quartz and calcite
Fig. E.M. 154, 237

Montmoiron, France
quartz and some calcite
Fig. E.M. 155

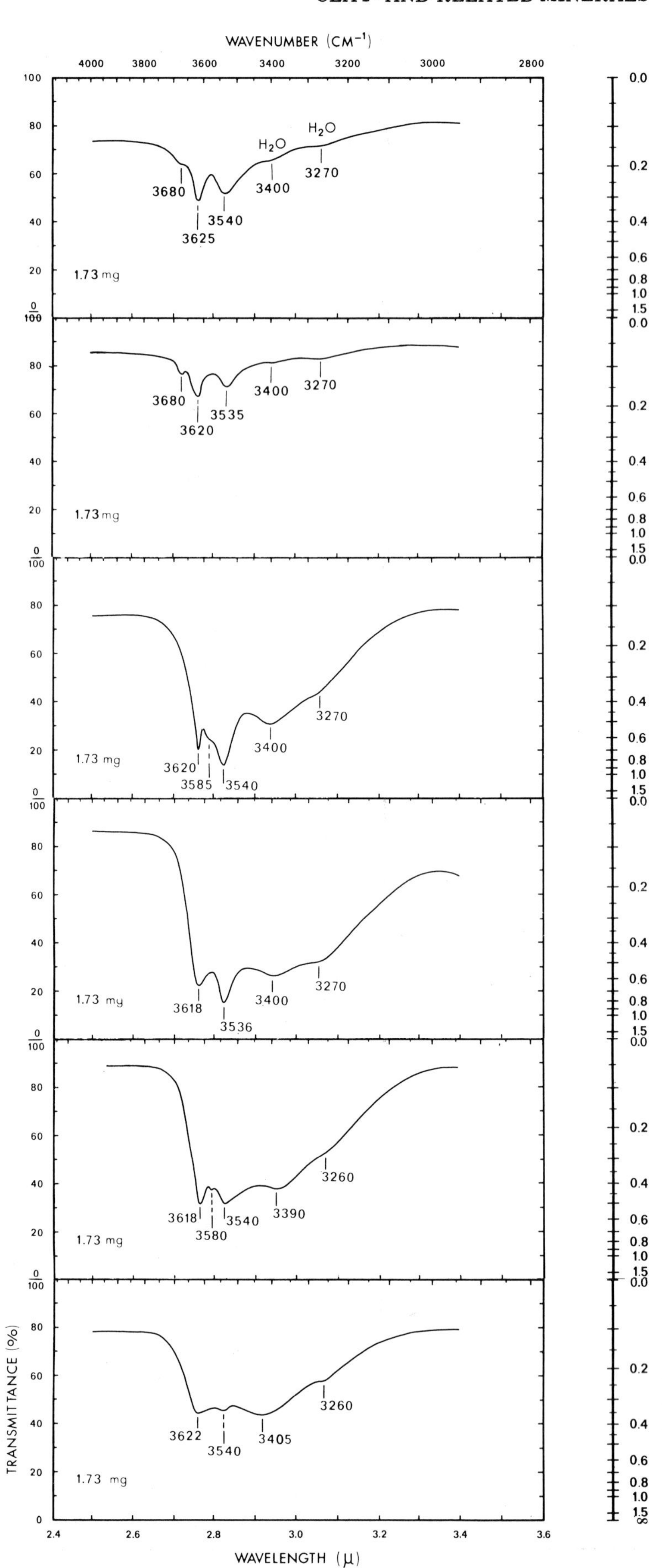

Fig. 6.60. Infrared spectra of attapulgite (palygorskite) from different origin.

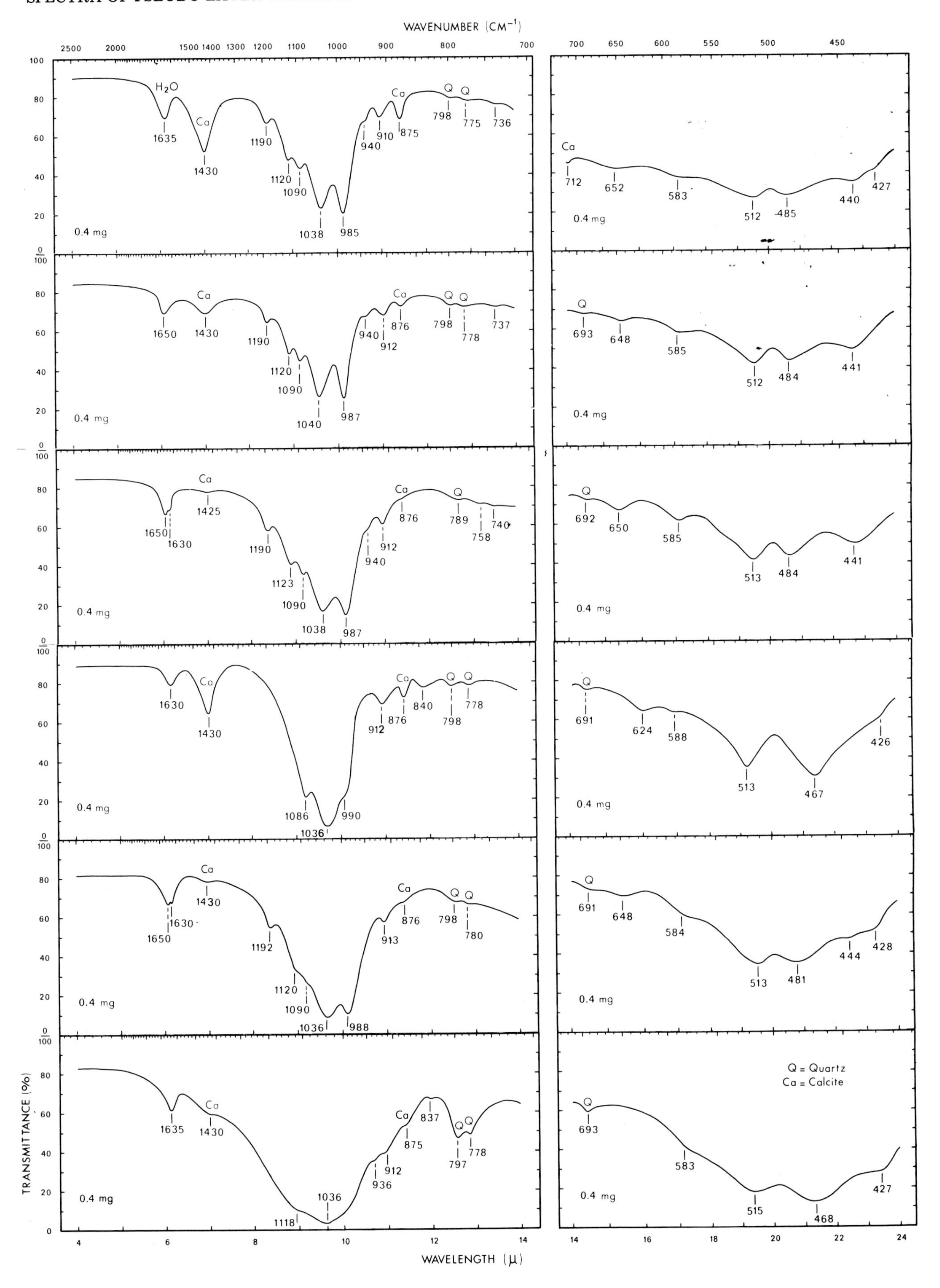
WAVENUMBER (CM⁻¹)
TRANSMITTANCE (%)
WAVELENGTH (μ)
Q = Quartz
Ca = Calcite
0.4 mg

Sepiolite

Ampandandrava, Malagasie
pure
Fig. E.M. 159

Salinelles, France
pure
Fig. E.M. 160

Hawthorne, Nev., U.S.A.
pure
Fig. E.M. 158

Vacia, Spain
pure

Vallecas, Spain
pure
Fig. E.M. 157

Rivola, Italy
pure

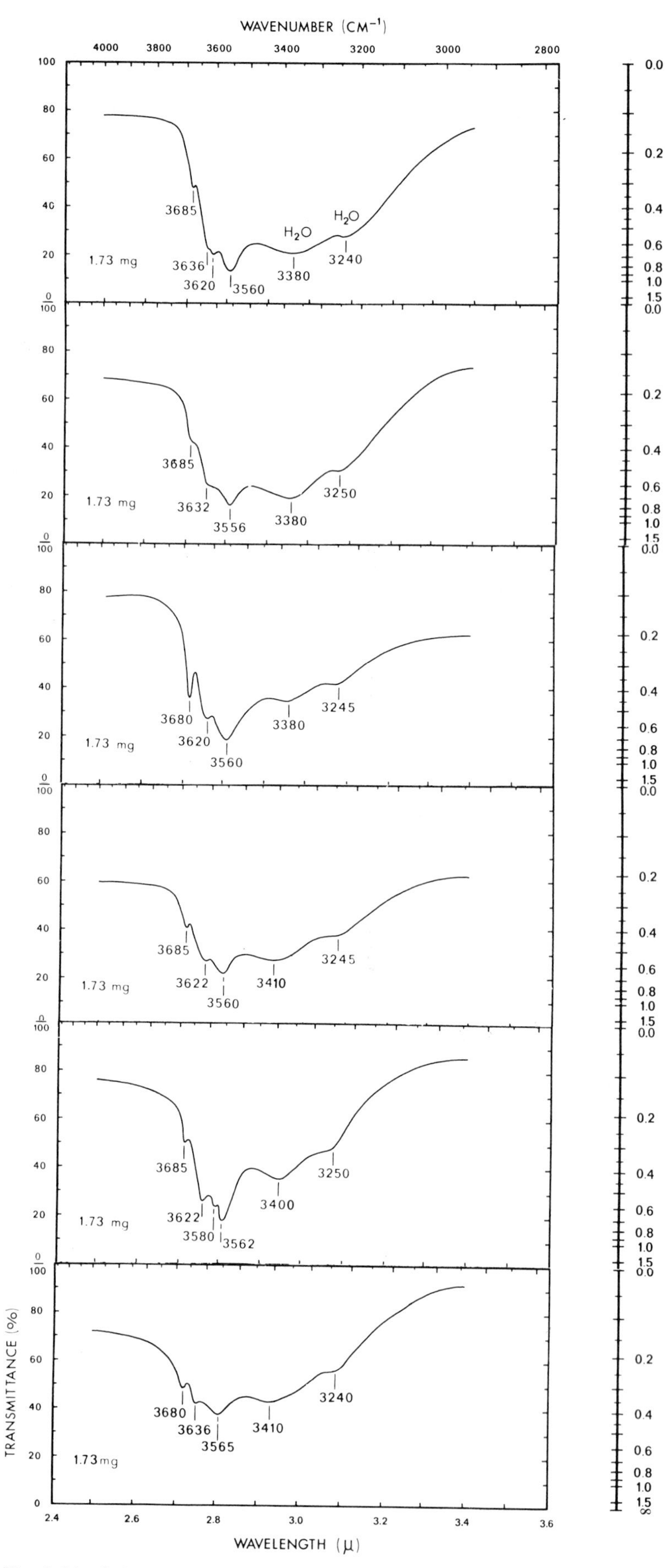

Fig. 6.61. Infrared spectra of sepiolite from different origin.

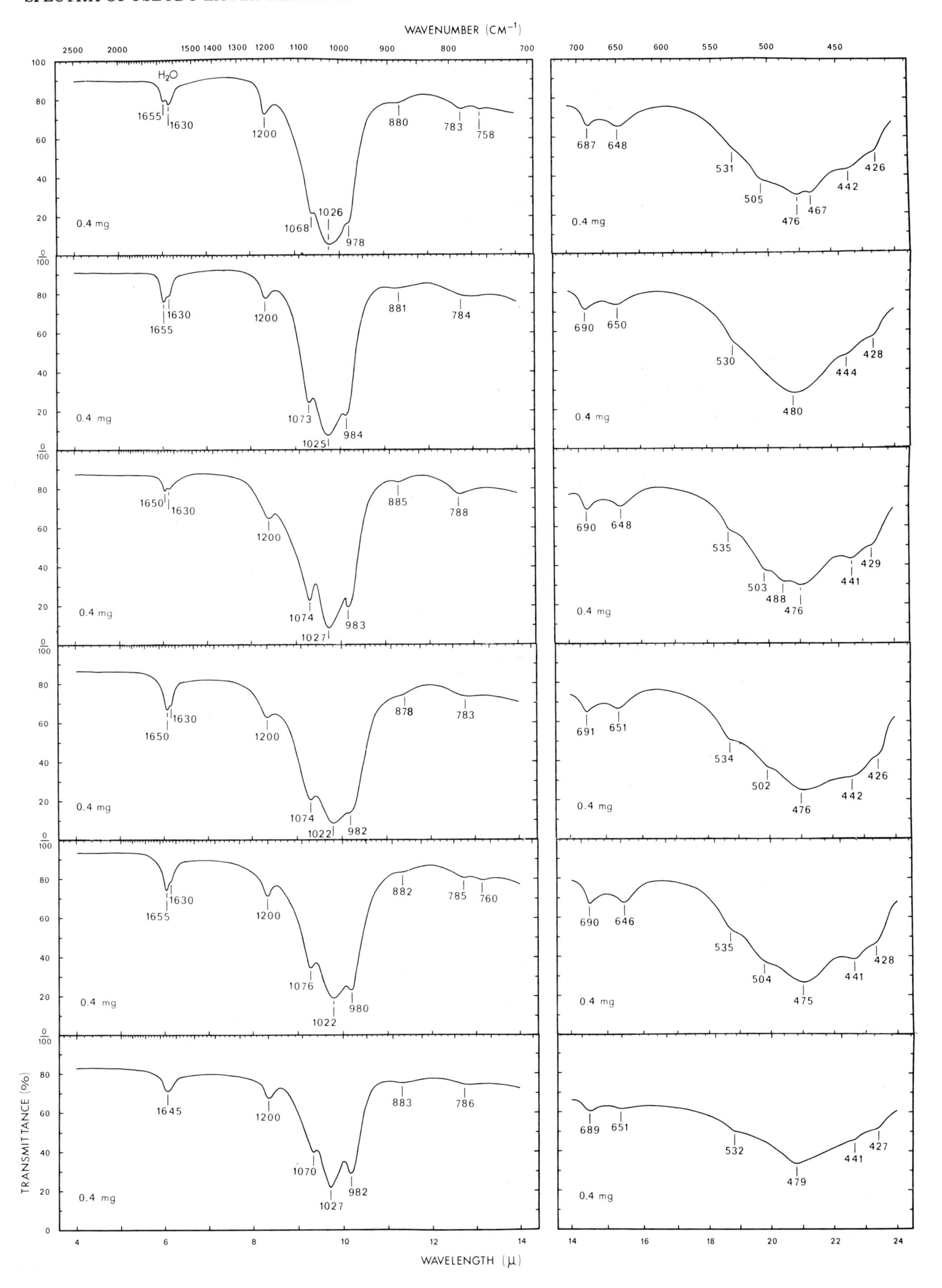

WAVENUMBER (CM⁻¹)
WAVELENGTH (μ)
TRANSMITTANCE (%)
H₂O
0.4 mg
1655 1630 1200 1068 1026 978 880 783 758
687 648 531 505 476 467 442 426
1655 1630 1200 1073 1025 984 881 784
690 650 530 480 444 428
1650 1630 1200 1074 1027 983 885 788
690 648 535 503 488 476 441 429
1650 1630 1200 1074 1022 982 878 783
691 651 534 502 476 442 426
1655 1630 1200 1076 1022 980 882 785 760
690 646 535 504 475 441 428
1645 1200 1070 1027 982 883 786
689 651 532 479 441 427

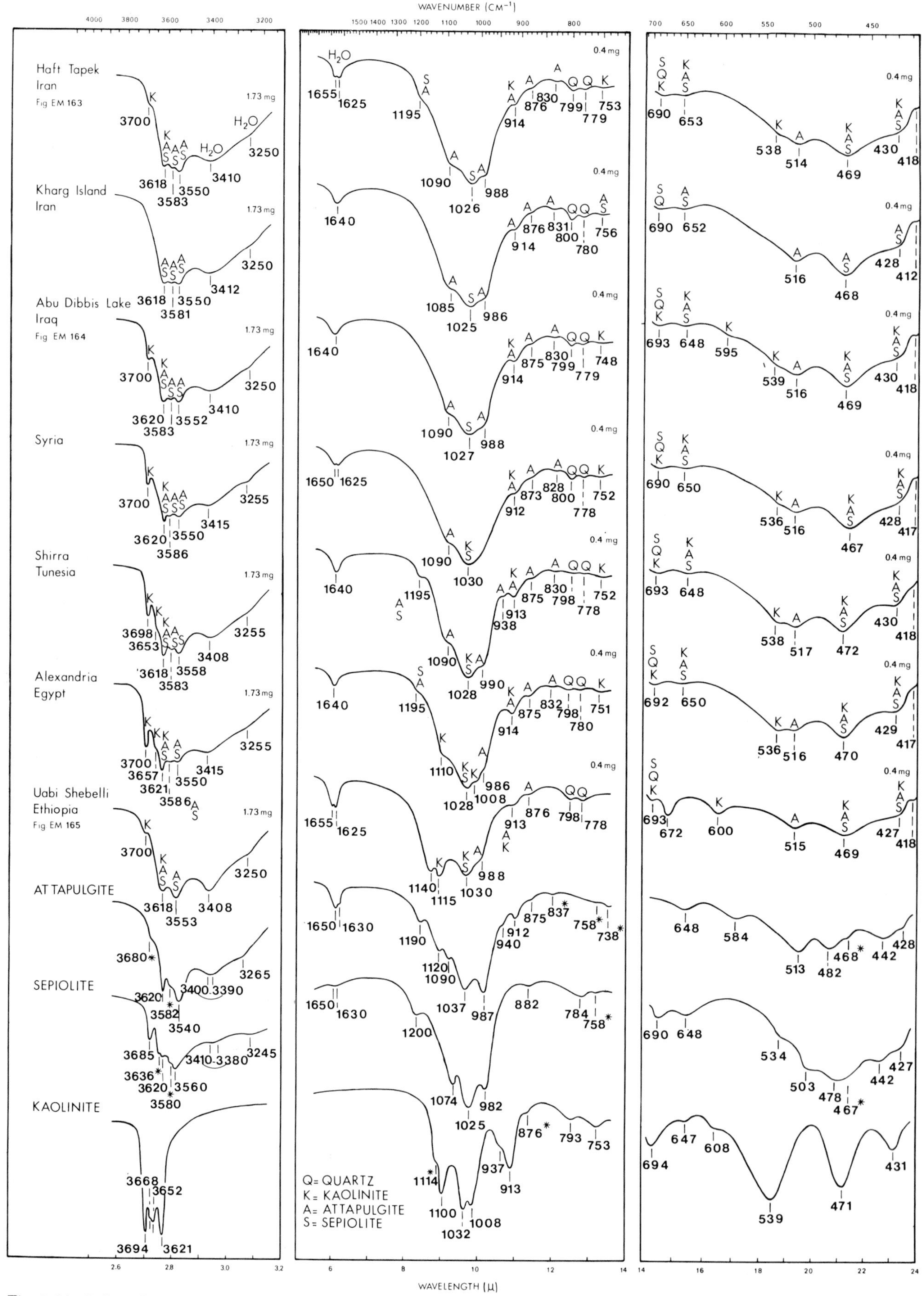

Fig. 6.62. Infrared spectra of sediments < 2 μm from different origin which contain attapulgite and sepiolite. Chlorite and feldspar of Iran sample overlapped.

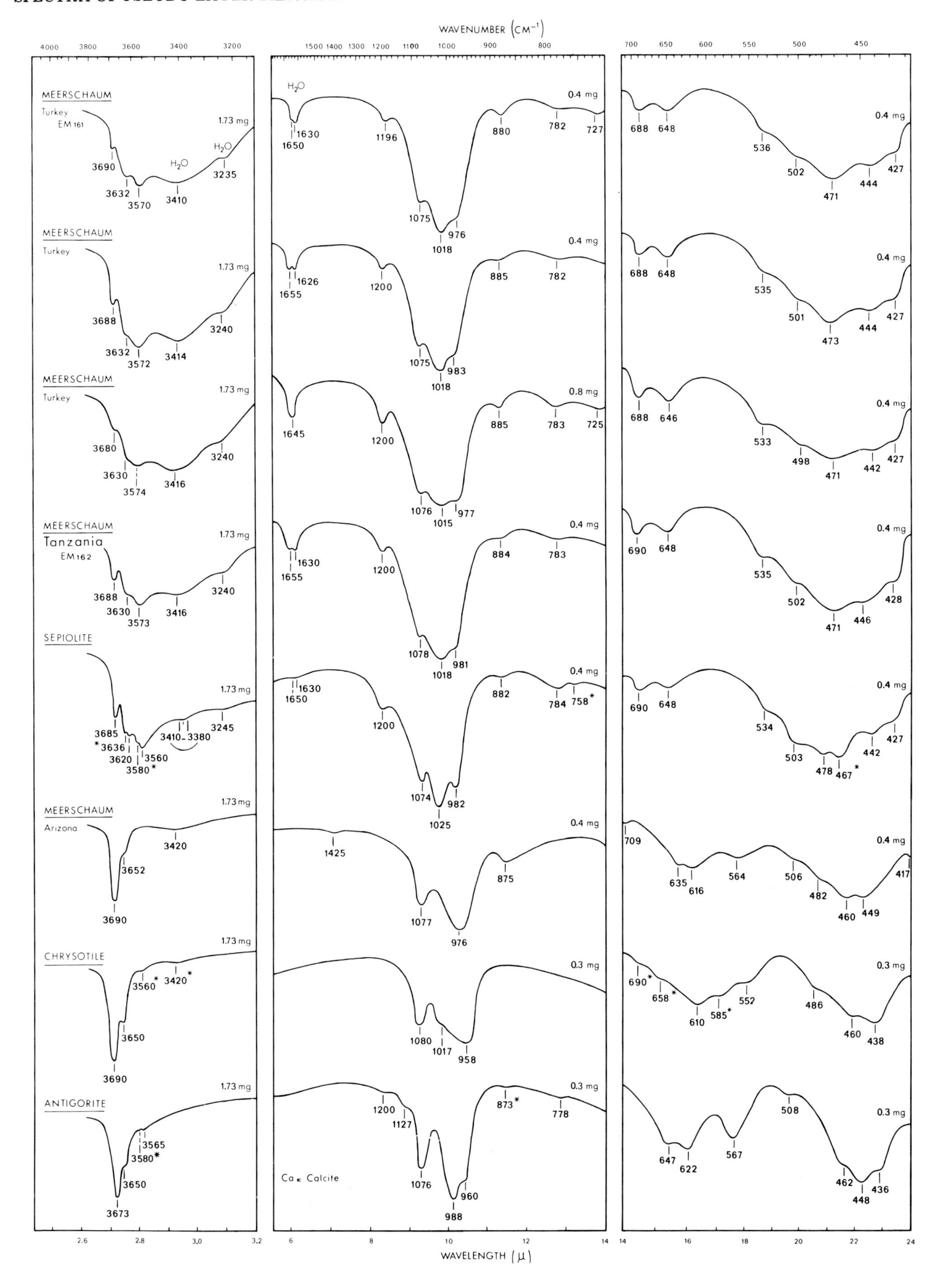

Fig. 6.63. Infrared spectra of meerschaum from different origin.

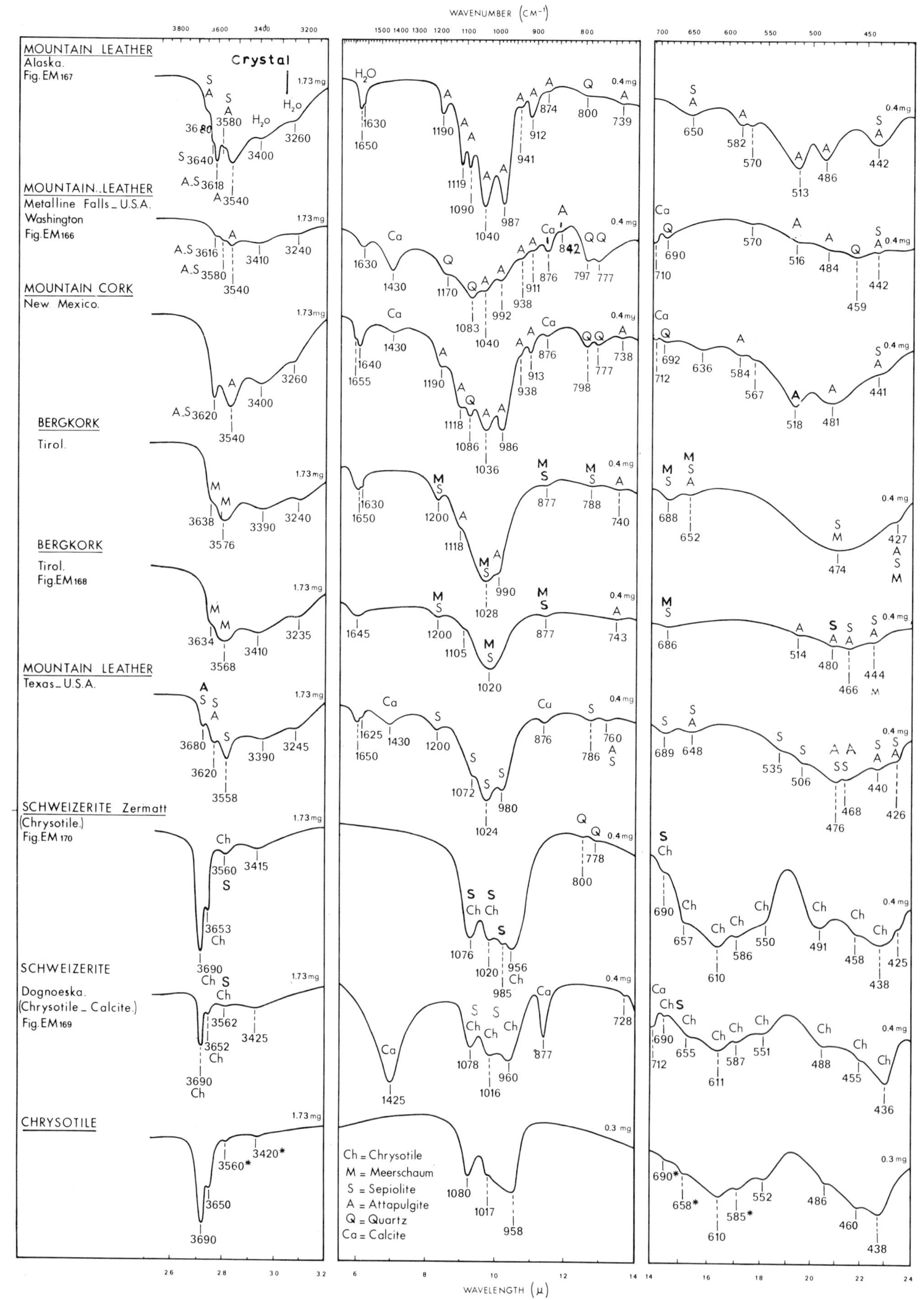

Fig. 6.64. Infrared spectra of Mountain leather (Mountain Cork, Bergkork) and Schweizerite from different origin.

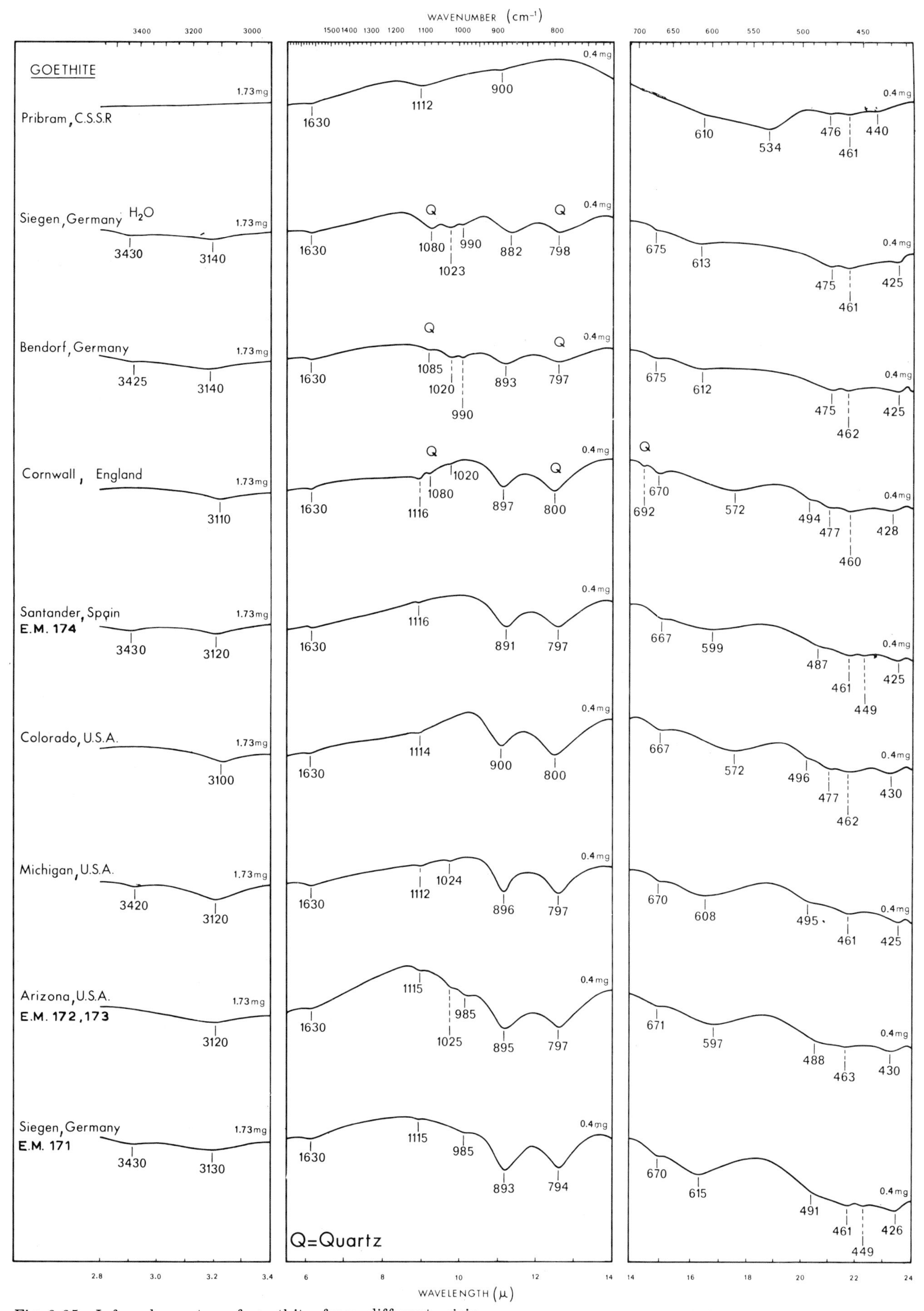

Fig. 6.65. Infrared spectra of goethite from different origin.

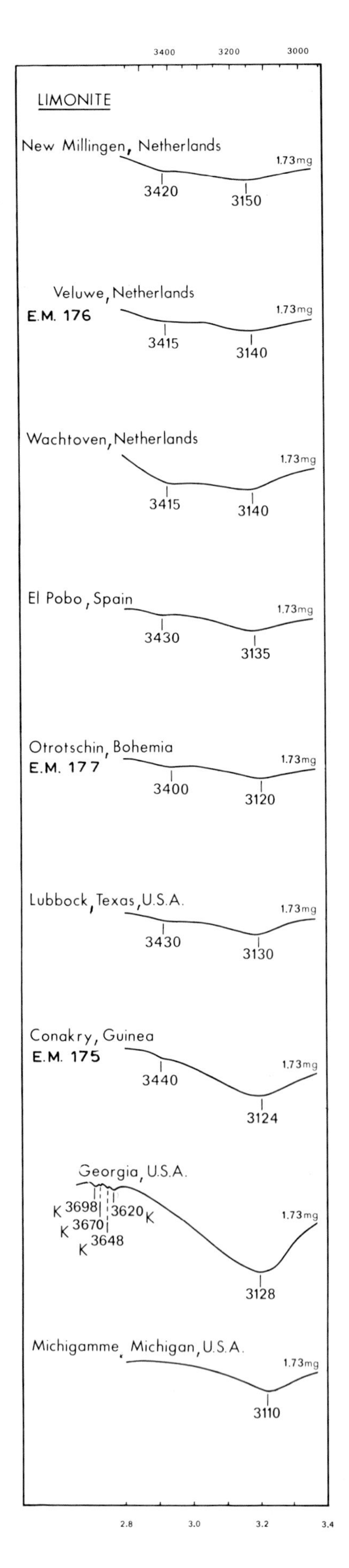

Fig. 6.66. Infrared spectra of limonite from different origin.

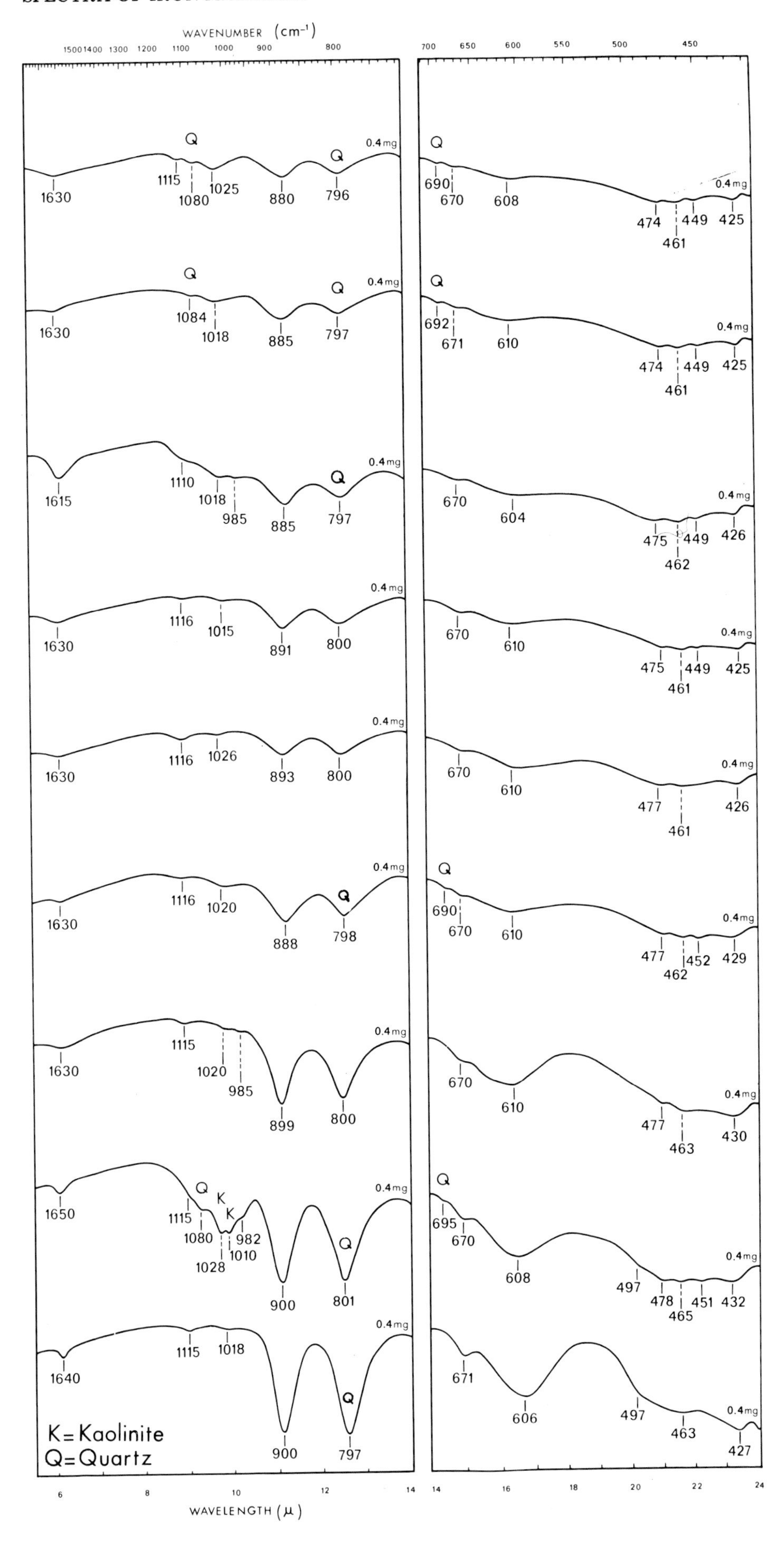
WAVENUMBER (cm⁻¹)
WAVELENGTH (μ)
K=Kaolinite
Q=Quartz
0.4 mg

Lepidocrocite concretions
Easton, Pa., U.S.A. . . . . (1)
Fig. E.M. 179, 180, 181

Lepidocrocite
Bieber, Germany. . . . . (2)

Koblenz, Germany . . . . (3)
some quartz

Fellinghausen, Germany , ... (4)

Burbach, Germany . . . . (5)
some quartz

Siegen, Germany . . . . (6)
some quartz

Milan, Italy . . . . (7)
some quartz

Iron rust
some maghemite and quartz . . . . (8)

some quartz and kaolinite . . . . (9)

some quartz . . . . (10)

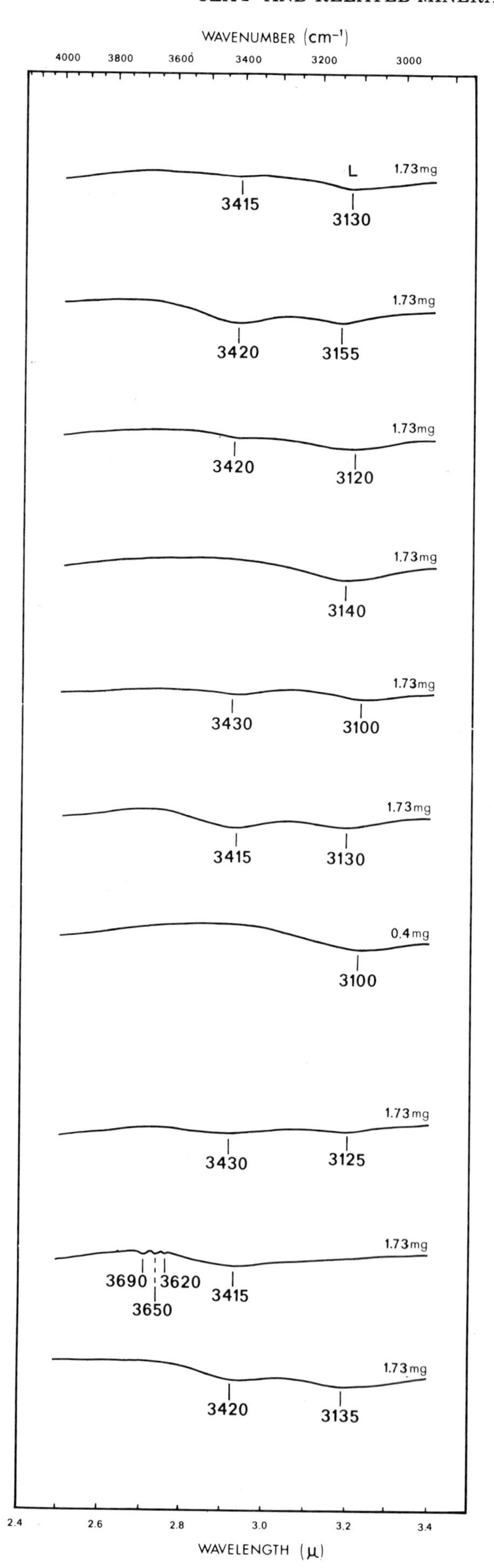

Fig. 6.67. Infrared spectra of lepidocrocite and iron rust from different origin. The samples contain after X-ray and I.R. analysis limonite as an impurity (3100—3155 cm^{-1}).

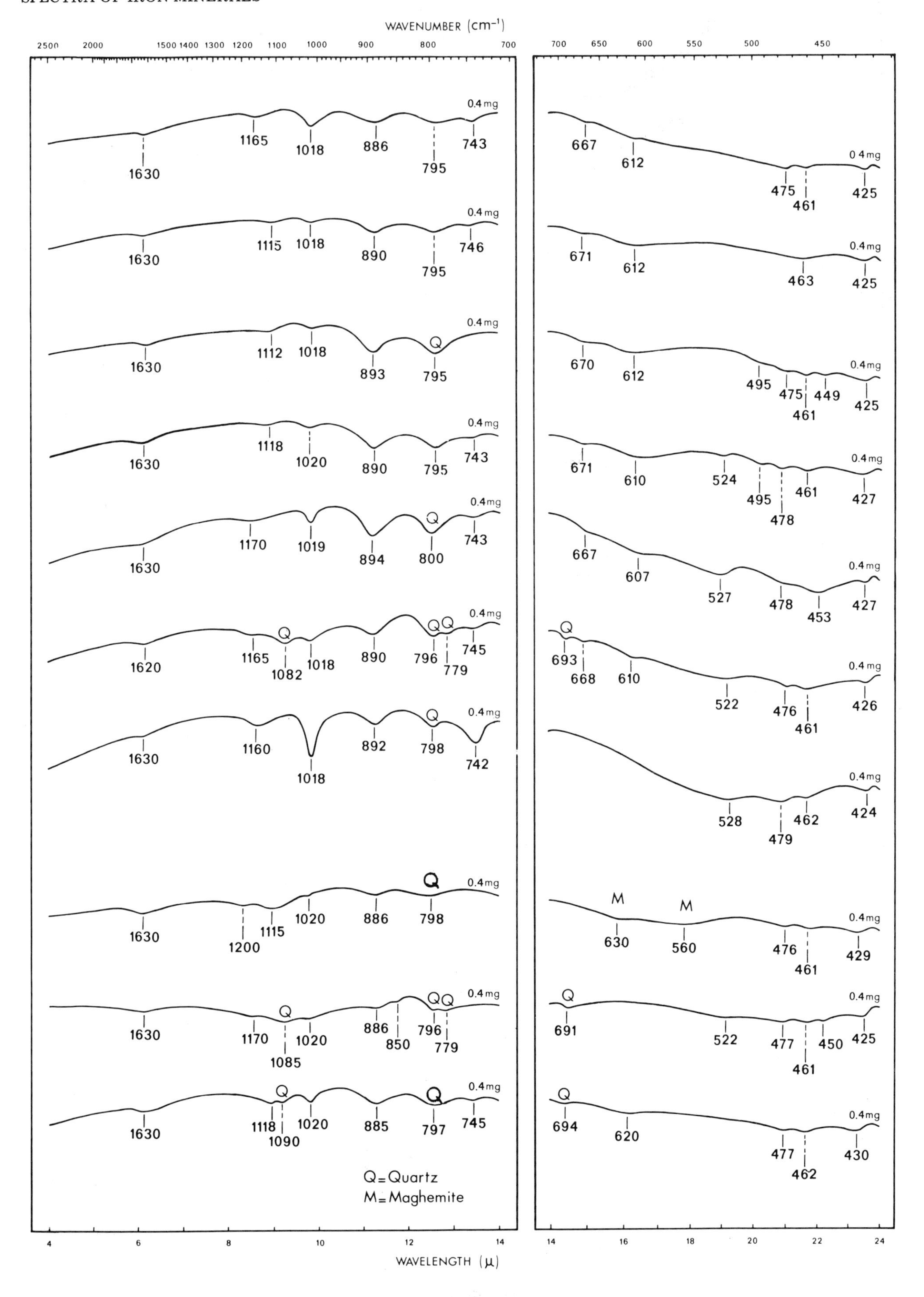
WAVENUMBER (cm⁻¹)
2500 2000 1500 1400 1300 1200 1100 1000 900 800 700
700 650 600 550 500 450
0.4 mg
1630 1165 1018 886 795 743
667 612 475 461 425
1630 1115 1018 890 795 746
671 612 463 425
1630 1112 1018 893 Q 795
670 612 495 475 461 449 425
1630 1118 1020 890 795 743
671 610 524 495 478 461 427
1630 1170 1019 894 Q 800 743
667 607 527 478 453 427
1620 1165 Q 1082 1018 890 Q Q 796 779 745
Q 693 668 610 522 476 461 426
1630 1160 1018 892 Q 798 742
528 479 462 424
1630 1200 1115 1020 886 Q 798
M 630 M 560 476 461 429
1630 1170 Q 1085 1020 886 850 Q Q 796 779
Q 691 522 477 461 450 425
1630 1118 Q 1090 1020 885 Q 797 745
Q 694 620 477 462 430
Q = Quartz
M = Maghemite
4 6 8 10 12 14
14 16 18 20 22 24
WAVELENGTH (μ)

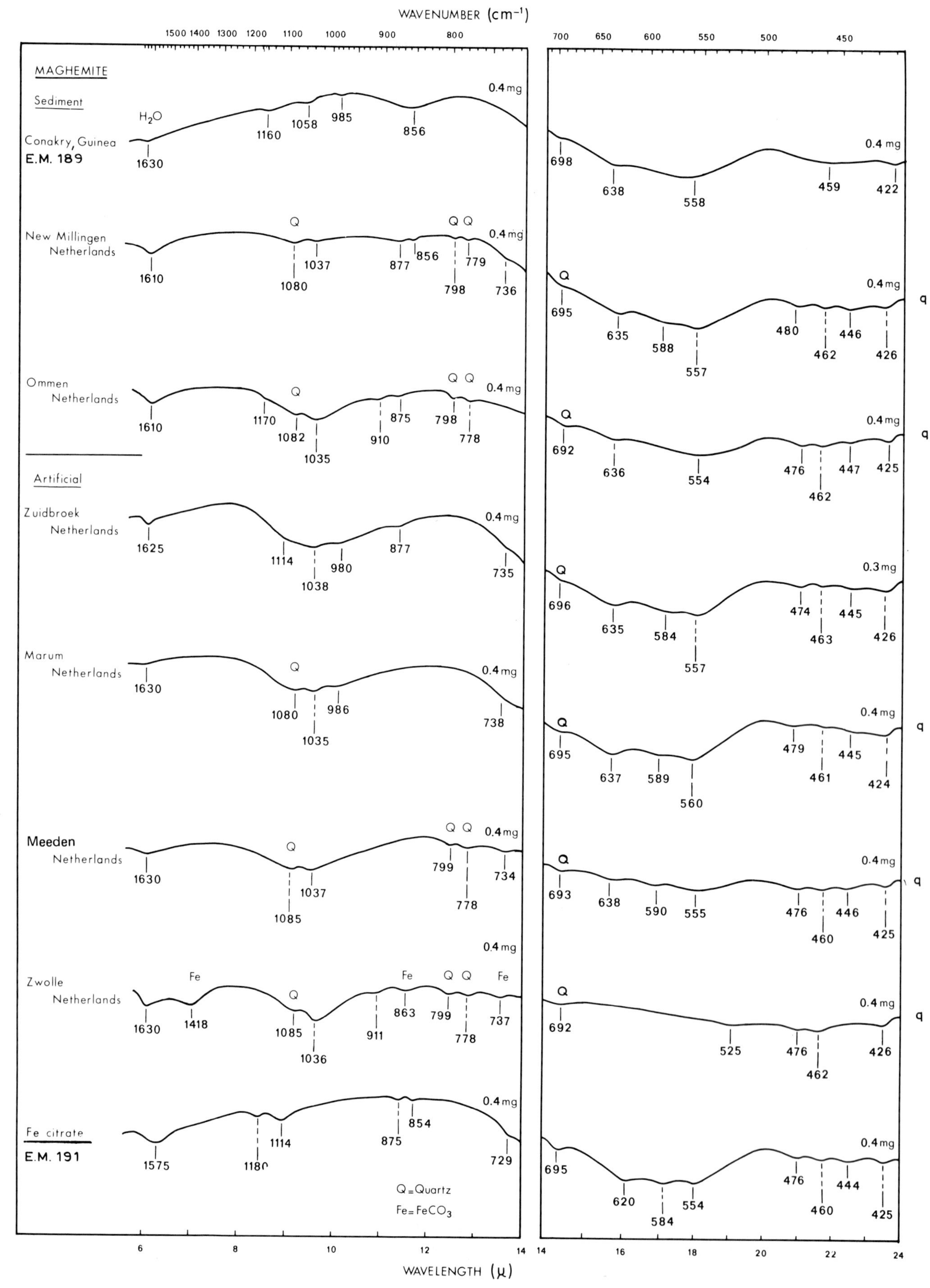

Fig. 6.68. Infrared spectra of maghemite from different origin.

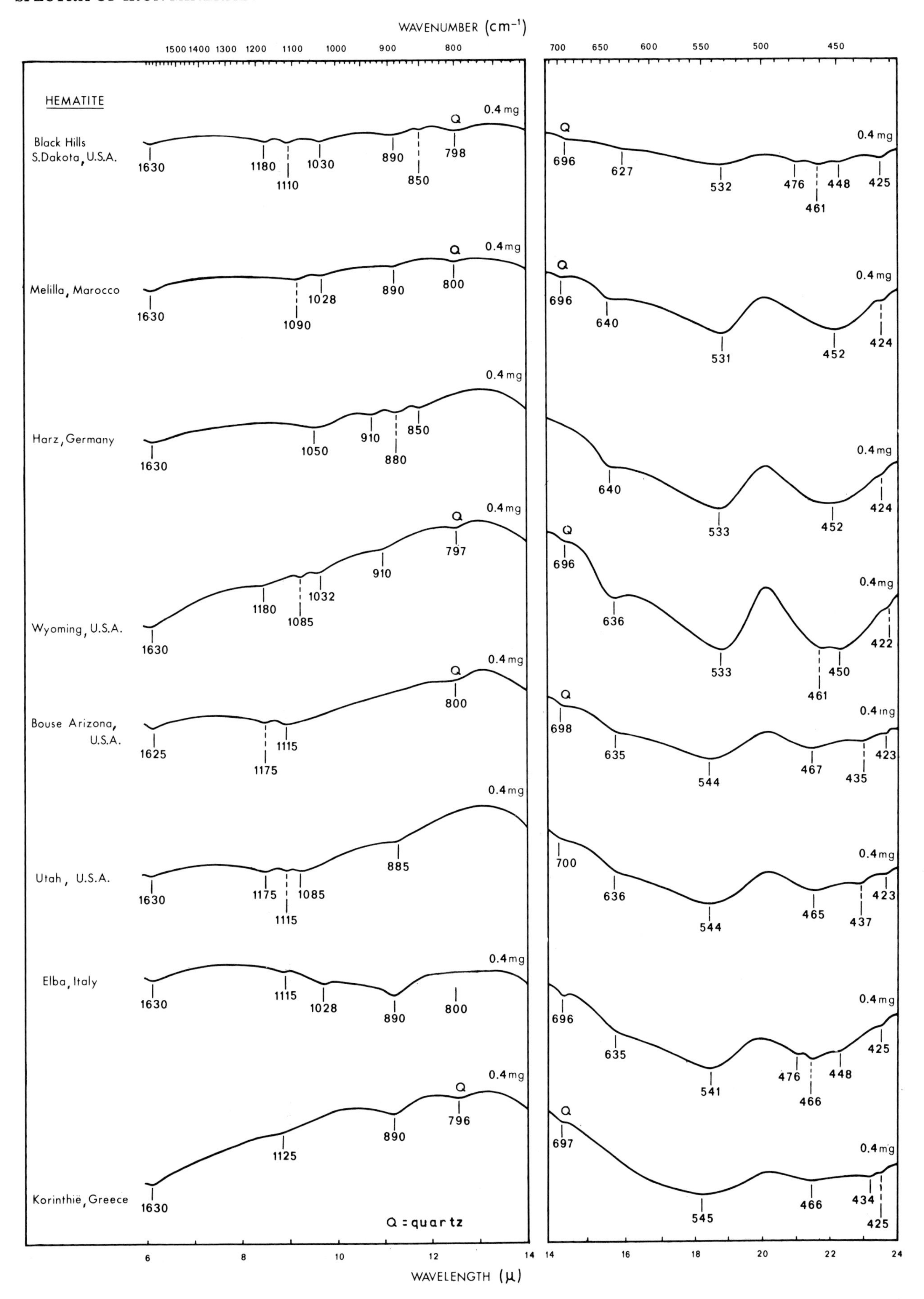

Fig. 6.69. Infrared spectra of hematite from different origin.

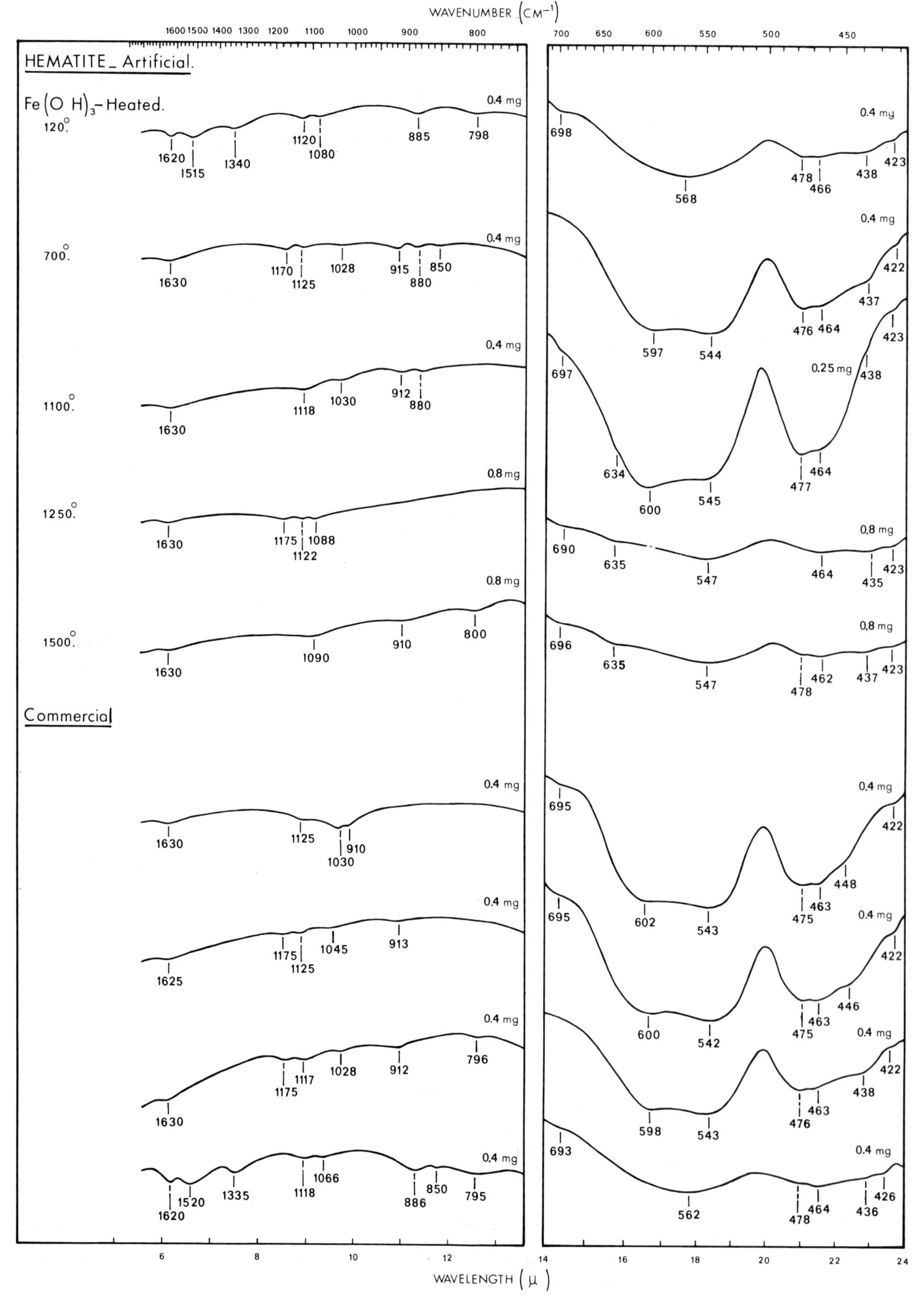

Fig. 6.70. Infrared spectra of artificial hematite from different origin.

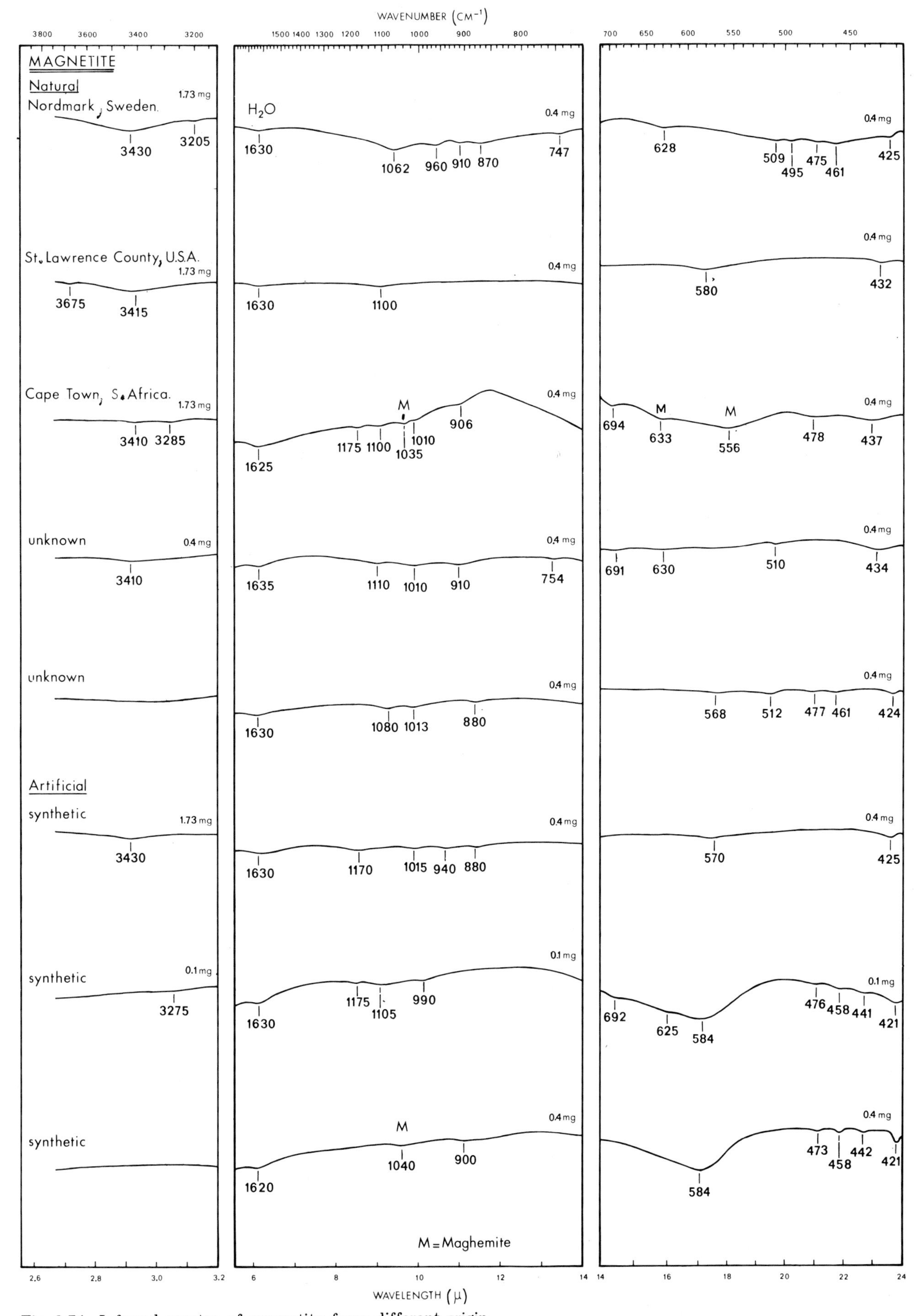

Fig. 6.71. Infrared spectra of magnetite from different origin.

Hydrargillite

Bintam, Indonesia
some kaolinite and quartz

Surinam
some kaolinite and quartz
Fig. E.M. 193, 194

Angola
some kaolinite, boehmite and quartz

Szöc, Hungary
some kaolinite and quartz
Fig. E.M. 195

British Guyana
some kaolinite and quartz

New Caledonia
pure
Fig. E.M. 196

Papoea, Indonesia
some boehmite

Nordstrandite
Serawak, Borneo

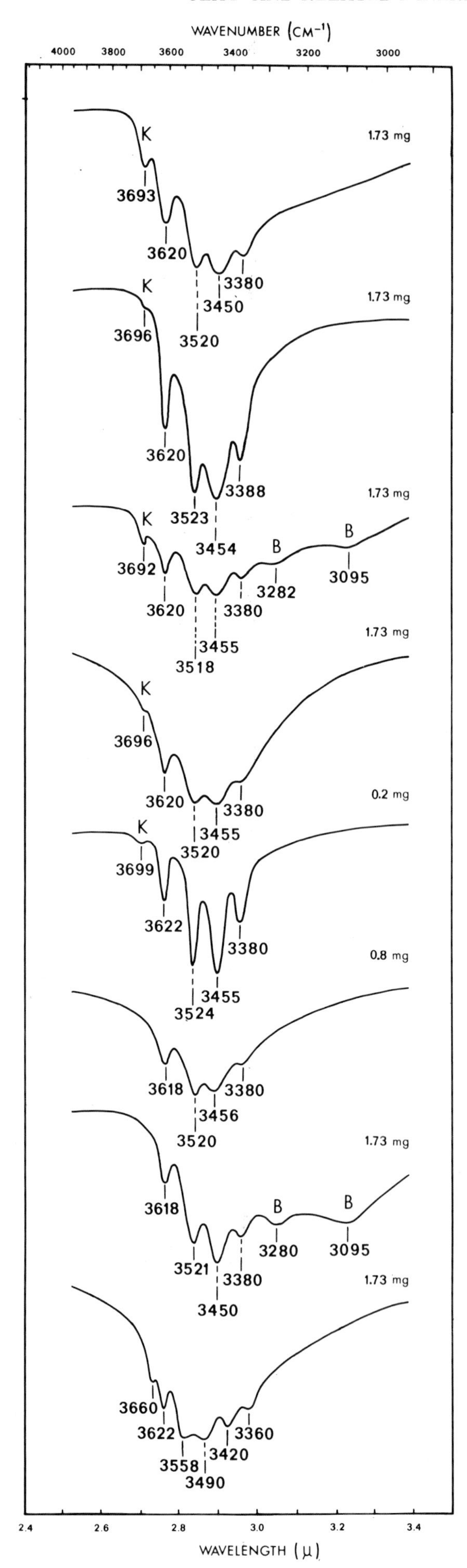

Fig. 6.72. Infrared spectra of hydrargillite from different origin and of nordstrandite.

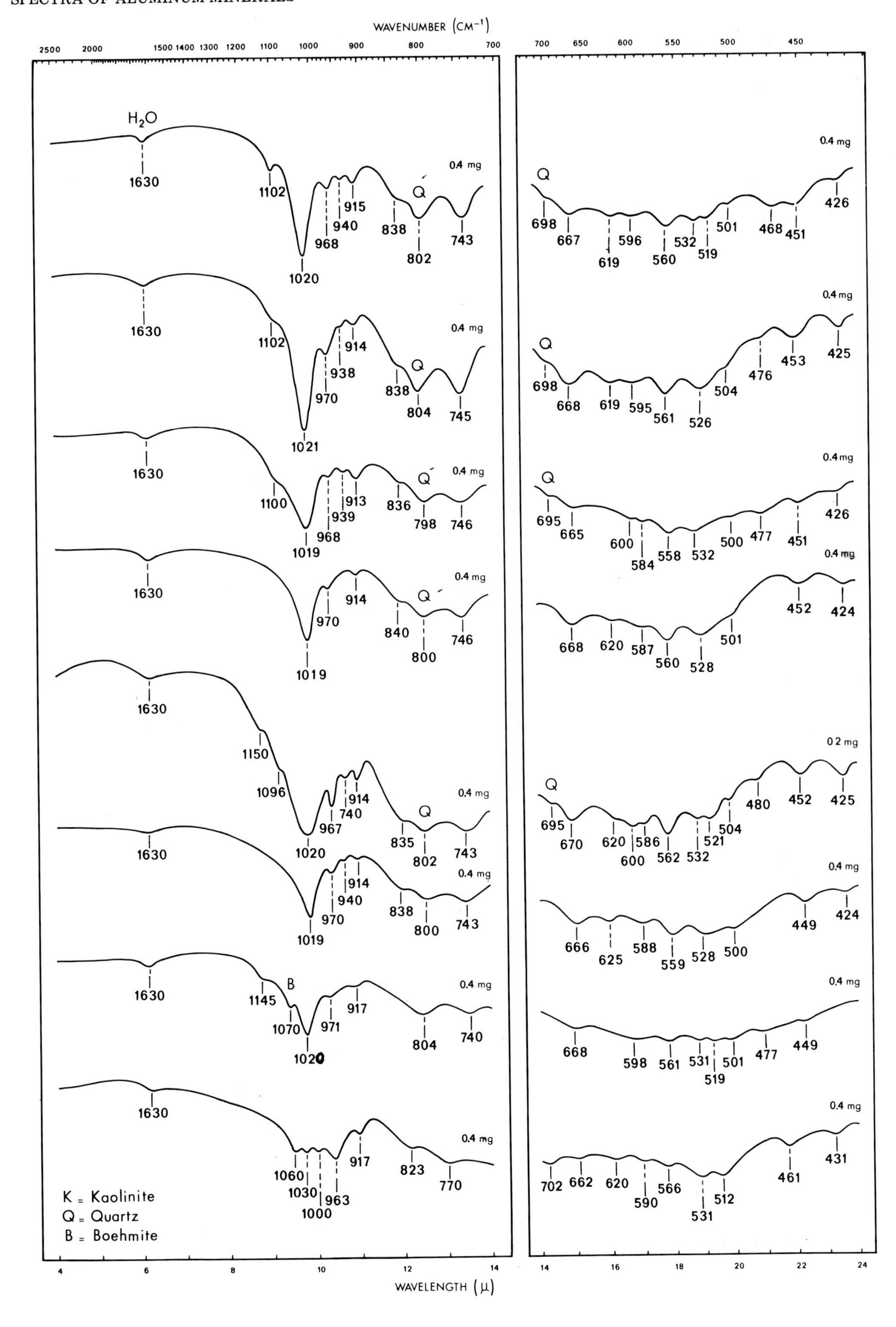
WAVENUMBER (CM⁻¹)
WAVELENGTH (μ)
H2O
0.4 mg
0.2 mg
K = Kaolinite
Q = Quartz
B = Boehmite

Hydrargillite (artificial)
some boehmite . . . . (1)

pure . . . . (2)

pure . . . . (3)

some boehmite . . . . (4)

Degussa (commercial) . . . . (5)

Bayerite (artificial)
Fig. E.M. 199 . . . . (6)

Fig. E.M. 200 . . . . (7)
some boehmite

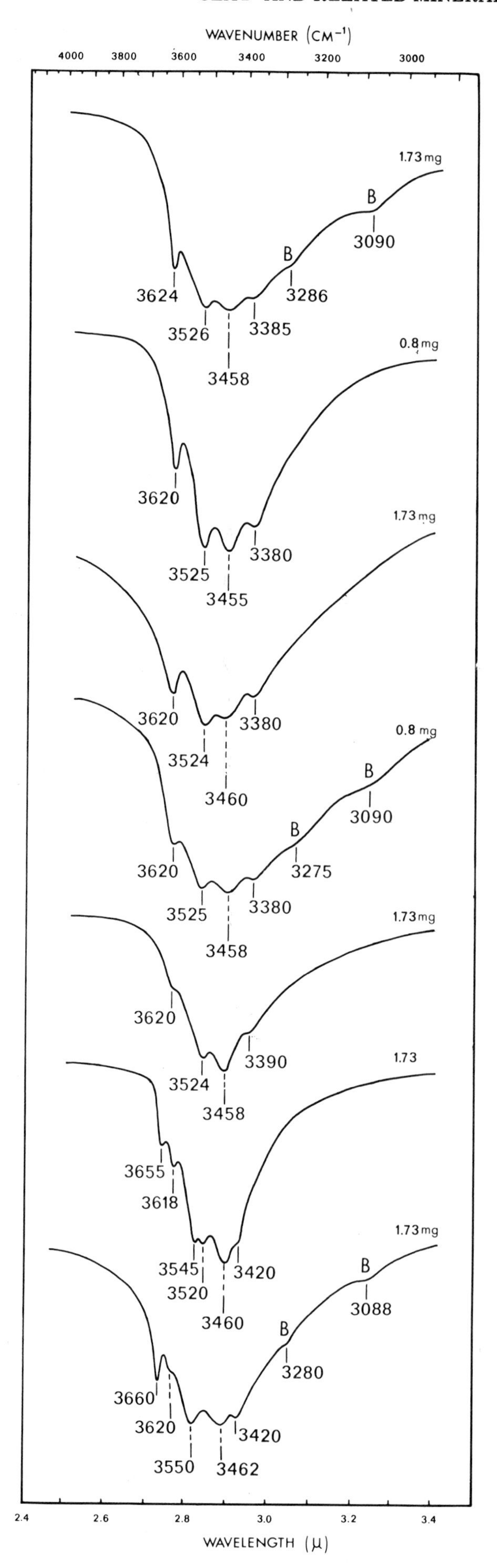

Fig. 6.73. Infrared spectra of artificial hydrargillite and artificial bayerite from different origin.

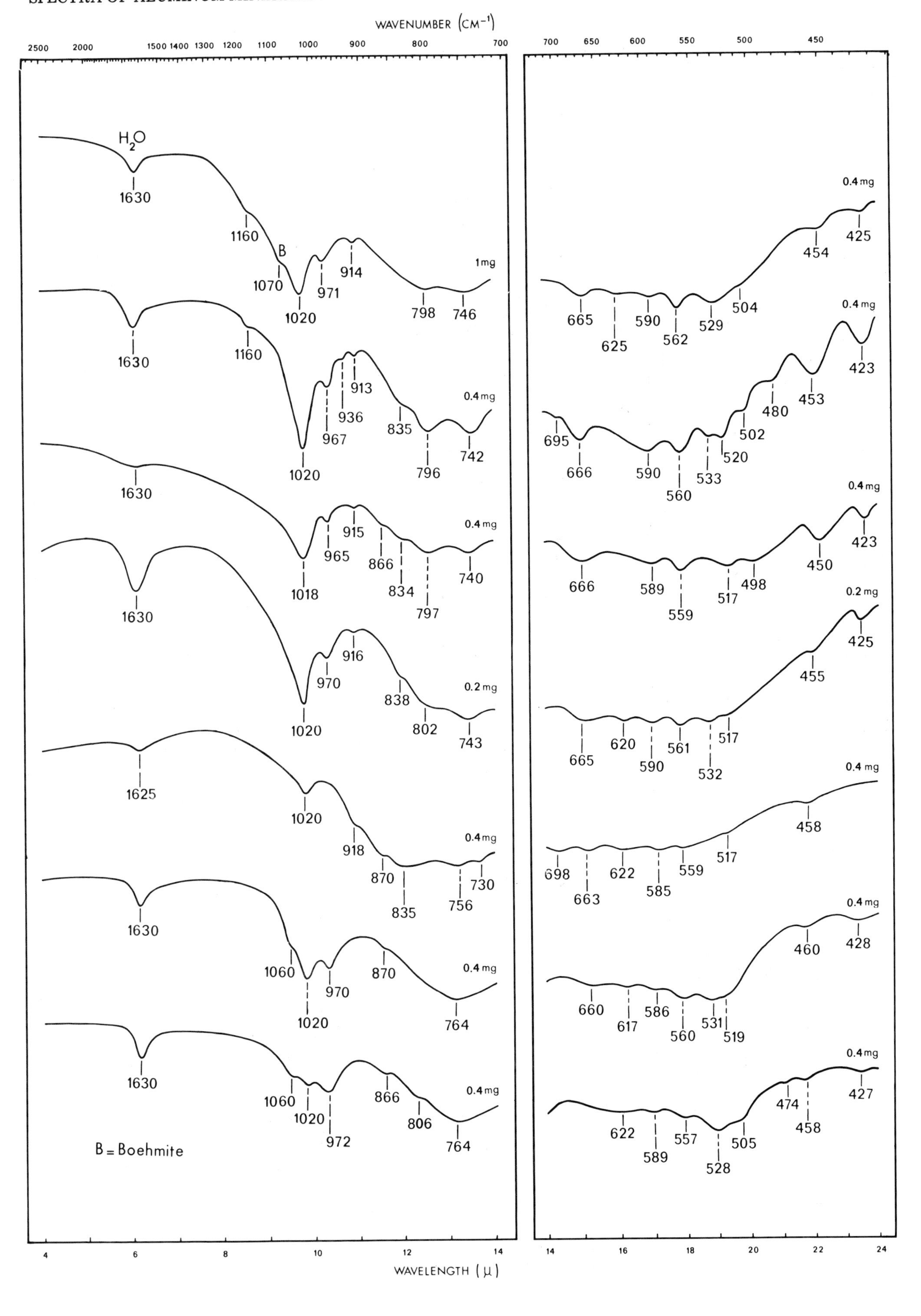
WAVENUMBER (CM-1)
2500
2000
1500
1400
1300
1200
1100
1000
900
800
700
650
600
550
500
450
H2O
1630
1160
B
1070
1020
971
914
798
746
1 mg
665
625
590
562
529
504
454
425
0.4 mg
1160
967
936
913
835
796
742
1018
965
915
866
834
797
740
1625
918
870
756
730
1060
764
972
806
B = Boehmite
WAVELENGTH (μ)

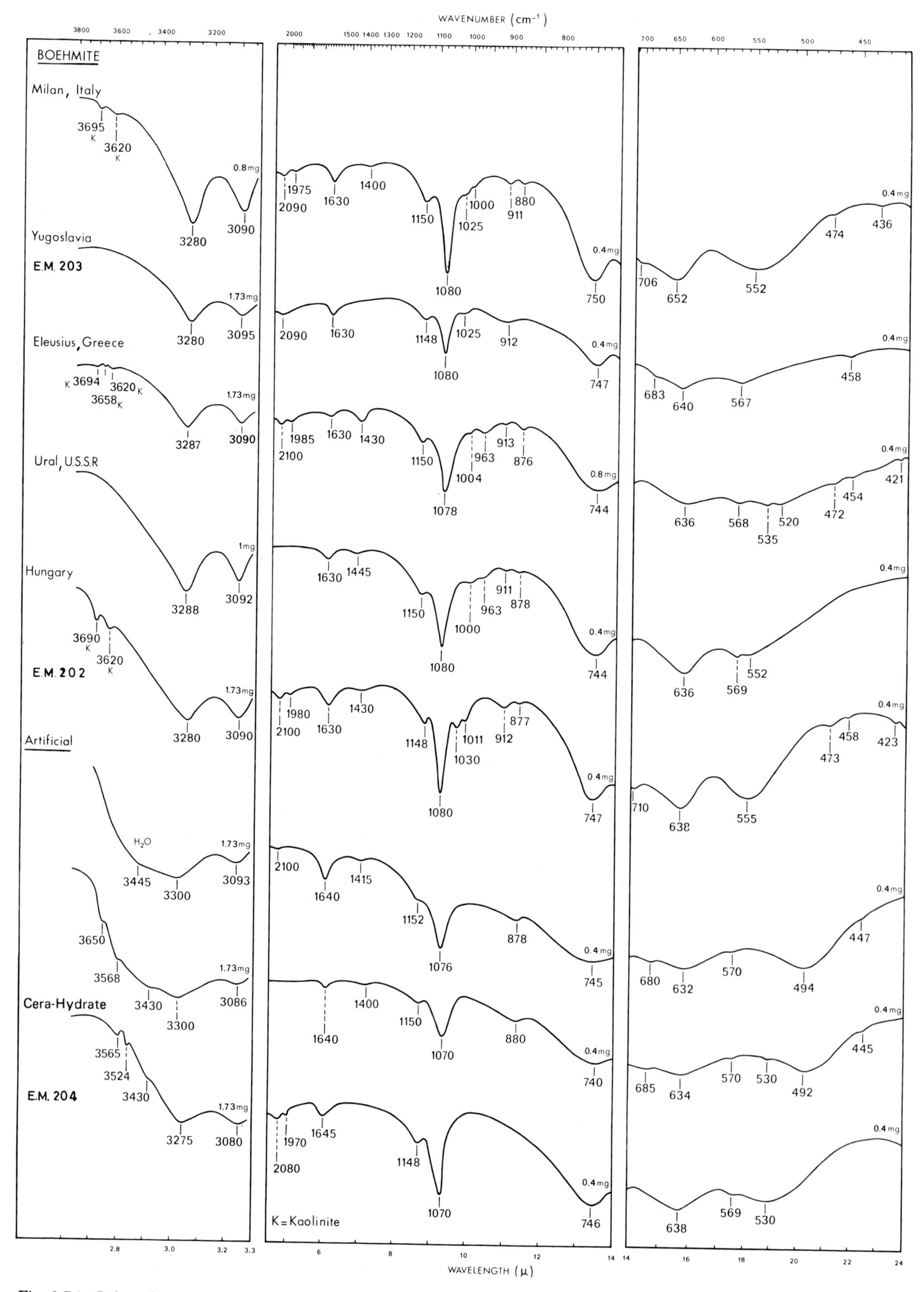

Fig. 6.74. Infrared spectra of boehmite and artificial boehmite from different origin.

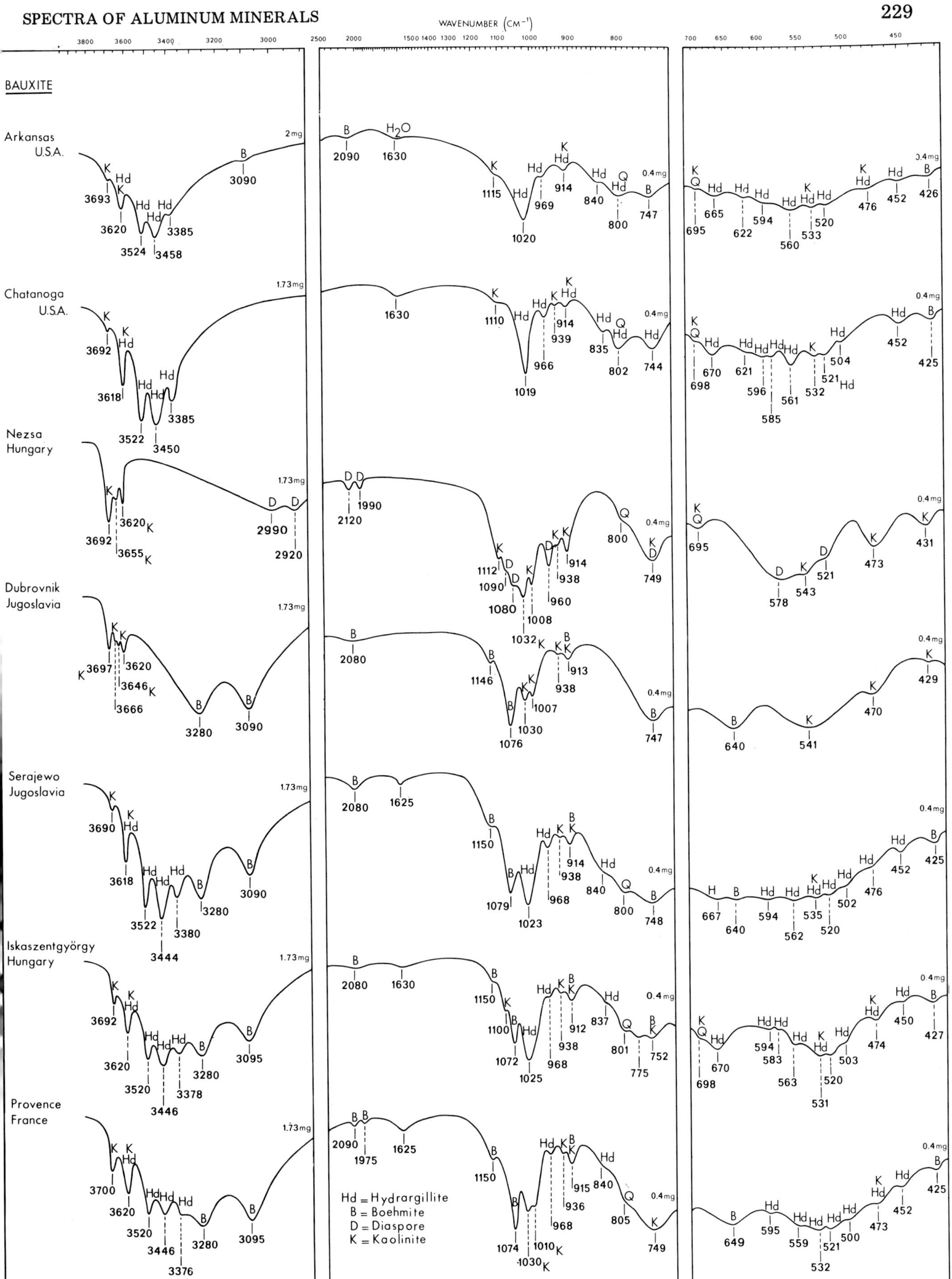

Fig. 6.75. Infrared spectra of bauxite from different deposits.

Diaspore

Chester, Vermont, U.S.A.
some kaolinite and pyrophyllite (overlapped)

Ural, U.S.S.R.
pure

Mn-diaspore
Cape Provence, South Africa
some kaolinite

Diaspore impure
Hungaria
kaolinite

Pyrophyllite
Rosebud, Miss., U.S.A.
chlorite, kaolinite and mica
Fig. E.M. 205

Dover Mine, Nev., U.S.A.
pyrophyllite and some kaolinite

Kaolinite

Pyrophyllite

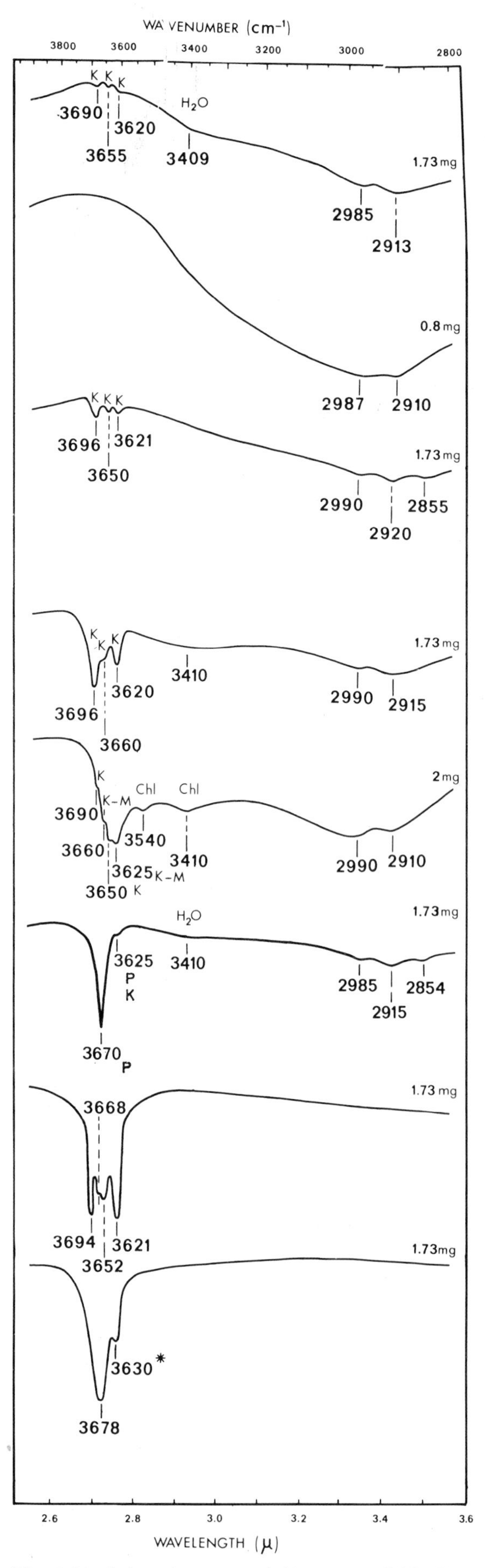

Fig. 6.76. Infrared spectra of diaspore and Mn-diaspore.

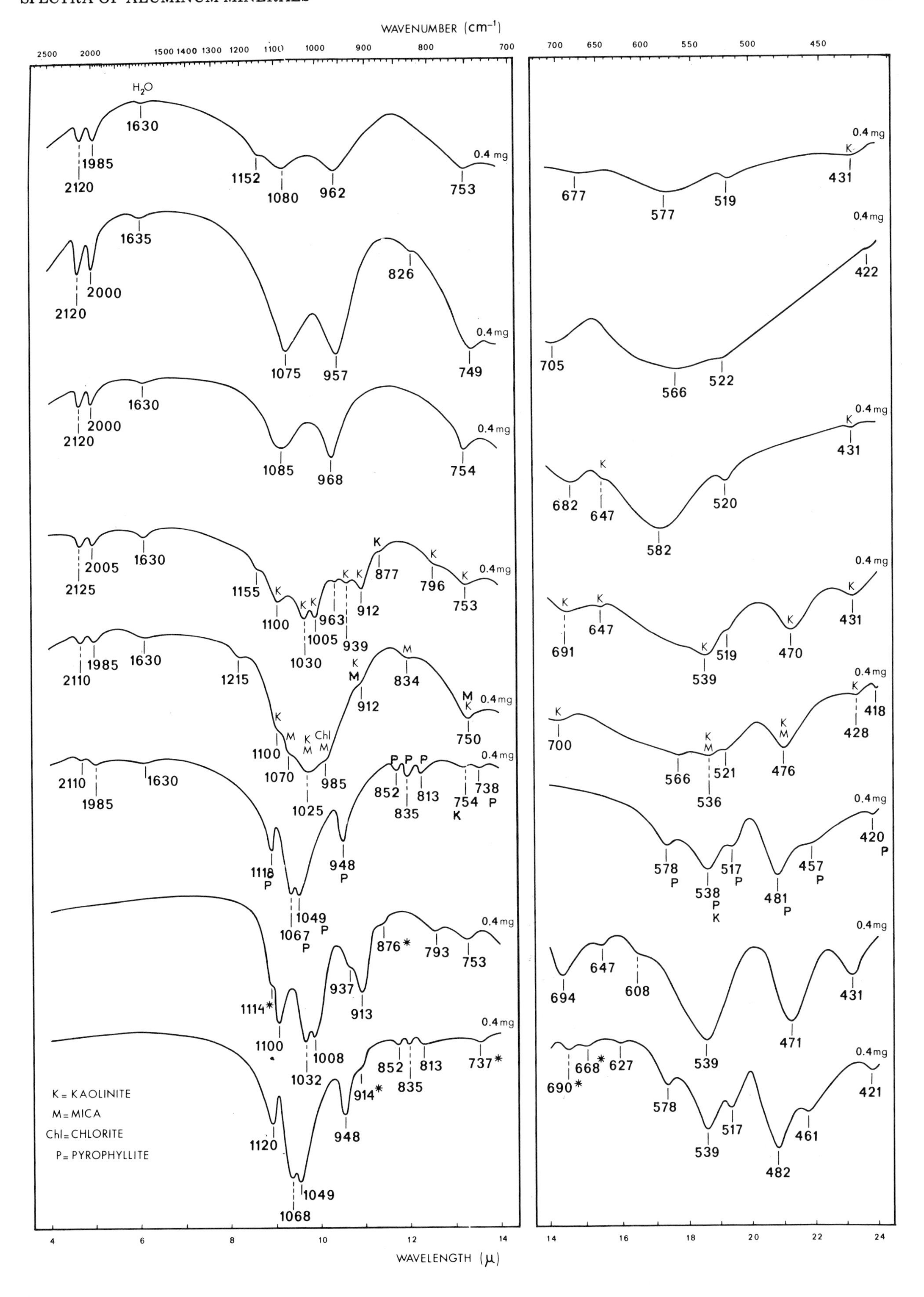
WAVENUMBER (cm⁻¹)
WAVELENGTH (μ)
K = KAOLINITE
M = MICA
Chl = CHLORITE
P = PYROPHYLLITE
0.4 mg

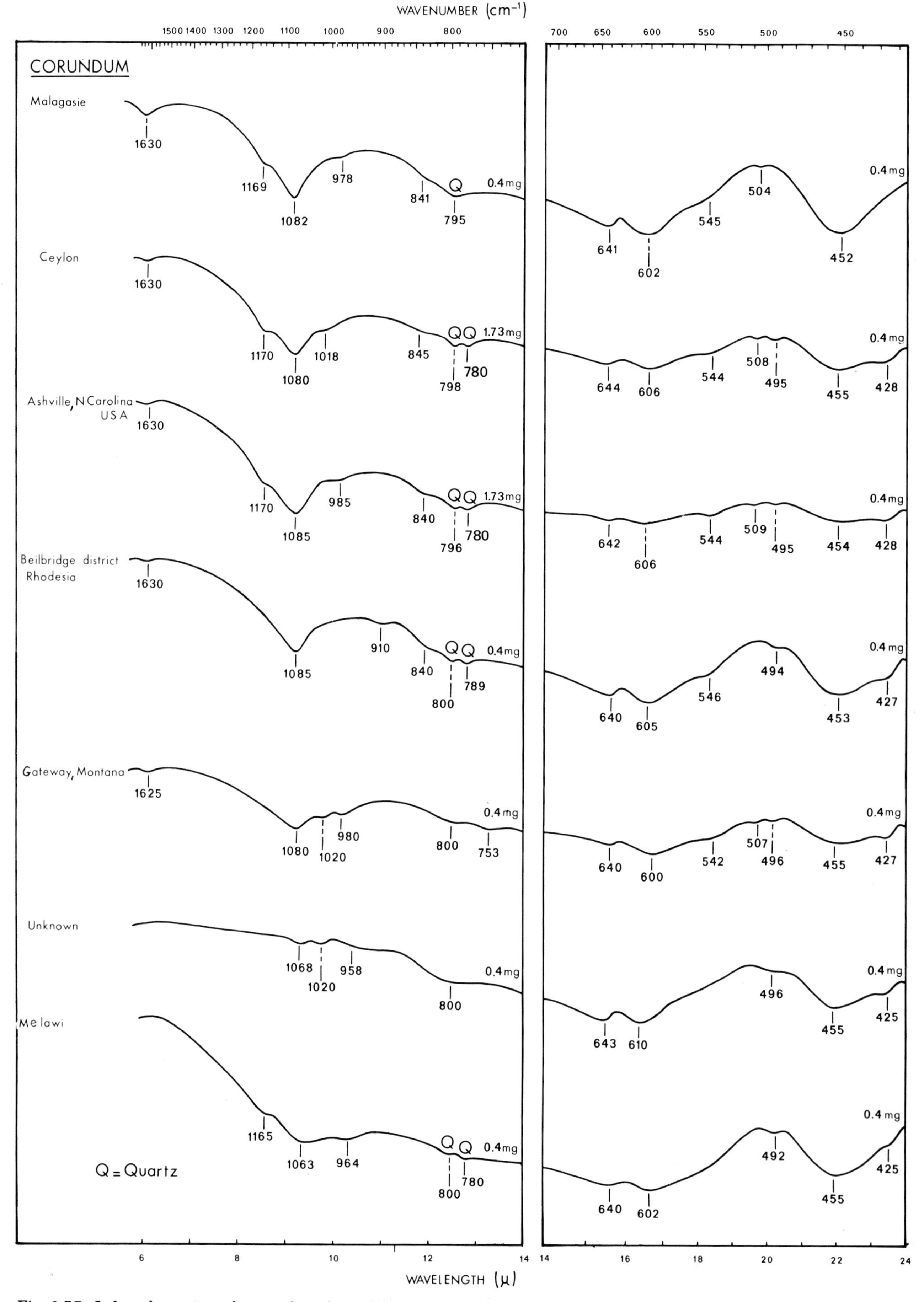

Fig. 6.77. Infrared spectra of corundum from different origin.

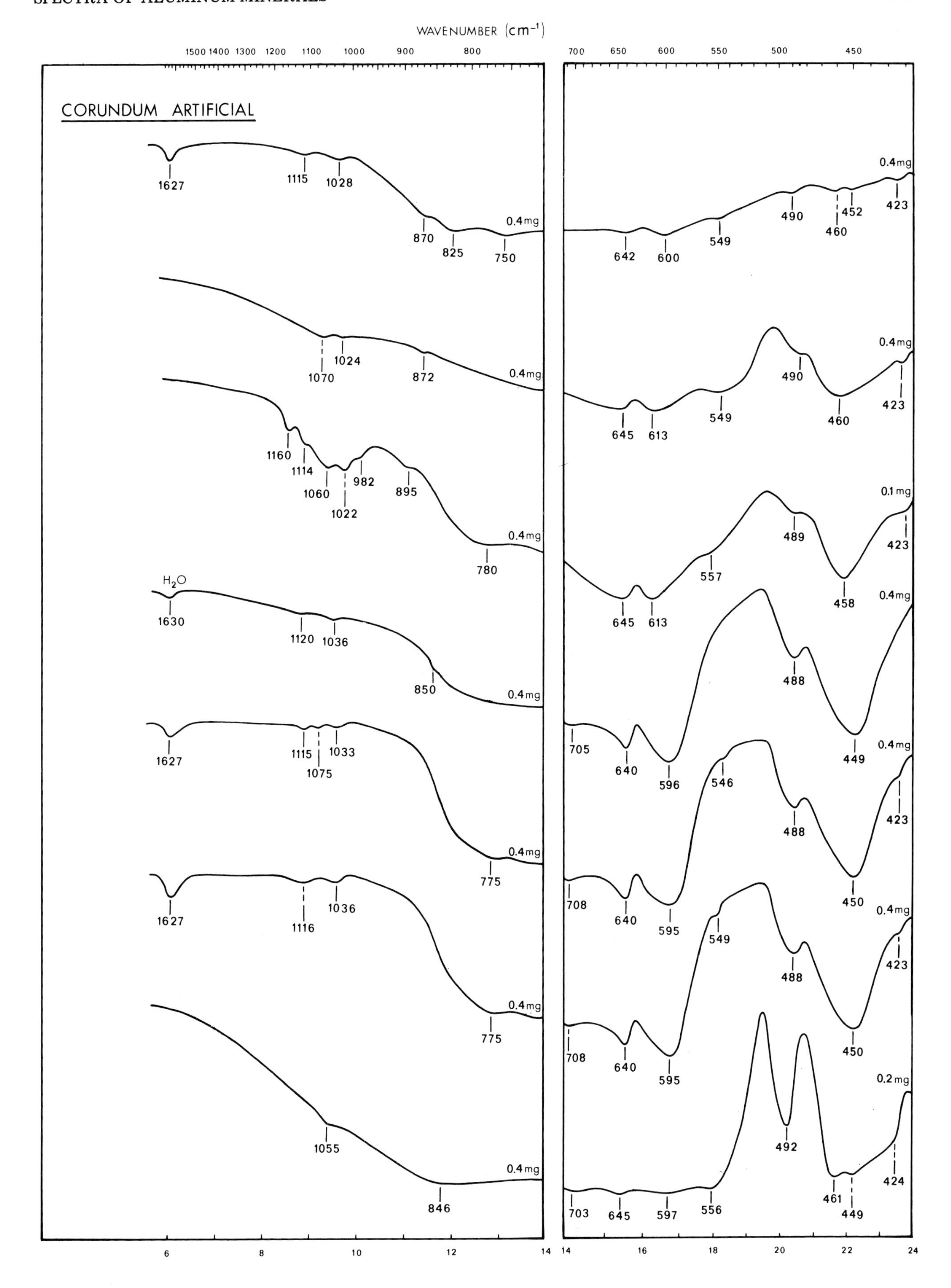

Fig. 6.78. Infrared spectra of artificial corundum.

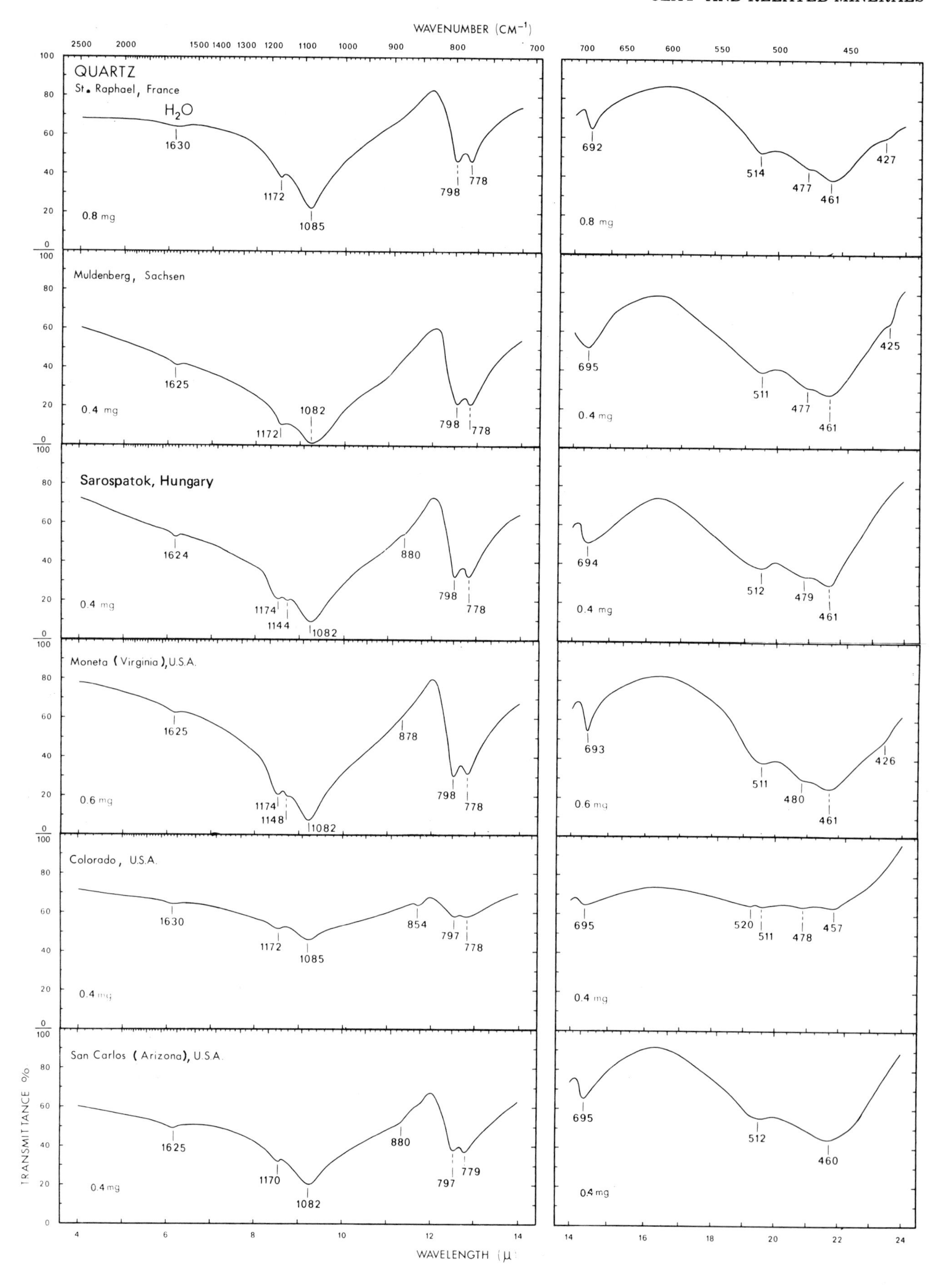

Fig. 6.79. Infrared spectra of quartz from different origin.

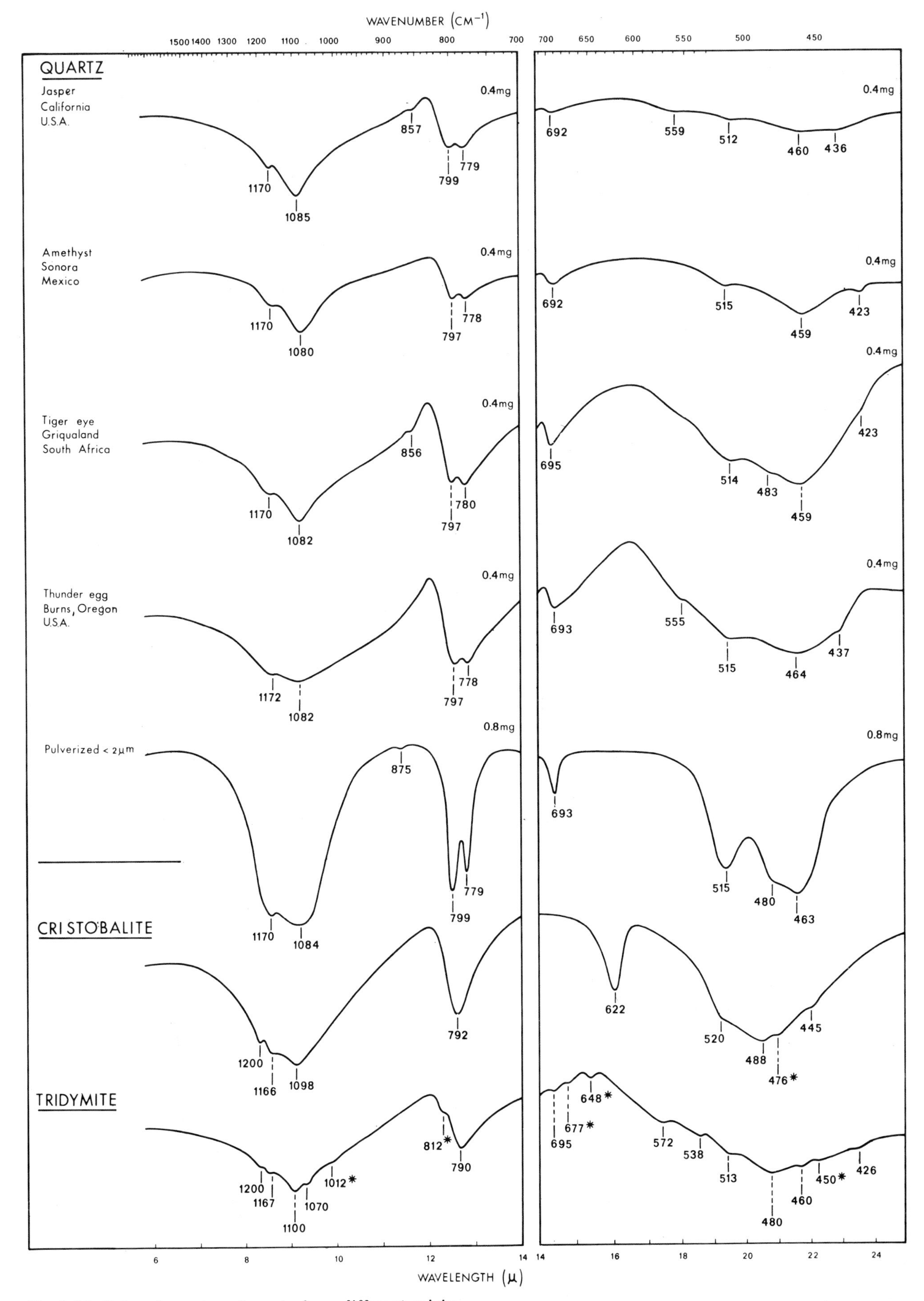

Fig. 6.80. Infrared spectra of quartz from different origin.

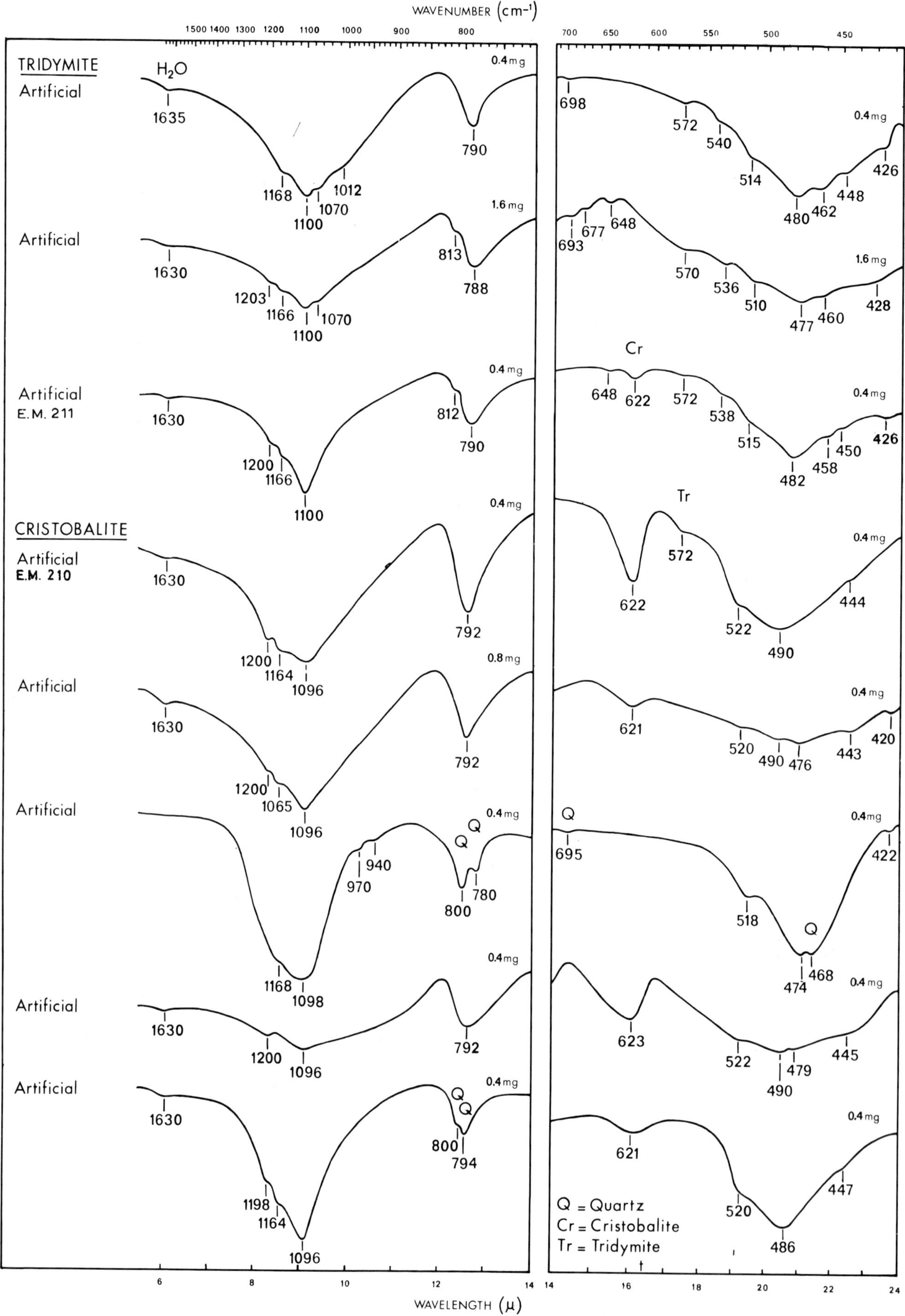

Fig. 6.81. Infrared spectra of tridymite and cristobalite from different origin.

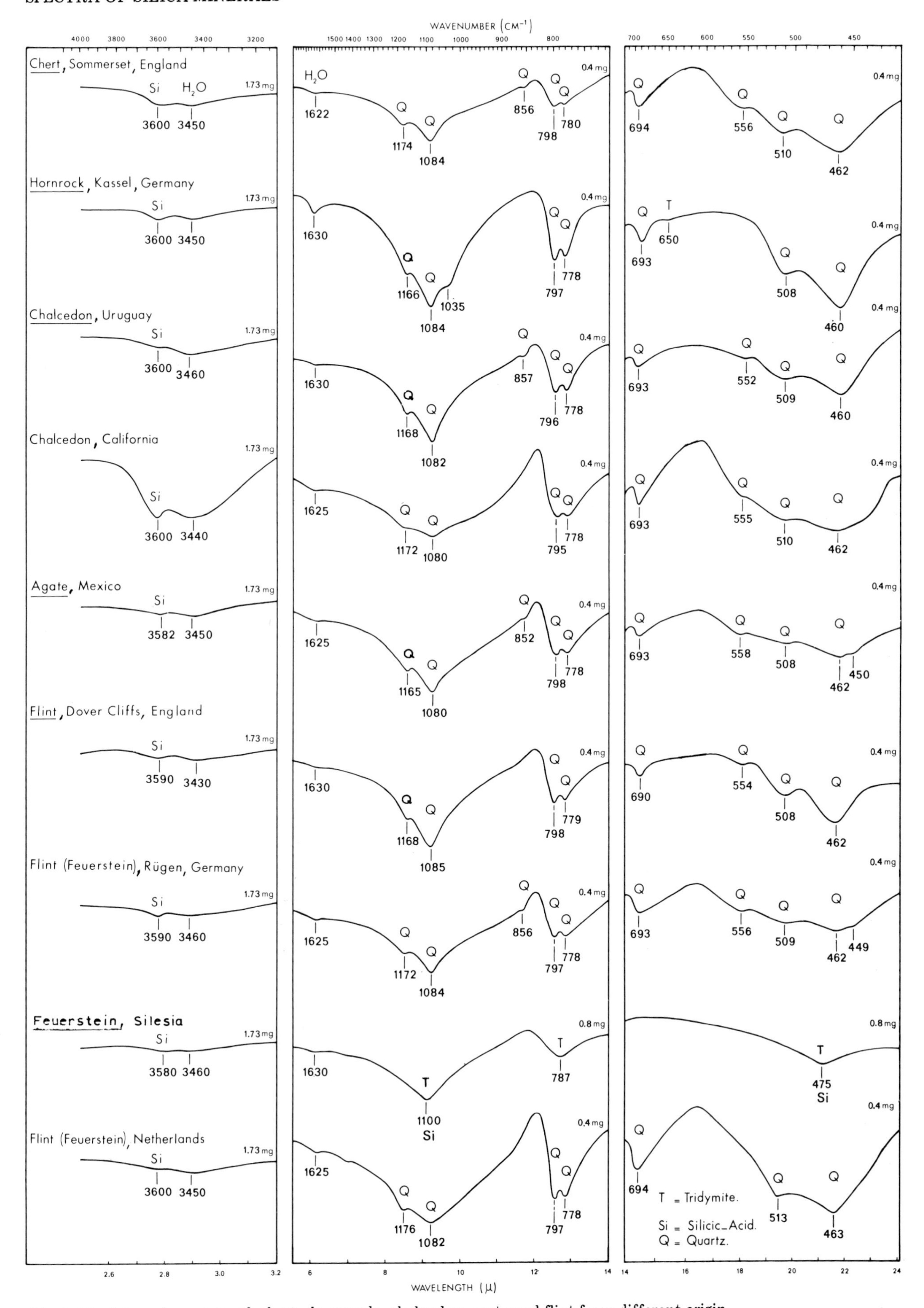

Fig. 6.82. Infrared spectra of chert, hornrock, chalcedon, agate and flint from different origin.

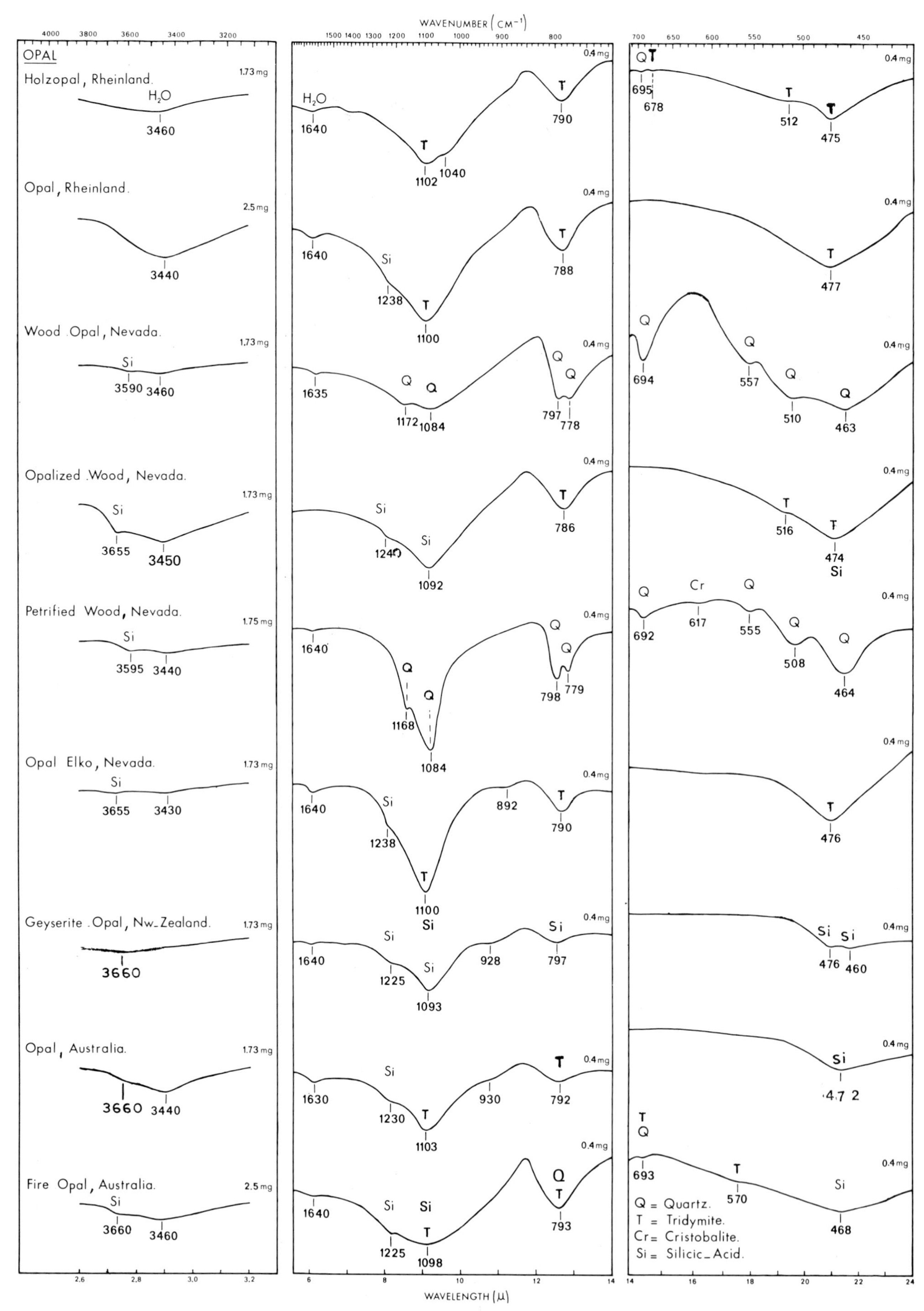

Fig. 6.83. Infrared spectra of opal and wood opal from different origin.

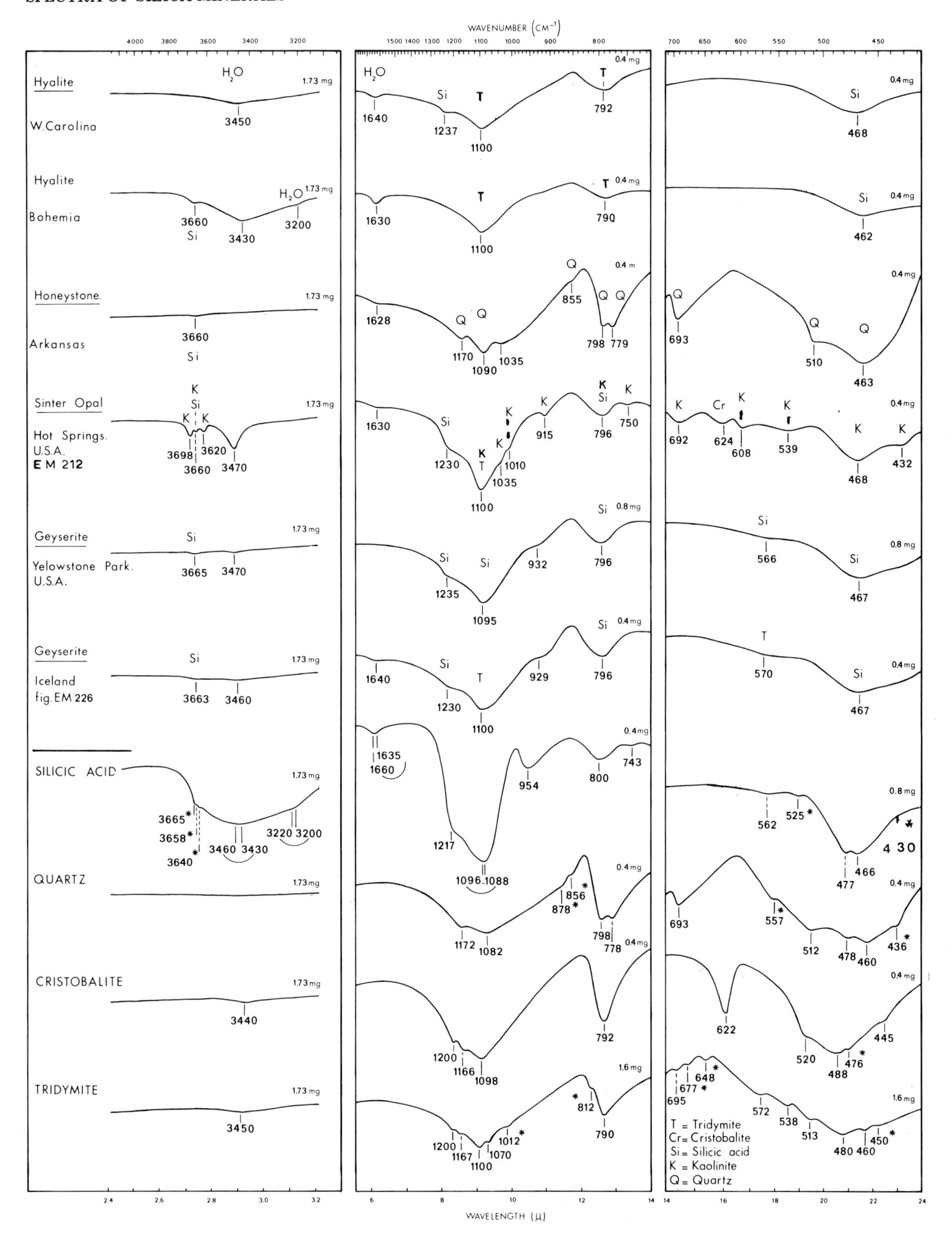

Fig. 6.84. Infrared spectra of hyalite, honeystone, sinter opal, and geyserite from different origin.

7. ADMIXTURES

Carbonates

Minerals of the carbonate group consisting of single or double cations of Ca, Mg, Sr are widespread in the world. They are not only found in rock fissures, rock veins, skeletons of foramifera, algae, precipitates etc., but also concentrated to huge amounts as massive rocks: e.g. calcite (Ca), dolomite (Ca,Mg) and magnesite (Mg); siderite (Fe) is less common. Hydrocarbonate, hydromagnesite, trona and Gaylussite are common components of efflorescences in salt lakes and salt marshes.

Calcite and aragonite; formula: $CaCO_3$

Aragonite is the dense metastable orthorhombic modification of calcite (calcite = trigonal).

Occurrence. Many sediments contain in their finer fractions calcite as a weathering or abrasion product of massive calcite rocks, or as a chemical precipitate from carbonaceous solutions e.g. seawater etc. Aragonite is found in fissures of basic magmatic rocks and at hot springs.
Shells mostly consist of calcite. But there are also shells which consist of aragonite only (Cardium edule) or of calcite only (Ostrea), or which have both. In that case there is a broadening of the ν_3 vibration at the smaller amounts. The amounts depend on environmental conditions during growth. Also algae and foramifera may contain considerable amounts of aragonite.

Infrared spectra (Fig. 7.1; Table I). Theoretically $3N-6$(N = number of atoms) = 6 modes of vibrations are possible for the free moving carbonate radical CO_3^{2-} group. Two of them are doubly degenerate e.g., ν_3 (asymmetric stretching) and ν_4 (planar bending). One is non-degenerate = ν_2 (out of plane bending). Another, ν_1 (symmetric stretching) is non-degenerate and weak; especially for the highly symmetrical hexagonal carbonate minerals. Splitting of the double-degenerate bands ν_3 and ν_4 and of the ν_2 and ν_4 band may be caused by interaction of neighbouring molecules or by Fermi resonance.
Infrared vibrations of large intensity are for calcite in increasing order of intensity: 712 cm^{-1} (ν_4), 876 cm^{-1} (ν_2) and 1422 cm^{-1} (ν_3). For aragonite they are 700, 712 cm^{-1} (ν_4), 858 cm^{-1} (ν_2) and 1473 cm^{-1} (ν_3).
Oyster shells (calcite) have the same spectrum as calcite. The precipitated fine calcite mineral has broad bands which are caused by disorder. Such in contrast to the X-ray spectrum which show sharp reflections. Aragonite can be distinguished from calcite (between brackets) by the following bands 1473 (1422), 1084 (1060), 858 (876).
Artificial aragonite, which as usually has been obtained in the heat, has a broad band at 1476 cm^{-1}. Artificial calcite which as usually is obtained in the cold has a poor spectrum.

Remarks. Calcite is an important commercial product serving many purposes, e.g. calcite is a source of CO_2 gas which is widely used for many purposes. Ground calcite (limestone = impure) is applied as an amelioration product to increase the permeability of sticky soils or as a fertilizer to decrease the acidity of soils. Limestone is used in the building industry. Lime is used in plastery and as a flux in steel furnaces to flux off the impurities. Fine precipitated $CaCO_3$ is a coater in paper industry. Calcite of large transparency, purity and perfection of crystallization is called iceland spar after Iceland where the best optical material has been gained for long years for the fabrication of Nicol prisms (polarizing microscopes, polarimeters, etc.).
The name calcite is derived from the Greek calx which means "reduce of a solid material to a powder by heat". Aragonite is named after the type locality Molina de Aragon in Spain.

Magnesite, gelmagnesite; formula: $MgCO_3$

Gelmagnesite is said to be a badly crystallized magnesite.

Occurrence. Magnesite is found in veins in serpentine. Magnesite rocks are not widespread. They resulted by replacement of limestones or dolomite through hydrothermal solutions. Gelmagnesite is only found incidentally. Many skeletons of coralline algae contain smaller or larger amounts of magnesite. The latter increases with growth rate e.g., temperature, light, food. Some sediments contain in their fractions magnesite as a weathering- or abrasion product.

Infrared spectra (Fig. 7.2; Table I). Magnesite has three strong bands at 1445, 886, 748 cm^{-1}. It can easily be distinguished from calcite (1422, 876, 712 cm^{-1}).
Gelmagnesite has the same bands but there are in addition some OH bands in the 2920—3700 cm^{-1} region.

Remarks. Magnesite is a source for the manufacture of metallic magnesium, the latter is a well known light metal. Burnt magnesite at ca. 1200° C is used in refractory bricks and together

TABLE I

CO_3 vibrations of various minerals, in relation to the kind of the cation.

ADMIXTURES

No.	Origin	ν_1 Weak	ν_2 Weak	ν_3 Weak	ν_4
A) Ditrigonal					
Magnesite					
1609	Nye Co. Nevada		854 887	1444	748
1445	Vaco mine California		856 886	1445	748
1117	Stevens Co., Calif		855 886	1444	748
411	Steiermark, Austr.		856 886	1446	748
1610	Utah		856 886	1485 1445	747
1885	S. Benito, Calif.		856 887	1485 1445	748
1614	S. Benardino, Calif.		856 885	1445	748
1884	St. Clara, Calif.		856 886	1485 1449	748
1611	Izobal Guatemala		855 887	1485 1445	747
	Average		856 886	1485 1445	748
Co = carbonate					
1764	Artificial	1065	838 866	1421	743
Smitsonite (Zn)					
1674	Kelly, Nw. Mexico		870	1421	743
1710	Tsumeb, Afr.		870	1422	743
1897	Grootfontein, Afr.		839 871	1422	743
1592	Durango, Mex.	1057	869	1422	743
1895	Rapid Bay, Austr.	1058	870	1421	743
18[illegible]	Cerro-Gordo, Calif.		870	1422	743
	Average	1058	839 870	1422	743
Siderite (Fe)					
1591	Maas plain, Netherl.	1032 1062	867	1419	736
1747	Steiermark, Austr.	1028 1060	867	1416	737
537	Siegen.Germ.	1030 1060	868	1417	737
2050	Roxburry, Conn.	1030 1060	868	1417	737
1204	Braunschweig, Germ.	1028	868	1416	736
	Average	1028 1060	868	1418	737
Rhodochrosite (Mn)					
850	Colorado	1070	837 867	1421	725
1888	Sonora, Mex.	1065	867	1422	725
851	Freiburg. Germ.	1065	867	1421	726
1593	Butte, Mont.	1070	867	1422	725
1889	Catamarca, Arg.	1066	837 867	1421	725
	Average	1066	837 867	1421	725
Cd Carbonate					
2247	Artificial	1075	861	1414	723
Calcite					
1411	Marmor album		847 876	1422	712
1043	Travertine, Rome.	1060	848 877	1421	711
1521	Crestmore, Calif.	1062	847 876	1420	712
1325	Chihuahua, Mex.	1058	847 876	1423	712
391	Serajewo, Jugosl.	1062	848 876	1423	712
1618	Japlin, Miss.	1058	847 876	1422	712
1427	Livingstone, Mont.	1058	847 876	1421	712
1535	Flagstaff, Ariz.		847 876	1420	712
	Average	1060	848 876	1422	712
1339	Ostrea shell	1062	848 877	1422	712

No.		ν_1 Weak	ν_2 Weak	ν_3 Weak	ν_4 Weak
B) Trigonal					
Dolomite $CaMg(CO_3)_2$					
1995	Teruel, Spain		850 882	1430	728
1604	Tyrone, Nw.Mex.		850 880	1430	727
1603	Vermont		850 882	1432	728
1601	Thornwood, N. York		850 882	1434	728
1496	Snarun, Norway		852 882	1434	728
1433	Ontario, Canada		881	1430	728
1130	Missouri		852 882	1428	728
1129	Nue Co., Nevada		852 882	1432	728
	Average		851 882	1432	728
C) Rhombic					
Cerusite Pb					
1868	Pinal Co., Arizone	1051	838	1389 1426	674
1869	Broken Hill, N.S.Wales	1051	838	1389 1430	676
1597	Vanadia, Nw. Mex.	1051	838	1388 1426	676
1866	Tsumeb, Afr.	1051	838	1388 1428	676
1520	Kellog, Idaho	1051	838	1390 1428	676
	Average	1051	838	1389 1428	676
Aragonite (Ca)					
1437	Pinal Co., Ariz.	1084	842 858	1473	700 711
1150	Corocoro, Boliv.	1084	843 858	1472	700 712
1160	Aragon, Spain	1084	843 858	1473	700 712
1333	Somersett Engl.	1084	843 858	1473	700 712
1620	Oeral U.S.S.R.	1085	844 858	1473	700 712
1113	Billn, Böhm	1084	843 858	1473	700 712
1600	Texas	1084	843 858	1473	700 712
1338	Carlsbad, C.S.S.R.	1084	843 858	1474	700 712
	Average	1084	843 858	1473	700 712
Shells					
534	Hydro vulva	1084	843 863	1475	700 713
291	Cerithium	1084	843 863	1475	700 713
1491	Cardium, edule	1084	843 862	1475	700 712
	Average	1084	843 863	1475	700 713
Strontianite (Sr)					
1771	Artificial	1073	843 860	1459	697 704
1873	Bruck, Austria	1072	842 859	1453	697 703
1867	Strontian, Scotl.	1073	842 858	1446	697 704
854	Ahlen, Germ.	1074	859	1460	697 704
	Average	1073	842 859	1458	697 704
Witherite (Ba)					
1853	Elisab.Town, Ill.	1060	840 859	1428	691
1891	Durham, Engl.	1059	859	1428	691
1892	Cumberland, Engl.	1060	839 859	1428	691
	Average	1060	839 859	1428	691

Cation	Ca	Sr	Pb	Ba	Mg	Co	Zn	Fe^{2+}	Mn	Cd	Ca	(Ca, Mg)
Radius Å	0.99	1.12	1.20	1.34	0.66	0.72	0.74	0.74	0.80	0.97	0.99	0.82
Mass g	40.07	87.63	207.21	137.36	24.32	58.97	65.37	55.84	54.93	112.4	40.07	32.20
Electronegat. eV	1.0	1.0	1.6	0.9	1.2	.17	1.5	1.7	1.4	1.5	1.0	1.1

v_1 = symmetric stretching; v_2 = out of plane bending; v_3 = doubly degenerate asymmetric stretching; v_4 = double degenerate planar bedding.

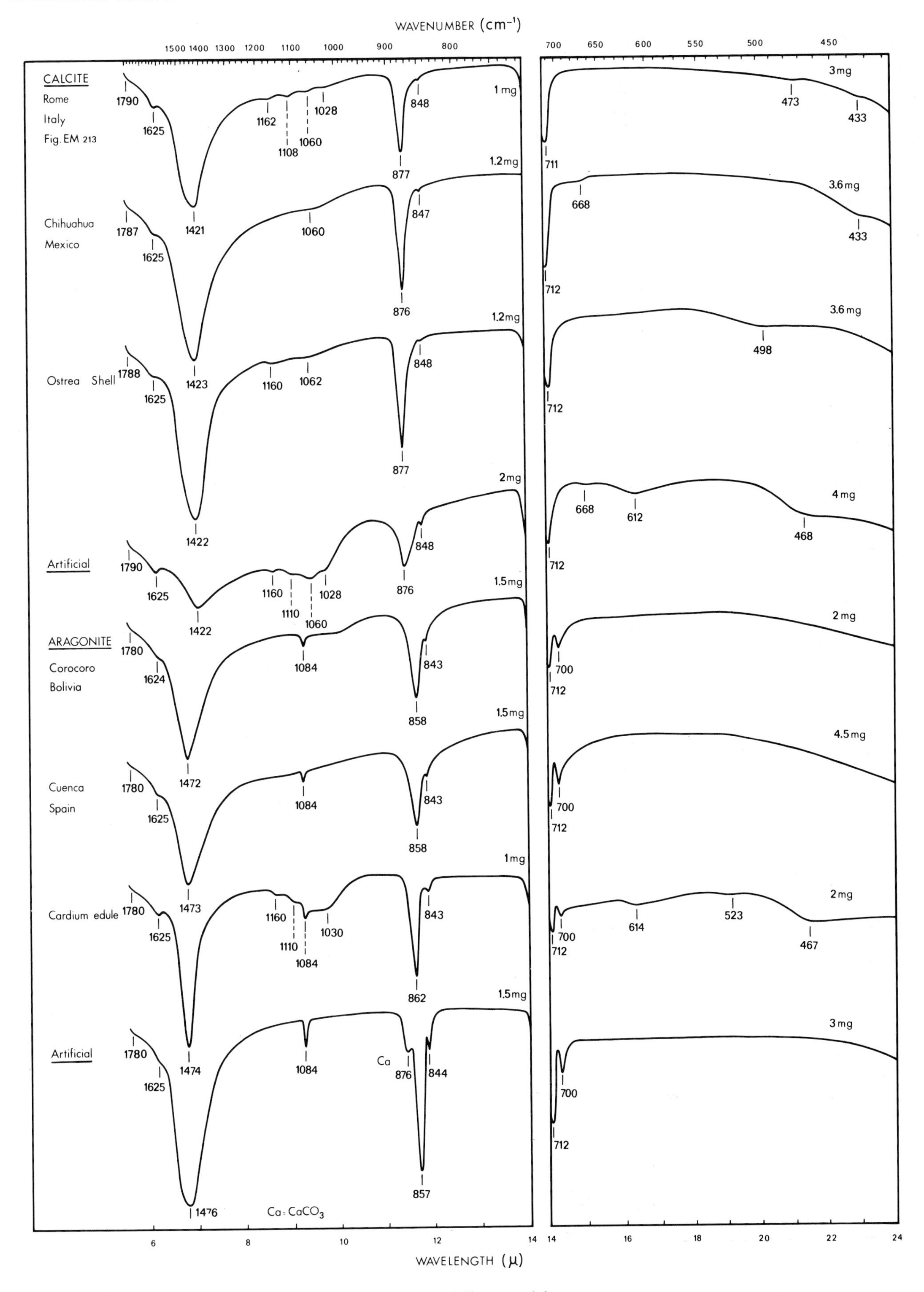

Fig. 7.1. Infrared spectra of calcite and aragonite from different origin.

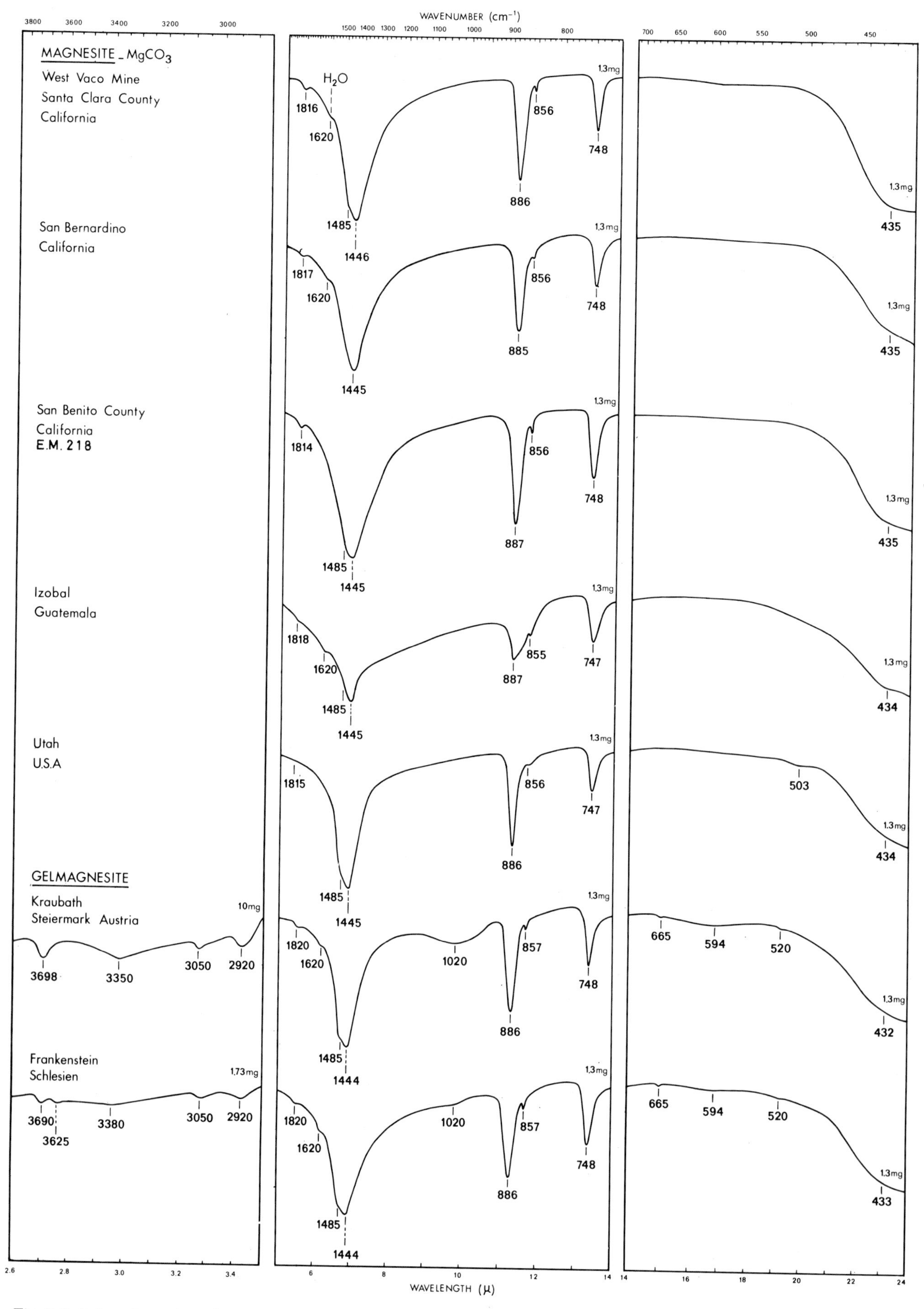

Fig. 7.2. Infrared spectra of magnesite and gelmagnesite from different origin.

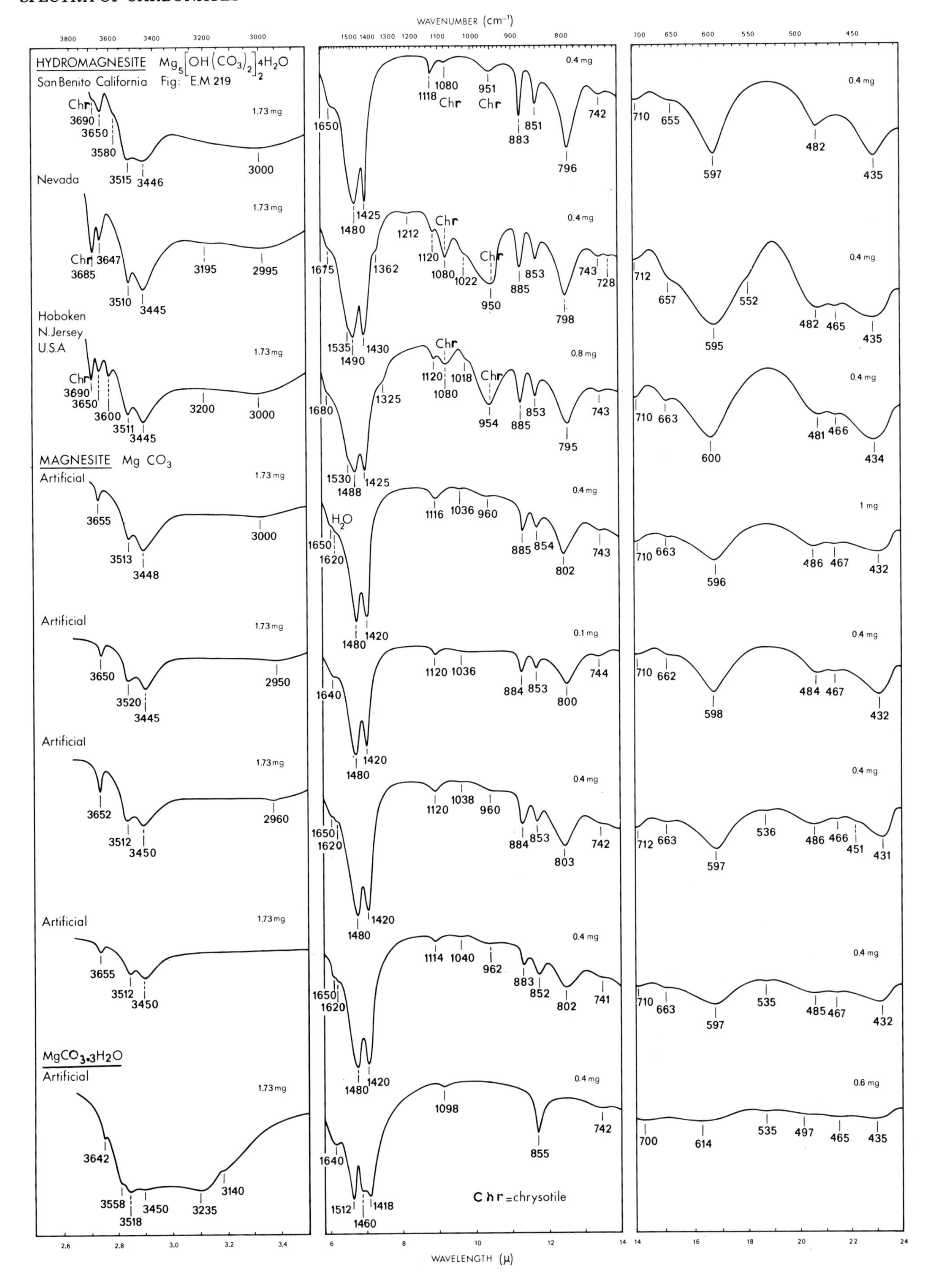

Fig. 7.3. Infrared spectra of hydromagnesite and artificial magnesite from different origin.

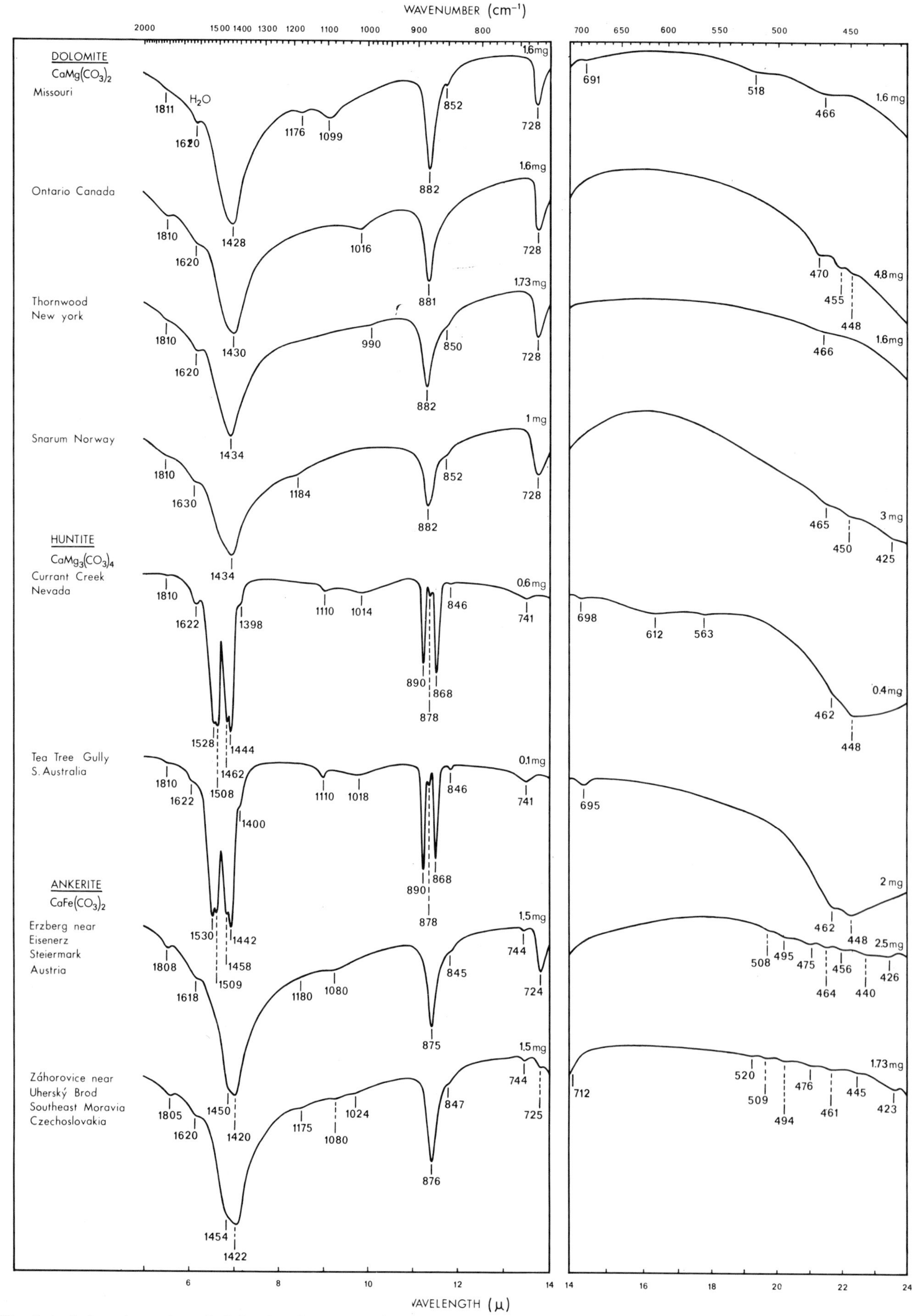

Fig. 7.4. Infrared spectra of dolomite, huntite and ankerite from different origin.

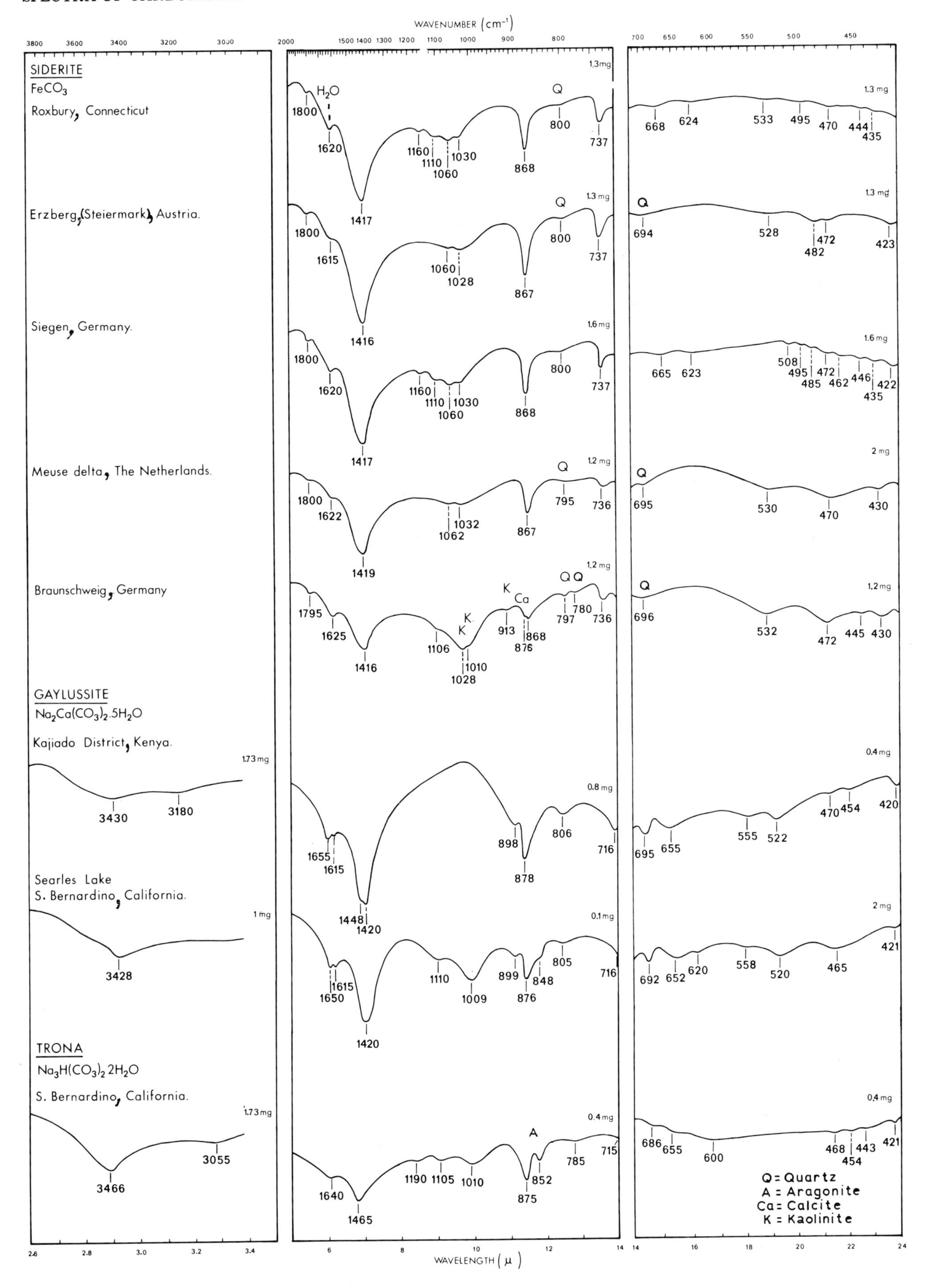

Fig. 7.5. Infrared spectra of siderite, gaylussite and trona from different origin.

with $MgCl_2$ in Sorel cement for floorings and stuccos. When mixed with asbestos, magnesite serves as a heat-isolation material. When the carbonate is heated at ca. 1500°C the magnesite is converted into periclase which is used in the manufacture of magnesite brick in steel furnaces.

Hydromagnesite and artificial magnesites; formulae:
$Mg_5[OH|(CO_3)_2]_2 4H_2O$ and $MgCO_3 \cdot 3H_2O$

Occurrence. Hydromagnesite is found in caves and caverns as an alteration product of serpentines. The $MgCO_3 \cdot 3H_2O$ investigated here, is an artificial product (natural mineral = nesquehonite).

Infrared spectra (Fig. 7.3). The spectra show that artificial magnesites made from $MgCl_2$ and Na_2CO_3 are alike natural occurring hydromagnesites. Characteristic bands are at: (artificial magnesite between brackets) 3650 (3652), 3511 (3512), 3445 (3448), 1485 (1480), 1428 (1420), 885 (884), 853 (853), 796 (802), 597 (597), 482 (485), 466 (467) cm^{-1}.
The $MgCO_3 \cdot 3H_2O$ mineral is somewhat different. It has a strong 3235 cm^{-1} band which points to H_2O molecules strongly bonded to Mg. Other characteristics for its identification are the 1512 and 1460 cm^{-1} bands and the absence of the 597 (597) vibrations of the hydromagnesites.

Remarks. Nesquehonite is named after the type locality Nesquehoning Pennsylvania where the mineral is found as small stalactites in a coal mine.

Dolomite, huntite, ankerite; formulae: $CaMg(CO_3)_2$, $CaMg_3(CO_3)_4$, $CaFe(CO_3)_2$

Dolomite and huntite are respectively 1 : 1 and 1 : 3 regular interstratified minerals consisting of calcite-like and magnesite-like layers.
Ankerite is a 1 : 1 regular interstratified mineral of calcite and siderite.

Occurrence. Dolomite rocks are found all over the world. Huntite, ankerite and many other polycomponent isomorphous systems (Ba-Mg, Pb-Mn, Ca-Mn, etc.) of the carbonates are all rare.

Infrared spectra (Fig. 7.4; Table I). Characteristic bands of large intensity for dolomite are in increasing order 728, 882 and 1432 cm^{-1}. Huntite can easily be recognized by the four bands in the 1443—1529 cm^{-1} and the three bands in the 868—890 cm^{-1} region. Characteristic bands for ankerite are at 1807, 1620, 1452, 1421, 876 cm^{-1}.

Remarks. Dolomite is used as a building material in civil engineering. Its resistance against weathering and abrasion is larger than that of calcite. When burnt it is used for the manufacture of magnesian brick for basic open-hearth furnaces.

Siderite (ferrocarbonate); formula: $FeCO_3$

Occurrence. Siderite may be found in the neighbourhood of peats. It resulted from loss of CO_2 from ferrobicarbonate solutions from the peats. It is commonly accompanied by limonite ore, an oxydation product of ferrocarbonate.

Infrared spectra (Fig. 7.5; Table I). Characteristic bands of large intensity for ferrocarbonate are in increasing order of intensity: 1618, 1800, 737, 868, 1418 cm^{-1}.

Remarks. Hard ferrocarbonate banks of large extent occurring incidentally near peats may be serious obstacles in the construction of roads, canals, etc.

Gaylussite, trona; formulae: $Na_2Ca(CO_3)_2 \cdot 5H_2O$, $Na_3H(CO_3)_2.2H_2O$

Occurrence. The above minerals are common substances in saline (solontchaks) and alkaline (solonetz) soils (marshes) which cover large parts of the world in the drier areas. Furthermore in salt lakes.

Infrared spectra (Fig. 7.5). Characteristic bands for gaylussite are at 1420 and 876 cm^{-1} which are also found for ankerite (1421, 876 cm^{-1}) and calcite (1422, 876 cm^{-1}). They can be distinguished from each other by the bands at 1178, 1080, 509, 494, 475, 442 cm^{-1} for ankerite and at 3429, 1652, 898, 805, 693, 654, 556, 521 cm^{-1} for gaylussite, and 1788, 1060 for calcite.
The spectrum of trona is very poorly developed. It has bands at 3466, 3055, and 1465 cm^{-1} which are not found for the other carbonates.

Remarks. The presence of salts in soils is very harmful for the cultivation of crops. Their removal is very expensive and not always justified economically. Millions of acres lie fallow for this reason and wait for reclamation.
Gaylussite is named after the French chemist L.J. Gay-Lussac (1778—1850) and strontianite after Strontian (Scotland).

Sulfur, sulfates and sulfides

The above minerals are widespread in the world; especially in sediments. They occur as films (sulfur), coatings, efflorescences (sulfates), banks (sulfides). But also rocks exist such as for gypsum and anhydrite.

Sulfur; formula: S_8

Occurrence. Rhombic S is found in craters (fumaroles) as a sublimation product of volcanic gases and as an oxidation product of H_2S by bacterial action (Beggiatoa).
The S accumulated in the cells as fine amorphous grains = μ sulfur and it is non-soluble in CS_2. Sulfur is also found as a thin white film floating on the water in marine estuaries as a result of bacterial action (Microspira desulfuricans, Desulfovibrio desulfuricans) in a reducing environment, with a redox coefficient (r_H) = < 10 (well-aerated soils = 26—30).
Another important source of sulfur is provided by the fossil banks near salt domes. In this case very large amounts may be concentrated which may easily be won (melting process).

Infrared spectra (Fig. 7.6). The spectra of sulfur do not show, except some contamination with $CaCO_3$, any appreciable vibrational activity. This is in contrast with their X-ray spectra, which all have well-defined reflections. The poor infrared spectrum is caused by the symmetric arrangement of the S_8 ring molecules in the crystal, which does not permit an appreciable change in their dipol moment when the atoms vibrate.

Remarks. Sulfur is a widely used raw material for the manufacture of all kinds of products: sulfuric acid, hard rubber, fungicides etc.

Sulfates; formulae:
$(K,Na,Ca,Mg,Fe,Al)[(OH,F,Cl)|(SO_4)]nH_2O$

Occurrence. Most sulfate minerals occur as films, coatings and efflorescences, e.g., mirabilite-$Na_2SO_4 \cdot 10H_2O$, glauberite-$CaNa_2[SO_4]_2$, polyhalite-$K_2Ca_2Mg[SO_4]_4 . 2H_2O$, melanterite-$FeSO_4 . 7H_2O$, jarosite-$K,Fe_3^{3+}[(OH)_6|(SO_4)_2]$, natrojarosite-$NaFe_3^{3+}[(OH)_6|(SO_4)_2)]$, ettringite-$Ca_6Al_2[(OH)_4|(SO_4)]_3 \cdot 30\ H_2O$) etc.
Gypsum $CaSO_4 \cdot 2H_2O$ and anhydrite ($CaSO_4$) are also found as massive rocks of large extension. Selenite is a colourless transparent variety of gypsum.
Sulfate solutions may be formed by action of S oxydizing bacteria e.g. Bacillus thiooxidans, Beggiatoa, etc.

Efflorescences of gypsum when formed at the surface of roofing slates and which resulted from reaction of the $CaCO_3$ in the slates with SO_2 fumes form neighbouring plants may cause large damages. Its volume is twice that of the single components. The same holds for ettringite which crystallizes with 30 molecules H_2O and which is formed when for instance sulfate solutions in the ground water react with mortar ("cement bacillus").

Infrared spectra (Fig. 7.7; Appendix 3). The free polyatomic sulfate molecule SO_4^{-2} has in its ideal tetrahedral configuration $3N$—6 = 9 normal vibrational modes, e.g., two triply degenerate vibrations ν_3 (ca. 1100 cm^{-1}) and ν_4 (ca. 620 cm^{-1}), one weak doubly vibration ν_2 (ca. 450 cm^{-1}) and one weak non-degenerate vibration ν_1 (ca. 1000 cm^{-1}), All four are Raman-active but only ν_3 and ν_4 are IR-active. However, in a crystal the symmetry may be changed from which results that also ν_1 and ν_2 become IR active.
The introduction of appreciable amounts of cations of different mass, size and charge in the sulfate crystal decreases the environmental symmetry of the SO_4 molecule. This holds in particular for the presence of two or more different cations and anions such as in various complicated sulfate minerals.
The lower the symmetry in the sulfate mineral, the more the triply and doubly degenerate vibrations are split into their single components and the stronger the weak ν_1 (and ν_2) vibrations will appear in the spectrum. Mass, radius and electronegativity of the surrounding ions are all factors which determine the wavelength of a certain vibration of the sulfate molecule. Consequently the spectra of the various sulfate minerals may show a wide variation and activity of their vibrations. In general those with the largest activity also have the most lowered structural symmetry.
Crystal water of the sulfate mineral is represented by a band at ca. 3200 cm^{-1}. But not all minerals with crystal water have this band so pronounced. The crystal water of the mineral may be so loosely bound that its O-H vibration is very near to that of common water adsorbed on the mineral surface (see mirabilite-$Na_2SO_4 \cdot 10H_2O$).
Gypsum, hemihydrate and anhydrite, three important raw materials, can easily be distinguished from each other e.g., gypsum: 3543, 3240, 1685, 1162, 668 cm^{-1}; hemihydrate: 3608, 3554 cm^{-1}; anhydrite: 1155, 677, 618 cm^{-1}.
Ettringite has broad bands of low intensity. Such because of its complicated structure. The water band at 3410 cm^{-1} is very strong.

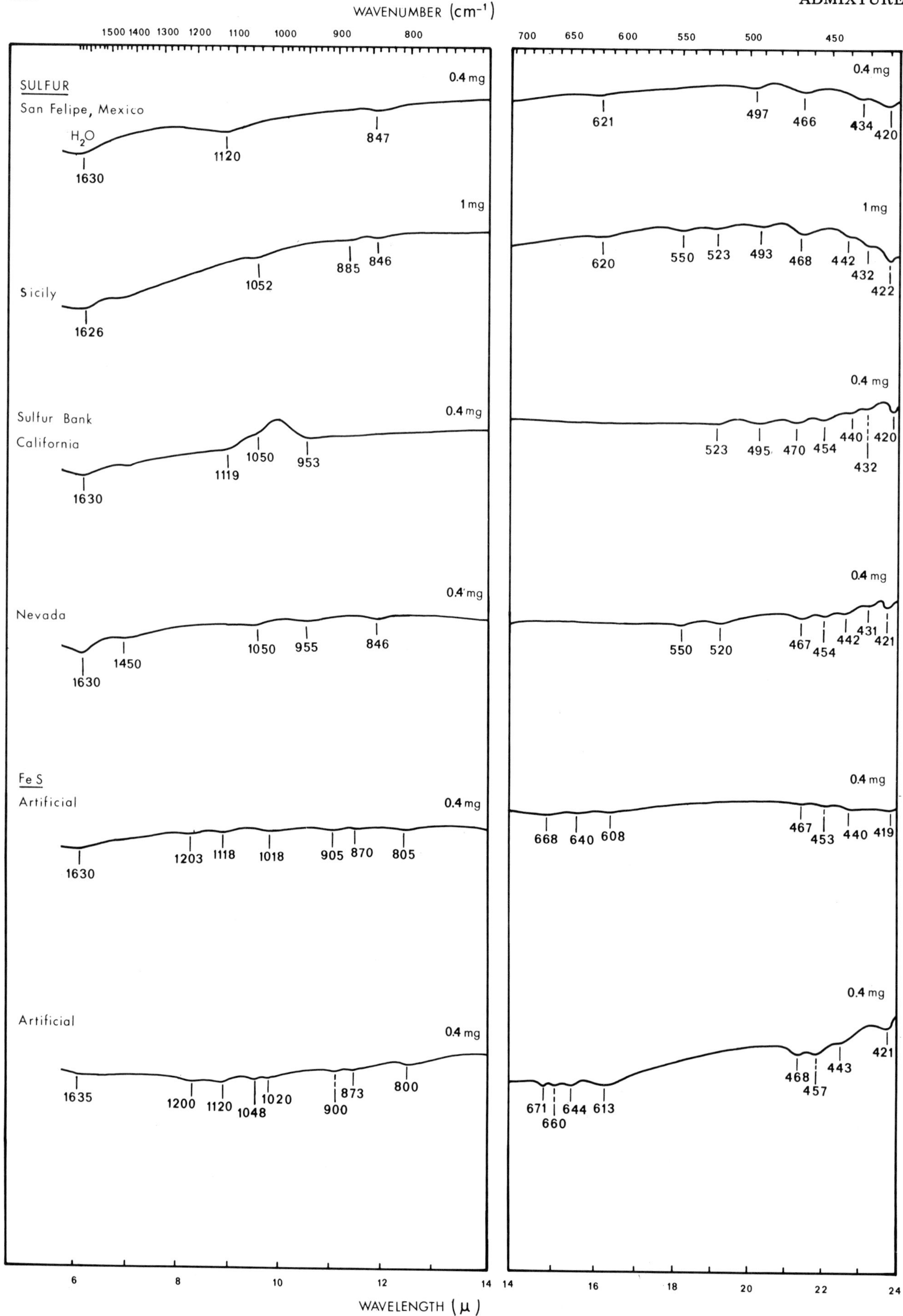

Fig. 7.6. Infrared spectra of sulfur and ferrosulfide from different origin.

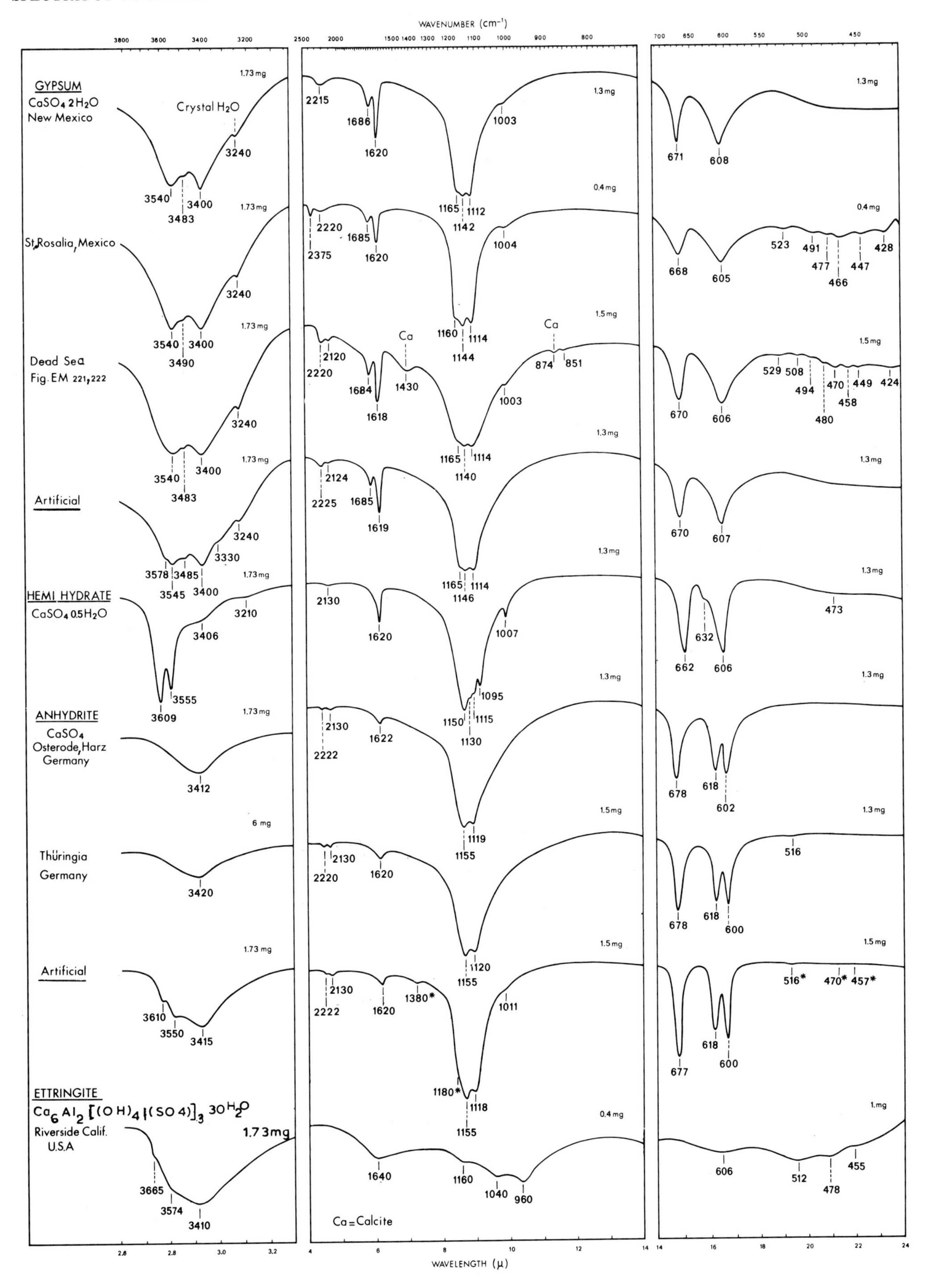

Fig. 7.7. Infrared spectra of gypsum, hemi-hydrate, and anhydrite from different origin, and of ettringite.

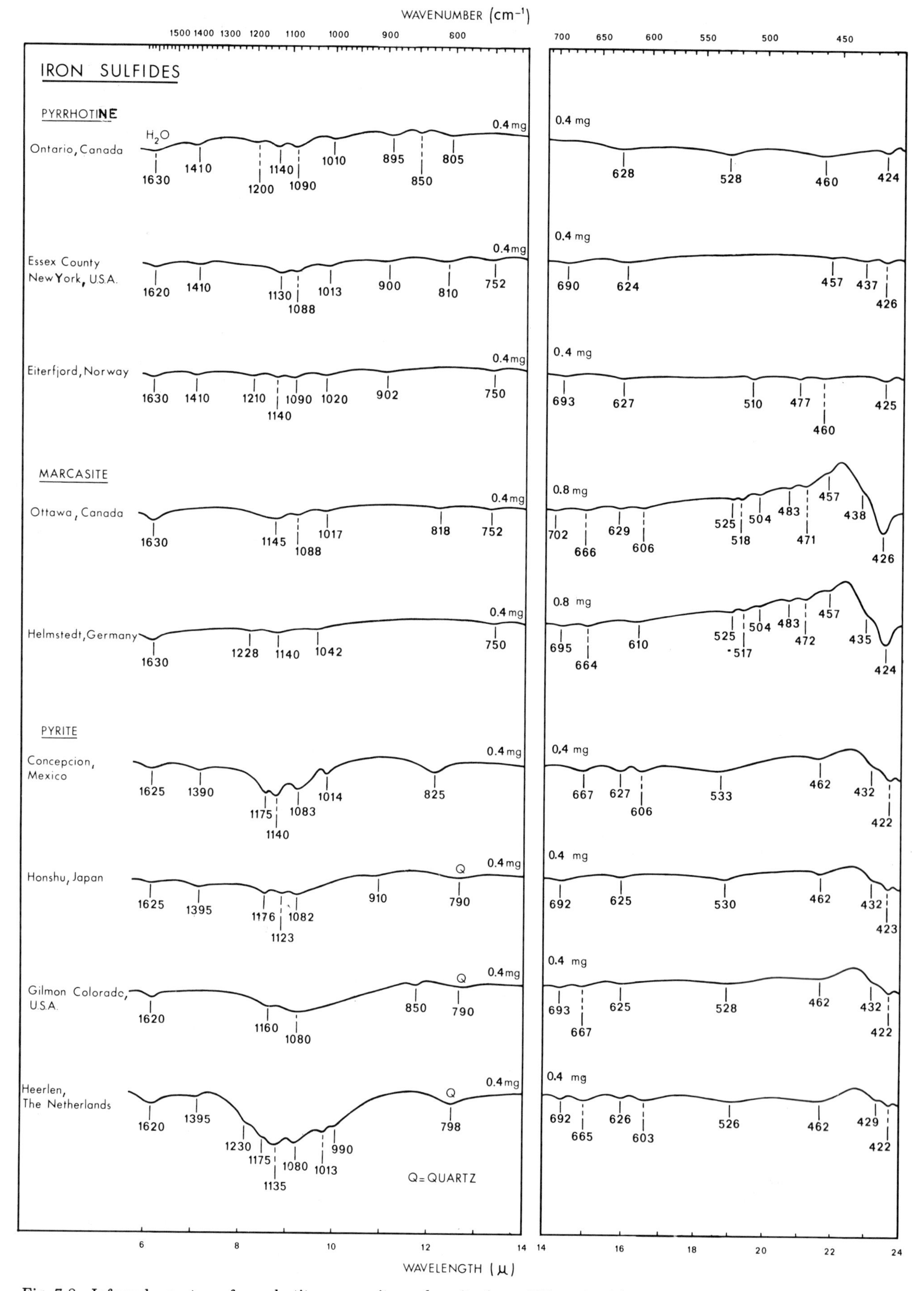

Fig. 7.8. Infrared spectra of pyrrhotite, marcasite and pyrite from different origin.

Remarks. Gypsum is of great importance in the manufacture of hemihydrate, also called Plaster of Paris = $CaSO_4 . ½H_2O$. which is used for many purposes in the building industry. It is obtained by heating gypsum to ca. 130°C (Kettle process). When the mineral is heated to over ca. 300°C, $CaSO_4$ is formed which is like the natural mineral anhydrite. It will badly take up H_2O (dead burnt gypsum) and therefore is worthless. Another application of gypsum is as an amelioration product for saline or badly permeable soils.
Barite ($BaSO_4$) is used in drilling muds to increase their specific gravity. Ettringite is named after the type locality Ettringer, Bavaria (W. Germany).

Sulfides: pyrrhotite, marcasite, pyrite; formulae: Fe_7S_8 to FeS (monocline to hexagonal), FeS_2 (rhombic) and FeS_2 (cubic)

Occurrence. The above minerals, when occurring with colloform, globular texture and concentrated to smaller or larger amounts (banks) in sediments, are mostly formed by action of sulfate reducing bacteria: Desulfovibrio desulfuricans, Microspira desulfuricans, etc. In the first instance they produce H_2S and S from sulfate solutions which further react with iron-hydroxide from the soil. There is a preference for pyrite to be formed in an alkalic solution and of marcasite in an acidic environment.
Black-coloured hydrotroilite an amorphous water-rich iron monosulfide ($FeS \cdot nH_2O$) is an intermediate product and melnikovit ($FeS_2 \cdot nH_2O$) a cryptocrystalline form of pyrite.

Infrared spectra (Fig. 7.8). The spectra of pyrrhotite and marcasite are all weak, non-constant and hardly different, except for the lowest wave numbers. The spectra of pyrite are only slightly different from each other. This is in contrast with their X-ray spectra, which show well-defined differences between the several minerals.

Remarks. The above sulfide minerals, when concentrated in deposits, are an important raw material for the manufacture of sulphuric acid and iron. For that purpose they are first roasted.

Amphiboles and pyroxenes

A. Amphibole, actinolite, tremolite (amphibole group); formulae [1]:
Amphibole: $(Ca,Na,K)_{2-3}$ $(Mg,Fe^{2+},Fe^{3+},Al)_{5}$-$[(OH,F)_2|(Si,Al)_2Si_6O_{22}]$
Actinolite: $Ca_2(Mg,Fe^{2+})_5[(OH,F)|Si_4O_{11}]_2$
Tremolite: $Ca_2Mg[(OH,F)|Si_4O_{11}]_2$

The structure consists of double chains of SiO_4 tetrahedrons linked together to $(Si_4O_{11})^{6-}$ ribbons and the negative charges compensated by Na,K,Ca,Mg,Fe and some OH bonded by Mg and Fe. Kaersutite is a titanium-rich variety of amphibole. The distinction between amphibole- and augite group minerals is sometimes very difficult because there are forms of intermediate composition.
Also the samples of each variety are not of constant composition. Thereby transitions from one to another occur.

Occurrence. Amphibole and to a lesser degree actinolite, are common minerals of basic volcanic rocks. They weather easily, although not so fast as the minerals of the pyroxene group. Consequently they are rarely found in common sediments. But in deep sea sediments where weathering is small, they are found in the finer fractions of basic volcanic rocks e.g., in the Pacific.
Tremolite is rarer than amphibole and actinolite.

Infrared spectra (Fig. 7.9). The spectra of the above minerals are well developed, although the bands are broad, and there are hardly any significant differences between the three members of the amphibole group. Kaersutite, the titaniferous amphibole mineral, has a very poor spectrum because of the introduction of an appreciable amount of a quite different atom in the crystal structure.

Remarks. At some localities amphibole and tremolite yield a fibrous material which is explored and sold as asbestos.

B. Augite, hypersthene, enstatite (pyroxene group); formulae[1]:
Augite: $(Ca,Mg,Fe^{2+},Fe^{3+},Ti,Al)_2[(Si,Al)_2O_6]$
Hypersthene: $(Fe,Mg)_2[Si_2O_6]$
Enstatite: $Mg_2[Si_2O_6]$

The structure consists of SiO_4 tetrahedrons linked together to single $(SiO_3)^{2-}$ chains and the negative charges compensated by Ca,Na,Fe,Mg.

Occurrence. Augite and hypersthene are common minerals of basic volcanic rocks. Because they weather easily, they are rarely found in the clay- or silt fraction of soils. But where weathering is slow as e.g., in deep-sea sediments of the Pacific, they are a common mineral in the finer fractions of basic volcanic sediments. Enstatite is also found in basic rocks but rarely.

Infrared spectra (Fig. 7.10). The spectra of the above minerals can easily be distinguished from those of the amphibole group by the failing of strong O-H stretching vibrations. There is only a small band left at 3650 cm^{-1} for hypersthene and at 3650 and 3683 cm^{-1} for enstatite which is caused by O-H vibrations where some O in the $(SiO_3)^{2-}$ chains have been replaced by OH. Another difference is the strong 960—966 cm^{-1} band of the pyroxene group minerals against the 980—994 cm^{-1} band of the amphibole group minerals. There is also a large difference between the shape of the spectra for the pyroxene- and the amphibole group minerals.
The spectra of hypersthene and enstatite are nearly alike. Augite may be distinguished from the foregoing by the absence of several weak, non-constant, occurring bands.

Remarks. When talc, saponite, hectorite, sepiolite and attapulgite, all Mg-rich but Fe-poor minerals are heated at ca. 1000°C, enstatite is formed. Badly ordered enstatite-like minerals are known to occur also in this process. When the sample is further heated, cristobalite appears.

[1] According to Strunz, 1970: Mineralogische Tabellen, Akad. Verlagsgesellsch., 621 pp.

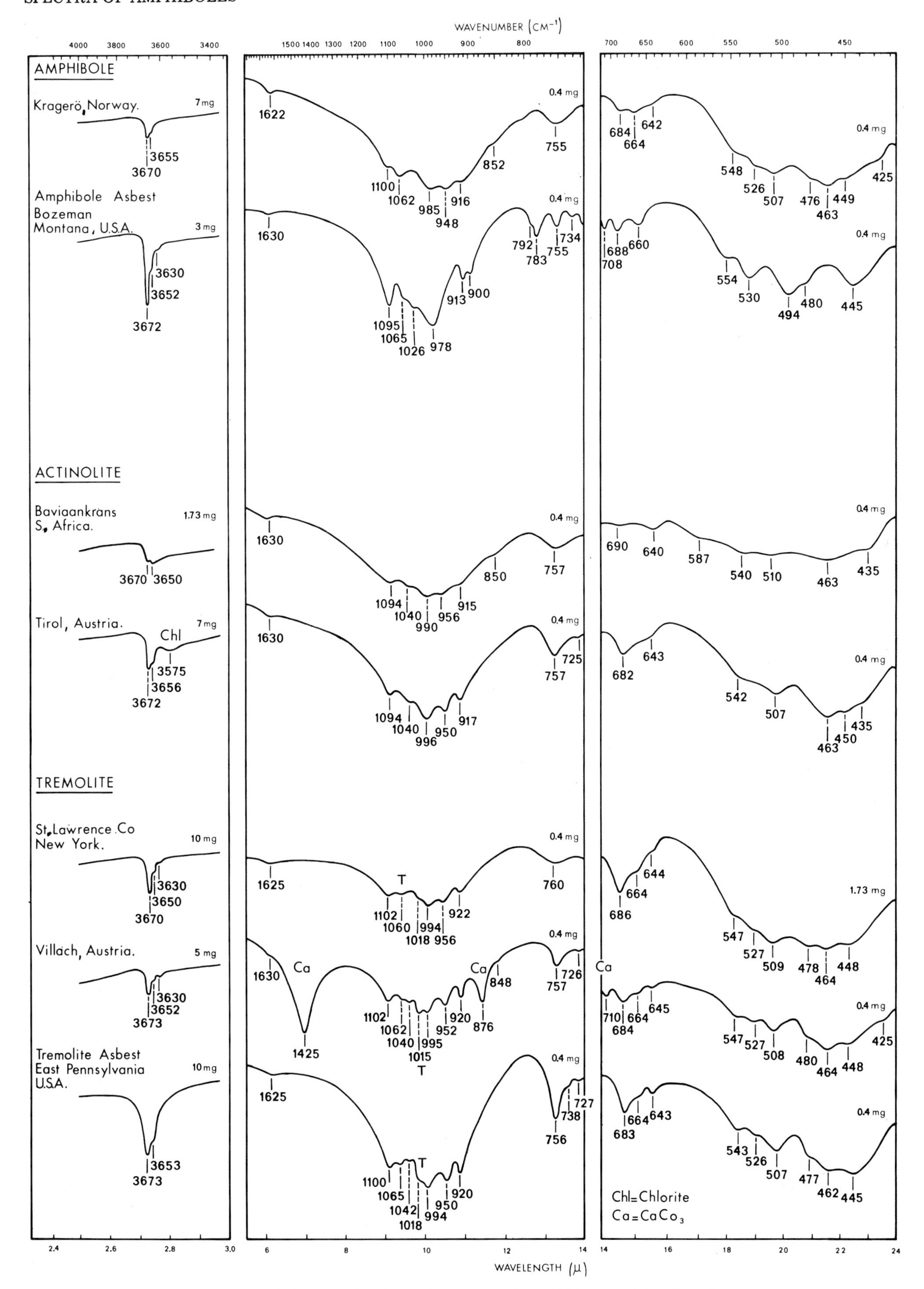

Fig. 7.9. Infrared spectra of amphibole, actinolite and tremolite from different origin. T = talc.

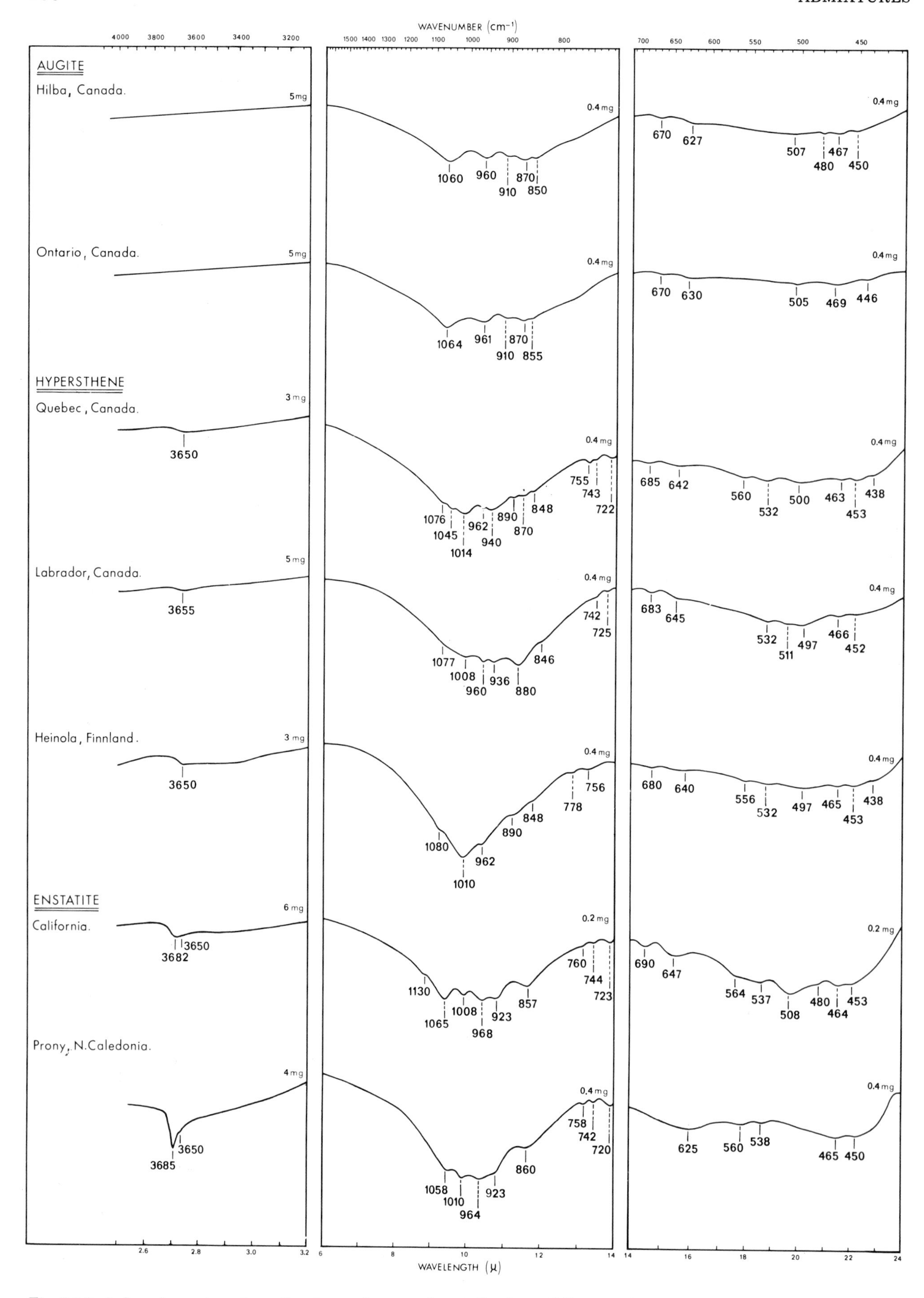

Fig. 7.10. Infrared spectra of augite, hypersthene and enstatite from different origin.

Titano minerals

Anatase, rutile, brookite, titanite(sphene), ilmenite; formulae: anatase, rutile, brookite = TiO_2; titanite = $CaTi[O|SiO_4]$; ilmenite = $FeTiO_3$.

Anatase (octahedrite) and rutile are tetragonal. The first has an oxygen lattice which is nearly cubically close-packed and the latter is nearly hexagonally close-packed. Rutile is the most stable phase of both. Brookite is orthorhombic. Titanite = monocline and ilmenite = hexagonal.

Occurrence. All the above minerals are found in small concentrations in basic magmatic and metamorphic rocks. They are all very resistant. Owing to weathering they may be concentrated in the clay ($< 2\ \mu m$) or silt ($< 16\ \mu m$) separate from various sediments, e.g. anatase in kaolinite and bauxite deposits to a maximum of even ca. 30%. Ilmenite deposits are found in U.S.A. and Finland. Another source of titanite are the black ilmenite-rich (50—70%) beach sands in India and Florida and the rutile-ilmenite beach sands in Australia.

Infrared spectra (Fig. 7.11, 7.12). The above minerals have, with the exception of the artificial products, only small weak bands. Probably, this is an effect on the large particle size of the natural minerals, even after grinding which is accentuated by the high refractive index. The scattering causes a decrease of the band intensities. Anatase can be distinguished from the others by its band at 811 cm^{-1}, rutile from the others by its band at 603 cm^{-1}, brookite from the others by its band at 1165 cm^{-1}, titanite from the others by its band at 850 and 565 cm^{-1}, and ilmenite from the others by its bands at 1572 and 1040* cm^{-1}.
Artificial anatase and rutile have large broad bands in the 750—530 cm^{-1} region, which mask their characteristic bands at 810 and 603 cm^{-1}, respectively.

Remarks. Titanium is widely used in metallurgy as an increaser of tensile strength of steel. The high indices of refraction of the TiO_2 minerals (anatase: ω = 2.554, ϵ = 2.493; brookite: α = 2.583, β = 2.586, γ = 2.741; rutile: ω = 2.616, ϵ = 2.903) are very useful in the manufacture of paints (titanwhite), printing inks, linoleum and as an opacifier in enamel-ware. The effect for particles of the same diameter is in this case twice that of ZnO and three times that of PbO. Fine (particle size = ca. 0.3 μm) artificial anatase is obtained from ilmenite, titanite ores by the sulfuric acid or hydrochloric acid process.
Artificial rutile, the most stable phase and therefore also the one that is the most wanted for certain purposes (paints, ceramics, etc.), is obtained by heating artificial anatase at ca. 900° C. Traces of ferric oxide have a strong catalytic effect on this transformation.
The name ilmenite is derived from the type locality in the Ilmen Mts., Ural.

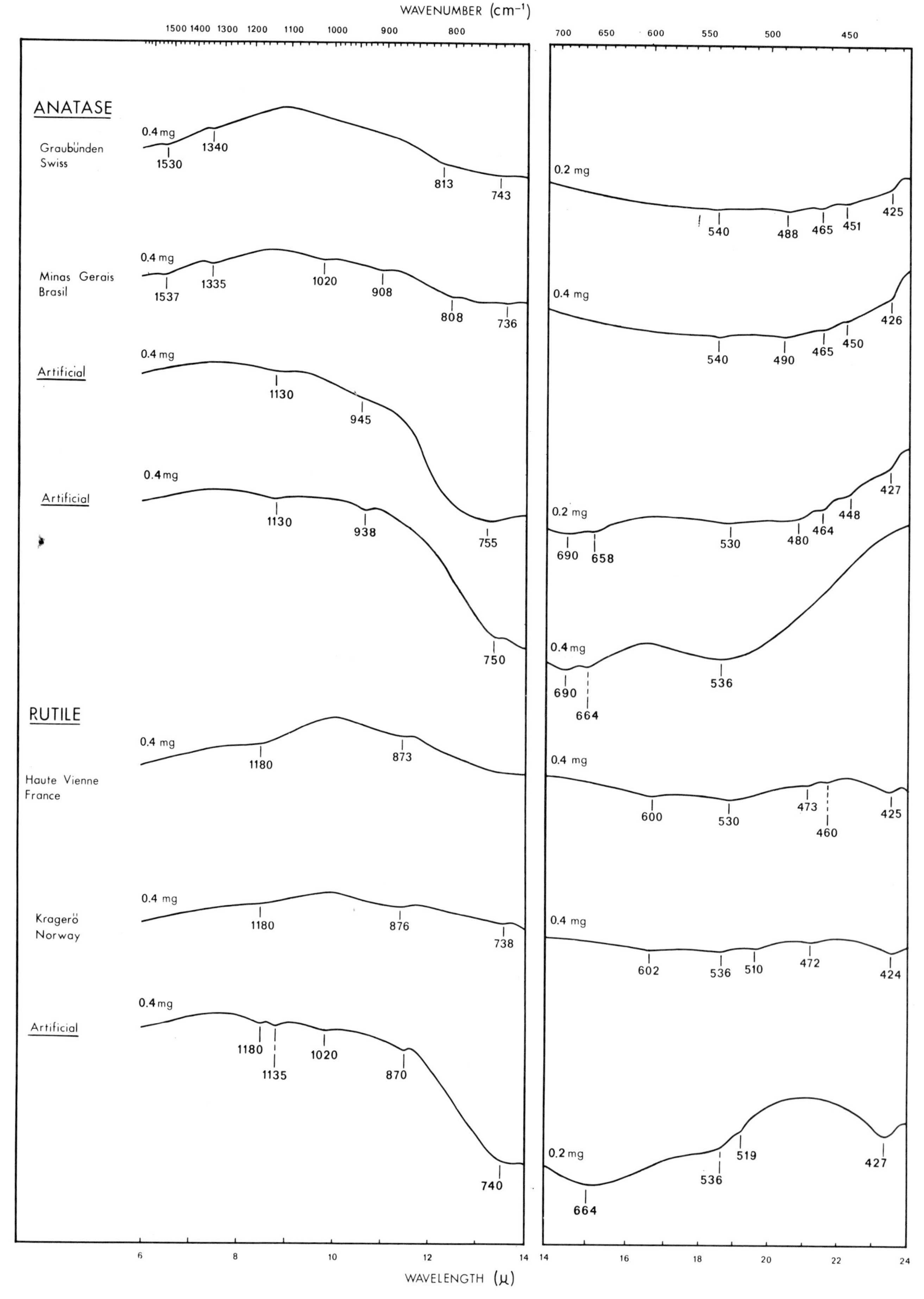

Fig. 7.11. Infrared spectra of anatase and rutile from different origin.

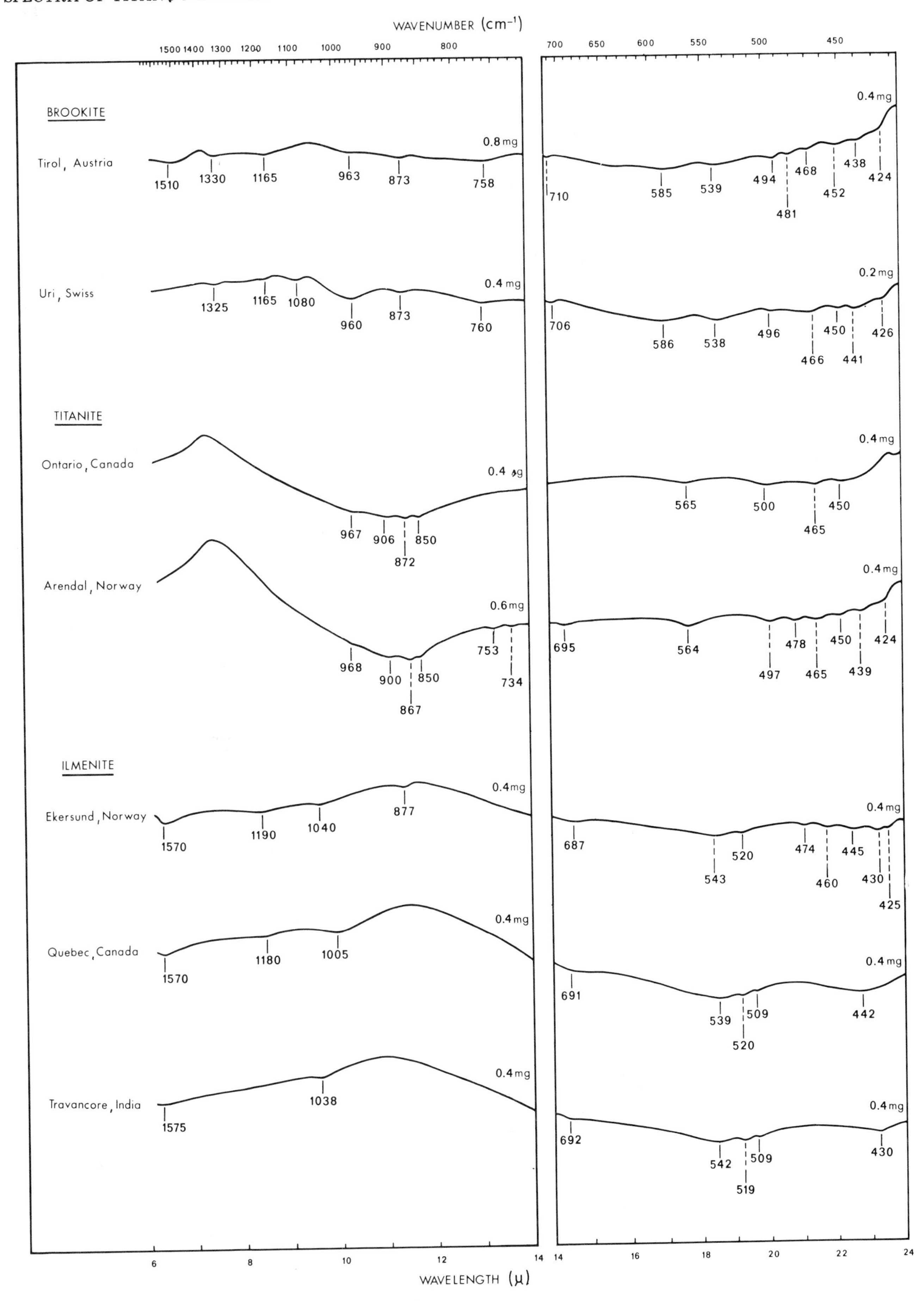

Fig. 7.12. Infrared spectra of brookite, titanite and ilmenite from different origin.

Feldspars

Feldspars consist of a three-dimensional framework of Si, Al-oxygen tetrahedrons which form rings consisting of $4(Al,Si)O_4$ tetrahedrons bridged by oxygens. The rings are linked together in such a way as to form a kind of honeycomb. Each oxygen atom belongs to two tetrahedrons and is furthermore linked to K, Na or Ca, which fill up the large interstices between the rings.

The structure of the feldspars depends on temperature of crystallization and subsequent thermal history. A feldspar which has retained its high temperature, Al/Si disordered (Al randomly distributed in the tetraeders) phase is called a high-temperature feldspar. Other phases are intermediate- and low-temperature. Triclinicity decreases and monoclinicity increases in the plagioclases when Al/Si disorder increases. The feldspars have caused many difficulties in their identification. Although their structure is very similar, there is a wide variability in details, because of the existence of series from one end member mineral to the other, differences in chemical composition, grade of Al/Si ordering, intergrowths and the nonhomogeneity of a sample. Thus the Rapakivi granites (Finland) contain K-feldspar phenocrysts, mantled by plagioclase, which resulted from magmatic differentiation. X-ray analysis, I.R. analysis, thermal analysis, optic methods (refractive index, axial angle, optic sign) and chemical analysis are used for the identification of the various feldspar minerals; each method having its own merits.

Occurrence. The feldspar minerals are very widespread; particularly in syenites, granodiorites and their volcanic equivalents. They are not only found in various kinds of rocks, but also in the $< 2\ \mu m$ separate of different sediments.

Resistancy against weathering decreases when Ca contents of the plagioclases increases. Anorthite is easily attacked. The K-feldspars, and especially albite, are very resistant. In volcanic deep-sea sediments (Japan, Pacific) and in glacial sediments where weathering action is slow, the amount of feldspar may be as high as 30%.

Na-Ca feldspars (plagioclases) - triclinic

Formula. These minerals form a discontinuous isomorphous series from $Na[AlSi_3O_8]$ = Ab to $Ca[Al_2Si_2O_8]$ = An by substitution of Na and Si by Ca and Al.

However, 100% pure endmembers are absent. They usually contain also some $K[AlS_3O_8]$.

The plagioclases are classified according to their molecular Ab/An percentage.

Albite $Na[AlSi_3O_8]$	An = 0—10%	Na, Si
Oligoclase	An = 10—30%	↓ ↑
Andesine	An = 30—50%	
Labradorite	An = 50—70%	
Bytownite	An = 70—90%	
Anorthite $Ca[Al_2Si_2O_8]$	An = 90—100%	Ca,Al

In the above series anorthite is the least resistant mineral against weathering and albite the most resistant.

Alkali feldspars

Also in this group there exists an isomorphous series between K-Na feldspars (sanidine) and K-feldspars (orthoclase, microcline). The latter are not Na-free, but always contain some %.

Many samples are not homogenous, but contain separate K-rich and Na-rich parts. The Ca members are only connected to low K contents — see also the large gap in the triangular diagram of the $Na[AlSi_3O_8]$-$K[AlSi_3O_8]$-$Ca[Al_2Si_2O_8]$ system.

Sanidine: $(K,Na)\ [(Si,Al)_4O_8]$-monocline, high-temperature phase with 6 to 7% Na_2O.

Microcline: $K[AlSi_3O_8]$-tricline low-temperature phase

Orthoclase: $K[AlSi_3O_8]$-monocline, intermediate-temperature phase.

Perthites

When two or more different structural minerals or phases of Al/Si ordering occur together intergrown (interlamellated) in the same sample, the mineral is called a perthite[1] (ultramicroscopic = cryptoperthite, microscopic = microperthite, and megascopic = perthite). The mineral results from the unmixing of an original homogeneous high-temperature alkali feldspar or by simultaneous crystallization of a K-rich and a Na-rich feldspar. To this group belong:

anorthoclase: triclinic or monoclinic albite-microline cryptoperthite

adularia: monoclinic orthoclase-microcline-sanidine cryptoperthite.

Moonstone

This mineral should consist of a lamellar micro- or crypto-intergrowth of albite (sanidine, orthoclase or microcline) with minute amounts of other feldspars. The milky iridiscent (opalescent) colour of this mineral is caused by interference effects.

[1] The name is derived from Perth, a town in the province of Quebec, where this type of feldspar mineral was first found by Thompson (1843).

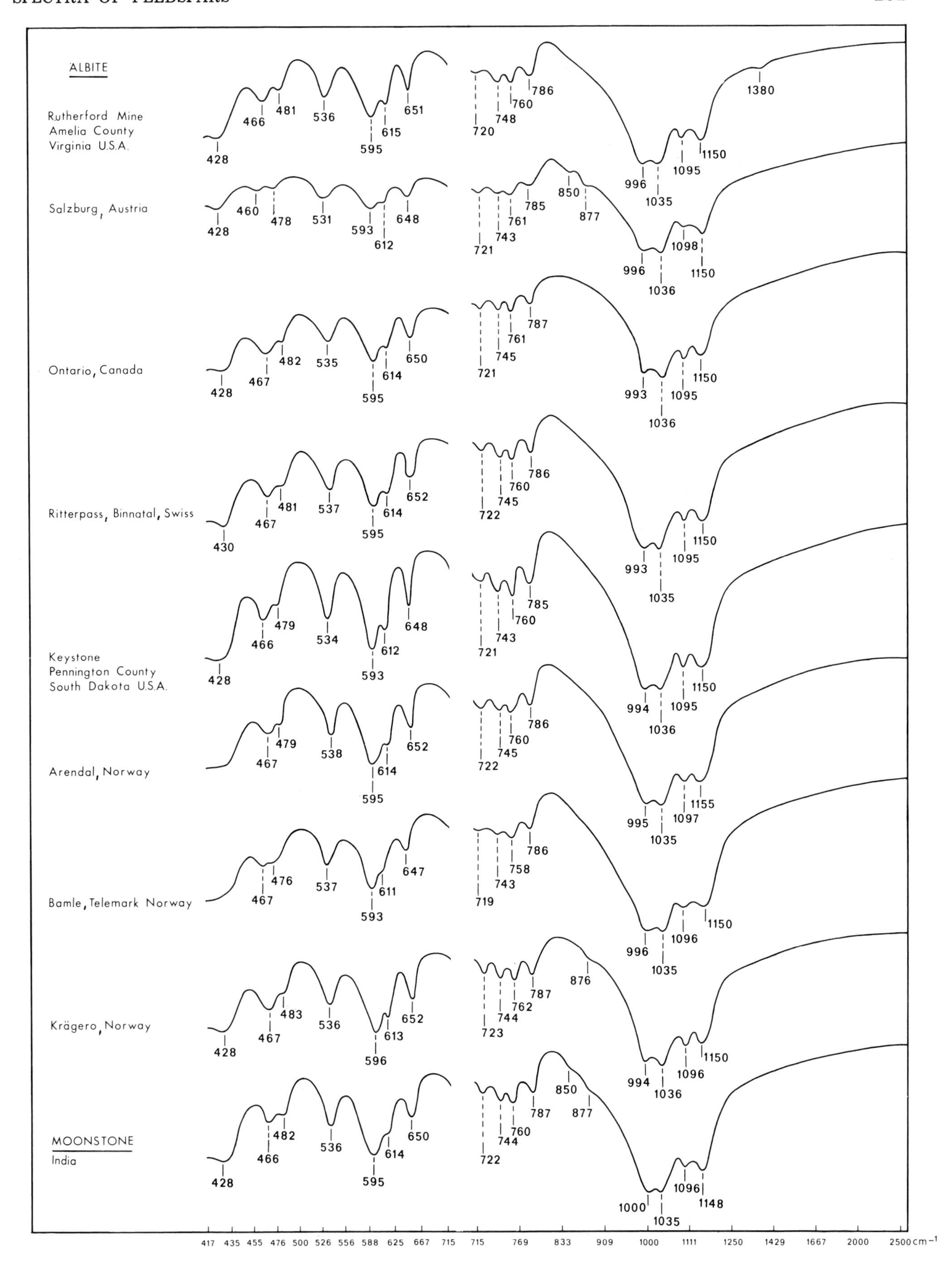

Fig. 7.13. Infrared spectra of albite from different origin (pellets: 0.4 mg on 300 mg KBr).

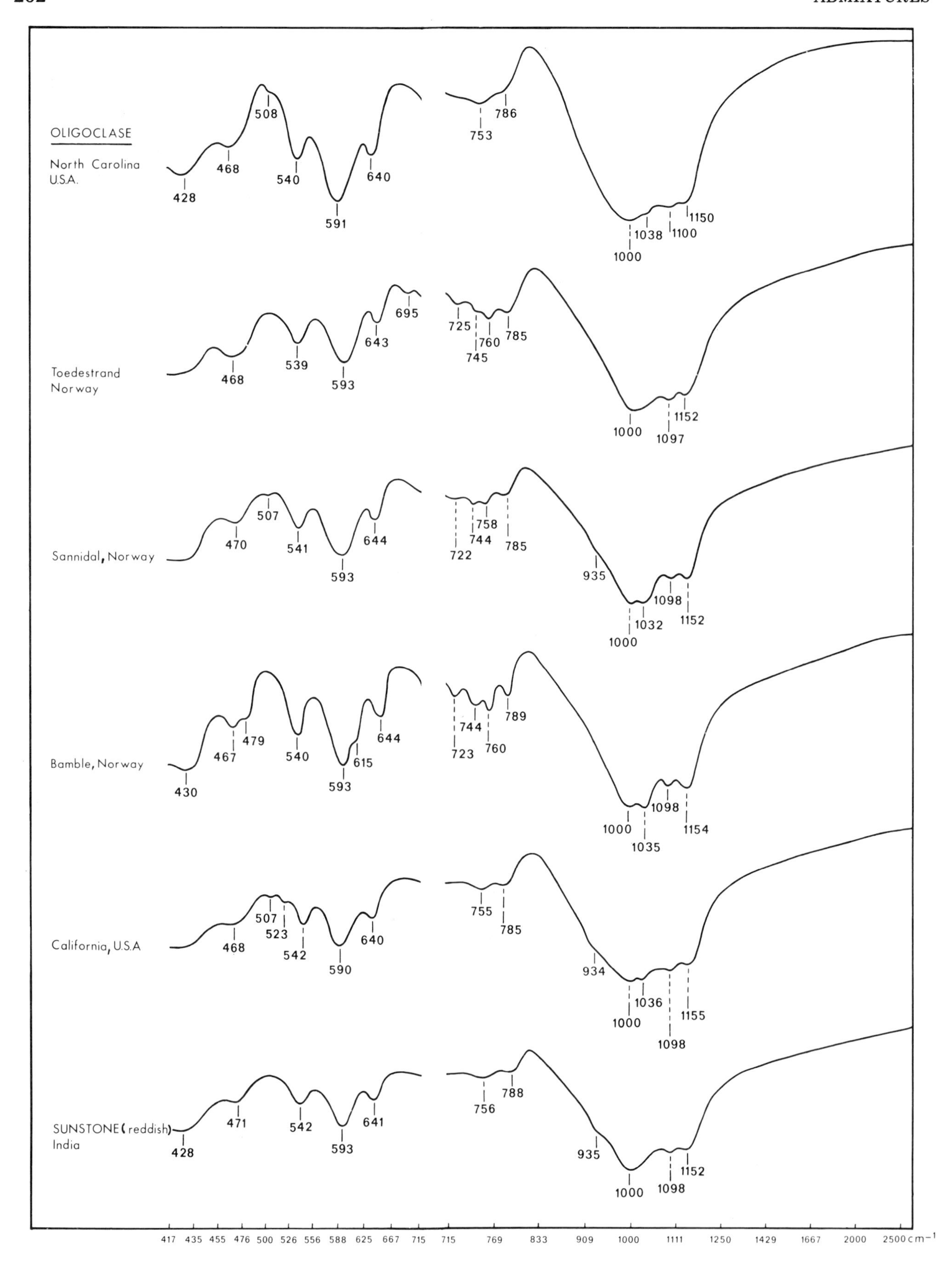

Fig. 7.14. Infrared spectra of oligoclase from different origin (pellets: 0.4 mg on 300 mg KBr).

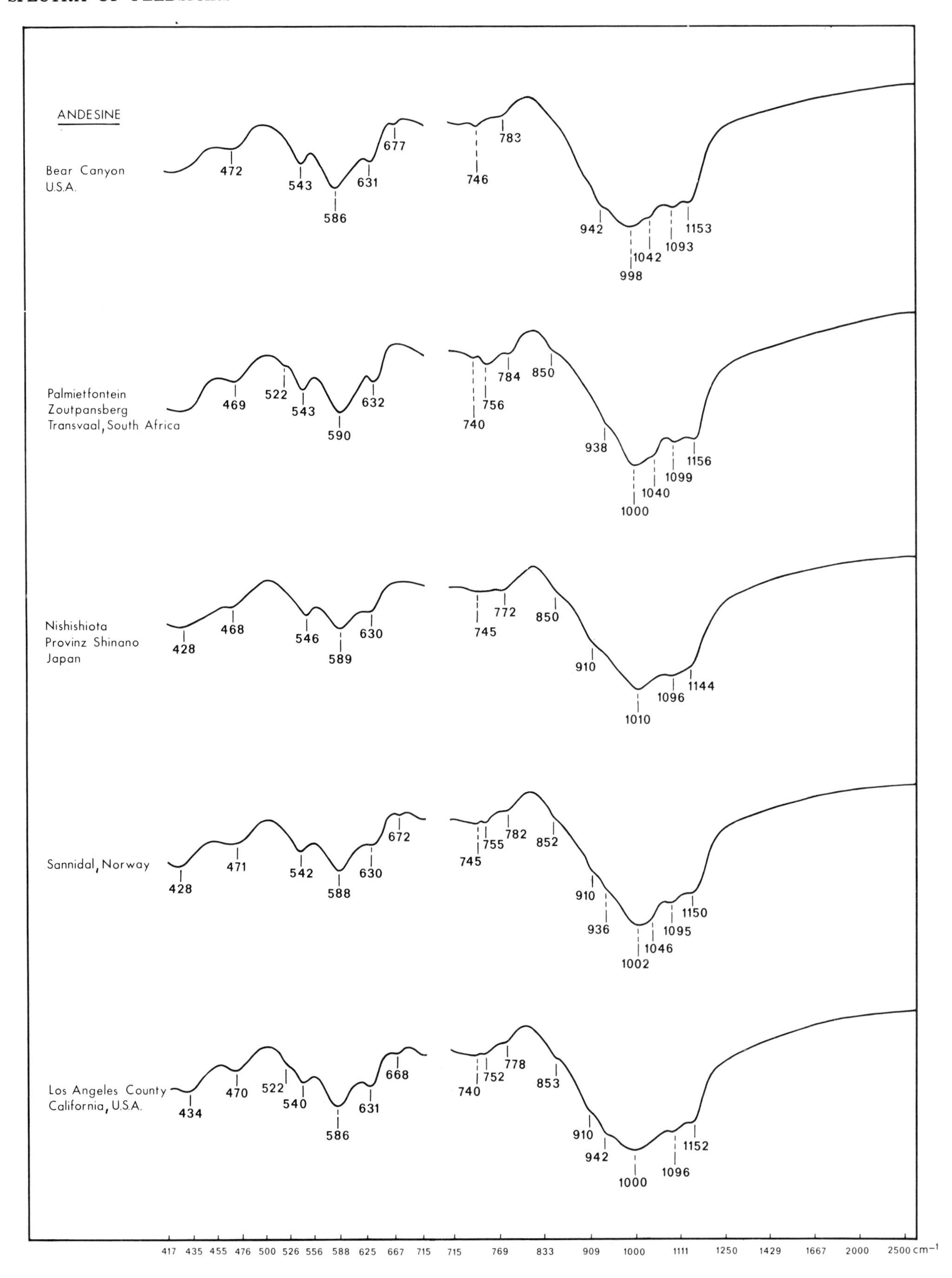

Fig. 7.15. Infrared spectra of andesine from different origin (pellets: 0.4 mg on 300 mg KBr).

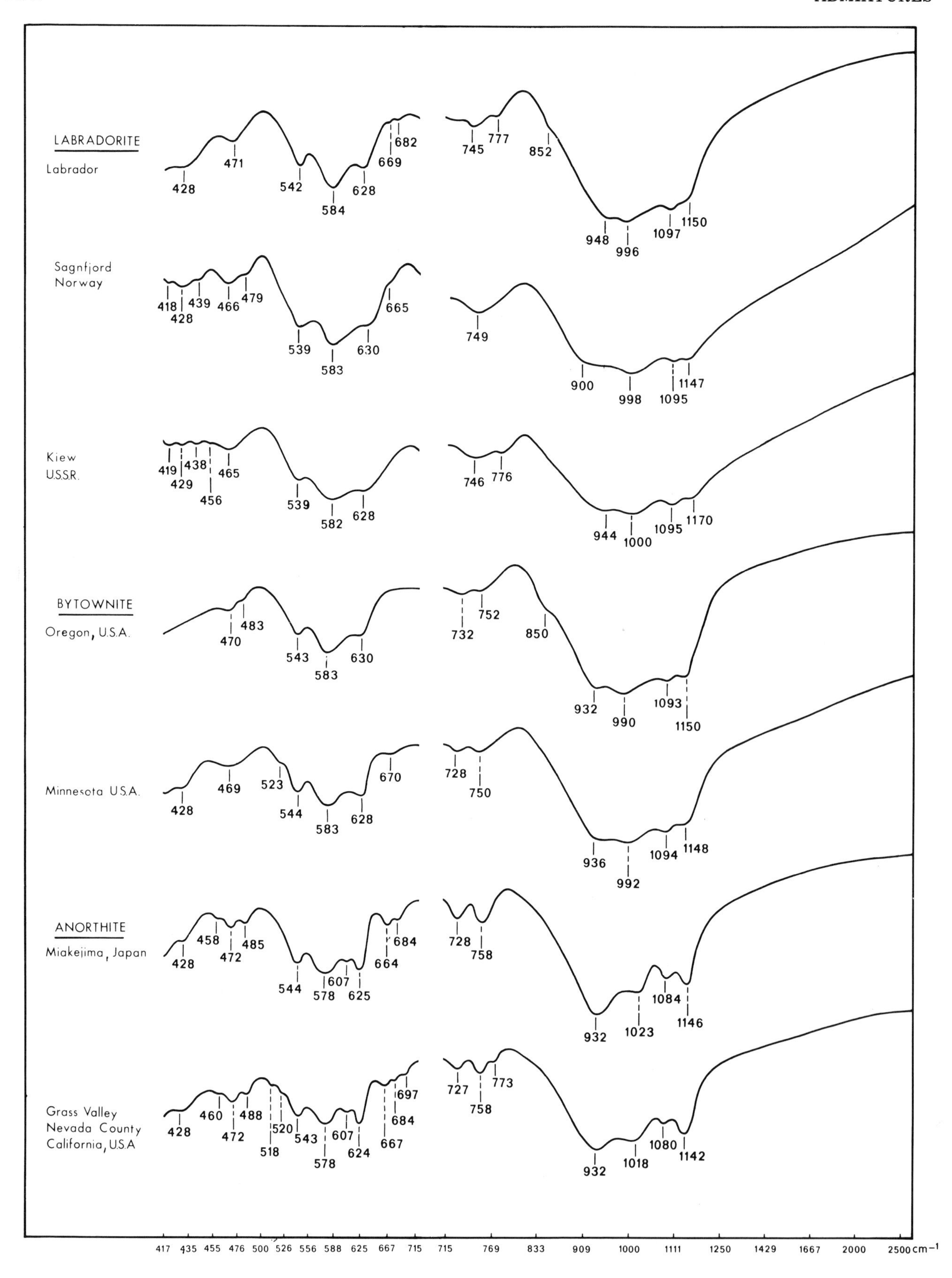

Fig. 7.16. Infrared spectra of labradorite, bytownite and anorthite from different origin (pellets: 0.4 mg on 300 mg

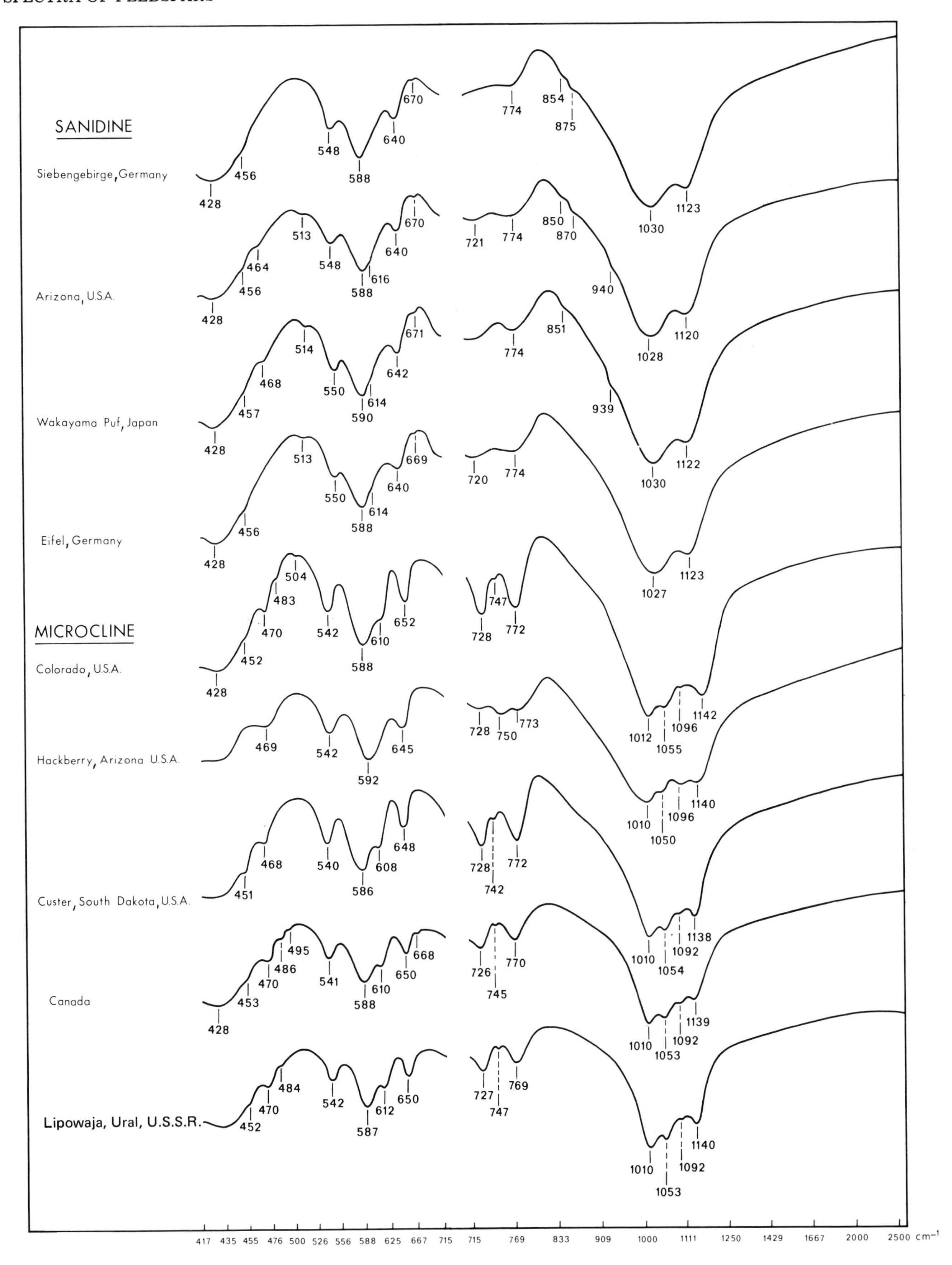

Fig. 7.17. Infrared spectra of sanidine and microcline from different origin (pellets: 0.4 mg on 300 mg KBr).

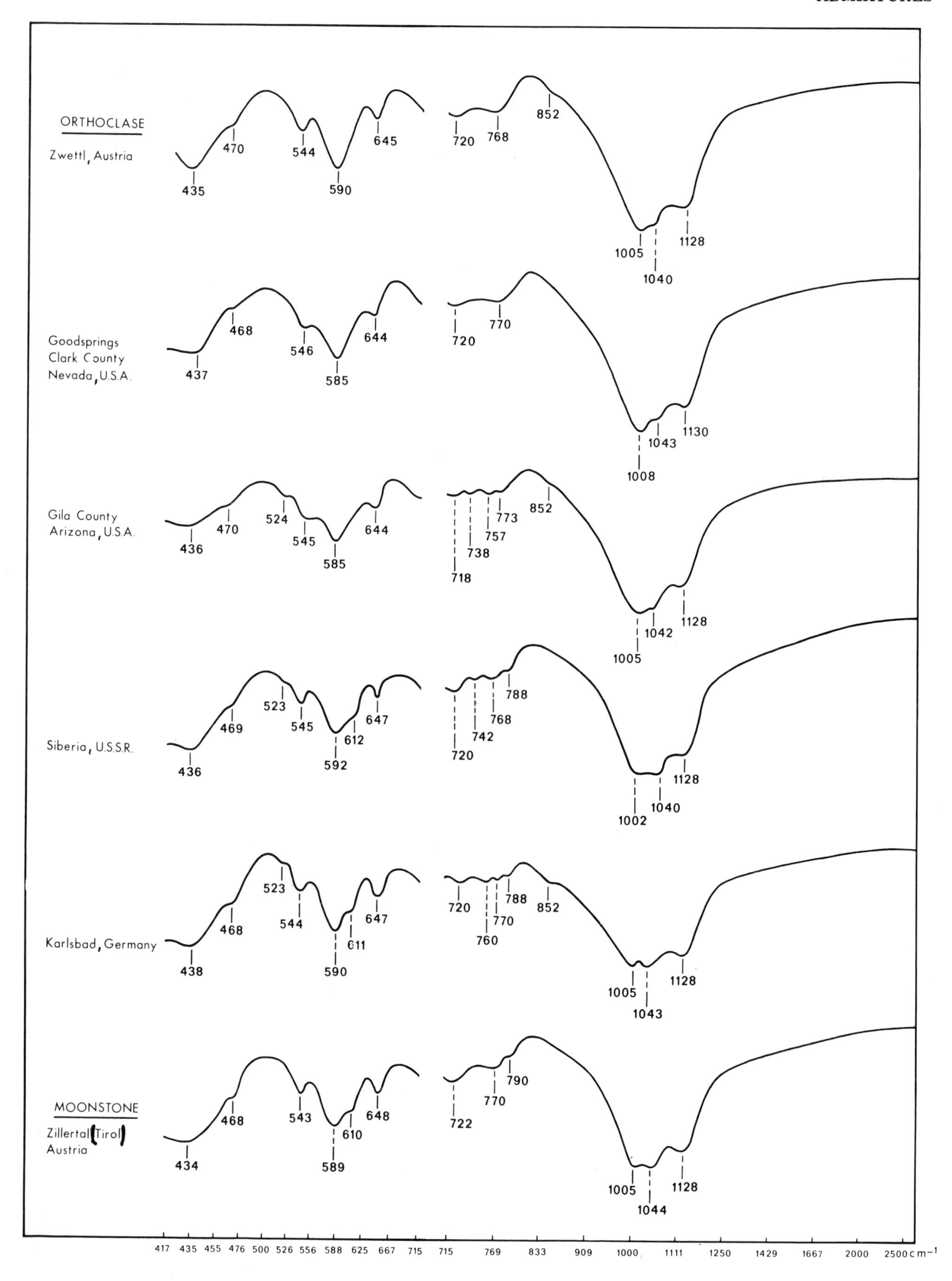

Fig. 7.18. Infrared spectra of orthoclase from different origin (pellets: 0.4 mg on 300 mg KBr).

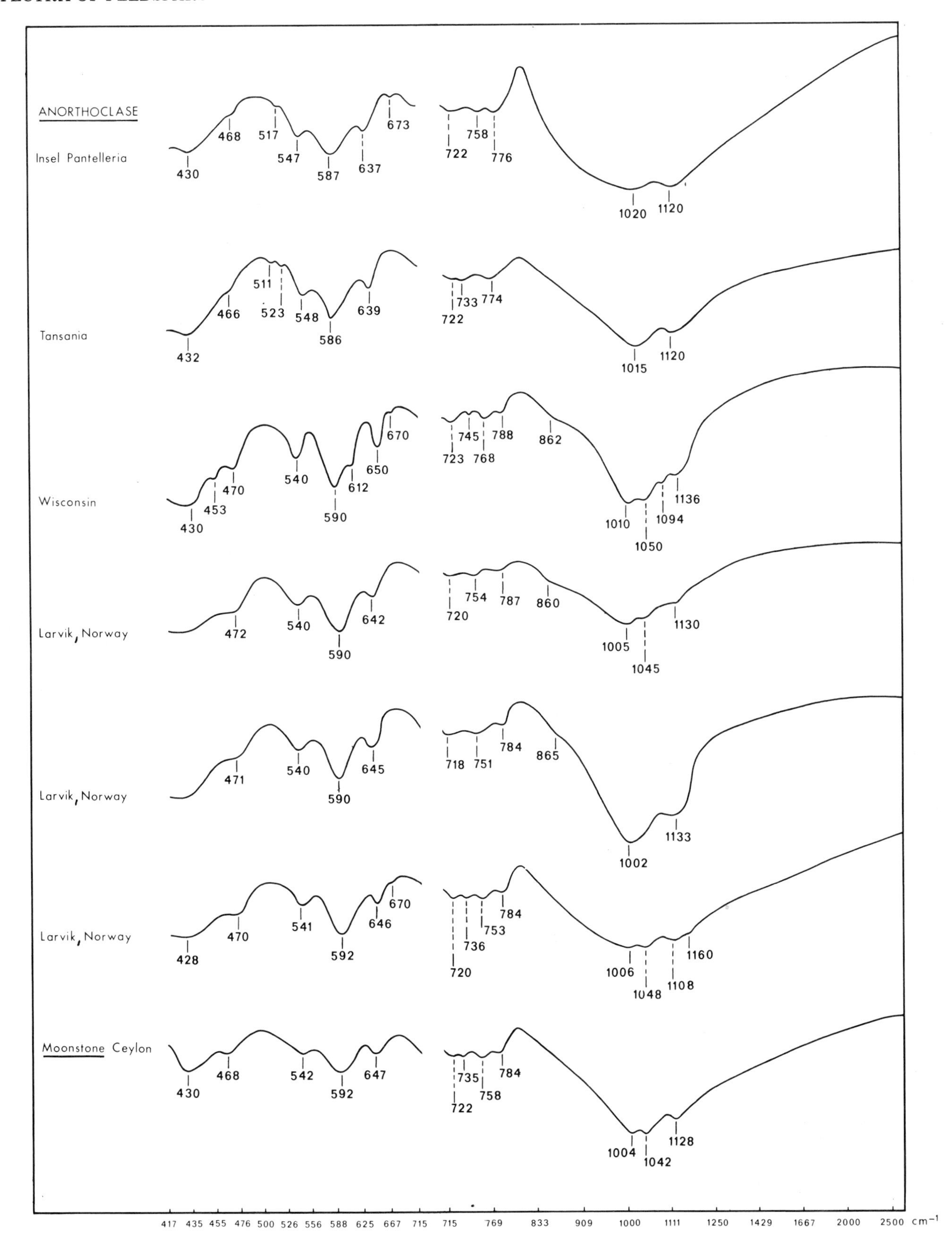

Fig. 7.19. Infrared spectra of anorthoclase from different origin (pellets: 0.4 mg on 300 mg KBr).

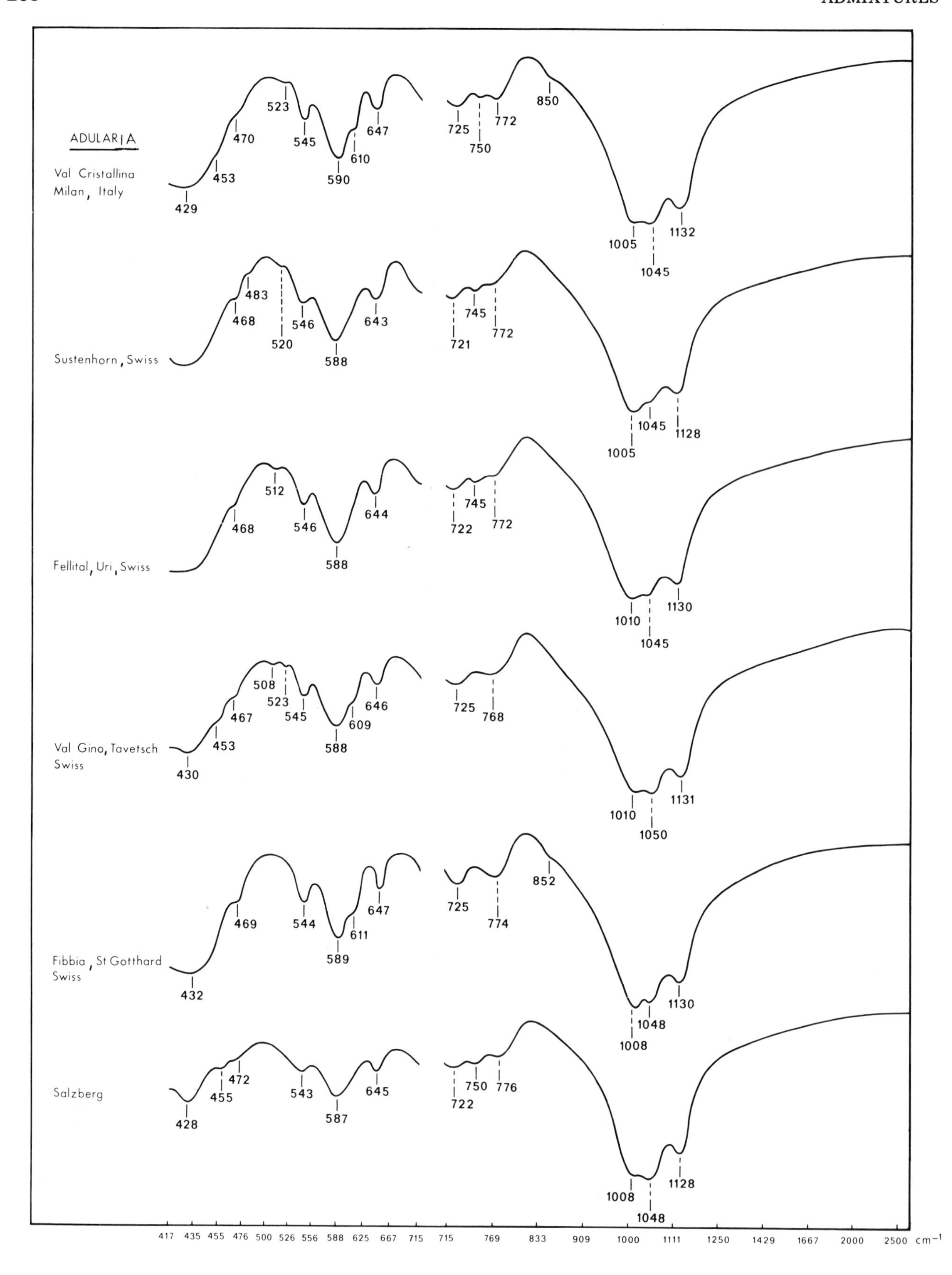

Fig. 7.20. Infrared spectra of adularia from different origin (pellets: 0.4 mg on 300 mg KBr).

Ba-feldspars

There also exists an isomorphous series between K feldspar (orthoclase) and Ba feldspar (celsian) = $Ba[Al_2Si_2O_8]$ e.g., hyalophane = $(K,Ba)[Al(Al,Si)Si_2O_8]$.
Another series is from Na-feldspar (albite) to Ba-Na feldspar, e.g. banalsite = $Ba,Na_2[Al_2Si_2O_8]$.

Infrared spectra (Fig. 7.13—7.20). The spectra of the Na-Ca feldspar series are quite similar. There is an increase of broadening in the 500—670 cm^{-1} and 900—1150 cm^{-1} regions, which is caused by the introduction of the larger Al(r = 0.45 Å) in the crystals substituting the smaller Si (r = 0.38 Å).
When going from albite to anorthite the 594 cm^{-1} band decreases to 578 cm^{-1} and the 650 cm^{-1} band to 625 cm^{-1}. There is a discontinuity at 30—50% albite (see figure on p. 54). The spectra of the three members of the alkali feldspars show some differences which can be used for their identification, e.g., 1029 (sanidine), 1053 + 1012 (microcline), 1042 + 1005 (orthoclase).
The spectra of the anorthoclase are very variable — see the many non-constant occurring bands. This is caused by the intergrowth of different sorts of feldspars.
Also the spectra of the adularia samples show several non-constant occurring bands which largely hinders their identification. But also by X-ray analysis the identification of feldspar minerals is difficult and not always possible. Preheated samples at 800—1400°C give in some cases relief.

Remarks. The name feldspar is derived from the Swedish "feldtspat" and given by Tilas (1740). It should refer to the presence of the "spar" (spath) in fields (feldts or fälts) overlying granite. Orthoclase is derived from "orthos" (upright) and "klasis" (fracture) because of its prominent cleavages at right angles. Plagioclase from "plagios" (oblique) because of its oblique [001] and [010] cleavages. Microcline from "mikros" (small) and "klineen" (to incline). Anorthoclase from the negative prefix "an" and orthos (upright) and "klasis" (fracture). Sanidine from "sanis" (tablet) and "idos" (appearance). Albite from "alba" (white). Oligoclase from "oligos" (little) and "klasis" (fracture). Anorthite from the negative prefix "an" together with "orthos" (upright). Hyalophane from "hualos" (glassy) and "phanes" (appearance). Andesine is named after a locality in the Andes mountains with its chiefly andesite rock type. Labradorite after the type locality = the coast of Labrador. Bytownite after the type locality Bytown (Canada), now Ottawa. Adularia after the type locality in the "Adular" mountains, Switzerland. Perthite after "Perth" (Quebec) an early locality where the mineral was found. Celsian in honour of the Swedish naturist A. Celsius: see for further details about the above names: Zenzén (1925); Spencer (1937).
Feldspars are widely used in glass and pottery industry and the manufacture of enamel. Furthermore in dentistry for the manufacture of artificial teeth. Plagioclases high in Ca are of small commercial importance.

References

Spencer, L.J., 1937. Some mineral names. Am. Mineralogist, 22: 682.
Tilas, D., 1740. Akad. Handl. Stockholm, I: 199.
Zenzén, N., 1925. On the first use of the term feldspat = feldspar. Geol. För. Förh. Stockholm, 47: 390.

Manganese minerals

Formulae:

Pyrolusite (polianite) - β MnO_2 - tetragonal
Ramsdellite - MnO_2 γ rhombic
Manganese dioxide - MnO_2 - commercial
Manganite - γ MnOOH - monocline
Groutite - α MnOOH - rhombic
Pyrochroite - $Mn(OH)_2$ - ditrigonal
Hausmannite - $MnMn_2O_4$ - tetragonal
Psilomelane $(Ba,H_2O)_2Mn_5O_{10}$ - monocline
Braunite - M_n^{2+} M_{n6}^{4+} $[O_8|SiO_4]$ - tetragonal
Manganogel - variable composition and amorphous

Occurrences. Manganese minerals are widespread. Intergrowths are manifold. The oxides may occur as ores in sediments, where they resulted from oxidation of the hydroxides, which is partly a biological process. Manganese minerals in sediments concentrated to hardpans are mostly associated with iron minerals. Large amounts of Mn nodules of biologic origin are found in deep sea sediments, but their exploration is not yet profitable.
Pyrolusite is the main Mn ore mineral (60% Mn) which has been explored. Others are manganite (ca. 40% Mn) and psilomelane (ca. 50% Mn).

Infrared spectra (Fig. 7.21, 7.22, 7.23). The spectra of most of the Mn minerals are poorly developed.
In consequence the distinction between the various minerals is not so easy. But also X-ray and in particular thermal analysis give poor results. The MnO_2 minerals pyrolussite, ramsdellite and manganese dioxide (commercial) may be distinguished by their shape. The first has a small amount of broad bands. Ramsdellite and the manganese dioxide have a larger amount of weak bands. Artificial MnO_2 is said to contain several modifications, called α, β, β', ϵ_1, ϵ_2 etc. depending on the way the sample was produced. Both MnOOH minerals manganite and groutite are characterized by their bands (groutite between brackets) at 1152 (1150), 1114 (1113), 1084 (1085) cm^{-1}. The latter may be distinguished from the former by the absence of a 1000 cm^{-1} band.
Pyrochroite the $Mn(OH)_2$ variety can easily be distinguished by its OH vibration at 3690, 3660 (3680) cm^{-1}; the poor bands in the 800—1200 cm^{-1} region and the two broad bands at 611 and 500 cm^{-1}.
The complicated Mn mineral hausmannite has only some broad bands.
For psilomelane which has Ba in the crystal the spectrum is also very poor. The same holds for braunite. The Mn silicate mineral, manganogel, gives a very bad broad spectrum, which was to be expected from its name.

Remarks. Manganogel has a radiating colloform structure pointing to a colloidal origin. Manganese is used in the manufacture of steels to get a hard steel. At the same time it acts as a deoxidizing- and desulfurizing agent, thus giving a clean metal.
Pyrolusite is used in the production of chlorine (and bromine) and as a colorizing material in bricks, roofs and pigments. The name pyrolussite is derived from pur (fire) and luwo (to wash), which is connected with its use in the manufacture of glass ware to prevent brown and green tints. In France it is called "le savon des verrièrs" which has the same meaning.
The name psilomelane is derived from psulos (smooth) and melas (black).

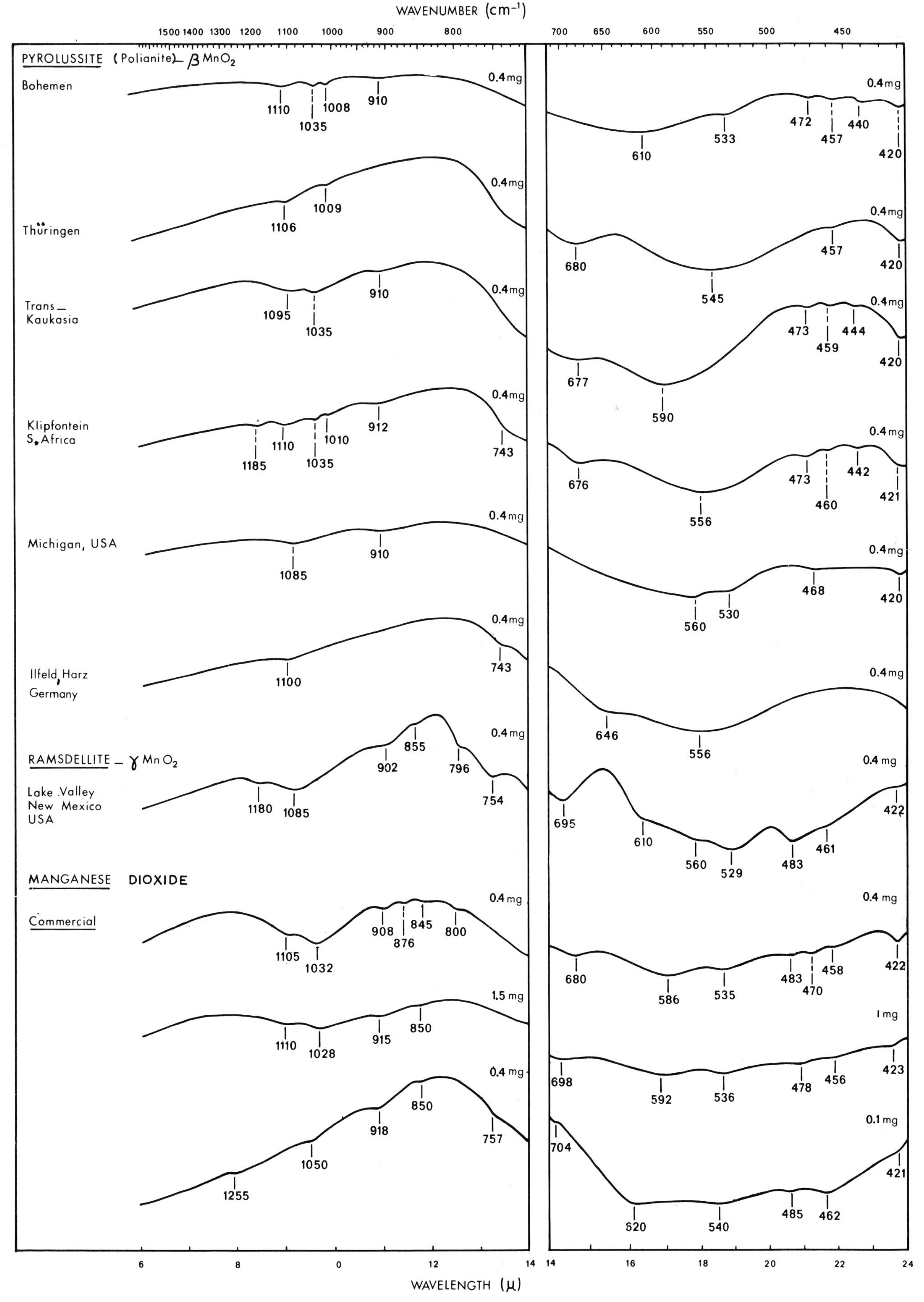

Fig. 7.21. Infrared spectra of pyrolusite, ramsdellite and commercial manganese dioxide from different origin.

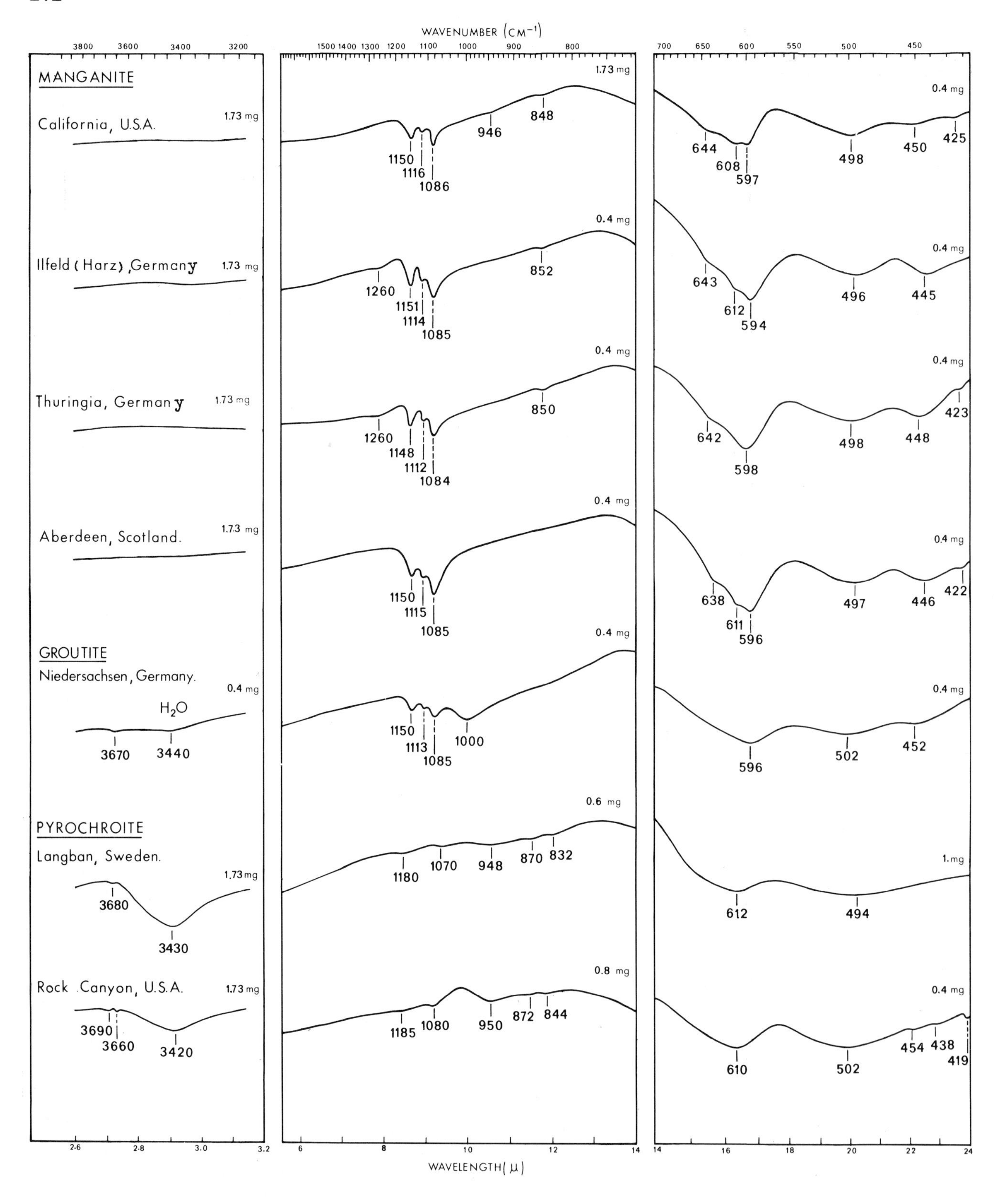

Fig. 7.22. Infrared spectra of manganite, groutite and pyrochroite from different origin.

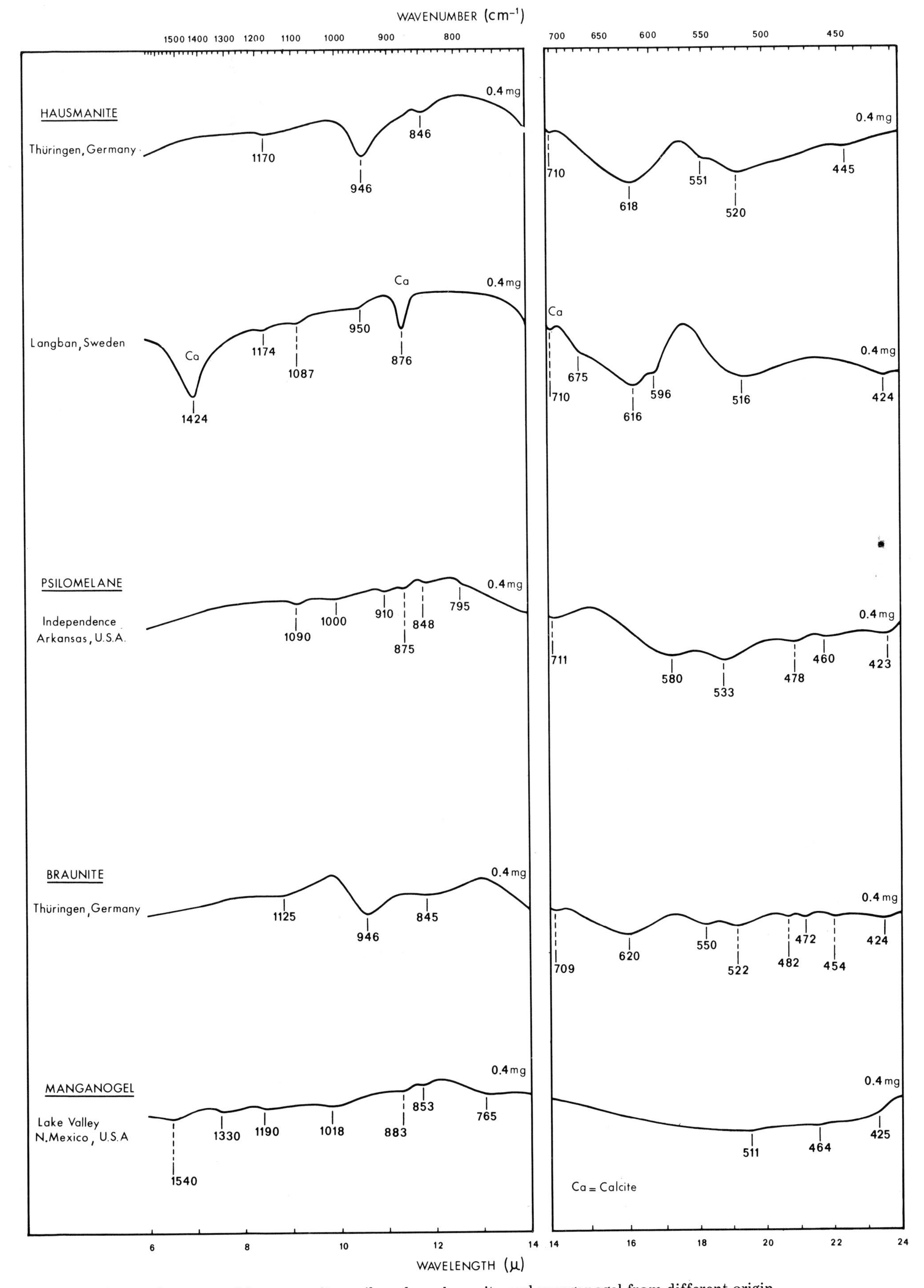

Fig. 7.23. Infrared spectra of hausmannite, psilomelane, braunite and manganogel from different origin.

8. AMORPHOUS MATERIALS

The term amorphous is somewhat vague because it depends on the method which is used to identify the mineral. For instance, a material which is planar badly ordered can be better detected by electron rays (λ = ca. 0.05 Å - electron diffraction microscopy) than by röntgen rays (λ ca. 1.6 Å-röntgen diffractometer). However, badly ordered bonds between atoms or atomgroups are more easily recognized by long-wave infrared rays (λ = ca. 3 to 25 μ) than by the shorter electron- or röntgen rays. But in the following, because of tradition, the term amorphous is maintained for all materials that give very bad X-ray spectra.

Silicic acid; formula: $H_2SiO_3 \cdot nH_2O$

It consists of a non-ordered (deformed) three-dimensional network of hydrogen-bonded monomeric-, dimeric- or polymeric $Si(OH)_4$ groups and H_2O molecules.

Occurrence. Amorphous silicic acid gel occurs chiefly in weathered sediments (red earths, latosols, gley soils). It is abundant in salt-affected soils in semi arid regions (solontchak and solonetz). The solubility of the undissociated monomers of $Si(OH)_4$ at ca. 25°C = 100–150 mg SiO_2 /l in the pH 2.5–8.5 traject. At higher pH its solubility is largely increased. In river waters and in the sea only ca. 30 and 10 mg is found respectively. The solubility of $Si(OH)_4$ is largely decreased by Al and Fe ions. Thus 20 p.p.m. of Al^{3+} reduce its solubility to 15 p.p.m. SiO_2. Also the temperature effect is large 0°C = 60–80 p.p.m.; 90°C = 315–350 p.p.m.
Silicic acid is also produced artificially by precipitation of a sodiumsilicate solution with HCl trade marks: Hisil (Columbia Southern Chemical Corp.), Metasilicic acid, MSBBS (Mallinckrodt).
Amorphous, high-graded SiO_2 of light weight (vol. wt. = ca. 0.060) is an artificial product which is obtained by hydrolysis of silicum tetrachloride with steam in the heat. Trade marks: Aerosil (Degussa), Ludox (Du Pont de Nemours), Cab-O-Sil (Cabot Inc., Boston). The product is very fine ca. 0.010 to 0.040 μm and still contains some silanolgroups.

Infrared spectra (Fig. 8.1). Highly dehydrated, high-graded Cab- O-Sil gives a sharp band of high intensity at 3750 cm^{-1}. For less hydrated samples, it is ca. 3665 and 3640 cm^{-1}, appearing as weak bands. Silicic acid contains strongly bonded H_2O molecules; see the O-H stretching band at 3220 to 3200 cm^{-1} and the O-H bending vibration even to as high as 1660 cm^{-1}.
X-ray analysis of silicic acid gives only a broad hump of large intensity between 5.5 to 2.7 Å in which only some badly defined tridymite and cristobalite reflections can be recognized.
The spectra of silicic acid have several bands in common or nearly in common with those of quartz (between brackets) e.g., 477 (478), 466 (460) cm^{-1}, tridymite (between brackets) 1217 (1200), 800 (790), 477 (480), 466 (460) cm^{-1}, cristobalite (between brackets) 1217 (1200), 1096–1088 (1098), 800 (792), 525* (520) cm^{-1}.
Silicic acid can easily be distinguished from quartz, tridymite and cristobalite by its O-H stretching vibrations at 3665 and 3640 cm^{-1} and the bands at 954 and 800 cm^{-1}.

Remarks. Silicic acid has a large surface = 400–500 m^2/g. It is used as a cracking catalyst or as a supporter for metal (Cu, Ni, Pa) catalysts. Also as an adsorbens for water and gases.
Aerosil etc. serves many purposes: reinforcing filler in rubber industry, filler in pharmaceutics and cosmetics, increaser of viscosity and thixotropy in paints, stabilizer for various kinds of emulsions, supporter for herbicides and fungicides, thermal insulator, etc.

Biogeneous silicious material (diatomaceous earth, sponge spicules); formula: $H_2SiO_3 \cdot nH_2O$

Occurrence. Diatoms (and radiolarians) are characteristic compounds of marine and lake sediments. Under specific conditions essential to life (pure water of ca. 30°C with a high content of SiO_2), they may breed rapidly and leave large amounts of silicious skeletons after their death. At various localities in the world are found large deposits of diatomaceous earth, also called infusorial earth, diatomaceous silica, Kieselguhr - Whitehills and Mammoth (Arizona), Santa Barbara (California), Oran (Algeria), Skye (Scotland), Isles of Fur and Mörs (Denmark), Luneburger heath (Germany), Toba Lake (N. Sumatra).
Spicules of sponges and phytoliths (plant opal in grasses) are another amorphous silicious material.

Infrared spectra (Fig. 8.2). The spectra of the above silicious materials come close to those of silicic acid. However, the 478 and 460 cm^{-1} band of quartz, the 480 and 460 cm^{-1} band of tridymite, the 488 and 476* cm^{-1} band of cristobalite, and the 477 and 466 bands of silicic acid have become a broad band at 468 cm^{-1}.

The diatoms have O-H stretching vibrations at 3625 cm^{-1}. For the sponge spicules they are at 3670 cm^{-1}. The X-ray spectra give only a broad hump of large intensity between 5.5 to 2.7 Å, in which well-defined reflexions of cristobalite and tridymite can only be recognized in samples of Scotland, the North Sea, Santa Barbara and the Mediterranean.

Remarks. Diatomaceous earth (infusorial earth, diatomite, kieselguhr) is used as an abrasive in metal polishers; as an important component of glasses and enamels, cosmetics and medicines. Also as filtermaterial for mineral oils, sugars, beers, wine.
The raw material is first heated to destroy the organic matter before it is handled. It then contains mainly cristobalite.

Obsidian, perlite, windowglass

The three above listed high-siliceous materials do not have symmetry and periodicity in the arrangement of their atoms as in crystals. They consist of a nonordered (deformed) three-dimensional network of Si and Al tetrahedra connected by oxygen linkages. The holes contain the cations which neutralize the deficit charges caused by Si-O/Al-O replacements. Hydrogen in obsidians occurs as randomly distributed OH groups bonded to Si and as H_2O molecules bonded to cations or adsorbed in small pores.
Their total amount (loss on ignition) is for obsidians up to ca. 1%. The perlites have up to ca. 5%. The adsorbed water is lost at ca. 300°C and the more strongly bonded water at 900—1000°C explosively. Due to the sudden loss of OH or H_2O molecules, the material may expand 8 to 10 × its original volume.

Occurrence. Obsidian and perlite are found in volcanic or post-volcanic regions. Their composition is variable.

Infrared spectra (Fig. 8.3). The spectra of these X-ray amorphous materials show well-defined bands, although they are broad, especially those of the window glass. Strongly bonded H_2O molecules are marked by a ca. 3200 cm^{-1} band. The O-H stretching vibrations are at 3600—3640 and 3670—3675 cm^{-1}. The spectra show that obsidian and perlite are the most like silicic acid. However, the ca. 1096—1088 cm^{-1} band of the latter has decreased to 1070—1058 cm^{-1}, and for the window glass even to 1042 cm^{-1}. X-ray analysis gives only a broad hump of large intensity between 5.5 and 2.7 Å. Small amounts of cristobalite and tridymite or ditto-like high-temperature minerals as present in some of the samples can, due to overlapping effects, only be found by the X-ray and not by the IR method. Feldspars (obsidian, Italy) can only be found by the IR method when large amounts occur.

Remarks. Artificial glass is used for various purposes: windows, household ware, instruments, etc. It consists of SiO_2 to which various substances have been added depending on the purpose for which the glass is used e.g., the introduction of Ca in the melt increases its resistance against corrosion. Lead silicate glasses have the largest transmittance and brillancy (flint glass). Crystal glass which has a musical tone is a potash lead silicate glass. Coloured glasses contain metallic gold (ruby), selenium (red), chromium and copper (green), cobalt (blue), platina (yellow to orange), cadmium and uranium (yellow), manganese (violet), iron oxide (brown), calciumfluoride and tinoxide (opal). Common window glasses are soda-lime glasses. By the introduction of the metal atoms, the Si-O-Si bonds in the disordered network of SiO_4 tetraeder are broken up. OH groups in the glasses are hydrogen bonded to oxygen atoms.
Devitrification, brittleness and cracking when the glass is cooled is caused by "stones" of minute crystals of quartz (Li_2O, MgO are accelerators), cristobalite (MgO,CaO,SrO,BaO are accelerators), tridymite (Li_2O,Na_2O,K_2O are accelerators) or a mixture of them. The crystals of the latter are the coarsest.
Heated expanded perlite is used as a light weight material in prefabricated constructions of houses. When finely ground it is used as a polishing material.

Allophane; formula: $X\,Al_2O_3\;Y\,SiO_2\,n\,H_2O$

Allophane is a gel-like, naturally occurring amorphous hydrous alumino silicate product of widely varying chemical composition. When Fe_2O_3 is high (30—40%) the material is called hisingerite. An intermediate product is called iron allophane.
By imogolite [1] is indicated an allophane-like mineral with some degree of order, e.g., distorted chains of Al-octahedra are in this case linked by isolated Si_2O_7-groups. It further consists of very

[1] Named after "Imogo" a brownish soil layer intervening between dark-coloured layers in ando soils of Japan: Yoshinaga, N., and Aomine, S., 1962. Allophane in some ando soils. Soil Sci. Plant Nutr. (Tokyo), 8: 114—121. Russell, J.D., Mac Hardy, W.J. and Frazer, A.R., 1969. Imogolites: A unique alumino silicate. Clay Minerals, 8: 87—99.

thin threads (ϕ = 50 to 100 Å) instead of globules or of particles with undefined morphology such as common allophane.

Occurrence. Allophane and imogolite are commonly found in volcanic ashes and pumices, where they form an intermediate stage to halloysite. Hisingerite is found as fillings in small fissures of volcanic Fe-rich (pyroxene) rocks or near iron ores (pyrrhotite, pyrite) from which they resulted.

Infrared spectra (Fig. 8.4.). The samples of the above-mentioned X-ray amorphous materials show broad bands which point to a large disorder in the structure. Vibrations of O-H are from 3570 to 3666 cm^{-1}. Several samples contain, although they were dried before at 110°C, appreciable amounts of adsorbed H_2O (1625—1635 cm^{-1}, and 3400—3470 cm^{-1}). Also strongly adsorbed H_2O (1650 and 3200—3280 cm^{-1}).
There is a wide variation in the position of the bands which is caused by the variable composition.
Small amounts of quartz (Imogo, Ragton, Sirau), tridymite (Ragton), cristobalite (Ragton, Imogo, Choyo volcano, Sirau volcano) and feldspar (Sirau volcano) cannot be identified by the IR method, only by X-ray analysis.

Remarks. Several samples from mineral museums or shops indicated as amorphous allophane proved to be in reality kaolinite, gypsum, mica, etc.
If allophane gel is dried at 110°C or even at room temperature a water-irreversible sandy material is formed. The reversional uptake of water, which reaction proceeds very slowly, may give an after-effect well-known in many dry regions with seasonal rainfall. Thus strongly permeable sediments of low plasticity and coarse particle size may change into a finely graded impermeable plastic clay. Its volume may, moreover, increase considerably and thus may lead to dambursts when used as a building material for dams.
Several chemicals have been recommended to remove crusts cementing the particles together (Na dithionite, H_2O_2, HCl). However, in most cases only strong reagents are effective, but they rather lead to partial decomposition of the minerals (figure EM. 234, 235, 236, 237 and 238).
Intercalation of allophane, silicic acid, Al-hydroxide, Fe-hydroxide between the layers of swelling minerals leads to pseudochlorites.
Artificial amorphous silica-alumina compounds are used as catalysts for many purposes.

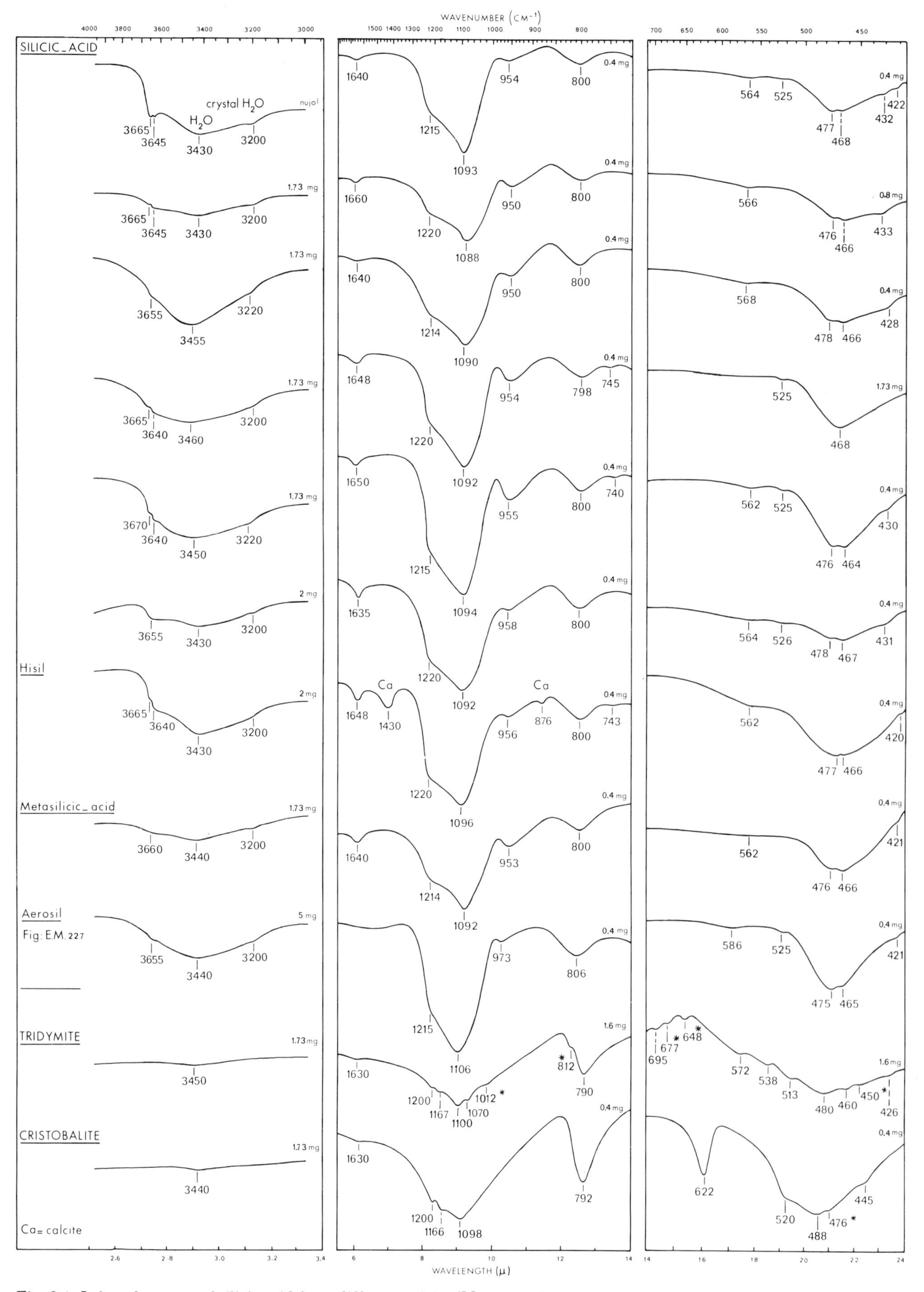

Fig. 8.1. Infrared spectra of silicic acid from different origin. (Most samples contain some cristobalite and tridymite which are overlapped)

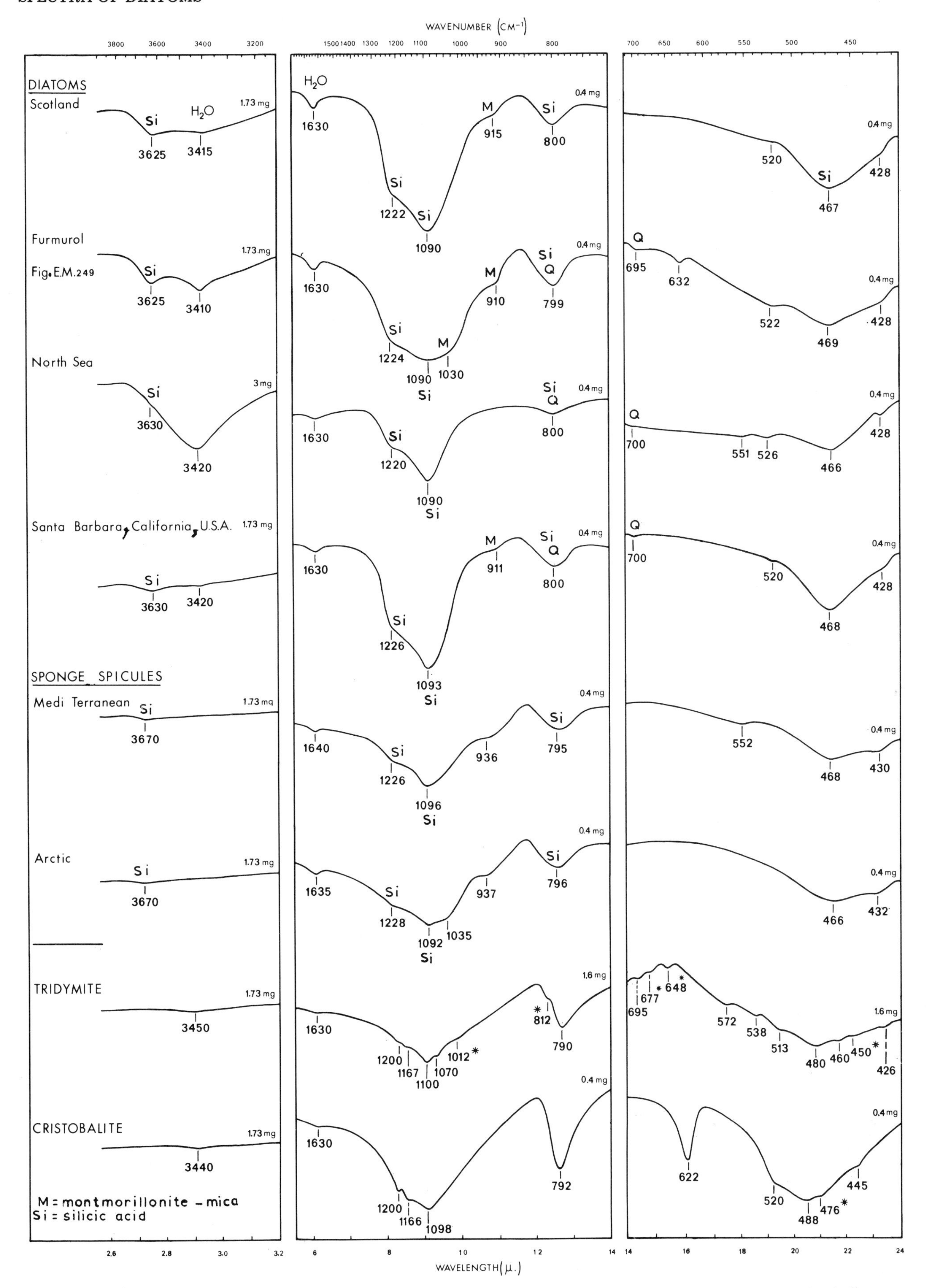

Fig. 8.2. Infrared spectra of diatoms and sponge-spicules from different origin. (Most samples contain some cristobalite and tridymite which are overlapped.)

Obsidian

Chihuahua, Mexico

Krabla, Iceland

Virgin Valley, Nev., U.S.A.

Oregon, U.S.A.
some tridymite and cristobalite (both overlapped)

Ischia, Italy
feldspar, some mica, and quartz (all overlapped)

G. Goentoer, Java

Melos, Greece
some cristobalite (overlapped)

Meteoric glass, tektite, Thailand

Perlite

Superior, Ariz., U.S.A.
some cristobalite (overlapped)

Mikawa, Japan
some cristobalite, mica and quartz (all overlapped)

Window glass
some tridymite (overlapped)

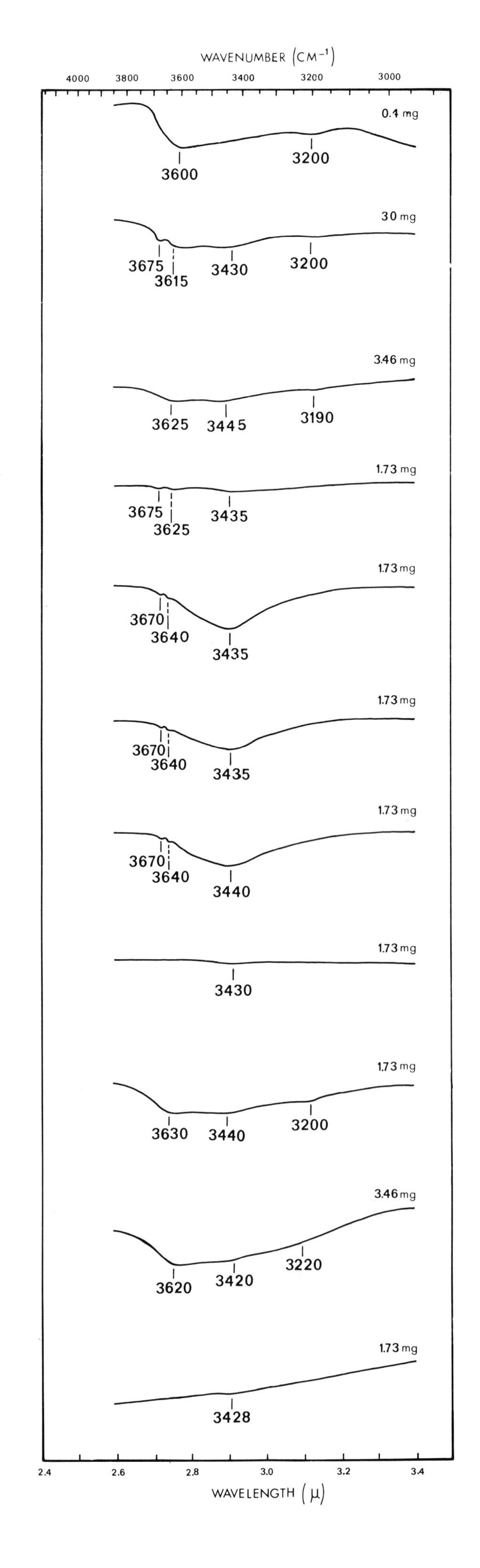

Fig. 8.3. Infrared spectra of obsidian and perlite from different origin and of window glass. (Some quartz, cristobalite and tridymite overlapped.)

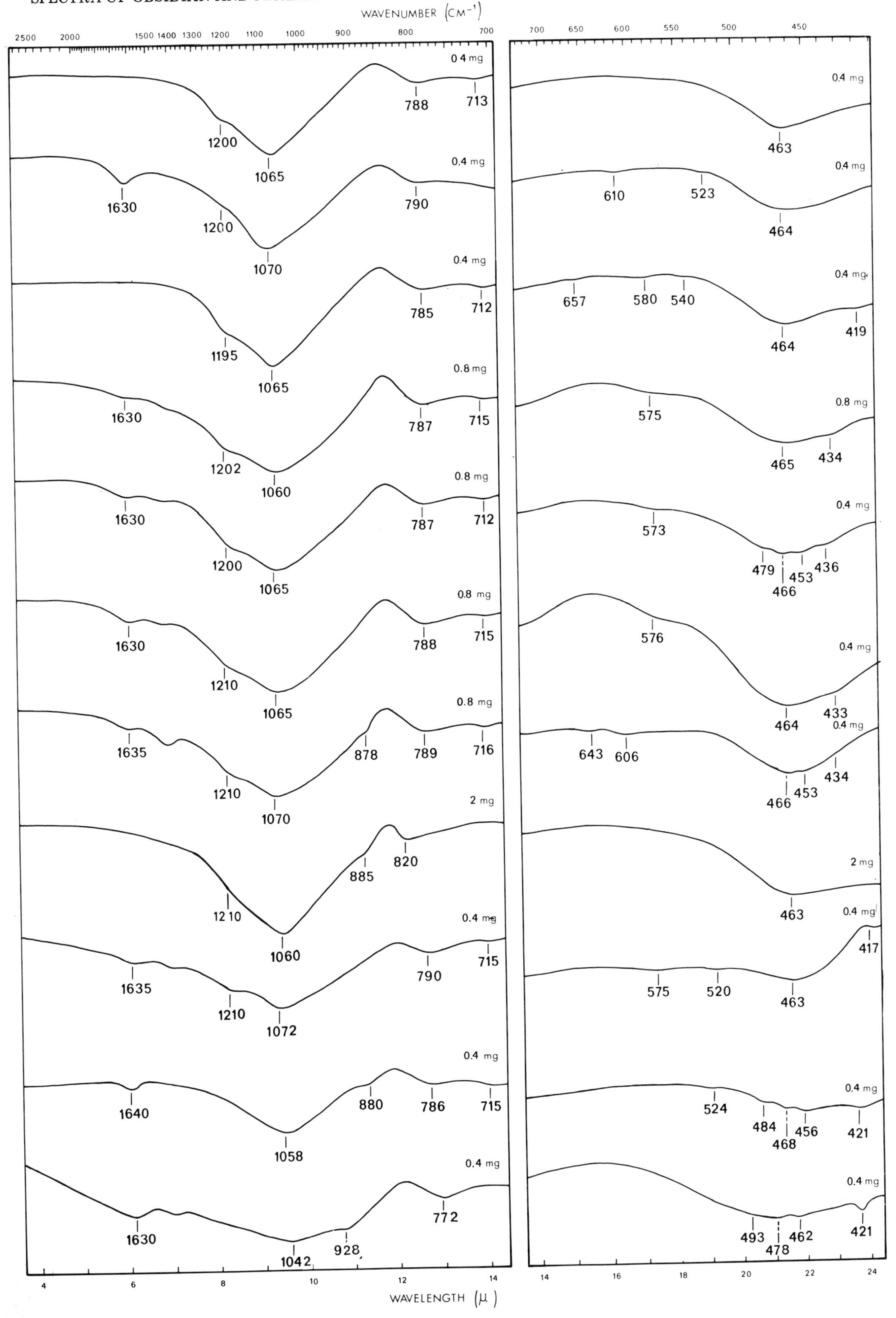
WAVENUMBER (CM⁻¹)
2500 2000 1500 1400 1300 1200 1100 1000 900 800 700
700 650 600 550 500 450
0.4 mg
1200 1065 788 713
463
0.4 mg
1630 1200 1070 790
610 523 464
0.4 mg
1195 1065 785 712
657 580 540 464 419
0.8 mg
1630 1202 1060 787 715
575 465 434
0.8 mg
1630 1200 1065 787 712
573 479 466 453 436
0.8 mg
1630 1210 1065 788 715
576 464 433
0.8 mg
1635 1210 1070 878 789 716
643 606 466 453 434
2 mg
1210 1060 885 820
463
0.4 mg
1635 1210 1072 790 715
575 520 463 417
0.4 mg
1640 1058 880 786 715
524 484 468 456 421
0.4 mg
1630 1042 928 772
493 478 462 421
4 6 8 10 12 14
14 16 18 20 22 24
WAVELENGTH (μ)

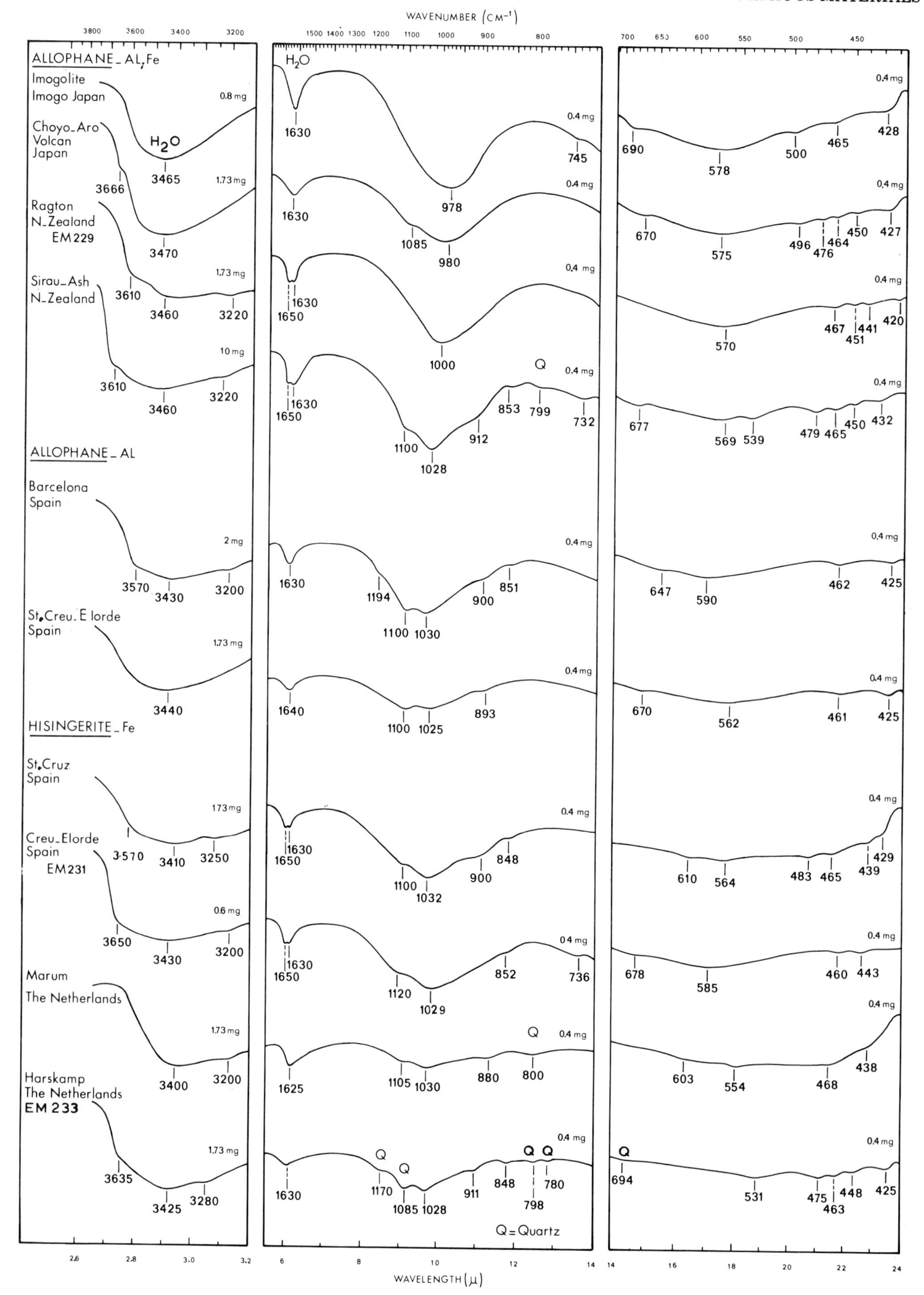

Fig. 8.4. Infrared spectra of allophane (imogolite) and hisingerite from different origin. (Some feldspar, cristobalite and tridymite overlapped.)

9. MISCELLANY

Vivianite; formula: $Fe_3{}^{2+}[(PO_4)_2]\ .\ 8H_2O$

Occurrence. This greenish ferro-iron phosphate mineral is found as prismatic crystals in iron ore. But also as an earthy mass in sediments; especially in peaty clays where it was formed under reducing conditions.

Infrared spectra (Fig. 9.1). Bands of high intensity are at: 1046, 1014, 976, 941 cm^{-1}.
According to the O-H band at 3380 cm^{-1}, the mineral may probably also contain $(OH)_4^{4-}$ groups besides adsorbed H_2O molecules (3440 cm^{-1}) and crystal water molecules (3175 cm^{-1}).

Substitution of $(PO_4)^{3-}$ for $(OH)_4^{4-}$ is suggested for griphite: $(Na,Al,Ca,Fe)_3(Al,Mn)_2[(PO_4)_{3-x}, (H_4O_4)_x]$ by Mc Connell (1942) and for hydroxyapatite: $Ca_{10}[(PO_4)_{4-x}, (H_4O_4)_x]$ by Mc Connell (1960).

Remarks. The mineral is rare and of small importance.

References

McConnell, D., 1942. Griphite, a hydrophosphate garnetoid. Am. Mineralogist, 27: 452—461.

McConnell, D., 1960. The stoichiometry of hydroxyapatite. Naturwissenschaften, 47: 227.

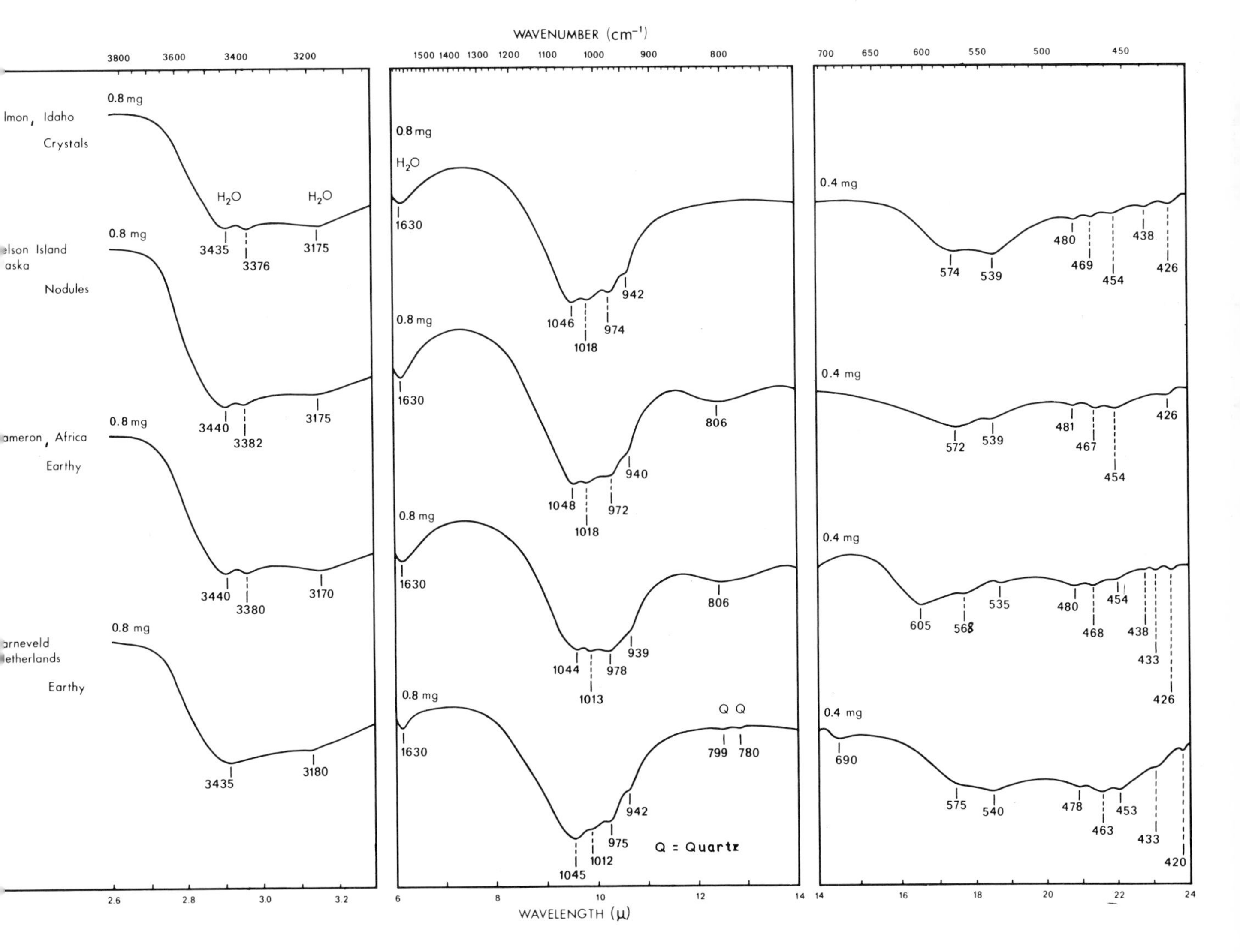

Fig. 9.1. Infrared spectra of vivianite from different origin.

Olivine and forsterite; formula: olivine $(MgFe)_2SiO_4$; forsterite Mg_2SiO_4

Fayalite = $Fe_2(SiO_4)$ is the Fe end member in this series. The above minerals are all neosilicates with island structure.

Occurrence. Olivine is found as grains in basic igneous rocks. When chlorite or serpentine is heated olivine is formed. High Al content of the chlorite used retards the transformation. Forsterite is found in small amounts in volcanic rocks or ejections. When chrysotile $(Mg_3(OH)_4 | Si_2O_5)$ is heated at ca. 600°C, forsterite and amorphous SiO_2 are formed.

Fayalite is a rare mineral.

Infrared spectra (Fig. 9.2). The bands are alike those of forsterite. Characteristics of high intensity are at: 986, 885, 839, 607, 503 cm^{-1}.
Neither by X-ray analysis can the two minerals be distinguished. The olivine samples were defiled with very small amounts of kaolinite that were only detectable by the infrared (3692, 3669, 3650, 3622 cm^{-1}) and not by the X-ray method.

Remarks. A transparent, pale, yellowish-green variety of olivine called peridot is sold in jewelry.

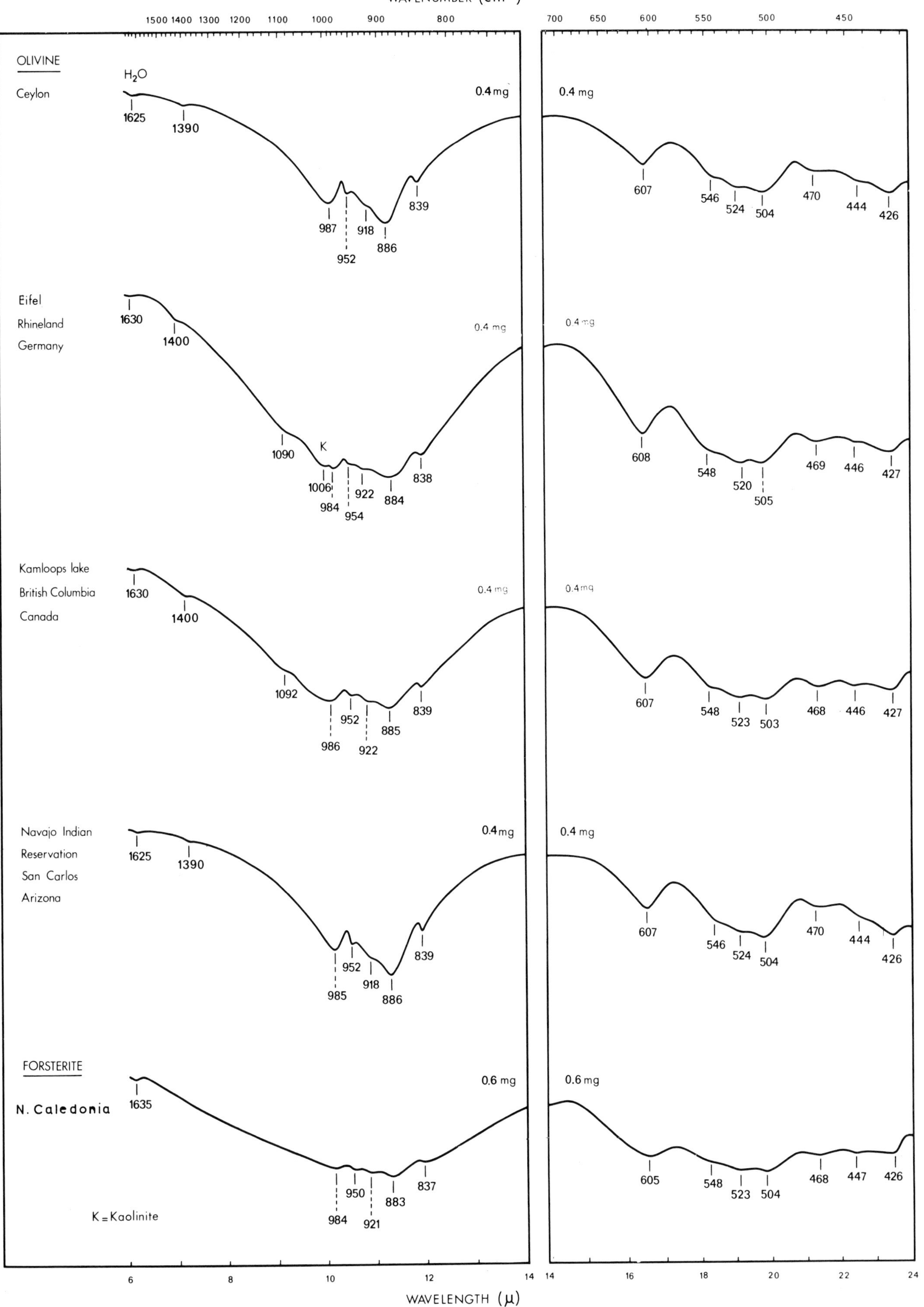

Fig. 9.2. Infrared spectra of olivine and forsterite from different origin.

Mullite and sillimanite; formulae:

Sillimanite: $Al_2O_3 . SiO_2$:$Al^6Al^4[O|SiO_4]$
mullite: $3Al_2O_3 . 2SiO_2$:$Al^6_4Al^4_4[O_3(O_{0.5},OH,F)$-$|Si_3AlO_{16}]$
β-mullite(praguite): $2Al_2O_3 . SiO_2$: Agrell and Smith (1960).

The structure of mullite, although its chemical composition is somewhat different (mullite = 72—78% and sillimanite = 62—66% Al_2O_3) is essentially the same as that of sillimanite = Al-O octahedral chains crosslinked by tetrahedral Si and Al.
They are both rhombic and their cell dimensions differ only slightly. a_0 = 7.50 Å; b_0 = 7.65 Å; c_0 = 5.75 Å; and a_0 = 7.44 Å; b_0 = 7.60 Å; c_0 = 5.75 Å, respectively.
Samples with cell dimensions intermediate between mullite and sillimanite are also found: Aramoki and Roy (1963).
There exists an isomorphous series between sillimanite and 2 : 1 mullite: Durovič (1962).

Occurrence. Mullite is but rarely found as small crystals in metamorphic rocks or in igneous intrusions. It is a component of porcelain occurring as needles, aggregates or globules, being a fusion (± 1450°C) product of kaolinite. At ca. 650°C metakaolinite is formed. At ca. 1000°C Al-Si spinel and at 1100—1200°C mullite and cristobalite. β-mullite is only known as an artificial product.
Also by heating pyrophyllite, montmorillonite, muscovite and illite at ca. 1000°C, mullite is formed.

Sillimanite is commonly found in schists, gneises and other metamorphic rocks, where it occurs in fibriolitic aggregates or as large prisms.

Infrared spectra (Fig. 9.3). Characteristic bands: mullite: 1175, 1120, 905, 815 cm^{-1}; sillimanite: 1195*, 1177, 960, 898, 819, 747, 687, 634, 486, 450 cm^{-1}.
The infrared spectra of sillimanite are better pronounced than those of artificial mullite.
As a contrast to X-ray analysis, mullite and sillimanite can be distinguished better from each other by the infrared method.

Remarks. It is named after the type locality the isle of Mull. The mineral gives excellent refractories withstanding high temperatures (to 1800°C) and having small expansion. It is a good insulator and resists corrosion. Mullite is obtained by heating Al-hydroxide + quartz, sillimanite, andalusite or kyanite at ca. 1500°C. Kyanite is the least suited because it increases in volume upon change into mullite.

References

Agrell, S.O. and Smith, J.V., 1960. Cell dimensions, solid solution, polymorphism and identification of mullite and sillimanite. J. Am. Ceram. Soc., 43: 69—76.
Aramoki, S. and Roy, R., 1963. A new polymorph of Al_2SiO_5 and further studies in the system Al_2O_3-SiO_2-H_2O. Am. Mineralogist, 48: 1322—1347.
Durovič, S., 1962. Isomorphism between sillimanite and mullite. J. Am. Ceram. Soc., 45: 157—161.

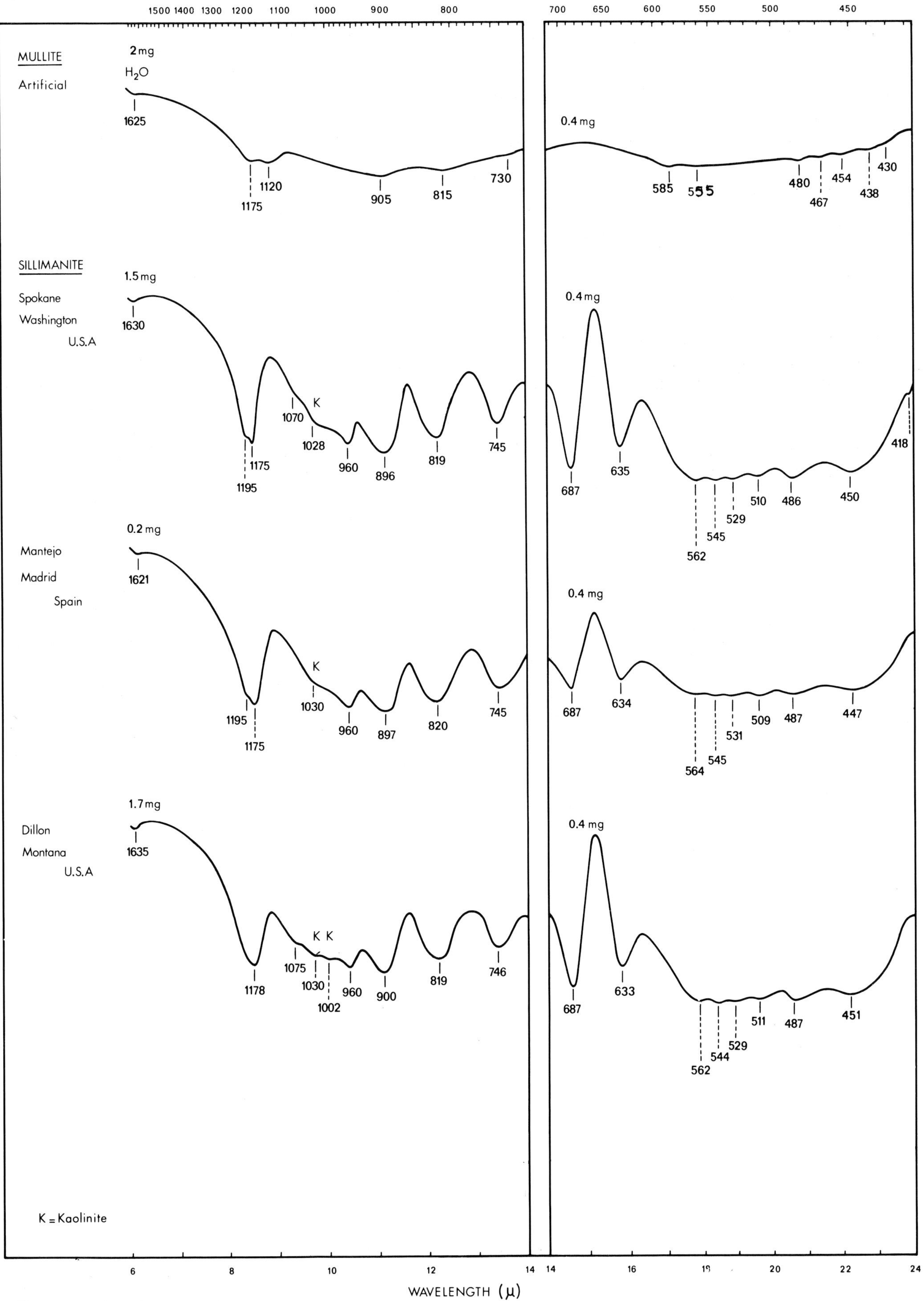

Fig. 9.3. Infrared spectra of mullite and sillimanite from different origin. Mica, quartz and feldspar impurities of sillimanite samples are overlapped.

Chloritoid, ottrelite; formula:
$Fe^{2+}AlAl_3[(OH)_4|O_2|(SiO_4)_2]$

The above mineral is blackish-green, brittle and resembles chlorite. Its structure may be considered to consist of an alternation of layers of brucite type and corundum type linked by Si-atoms in fourfold coordination and by hydrogen bonds. The SiO_4 tetrahedra, the vertices of which point inwards, do not form continuous silicate sheets as in the mica's, but occur as islands (Harrison and Brindley, 1957).
Ottrelite is supposed to be a Mn(Fe)-rich variety.

Occurrence. Chloritoid is found in metamorphic rocks and schists where it resulted from the decomposition of pyroxenes and amphiboles.

Infrared spectra (Fig. 9.4). The spectra from different localities are not uniform, especially not the relative intensities of their bands. Characteristic bands are at: 3448, 1102, 900, 850, 742, 609, 588, 548, 445 cm^{-1}. Chloritoid can be easily distinguished from ottrelite by the bands at (ottrelite between brackets): 1102 (absent), 900 (912), 850 (874), absent (832), 668 (absent), 609 (absent), 588 (absent), 548 (absent), 518 (530), 445 (absent).

Remarks. Ottrelite is named after its type locality Ottre, Belgium. The structure of chloritoid needs further investigation.
Monoclinic, triclinic varieties and intergrowths are found (Snelling, 1957). Also the X-ray spectra show large differences.

References

Harrison, F.W. and Brindley, G.W., 1957. The crystal structure of chloritoid. Acta Crystall., 10: 77—82.

Snelling, N.J., 1957. A contribution to the mineralogy of chloritoid. Mineral. Mag., 31: 469—475.

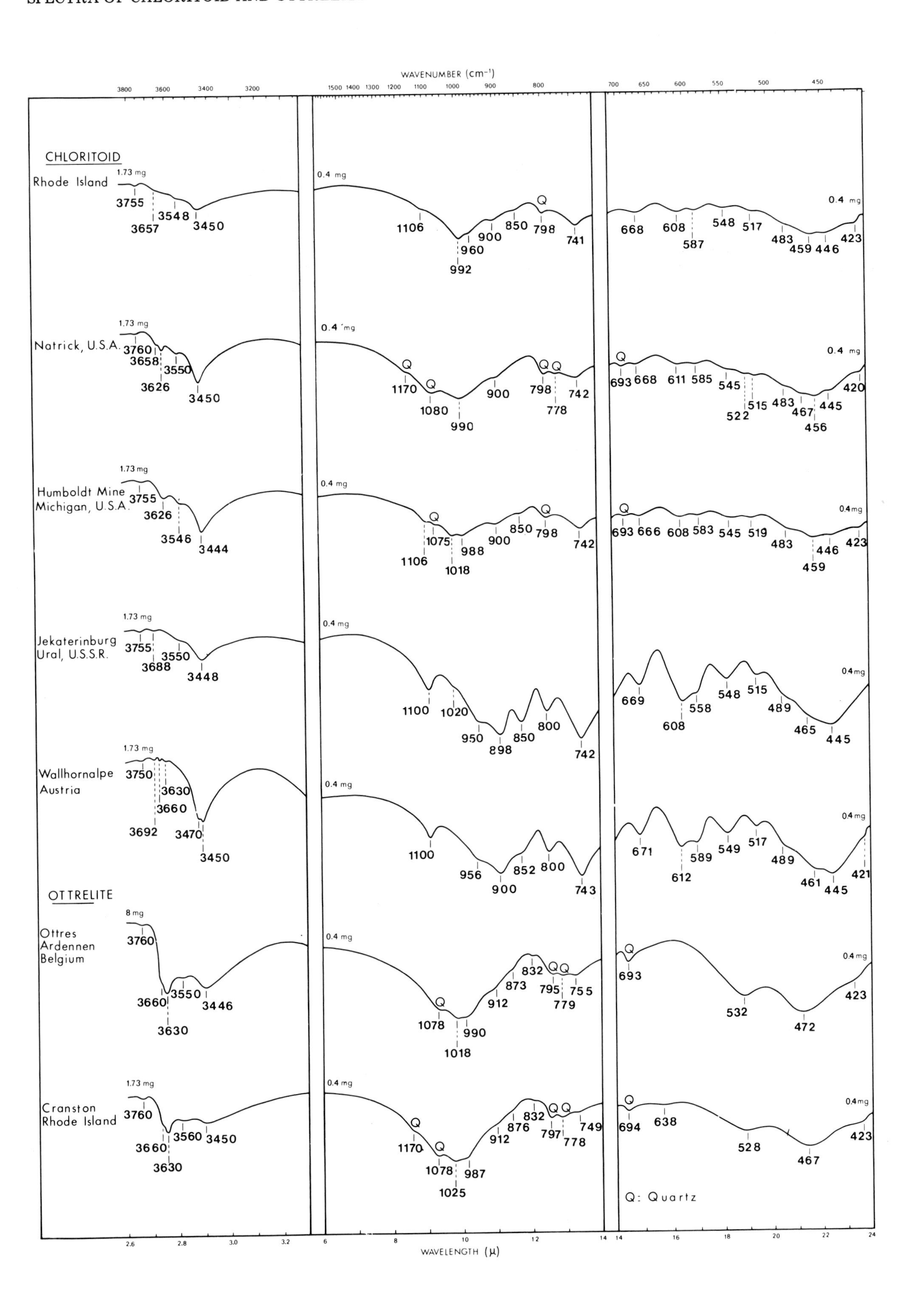

Fig. 9.4. Infrared spectra of chloritoid and ottrelite from different origin.

Stilpnomelane; formula:

$$\left[\begin{matrix}(Fe^{2+},Fe^{3+},Mg,Al)_3(OH)_2 \mid Si_4O_{10} \\ (K,H_2O)\end{matrix}\right](H_2O)_2$$

The above iron-rich ($FeO + Fe_2O_3$ = 25—30%) layer silicate mineral may readily be mistaken for biotite, but it is more brittle.
According to X-ray analysis the mineral has a 12.4 Å basal spacing, which can be followed to its 7th order. When heated at 550° for 2.5 h there is no change. The (060) reflection at 1.52 Å points to a trioctahedral mineral.

Occurrence. It is commonly found together with iron ores; and also in schists together with chlorite.

Infrared spectra (Fig. 9.5). Bands at: 3548, 1035, 660, 595, 476, 463 cm^{-1}. Ferro stilpnomelane (California) cannot be distinguished from common ferri stilpnomelane. The band at 3550 cm^{-1} points to a chloritic structure. It has also been observed that stilpnomelane develops from chlorite. So it may be a chlorite-biotite intermediate.

Remarks. The structure of stilpnomelane needs further examination. Well-selected specimens should be used.

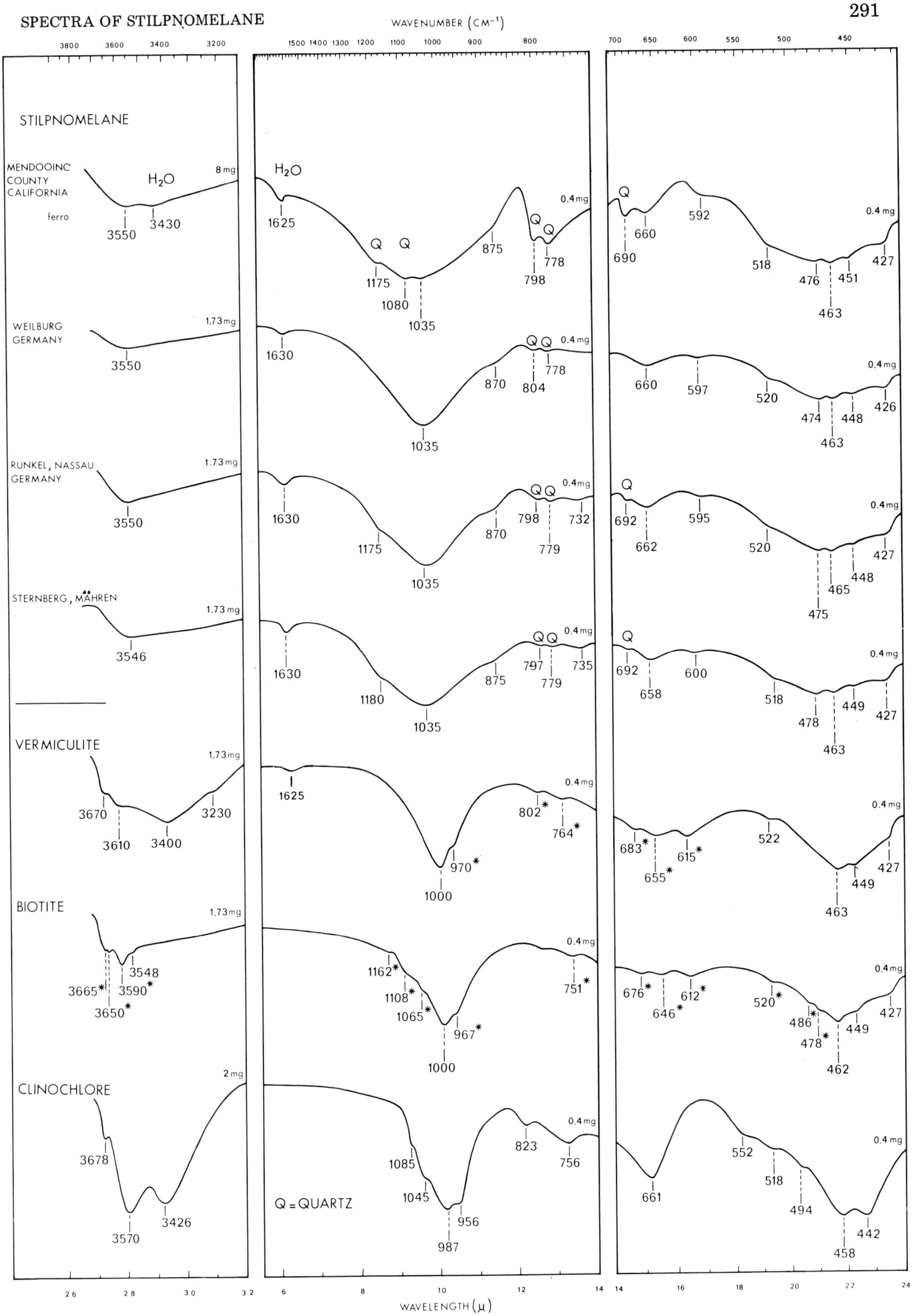

WAVENUMBER (CM^{-1})

3800 3600 3400 3200 1500 1400 1300 1200 1100 1000 900 800 700 650 600 550 500 450

STILPNOMELANE

MENDOOINC COUNTY CALIFORNIA ferro
8 mg H_2O 3550 3430
H_2O 0.4 mg 1625 Q Q 1175 1080 1035 875 Q Q 798 778
Q 690 660 592 0.4 mg 518 476 463 451 427

WEILBURG GERMANY
1.73 mg 3550
1630 Q Q 0.4 mg 1035 870 804 778
660 597 0.4 mg 520 474 463 448 426

RUNKEL, NASSAU GERMANY
1.73 mg 3550
1630 1175 1035 870 Q Q 0.4 mg 798 779 732
Q 692 662 595 0.4 mg 520 475 465 448 427

STERNBERG, MÄHREN
1.73 mg 3546
1630 1180 1035 875 Q Q 0.4 mg 797 779 735
Q 692 658 600 0.4 mg 518 478 463 449 427

VERMICULITE
1.73 mg 3670 3610 3400 3230
1625 1000 970* 0.4 mg 802* 764*
0.4 mg 683* 655* 615* 522 463 449 427

BIOTITE
1.73 mg 3665* 3650* 3590* 3548
0.4 mg 1162* 1108* 1065* 1000 967* 751*
0.4 mg 676* 646* 612* 520* 486* 478* 462 449 427

CLINOCHLORE
2 mg 3678 3570 3426
0.4 mg 1085 1045 987 956 823 756
0.4 mg 661 552 518 494 458 442

Q = QUARTZ

26 28 30 32 6 8 10 12 14 14 16 18 20 22 24

WAVELENGTH (μ)

Fig. 9.5. Infrared spectra of stilpnomelane from different origin.

Zeolites

The following groups can be distinguished (classification after Strunz, 1970):

Natrolite:
Natrolite: $Na_2[Al_2Si_3O_{10}]2H_2O$
Scolecite: $Ca[Al_2Si_3O_{10}]3H_2O$
Mesolite: $Na_2Ca_2[Al_2Si_3O_{10}]_3 8H_2O$
Thomsonite: $NaCa_2[Al_2(Al,Si)|Si_2O_{10}]_2 6H_2O$
Gonnardite: $(Ca,Na)_3[(Al,Si)_5O_{10}]6H_2O$

Heulandite:
Heulandite: $Ca[Al_2Si_7O_{18}]6H_2O$
Stilbite (desmine): $Ca[Al_2Si_7O_{18}]7H_2O$

Phillipsite:
Phillipsite: $KCa[Al_3Si_5O_{16}]6H_2O$

Chabasite:
Chabasite: $(Ca,Na_2)[Al_2Si_4O_{12}]6H_2O$

Mordenite:
Mordenite (ptilolite): $(Ca,K_2,Na_2)[AlSi_5O_{12}]_2$
$6H_2O$
Laumontite: $Ca[AlSi_2O_6]_2 4H_2O$

The above minerals consist like the feldspars, of a three-dimensional frame work of SiO_4 and AlO_4 tetraeders which are connected by O atoms. Four, six and even twelve oxygen rings are possible. Even numbered rings are more usual than odd. The negative charges produced by the Si/Al substitutions are balanced by K, Na, Ca, Ba cations with which H_2O molecules are associated.
The zeolites differ from feldspars in having more open structure and containing loosely adsorbed water and water more tightly bonded. When the cavities and channels filled with water, which occupy to ca. 50% of its volume in the most porous zeolites, are emptied by heating (ca. 350°C) the mineral does not disintegrate as e.g., gypsum, vermiculite, etc.
The cations in the continuous intersecting channels may be exchanged by other cations depending on their size and that of the narrower windows in the channels. The largest have a diameter of even 13 Å, (faujasite).
Also for the adsorption of molecules there is a limiting size above which no adsorption takes place, e.g. chabasite readily adsorbs He, H_2, O_2, N_2, CO_2, NH_3, methylalcohol, ethylalcohol, but not aceton, ether and the higher alcohols.

Occurrence. Zeolites are commonly found in cracks and cavities of basic volcanic rocks where they resulted from hydrothermal activities. But phillipsite and in particular clinoptilolite a Si,-Na,K-rich variety of heulandite (Ca) may also be of authigenic origin.
For instance, phillipsite is found in pelagic sediments in the Pacific, where it resulted from action of the salt water on the sedimentary materials. Clinoptilolite is found also in tuffaceous rocks in the Mojave desert, in bentonite beds in northern Alaska and Patagonia, in Jurassic and Cretaceous sediments in southeast England etc.

Infrared spectra (Fig. 9.6, 9.7, 9.8, 9.9). Natrolite, scolecite, mesolite and thomsonite can easily be distinguished from each other. In particular by the shape of their bands at the highest wavelengths. Such in contrast with X-ray analysis. But also the other zeolite minerals can be distinguished from each other: Characteristic bands are at:
Natrolite: 3546, 1000*, 982, 967 cm^{-1}
Scolecite: 3507, 1105, 987, 928, 498 cm^{-1}
Mesolite: 3540, 980, 963 cm^{-1}
Thomsonite: 1060, 992 cm^{-1}
Heulandite: 3606, 1200, 1050*, 603 cm^{-1}
Stilbite: 3592, 560 cm^{-1}
Phillipsite: 1004, 585 cm^{-1}
Chabasite: 1022 cm^{-1}
Mordenite: 3610, 1221 cm^{-1}
Laumontite: 3560, 3288, 1130, 995, 763 cm^{-1}.

Remarks. The name zeolite is derived from zeo (= to boil) and lithos (= stone) which is connected with the property of the mineral to boil when heated because of a severe loss of H_2O molecules = ca. 10—15%. Phillipsite is named after the well-known English mineralogist W. Phillips. Because of the large internal surface, the high cation-exchange capacity and the ultimate selective properties for molecules and cations, the zeolites are of large importance as molecular and cation sieves in separations, e.g., *n*-octane from iso-octane, thiophene from benzene, ethylene from ethene, aromatic from aliphatic hydrocarbons. Further in the polymerization of olefines of low molecular weight to gasoline, the storage and separation of gases: the more the molecule is polar the more tightly it is bound. Further as cation sieves, e.g., in the removal of Cs from high-graded radioactive wastes and in their storage. However, their use is severely limited by being unstable against acids and alkalis.

Natrolite

Puy de Dôme, France

Springfield, Ore., U.S.A.

Poona, India

Bohemia, Czechoslovakia

Scolecite

Lana Co., Ore., U.S.A.

Rio Grande do Sul, Brazil

Mesolite
Corvallis, U.S.A.

Thomsonite

Table Mountain, Colo., U.S.A.

Michigan, U.S.A.

Bohemia, Czechoslovakia

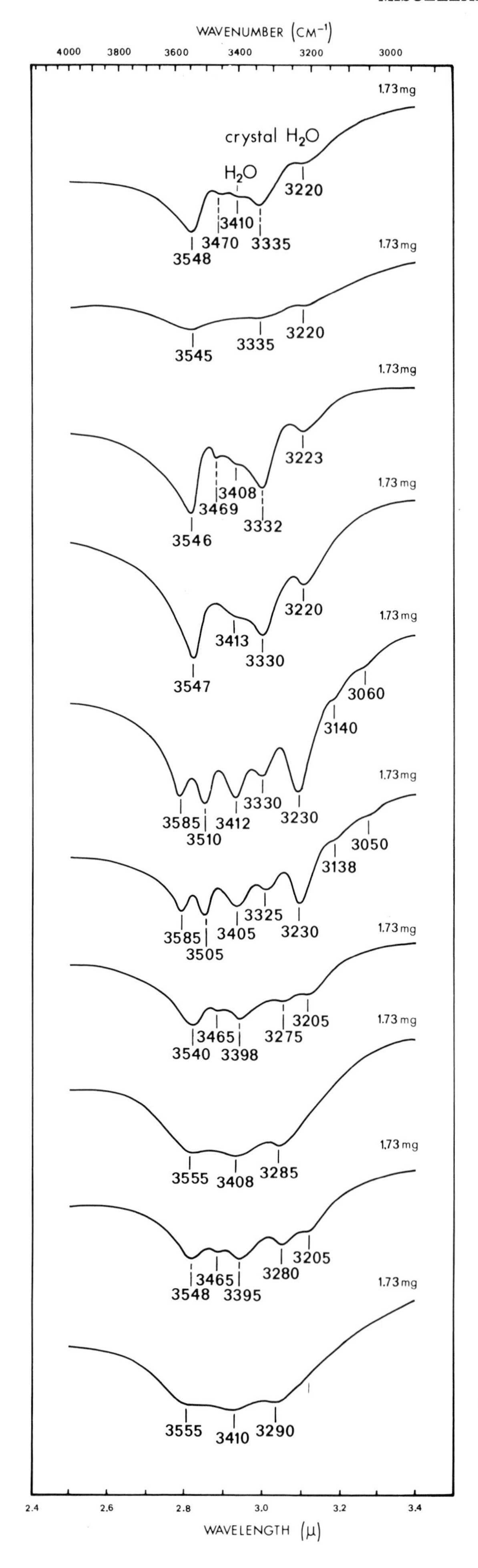

Fig. 9.6 Infrared spectra of minerals of the natrolite group from different origin.

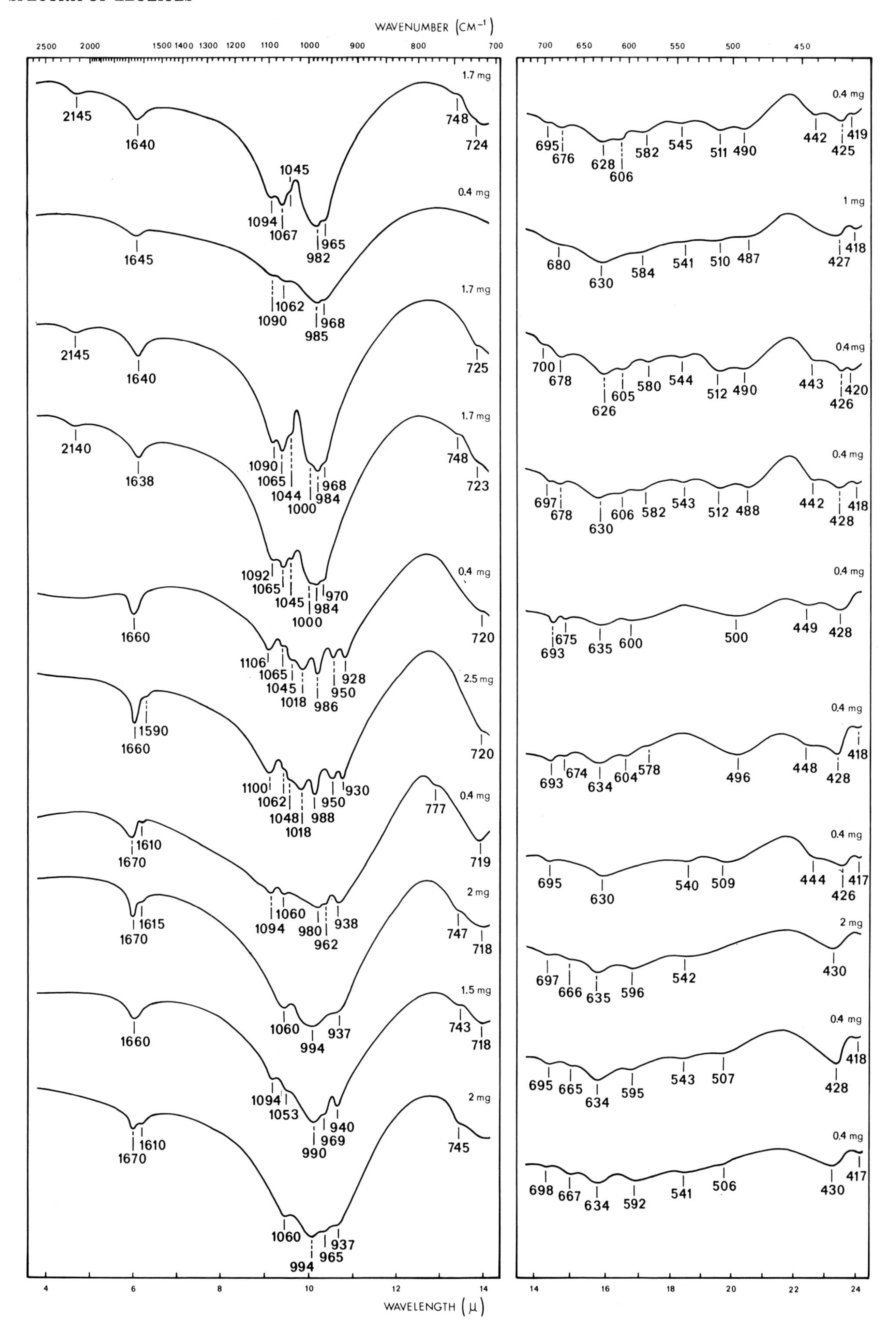
WAVENUMBER (CM⁻¹)
2500 2000 1500 1400 1300 1200 1100 1000 900 800 700
700 650 600 550 500 450
1.7 mg
2145 1640 1045 1094 1067 982 965 748 724
0.4 mg
695 676 628 606 582 545 511 490 442 425 419
0.4 mg
1645 1090 1062 985 968
1 mg
680 630 584 541 510 487 427 418
1.7 mg
2145 1640 725
0.4mg
700 678 626 605 580 544 512 490 443 426 420
1.7 mg
2140 1638 1090 1065 1044 1000 984 968 748 723
0.4 mg
697 678 630 606 582 543 512 488 442 428 418
1092 1065 1045 1000 984 970
0.4 mg
1660 1106 1065 1045 1018 986 950 928 720
0.4 mg
693 675 635 600 500 449 428
2.5 mg
1660 1590 1100 1062 1048 1018 988 950 930 720
0.4 mg
693 674 634 604 578 496 448 428 418
0.4 mg
1670 1610 1094 1060 980 962 938 777 719
0.4 mg
695 630 540 509 444 426 417
2 mg
1670 1615 1060 994 937 747 718
2 mg
697 666 635 596 542 430
1.5 mg
1660 1094 1053 990 969 940 743 718
0.4 mg
695 665 634 595 543 507 428 418
2 mg
1670 1610 1060 994 965 937 745
0.4 mg
698 667 634 592 541 506 430 417
4 6 8 10 12 14
14 16 18 20 22 24
WAVELENGTH (μ)

Heulandite (Ca)

Theigarhörn, Iceland

Fassathal, Tirol

Fall Creak Dam, Ore., U.S.A.

Clinoptilolite (Na-heulandite)
Hector, Calif., U.S.A.

Stilbite (desmine)

Banat Csiklova, Roumania

Harz, West Germany

Iceland

Faröer Islands

Poona, India

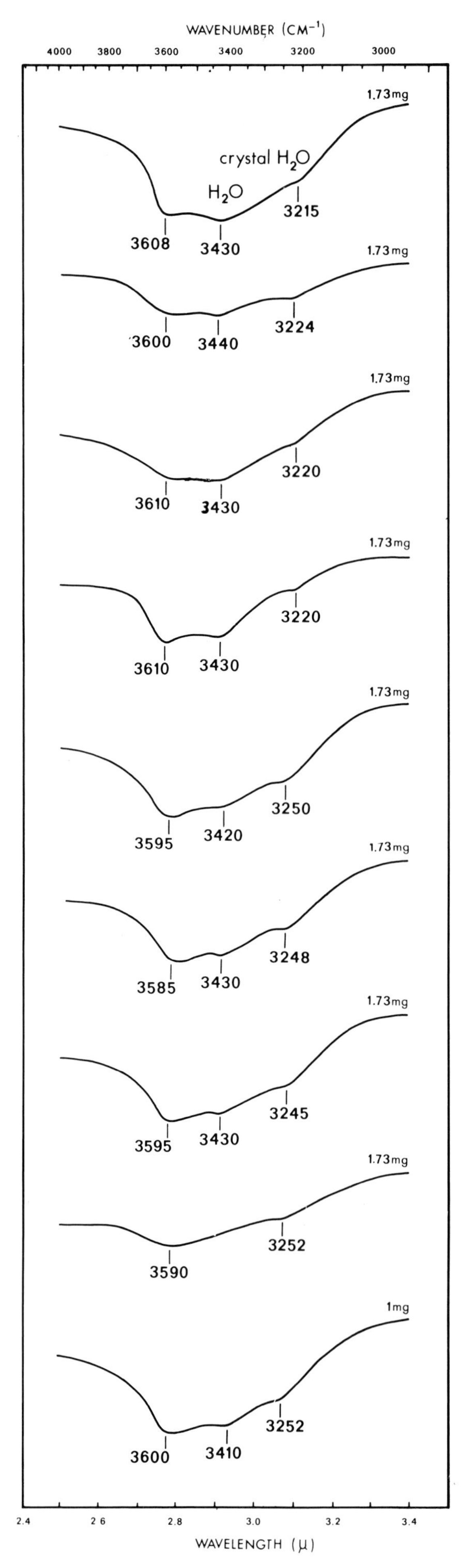

Fig. 9.7. Infrared spectra of heulandite, clinoptilolite and stilbite (desmine) from different origin.

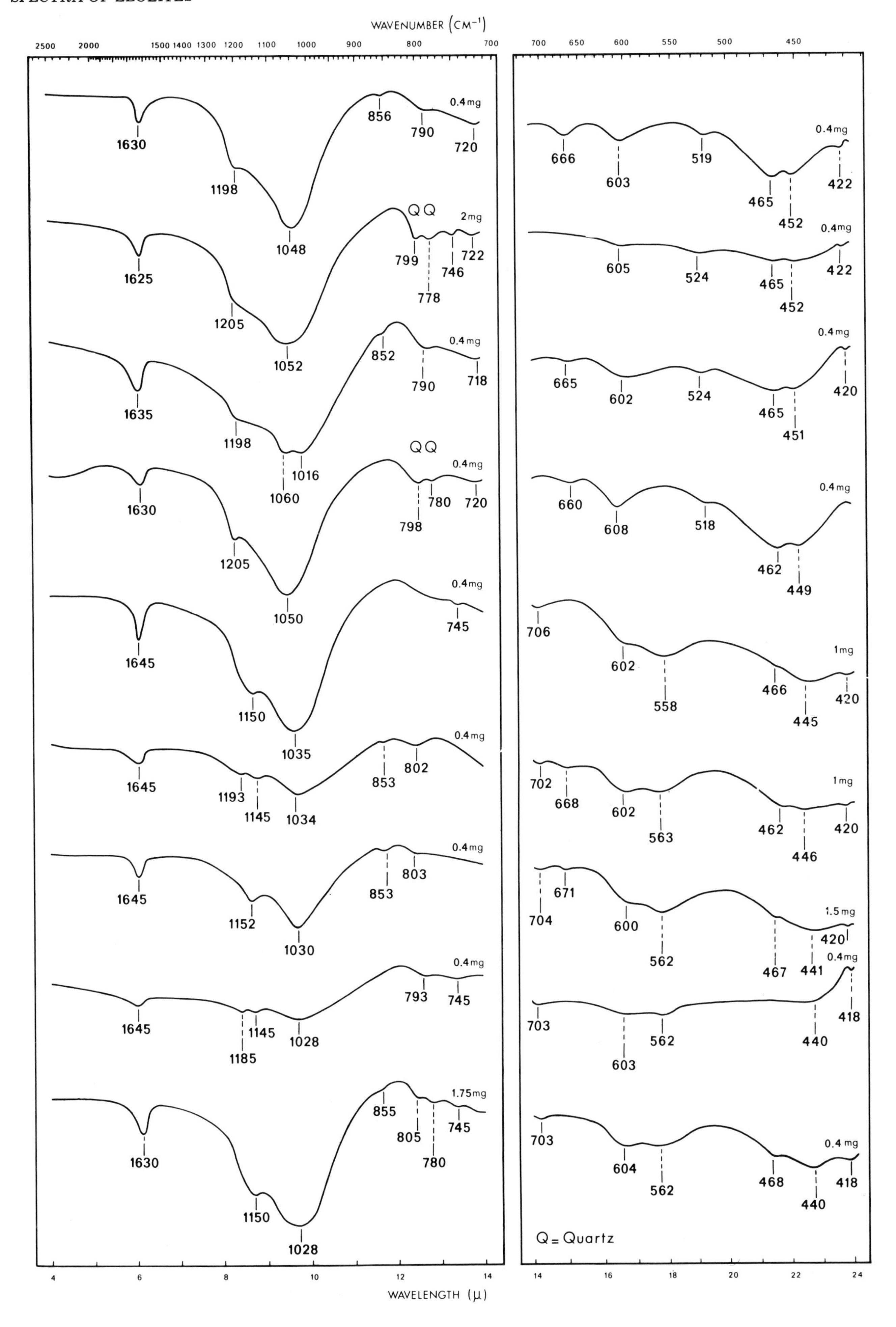
WAVENUMBER (CM⁻¹)
2500 2000 1500 1400 1300 1200 1100 1000 900 800 700
700 650 600 550 500 450
0.4mg 1630 1198 1048 856 790 720
666 603 519 465 452 422 0.4mg
Q Q 2mg 1625 1205 1052 799 778 746 722
605 524 465 452 422 0.4mg
0.4mg 1635 1198 1060 1016 852 790 718
665 602 524 465 451 420 0.4mg
Q Q 0.4mg 1630 1205 1050 798 780 720
660 608 518 462 449 0.4mg
0.4mg 1645 1150 1035 745
706 602 558 466 445 420 1mg
0.4mg 1645 1193 1145 1034 853 802
702 668 602 563 462 446 420 1mg
0.4mg 1645 1152 1030 853 803
704 671 600 562 467 441 420 1.5mg
0.4mg 1645 1185 1145 1028 793 745
703 603 562 440 418 0.4mg
1.75mg 1630 1150 1028 855 805 780 745
703 604 562 468 440 418 0.4 mg
Q = Quartz
4 6 8 10 12 14
14 16 18 20 22 24
WAVELENGTH (μ)

Phillipsite

Fichtelberg, G.D.R.

Radkersburg, Austria

Italy

Aqua Acetosa, Italy
(contains much amorphous material)

Chabasite
Hungary

Faröer Islands

Oregon, U.S.A.

Bohemia, Czechoslovakia

Nova Scotia, U.S.A.

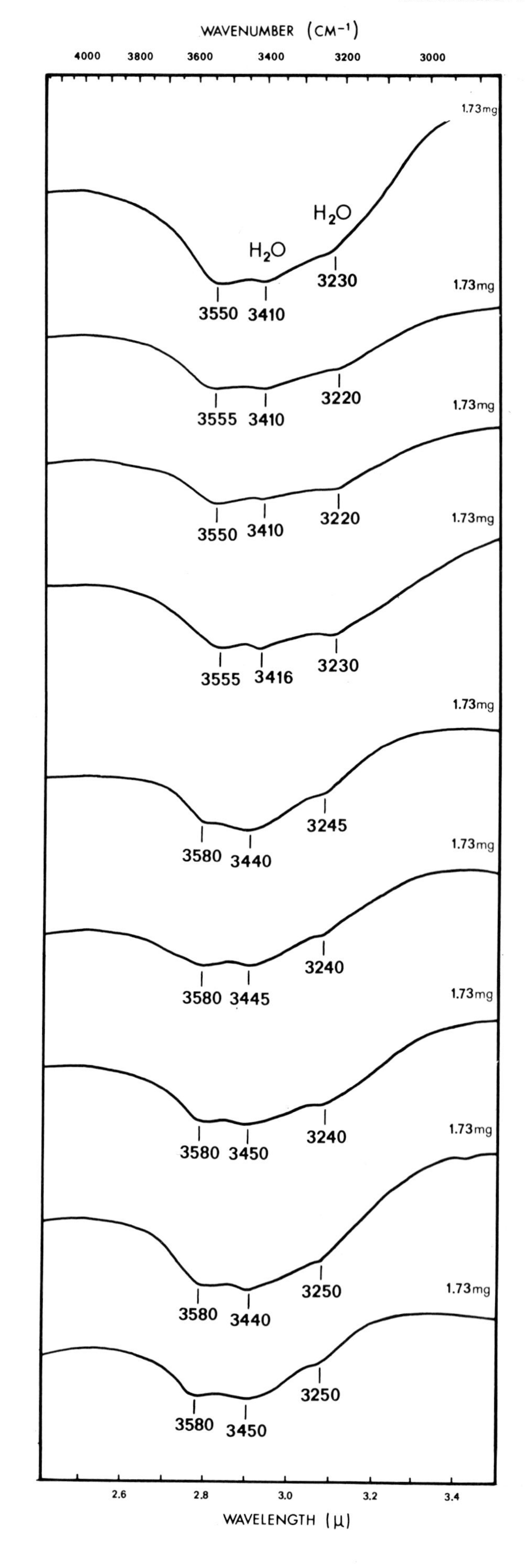

Fig. 9.8. Infrared spectra of phillipsite and chabasite from different origin.

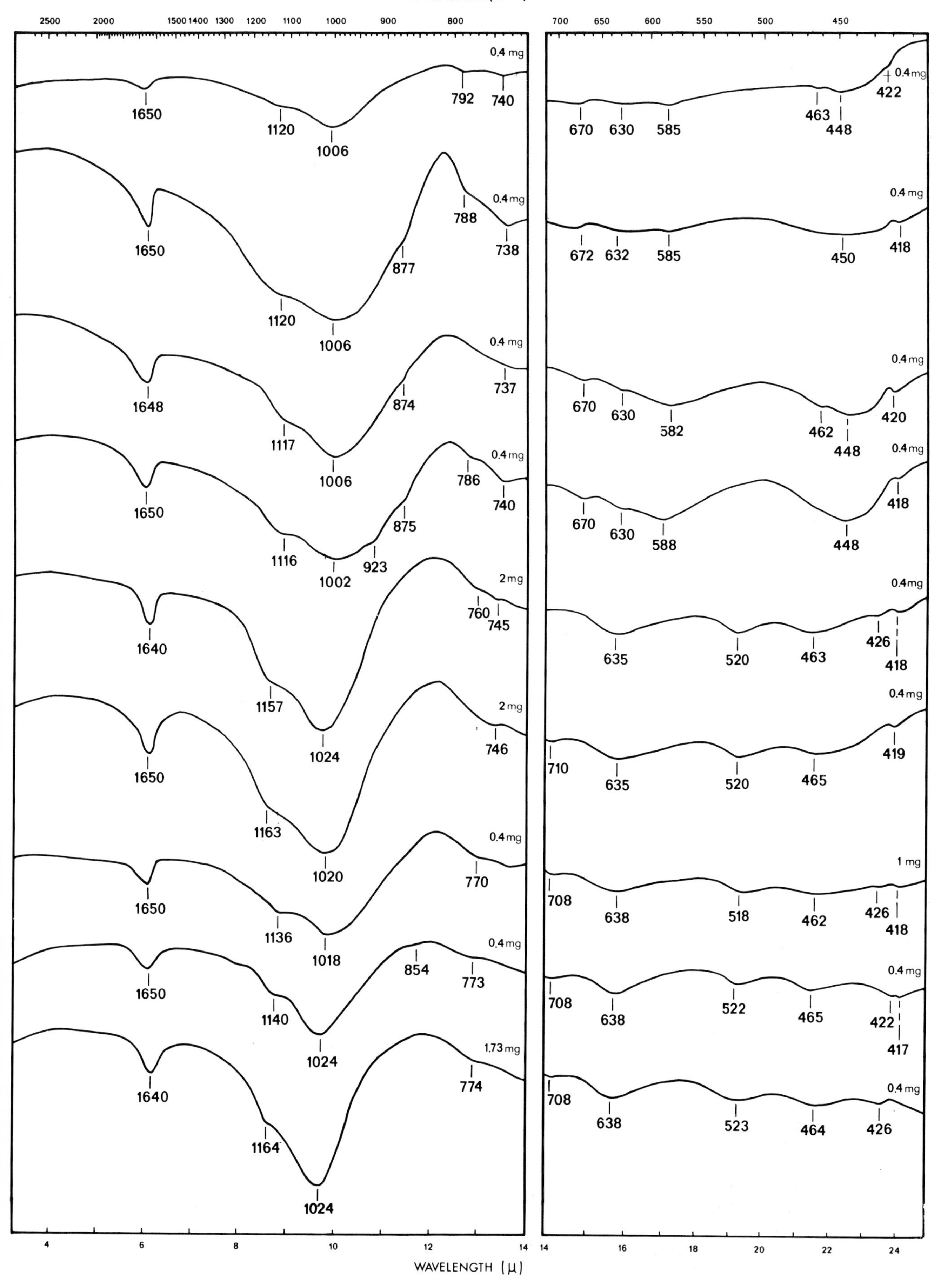

WAVENUMBER (CM⁻¹)
2500 2000 1500 1400 1300 1200 1100 1000 900 800
700 650 600 550 500 450
0.4 mg
1650 1120 1006 792 740
670 630 585 463 448 422
0.4 mg
1650 1120 1006 877 788 738
672 632 585 450 418
0.4 mg
1648 1117 1006 874 737
670 630 582 462 448 420
0.4 mg
1650 1116 1002 923 875 786 740
670 630 588 448 418
2 mg
1640 1157 1024 760 745
635 520 463 426 418
2 mg
1650 1163 1020 746
710 635 520 465 419
0.4 mg
1650 1136 1018 770
1 mg
708 638 518 462 426 418
0.4 mg
1650 1140 1024 854 773
708 638 522 465 422 417
1.73 mg
1640 1164 1024 774
708 638 523 464 426
4 6 8 10 12 14
14 16 18 20 22 24
WAVELENGTH (μ)

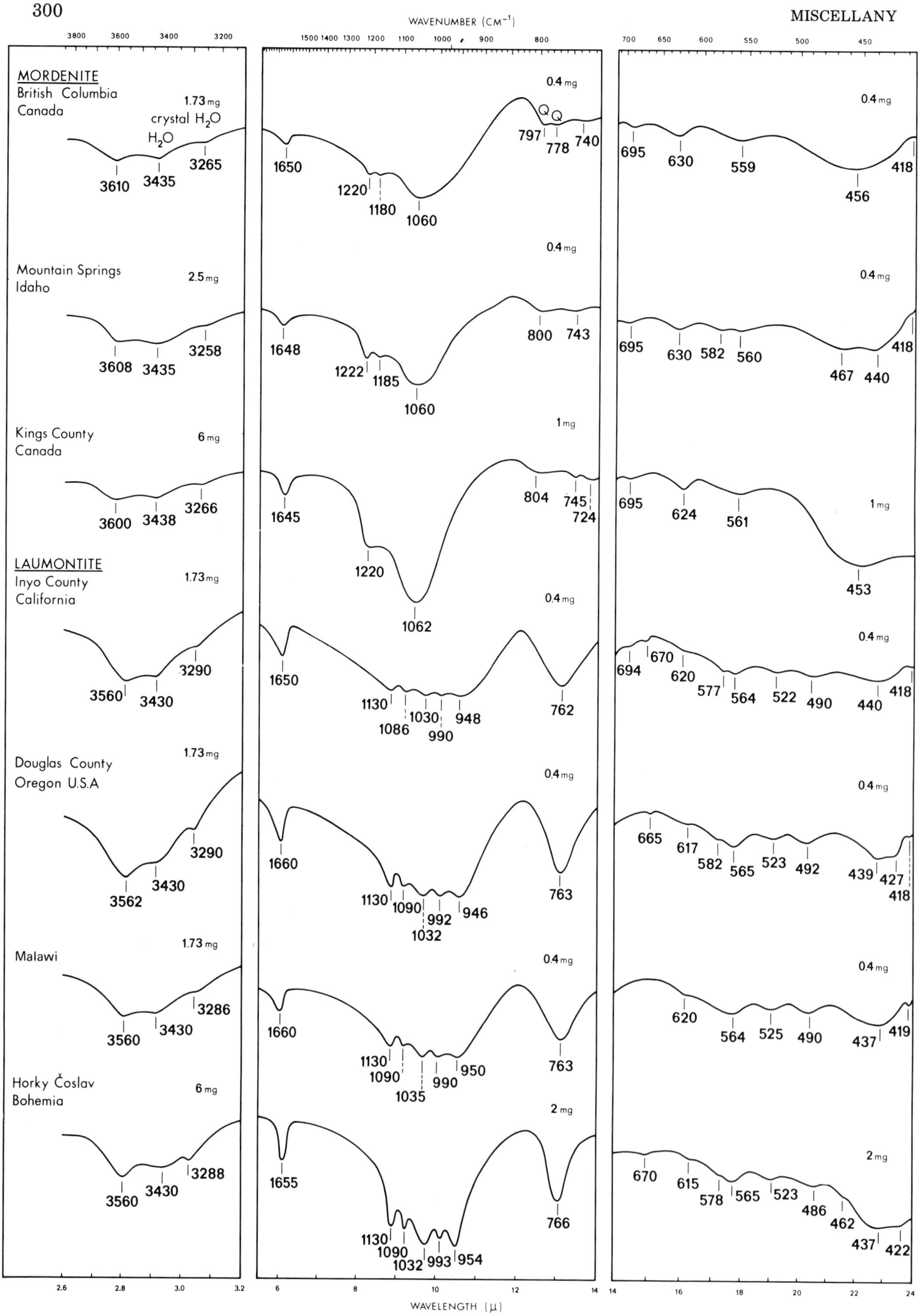

WAVENUMBER (CM^{-1})

MORDENITE
British Columbia Canada

Mountain Springs Idaho

Kings County Canada

LAUMONTITE
Inyo County California

Douglas County Oregon U.S.A

Malawi

Horky Čoslav Bohemia

WAVELENGTH (μ)

Fig. 9.9. Infrared spectra of mordenite and laumontite from different origin.

Cation and anion exchangers; formula:

(M^+) X Al_2O_3 Y SiO_2 n H_2O (M^+ = exchangeable cation)

Inorganic amorphous products with a large surface and high cation-exchange capacity are called permutites. They can be obtained by pouring hot solutions of Al, Mg, Fe or Ni chloride into a hot solution of Na-silicate in adequate proportions (Si/Al molair = 1). Another way is to melt SiO_2, Al_2O_3 and NaOH. The precipitate or melt is pulverized, dialyzed and dried.
The cation-exchange capacity of the permutites is ca. 200 mequiv./100 g as compared with ca. 100—150 mequiv./100 g for bentonite cation exchangers and natural zeolites.
Organic cation and anion exchangers consist of resins with crosslinked polystyrene matrix. The cation exchangers (Zeo Karb 225, Amberlite 200, etc.) have sulfonic groups and the anion exchangers (Amberlite 410, etc.) quaternary ethanol ammonium groups $[N(C_2H_5]_4]^+$ substituted in the benzene ring.
Cation-exchange capacity is 350—400 mequiv. / 100 g. Moreover, they are more resistant to acid than the natural (mineral) and the artificial ion exchangers.

Infrared spectra (Fig. 9.10). The samples show bands which point to some ordering of the atoms. As a constrast X-ray analyses show almost amorphous material.
For the samples heated in an autoclave the ordering of the atoms has somewhat increased according to I.R. They even show a montmorillonite-like picture (also after X-ray analysis) and suggest that the kind of cation used in the preparation of the permutite determines the place of some of the bands among which that of the most intensive ones (heated between brackets): Al: 3557 (3550), 1037 (1036), Mg: 3680 (3682), 1018 (1018); Ni: 3635 (3630), 1025 (1028).
The infrared spectra of the Zeo Karb 225 and Amberlite 200 polystyrene cation exchange resins, show absorption bands of polymer chain aliphates and aromatics at 3060, 3030 and 2980 cm^{-1} (νC-H).
Other characteristic bands of polystyrene occur at 624, 584 and 463 cm^{-1}.
The phenyl nucleus is represented at 1645 (ν C=C or aromatics) and at 836 and 777 cm^{-1} (bending vibration of two adjacent H atoms of the nucleus). Further several bands occur which point to sulfonic acid (R-SO_2-OH) substitution, e.g., at 1225, 1187, 1126 (ν_{as} SO_2), 1040, 1010 (ν_s SO_2). Hydroxyl vibrations of water (ν_s O-H) are at 3440 cm^{-1}.

Remarks. Inorganic and organic cation exchangers are used in industry to concentrate trace ions from diluted solution. Further as water softeners in laundries, purifiers of water, etc. When saturated with a certain ion, they can be regenerated and then used again.
The silica-alumina amorphs are also used as catalysts for various purposes.

Ion exchangers
Al-permutite
Inorganic:

Fig. E.M. 105, 239

Al-permutite heated*

Mg-permutite

Mg-permutite heated*

Ni-permutite

Ni-permutite heated*

Organic
Zeo-Karb 225 (cation)

Amberlite 200 (cation)

*Heated in an autoclave for 1 week at ca. 200°C and 300 kg/cm².

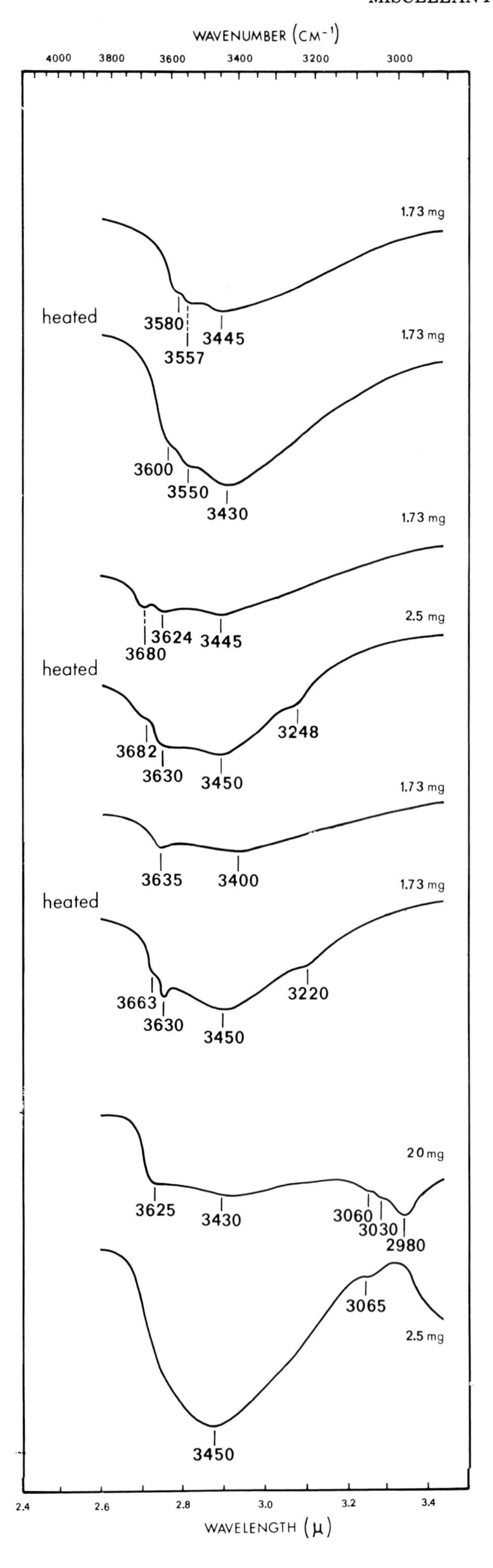

Fig. 9.10. Infrared spectra of inorganic and organic ion exchangers.

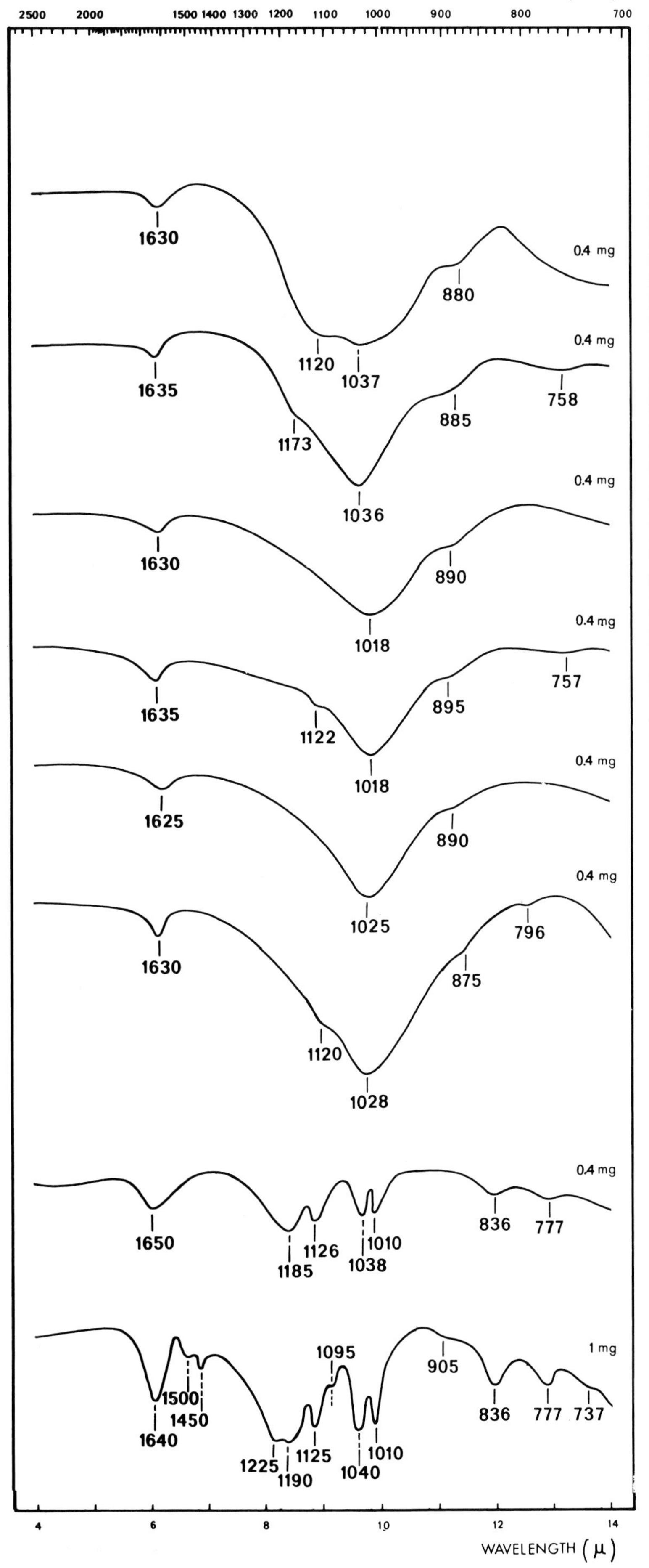
WAVENUMBER (CM⁻¹)
2500 2000 1500 1400 1300 1200 1100 1000 900 800 700
1630 880 0.4 mg 1120 1037
1635 0.4 mg 1173 885 758 1036
1630 0.4 mg 890 1018
1635 0.4 mg 1122 895 757 1018
1625 0.4 mg 890 1025
1630 0.4 mg 1120 875 796 1028
1650 0.4 mg 1185 1126 1038 1010 836 777
1640 1500 1450 1 mg 1225 1190 1125 1095 1040 1010 905 836 777 737
4 6 8 10 12 14
WAVELENGTH (μ)

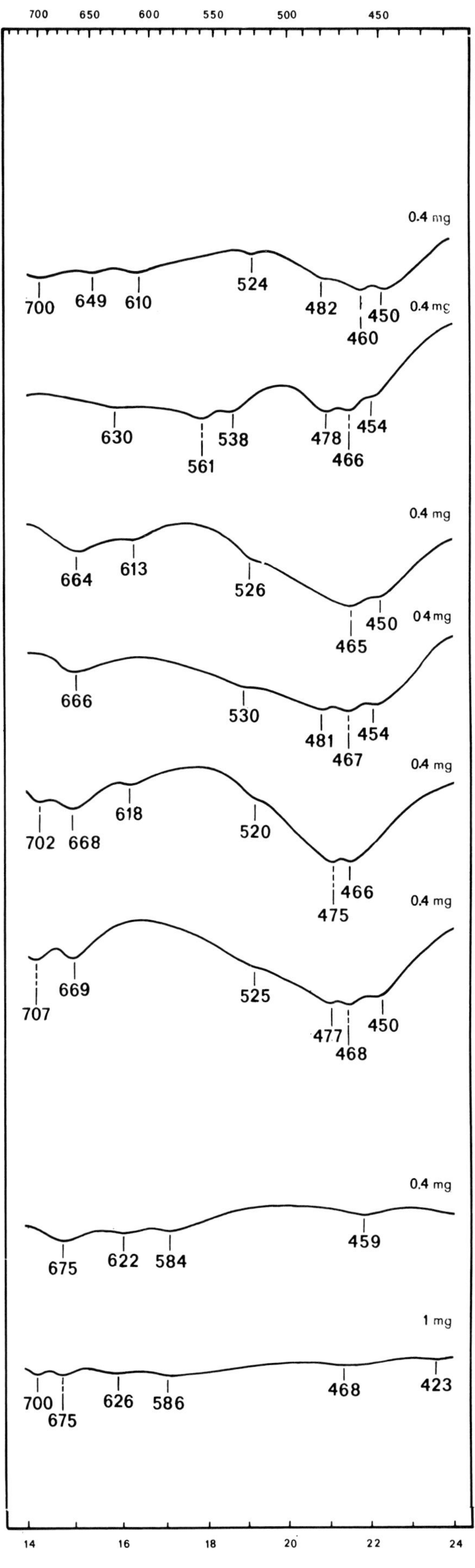
700 650 600 550 500 450
0.4 mg 700 649 610 524 482 460 450
0.4 mg 630 561 538 478 466 454
0.4 mg 664 613 526 465 450
0.4 mg 666 530 481 467 454
0.4 mg 702 668 618 520 475 466
0.4 mg 707 669 525 477 468 450
0.4 mg 675 622 584 459
1 mg 700 675 626 586 468 423
14 16 18 20 22 24

10. ORGANIC MATTER

General

Large amounts of coalification products like peat, brown-coal, anthracite and other types of the coalification series are found in nature, often accumulated in large deposits during geologic periods.

The composition of the organic debris from plants and animals is very complicated and variable; the most predominant substances of the organic-matter fraction of soils are cellulose, lignin, proteins and mixtures of high-molecular, dark-coloured substances of variable composition depending on climatic conditions and kind of soil. The dark-coloured soil organic matter (humic substances) mainly consists of humic acids and fulvic acids. Humic substances are an essential factor influencing soil fertility. On account of its favourable physical properties, it has a large capacity to retain water and to exchange cations, among those which are indispensable for plant growth (for details, see Flaig et al., 1974).

The main absorption bands in the infrared spectra of organic substances occur with only few exceptions in the range between 4000—700 cm^{-1}. According to the energetic relation between molecular structure and absorption properties of radiation in the infrared range of light, characteristic absorption bands of unique parts of the molecules or functional groups occur in equal ranges of infrared spectra. Therefore, in the range of high wavenumbers of the infrared spectra of organic molecules we find the valency vibrations, which are mostly related to the different linkages between carbon and hydrogen, or oxygen and hydrogen (3000—1600 cm^{-1}). This is followed by the range of absorptions caused by double linkages between adjacent carbon atoms, or carbon and oxygen (1800—1600 cm^{-1}), followed by the various types of single bonds between adjacent carbon atoms, or carbon and heteroatoms, like oxygen, nitrogen, sulphur and others. The correct assignment in the "fingerprint" region between 1250 and 700 cm^{-1} is difficult on account of the various possibilities of structural arrangements of the organic molecule functions.

As natural occurring substances are mostly mixtures of complex nature with different types of linkages and functional groups, this leads to an overlapping of adsorption bands. Moreover, structural environment influences strongly the vibration frequencies of single components. For instance the carbonyl vibration of 1.8-dihydroxyanthraquinone is lowered from about 1700 to about 1400 cm^{-1} because of simultaneous bonding of the carbonyl oxygen with two hydroxyl groups. When the pellet technique is used, the time of grinding and evacuation is a further disturbing factor. The table of assignment of the main absorption bands is a summary of a few generalizations which may serve as a guide for first interpretations of unknown infrared data.

References

Flaig, W., Beutelspacher, H. and Rietz, E., 1975. Chemical composition and physical properties of humic substances. In: J.E. Gieseking (Editor), Soil Components, 1. Springer-Verlag, New York, N.Y.

Peat, brown coal, lignite, hard coal, anthracite, coke; formula: C, O, H, N.

The above organic materials contain several kinds of functional groups; mainly hydroxyl (R-OH), alkanes (R-$(CH_2)_n$-R′), carboxyl (R-COOH), carbonyl (R-CO-R′), esters (R-CO-O-R′), linear- and cyclic-ether groups (R-O-R′) in variable amounts.

Peat stands at the beginning of the coalification process and anthracite at the end. During coalification the content of carbon increases and that of oxygen, hydrogen and nitrogen decreases.

Gray peat = ca. 58%; black peat = ca. 62%; brown coal = ca. 65%; lignite = ca. 75%; hard coal = ca. 85%; anthracite = ca. 95%; coke = ca. 97% carbon.

Occurrence. Very large deposits are widespread found all over the world. The organic matter was accumulated under favourable conditions at mild temperature and high rainfall.

The progressive change from peat to anthracite is caused by chemical action and increase of pressure and temperature by overburdening or folding during geologic processes. This metamorphic transformation lasts over long periods of time.

Infrared spectra (Fig. 10.1; Table II). The infrared spectra of the several products of coalification show mixtures of different organic compounds and residues of plant materials. There is an overlapping of the bands with a large decrease in the wave numbers and the intensities of absorption when coalification process proceeds: especially for ca. 2950 cm^{-1} ($\nu_{as}CH_3$); ca. 2920 cm^{-1} ($\nu_{as}CH_2$); ca. 2850 cm^{-1} ($\nu_s CH_2$); ca. 1700 cm^{-1} ($\nu C = O$ carbonic acids, ketones); ca. 1622 cm^{-1} ($\nu C = C$ phenyl nucleus); ca. 1420 cm^{-1} (νOH alcohols); ca. 1160, 1058, 1030 cm^{-1}

(νC-O bands of equatorial and axial stretching of ethereal and hydroxylic groups of cellulose).

Remarks. The above materials are mainly used for the production of heat and energy.
Coke is produced by anaerobic distillation of coal at ca. 1000°C. Important by-products are obtained in this case, c.q. light oils, ammonia, base products for plastics etc.

Lignin; formula: C, H, O

Lignin is a well-known polymer and copolymer of different phenylpropenyl alcohols. They contain hydroxyl-, carbonyl-, methoxyl-, ethyl- and aromatic groups. Hydroxy-benzaldehyd, phtalic acid, protocatechuic acid, vanillic acid, syringic acid, ferulic acid, gallic acid, and guaiacylic acid all are products which can be isolated from lignin by several procedures. The amount of these lignin-related compounds in the lignin molecule depends on the modes applied and the kind of the material from which the lignin was isolated. Thus conifer lignin is mainly built up of coniferyl alcohol, while graminae mainly consist of p-coumaryl alcohol, etc.

Occurrence. The cell-wall of plants contains lignin. Milled wood has about 15—30% lignin. Grasses have only some percents. Lignin can be extracted from the above materials, e.g. by the method of Björkman (1954) (extraction with dioxane, precipitation with ethyl ether). The methoxyl content of the final product is ca. 15%.
Lignin is assumed to be the main source in the formation of natural humic acids in soils by weathering action or microbiological attack.

Infrared spectra (Fig. 10.2; Table II). According to the complex nature of the lignin molecules which may be assumed to be built up of a mixture of p-hydroxybenzaldehyde, vanillic-ferulic- and syringic aldehyde, strong similarity, but not identity, exists between lignin from straw and the above mentioned compounds, especially that with ferulic aldehyde.
Functional groups pointing to different types of aliphatic C-H vibrations occur at 2967 ($\nu_{as}CH_3$); 2938 ($\nu_{as}CH_2$); 1660 (νC=C); 1460 ($\delta_{as}CH_3$,-δCH_2); 1362 ($\delta_s CH_3$) cm^{-1}.
Vibrations of the aromatic skeleton occur at 1620, (νC=C phenyl conjugated alkanes); 1600 and 1505 (νC=C phenyl nucleus) in combination with those at 1460 ($\delta_{as}CH_3$, CH_2 scissoring) and 1427 cm^{-1} (δO-H in plane).
Vibrations of the various C = O functions of ketones, aldehydes, acids and esters are determined by absorption bands at 1740 (νC=O), 1333 (δO-H or νC=O in carbonyl), 1265 (ν_{as}=C-O-C) of ethers like $-OCH_3$ together with those at 1235 (νC-O in plane), 1180 and 1125 (νC-O), 1086 (νC-O sec. alcohols), 1048 (νC-O primary alcohols) and 1030 cm^{-1} (= C-O-C symmetr. stretching).

Remarks. The name is derived from lignum = wood, the main source of lignin. Lignin is a waste product in the manufacture of cellulose. It is used as a fuel and as an adsorbent of water in agriculture. It is assumed that lignin is an important initial material in the formation of humic acids in soils.

References

Björkman, A., 1954. Isolation of lignin from finally divided wood with neutral solvents. Nature, 174: 1057—1058.

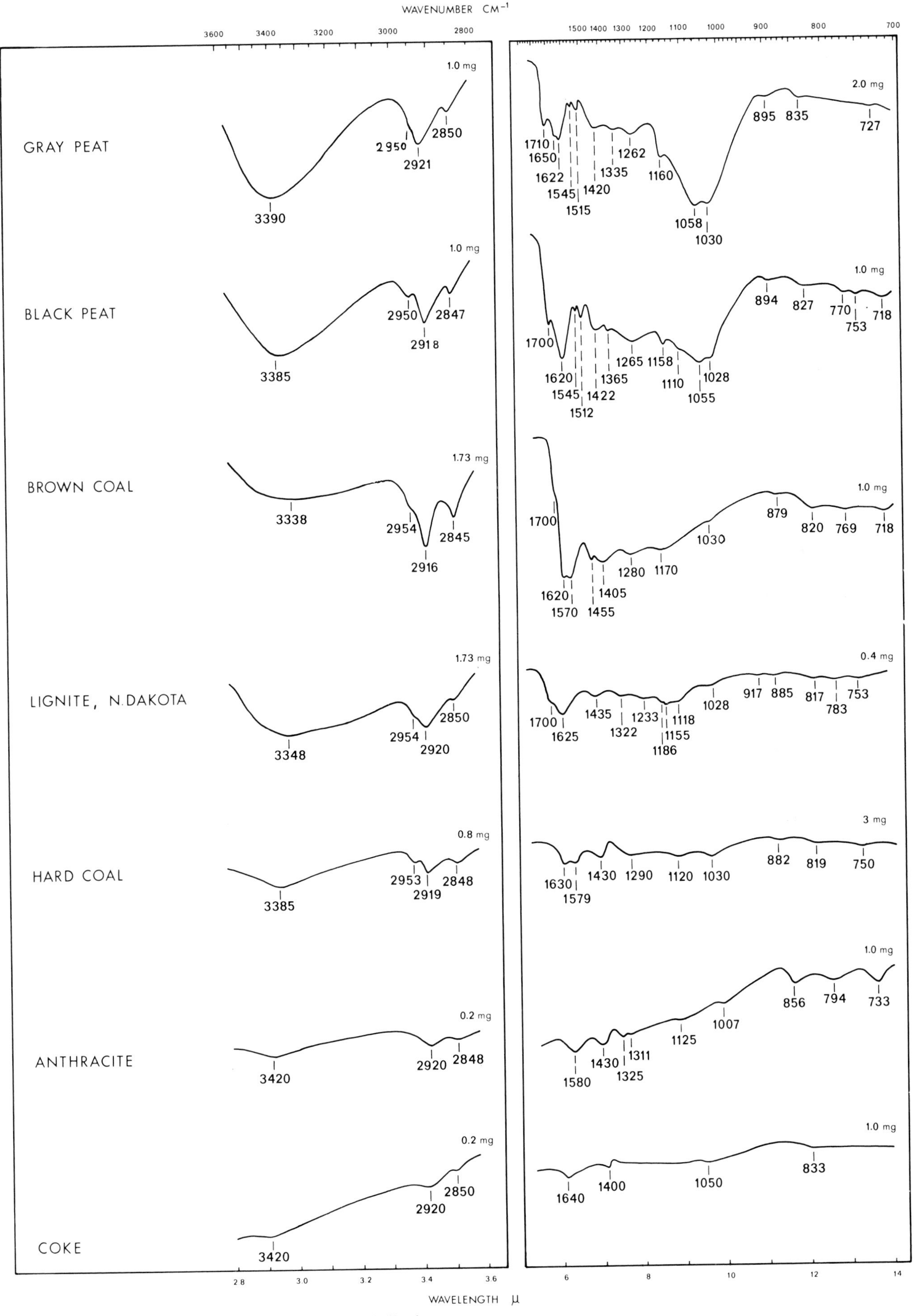

Fig. 10.1. Infrared spectra of peat, brown coal, lignite, hard coal, anthracite and coke.

TABLE II

Wavenumber (cm^{-1}) and assignment of most predominant vibrations which may occur in organic matter of soils according to literature and own observations

Groups	Bands (cm^{-1})	Intens.*	Assignment	Remarks
ALKANES	3000—2800	s	ν C-H	2 or 3 bands
$R-CH_3$	1470—1430	w	δ C-H	
	1380—1355	m	δ C-H	
	1255—1145	m	skeletal	
$R, R' > CH_2$	3000—2800	s	ν C-H	
	1470—1430	m	δ C-H	
	725—720	w	rocking	$R-(CH_2)_n-R'$ in paraffines
$R, R', R'' > CH$	2860	w	ν C-H	
	1340	w	δ C-H	
ALKENES	3100—3000	m	ν C-H	
$RCH=CH_2$; $RCH=CHR'$	1670—1600	v	ν C=C	
	1420—1400	m	δ C-H (in-plane)	
	1000—670	s	δ C-H (out-of-plane)	
AROMATICS	3100—3000	w	ν C-H	ϕ-H ($\phi = C_6H_5$)
	2000—1660	w	overtone	phenyl nucleus
	1600—1450	m	ν C=C	
	1225—950	w	δ C-H (in-plane)	
	900—690	s	δ C-H (out-of-plane)	
HYDROXYL	3640—3610	v	ν O-H	free O-H
-OH	3600—3500	s	ν O-H	dimeric OH, OH ... OH
	3400—3200	s	ν O-H	polymeric O-H
	3200—2500	v	ν O-H	OH ... O=C
ALCOHOLS, PHENOLS	1400—1250	s	δ O-H	
$R-CH_2-OH$; R-OH	1200—1000	s	ν C-O	
ETHERS and RELAT. COMP.	2850—2780	m	ν C-H	$R-O-CH_3$
R-O-R′	1200—1040	s	ν C-O-C	saturated ethers, ketols,
	1275—1200	s	ν_{as}=C-O-C	vinyl ethers, arom. ethers,
	1075—1020	s	ν_s=C-O-C	vinyl ethers, arom. ethers
ACIDS	3550—3500	w	ν O-H	vibration frequencies of acids
R-COOH	3000—2500	w	ν O-H	are mostly derived from dimers
	2700—2500	w	ν O-H	
	1740—1650	s	ν C=O	by conversion to inorganic salts,
	1440—1395	m	δ O-H	the characteristic frequencies
	1315—1280	m	ν C-O	are replaced by the two antisym.
	960—880	m	δ OH ... O (out-of-plane)	and sym. stretching vibrations of the carboxylate ion
CARBOXYLATES	1640—1550	s	ν_{as}C-O	COO^-
$R-COO^-$	1440—1340	m	ν_sC-O	COO^-
	680—400	m	bending and rocking	
ESTERS	1745—1735	s	ν C=O	
R-CO-O-R′	1330—1050	s	ν (O-C), and ν (OC-O)	
ALDEHYDES	2900—2680	w	ν C-H	2 bands
	1760—1645	s	ν C=O	
R-CHO	1440—1325	m	ν C-C	aliphatic aldehydes
	1420—1160	m	δ C-C	arom. aldehydes (3 bands)
	950—770	m	δ C-H	
KETONES	1780—1630	s	ν C=O	
R-CO-R′	1200—1100	s	ν C-O	aliphatic ketones
	1320—1220	s	ν C-O	aromatic ketones

* s = strong; m = medium; w = weak; v = variable.

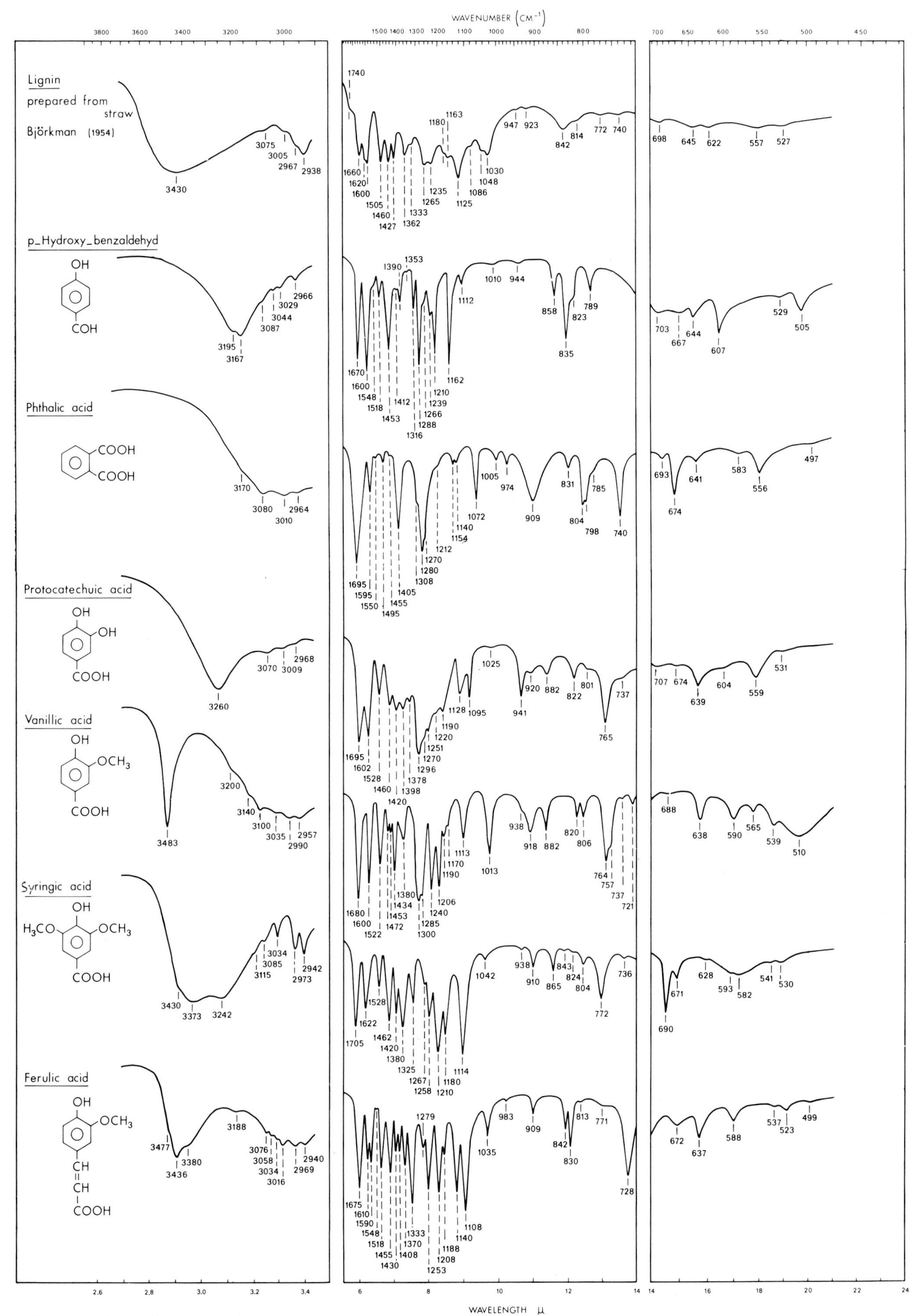

Fig. 10.2. Infrared spectra of lignin and some related compounds (pellets: 0.5 mg, lignin = 1.1 mg on 300 mg KBr).

Humic acids

Humic acids natural; formula: C, H, O, N (S)

To this group of acids belong humic materials from soils which are extracted by alkaline solutions and then precipitated by HCl. Calcareous soils are pretreated with HCl before alkaline extraction.
Humic acids are mixtures of several kinds of molecules with no defined structure and in different amounts. The main functional groups are: alkanes, hydroxyl-, phenolic-, carboxyl-, carbonyl- and methoxyl groups. The apparent molecular weight obtained by different methods varies from 1000 to 100 000. The amount of carbon of these polydisperse substances = 55—60%, hydrogen contents = 3—5%. Oxygen contents = 30—40% and nitrogen contents = 2—6%.

Occurrence. Depending on climatic and environmental conditions, humic acids are accumulated in soils in different amounts. They may even accumulate to ca. 20% in tschernozems.
Several fungi (*Stachybotrys chartarum*, *Epicoccum nigrum*) produce dark brown humic-acid-like products. Phenols formed by the fungus from carbohydrates occur as an intermediate stage in this process. Derivatives of amino acids are also a by-product.
Dark-coloured gel-like products which are found in peats are called dopplerite after its discoverer Christian Doppler (1849).

Infrared spectra (Fig. 10.3, 10.4; Table II). The spectrum shows that the humic acids from various origins are similar but not identical. The variations are caused by slight differences in their composition.
The most characteristic feature of humic acids is given by the strong but relatively broad absorptions of the C = O stretching vibration of several functional groups which may occur in preparations of humic substances, e.g. carboxylic acids, cyclic and acyclic aldehydes and ketones at 1709 to 1715 cm^{-1}, C = C vibrations of double bonds in aromatic compounds at 1600 to 1613 cm^{-1}. The bands at 1698 to 1701 and at ca. 1400 cm^{-1} point to the presence of carboxyls and carbonyl groups. As a contrast to those of the other humic acid samples, the spectrum of the humic acid from tschernozem (USSR) and from marsh soils do not contain alkanes ($-CH_3$, $>CH_2$, $\geq CH$) vibrations at ca. 2925 and ca. 2845 cm^{-1}. The spectrum of the commercial humic acids (Kasseler brown) resembles that of the dopplerite samples but is not identical. The infrared spectrum of the mycel humic acid is strongly differentiated. The most characteristic bands are at 2924 cm^{-1} (alkanes), 1715 cm^{-1} (C = O in ketones and carboxylic acids), 1656 cm^{-1} (amino acids); 1616 cm^{-1} (C = C in aromatics and COO^- of carboxylates), 1464 cm^{-1} (alkanes), 1081 cm^{-1} (ethers) and 772 cm^{-1} (COO^--scissoring). The last mentioned mostly occurs in natural humic and fulvic acids.

Remarks. Humic acids have a large cation exchange capacity (CEC) = ca. 200—400 mequiv./100 g. It is mainly caused by carboxyl groups. At higher pH values especially phenol groups become active. Because of their large capacity to retain cations by exchange reactions, humic acids are of great importance for soil fertility.
Commercial humic acids are manufactured from brown coal (trade mark: Kasseler brown) or from lignite (trade mark: Baroid Carbonox) and by extraction with NaOH and afterwards precipitation with HCl and washing out the excess of salts and acids.

Humic acids from hydroquinones

Artificial humic acid-like substances may be formed from hydroquinone ($C_6H_4(OH)_2$) or other phenoles by oxidative polymerization at a pH = 8 in alkaline medium.
The dark-brown substances are flocculated with HCl and afterwards purified from inorganic components and low polymer substances by dialysis.

Infrared spectra (Fig. 10.3). The infrared spectrum shows a large decrease in the number of absorption bands when oxidation proceeds. Also their frequencies are changed. The spectrum of the wholly oxidized samples resembles that of the natural humic acids but is not identical.

Fulvic acids; formula: C, H, O, N (S)

To this group of acids belong weak yellow to brown yellow-coloured organic substances from soils which are dissolved by alkaline solutions. In contrast to humic acids they do not precipitate when mineral acids are added. There is a gradual transition of the one into the other. The chemical composition of fulvic acids reveals a small amount of carbon, but a larger amount of hydrogen and oxygen. Structural investigations lead to the assumption that, when compared with humic acids, the content of functional groups containing double bonds, aliphatic side chains and aromatics is smaller while that of the carboxylic groups is larger.

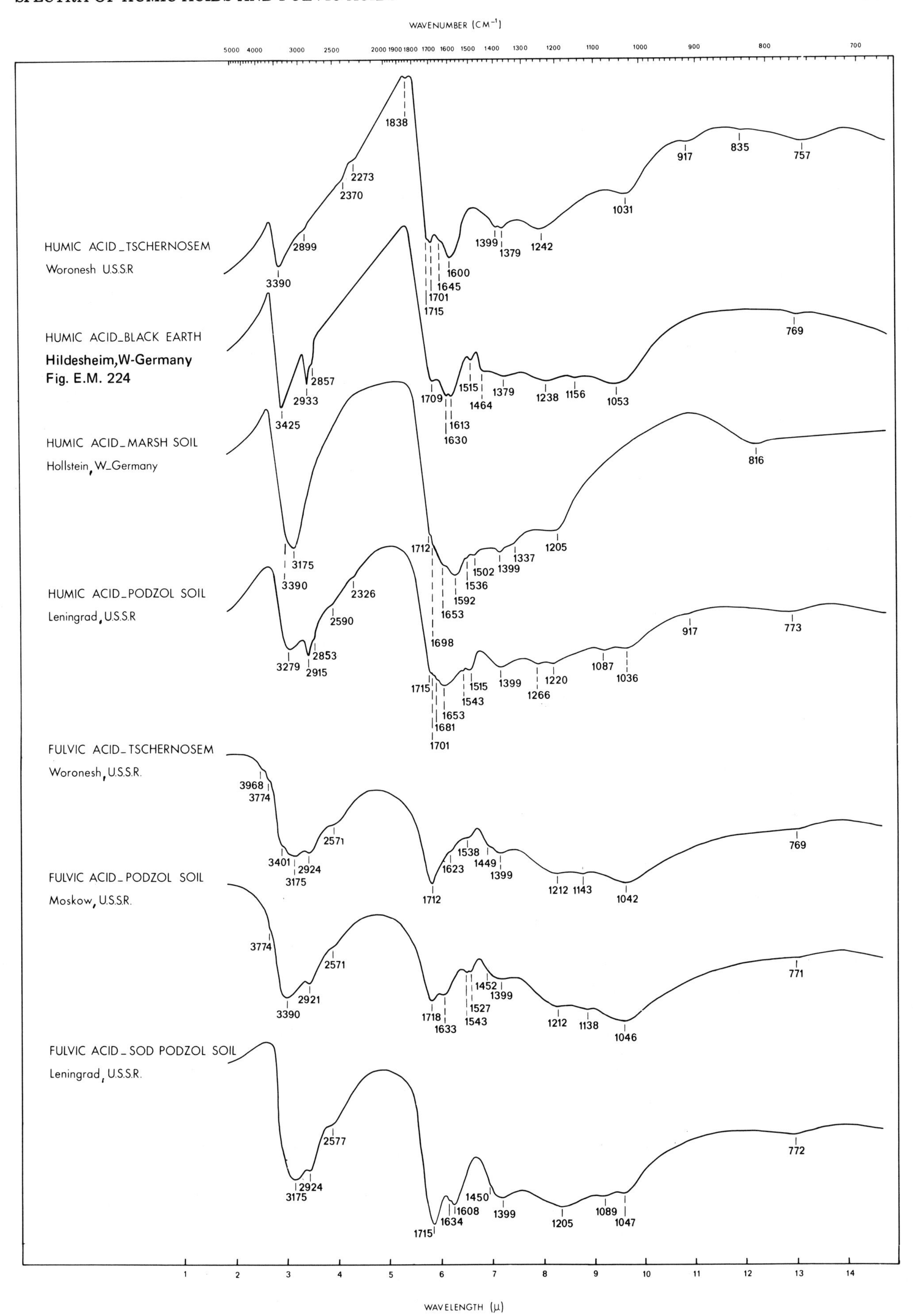

Fig. 10.3. Infrared spectra of humic acids and fulvic acids from different origin (pellets: 0.7 mg on 300 mg KBr).

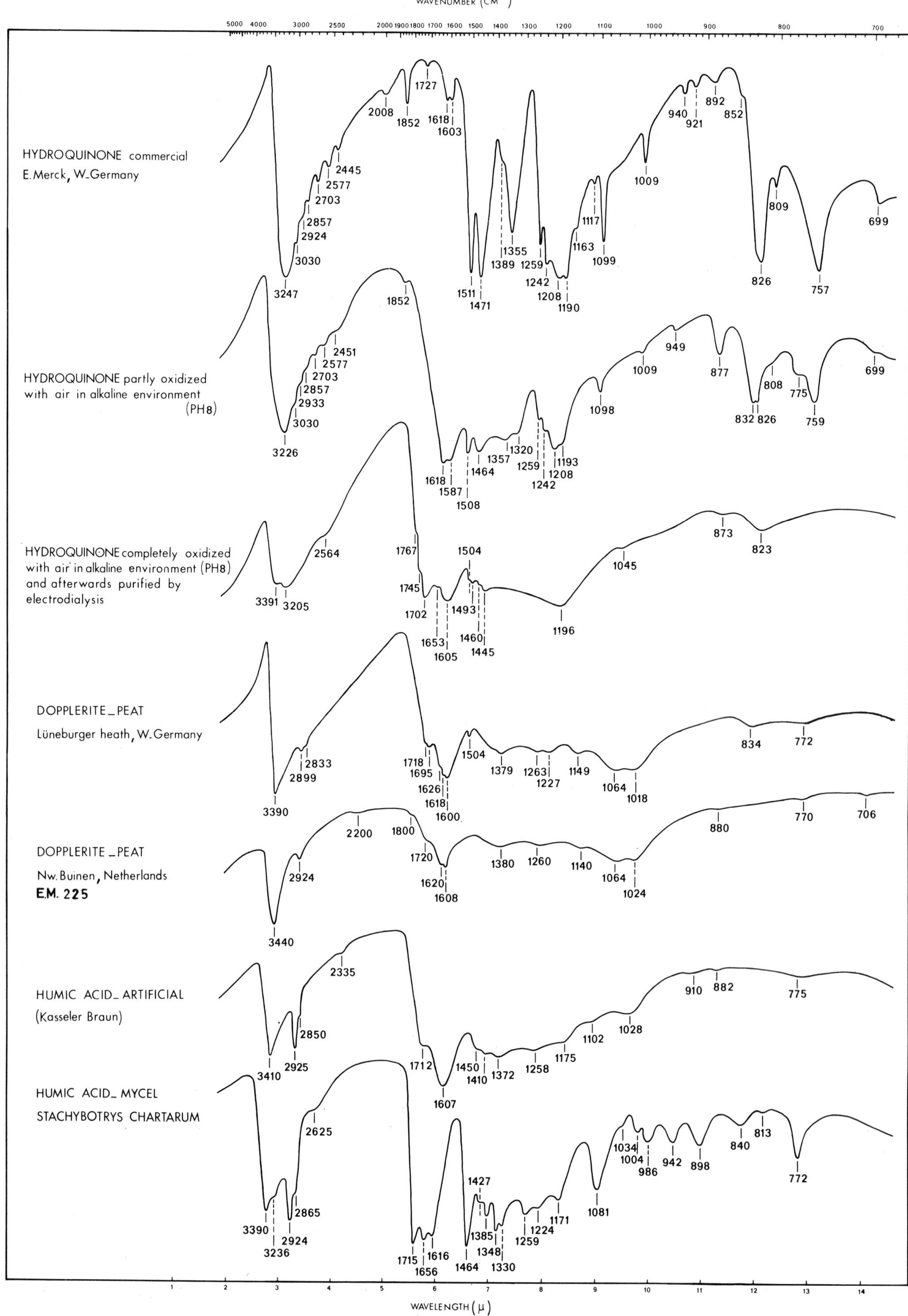

Fig. 10.4. Infrared spectra of hydroquinone, dopplerite and humic-acids from different origin. (pellets: 0.8 mg, dopplerite = 1.1 mg on 300 mg KBr).

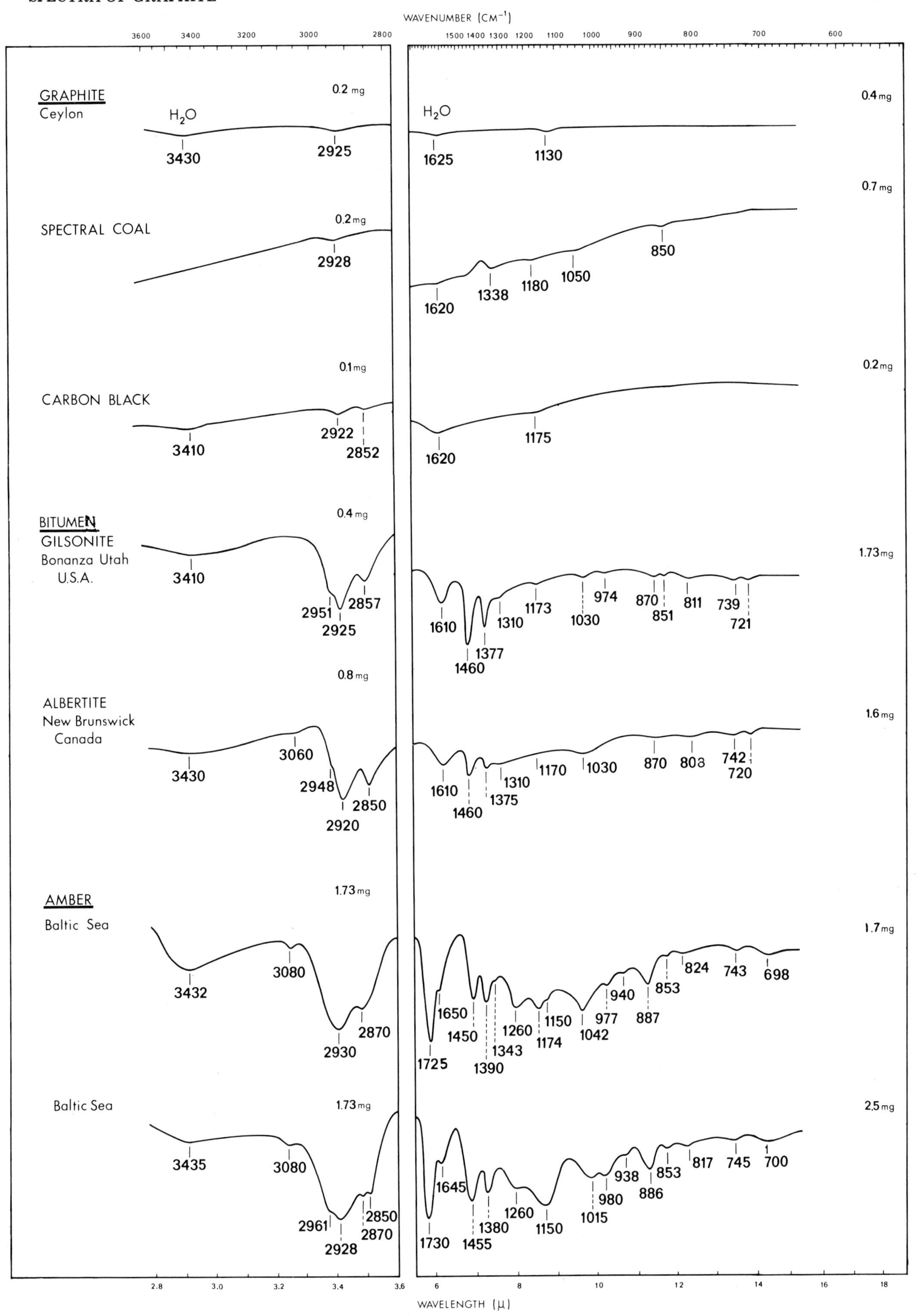

Fig. 10.5. Infrared spectra of graphite, coal, carbon black, bitumen and amber from different origin.

Occurrence. The humic matter of black earth, brown earth and podzols contains ca. 15, ca. 45 and ca. 70% of fulvic acids, respectively. Extracts from the B horizon of podzols even mainly consist of fulvic acids.

Infrared spectrum (Fig. 10.3; Table II). Like humic acids, the spectra of fulvic acids from different origin show very similar but not identical features. As compared with those of humic acids, the content of alkanes (2924 $\nu_{as}CH_2$) or that of carboxylic and carbonyl groups 1715 (νC=O), 1205—1212 (νC-O of carboxyls and phenols), 1399 (δ-OH), 772 (COO^--scissoring) cm^{-1}, has increased. On the other hand the amount of aromatics (1592—1613 cm^{-1}) has decreased. The predominant acidic character of the fulvic acids is furthermore demonstrated by the occurrence of the shoulder of the ν-OH vibration of the carboxyl group at 2570 cm^{-1}. Vibrations of ethers may be derived from the absorption band at 1047 cm^{-1}.

Remarks. Fulvic acids are to be considered as a precursor of humic acids. They are altered to humic acids by condensation and polymerization.

Graphite (spectral coal, carbon black); formula: C

In the graphite-group structure the carbon atoms are located at the corners of regular hexagons, which form two-dimensional layers of indefinite size. The layers run parallel to each other at a distance of 3.33 Å. The large cleavage and softness of this mineral is inherent to weak Van der Waal's forces holding the layers together.

Occurrence. Graphite is mainly found in metamorphic rocks. Black limestones may yield white marbles with disseminated graphite. By metamorphic action coal beds may change into graphite: Sonora (Mexico). Graphite also occurs in pegmatite because of hydrothermal action (Ceylon).
Artificial graphite is made by heating amorphous anthracite or coal in electrically heated retorts at ca. 1500° C. In contrast to natural graphite, artificial graphite is very hard, because in this case the crystallites are not ordered in equidistantial layers but are intergrown haphazardly together to a dense mass.

Infrared spectra (Fig. 10.5; Table II). Graphite spectra do not show infrared bands because there is no change in dipolmoment when the atoms vibrate in this symmetric crystal built up of equal atoms.

The bands at 3430 and 1625 cm^{-1} are due to some adsorbed H_2O, that at 1130 cm^{-1} to some νC-O impurity and that at 2925 cm^{-1} to some νCH_2 of aliphates. The spectral coal has thereby also a small C-O impurity at 1180 cm^{-1}. Its bands at 1338 and 850 cm^{-1} are caused by NO_3 stretching and O-N-O bending vibrations from nitrate which was added during its fabrication. The carbon black spectrum is poor in absorption bands. The band at 1175 cm^{-1} is caused by some C-O impurity, that at 3410 and 1620 cm^{-1} by some adsorbed H_2O and that at 2922 and 2852 cm^{-1} by a weak methylene doublet ν_{as} CH_2 and ν_s CH_2, respectively.

Remarks. The name graphite is derived from the greek grapho which means "I write". Graphite is very resistant to temperature (melting point = ca. 3000° C). It is a good conductor of electricity and its resistance to chemicals is high. Graphite is widely used mixed with oil for lubricants, as electrodes in arc dischargers, as polisher for stoves, dynamo brushes and in the manufacture of pencils, as a moderator to hamper neutron velocity, etc.

Activite (trade name) is pretreated coal with sulfuric acid. It is widely used as a cation exchanger for the purification of water.
Activated coal (medical coal) is manufactured by slow heating of animal wastes (blood, bones) or of sugar. Some $ZnCl_2$ is added to prevent sintering. Afterwards the material is treated with steam or air to increase its surface being of the order of ca. 800 m^2/g. Activated coal is used for adsorption of all kinds of (harmful) gases and liquids.
Saran carbons are molecular sieves of carbon which are resistant to heat and acids and which are produced, e.g. by heating polyvinyl chloride at ca. 100° C.
Spectral coal is artificial graphite, which with some oil is pressed to rods. It contains some nitrate which is added to get better burning. Because of the absence of disturbing emission lines, the product is widely used in flamephotometry.

Carbon black is a very fine (ϕ = 100—600 Å, specif. surface = 40—1000 m^2/g) X-ray amorphous artificial porous product with randomly arranged platelets. It is obtained by burning natural gas in tiny flames falling on a cooled plate with just the correct amount of air. Carbon black is widely used as reinforcing agent in rubber industry, printing inks and lacquers.

Bitumen; formula: C, H, O (N,S)

Bitumen contains mainly long-chain aliphatic components (CH, CH_2, CH_3) coloured by black pigments.

Occurrence. Bitumen is a restproduct of the oxidation or evaporation of petroleum. Bitumen of high purity called gilsonite and albertite are found in Utah, Alberta, New Brunswick, Canada, etc.

Infrared spectra (Fig. 10.5; Table II). The infrared spectra of the natural occurring gilsonite and albertite samples show very similar properties. Both samples show valency vibrations of hydroxyl groups with strong intramolecular hydrogen bonding at 3430 and 3410 cm^{-1} (ν-OH). Absorption of the different types of aliphates occur at the wave numbers 2950 ($\nu_{as}CH_3$), 2922 ($\nu_{as}CH_2$) and 2853 (ν_sCH_2), 1460 ($\delta_{as}CH_3$ and CH_2 scissoring), and 1376 cm^{-1} (δ_sCH_3), 740 (C-H wagging). The absorption at 720 cm^{-1} is caused by a -(CH_2)n-rocking vibration of long chains in paraffines. The broad absorption with a maximum at 1610 cm^{-1} is due to intramolecular hydrogen linkages of the stretching vibration of the phenyl nucleus.
Albertite can be distinguished from gilsonite by the absence of naphtenes (974 and 851 cm^{-1}) and the presence of alkenes (3060 cm^{-1}).

Remarks. Gilsonite (named after S.H. Gilson) and albertite, named after Alberta, (Canada), have found many applications in arts. It is also used in varnishes, stains, inks, paints, etc.
Asphalt is the impure product of bitumen mixed with fine silicate minerals. It is also found in large deposits: Pich Lake (Trinidad), Venezuela, Iraq, Dead Sea, etc.
Asphalt is widely used in roofing, paving, briquetting, pipe coatings.

Amber (bernstein, succinite); formula: ca. 78% C, 10% H, 11% O

Structure. The material consists of a mixture of several kinds of resins, succinic acid (3—8%) and etheric oils. Amber melts at ca. 375°C, and it is isomorphous to X-rays.

Occurrence. Amber is a fossil resin from pine trees of tertiary period. The best-known occurrence is the Baltic Coast. Other regions are the coast of Denmark, Sweden, Sicily, Roumania and Portugal.

Infrared spectra (Fig. 10.5; Table II). As the etheric oils are mainly derived from terpenes, we find besides adsorption of the aliphatic groups at 2961* ($\nu_{as}CH_3$); 2870 (ν_sCH_3) methyl doublet; 2928 ($\nu_{as}CH_2$) and 2850* (ν_sCH_2) methylene doublet; 1453 ($\delta_{as}CH_3$, CH_2 scissoring); 1385 (δ_sCH_3) 1174* and 1150 ($(CH_3)_2CH$—); 744 (C-H wagging). That of alkenes and unsaturated cyclic compounds at 3080 (ν_{as}=CH_2), 1648 (νC=C); 978 (R-CH=CH_2), 886 (RR'C=CH_2) and 700 cm^{-1} (RCH=CHR'). Carbonyl vibrations of aliphatic carbonic acids (succinic acid), esters and ethers are found at 1728 (νC=O); 1260 (νC-O); 1042* (νO-C); 1015* (νOC-O) and hydroxyl vibrations at 3433, (νOH); 1385 (δOH overlapping with δ_sCH_3); and 939 cm^{-1} (OH...H out-of-plane).
By oxidation the intensity of the 886 cm^{-1} band (δCH of olefin) decreases and that of the 1260, 1174*, 1150 band increases: compare Beck et al. (1965).

Remarks. Amber is used for personal adornment (necklaces, bracelets, gemstones). Because of its high electrical susceptibility (when rubbed, amber is electrically charged), amber is called elektrum.
The mineral contains some succinic acid which sublimates when it is heated and which is produced in this way on a commercial scale. Wastes of amber are distilled with water to obtain the etheric oils which are applied in lacquers.

Reference

Beck, C., Wilbur, E., Meret, S., Kossove, D. and Kermani, K., 1965. The infrared spectra of amber and the identification of Baltic amber. Archaeometry, 8: 96—109.

11. INORGANIC (CLAY) — ORGANIC COMPLEXES

The study of clay-organic complexes by infrared spectroscopy has largely increased in recent years. Many researches have been published in a short time. Montmorillonite has been mostly used in the experiments and to a lesser degree vermiculite. In this case the uptake of organic molecules can easily be followed by measuring the distances between the layers by X-ray analysis. Addition of glycerol or aethylene glycol has for many years even been used to identify swelling minerals. Artificially prepared fine rutile, anatase, silica, alumina and silica-alumina are well known catalysts which adsorb carbondioxide, ammonia, nitrogen, oxygen, benzene, etc.

Organic—inorganic complexes may cause a shift in the wave numbers and intensities of the original components. It is caused by strong electrostatic forces (metal-organic complexes) or by O-H interaction (hydrogen bonding in mineral-organic complexes). Moreover by cation-dipole interactions between mineral and organic molecules, their strength (heat of adsorption) depending on the kind of cation (polarizability) and the kind of organic molecule (ionization potential, dielectric constant). Clay minerals may act as electron acceptors and donors. Electron acceptor sites are Al at the crystal edges (Lewis acids) or H from adsorbed H_2O (Brönsted acids). Electron donors are metals in the lower valency state, e.g., ferro iron.

Attempts have been made to get Si-C organo clay derivatives by Grignard (Mg catalysator) or Friedel-Crafts ($AlCl_3$ catalysator) reaction with Si-Cl minerals. The latter were obtained by refluxing on a waterbath (Si-OH) groups of minerals with thionylchloride ($SOCl_2$) or phosphortrichloride (PCl_3) in benzene (15 h at ca. 80°C) by which it is supposed that OH of the minerals is replaced by C of the organic molecule:

$$\equiv\text{Si-Cl} + C_2H_5MgJ \xrightarrow{\text{6 h at ca. 40°C in ether}} \equiv\text{Si-}C_2H_5 + \text{MgJCl (Grignard)}$$

$$\equiv\text{Si-Cl} + C_6H_6 \xrightarrow{\text{6 h at ca. 80°C}} \equiv\text{Si-}C_6H_5 + \text{HCl (Friedel-Crafts)}$$

The spectrum of "phenyl" montmorillonite non compensated and compensated with montmorillonite has only some small bands in the 1600—1400 cm^{-1} regions (Fig. 11.1).[1]

They have some resemblance to those of phenol considering the possibility that the organo-mineral complexes may have caused some changes in their wave numbers and intensities.

However, by chemical analysis the product and also that of other clay minerals (hectorite, saponite, etc.) and minerals (Al-hydroxide, permutite, limonite) treated in the same way contained some percent of total sulfur.

Probably only sulfoxide compounds were formed by treating the mineral with thionylchloride which afterwards have reacted with the metal catalyzed Friedel-Crafts solution, giving phenyl sulfur complexes of complicated nature.

Remarks

Clay minerals are of large importance as catalysts for many kinds of reactions including polymerizations. They may protect organic molecules against weathering agents and biologic activities. The growth of micro-organisms may be stimulated by clay minerals because of absorption of nutritive components and toxic secretions. Clay organic interactions have been suggested to explain the formation of oils or even of the origin of life.

Of great practical importance is the adsorption of organic herbicides and fungicides on clay minerals as supporters. For soil mechanics it is the

[1] By courtesy of E. Rietz, Braunschweig (West Germany).

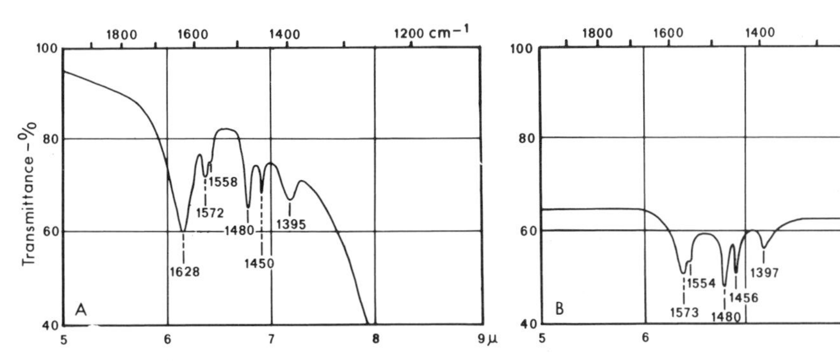

Fig. 11.1. Montmorillonite treated after the $SOCl_2$ Friedel-Crafts procedure (A). B. The same but compensated with montmorillonite. (Courtesy E. Rietz.)

change of fine highly swelling non-permeable hydrophilic clays of low aggregate stabilization into coarse non-swelling permeable hydrophobic clays of high shear strength.

It is caused by a bridging to carbonyl-, carboxyl-, amide-, amino groups connected to long-chain polymeric hydrophobic, aliphatic or aromatic groups. The existence of stable clay organic compounds has proved to be of practical use for the construction of large storage basins for crude oil to overcome seasonal demands. These oils mostly contain small amounts of olefines = unsaturated (double bonds) hydrocarbons. The Russian oils have appreciable amounts of naphthenes (unsaturated hydrocarbons with cyclic bonds). These kinds of organics are strongly adsorbed by montmorillonite. Butadiene (C_4H_6) even polymerizes spontaneously without additional catalyst when mixed with montmorillonite.

The strong adsorption is caused by the electron seaking (electrophylitic) property of the double bond: see also the strong addition of Br_2, Cl_2, J_2. As a result the pores of the clay are clogged at the contact with the oil. Thus a layer of montmorillonitic material to a thickness of only 20—25 cm is sufficient to prevent loss of crude oil by seepage for a height of the oil column of ca. 100 m. Costs of these earthen storage reservoirs (Venezuela) with a capacity of ca. 2 million m^3 is about 15% of that of steel tankages.

CONCLUDING REMARKS

The spectra clearly demonstrate that the infrared spectroscope is a suitable tool in the identification of clay minerals and their admixtures especially to give details of their structure.

Ordering of the atoms can be better recognized in an earlier stage by infrared spectroscopy than by electron- or X-ray diffraction. The infrared method proceeds very fast (ca. 20 min) and only some mg of the mineral are needed for an analysis.

However, also in this Atlas as in the preceding about electron microscopy, a number of minerals with non-characteristic features remain open. Fortunately, most of them are not the same in both cases, and if not so they may be identified by X-ray diffraction or thermal analysis. Atlases about these techniques will appear in due time.

APPENDIX 1. Conversion of wavelengths (μm) to wavenumbers (cm^{-1}).

	0	1	2	3	4	5	6	7	8	9
1.0	10000	9901	9804	9709	9615	9524	9434	9346	9259	9174
1.1	9091	9009	8929	8850	8772	8696	8621	8547	8475	8403
1.2	8333	8264	8197	8130	8065	8000	7937	7874	7813	7752
1.3	7692	7634	7576	7519	7463	7407	7353	7299	7246	7194
1.4	7143	7092	7042	6993	6944	6897	6849	6803	6757	6711
1.5	6667	6623	6579	6536	6494	6452	6410	6369	6329	6289
1.6	6250	6211	6173	6135	6098	6061	6024	5988	5952	5917
1.7	5882	5848	5814	5780	5747	5714	5682	5650	5618	5587
1.8	5556	5525	5495	5464	5435	5405	5376	5348	5319	5291
1.9	5263	5236	5208	5181	5155	5128	5102	5076	5051	5025
2.0	5000	4975	4950	4926	4902	4878	4854	4831	4808	4785
2.1	4762	4739	4717	4695	4673	4651	4630	4608	4587	4566
2.2	4545	4525	4505	4484	4464	4444	4425	4405	4386	4367
2.3	4348	4329	4310	4292	4274	4255	4237	4219	4202	4184
2.4	4167	4149	4132	4115	4098	4082	4065	4049	4032	4016
2.5	4000	3984	3968	3953	3937	3922	3906	3891	3876	3861
2.6	3846	3831	3817	3802	3788	3774	3759	3745	3731	3717
2.7	3704	3690	3676	3663	3650	3636	3623	3610	3597	3584
2.8	3571	3559	3546	3534	3521	3509	3497	3484	3472	3460
2.9	3448	3436	3425	3413	3401	3390	3378	3367	3356	3344
3.0	3333	3322	3311	3300	3289	3279	3268	3257	3247	3236
3.1	3226	3215	3205	3195	3185	3175	3165	3155	3145	3135
3.2	3125	3115	3106	3096	3086	3077	3067	3058	3049	3040
3.3	3030	3021	3012	3003	2994	2985	2976	2967	2959	2950
3.4	2941	2933	2924	2915	2907	2899	2890	2882	2874	2865
3.5	2857	2849	2841	2833	2825	2817	2809	2801	2793	2786
3.6	2778	2770	2762	2755	2747	2740	2732	2725	2717	2710
3.7	2703	2695	2688	2681	2674	2667	2660	2653	2646	2639
3.8	2632	2625	2618	2611	2604	2597	2591	2584	2577	2571
3.9	2564	2558	2551	2545	2538	2532	2525	2519	2513	2506
4.0	2500	2494	2488	2481	2475	2469	2463	2457	2451	2445
4.1	2439	2433	2427	2421	2415	2410	2404	2398	2392	2387
4.2	2381	2375	2370	2364	2358	2353	2347	2342	2336	2331
4.3	2326	2320	2315	2309	2304	2299	2294	2288	2283	2278
4.4	2273	2268	2262	2257	2252	2247	2242	2237	2232	2227
4.5	2222	2217	2212	2208	2203	2198	2193	2188	2183	2179
4.6	2174	2169	2165	2160	2155	2151	2146	2141	2137	2132
4.7	2128	2123	2119	2114	2110	2105	2101	2096	2092	2088
4.8	2083	2079	2075	2070	2066	2062	2058	2053	2049	2045
4.9	2041	2037	2033	2028	2024	2020	2016	2012	2008	2004
5.0	2000	1996	1992	1988	1984	1980	1976	1972	1969	1965
5.1	1961	1957	1953	1949	1946	1942	1938	1934	1931	1927
5.2	1923	1919	1916	1912	1908	1905	1901	1898	1894	1890
5.3	1887	1883	1880	1876	1873	1869	1866	1862	1859	1855
5.4	1852	1848	1845	1842	1838	1835	1832	1828	1825	1821

	0	1	2	3	4	5	6	7	8	9
5.5	1818	1815	1812	1808	1805	1802	1799	1795	1792	1789
5.6	1786	1783	1779	1776	1773	1770	1767	1764	1761	1757
5.7	1754	1751	1748	1745	1742	1739	1736	1733	1730	1727
5.8	1724	1721	1718	1715	1712	1709	1706	1704	1701	1698
5.9	1695	1692	1689	1686	1684	1681	1678	1675	1672	1669
6.0	1667	1664	1661	1658	1656	1653	1650	1647	1645	1642
6.1	1639	1637	1634	1631	1629	1626	1623	1621	1618	1616
6.2	1613	1610	1608	1605	1603	1600	1597	1595	1592	1590
6.3	1587	1585	1582	1580	1577	1575	1572	1570	1567	1565
6.4	1563	1560	1558	1555	1553	1550	1548	1546	1543	1541
6.5	1538	1536	1534	1531	1529	1527	1524	1522	1520	1517
6.6	1515	1513	1511	1508	1506	1504	1502	1499	1497	1495
6.7	1493	1490	1488	1486	1484	1481	1479	1477	1475	1473
6.8	1471	1468	1466	1464	1462	1460	1458	1456	1453	1451
6.9	1449	1447	1445	1443	1441	1439	1437	1435	1433	1431
7.0	1429	1427	1425	1422	1420	1418	1416	1414	1412	1410
7.1	1408	1406	1404	1403	1401	1399	1397	1395	1393	1391
7.2	1389	1387	1385	1383	1381	1379	1377	1376	1374	1372
7.3	1370	1368	1366	1364	1362	1361	1359	1357	1355	1353
7.4	1351	1350	1348	1346	1344	1342	1340	1339	1337	1335
7.5	1333	1332	1330	1328	1326	1325	1323	1321	1319	1318
7.6	1316	1314	1312	1311	1309	1307	1305	1304	1302	1300
7.7	1299	1297	1295	1294	1292	1290	1289	1287	1285	1284
7.8	1282	1280	1279	1277	1276	1274	1272	1271	1269	1267
7.9	1266	1264	1263	1261	1259	1258	1256	1255	1253	1252
8.0	1250	1248	1247	1245	1244	1242	1241	1239	1238	1236
8.1	1235	1233	1232	1230	1229	1227	1225	1224	1222	1221
8.2	1220	1218	1217	1215	1214	1212	1211	1209	1208	1206
8.3	1205	1203	1202	1200	1199	1198	1196	1195	1193	1192
8.4	1190	1189	1188	1186	1185	1183	1182	1181	1179	1178
8.5	1176	1175	1174	1172	1171	1170	1168	1167	1166	1164
8.6	1163	1161	1160	1159	1157	1156	1155	1153	1152	1151
8.7	1149	1148	1147	1145	1144	1143	1142	1140	1139	1138
8.8	1136	1135	1134	1133	1131	1130	1129	1127	1126	1125
8.9	1124	1122	1121	1120	1119	1117	1116	1115	1114	1112
9.0	1111	1110	1109	1107	1106	1105	1104	1103	1101	1100
9.1	1099	1098	1096	1095	1094	1093	1092	1091	1089	1088
9.2	1087	1086	1085	1083	1082	1081	1080	1079	1078	1076
9.3	1075	1074	1073	1072	1071	1070	1068	1067	1066	1065
9.4	1064	1063	1062	1060	1059	1058	1057	1056	1055	1054
9.5	1053	1052	1050	1049	1048	1047	1046	1045	1044	1043
9.6	1042	1041	1040	1038	1037	1036	1035	1034	1033	1032
9.7	1031	1030	1029	1028	1027	1026	1025	1024	1022	1021
9.8	1020	1019	1018	1017	1016	1015	1014	1013	1012	1011
9.9	1010	1009	1008	1007	1006	1005	1004	1003	1002	1001

APPENDIX 2. The most characteristic bands of clay minerals and their admixtures (▮ = strong, ı = moderate, weak not mentioned).

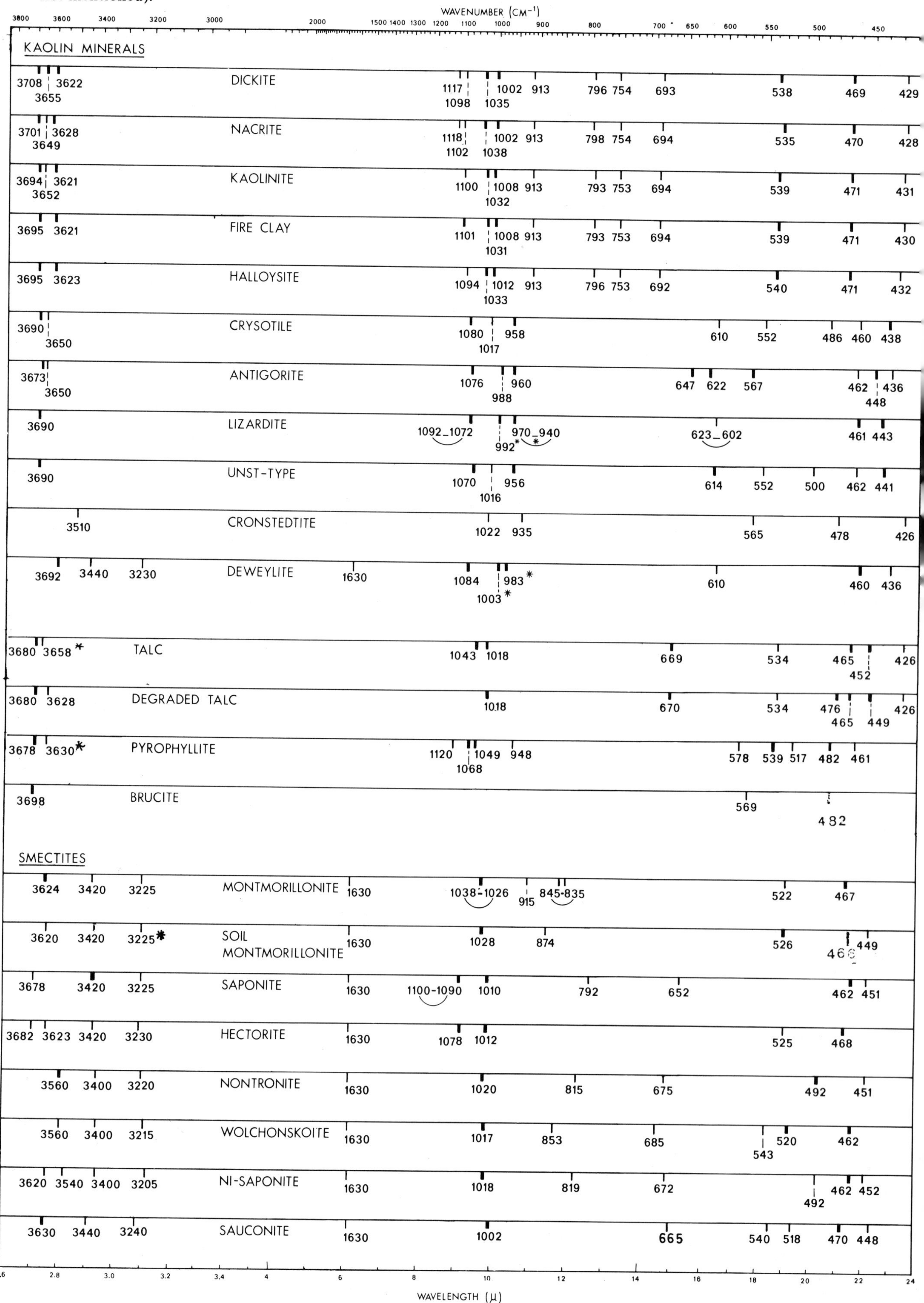

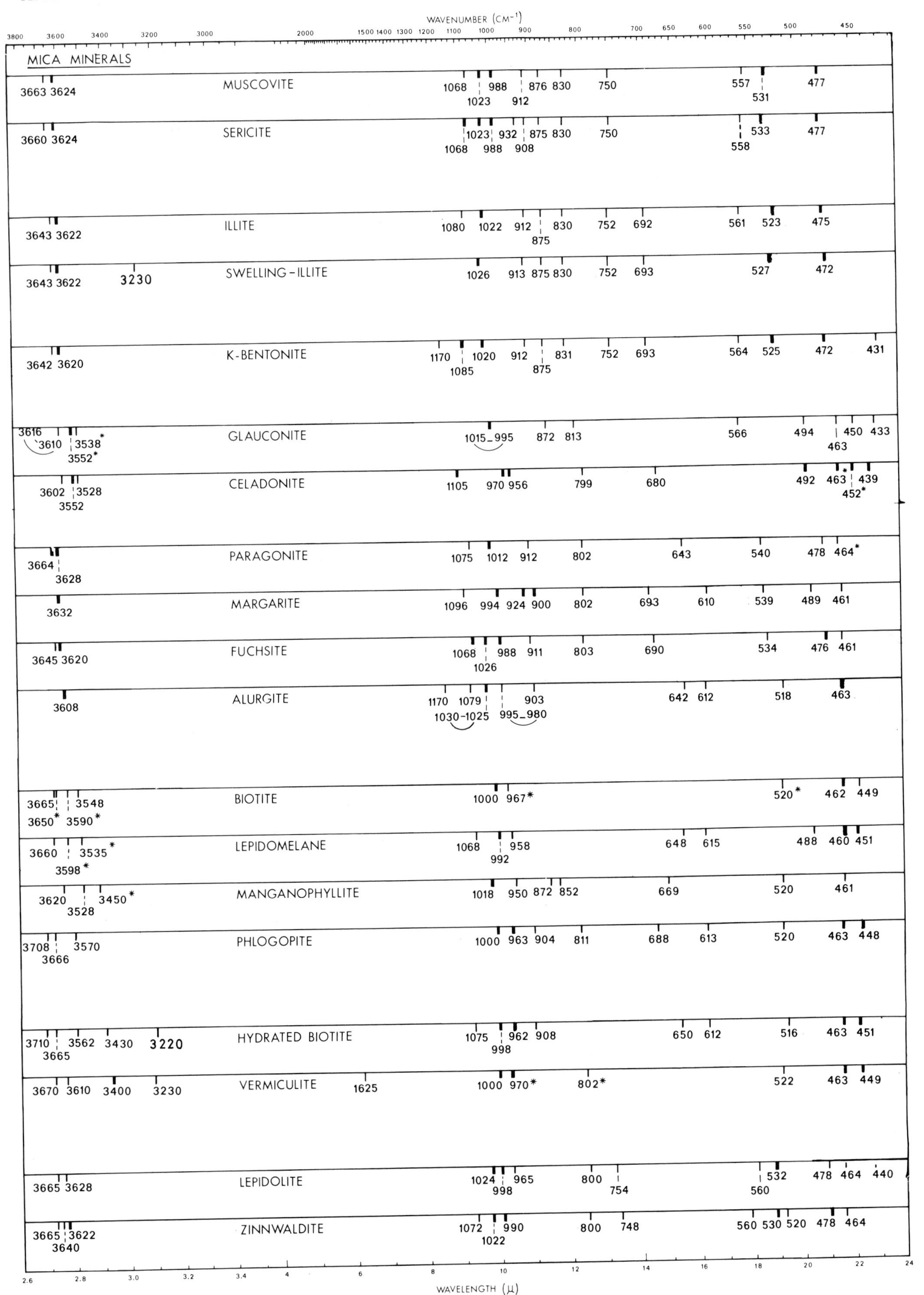
WAVENUMBER (CM-1)
WAVELENGTH (μ)
MICA MINERALS
MUSCOVITE
SERICITE
ILLITE
SWELLING-ILLITE
K-BENTONITE
GLAUCONITE
CELADONITE
PARAGONITE
MARGARITE
FUCHSITE
ALURGITE
BIOTITE
LEPIDOMELANE
MANGANOPHYLLITE
PHLOGOPITE
HYDRATED BIOTITE
VERMICULITE
LEPIDOLITE
ZINNWALDITE

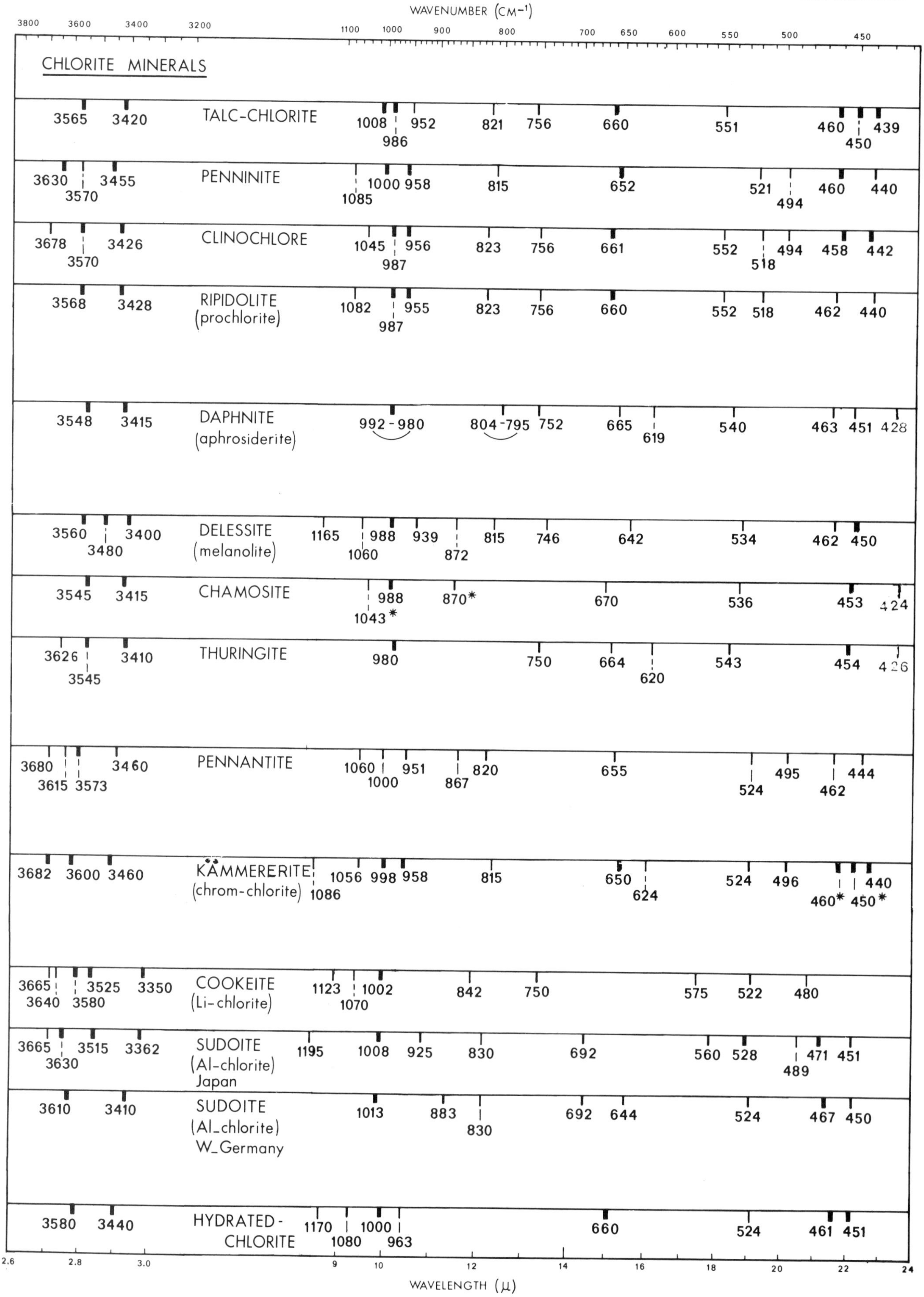

WAVENUMBER (cm^{-1})

3800 3600 3400 3200 1100 1000 900 800 700 650 600 550 500 450

CHLORITE MINERALS

Mineral	Bands (cm^{-1})
TALC-CHLORITE	3565, 3420, 1008, 986, 952, 821, 756, 660, 551, 460, 450, 439
PENNINITE	3630, 3570, 3455, 1085, 1000, 958, 815, 652, 521, 494, 460, 440
CLINOCHLORE	3678, 3570, 3426, 1045, 987, 956, 823, 756, 661, 552, 518, 494, 458, 442
RIPIDOLITE (prochlorite)	3568, 3428, 1082, 987, 955, 823, 756, 660, 552, 518, 462, 440
DAPHNITE (aphrosiderite)	3548, 3415, 992 - 980, 804 - 795, 752, 665, 619, 540, 463, 451, 428
DELESSITE (melanolite)	3560, 3480, 3400, 1165, 1060, 988, 939, 872, 815, 746, 642, 534, 462, 450
CHAMOSITE	3545, 3415, 1043*, 988, 870*, 670, 536, 453, 424
THURINGITE	3626, 3545, 3410, 980, 750, 664, 620, 543, 454, 426
PENNANTITE	3680, 3615, 3573, 3460, 1060, 1000, 951, 867, 820, 655, 524, 495, 462, 444
KÄMMERERITE (chrom-chlorite)	3682, 3600, 3460, 1086, 1056, 998, 958, 815, 650, 624, 524, 496, 460*, 450*, 440
COOKEITE (Li-chlorite)	3665, 3640, 3580, 3525, 3350, 1123, 1070, 1002, 842, 750, 575, 522, 480
SUDOITE (Al-chlorite) Japan	3665, 3630, 3515, 3362, 1195, 1008, 925, 830, 692, 560, 528, 489, 471, 451
SUDOITE (Al-chlorite) W-Germany	3610, 3410, 1013, 883, 830, 692, 644, 524, 467, 450
HYDRATED-CHLORITE	3580, 3440, 1170, 1080, 1000, 963, 660, 524, 461, 451

2.6 2.8 3.0 9 10 12 14 16 18 20 22 24

WAVELENGTH (μ)

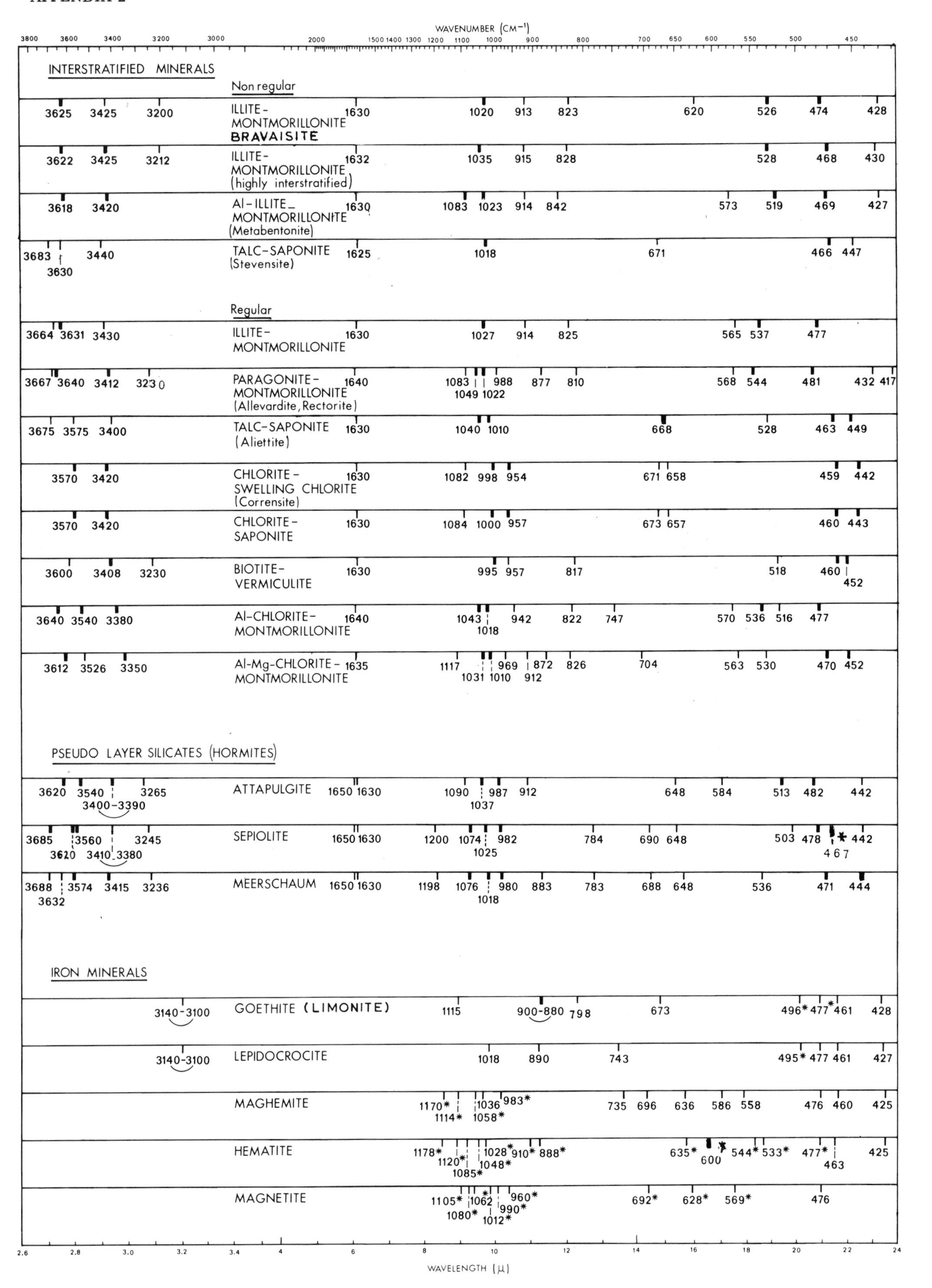
WAVENUMBER (CM⁻¹)
WAVELENGTH (μ)
INTERSTRATIFIED MINERALS
Non regular
ILLITE-MONTMORILLONITE BRAVAISITE
ILLITE-MONTMORILLONITE (highly interstratified)
Al-ILLITE-MONTMORILLONITE (Metabentonite)
TALC-SAPONITE (Stevensite)
Regular
ILLITE-MONTMORILLONITE
PARAGONITE-MONTMORILLONITE (Allevardite, Rectorite)
TALC-SAPONITE (Aliettite)
CHLORITE-SWELLING CHLORITE (Corrensite)
CHLORITE-SAPONITE
BIOTITE-VERMICULITE
Al-CHLORITE-MONTMORILLONITE
Al-Mg-CHLORITE-MONTMORILLONITE
PSEUDO LAYER SILICATES (HORMITES)
ATTAPULGITE
SEPIOLITE
MEERSCHAUM
IRON MINERALS
GOETHITE (LIMONITE)
LEPIDOCROCITE
MAGHEMITE
HEMATITE
MAGNETITE

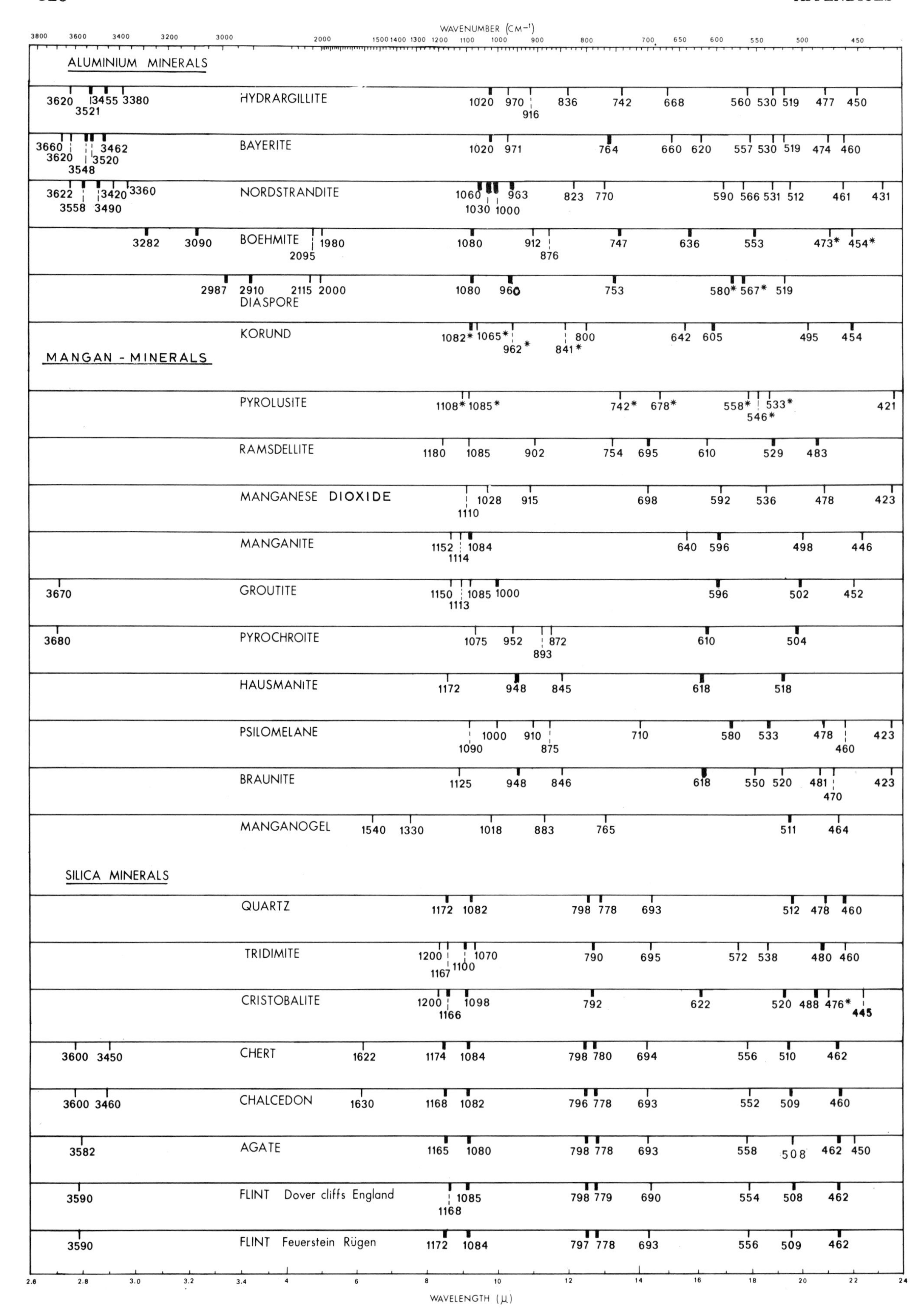
WAVENUMBER (CM⁻¹)
ALUMINIUM MINERALS
HYDRARGILLITE
BAYERITE
NORDSTRANDITE
BOEHMITE
DIASPORE
KORUND
MANGAN - MINERALS
PYROLUSITE
RAMSDELLITE
MANGANESE DIOXIDE
MANGANITE
GROUTITE
PYROCHROITE
HAUSMANITE
PSILOMELANE
BRAUNITE
MANGANOGEL
SILICA MINERALS
QUARTZ
TRIDIMITE
CRISTOBALITE
CHERT
CHALCEDON
AGATE
FLINT Dover cliffs England
FLINT Feuerstein Rügen
WAVELENGTH (μ)

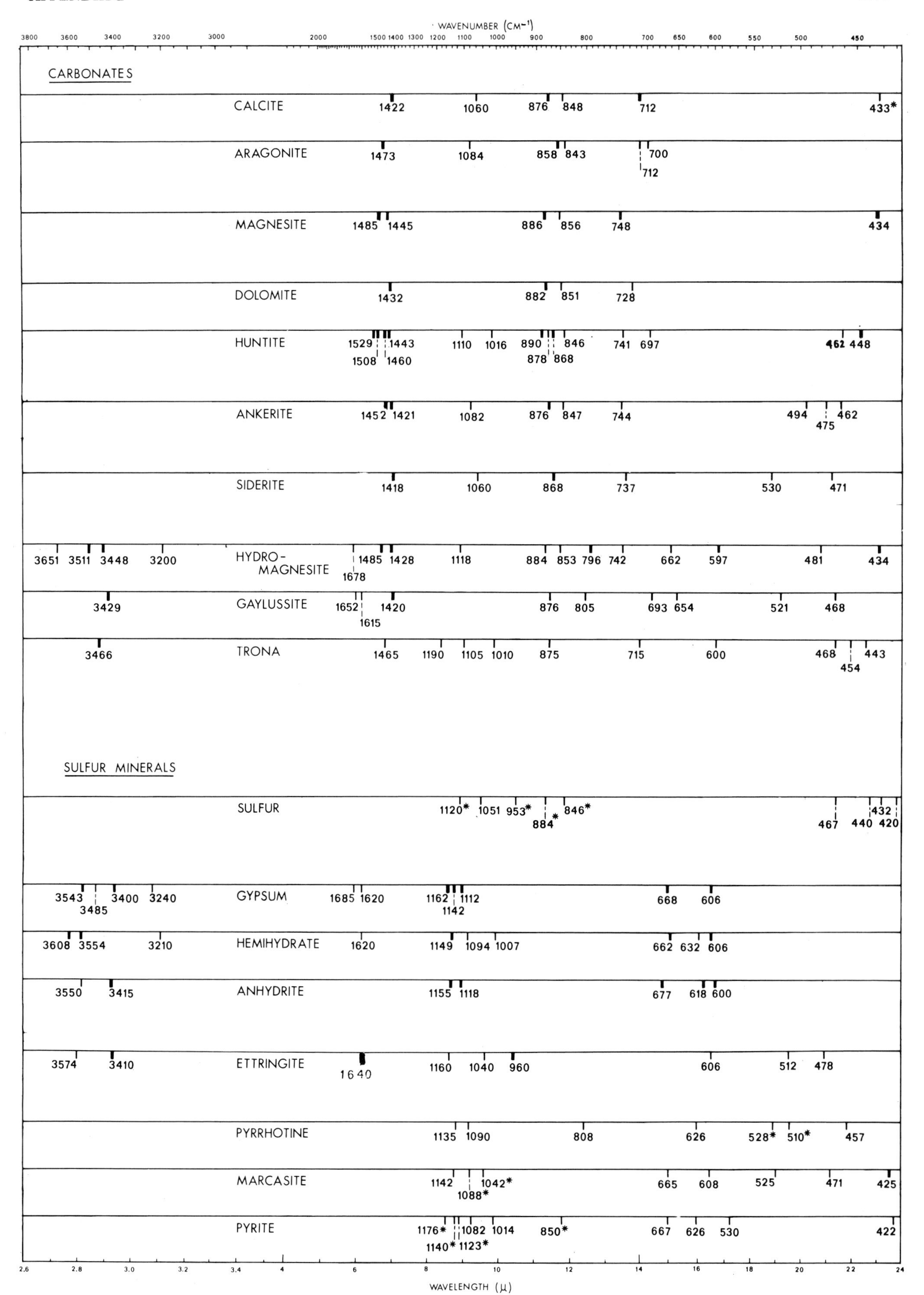
WAVENUMBER (CM⁻¹)
CARBONATES
CALCITE 1422 1060 876 848 712 433*
ARAGONITE 1473 1084 858 843 700 712
MAGNESITE 1485 1445 886 856 748 434
DOLOMITE 1432 882 851 728
HUNTITE 1529 1508 1443 1460 1110 1016 890 878 868 846 741 697 462 448
ANKERITE 1452 1421 1082 876 847 744 494 475 462
SIDERITE 1418 1060 868 737 530 471
HYDRO-MAGNESITE 3651 3511 3448 3200 1678 1485 1428 1118 884 853 796 742 662 597 481 434
GAYLUSSITE 3429 1652 1615 1420 876 805 693 654 521 468
TRONA 3466 1465 1190 1105 1010 875 715 600 468 454 443
SULFUR MINERALS
SULFUR 1120* 1051 953* 884* 846* 467 440 432 420
GYPSUM 3543 3485 3400 3240 1685 1620 1162 1142 1112 668 606
HEMIHYDRATE 3608 3554 3210 1620 1149 1094 1007 662 632 606
ANHYDRITE 3550 3415 1155 1118 677 618 600
ETTRINGITE 3574 3410 1640 1160 1040 960 606 512 478
PYRRHOTINE 1135 1090 808 626 528* 510* 457
MARCASITE 1142 1088* 1042* 665 608 525 471 425
PYRITE 1176* 1140* 1123* 1082 1014 850* 667 626 530 422
WAVELENGTH (μ)

WAVENUMBER (CM⁻¹): 3800 3600 3400 3200 3000 2000 1500 1400 1300 1200 1100 1000 900 800 700 650 600 550 500 450

AMPHIBOLES

Mineral	Absorption bands (cm⁻¹)
AMPHIBOLE	3672 3653 3628* 1097 1060 980 948 915 756 686 660 530 506 478 464 448
ACTINOLITE	3670 3653 1092 1038 992 953 915 757 688 643 530* 508* 463
TREMOLITE	3675 3653 1102 1018 994 952 920 756 684 663 643 527 509 478 463 445
AUGITE	1062 960 910 870 852 670 628 507 468 448
HYPERSTHENE	3650 1076 1010 962 885 846 683 642 532 497 464 453
ENSTATITE	3683 3650 1060 1009 966 923 859 743 647* 562 537 464 452

TITANO MINERALS

Mineral	Absorption bands (cm⁻¹)
ANATASE	1338* 1020* 811 738 540 488 465
RUTILE	874 738 603 520* 494* 473* 458*
BROOKITE	1328 1165 962 873 758 539 495 480 450* 438
TITANITE	967 870 850 752* 735* 565 497 465 438*
ILMENITE	1572 1185 1040 690 541 520 460 432

FELDSPARS

Mineral	Absorption bands (cm⁻¹)
ALBITE	1150 1096 1035 995 786 760 744 650 594 536 467 428
OLIGOCLASE	1151 1098 1036 998 946 786 758 641 592 540 469 428*
ANDESINE	1150 1097 1042* 1000 780 742 631 588 542 470 428*
LABRADORITE	1149 1096 998 946 776 746 630 584 540 468 428
BYTOWNITE	1149 1094 990 935 751 729 628 583 544 483 470 428
ANORTHITE	1143 1082 1020 932 757 727 625 578 543 488 472 428
SANIDINE	1122 1029 775 640 588 548 428
MICROCLINE	1140 1092 1053 1012 770 728 648 609 587 540 468 428*
ORTHOCLASE	1128 1042 1005 770 646 589 545 469 437
ANORTHOCLASE	1160*–1094* 1048*–1006* 720 590 470 428
ADULARIA	1130 1048 1006 773 722 645 588 543 468 430*

WAVELENGTH (μ): 2.6 2.8 3.0 3.2 3.4 4 6 8 10 12 14 16 18 20 22 24

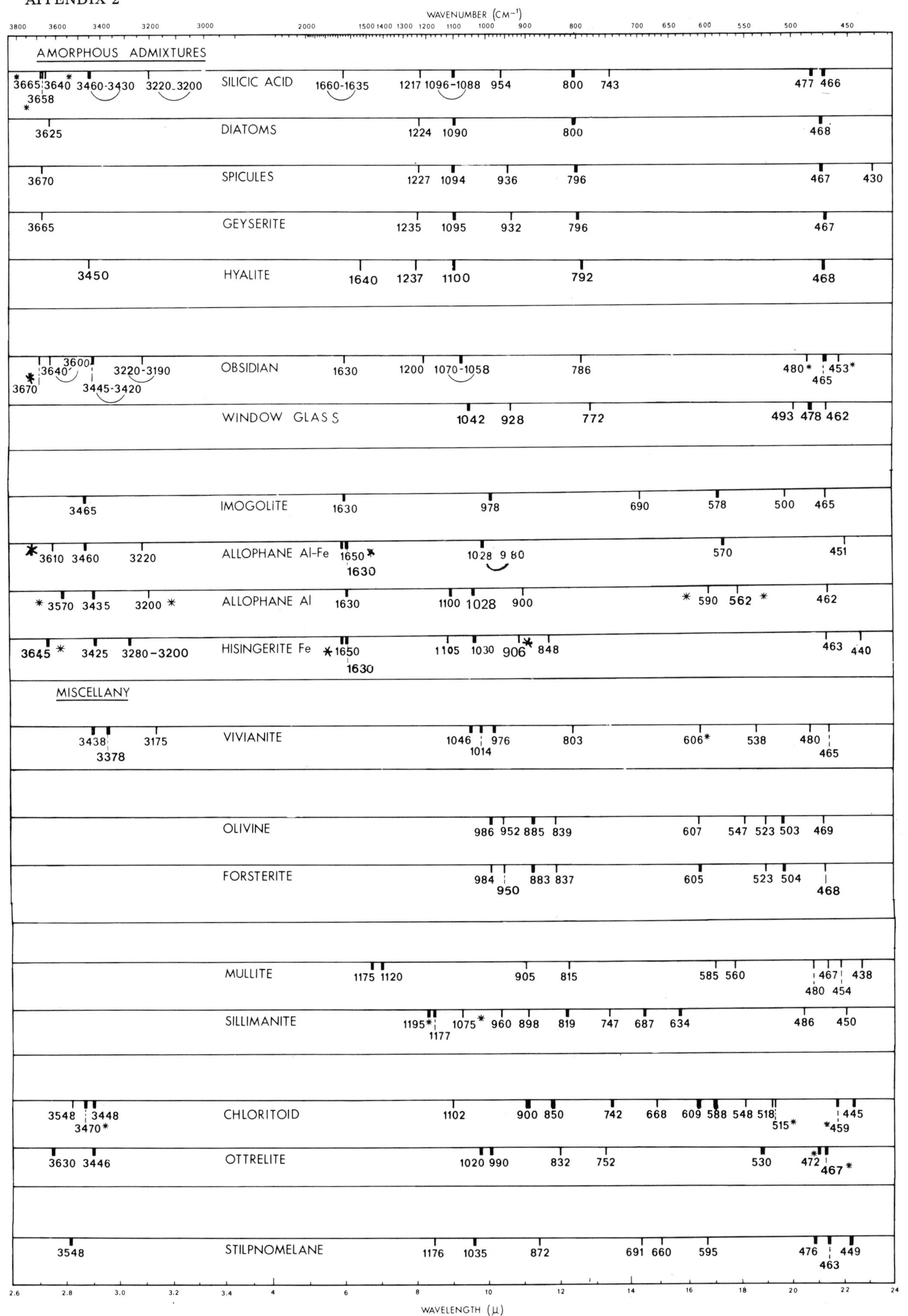

WAVENUMBER (CM⁻¹)
AMORPHOUS ADMIXTURES
SILICIC ACID
DIATOMS
SPICULES
GEYSERITE
HYALITE
OBSIDIAN
WINDOW GLASS
IMOGOLITE
ALLOPHANE Al-Fe
ALLOPHANE Al
HISINGERITE Fe
MISCELLANY
VIVIANITE
OLIVINE
FORSTERITE
MULLITE
SILLIMANITE
CHLORITOID
OTTRELITE
STILPNOMELANE
WAVELENGTH (μ)

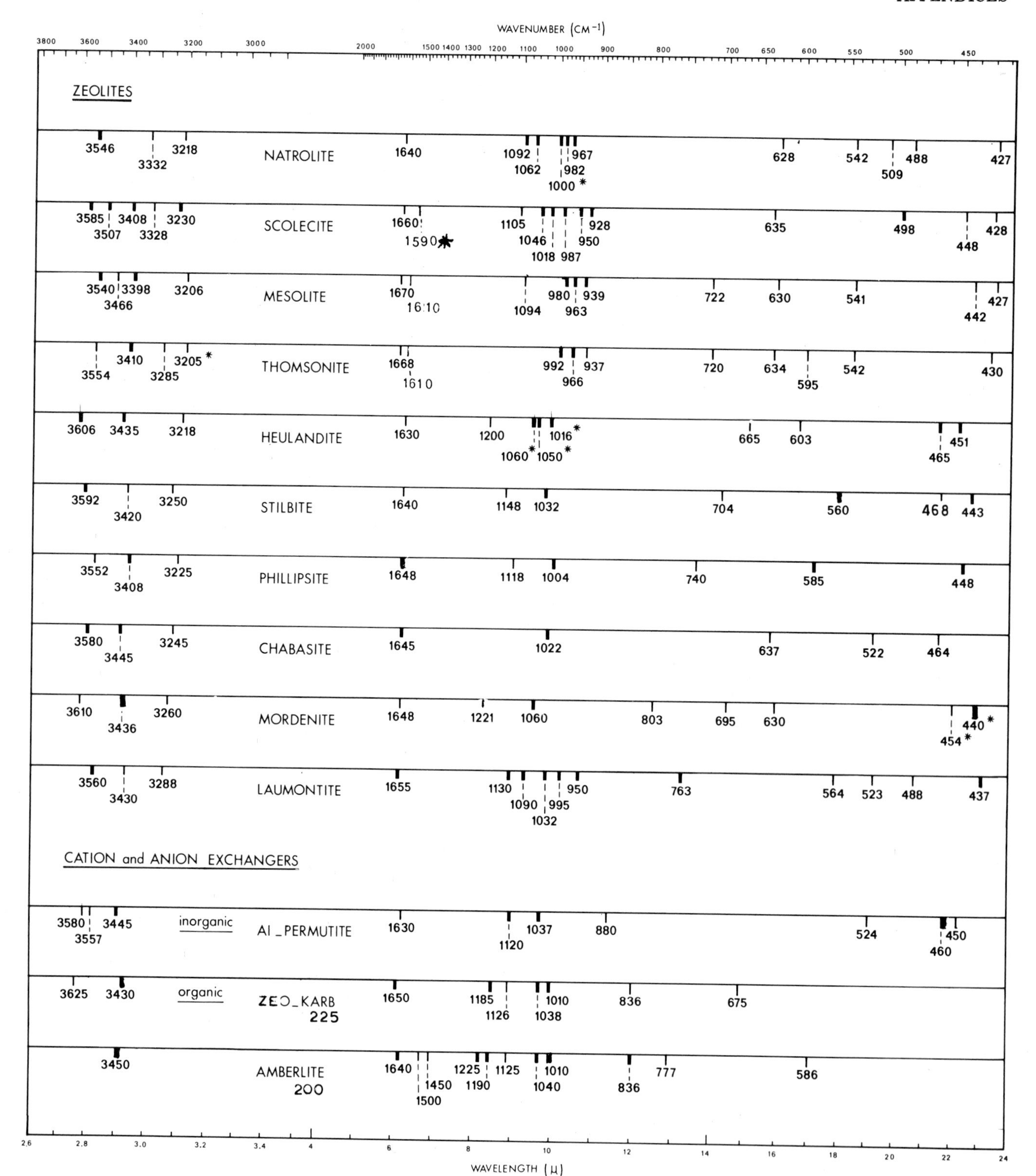
WAVENUMBER (CM⁻¹)
ZEOLITES
NATROLITE
3546 3332 3218 1640 1092 1062 967 982 1000 * 628 542 509 488 427
SCOLECITE
3585 3507 3408 3328 3230 1660 1590 1105 1046 1018 987 928 950 635 498 448 428
MESOLITE
3540 3466 3398 3206 1670 1610 1094 980 963 939 722 630 541 442 427
THOMSONITE
3554 3410 3285 3205 * 1668 1610 992 966 937 720 634 595 542 430
HEULANDITE
3606 3435 3218 1630 1200 1060 * 1050 * 1016 * 665 603 465 451
STILBITE
3592 3420 3250 1640 1148 1032 704 560 468 443
PHILLIPSITE
3552 3408 3225 1648 1118 1004 740 585 448
CHABASITE
3580 3445 3245 1645 1022 637 522 464
MORDENITE
3610 3436 3260 1648 1221 1060 803 695 630 454 * 440 *
LAUMONTITE
3560 3430 3288 1655 1130 1090 1032 995 950 763 564 523 488 437
CATION and ANION EXCHANGERS
inorganic
Al _PERMUTITE
3580 3557 3445 1630 1120 1037 880 524 460 450
organic
ZEO_KARB 225
3625 3430 1650 1185 1126 1038 1010 836 675
AMBERLITE 200
3450 1640 1500 1450 1225 1190 1125 1040 1010 836 777 586
WAVELENGTH (μ)

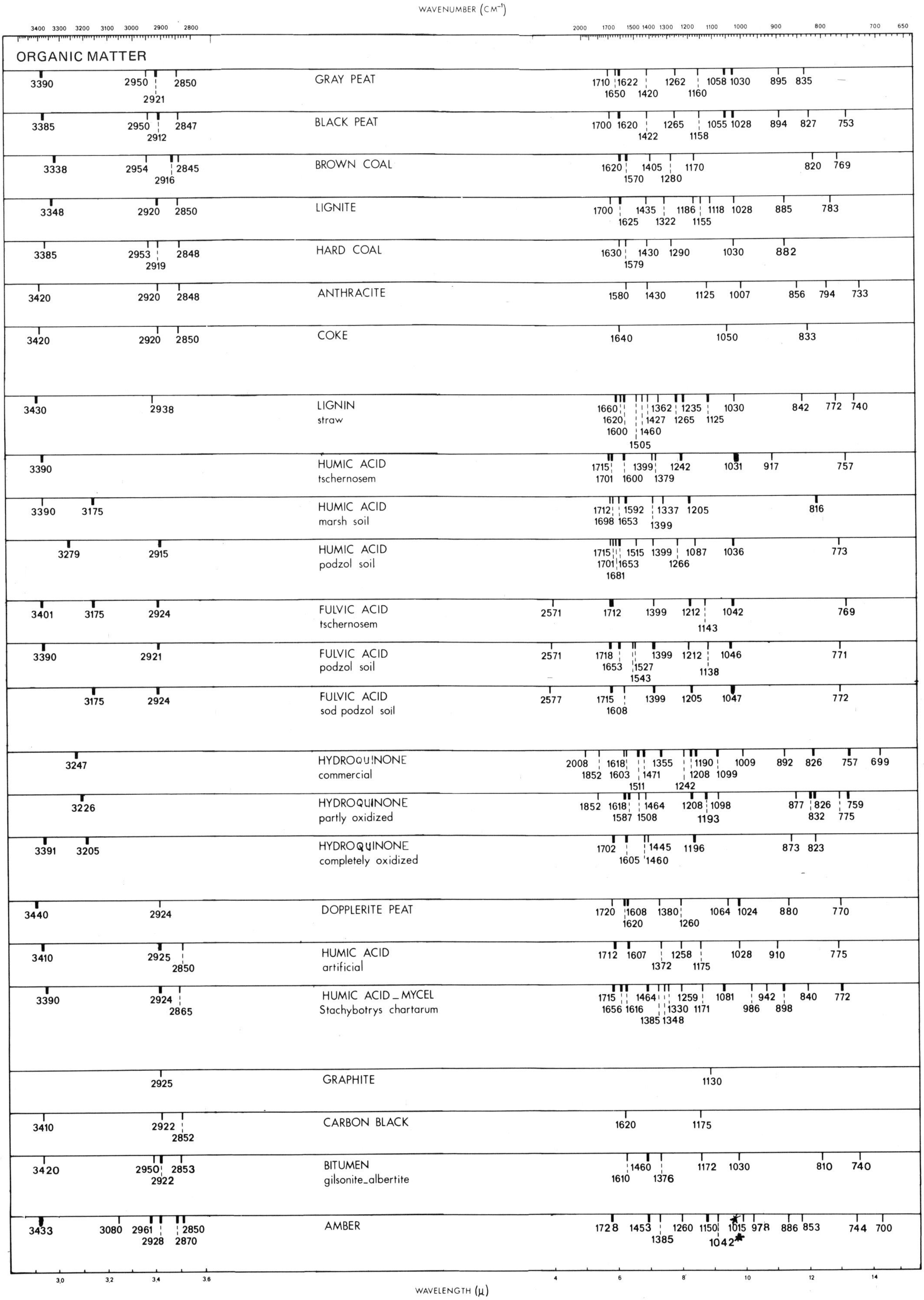

WAVENUMBER (CM⁻¹)

3400 3300 3200 3100 3000 2900 2800 — 2000 1700 1500 1400 1300 1200 1100 1000 900 800 700 650

ORGANIC MATTER

Sample	Band positions (cm⁻¹)
GRAY PEAT	3390, 2950, 2921, 2850, 1710, 1650, 1622, 1420, 1262, 1160, 1058, 1030, 895, 835
BLACK PEAT	3385, 2950, 2912, 2847, 1700, 1620, 1422, 1265, 1158, 1055, 1028, 894, 827, 753
BROWN COAL	3338, 2954, 2916, 2845, 1620, 1570, 1405, 1280, 1170, 820, 769
LIGNITE	3348, 2920, 2850, 1700, 1625, 1435, 1322, 1186, 1155, 1118, 1028, 885, 783
HARD COAL	3385, 2953, 2919, 2848, 1630, 1579, 1430, 1290, 1030, 882
ANTHRACITE	3420, 2920, 2848, 1580, 1430, 1125, 1007, 856, 794, 733
COKE	3420, 2920, 2850, 1640, 1050, 833
LIGNIN straw	3430, 2938, 1660, 1620, 1600, 1505, 1460, 1427, 1362, 1265, 1235, 1125, 1030, 842, 772, 740
HUMIC ACID tschernosem	3390, 1715, 1701, 1600, 1399, 1379, 1242, 1031, 917, 757
HUMIC ACID marsh soil	3390, 3175, 1712, 1698, 1653, 1592, 1399, 1337, 1205, 816
HUMIC ACID podzol soil	3279, 2915, 1715, 1701, 1681, 1653, 1515, 1399, 1266, 1087, 1036, 773
FULVIC ACID tschernosem	3401, 3175, 2924, 2571, 1712, 1399, 1212, 1143, 1042, 769
FULVIC ACID podzol soil	3390, 2921, 2571, 1718, 1653, 1543, 1527, 1399, 1212, 1138, 1046, 771
FULVIC ACID sod podzol soil	3175, 2924, 2577, 1715, 1608, 1399, 1205, 1047, 772
HYDROQUINONE commercial	3247, 2008, 1852, 1618, 1603, 1511, 1471, 1355, 1242, 1208, 1190, 1099, 1009, 892, 826, 757, 699
HYDROQUINONE partly oxidized	3226, 1852, 1618, 1587, 1508, 1464, 1208, 1193, 1098, 877, 832, 826, 775, 759
HYDROQUINONE completely oxidized	3391, 3205, 1702, 1605, 1460, 1445, 1196, 873, 823
DOPPLERITE PEAT	3440, 2924, 1720, 1620, 1608, 1380, 1260, 1064, 1024, 880, 770
HUMIC ACID artificial	3410, 2925, 2850, 1712, 1607, 1372, 1258, 1175, 1028, 910, 775
HUMIC ACID _ MYCEL Stachybotrys chartarum	3390, 2924, 2865, 1715, 1656, 1616, 1464, 1385, 1348, 1330, 1259, 1171, 1081, 986, 942, 898, 840, 772
GRAPHITE	2925, 1130
CARBON BLACK	3410, 2922, 2852, 1620, 1175
BITUMEN gilsonite_albertite	3420, 2950, 2922, 2853, 1610, 1460, 1376, 1172, 1030, 810, 740
AMBER	3433, 3080, 2961, 2928, 2870, 2850, 1728, 1453, 1385, 1260, 1150, 1042*, 1015*, 978, 886, 853, 744, 700

3.0 3.2 3.4 3.6 — 4 6 8 10 12 14

WAVELENGTH (μ)

APPENDIX 3. Infrared spectra of various sulfate minerals from different origin (after Struntz, 1970).

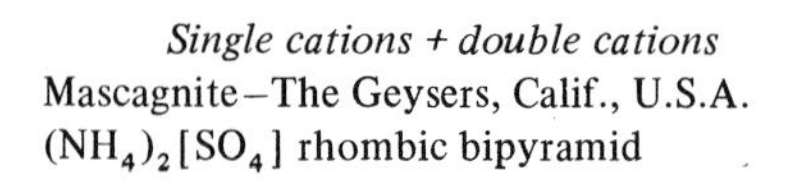

Single cations + double cations

Mascagnite–The Geysers, Calif., U.S.A.
$(NH_4)_2[SO_4]$ rhombic bipyramid

Themardite–Camp Verde, Ariz., U.S.A.
α-$Na_2[SO_4]$ rhombic bipyramid

Anhydrite–Osterode, W. Germany
$Ca[SO_4]$ rhombic bipyramid

Celestite (coelestine)–Clay Center, Ohio, U.S.A.
$Sr[SO_4]$ rhombic bipyramid

Baryte–Bingham, New Mex., U.S.A.
$Ba[SO_4]$ rhombic bipyramid

Anglesite–Rosiclare, Ill., U.S.A.
$Pb[SO_4]$ rhombic bipyramid

Langbeinite–Carlsbad, New Mex., U.S.A.
$K_2Mg_2[SO_4]_3$ cubic tetarthohedron

Glauberite–Salton Sea, Calif., U.S.A.
$CaNa_2[SO_4]_2$ monoclinic prismatic

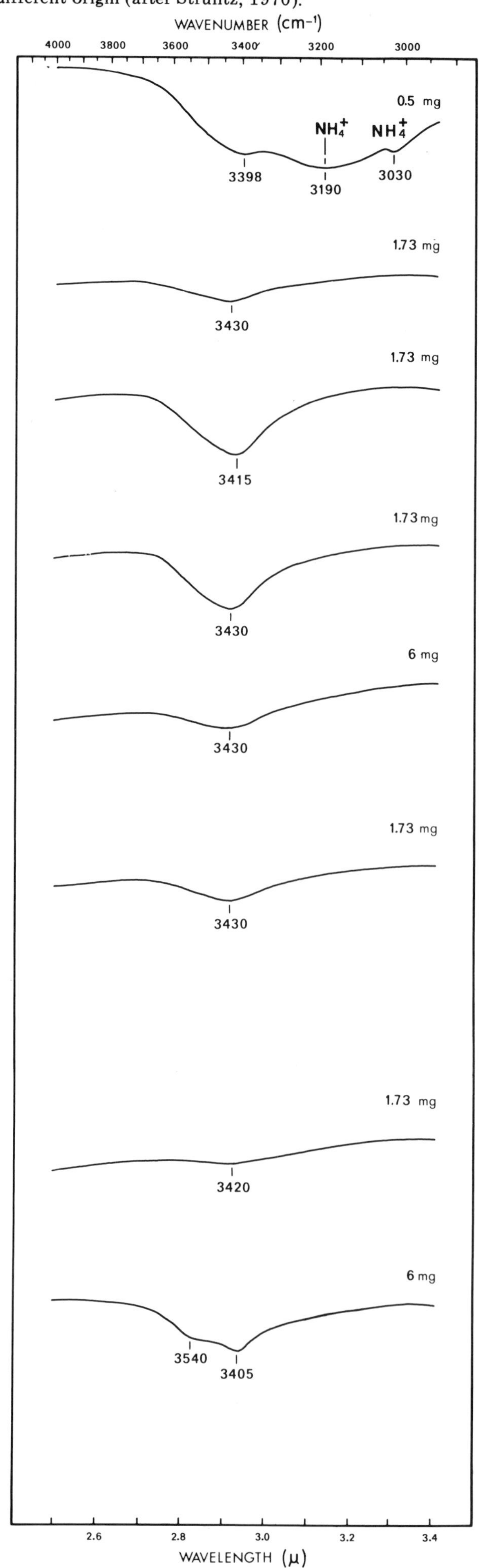

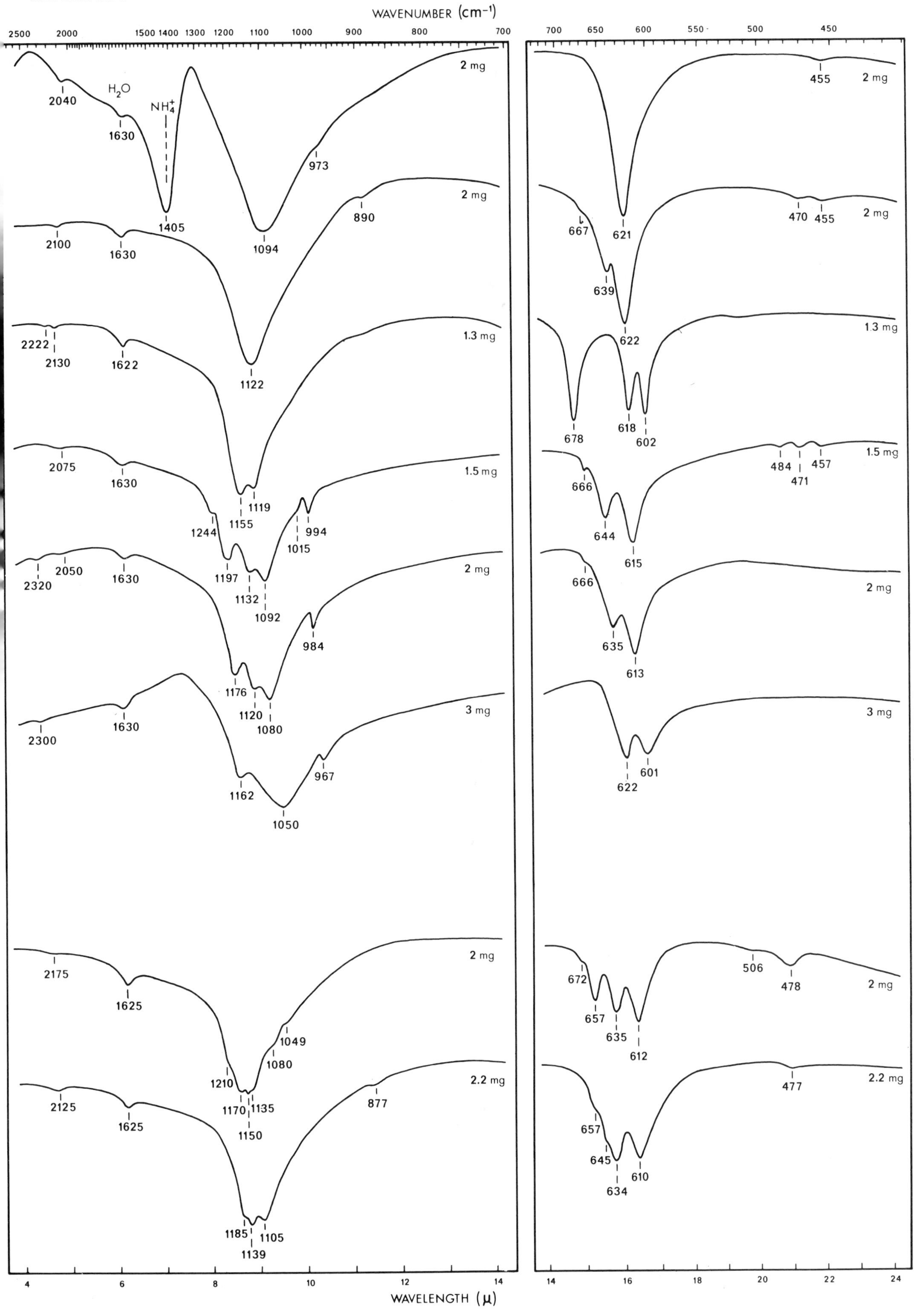

WAVENUMBER (cm⁻¹)
2500
2000
1500 1400 1300 1200 1100 1000 900 800 700
700 650 600 550 500 450
H2O
NH4+
2 mg
2040
1630
1405
1094
973
621
455
2 mg
2100
1630
890
667
639
622
470 455
1.3 mg
2222
2130
1622
1122
678
618
602
1.5 mg
2075
1630
1244
1155
1119
1015
994
666
644
615
484 471 457
2320
2050
1630
1197
1132
1092
984
666
635
613
3 mg
2300
1630
1176
1120
1080
1162
1050
967
622
601
2175
1625
1210
1049
1080
1170 1135
1150
672
657
635
612
506
478
2.2 mg
2125
1625
877
1185
1139
1105
657
645
634
610
477
4 6 8 10 12 14
14 16 18 20 22 24
WAVELENGTH (μ)

Single cations + crystalwater

Mirabilite–Dauria, U.S.S.R.
$Na_2[SO_4] \cdot 10H_2O$ monoclinic prismatic

Kieserite–Stassfurt, G.D.R.
$Mg[SO_4]$ monoclinic prismatic

Epsomite–Calatayud, Spain
$Mg[SO_4] \cdot 7H_2O$ rhombic bisphenoid (pseudo-tetragonal)

Anulogen–Manhattan Mine, Calif., U.S.A.
$Al_2[SO_4] \cdot 16H_2O$ triclinic pinacoid

Melanterite–Ominato, Japan
$Fe[SO_4] \cdot 7H_2O$ monoclinic prismatic

Kornelite–Coso Hot Springs, Calif., U.S.A.
$Fe_2^{3+}[SO_4]_3 \cdot 7\frac{1}{2}H_2O$ monoclinic prismatic

Coquimbite–Calf Mesa, Utah, U.S.A.
$Fe_2^{3+}[SO_4]_3 \cdot 9H_2O$ hexagonal

Bieberite–Kitwe, N. Rhodesia
$Co[SO_4] \cdot 7H_2O$ monoclinic prismatic

Chalcantite–Gila Cy, Ariz., U.S.A.
$Cu[SO_4] \cdot 5H_2O$ triclinic pinacoid

Goslarite–Dutte, Mont., U.S.A.
$Zn[SO_4] \cdot 7H_2O$ rhombic bisphenoid (pseudo-tetragonal)

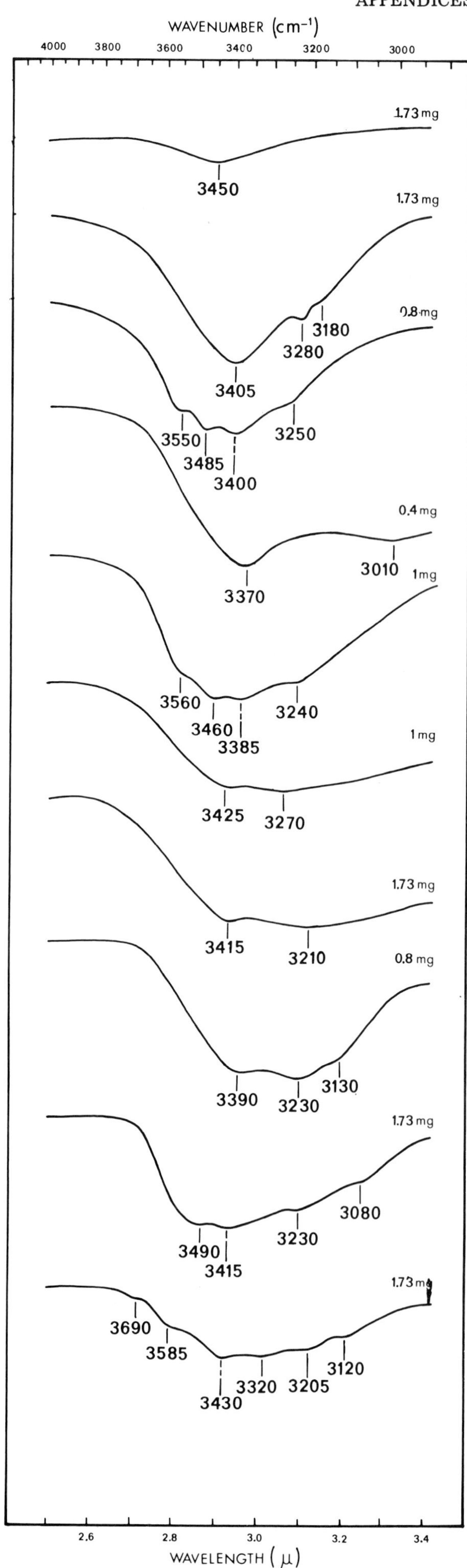

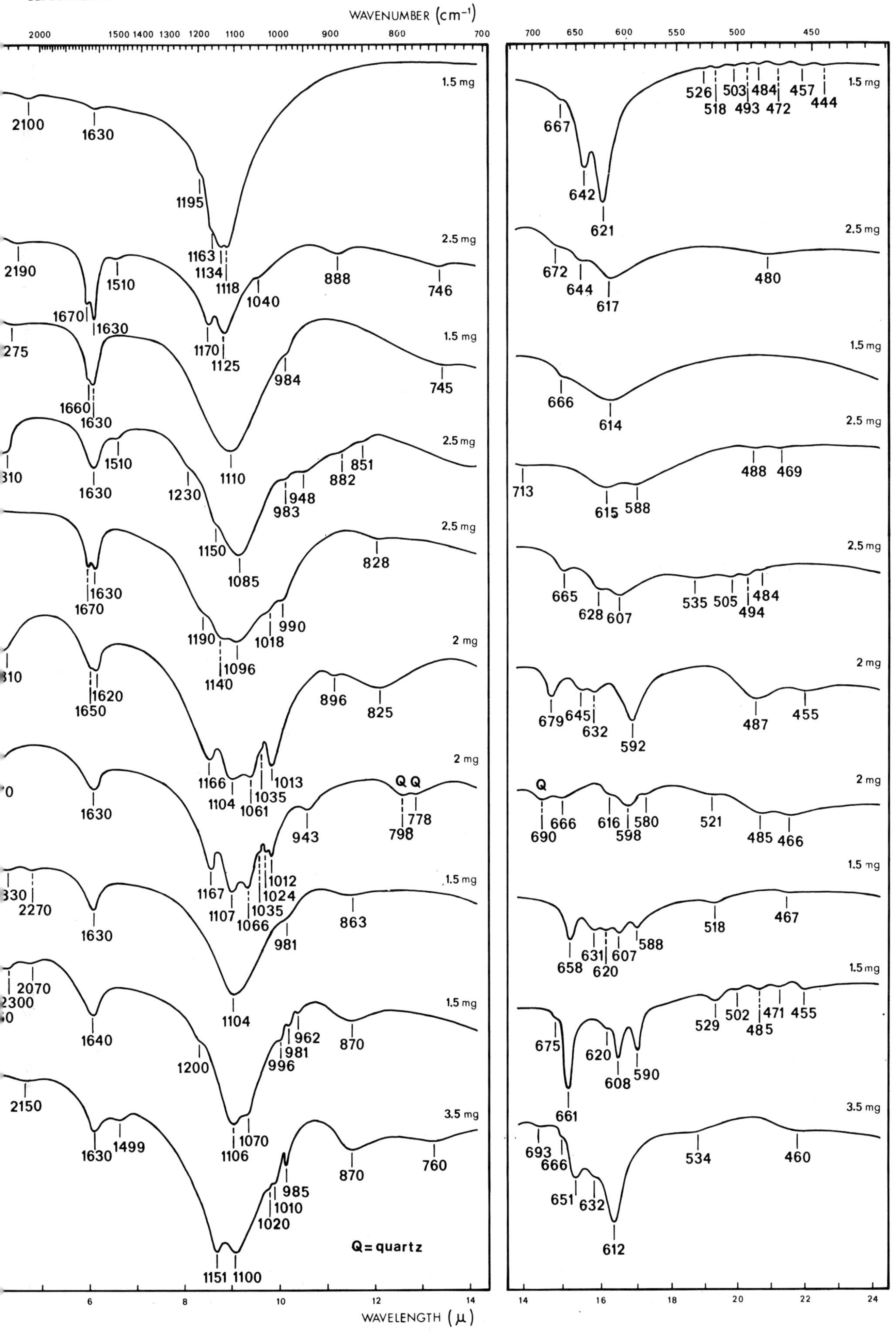

WAVENUMBER (cm⁻¹)
WAVELENGTH (μ)
Q = quartz
1.5 mg
2.5 mg
2 mg
3.5 mg

Double cations + crystalwater

Bousingaultite–South Mountain, Calif., U.S.A.
$(NH_4)_2Mg[SO_4]_2 \cdot 6H_2O$ monoclinic prismatic

Bloedite–Bertram Siding, Calif., U.S.A.
$Na_2Mg[SO_4]_2 \cdot 4H_2O$ monoclinic prismatic

Tamarugite–Cerro Pintodas, Calif., U.S.A.
$NaAl[SO_4]_2 \cdot 6H_2O$ monoclinic prismatic

Mendozite–Gold Bottom Mine, Calif., U.S.A.
$NaAl[SO_4]_2 \cdot 11H_2O$ monoclinic, fibrous

Pickeringite–Tucumcari, New Mex., U.S.A.
$MgAl_2[SO_4]_4 \cdot 22H_2O$ monoclinic sphenoidon

Kalinite–San Felipe, Calif., U.S.A.
$KAl[SO_4]_2 \cdot 11H_2O$ monoclinic, fibrous

Potashalumite–Marysvale, Utah, U.S.A.
$KAl[SO_4]_2 \cdot 12H_2O$ cubic dyakisdodecahedric

Krausite–Borate, Calif., U.S.A.
$KFe^{3+}[SO_4]_2 \cdot H_2O$ monoclinic

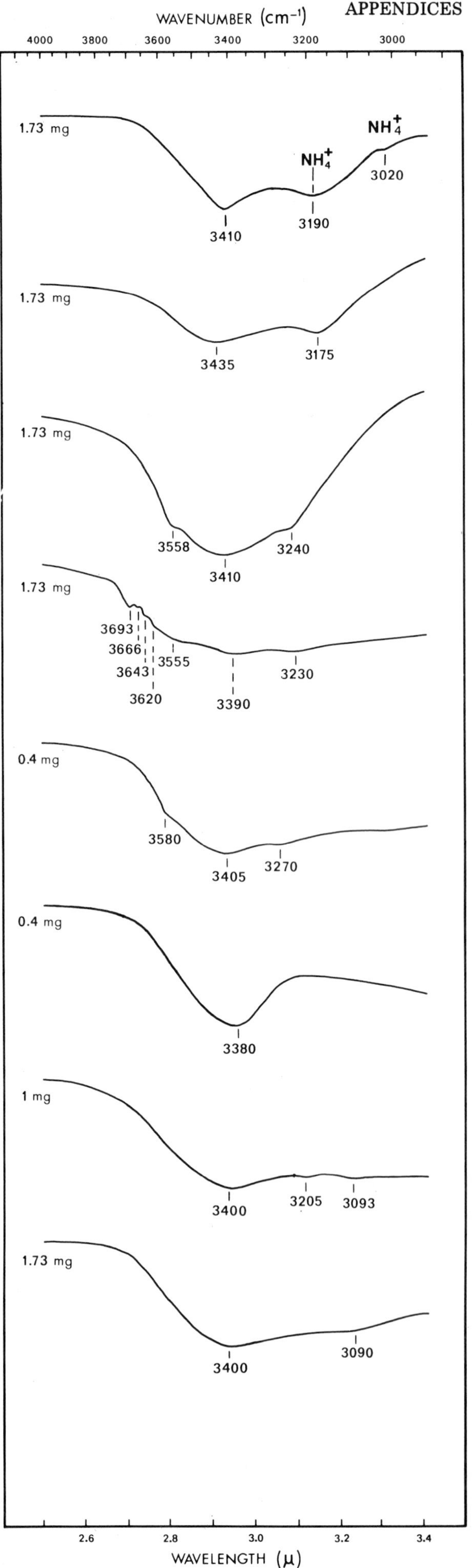

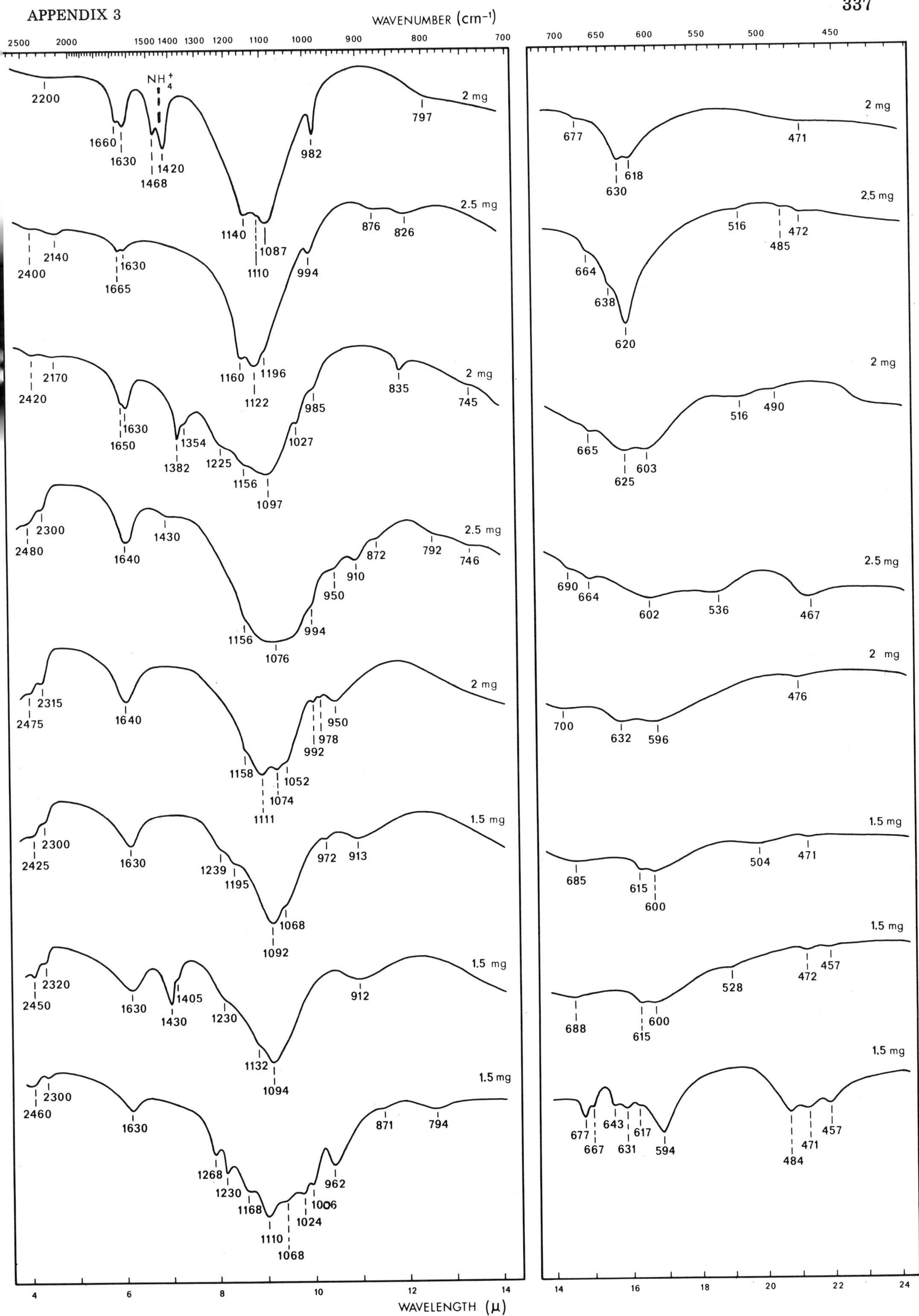
WAVENUMBER (cm⁻¹)
2500 2000 1500 1400 1300 1200 1100 1000 900 800 700
700 650 600 550 500 450
NH$_4^+$
2200 1660 1630 1468 1420 982 797 2 mg
677 630 618 471 2 mg
2400 2140 1665 1630 1140 1110 1087 994 876 826 2.5 mg
664 638 620 516 485 472 2.5 mg
2420 2170 1650 1630 1160 1122 1196 985 835 745 2 mg
665 625 603 516 490 2 mg
1382 1354 1225 1156 1097 1027
2480 2300 1640 1430 1156 1076 994 950 910 872 792 746 2.5 mg
690 664 602 536 467 2.5 mg
2475 2315 1640 1158 1111 1074 1052 992 978 950 2 mg
700 632 596 476 2 mg
2425 2300 1630 1239 1195 1092 1068 972 913 1.5 mg
685 615 600 504 471 1.5 mg
2450 2320 1630 1430 1405 1230 1132 1094 912 1.5 mg
688 615 600 528 472 457 1.5 mg
2460 2300 1630 1268 1230 1168 1110 1068 1024 1006 962 871 794 1.5 mg
677 667 643 631 617 594 484 471 457 1.5 mg
4 6 8 10 12 14
14 16 18 20 22 24
WAVELENGTH (μ)

Rhomboclase–Alcaparosa, Chile
$Fe^{3+}H[SO_4]_2 \cdot 4H_2O$ rhombic bipyramid

Halotrichite–Almeda, Calif., U.S.A.
$Fe^{2+}Al_2[SO_4]_4 \cdot 22H_2O$ monoclinic sphenoidon

Roemerite–Sierra Gorda, Chile
$Fe^{2+}Fe_2^{3+}[SO_4]_4 \cdot 14H_2O$ triclinic pinacoid

Kroehnkite–Chuquicamata, Chile
$Na_2Cu[SO_4]_2 \cdot 2H_2O$ monoclinic prismatic

Single cations + anions and ditto + crystalwater

Sulphohalite–Searles Lake, Calif., U.S.A.
$Na_6[(F,Cl) \mid (SO_4)_2]$ cubic

Antlerite–Chuquicamata, Chile
$Cu_3[(OH_4) \mid (SO_4)]$ rhombic bipyramid

Brochantite–Horn Silver Mine, Utah, U.S.A.
$Cu_4[(OH)_6 \mid (SO_4)]$ monoclinic prismatic

Aluminite–Halle-Saale, Germany
$Al_2[(OH)_4 \mid (SO_4)] \cdot 7H_2O$ monoclinic

Metahohmannite–Quetena, Chile
$Fe^{3+}[(OH) \mid (SO_4)] \cdot 1½H_2O$

Fibroferrite–Sierra Gorda, Chile
$Fe^{3+}[(OH) \mid (SO_4)] \cdot 4½H_2O$ rhombic

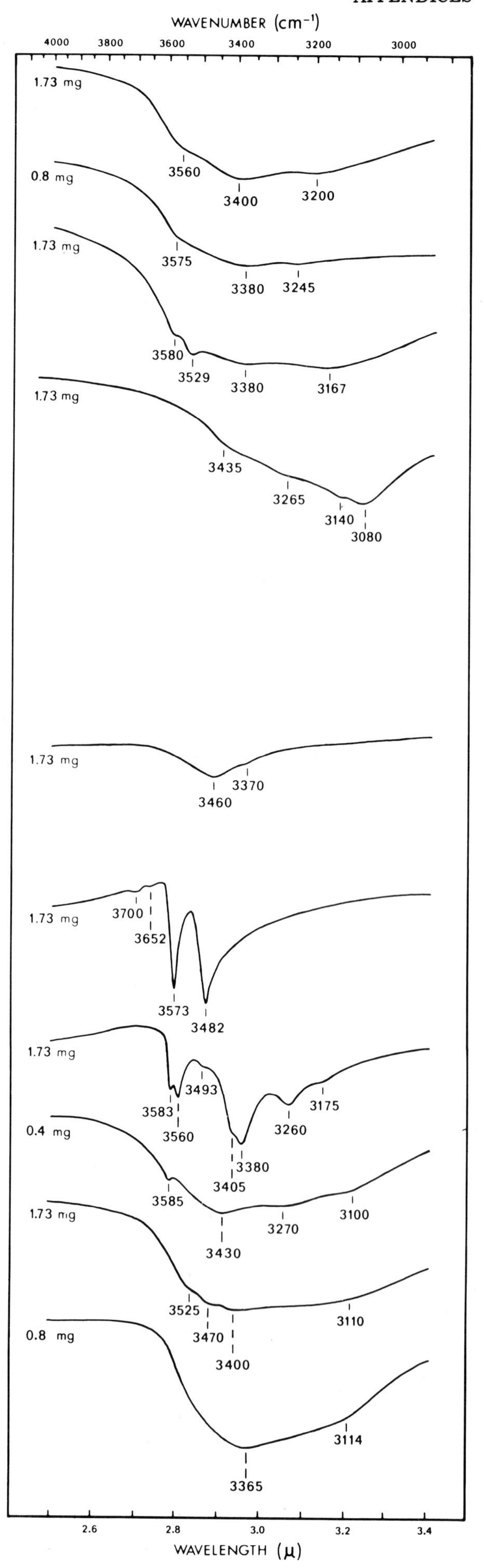

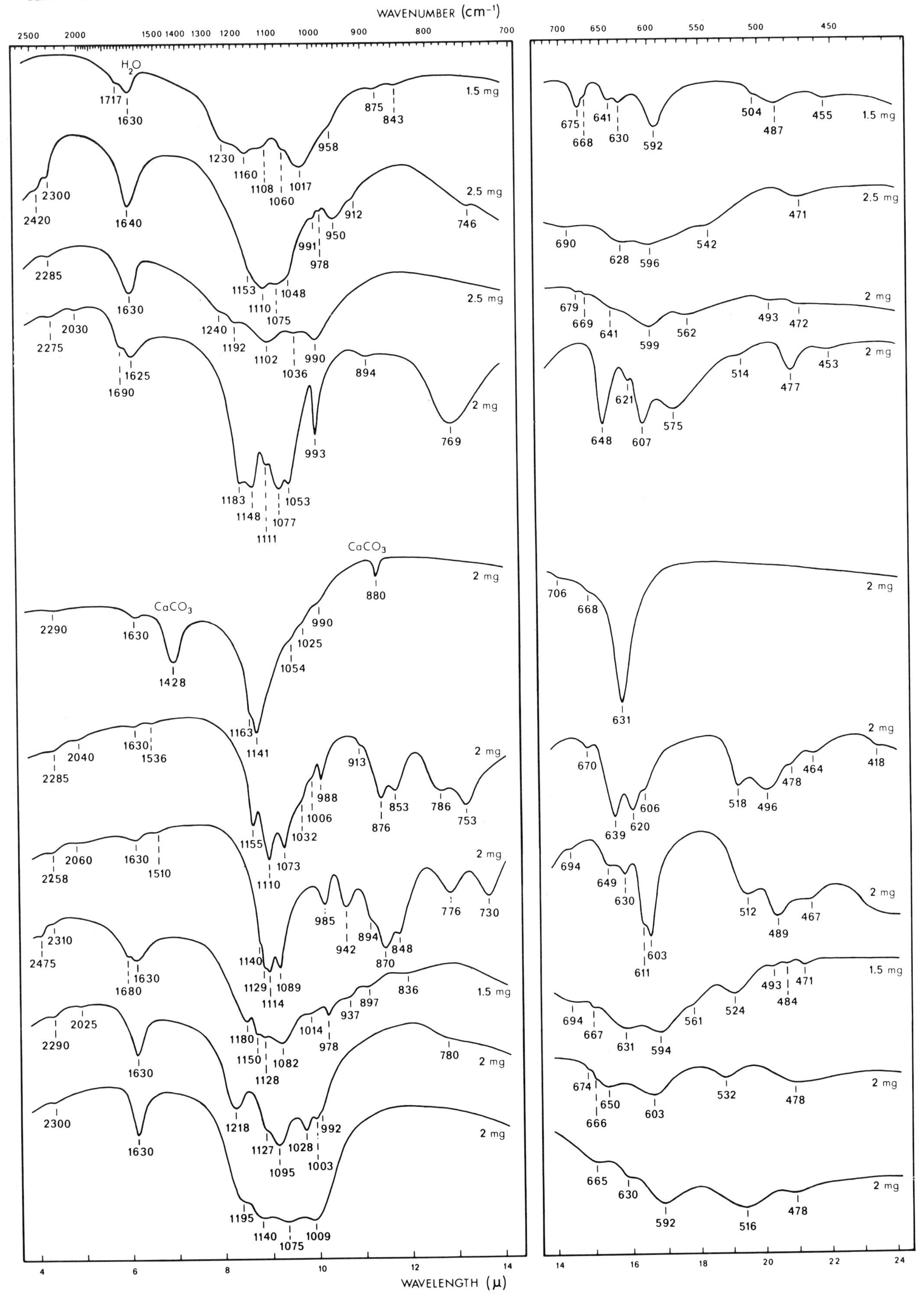
WAVENUMBER (cm⁻¹)
2500 2000 1500 1400 1300 1200 1100 1000 900 800 700
700 650 600 550 500 450
H2O
1717 1630 1230 1160 1108 1060 1017 958 875 843 1.5 mg
675 668 641 630 592 504 487 455 1.5 mg
2300 2420 1640 746 2.5 mg
471 2.5 mg
690 628 596 542
912 950 991 978
2285 1630 1153 1110 1075 1048 2.5 mg
679 669 641 599 562 493 472 2 mg
2030 2275 1625 1690 1240 1192 1102 1036 990 894 2 mg
514 477 453 2 mg
621 648 607 575
769 993 1183 1148 1111 1077 1053
CaCO3
880 2 mg
706 668 631 2 mg
2290 1630 CaCO3 1428 990 1025 1054 1163 1141
2040 2285 1630 1536 913 988 1006 1032 853 876 786 753 1155 1110 1073 2 mg
670 639 620 606 518 496 478 464 418 2 mg
2060 2258 1630 1510 2 mg
694 649 630 603 611 512 489 467 2 mg
985 942 894 870 848 776 730
2310 2475 1680 1630 1140 1129 1114 1089 836 1.5 mg
694 667 631 594 561 524 493 484 471 1.5 mg
2025 2290 1630 1180 1150 1128 1082 1014 978 937 897 780 2 mg
674 650 666 603 532 478 2 mg
2300 1630 1218 1127 1095 1028 1003 992 2 mg
1195 1140 1075 1009
665 630 592 516 478 2 mg
4 6 8 10 12 14
14 16 18 20 22 24
WAVELENGTH (μ)

Double cations + anions

Natrojarosite–Bool Coomatta, Australia
$NaFe_3^{3+}[(OH)_6 \mid (SO_4)_2]$ ditrigonal bipyramid

Alunite–Marysvale, Utah, U.S.A.
$KAl_3[(OH)_6 \mid (SO_4)_2]$ ditrigonal bipyramid

Jarosite–Copiaco, New Mex., U.S.A.
$KFe_3^{3+}[(OH)_6 \mid (SO_4)_2]$ ditrigonal bipyramid

Plumbojarosite–Darwin, Calif., U.S.A.
$Pb^2Fe_6^{3+}[(OH)_6 \mid (SO_4)_2]_2$ ditrigonal scalenohedron

Linarite–Tecopa, Calif., U.S.A.
$PbCu[(OH) \mid (SO_4)]$ monoclinic prismatic

Double cations + anions + crystalwater

Metasideronatrite–Sierra Gorda, Chile
$Na_2Fe^{3+}[(OH) \mid (SO_4)_2] \cdot 1\frac{1}{2}H_2O$ rhombic bipyramid

Sideronatrite–Sierra Gorda, Chile
$Na_2Fe^{3+}[(OH) \mid (SO_4)_2] \cdot 3H_2O$ rhombic

Natrochalcite–Chuquicamata, Chile
$NaCu_2[(OH) \mid (SO_4)_2] \cdot H_2O$ monoclinic prismatic

Botryogen–Knoxville, Calif., U.S.A.
$MgFe^{3+}[(OH) \mid (SO_4)_2] \cdot 7H_2O$ monoclinic prismatic

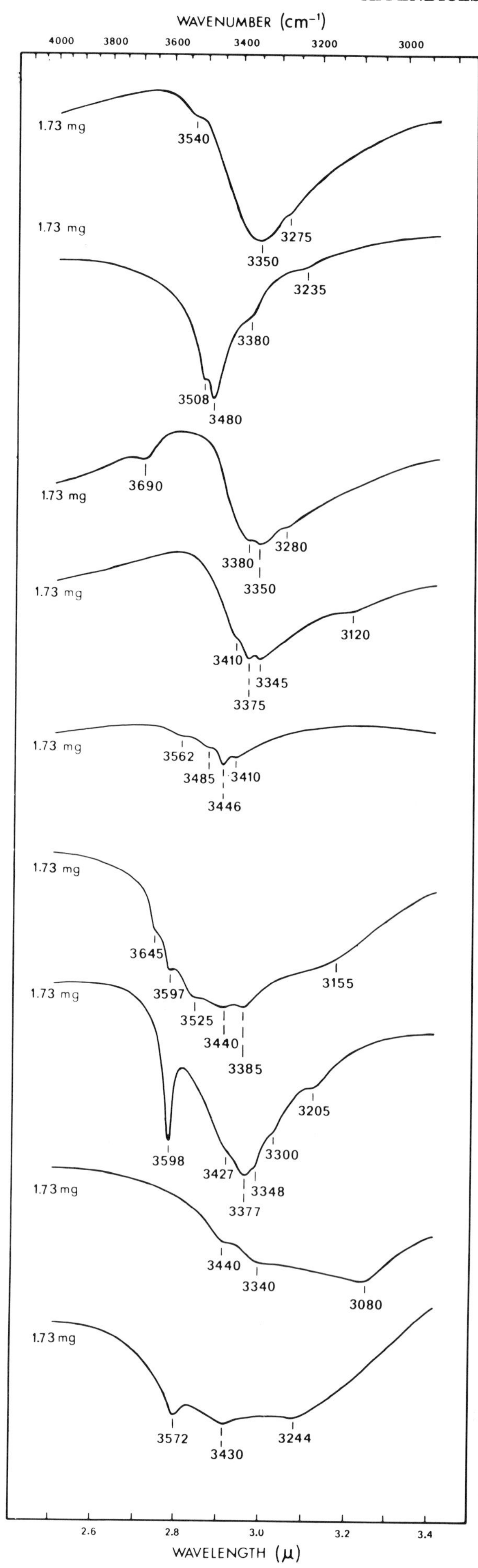

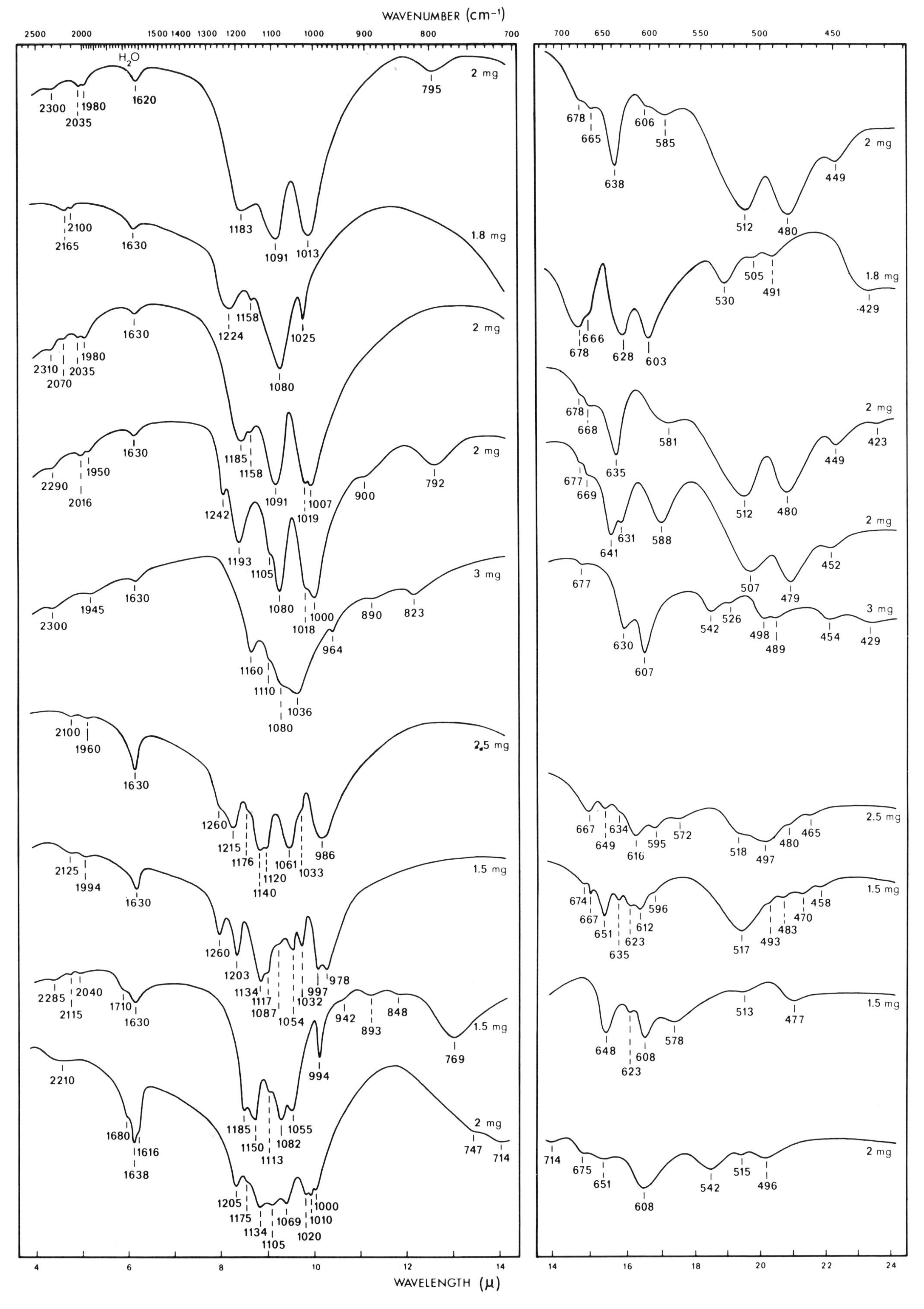

WAVENUMBER (cm⁻¹)
WAVELENGTH (μ)
H2O
2 mg
1.8 mg
2 mg
2 mg
3 mg
2.5 mg
1.5 mg
1.5 mg
2 mg

Metavoltine–Vulcana, Italy
α-$K_5Fe_3^{3+}[(OH) | (SO_4)_3]_2 \cdot 8H_2O$ hexagonal bipyramid

Cyanotrichite–Grand Canyon, Ariz., U.S.A.
$Cu_4Al_2[(OH)_{12} | (SO_4)] \cdot 2H_2O$ rhombic

Triple cations + crystalwater

Polyhalite–Carlsbad, New Mex., U.S.A.
$K_2Ca_2Mg[SO_4]_4 \cdot 2H_2O$ triclinic pinacoid, pseudorhombic

Leightonite–Chuquicamata, Chile
$K_2Ca_2Cu[SO_4]_4 \cdot 2H_2O$ triclinic pinacoid pseudorhombic

Voltaite–Alcaparosa, Chile
$K_2Fe_5^{2+}Fe_4^{3+}[SO_4]_{12} \cdot 18H_2O$ cubic hexoctahedron

Triple cations + anions + crystalwater

Ungemachite–Sierra Gorda, Chile
$K_3Na_9Fe^{3+}[(OH) | (SO_4)_2]_3 \cdot 9H_2O$ trigonal rhombohedric

Copiapite–Borate, Calif., U.S.A.
$(Fe^{2+},Mg)Fe_4^{3+}[(OH) | (SO_4)_3]_2 \cdot 20H_2O$ triclinic pinacoid

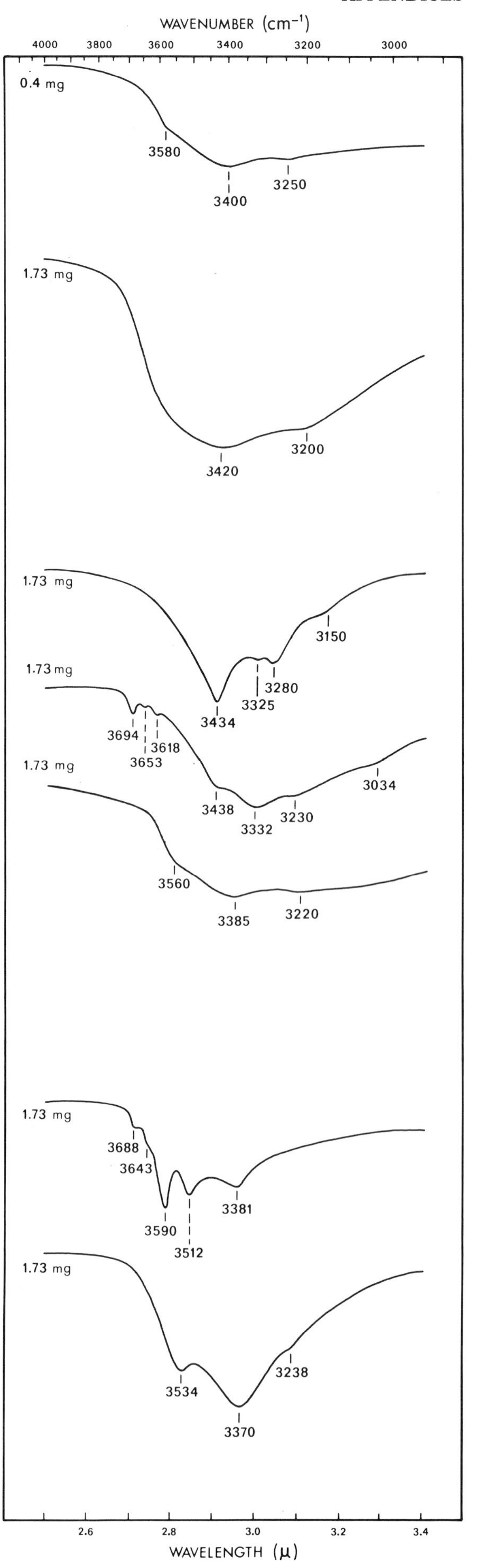

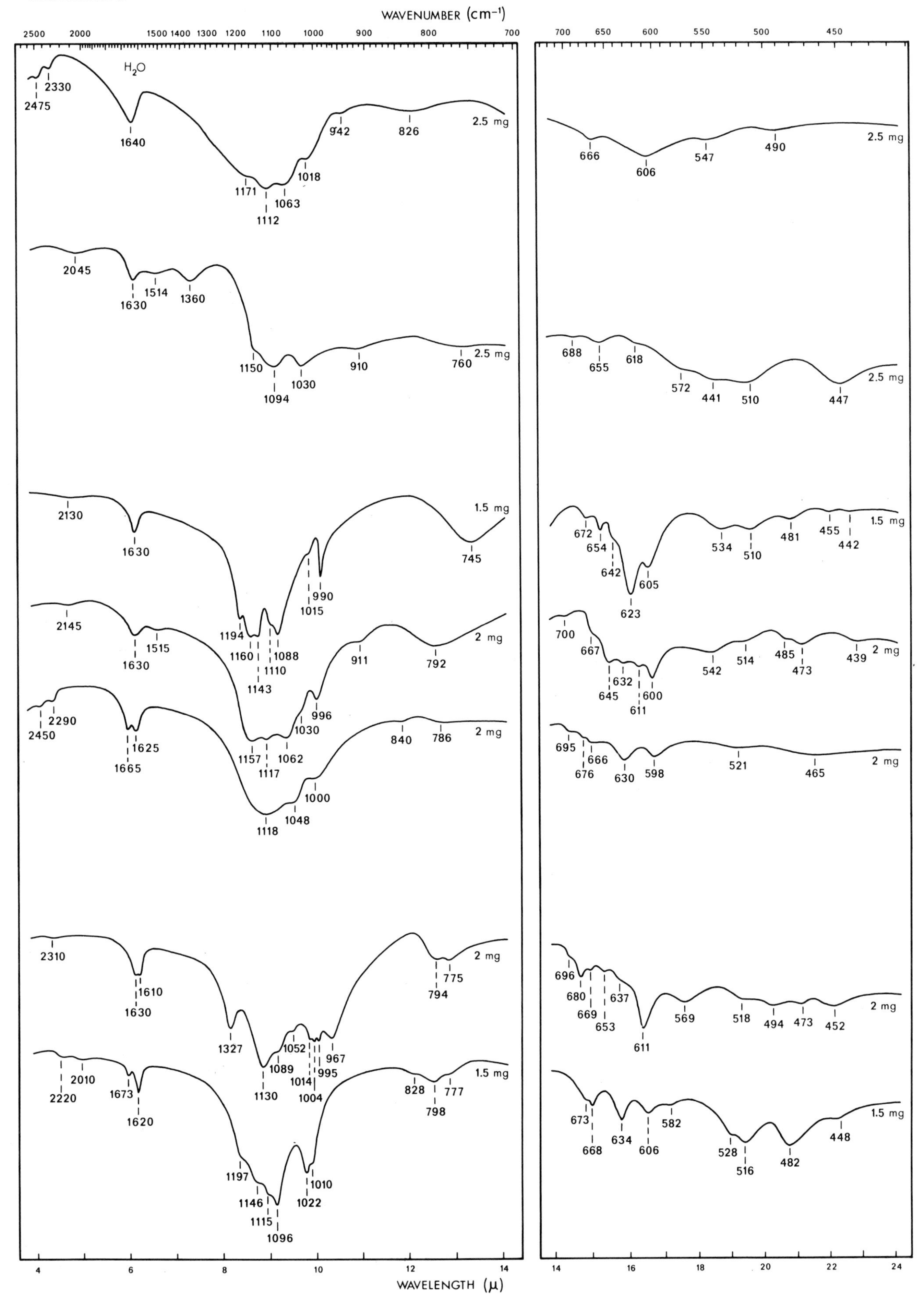
WAVENUMBER (cm⁻¹)
2500 2000 1500 1400 1300 1200 1100 1000 900 800 700
700 650 600 550 500 450
H2O
2475 2330 1640 1171 1112 1063 1018 942 826 2.5 mg
666 606 547 490 2.5 mg
2045 1630 1514 1360 1150 1094 1030 910 760 2.5 mg
688 655 618 572 441 510 447 2.5 mg
2130 1630 1194 1160 1143 1110 1088 1015 990 745 1.5 mg
672 654 642 623 605 534 510 481 455 442 1.5 mg
2145 1630 1515 1157 1117 1062 1030 996 911 792 2 mg
700 667 645 632 611 600 542 514 485 473 439 2 mg
2450 2290 1665 1625 1118 1048 1000 840 786 2 mg
695 676 666 630 598 521 465 2 mg
2310 1630 1610 1327 1130 1089 1052 1014 1004 995 967 794 775 2 mg
696 680 669 653 637 611 569 518 494 473 452 2 mg
2220 2010 1673 1620 1197 1146 1115 1096 1022 1010 828 798 777 1.5 mg
673 668 634 606 582 528 516 482 448 1.5 mg
4 6 8 10 12 14
14 16 18 20 22 24
WAVELENGTH (μ)

APPENDIX 4. The most characteristic bands of sulfates (▮ = strong, | = moderate, weak not mentioned).

WAVENUMBER (CM^{-1}): 3600 3500 3400 3300 3200 3100 3000 2900 … 2500 2000 1800 1600 1400 1300 1200 1100 1000 900 800 700 650 600 550 500 450

WAVELENGTH (μ): 2.8 3.0 3.2 3.4 3.6 … 4 6 8 10 12 14 16 18 20 22

Mineral	Bands (cm^{-1})
SINGLE CATIONS + DOUBLE CATIONS	
MASCAGNITE	3190, 3030, 2040, 1405, 1094, 973, 621, 455
THENARDITE	2100, 1122, 890, 639, 622
ANHYDRITE	2222, 2130, 1155, 1119, 678, 618, 602
CELESTITE	2075, 1197, 1132, 1092, 994, 644, 615, 484, 471, 457
BARYTE	2320, 2050, 1176, 1120, 1080, 984, 635, 613
ANGLESITE	2300, 1162, 1050, 967, 622, 601
LANGBEINITE	1170, 1150, 1135, 657, 635, 612, 478
GLAUBERITE	3540, 2125, 1185, 1139, 1105, 877, 657, 634, 610, 477
SINGLE CATIONS + CRYSTALWATER	
MIRABILITE *	2100, 1163, 1134, 1118, 667, 642, 621, 472, 457
KIESERITE	3405, 3280, 2190, 1670, 1630, 1170, 1125, 1040, 888, 746, 672, 644, 617, 480
EPSOMITE	3550, 3485, 3400, 3250, 2275, 1660, 1630, 1110, 745, 666, 614
ALUNOGEN	3370, 3010, 1630, 1150, 1085, 948, 615, 588, 488
MELANTERITE	3560, 3460, 3385, 3240, 1670, 1630, 1140, 1096, 1018, 990, 665, 628, 607, 494, 484
KORNELITE	3425, 3270, 1650, 1620, 1166, 1104, 1061, 1013, 896, 825, 679, 632, 592, 487
COQUIMBITE	3415, 3210, 1630, 1167, 1107, 1066, 943, 666, 616, 598, 485, 466
BIEBERITE	3390, 3230, 2330, 2270, 1630, 1104, 981, 863, 658, 631, 620, 607, 588, 518
CHALCANTITE	3490, 3415, 3230, 3080, 2450, 2300, 2070, 1640, 1106, 1070, 870, 661, 608, 590, 529, 485, 455
GOSLARITE	3585, 3430, 3320, 3205, 3120, 2150, 1630, 1499, 1151, 1100, 985, 870, 760, 651, 612, 534, 460
DOUBLE CATIONS + CRYSTALWATER	
BOUSINGAULTITE	3410, 3190, 3020, 2200, 1660, 1630, 1468, 1420, 1140, 1087, 982, 797, 630, 618
BLOEDITE	3435, 3175, 2400, 2140, 1665, 1630, 1160, 1122, 994, 876, 826, 638, 620, 472
TAMARUGITE	3558, 3410, 3240, 2420, 2170, 1650, 1630, 1382, 1097, 1027, 835, 625, 603, 516
MENDOZITE	3390, 3230, 2480, 2300, 1640, 1076, 950, 910, 664, 602, 536, 467
PICKERINGITE	3580, 3405, 3270, 2475, 2315, 1640, 1111, 1074, 1052, 950, 632, 596, 476
KALINITE	3380, 2425, 2300, 1630, 1092, 972, 913, 685, 615, 600, 504
POTASHALUMITE	3400, 3205, 3093, 2450, 2320, 1630, 1430, 1405, 1094, 912, 688, 615, 600, 472
KRAUSITE	3400, 3090, 2460, 2300, 1630, 1268, 1230, 1168, 1110, 1024, 1006, 962, 794, 677, 594, 484, 471, 457

* Crystalwater lost by pressing with KBr.

WAVENUMBER (CM^{-1})

Mineral	Bands (wavenumber, cm^{-1})
DOUBLE CATIONS + CRYSTALWATER	
RHOMBOCLASE	3560 3400 3200 1717 1630 1230 1160 1108 1017 875 675 641 630 592 487 455
HALOTRICHITE	3575 3380 3245 2420 2300 1640 1110 1075 1048 950 690 628 596 542 471
ROEMERITE	3580 3529 3380 3167 2285 1630 1192 1102 990 669 599 562 493
KROEHNKITE	3435 3080 2275 2030 1690 1625 1183 1148 1077 1053 993 894 769 648 607 575 477 453
SINGLE CATIONS + ANIONS and DITTO + CRYSTALWATER	
SULFOHALITE	2290 1163 1141 1054 990 631
ANTLERITE	3573 3482 2285 2040 1155 1110 1073 988 876 853 786 753 639 620 606 518 496 464
BROCHANTITE	3583 3560 3405 3380 3260 3175 2258 2060 1129 1114 1089 985 942 870 848 776 730 649 630 611 603 512 489 467
ALUMINITE	3585 3430 3270 3100 2475 2310 1680 1630 1180 1150 1128 1082 978 897 631 594 524 493 471
METAHOHMANNITE	3525 3400 3110 2290 2025 1630 1218 1127 1095 1028 1003 650 603 532 478
FIBROFERRITE	3365 2300 1630 1195 1140 1075 1009 665 630 592 516 478
DOUBLE CATIONS + ANIONS	
NATROJAROSITE	3540 3350 2300 2035 1980 1183 1091 1013 795 678 638 585 512 480 449
ALUNITE	3508 3480 3380 2165 2100 1224 1080 1025 678 628 603 530 491 429
JAROSITE	3380 3350 3280 2310 2070 1980 1185 1091 1019 1007 668 635 581 512 480 449
PLUMBOJAROSITE	3410 3375 3345 3120 2290 2016 1950 1242 1193 1080 1018 1000 792 641 631 588 507 479 452
LINARITE	3562 3446 2300 1945 1160 1080 1036 964 823 677 630 607 542 498 489 454
DOUBLE CATIONS + ANIONS + CRYSTALWATER	
METASIDERONATRITE	3597 3440 3385 3155 2100 1960 1630 1215 1140 1120 1061 986 667 649 616 595 518 497 465
SIDERONATRITE	3598 3377 3205 2125 1994 1630 1260 1203 1134 1054 1032 997 978 667 651 623 612 517 470
NATROCHALCITE	3440 3340 3080 2285 2115 1710 1630 1185 1150 1082 1055 994 893 769 648 608 578 513 477
BOTRYOGEN	3572 3430 3244 2210 1680 1638 1616 1205 1134 1105 1069 1020 1010 1000 675 651 608 542 515 496
METAVOLTINE	3580 3400 3250 2475 2330 1640 1171 1112 1063 1018 826 666 606 547 490
CYANOTRICHITE	3420 3200 2045 1630 1360 1150 1094 1030 910 760 655 572 510 447 441
TRIPLE CATIONS + CRYSTALWATER	
POLYHALITE	3434 3280 3150 1630 1194 1160 1143 1088 990 745 623 605 534 510 481
LEIGHTONITE	3438 3332 3230 2145 1630 1157 1117 1062 996 792 645 632 600 542 473 439
VOLTAITE	3560 3385 3220 2450 2290 1665 1625 1118 1048 1000 840 786 666 630 598 521 465
TRIPLE CATIONS + ANIONS + CRYSTALWATER	
UNGEMACHITE	3590 3512 3381 1630 1610 1327 1130 1089 967 794 775 680 611 569 518 494 452
COPIAPITE	3534 3370 3238 2220 2010 1673 1620 1197 1096 1022 1010 798 668 634 606 528 516 482 448

WAVELENGTH (μ)

BIBLIOGRAPHY

Handbooks

Afremow, L.C., 1970. I.R. Spectroscopy: its Use in the Coatings Industry. Federation of Societies for Paint Technology, Philadelphia, 456 pp.

Allen, H.C. Jr. and Cross, P.C., 1960. Molecular Vibrators: the Theory and Interpretation of High-Resolution Infrared Spectra. Wiley, London, 384 pp.

Alpert, N.L., Keiser, W.E. and Szymanski, H.A., 1970. I.R.—Theory and Practice of Infrared Spectroscopy. Plenum Press, New York, 2nd ed., 380 pp.

Barrow, G.M., 1962. Introduction to Molecular Spectroscopy. McGraw-Hill, New York, 318 pp.

Bauman, R.P., 1962. Absorption Spectroscopy. Wiley, New York, 611 pp.

Beaven, G.H., Johnson, E.A., Willis, H.A. and Miller, R.G.J., 1962. Molecular Spectroscopy. McMillan, New York, 336 pp.

Bellamy, L.J., 1964. The Infrared Spectra of Complex Molecules. Associated Book Publ., London, 448 pp.

Bellamy, L.J., 1966. Ultrarot Spektrum und Chemische Konstitution. Dietrich Steinkopf, Darmstadt, 325 pp.

Brügel, W., 1962. An Introduction to Infrared Spectroscopy (translated by A.R. and A.J.D. Katritzky). Wiley, New York, 419 pp.

Brügel, W., 1969. Einführung in die Ultraspektroskopie. Dietrich Steinkopf, Darmstadt, 426 pp.

Colthup, N.B., Daly, L.H. and Wiberley, S.E., 1964. Introduction to Infrared and Raman Spectroscopy. Acad. Press, New York-London, 511 pp.

Conley, R.T., 1966. Infrared Spectroscopy. Allyn and Bacon, Boston, 293 pp.

Conn, G.K.T. and Avery, D.G., 1960. Infrared Methods. Monograph 7, Physical Chemistry. Acad. Press, New York, 203 pp.

Cross, A.D. and Jones, R.A., 1969. An Introduction to Practical Infrared Spectroscopy. Butterworths, London, 3rd ed., 102 pp.

Davies, M., 1963. Infrared Spectroscopy and Molecular Structures. Elsevier, Amsterdam, 468 pp.

Derkosch, J., 1967. Absorptionspektral-analyse im ultravioletten, sichtbaren und infraroten Gebiet. Akadem. Verlaggesellschaft, Frankfurt a. Main, 427 pp.

Farmer, V.C., 1974. The Infrared Spectra of Minerals (Mineral. Soc. Monogr., 4). Adlard, Dorking, 539 pp.

Hair, M.L. 1967. Infrared Spectroscopy in Surface Chemistry. M. Dekker, New York, 315 pp.

Halban, H.E., 1963. Infrared Spectroscopy and Molecular Structure. Elsevier, New York, 418 pp.

Kackforth, H.I., 1960. Infrared Radiation. MacGraw-Hill, New York, 303 pp.

Kendall, D.N., 1966. Applied Infrared Spectroscopy. Reinhold, New York and Chapman Hall, London, 560 pp.

Kössler, I., 1966. Methoden der Infrarot-Spektroskopie in der chemischen Analyse. Akad. Verlagsges. Geest Portig, Leipzig, 227 pp.

Krimm, S., 1963. Infrared Spectroscopy and Molecular Structure. Elsevier, New York, 287 pp.

Lecomte, J., 1949a. Le Rayonnement infrarouge, 1. Applications biologiques, physiques et techniques. Gauthier-Villards, Paris, 391 pp.

Lecomte, J., 1949b. Le Rayonnement infrarouge, 2. La Spectrométrie infrarouge et ses Applications physicochimiques. Gauthier-Villards, Paris, 764 pp.

Lecomte, J., 1958. Spectroscopie dans l'infrarouge. In: S. Flügge (Editor), Handbuch der Physik, 26. Licht und Materie. Springer, Berlin-Göttingen-Heidelberg, pp. 244—937.

Lothian, G.F., 1958. Absorption Spectrophotometry. Hilger and Watts, London, 2nd ed., 246 pp.

Mangini, A. (Editor), 1962. Advances in Molecular Spectroscopy. (Proc. 4th International Meeting on Molecular Spectroscopy.) Pergamon, London-New York-Paris, 1393 pp.

Mayer, F.X. and Luszczak, A., 1951. Absorptions-Spektranalyse. Walter de Gruyter, Berlin, 238 pp.

Mellon, M.G., 1950. Analytical Absorption Spectroscopy. Wiley, New York, 618 pp.

Meloan, C.E., 1963. Elementary Infrared Spectroscopy. MacMillan, New York, 193 pp.

Nakanishi, K., 1962. Practical Infrared Absorption Spectroscopy. Holden-Day, San Francisco, 242 pp.

Pimentel, G.C. and McClellan, C., 1960. The Hydrogen Bond. Reinhold, San Francisco, 475 pp.

Rao C.N.R., 1963. Chemical Applications of Infrared Spectroscopy. Acad. Press, New York, 683 pp.

Simon, I., 1966. Infrared Radiation. D. van Nostrand, New York, 119 pp.

Smith, R.A., Jones, F.E. and Chasmar, R.P., 1968. The Detection and Measurement of Infrared Radiation. Clarendon Press, Oxford, 2nd. ed., 522 pp.

Strouts, C.R.N., Gilfillon, J.H. and Wilson, H.N., 1962. Infrared Absorption Spectrophotometry. Chemical Analysis, 2: 264—361.

Szymanski, H.A., 1962, 1964. Progress in Infrared Spectroscopy, 1,2. Plenum Press, New York, 1:452 pp; 2:272 pp.

Szymanski, H.A., 1964. Theory and Practice of Infrared Spectroscopy. Plenum Press, New York, 389 pp.

Ulrich, W.F., 1963. Bibliography of Infrared Applications. Bull. 7037, Beckman Corp., 96 pp., 3332 ref.

Van der Maas, J.H., 1972. Basic Infrared Spectroscopy. Heyden, London, 109 pp.

Volkmann, H., 1972. Handbuch der Infrarot Spektroskopie. Verlag Chemie Weinheim, 505 pp.

West, W., 1964. Chemical Applications of Spectroscopy. Interscience, London-New York, 787 pp.

White, R.J., 1964. Handbook of Industrial Infrared-Analysis. Plenum Press, New York, 440 pp.

Wilson, E.B. Jr., Decius, J.C. and Cross, P.C., 1955. Molecular Vibrations; the Theory of Infrared and Raman Vibrational Spectra. Mc.Graw-Hill, New York, 388 pp.

Documentation (mainly organics)

A.S.T.M., 1964. Identification of material by absorption spectroscopy, using the Wyandotte-A.S.T.M. (Kuentzel) punched card index. A.S.T.M.-Stand., 31: 579—587

A.P.I. Am. Petrol. Inst., 1950. Infrared and ultraviolet spectral data. Am. Petrol. Inst., Res. Proj. 44. Carnegie Inst. Technol., Pittsburgh.

A.S.T.M. Infrared Optical Coincidence Index, S.T.P. 365, with the "peek-a-boo" data retrieval system.

A.S.T.M. Wyandotte Infrared Spectral Index Cards. Am. Soc. Test. Mater., Philadelphia, 1963.

Ayton, M.W., Derby, T.J., Bateler, V.H. and Brown, C.R., 1957. Infrared. A Bibliography, 2. Library of Congress, Techn. Inf. Division. PB 121998, 150 pp.

Brown, C.R., Ayton, M.W., Goodwin, T.C. and Derby, T.J. 1954. Infrared Bibliography 1,2. Off. Tech. Serv. U.S. Dept. Commerce. 1:374 pp., 2:150 pp.

Butterworth's D.M.S. system. Documentation of Molecular Spectroscopy. 1956 D.M.S. - I cards. Butterworth's Sci. Publ., London, and Verlag Chem. West Germany.

Coblentz, W.W., 1959. Society Spectra. The Sadtler Research Laboratories, Philadelphia, 160 pp. + supplements.

Coblentz, W.W., 1962. Investigations of infrared spectra. (Reprinted by the Coblentz Soc. Inc. c/o the Perkin Elmer Corpor., Norwalk, 660 pp.

Haase, H., 1956. Infrarot Bibliographie. Forschungsber. Dtsch. Wirtsch. Verkehrsminist. Nordrhein - Westfalen. Westdeutscher Verlag, Köln - Opladen, 233: 80 pp.

Hershenson, H.M., 1959, 1964. Infrared Absorption Spectra Index for 1945-1957 (111 pp. ed. 1959) and Index for 1958—1962 (153 pp. ed. 1964) Academic Press, New York.

Hirayama, K., 1958. The data arrangement of infrared spectra; the DMS system. Kagako no Ryoiki Zokan, 31: 113—143.

Mecke, R. and Langenbucher, F., 1965. Infrared Spectra of Selected Chemical Compounds. Heyden, London, 8 volumes, with indices.

Ministry of Aviation Technical Information and Library Services, 1960. An Atlas of Published Infrared Spectra, I, II, and III. Her Majesty's Stationary Office, London.

NRC - NBS. Infrared spectral absorption data. Natl. Res. Council, Natl. Bur. Stand., Wash.

Otting, W., 1963. Spektrale Zuordnungtafel der IR-Absorptionbanden. Springer, Berlin, 18 pp.

Randall, H.M., Fowler, R.G., Fuson, N. and Dange, J.R., 1949. Infrared Determination of Organic Structure. Van Nostrand, New York, 239 pp.

Sadtler Research Laboratories' Spectra, 1960. S.P. Sadtler Res. Lab., Philadelphia. Heyden, London (+ suppl. and "spec finder" to locate the spectra).

Spectra Index and Scientific Documentation Centre, 1963 Lindsay & Co., Dunfermline, England.

Szymanski, H.A., 1964, 1966, 1967. Interpreted Infrared Spectra. Plenum Press, New Jersey, 1: 293 pp., 2: 304 pp., 3: 275 pp.

Szymanski, H.A., 1970. Infrared Band Handbook. Plenum Press, New York, (anorganic + organic) revised ed., ca. 1350 pp.

1. Basic considerations about electromagnetic radiation

Badger, R.M. J. Chem. Phys., 8 (1940) 288.
Barrow, G.M. J. Phys. Chem., 59 (1955) 1129.
Bellamy, L.J. Spectrochim. Acta, 14 (1959) 192.
Bovey, L.H.F. and Sutherland, G.B.B.M. J. Chem. Phys., 17 (1949) 843.
Brown, G.E. and Gibbs, G.V. Am. Mineralogist, 54 (1969) 1528.
Burstein, E. Phys. Rev., 97 (1955) 39.
Crawford, B. J. Chem. Phys., 29 (1958) 1042.
Flynn, T.D. Werner, R.L. and Graham, B.M. Austr. J. Chem., 12 (1959) 575.
Gilfert, J.C. and Williams, D. J. Opt. Soc. Am., 48 (1958) 765.
Gilfert, J.C. and Williams, D. J. Opt. Soc. Am., 49 (1959) 212.
Girin, O.P. Opt. i Spektr., 9 (1960) 673.
Gordy, W. J. Chem. Phys., 14 (1946) 305.
Gribov, L.A. Opt. i Spektr., 11 (1961) 146.
Gurevich, L.E. and Uritskii, Z.I. Fiz. Tverdogo Tela (Solid State Physica), 2 (1960) 1239.
Haas, C. and Ketelaar, J.A.A. Physica, 22 (1956) 1286.
Hornig, D.F. J. Chem. Phys., 16 (1948) 1063.
Hrostowski, H.J. and Pimentel, G.C. J. Chem. Phys., 19 (1951) 661.
Huggins, M.L. J. Am. Chem. Soc., 75 (1953) 4126.
Hughes, R.H., Martin, R.J. and Coggeshall, N.D. J. Chem. Phys., 24 (1956) 489.
Jones, G.O., Martin, D.H., Mawer, P.A. and Perry, C.H. Proc. R. Soc., A 261 (1961) 10.
Kuratani, K. Kagako no Ryoiki Zokan, 21 (1956) 27.
Lax, M. and Burstein, E. Phys. Rev., 97 (1955) 39.
Lippincott, E.R., Weir, C.E., Van Valkenburg, A. and Bunting, E.N. Spectrochim. Acta, 16 (1960) 58.
Lüttke, W. and Mecke, R. Z. Elektrochem., 53 (1949) 241.
Matossi, F. J. Chem. Phys., 17 (1949) 679.
Mecke, R. J. Chem. Phys., 20 (1952) 1935.
Norrish, R.S. Nature, 187 (1960) 142.
Ovander, L.N. Opt. i Spektr., 11 (1961) 129.
Patty, R.R. and Williams, D. J. Opt. Soc. Am., 51 (1961) 1351.
Petrash, G.G. Opt. i Spektr., 9 (1960) 121.
Prima, A.M. Tr. Inst. Fiz. Mat., Akad. Nauk. Bolerus. S.S.S.R., 2 (1957) 124.
Pullin, A.D.E. Proc. R. Soc., A 255 (1960) 39.
Rocchiccioli, C. Bull. Soc. Fr. Ceram., 68 (1965) 39.
Shrinivasacharya, K.G. and Santhamma, C.J. Molecul. Spectr., 15 (1965) 435.
Smith, R.P. J. Phys. Chem., 60 (1956) 1293;
Sterzel, W. Naturwissenschaften, 51 (1964) 505.
Verdier, P.H. and Wilson, E.B. J. Chem. Phys., 30 (1959) 1372.
Vincent-Geisse, J. J. Phys. Radium, 17 (1956) 63.
White, W.B. Mater. Res. Bull., 2 (1967) 381.

2. Instrumental

Monochromators

Acquista, N. and Plyler, E.K. J. Opt. Soc. Am., 43 (1953) 977.
Afremow, L.C. and Vandeberg, J.T. J. Paint Technol., 38 (1966) 169.
Blout, E.R., Corley, R.S. and Snow, P.L. J. Opt. Soc. Am., 40 (1950) 415.
Coates, V.J. Spectrochim. Acta, 15 (1959) 820.
Duncanson, L.A., Eddell, J.W., Lloyd, M.B. and Moore, W.T. Spectrochim. Acta, 15 (1959) 64.
Fastie, W.G. J. Opt. Soc. Am., 42 (1952) 647.
Gottlieb, M. J. Opt. Soc. Am., 50 (1960) 343.
Greenler, R.G. J. Opt. Soc. Am., 47 (1957) 130.
Ham, N.S., Walsh, A. and Willis, J.B. J. Opt. Soc. Am., 42 (1952) 496.
Harrison, G.R. J. Opt. Soc. Am., 39 (1949) 522.
Hatcher, R.D. and Rohrbaugh, J.H. J. Opt. Soc. Am., 46 (1956) 104.
Heavens, O.S., Ring, J. and Smith, S.D. Spectrochim. Acta, 10 (1957) 179.
Hepner, G. J. Phys. Radium, 14 (1953) 717.
Kendrick, E. and Schnurmann, R. J. Opt. Soc. Am., 44 (1954) 501.
Pirlot, G. Bull. Soc. Chim. Belg., 59 (1950) 352.
Plyler, E.K. and Acquista, N. J. Res. Natl. Bur. Stand., 49 (1952) 61.
Plyler, E.K. and Acquista, N. J. Opt. Soc. Am., 48 (1958) 668.
Plyler, E.K. and Ball, J.J. J. Opt. Soc. Am., 42 (1952) 266.
Plyler, E.K. and Blaine, L.R. J. Res. Bur. Stand., 60 (1958) 55.
Rank, D.H. J. Opt. Soc. Am., 50 (1960) 657.
Rank, D.H., Eastman, D.P., Birtley, W.B., Skorinko, G. and Wiggins, T.A. J. Opt. Soc. Am., 50 (1960) 821.
Rochester, J.C.O. and Martin, A.E. Nature, 168 (1951) 785.
Straat, H.W. J. Opt. Soc. Am., 43 (1953) 593.
Walsh, A. J. Opt. Soc. Am., 42 (1952) 94.
Yamada, Y. and Mitsuishi, A. Tech. Rep. Osaka Univ., 4 (1954) 191.

Sources of light

Crosswhite, H.M. Dieke, G.H. and Legagneur, C.S. J. Opt. Soc. Am., 45 (1955) 270.
Ebers, E.S. and Nielsen, H.H. Rev. Sci. Instr., 11 (1940) 429.
Friedel, R.A. and Sharkey, A.G. Rev. Sci. Instr., 18 (1947) 928.
Hisdal, B. J. Opt. Soc. Am., 52 (1962) 395.
Plyler, E.K., Yates, D.J.C. and Gebbie, H.A. J. Opt. Soc. Am., 52 (1962) 859.

Detectors

Boyle, W.S. and Rodgers, K.F. J. Opt. Soc. Am., 49 (1959) 66.
Bratt, P., Engeler, W., Levinstein, H., MacRae, A. and Pehek, J. Infrared Phys., 1 (1961) 27.
Chasmar, R.P., Mitchell, W.H. and Rennie, A. J. Opt. Soc. Am., 46 (1956) 469.
Černý, M., Kofink, W. and Lippert, W. Ann. Phys., 8 (1951) 65.
Hanel, R.A. J. Opt. Soc. Am., 51 (1961) 220.
Hyde, W.L. Spectrochim. Acta, 6 (1953) 9.
Jones, R.C. J. Opt. Soc. Am., 43 (1953) 1.
Jones, R.C. J. Opt. Soc. Am., 43 (1953) 1008.
Lasser, M.E., Cholet, P. and Wurst, E.C. J. Opt. Soc. Am., 48 (1958) 468.
Leo, W. and Hübner, W. Z. Angew. Phys., 2 (1950) 454.
Low, F.J. J. Opt. Soc. Am., 51 (1961) 1300.
Moss, T.S. J. Opt. Soc. Am., 40 (1950) 603.
Stubbs, H.E. and Phillips, R.G. Rev. Sci. Instr., 31 (1960) 115.
Vogl, T.P., Shifrin, G.A. and Leon, B.J. J. Opt. Soc. Am., 52 (1962) 957.
Weber, P.E. Optik, 6 (1950) 152.
Wormser, E.M. J. Opt. Soc. Am., 43 (1953) 15.
Wotring, A.W., Wall, R.F. and Zinn, T.L. Anal. Chem., 28 (1956) 1396.
Young, A.S. J. Sci. Instr., 32 (1955) 142.

Calibration

Acquista, N. and Plyler, E.K. J. Res. Natl. Bur. Stand., 49 (1952) 13.
Benedict, W.S. and Plyler, E.K. J. Res. Natl. Bur. Stand., 46 (1951) 246.
Bethke, G.W. J. Opt. Soc. Am., 46 (1956) 560.
Coleman, B. and Meggers, W.F. Natl. Bur. Stand., Monogr. 3 (1960).
Commission of Molecular Structure and Spectroscopy. Pure Appl. Chem., 1 (1961) 537 and 603.
Downie, A.R., Magoon, M.C., Purcell, T. and Crawford, B. J. Opt. Soc. Am., 43 (1953) 941.
Dows, D.A. J. Opt. Soc. Am., 48 (1958) 73.
Edlèn, B. J. Opt. Soc. Am., 43 (1953) 339.
Feigel'son, E.M. Izv. Akad. Nauk. S.S.S.R., Ser. Geofiz. (1955) 69.
Filler, A.S. and Indyk, L. J. Opt. Soc. Am., 51 (1961) 572.
Guy, W. and Towler, J.H. J. Sci. Instr., 28 (1951) 103.
Humphreys, C.J. J. Opt. Soc. Am., 43 (1953) 1027.
Jones, L.H. J. Chem. Phys., 24 (1956) 1250.
Jones, R.N. Faure, P.K. and Zaharias, W. Rev. Univ. Mines, 9th Ser., 15 (1959) 417.
Jones, R.N. Jonathan, N.B.W., Mackenzie, M.A. and Nadeau, A. Spectrochim. Acta, 17 (1961) 77.
Jones, R.N. and Nadeau, A. Spectrochim. Acta, 20 (1964) 1175.
Kessler, F.R. and Hauser, H. Z. Angew. Phys., 17 (1964) 280.
Kiess, C.C. J. Res. Natl. Bur. Stand., 48 (1952) 377.
König, R. and Malewski, G. Jenaer Rundsch., 12 (1967) 132.
Marrison, L.W. J. Sci. Instr., 29 (1952) 233.
Martin, A.E. J. Opt. Soc. Am., 41 (1951) 56.
McKinney, D.S. and Friedel, R.A. J. Opt. Soc. Am., 38 (1948) 222.
Meggers, W.F. and Westfall, F.O. J. Res. Natl. Bur. Stand., 44 (1950) 447.
Mills, I.M., Scherer, J.R. Crawford, B. and Youngquist, M. J. Opt. Soc. Am., 45 (1955) 785.
Penndorf, R. J. Opt. Soc. Am., 47 (1957) 176.
Plyler, E.K. and Peters, C.W. J. Res. Natl. Bur. Stand., 45 (1950) 279 and 462.
Plyler, E.K., Blaine, L.R. and Nowak, M. J. Res. Natl. Bur. Stand., 58 (1957) 195.
Plyler, E.K., Blaine, L.R. and Tidwell, E.D. J. Res. Natl. Bur. Stand., 55 (1955) 279.
Plyler, E.K., Danti, A., Blaine, L.R. and Tidwell, E.D. J. Res. Nat. Bur. Stand., 64A (1960) 29.
Plyler, E.K., Gailar, N.M. and Wiggins, T.A. J. Res. Natl. Bur. Stand., 48 (1952) 221.
Rank, D.H., Bennet, J.M. and Bennett, H.E. J. Opt. Soc. Am., 46 (1956) 477.
Rank, D.H., Birtley, W.B., Eastman, D.P., Rave, B.S. and Wiggins, T.A. J. Opt. Soc. Am., 50 (1960) 1275.
Rank, D.H., Skorinko, G., Eastman, D.P. and Wiggins, T.A. J. Molecul. Spectrosc., 4 (1960) 518.
Rao, K.N., Brim, W.W., Sinnett, V.L. and Wilson, R.H. J. Opt. Soc. Am., 52 (1962) 862.
Rao, K.N., Ryan, L.R. and Nielsen, H.H. J. Opt. Soc. Am., 49 (1959) 216.
Ross, W.L. and Little, D.E. J. Opt. Soc. Am., 41 (1951) 1006.
Schubert, M. Exp. Tech. Phys., 6 (1958) 203.

Tabačik, V. and Horák, M. Jenaer Rundsch., 9 (1964) 202.
Towler, J.H. and Guy, W. J. Sci. Instr., 29 (1952) 393.
Walker, I.K. and Todd, H.J. Anal. Chem., 31 (1959) 1603.

Spectrophotometers

Adams, D.M. Spectrochim. Acta, 18 (1962) 1039.
Bergmann, G. Z. Anal. Chem., 170 (1959) 66.
Blout, E.R. and Abbate, M.J. J. Opt. Soc. Am., 45 (1955) 1028.
Brodersen, S. J. Opt. Soc. Am., 43 (1953) 877.
Cahn, L. and Henderson, B.D. J. Opt. Soc. Am., 48 (1958) 380.
Childers, E. and Struthers, G.W. Anal. Chem., 25 (1953) 1311.
Cole, A.R.H. J. Opt. Soc. Am., 43 (1953) 807.
Daly, E.F. Nature, 166 (1950) 1072.
Dmitrievskii, O.D., Neporent, V.S. and Nikiten, V.A. Opt. i Spektr., 3 (1957) 180.
Eggers, D.F. and Emerson, M.T. J. Opt. Soc. Am., 50 (1960) 11.
Ford, M.A., Price, W.C. and Wilkinson, G.R. J. Sci. Instr., 35 (1958) 55.
Funck, E. and Beckmann, L. Chem. Ing. Tech., 31 (1959) 711.
Görlich, P. and Kortum, H. Feingerätetechnik, 8 (1959) 152 and 195.
Hales, J.L. J. Sci. Instr., 36 (1959) 264.
Herscher, L.W. Spectrochim. Acta, 15 (1959) 901.
Herscher, L.W., Ruhl, H.D. and Wright, N. J. Opt. Soc. Am., 48 (1958) 36.
Kaye, W. Spectrochim. Acta, 7 (1955) 181.
Kneubühl, F.K., Moser, J.F. and Steffen, H. J. Opt. Soc. Am., 56 (1966) 760.
Liston, M.D. and White, J.U. J. Opt. Soc. Am., 40 (1950) 36.
Lord, R.C. and McCubbin, T.K. J. Opt. Soc. Am., 47 (1957) 689.
Lyon, R.J.P., in: J. Zussmann. Physical Methods in Determinative Mineralogy. Academic Press, London-New York, (1967) p.371.
Martin, A.E. J. Opt. Soc. Am., 38 (1948) 70.
Mc Cubbin, T.K. and Sinton, W.M. J. Opt. Soc. Am., 42 (1952) 113.
National Bureau of Standards, Washington D.C., Appl. Spectr., 12 (1958) 175.
Neporent, B.S. Opt. i Spektr., 3 (1957) 289.
Oetjen, R.A., Haynie, W.H., Ward, W.M., Hansler, R.L., Schanwecker, H.E. and Bell, E.E. J. Opt. Soc. Am., 42 (1952) 559.
Parker, C.J. and Nordberg, M.E. J. Opt. Soc. Am., 49 (1959) 856.
Philpotts, A.R., Thain, W. and Smith, P.G. Anal. Chem., 23 (1951) 268.
Plyler, E.K. and Blaine, L.R. J. Res. Natl. Bur. Stand., 62 (1959) 7.
Porter, J.F. J. Opt. Soc. Am., 51 (1961) 789.
Robinson, D.W. J. Opt. Soc. Am., 49 (1959) 966.
Rochester, J.C.O. and Martin, A.E. Nature, 168 (1951) 785.
Rugg, F.M., Calvert, W.L. and Smith, J.J. J. Opt. Soc. Am., 41 (1951) 32.
Schnurmann, R. and Kendrick, E. Anal. Chem., 26 (1954) 1263.
Schnurmann, R. and Luther, H. Chem. Ing. Tech., 22 (1950) 409.
Siebert, W. Z. Elektrochem., 54 (1950) 521.
Steger, E. Chem. Tech., 12 (1960) 216.
Steger, E. Jena Nachr., 9 (1961) 23.
Suchy, J. and Vasatkova, M. Jenaer Rundsch., 9 (1964) 117.
Tarbet, C.S.C. and Daly, E.F. J. Opt. Soc. Am., 49 (1959) 603.
Tetlow, K.S., Mc Auslan, J., Brimley, K.J. and Price, W.C. J. Sci. Instr., 28 (1951) 161.
Volkmann, H. Sprechsaal Keram., Glas, Email, 94 (1961) 363.
White, J.U., Alpert, N.L., De Bell, A.G. and Chapman, R.M. J. Opt. Soc. Am., 47 (1957) 358.
White, J.U. and Liston, M.D. J. Opt. Soc. Am., 40 (1950) 29 and 93.
Williams, V.Z. Rev. Sci. Instr., 19 (1948) 135.
Wright, N. and Herscher, L.W. J. Opt. Soc. Am., 37 (1947) 211.
Yates, K.P. and Buhl, R.F. J. Opt. Soc. Am., 45 (1955) 192.
Yoshinaga, H. and Yamada, Y. J. Phys. Soc. Japan, 7 (1952) 223.

3. Preparation of the samples

Scattering

Bambauer, H.U., Brunner, G.O. and Laves, F. Am. Mineralogist, 54 (1969) 718.
Phillippi, C.M. A.F.M.L. - Tech. Rep., 67× (1968) 437.
Rank, D.H. and Douglas, A.E. J. Opt. Soc. Am., 38 (1948) 966.
Stewart, J.E. J. Res. Natl. Bur. Stand., 54 (1955) 41.
Valasék, J. Theoretical and Experimental Optics. Wiley, London, 1960, 454 pp.
Zimm, B.H. and Dandliker, W.B. J. Phys. Chem., 58 (1954) 644.

Christiansen effect

Barnes, R.B. and Bonner, L.G. Phys. Rev., 49 (1936) 732.
Christiansen, C. Wied. Ann. Phys., 23 (1884) 298.
Price, W.C. and Tetlow, K.S. J. Chem. Phys., 16 (1948) 1157.
Redfield, D. and Baum, R.L. J. Opt. Soc. Am., 51 (1961) 184.

Grinding

Ard, J.S. Anal. Chem., 25 (1953) 1780.
Bonhomme, J. and Duyckaerts, G. Ind. Chim. Belg., 20 (1955) 145.
Farmer, V.C. Spectrochim. Acta, 8 (1957) 374.
Farmer, V.C. and Russell, J.D. Spectrochim. Acta, 22 (1966) 389.
Hayashi, H., Koshi, K., Hamada, A. and Sakabe, H. Clay Sci., 1 (1962) 99.
Kaufmann, W. Chem. Z., 81 (1957) 8 and 43.
Mackenzie, R.C., Meldau, R. and Farmer, V.C. Ber. Dtsch. Keram. Ges., 33 (1956) 222.
Miller, J.G. and Oulton, T.D. Clays Clay Minerals, 18 (1970) 313.
Rouxhet, P.G. and Brindley, G.W. Clay Minerals, 6 (1966) 219.

Powder film, mull, and pellet method

Ard, J.S. Anal. Chem., 25 (1953) 1743.
Bak, B. and Christensen, D. Acta Chem. Scand., 10 (1956) 692.
Bak, B. and Christensen, D. Z. Anal. Chem., 156 (1957) 115.
Baudet, P., Otten Cl. and Cherbuliez, E. Helv. Chim. Acta, 47 (1964) 2430.
Boldin, A.A. and Vasil'ev, R.F. Zav. Lab., 27 (1961) 819.
Bonhomme, J. and Duyckaerts, G. Ind. Chim. Belg., 20 (1955) 145.
Braunbeck, J. Angew. Chem., 72 (1960) 31.
Clauson-Kaas, N., Nedenskov, P., Bak, B. and Andersen, J.R. Acta Chem. Scand., 8 (1954) 1088.
Crocket, D.S. and Haendler, H.M. Anal. Chem., 31 (1959) 626.
Deodhar, P.S. and Kargaonkar, K.S. Rev. Sci. Instr., 30 (1959) 138.
Dunken, H. and Fink, P. Z. Chemie, 3 (1963) 232.
Durie, R.A. and Szewczyk, J. Spectrochim. Acta, 15 (1959) 593.
Edwards, G.J. Appl. Spectr., 18 (1964) 94.
Elsey, R.D. and Haszeldine, R.N. Chem. Ind. (1954) 1177.
Fahr, E. and Neumann, W.P. Z. Anal. Chem., 149 (1956) 114.
Farmer, V.C. Chem. Ind. (London) (1955) 586.
Farmer, V.C. Spectrochim. Acta, 22 (1966) 1053.
Farmer, V.C. and Russell, J.D. Spectrochim. Acta, 22 (1966) 389.
Feinberg, J.G., Rapson, H.D.C. and Taylor, M.P. Nature, 181 (1958) 763.
Ford, M.A. and Wilkinson, G.R. J. Sci. Instr., 31 (1954) 338.
Garner, H.R. and Packer, H. Appl. Spectrosc., 22 (1968) 122.
Grendon, H.T. and Lovell, H.L. Anal. Chem., 32 (1960) 300.
Gusev, G.M. and Nikolaev, A.V. Dokl. Akad. Nauk. U.S.S.R., 182 (1968) 343.
Hacskaylo, M. Anal. Chem., 26 (1954) 1410.
Hales, J.S. and Kynaston, W. Analyst, 79 (1954) 702.
Hambleton, F.H., Hockey, J.A. and Taylor, J.A.G. Nature, 208 (1965) 138.
Hampel, B. Spectrochim. Acta, 19 (1963) 1276.
Hazebroek, H.F. Spectrochim. Acta, 15 (1959) 1.
Heintz, G. Jenaer Rundsch., 2 (1960) 71.
Hidalgo, A. and Serratosa, J.M. An. Edafol. Fisiol. Veg. (Madrid), 14 (1955) 269.
Higgins, H.G. Austr. J. Chem., 10 (1957) 496.
Hunt, J.M. and Turner, D.S. Anal. Chem., 25 (1953) 1169.
Ingerbrigtson, D.N. and Smith, A.L. Anal. Chem., 26 (1954) 1765.
Jensen, J.B. Dansk. Tidsskr. Farm., 32 (1958) 205.
Jensen, J.B. Dansk. Tidsskr. Farm., 33 (1959) 33.
Kirkland, J.J. Z. Anal. Chem., 151 (1956) 278.
Kopff, R., Rueff, R. and Benoit, H. Bull. Soc. Chim. Fr., 27 (1960) 1694.
Kutzelnigg, W., Nonnenmacher, G. and Mecke, R. Chem. Ber., 93 (1960) 1279.
Lane, J.L., Sen, D.N. and Quagliano, J.V. J. Chem. Phys., 22 (1954) 1855.
Lejeune, R. and Duyckaerts, G. Spectrochim. Acta, 6 (1954) 194.
Levitt, S.R. and Condrate, R.A. Am. Mineralogist, 55 (1970) 522.
Lohr, L.J. and Kaier, R.J. Anal. Chem., 32 (1960) 301.
Longworth, R. and Morawetz, H. Chem. Ind., (1955) 1470.
Lyon, R.J.P. Am. Mineralogist, 48 (1963) 1170.
Mason, W.B. Spectrochim. Acta, 15 (1959) 306.
McDevitt, N.T. and Baun, W.L. Appl. Spectrosc., 14 (1960) 135.
Meloche, V.W. and Kalbus, G.E. J. Inorg. Nucl. Chem., 7 (1958) 104.
Milkey, R.G. Anal. Chem., 30 (1958) 1931.
Morris, J. and Van der Walt, S.J. J. S. Afr. Chem. Inst., 14 (1961) 16.
Ogawa, M. Kagaku no Ryoiki Zokan, 38 (1959) 143.
Olsen, A.L. Anal. Chem., 31 (1959) 321.
Ponder, L.H. Appl. Spectrosc., 18 (1964) 191.
Primas, H. and Günthard, Hs.H. Helv. Chim. Acta, 37 (1954) 360.
Pytlewski, L.L. and Marchesani, V. Anal. Chem., 37 (1965) 618.
Rupke, H. and Neudert, W. Z. Anal. Chem., 170 (1959) 78.
Schiedt, U. and Reinwein, H. Z. Naturforsch., 7b (1952) 270.
Smallwood, S.E.F. and Hart, P.B. Spectrochim. Acta, 19 (1963) 285.
Smethorst, B. and Steele, D. Spectrochim. Acta, 20 (1964) 242.
Sterling, K.J. Anal. Chem., 38 (1966) 1804.
Stimson, M.M. and O'Donnell, M.J. J. Am. Chem. Soc., 74 (1952) 1805.
Svejda, H. Mikrochim. Acta, 3 (1962) 560.
Van der Maas, J.H. and Tolk, A. Spectrochim. Acta, 18 (1962) 235.
Von Dietrich, H. Z. Naturforsch., 118 (1956) 175.
Weinstein, B. Chem. Anal., 49 (1960) 29.
White, J.U., Weiner, S., Alpert, N.L. and Ward, W.M. Anal. Chem., 30 (1958) 1694.

Reflection method

Anderson, S. J. Am. Ceram. Soc., 33 (1950) 45.
Bent, R. and Ladner, W.R. Fuel, 44 (1965) 243.
Blyholder, G. and Richardson, E.A. J. Phys. Chem., 68 (1964) 3882.
Brügel, W. Z. Phys., 128 (1950) 255.

Burns, R.G. and Vaughan, D.J. Am. Mineralogist, 55 (1970) 1576.
Derksen, W.L., Monahan, T.I. and Lawes, A.J. J. Opt. Soc. Am., 47 (1957) 995.
Fahrenfort, J. and Visser, W.M. Spectrochim. Acta, 18 (1962) 1103.
Fahrenfort, J., in: E.R. Lippincott and M. Margoshes (Editors), Proc. 10th Coll. Spectrosc. Int., Univ. Maryland, Spartan Books (1962) 437.
Fahrenfort, J. and Visser, W.M. Spectrochim. Acta, 21 (1965) 1433.
Florinskaya, V.A. Proc. 3rd. All-Union Conf. Glassy State, Leningrad, 1959, (1960) 154.
Florinskaya, V.A. and Pechenkina, R.S. Dokl. Akad. Nauk. S.S.S.R., 85 (1952) 1265.
Florinskaya, V.A. and Pechenkina, R.S. Izv. Akad. Nauk. S.S.S.R., Ser. Fiz., 17 (1953) 649.
Flournoy, P.A. J. Chem. Phys., 39 (1963) 3156.
Ford, C.G. Nature, 212 (1966) 72.
Gilby, A.C. Burr, J. and Crawford, B., J. Phys. Chem., 70 (1966) 1520.
Gottlieb, K. and Schrader, B. Z. Anal. Chem., 216 (1966) 307.
Haas, C. and Ketelaar, J.A.A. Phys. Rev., 103 (1956) 564.
Hansen, W.N. Anal. Chem., 35 (1963) 765.
Hansen, W.N. and Horton, J.A. Anal. Chem., 36 (1964) 783.
Hansen, W.N. Spectrochim. Acta, 21 (1965) 815.
Harrick, N.J. J. Phys. Chem., 64 (1960) 1110.
Harrick, N.J. Anal. Chem., 36 (1964) 188.
Harrick, N.J. and Riederman, N.H. Spectrochim. Acta, 21 (1965) 2135.
Harrick, N.J. Internal Reflection Spectroscopy. Interscience, New York (1967) 327 pp.
Hass, G. J. Opt. Soc. Am., 45 (1955) 945.
Hass, G., Schroeder, H.H. and Turner, A.F. J. Opt. Soc. Am., 46 (1956) 31.
Hidalgo, A. and Serratosa, J.M. Anal. Real. Soc. Espan. Fis. Quimica, 57A (1961) 225.
Hovis, W.A. and Callahan, W.R. J. Opt. Soc. Am., 56 (1966) 639.
Jayme, G. and Rohmann, E.M. Papier, 19 (1965) 497.
Jayme, G. and Rohmann, E.M. Papier, 20 (1966) 1.
Krohmer, P. and Duelli, R. Tips Bodenseewerk, Perkin Elmer 35 UR (1967) 9 pp.
Kronstein, M., Kraushaar, R.J. and Deacle, R.E. J. Opt. Soc. Am., 53 (1963) 458.
Lindberg, J.D. and Snijder, D.G. Am. Mineralogist, 57 (1972) 485.
Lyon, R.J.P. and Burns, E.A. Econ. Geol., 58 (1963) 274.
Markin, E.P. and Sobolev, N.N. Opt. Spectr., 9 (1960) 309.
Miller, R.A. and Campbell, F.J. Rept. Naval Res. Lab. Progr. U.S. (1964) 11.
Pepperhoff, W. Z. Elektrochem., 58 (1954) 520.
Rawlins, T.G.R. Can. Spectr., 9 (1963) 12.
Rigault, G. and Aquilano, D. Period. Mineral (Rome), 33 (1964) 445.
Robinson, T.S. and Price, W.C. Proc. Phys. Soc. (London), 66 B (1953) 969.
Sevchenko, N.A. and Florinskaya, V.A. Dokl. Akad. Nauk. S.S.S.R., 109 (1956) 1115.
Sevchenko, N.A. and Florinskaya, V.A. Opt.i Spektr., 4 (1958) 261.
Sherman, B. Appl. Spectrosc., 18 (1964) 7.
Simon, I. J. Opt. Soc. Amer., 41 (1951) 336.
Simon, I. and Mc Mahon, H.O. J. Chem. Phys., 21 (1953) 23.
Simon, I. and Mc Mahon, H.O. J. Am. Ceram. Soc., 36 (1953) 160.
Su, G.J., Borrelli, N.F. and Miller, A.R. Phys. Chem. Glasses, 3 (1962) 167.
Vincent-Geisse, J. Meth. Phys. Anal. (1965) 108.
Wendlandt, W.W. and Hecht, H.G. Reflectance spectroscopy. Interscience, New York, (1966) 298 pp.
Wilks, P.A. and Hirschfeld, T. in: E.G. Brame (Editor), Applied Spectroscopy Reviews, Marcel Dekker, New York, 1 (1967) p.99.
Zolotarev, V.M. and Kislovskii, L.D. Prib. Tekhn. Eksper., 9 (1964) 175.
Zolotarev, V.M. and Kislovskii, L.D. Opt. i Spektr., 19 (1965) 623.

4. Assignment of infrared bands

Correlations

Adler, H.H. and Kerr, P.F. Am. Mineralogist, 48 (1963) 839.
Colthup, N.B. J. Opt. Soc. Am., 40 (1950) 397.
Dachille, F. and Roy, R. Z. Kristall., 111 (1959) 462.
Farmer, V.C. and Russell, J.D. Spectrochim. Acta, 22 (1966) 389.
Farmer, V.C., Russell, J.D. and Ahlrichs, J.L. Trans. 9th Int. Congr. Soil. Sci., 3 (1969) 101.
Hafner, St. and Laves, F. Z. Kristall., 109 (1957) 204.
Kleber, W. and Moenke, H. Ber. Dtsch. Akad. Wiss., Berlin, 6 (1964) 384, 628.
Kojima, K. Kagaku Ryoiki Zokan, 21 (1956) 49.
Laves, F. and Hafner, St. Z. Kristall., 108 (1956) 52.
Liese, H.C. Am. Mineralogist, 48 (1963) 980.
Rousseaux, J.M., Gomez, L.C., Nathan, Y. and Rouxhet, P.G. Int. Clay Conf. Madrid, I, (1972) 117.
Schwarzmann, E. Naturwissenschaften, 49 (1962) 103.
Stubičan, V. and Roy, R. Z. Kristall., 115 (1961) 200.
White, W.B. Am. Mineralogist, 51 (1966) 275.
Yoganarasimhan, S.R. and Rao, C.N.R. Chem. Anal., 51 (1962) 21.

Adsorbed water and crystal water

Adams, R.M. and Katz, J.J. J. Opt. Soc. Am., 46 (1956) 895.
Adams, R.V. and Douglas, R.W. J. Soc. Glass Tech., 43 (1959) 147.
Andreev, S.N. and Balicheva, T.G. Dokl. Akad. Nauk, 90 (1953) 149.
Benedict, W.S., Claassen, H.H. and Shaw, J.H. J. Res. Natl. Bur. Stand., 49 (1952) 91.
Benessi, H.A. and Jones, A.C. J. Phys. Chem., 63 (1959) 179.
Bernard, M.P. Compt. Rend., 254 (1962) 450.
Bertsch, L. and Halgood, H.W. J. Phys. Chem., 67 (1963) 1621.
Bethell, D.E. and Sheppard, N. J. Chem. Phys., 21 (1953) 1421.
Breger, I.A., Chandler, J.C. and Zubovic, P. Am. Mineralogist, 55 (1970) 825.
Bulanin, M.O. Opt.i Spektr., 2 (1957) 557.
Catalano, E. and Milligan, D.E. J. Chem. Phys., 30 (1959) 45.
Chidambaram, R. J. Chem. Phys., 36 (1962) 2361.
Cranquist, W.T. and Kennedy, J.V. Proc. 15th Natl. Conf. Clays Clay Miner., Pittsburgh, 1966, (1967) 103.
Curcio, J.A. and Petty, C.C. J. Opt. Soc. Am., 41 (1951) 302.
Curtis, N.F. Proc. Chem. Soc., (1960) 410.
Dalby, F.W. and Nielsen, H.H. J. Chem. Phys., 25 (1956) 934.
Draegert, D.A., Stone, N.W.B., Carnutte, B. and Williams, D. J. Opt. Soc. Am., 56 (1966) 64.
Drouard, E. Compt. Rend., 240 (1955) 1700.
Duval, C. and Lecomte, J. J. Chim. Phys., 50 (1953) 664.
Falk, M. and Giguère, P.A. Can. J. Chem., 35 (1957) 1195.
Falk, M. and Ford, T.A. Can. J. Chem., 44 (1966) 1699.
Farmer, V.C. and Russell, J.D. Tr. Faraday Soc., 67 (1971) 2737.
Ferriso, C.C. and Hornig, D.F. J. Am. Chem. Soc., 75 (1953) 4113.
Ferriso, C.C. and Hornig, D.F. J. Chem. Phys., 23 (1955) 1464.
Ford, T.A. and Falk, M. J. Chem., 46 (1968) 3579.
Forslind, E. Swed. Cement, Concrete Res. Inst. Stockholm (1948), Bull. 11: 20 pp.
Frank, H.S. Proc. R. Soc., A 247 (1958) 481.
Frohnsdorff, G.J.C. and Kington, G.L. Proc. R. Soc., A 247 (1958) 469.
Fujita, J., Nakamoto, K. and Kobayashi, M. J. Am. Chem. Soc., 78 (1956) 3963.
Gamo, I. Bull. Chem. Soc. Japan, 34 (1961) 760.
Gamo, I. Bull. Chem. Soc. Japan, 34 (1961) 764.
Giquère, P.A. and Harvey, K.B. Can. J. Chem., 34 (1956) 798.
Glemser, O. Nature, 183 (1959) 1476.
Glemser, O. Angew. Chem., 73 (1961) 785.
Glemser, O. and Hartert, E. Naturwissenschaften, 42 (1955) 534.
Gliemeroth, G. Tonind. Z. Keram. Rundsch., 87 (1963) 505, 529.
Götz, J. and Vozáhlová. Glastech. Ber., 41 (1968) 47.
Goulden, J.D.S. Spectrochim. Acta, 15 (1959) 657.
Graham, J. Rev. Pure Appl. Chem., 14 (1964) 81.
Greinacher, E. Lüttke, W. and Mecke, R. Z. Elektrochem., 59 (1955) 23.
Haas, C. and Hornig, D.F. J. Chem. Phys., 32 (1960) 1763.
Hartert, E. Naturwissenschaften, 43 (1956) 275.
Hendricks, S.B. and Jefferson, M.E. Am. Mineralogist, 23 (1938) 863.
Hornig, D.F., White, H.F. and Reding, F.P. Spectrochim. Acta, 12 (1958) 338.
Jere, G.V. and Patel, C.C. Nature, 194 (1962) 470.
Kalousek, G.L. and Roy, R. J. Am. Ceram. Soc., 40 (1957) 236.
Karyakin, A.V., Petrov, A.V., Gerlit, Yu.B. and Zubrilina, M.E. Teor. Eksp. Khim., 2 (1966) 494.
Kermarrec, Y. Compt. Rend., 258 (1964) 5836.
Keyworth, D.A. Talanta, 8 (1961) 461.
Kinsey, E.L. and Ellis, J.W. Phys. Rev., 49 (1936) 105.
Kiselev, A.V. and Lygin, V.I. Kolloid. Zh., 23 (1961) 157.
Kiselev, A.V. and Lygin, V.I., in: L.H. Little (Editor), Infrared Spectra of Absorbed Species. Acad. Press, London (1966), pp. 352—381.
Kislovskii, L.D. Opt. Spectrosc., 7 (1959) 201.
Lane, J.A. J. Opt. Soc. Am., 56 (1966) 1398.
Lecomte, J. J. Chem. Phys., 50 (1953) 53.
Lecomte, J. J. Chim. Anal., 36 (1954) 118.
Lippincott, E.R. Weir, C.E. and Van Valkenburg, A. J. Chem. Phys., 32 (1960) 612.
Lippincott, E.R., Stromberg, R.R. Grant, W.H. and Cessac, G.L. Science, 164 (1969) 1482.
Lord, R.C. and Merrifield, R.E. J. Chem. Phys., 21 (1953) 166.
Louisfert, J. J. Phys. Radium, 8 (1947) 45.
Low, P.F. and White, J.C. Clays Clay Miner., 18 (1970) 63.
Lucchesi, P.J. and Glasson, W.A. J. Am. Chem. Soc., 78 (1956) 1347.
Luck, W. Fortschr. Chem. Forsch., 4 (1964) 653.
Macey, H.H. Trans. Br. Ceram. Soc., 41 (1942) 73.
Matsumura, O. Mem. Fac. Sci. Kyusyu. Univ., 1 B (1951) 1.
Moulson, A.J. and Roberts, J.P. Trans. Farad. Soc., 57 (1961) 1208.
Murcray, D.G., Murcray, F.H., Williams, W.J. and Leslie, F.E. J. Opt. Soc. Am., 51 (1961) 186.
Nachod, F.C. and Martini, C.M. Appl. Spectrosc., 13 (1959) 45.
Ockman, N. Adv. Phys., 7 (1958) 199.
Ockman, N. and Sutherland, G.B.B.M. Proc. R. Soc., A 247 (1958) 434.
Palmer, C.H. J. Opt. Soc. Am., 50 (1960) 1232.
Parfitt, R.L. and Greenland, D.J. Clay Miner., 8 (1970) 317.
Plyler, E.K. and Acquista, N. J. Opt. Soc. Am., 44 (1954) 505.
Pobeguin, T. Compt. Rend., 248 (1959) 3585.
Price, W.C., Sherman, W.F. and Wilkinson, G.R. Proc. R. Soc. London, 247 (1958) 467.
Prost, R. Proc. Int. Clay Conf., Madrid, 2 (1972) 209.
Prost, R. and Chaussidon, J. Clay Miner., 8 (1969) 143.
Razouk, R.I., Salem, A.Sh. and Mik-

hail, R.Sh. J. Phys. Chem., 64 (1960) 1350.
Redington, R.L. and Milligan, D.E. J. Chem. Phys., 37 (1962) 2162.
Rocchiccioli, C. Chim. Anal. (Paris), 46 (1964) 452.
Rundle, R.E. and Parasol, M. J. Chem. Phys., 20 (1952) 1487.
Ryskin, Ya.I., Stavitskaya, G.P. and Toropov, N.A. Russ. J. Inorg. Chem., 5 (1960) 1315.
Sartori, G., Furlani, C. and Damiani, A. J. Inorg. Nuclear Chem., 8 (1958) 119.
Schiffer, J. and Hornig, D.F. J. Chem. Phys., 49 (1968) 4150.
Scholze, H. Glastech. Ber., 32 (1959) 81, 142, 278, 314, 381, 421.
Scholze, H. Naturwissenschaften, 47 (1960) 226.
Scholze, H. and Dietzel, A. Glastech. Ber., 28 (1955) 375.
Scholze, H. and Dietzel, A. Naturwissenschaften, 42 (1955) 342.
Senior, W.A. and Thompson, W.K. Nature, 205 (1965) 170.
Sidorov, A.N. Opt. i Spectr., 8 (1960) 424.
Stanevich, A.E. and Yaroslavski, N.G. Opt. i Spectr., 10 (1961) 278.
Staufer, F.R. and Walsh, T.E. J. Opt. Soc. Am., 56 (1966) 401.
Stekhanov, A.I. Vestn. Leningrad Univ. 12 (22) Ser. Fiz. Khim., 4 (1957) 62.
Stekhanov, A.I. Bull. Acad. Sci. U.S.S.R., Phys. Ser., 21 (1957) 319.
Swain, C.G. and Bader, R.F.W. Tetrahedron, 10 (1960) 182.
Szymanski, H.A. Stamires, D.N. and Lynch, G.R. J. Opt. Soc. Am., 50 (1960) 1323.
Taylor, J.H., Benedict, W.S. and Strong, J. J. Chem. Phys., 20 (1952) 1884.
Thompson, W.K. Proc. Br. Ceram. Soc., 5 (1965) 143.
Thompson, W.K. Senior, W.A. and Pethica, B.A., Nature, 211 (1966) 1086.
Van der Elsken, J. and Robinson, D.W. Spectrochim. Acta, 17 (1961) 1249.
Van der Marel, H.W. Z. Pflanzernähr., Düng. Bodenk., 114 (1966) 161.
Van Thiel, M., Becker, E.D. and Pimentel, G.C. J. Chem. Phys., 27 (1957) 486.
Vasilevskii, K.P. Opt. i Spectr., 11 (1961) 110.
Waggener, W.C. Anal. Chem., 30 (1958) 1569.
White, J.L. and Burns, A.F. Science, 141 (1963) 800.
White, W.B. Am. Mineralogist, 56 (1971) 46.
Williams, D. Nature, 210 (1966) 194.
Williams, D. and Millett, W. Phys. Rev., 66 (1944) 6.
Yukhnevich, G.V., Karyakin, A.V., Khitarov, N.I. and Senderov, E.E. Geochemistry, (1961) 937.
Zhdanov, S.P., Kiselev, A.V., Lygin, V.I. and Titova, T.I. Zh. Fiz. Khim., 38 (1964) 2408.
Zimmermann, P. and Pimentel, G.C. Adv. Molecul. Spectr., 2 (1962) 726.
Zolotarev, V.M. Opt.i. Spektr., 23 (1967) 816.

Hydrogen bonding (O-H)

Adler, H.H. Am. Mineral., 51 (1966) 258.
Ahlrichs, J.L. Clays and Clay Minerals, 16 (1968) 63.
Albert, N. and Badger, R.M. J. Chem. Phys., 29 (1958) 1193.
Angell, C.L. and Schaffer, P.C. J. Phys. Chem., 69 (1965) 3436.
Boehm, H.P. and Schneider, M. Z. Anorg. Allg. Chem., 301 (1959) 326.
Boutin, H. and Bassett, W. Am. Mineral., 48 (1963) 659.
Brockmann, H. and Franck, B. Naturwiss., 42 (1955) 45.
Burns, R.G. and Strens, R.G.J. Science, 153 (1966) 890.
Cannon, C.G. Spectroch. Acta, 10 (1958) 341.
Carter, J.L., Lucchesi, P.J. and Yates, D.J.C. J. Phys. Chem., 68 (1964) 1385.
Chaussidon, J. Clays Clay Minerals, 20 (1972) 59.
Coggeshall, N.D. J. Chem. Phys., 18 (1950) 978.
Coulson, C.A. Research, 10 (1957) 149.
Draegert, D.A., Stone, N.W.B., Curnutte, B. and Williams, D. J. Opt. Soc. Amer., 56 (1966) 64.
Errera, J. Physica, 4 (1937) 1097.
Farmer, V.C., Russell, J.D., Ahlrichs, J.L. and Velde, B. Bull. Groupe France Argiles, 19 (1967) 5.
Fernandez, M., Serratosa, J.M. and Johns, W.D. Proc. Reunion Hispano-Belga de Miner. Arcilla (Madrid) (1970) 163.
Fijal, J. and Zietkiewicz, J. Bull. Acad. Pol. Sci., Ser. Geol. Geogr., 17 (1969) 7.
Finch, J.N. and Lippincott, E.R. J. Chem. Phys., 24 (1956) 908.
Finch, J.N. and Lippincott, E.R. J. Phys. Chem., 61 (1957) 894.
Flett, M.St.C. Spectrochim. Acta, 10 (1957) 21.
Folman, M. and Yates, D.J.C. Proc. Roy. Soc. (London), A 246 (1958) 32.
Ford, T.A. and Falk, M. Can. J. Chem., 46 (1968) 3579.
Frank, H.S. Proc. Roy. Soc. (London), A 247 (1958) 481.
Glemser, O. Nature, 183 (1959) 943.
Glemser, O. and Hartert, E. Naturwiss., 40 (1953) 552.
Glemser, O. and Hartert, E. Naturwiss., 42 (1955) 534.
Glemser, O. and Hartert, E. Z. Anorg. Allg. Chem., 283 (1956) 111.
Gorman, M. J. Chem. Educ., 34 (1957) 304.
Götz, J. and Vosáhlová, Glastechn. Ber., 41 (1968) 47.
Habgood, H.W. J. Phys. Chem., 69 (1965) 1764.
Hartert, E. and Glemser, O. Naturwiss., 40 (1953) 199.
Hartert, E. and Glemser, O. Z. Elektroch., 60 (1956) 746.
Hayashi, H. and Oinuma, K. Am. Mineralogist, 52 (1967) 1206.
Hofacker, L. and Glemser, O. Naturw., 42 (1955) 369.
Huggins, Ch. and Pimentel, G.C. J. Phys. Chem., 60 (1956) 1615.
Kakitani, S. Kôbutsugaku-Zasshu, 3 (1956) 49.
Kats, A. Philips Res. Repts., 17 (1962) 133, 201.
Keeling, P.S. Trans. Brit. Ceram. Soc., 64 (1965) 137.
Keller, W.D. and Pickett, E.E. Am. J. Sci., 252 (1954) 87.
Kohl, R.A. and Taylor, S.A. Soil. Sci., 91 (1967) 223.
Ledoux, R.L. and White, J.L. Science, 143 (1964) 244.
Lippincott, E.R. and Schroeder, R. J. Chem. Phys., 23 (1955) 1099.
Lisachenko, A.A. and Filimonov, V.N. Dokl. Akad. Nauk S.S.S.R., 177 (1967) 391.
Lord, R.C. and Merrifield, R.E. J. Chem. Phys., 21 (1953) 166.
Low, P.F. and White, J.L. Clays Clay Miner., 18 (1970) 63.
Lygin, V.I. Vestn. Moskov Univ. Ser. Mat. Mekk. Astron. Fiz. Khim., 13 (1958) 223.
Nakamoto, K., Margoshes, M. and Rundle, R.E. J. Am. Chem. Soc., 77 (1955) 6480.
Pampuch, R. and Wilkos, K. Proc. 7th Conf. Silicate Ind., Budapest, 1963, (1965) 179.
Pimentel, G.C. and Sederholm, Ch.H. J. Chem. Phys., 24 (1956) 639.
Pimentel, G.C. and McClellan, A.L. Reinhold (1960) 475 pp.
Pobequin, T. Compt. Rend., 248 (1959) 3585.
Posner, A.S., Stutman, J.M. and Lippincott, E.R. Nature, 188 (1960) 486.
Radwan, M.K. J. Chem. Unit. Arab. Rep., 7 (1964) 55.
Romo, L.A. and Roy, R. Am. Mineralogist, 42 (1957) 165.

Rosenqvist, I.Th. Proc. 11th Natl. Conf. Clays Clay Miner., Ottawa, 1962, (1963) 117.
Rosenqvist, I.Th. and Jörgensen, P. Nature, 204 (1964) 176.
Rouxhet, P.G. Clay Miner., 8 (1970) 375.
Roy, D.M. and Roy, R. Geochim. Cosmochim. Acta, 11 (1957) 72.
Rundle, R.E. and Parasol, M. J. Chem. Phys., 20 (1952) 1487.
Russell, J.D. and Fraser, A.R. Clays Clay Miner., 19 (1971) 55.
Ryskin, Ia.I. Opt. i Spektr., 7 (1959) 177.
Ryskin, Ia.I. and Stavitskaya, G.P. Opt.i Spektr., 7 (1959) 834.
Ryskin, Ia.I. Opt. i Spektr., 12 (1962) 518.
Saumagne, P. and Josien, M.L. Bull. Soc. Chem. Fr., 25 (1958) 813.
Scholze, H. Naturwissenschaften, 47 (1960) 226.
Schwarzmann, E. Naturwissenschaften, 49 (1962) 103.
Schwarzmann, E. and Marsmann, H. Naturwissenschaften, 53 (1966) 349.
Serratosa, J.M. and Bradley, W.F. Nature, 181 (1958) 111.
Serratosa, J.M., Hidalgo, A. and Vinas, J.M. Nature, 195 (1962) 486.
Stekhanov, A.I. Izv. Akad. Nauk. Ser. Fiz., 21 (1957) 311.
Stekhanov, A.I. Izv. Akad. Nauk. Ser. Fiz., 22 (1958) 1109.
Stekhanov, A.I. and Gabrichidze, Z.A. Opt. i Spektr., 11 (1961) 359.
Stekhanov, A.I. and Popova, E.A. Opt. i Spektr., 25 (1968) 378.
Tsubomura, H. J. Chem. Phys., 23 (1955) 2130.
Tsubomura, H. J. Chem. Phys., 24 (1956) 927.
Uytterhoeven, J. and Fripiat, J.J. Bull. Soc. Chim. Fr., 29 (1962) 788.
Van der Elsken and Robinson, D.W. Spectrochim. Acta, 17 (1961) 1249.
Van Thiel, M., Becker, E.D. and Pimentel, G.C. J. Chem., Phys., 27 (1957) 486.
Vergnoux, A.M. J. Chim. Phys. Phys. Chim. Biol., 50 (1953) C75.
Weiss, H.G., Knight, J.A. and Shapiro, I. J. Am. Chem. Soc., 82 (1960) 1262.
West, R. and Baney, R.H. J. Phys. Chem., 64 (1960) 822.
White, J.L., Jelli, A.N., André, J.M. and Fripiat, J.J. Trans. Faraday Soc., 43 (1967) 461.
White, W.B. Am. Mineralogist, 56 (1971) 46.
Wilkins, R.W.T. Mineral. Mag., 36 (1967) 325.
Wolff, R.G. Am. Mineralogist, 50 (1965) 240.

Heating

Chaussidon, J. and Calvet, R. J. Phys. Chem., 69 (1965) 2265.
Dixon, J.B. and Jackson, M.L. Science, 129 (1959) 1616.
Egawa, T. Adv. Clay Sci., 3 (1961) 103.
Ernst, W.G. and Wai, C.M. Am. Mineralogist, 55 (1970) 1226.
Faust, G.T., Hathaway, J.C. and Millot, G. Am. Mineralogist, 44 (1959) 342.
Freund, F. Proc. Int. Clay Conf. Tokio, (1969) 121.
Fripiat, J.J. and Toussaint, F. Nature, 186 (1960) 627.
Fripiat, J.J., Chaussidon, J. and Touillaux, R. J. Phys. Chem., 64 (1960) 1234.
Grim. R.E. and Kulbicki, G. Am. Mineralogist, 46 (1961) 1329.
Hayashi, H. and Oinuma, K. Clay Sci., 1 (1963) 134.
Hayashi, H., Otsuka, R. and Imai, N. Am. Mineralogist, 54 (1969) 1613.
Heller, L., Farmer, V.C. Mackenzie, R.C., Mitchell, B.D. and Taylor, H.F.W. Clay Miner. Bull., 5 (1962) 56.
Hennicke, H.W. and Niesel, K. Tonind. Z. Keram. Rundsch., 89 (1965) 496.
Hidalgo, A., Serratosa, J.M. and Jubrias, M. An. Edaf. Fisiol. Veg., 15 (1956) 607.
Kato, C. Yogyo Kyokai Shi, 70 (1962) 81.
Kato, C. Yogyo Kyokai Sci, 70 (1962) 124.
Keller, W.D., Pickett, E.E. and Reesman, A.L. Proc. Int. Clay Conf., Jerusalem, 1 (1966) 75.
Kolterman, M. and Müller, K.P. Tonind. Z. Rundsch., 89 (1965) 406.
Miller, J.G. J. Phys. Chem., 65 (1961) 800.
Pampuch, R. Polska Akad. Nauk. Oddz. Krakowie, Kom. Nauk. Miner. Pr. Miner., 6 (1966) 53.
Pampuch, R. and Wilkos, K. Proc. 7th Conf. Silic. Ind., Budapest, 1963, (1965) 179.
Rouxhet, P.G. Am. Mineralogist, 55 (1970) 841.
Roy, R., Roy, D.M. and Francis, E.E. J. Am. Ceram. Soc., 38 (1955) 38.
Russell, J.D. and Farmer, V.C. Clay Miner. Bull., 5 (1964) 443.
Serratosa, J.M. Am. Mineralogist, 45 (1960) 1101.
Stubičan, V. Miner. Mag., 32 (1959) 38.
Stubičan, V. and Günthard, Hs.H. Nature, 179 (1957) 542.
Stubičan, V. and Roy, R. J. Phys. Chem., 65 (1961) 1348.
Tarasevich, Yu.I. Dopov. Akad. Nauk Ukr. R.S.R., Ser. B., 32 (1970) 938.
Tsuzuki, Y. and Nagasawa, K. Clay Sci., 3 (1969) 87.
Vedder, W. and Wilkins, R.W.T. Am. Mineralogist, 54 (1969) 482.
Zhdanov, S.P., Kiselev, A.V. Lygin, V.I. and Titova, T.I. Dokl. Akad. Nauk. S.S.S.R., 150 (1963) 584.

Deuteration

Adams, R.N. and Katz, J.J. J. Opt. Soc. Am., 46 (1956) 895.
Bain, O. and Giguère, P.A. Can. J. Chem., 33 (1955) 527.
Benedict, W.S., Gailar, N. and Plyler, E.K. J. Chem. Phys., 24 (1956) 1139.
Berglund-Larsson, U. Acta Chem. Scand., 10 (1956) 701.
Buchanan, R.A., Kinsey, E.L. and Caspers, H.H. Chem. Phys., 36 (1962) 2665.
Bulanin, M.O. Optik.i Spektr., 2 (1957) 557.
Busing, W.R. J. Chem. Phys., 23 (1955) 933.
Catalano, E. and Milligan, D.E. J. Chem. Phys., 30 (1959) 45.
Draegert, D.A., Stone, N.W.B., Curnutte, B. and Williams, D. J. Opt. Soc. Am., 56 (1966) 64.
Farmer, V.C. and Russell, J.D. Trans. Faraday Soc., 67 (1971) 2737.
Friedman, I. and Smith, R.L., Geochim. Cosmochim. Acta, 15 (1958) 218.
Fripiat, J.J. and Gastuche, M.C. Bull. Soc. Chim. Fr., 25 (1958) 626.
Fripiat, J.J., Gastuche, M.C. and Brichard, R. J. Phys. Chem., 66 (1962) 805.
Fry, D.L., Mohan, P.V. and Lee, R.W. J. Opt. Soc. Am., 50 (1960) 1321.
Gaunt, J. Analyst, 79 (1954) 580.
Giquère, P.A. and Harvey, K.B. Can. J. Chem., 34 (1956) 798.
Gribina, I.A. and Tarasevich, Y.I. Teor. Eksp. Khim., 8 (1972) 512.
Hexter, R.M. J. Chem. Phys., 34 (1961) 941.
Hexter, R.M. J. Chem. Phys., 38 (1963) 1024.
Hornig, D.F., White, H.F. and Reding, F.P. Spectrochim. Acta, 12 (1958) 338.
Hoyer, H. Monatsh. Chem., 90 (1959) 357.
Jones, L.H. J. Chem. Phys., 22 (1954) 217.
Jones, R.N. and MacKenzie, M.A. Talanta, 3 (1960) 356.
Jörgensen, P. and Rosenqvist, I.Th. Norsk Geol. Tidsskr., 43 (1963) 497.
Lecomte, J., Ceccaldi, M. and Roth, E. J. Chim. Phys., 50 (1953) 166.
Ledoux, R.L. and White, J.L. Science, 145 (1964) 47.
Leonard, R.A. Proc. Soil Sci. Am., 34 (1970) 339.

Lohman, J.B., Reding, F.P. and Hornig, D.F. J. Chem. Phys., 19 (1951) 252.
Lygin, V.I. and Kiselev, A.V. Kolloid. Zh., 23 (1961) 299.
Mann, J. and Marrinan, H.J. Trans. Faraday Soc., 52 (1956) 481.
Mortland, M.M., Fripiat, J.J., Chaussidon, J. and Uytterhoeven, J. J. Phys. Chem., 67 (1963) 248.
Nachod, F.C. and Martini, C.M. Appl. Spectr., 13 (1959) 45.
Newman, C., Polo, S.R. and Wilson, M.K. Spectrochim. Acta, 15 (1959) 793.
Nordman, C.E. and Lipscomb, W.N. J. Chem. Phys., 19 (1951) 1422.
Pliskin, W.A. and Eischens, R.P. Z. Phys. Chem., Neue Folge, 24 (1960) 11.
Reding, F.R. and Hornig, D.F. J. Chem. Phys., 19 (1951) 594.
Rocchiccioli, C. Compt. Rend., 253 (1961) 838.
Romo, L.A. J. Phys. Chem., 60 (1956) 987.
Rosenqvist, I.Th. and Jörgensen, P. Nature, 204 (1964) 176.
Rouxhet, P.G. Clay Miner., 8 (1970) 375.
Rouxhet, P.G. Am. Mineralogist, 55 (1970) 841.
Roy, D.M. and Roy, R. Proc. 4th Natl. Conf. Clays Clay Minerals, Pennsylv. State Univ., 1955, (1956) 82.
Roy, D.M. and Roy, R. Geochim. Cosmochim. Acta, 11 (1957) 72.
Russell, J.D., Farmer, V.C. and Velde, B. Mineral. Mag., 37 (1970) 869.
Saumagne, P. and Josien, M.L. Bull. Soc. Chim. Fr., 25 (1958) 813.
Stubičan, V. and Roy, R. J. Phys. Chem., 65 (1961) 1348.
Swain, C.G. and Bader, R.F.W. Tetrahedron, 10 (1960) 182.
Swenson, Ch.A. Spectrochim. Acta, 21 (1965) 987.
Szymanski, H.A., Stamires, D.N. and Lunch, G.R. J. Opt. Soc. Am., 50 (1960) 1323.
Tarasevich, Iu.I. and Ovcharenko, F.D. Dokl. Akad. Nauk S.S.S.R., 184 (1969) 142.
Tarasevich, Iu.I. and Ovcharenko, F.D. Dokl. Akad. Nauk S.S.S.R., 187 (1969) 372.
Tarasevich, Iu.I. Proc. Int. Clay Conf., Tokyo, 1 (1969) 261.
Van der Elsken, J. and Robinson, D.W. Spectrochim. Acta, 17 (1961) 1249.
Van Panthaleon van Eck, C.L., Mendel, H. and Fahrenfort, J. Proc. R. Soc. London, A 247 (1958) 472.
Wada, K. Soil Sci. Plant Nutr., 12 (1966) 176.
Wada, K. Clay Miner., 7 (1967) 51.
Waggener, W.C. Anal. Chem., 30 (1958) 1569.
Wagner, E.L. and Hornig, D.F. J. Chem. Phys., 18 (1950) 296.
Waldron, R.D. J. Chem. Phys., 26 (1957) 809.
Wei, Y.K. and Bernstein, R.B. J. Phys. Chem., 63 (1959) 738.
Yellin, W. and Courchene, W.L. Nature, 219 (1968) 852.
Zolotarev, V.M. Opt. i Spektr., 23 (1967) 816.

Electric vector and transition moment of the vibration

Barrow, G.M. J. Chem. Phys., 21 (1953) 219.
Bassett., W.A. Bull. Geol. Soc. Am., 71 (1960) 449.
Bird, G.R. and Parrish, M. J. Opt. Soc. Am., 50 (1960) 886.
Burns, R.G. Mineral. Mag., 35 (1966) 715.
Chaussidon, J. Clays Clay Miner., 20 (1972) 59.
Conn, G.K.T. and Eaton, G.K. J. Opt. Soc. Am., 44 (1954) 553.
Edwards, D.F. and Bruemmer, M.J. J. Opt. Soc. Am., 49 (1959) 860.
Elliott, A., Ambrose, E.J. and Temple, R.B. J. Chem. Phys., 16 (1948) 877.
Farmer, V.C. and Russell, J.D. Spectrochim. Acta, 20 (1964) 1149.
Farmer, V.C. and Russell, J.D. Trans. Faraday Soc., 67 (1971) 2737.
Faye, G.H., Manning, P.G. and Nickel, E.H. Am. Mineralogist, 53 (1968) 1174.
Grechushnikov, B.N. and Petrov, I.P. Opt.i Spektr., 14 (1963) 305.
Gribov, L.A. Opt. i Spektr., 9 (1960) 658.
Hanisch, K. N. Jahrb. Mineral., Monatsh. (1966) 109.
Hanisch, K. and Zemann, J. N. Jahrb. Mineral., Monatsh. (1966) 19.
Harrick, N.J. J. Opt. Soc. Am., 49 (1959) 815.
Hass, M. and Sutherland, G.B.B.M. Proc. R. Soc. London, A 236 (1956) 427.
Kats, A. Philips Res. Rep., 17 (1962) 135 and 201.
Lagemann, R.T. and Miller, T.G. J. Opt. Soc. Am., 41 (1951) 1063.
Louisfert, J. J. Phys. Radium, 8 (1947) 45.
Mann, J. and Thompson, H.W. Proc. R. Soc., A 211 (1952) 168.
Mara, R.T. and Sutherland, G.B.B.M. J. Opt. Soc. Am., 43 (1953) 1100.
Matthieu, J.P. J. Phys. Radium, 16 (1955) 219.
Meier, R. and Günthard, H.H. J. Opt. Soc. Am., 49 (1959) 1122.
Mitsuishi, A., Yamada, Y., Fujita, S. and Yoshinaga, H. J. Opt. Soc. Am., 50 (1960) 433.
Rigault, G. and Aguilano, D. Period. Miner. (Rome), 33 (1964) 445.
Ruprecht. G., Ginsberg, D.M. and Leslie, J.D. J. Opt. Soc. Am., 52 (1962) 665.
Schwarzmann, E. and Marsmann, H. Naturwissenschaften, 53 (1966) 349.
Serratosa, J.M. and Bradley, W.F. Nature, 181 (1958) 111.
Serratosa, J.M. and Bradley, W.F. J. Phys. Chem., 62 (1958) 1164.
Serratosa, J.M., Hidalgo, A. and Vinas, J.M. Nature, 195 (1962) 486.
Serratosa, J.M. and Hidalgo, A. Appl. Opt., 3 (1964) 315.
Serratosa, J.M. Nature, 208 (1965) 679.
Serratosa, J.M. Proc. 14th Natl. Conf. Clays Clay Miner., Berkeley, 1965, (1966) 385.
Serratosa, J.M. Clays Clay Miner., 16 (1968) 93.
Smakula, A. Opt. Acta, 9 (1962) 205.
Stewart. J.E. J. Chem. Phys., 23 (1955) 986.
Takahashi, S. J. Opt. Soc. Am., 51 (1961) 441.
Tsuboi, M. Bull. Chem. Soc. Japan, 23 (1950) 83.
Webber, D.S. Phys. Rev., 96 (1954) 846.
Weismiller, R.A., Ahlrichs, J.L. and White, J.L. Proc. Soil Sci. Soc. Am., 31 (1967) 459.
Wilkinson, G.R., Price, W.C. and Bradbury, E.M. Spectrochim. Acta, 14 (1959) 284.

5. Quantitative analysis

Anonymous. ASTM-Stand., 31 (1964) 426.

Aucott, J.W. and Marshall, M. Mineral. Mag., 37 (1969) 256.

Biernacka, T. Chem. Anal. (Warsawa), 10 (1965) 1075.

Bonhomme, J. Spectrochim. Acta, 7 (1955) 32.

Bonhomme, J. and Duyckaerts, G. Ind. Chem. Belg., 20 (1955) 145.

Bradley, K.B. and Potts, W.J. Appl. Spectrosc., 12 (1958) 77.

Brown, T.L. J. Phys. Chem., 61 (1957) 820.

Brown, T.L. and Rogers, M.T. J. Am. Chem. Soc., 79 (1957) 577.

Brown, T.L. Chem. Rev., 58 (1958) 581.

Browning, R.S., Wiberley, S.E. and Nachod, F.C. Anal. Chem., 27 (1955) 7.

Cannon, C.G. and Butterworth, I.S.C. Anal. Chem., 25 (1953) 168.

Chulanovskii, V.M. Usp. Fiz. Nauk, 68 (1959) 147.

Collier, G.L. and Singleton, F. J. Appl. Chem. (London), 6 (1956) 495.

Crawford, B.J. J. Chem. Phys., 29 (1958) 1042.

Diamond, W.J. Appl. Spectrosc., 12 (1958) 10.

Dobrowsky, A. Monatsh. Chem., 77 (1947) 185.

Duyckaerts, G. Bull. Soc. R. Sci., Liège, 21 (1952) 7.

Duyckaerts, G. Spectrochim. Acta, 7 (1955) 25.

Duyckaerts, G. Analist, 84 (1959) 201.

Foreman, R.W. and Jackson, W.J. Instruments, 22 (1949) 499.

Fraser, R.T.M. Anal. Chem., 31 (1959) 1602.

French, R.O., Wadsworth, M.E., Cook, M.A. and Cutler, I.B. J. Phys. Chem., 58 (1954) 805.

Friedel, R.A. and Queiser, I.A. Anal. Chem., 29 (1957) 1362.

Friedman, I. and Smith, Geochim. Cosmochim. Acta, 15 (1958) 218.

Fry, D.L., Mohan, P.V. and Lee, R.V. J. Opt. Soc. Am., 50 (1960) 1321.

Giese, A.T. and French, C.S. Appl. Spectrosc., 9 (1955) 78.

Gottschalk, G. and Dehmel, P. Z. Anal. Chem., 163 (1958) 330.

Götz, J. and Vosáhlová. Glastech. Ber., 41 (1968) 47.

Hammer, C.F. and Roe, H.R. Anal. Chem., 25 (1953) 668.

Hardy, A.C. and Mansfield Young, F. J. Opt. Soc. Am., 38 (1948) 854.

Hawkes, J.C. J. Appl. Chem., 7 (1957) 123.

Heighl, J.J., Bell, M.F. and White, J.U. Anal. Chem., 19 (1947) 293.

Hoffman, K. and Fischer, L. Chem. Ing. Tech., 27 (1955) 604.

Hollenberg, J.L. and Dows, D.A. J. Chem. Phys., 34 (1961) 1061.

Hunt, J.M. and Turner, D.S. Anal. Chem., 25 (1953) 1169.

Kirkland, J.J. Anal. Chem., 27 (1955) 1537.

Kirkland, J.J. Anal. Chem., 29 (1957) 1127.

Kössler, I. and Matyska, B. Chem. Zvesti, 6 (1952) 99.

Kössler, I. Chem. Listy, 49 (1955) 1244.

Kuentzel, L.E. Anal. Chem., 27 (1955) 301.

Lehmann, H. and Dutz, H. Tonind. Z. Keram. Rundsch., 83 (1959) 219.

Liebhafsky, H.A. and Pfeiffer, H.G. J. Chem. Educ., 30 (1953) 450.

Lipinski, D. and Schwiete, H.E. Glas Email Keram. Tech., 14 (1963) 325.

Lyon, R.J.P. and Tuddenham, W.M. Mineral. Eng., 11 (1959) 1233.

Lyon, R.J.P., Tuddenham, W.M. and Thompson, C.S. Econ. Geol., 54 (1959) 1047.

Manno, R.P., Paraskevopoulos, N. and Matsuguma, H.J. Appl. Spectrosc., 13 (1959) 57.

Martin, A.E. Trans. Faraday Soc., 47 (1951) 1182.

Martin, A.E. Nature, 180 (1957) 231.

Martin, A.E. Nature, 181 (1958) 1195.

Mc Donald, I.R.C. Nature, 174 (1954) 703.

Nichols, A.B. J. Am. Pharm. Assoc., 42 (1953) 223.

Nielsen, J.R., Thornton, V. and Brock Dale, E. Rev. Mod. Phys., 16 (1944) 307.

Oelert, H.H. Erdöl Kohle, 18 (1965) 876.

Otvos, J.W., Stone, H. and Haro, W.R. Spectrochim. Acta, 9 (1957) 148.

Pemsler, J.P. Rev. Sci. Instr., 28 (1957) 274.

Penner, S.S. and Weber, D. J. Chem. Phys., 19 (1951) 1351.

Penner, S.S. and Weber, D. J. Chem. Phys., 19 (1951) 1361.

Perkin, F.H. J. Opt. Soc. Am., 38 (1948) 72.

Perry, J.A., Bain, G.H. and Traver, W.B. Appl. Spectrosc., 10 (1956) 191.

Perry, J.A. and Bain, G.H. Anal. Chem., 29 (1957) 1123.

Philpotts, A.R. and Maddams, W.F. Proc. Int. Symp. Microchem., Birmingham Univ., 1958, (1959) 373.

Powell, H. J. Appl. Chem., 6 (1956) 488.

Robinson, D.Z. Anal. Chem., 23 (1951) 273.

Robinson, D.Z. Anal. Chem., 24 (1952) 619.

Rouxhet, P.G. Clay Miner., 8 (1970) 375.

Schnurmann, R. and Kendrick, E. Anal. Chem., 26 (1954) 1263.

Scholze, H. Fortschr. Mineral., 38 (1960) 122.

Schowtka, K.H. and Kriegsmann, H. Jena Nachr. VEB Carl Zeiss, 9 (1961) 3.

Schubert, K.D. Bergakademie, 17 (1965) 70.

Shrewsbury, D.D. Unicam Instr. Ltd., Arbury Works, Cambridge, 6 (1958) 1.

Steele, W.C. and Wilson, M.K. Am. Soc. Testing Materials, Spec. Tech. Publ., 269 (1960) 185.

Stewart, J.E. J. Res. Natl. Bur. Stand., 54 (1955) 41.

Strong, F.C. Anal. Chem., 24 (1952) 338.

Susi, H. and Rector, H.E. Anal. Chem., 30 (1958) 1933.

Swinehart, D.F. J. Chem. Ed., 39 (1962) 333.

Van der Marel, H.W. Silic. Ind., 25 (1960) 23 and 76.

Van der Marel, H.W. Acta Univ. Carolinae Geol. Suppl. 3, Prague, (1961) 23.

Van der Marel, H.W. Contrib. Mineral. Petrol., 12 (1966) 96.

Wada, K. and Greenland, D.J. Clay Miner., 8 (1970) 241.

Ward, W.R. and Philpotts, A.R. J. Appl. Chem., 8 (1958) 265.

Ward, W.R. J. Appl. Chem., 10 (1960) 277.

Washburn, W.H. and Mahoney, M.J. Anal. Chem., 30 (1958) 1053.

White, L. and Barrett, W.J. Anal. Chem., 28 (1956) 1538.

Wiberley, S.W., Sprague, J.W. and Campell, J.E. Anal. Chem., 29 (1957) 210.

Willard, H.H., Merritt, L.L. and Dean, J.A. Instrumental Methods of Analysis. D. van Nostrand, (1958) 626 pp.

Williams, V.Z., Coates, V.J. and Gaarde, F., Anal. Chem., 27 (1955) 2017

Wright, N. Ind. Eng. Chem. Anal. Ed., 13 (1941) 1.

6. Clay- and related minerals

General

Abbey, S., 1964. Mineralogical analysis of rocks. A possible application of infrared spectroscopy. Can. Spectrosc., 9: 62—65.

Adler, H.H., 1963. Some basic considerations in the application of infrared spectroscopy to mineral analysis. Econ. Geol., 58: 558—568.

Adler, H.H., Kerr, P.F., Bray, E.E., Stevens, N.P., Hunt, J.M., Keller, W.D. and Pickett, E.E., 1950. Infrared spectra of reference clay minerals. Am. Pet. Inst., Proj. 49, Prelim. Rep., 8: 146 pp.

Afremov, L.C. and Vandeberg, J.I., 1966. High-resolution spectra of inorganic pigments and extenders in the mid-infrared region from 1500 cm^{-1} to 200 cm^{-1}. J. Paint Technol., 36: 169—202.

Ahlrichs, J.L., 1968. Hydroxyl stretching frequencies of synthetic Ni-Al- and Mg-hydroxy interlayers in expanding clays. Clays Clay Miner., 16: 63—71.

Ahlrichs, J.L., Russell, J.D., Harter, R.D. and Weismiller, R.A., 1966. Infrared spectroscopy of clay mineral systems. Proc. Indiana Acad. Sci., 75: 247—255.

Akhmanova, M.V., 1959. Infrared absorption spectra of minerals. Usp. Khim., 28: 312—335.

Akhmanova, M.V., Karyakin, A.V. and Yukhnevich, G.V., 1963. Determination of hydroxyl groups in silicate minerals by infrared spectrophotometry. Geochemistry, 61: 596—600.

Aleshin, S.N., 1964. Properties of soil clay minerals, saturated with different cations (determined by infrared and data analysis). Fiz. Khim. Biol. Mineral. Pochv. (Akad. Nauk, S.S.S.R.) Dokl. K. 8 Mezhdunar. Kongr. Pochvoved., Bucharest: 219—228.

Alexanian, C., Morel, P. and Le Bouffant, L., 1966. The infrared absorption spectrum of natural minerals. Bull. Soc. Fr. Ceram., 71: 3—38.

Aronson, J.R., Emslie, A.G., Allen, R.V. and McLinden, H.G., 1967. Studies of the middle- and far- infrared spectra of mineral surfaces for application in remote compositional mapping of the moon and planets. J. Geophys. Res., 72: 687—703.

Beutelspacher, H., 1956. Infrarot Untersuchungen an Bodenkolloiden. 6th Int. Congr. Soil Sci. Paris, 1: 329—336.

Beutelspacher, H., 1956. Beiträge zur Ultrarotspektroskopie von Bodenkolloiden. Landwirtsch. Forsch., 7: 74—80.

Beutelspacher, H. and Van der Marel, H.W., 1961. Über die amorphen Stoffe in den Tonen verschiedener Böden. Acta Univ. Carolinae, Geol. Suppl., 1: 97—114.

Beutelspacher, H. and Fiedler, E., 1963. Einflusz verschiedener Vorbehandlungen auf aufweitbare Tonminerale und ihre Identifizierung. Landbauforsch. Völkenrode, 13: 85—98.

Beutelsbacher, H., Rietz, E. and Van der Marel, H.W., 1972. Bodenkunde. In: H. Volkman (Editor), Handbuch der Infrarot Spektroskopie. Verlag. Chemie, pp. 297—327.

Bishui, B.M. and Prasad. J., 1960. Infrared spectra of some clay minerals and related structures. Cent. Glass Ceram. Res. Inst. Bull., (India), 7: 97—109.

Bishui, B.M. and Prasad, J., 1965. Physiochemical properties of some Indian clays. IV: infrared absorption spectra. Centr. Glass Ceram. Res. Inst., Bull. (India), 12: 1—5.

Biswas, M.A. and Basak, A.K., 1961. Infrared absorption analysis of some East Pakistani clays. Pak. J. Sci. Ind. Res., 4: 118—120.

Bokii, G.B. and Plyusnina, I.I., 1958. Infrared absorption spectra of cyclic silicates in the 7-21 wavelength range. Nauchn. Dokl. Vyssh. Shk., Geol-Geogr., 1958: 116—123.

Bolydrev, A.I. and Komarov, I.A., 1964. Recording infrared spectra of mineral and humic constituents of soil absorption complexes. Dokl. Ross. S-kh. Akad., 99: 25—31.

Bonhomme, J., 1955. Contribution à l'analyse qualitative par les spectres d'absorption infrarouge des poudres. II. Etude experimentale. Spectrochim. Acta, 7: 32—44.

Bonhomme, J. and Duyckaerts, G., 1955. L'importance de la granulometrie et spectrometrie infrarouge des poudres. Ind. Chem. Belg., 20: 145—150.

Bradshaw, P.M.D., 1967. Measurement of the modal composition of granitic rock powder by point-counting, infrared spectroscopy and X-ray diffraction. Mineral. Mag., 36: 94—100.

Brown, G., 1954. A qualitative study of some gleyed soils from N.W. England. J. Soil. Sci., 5: 145—155.

Buswell, A.M., Krebs, K. and Rodebush, W.H., 1937. Absorption bands of hydrogels between 2.5 and 3.5 μ. J. Am. Chem. Soc., 59: 2603—2605.

Cabannes-Ott, Ch., 1960. Sur la constitution de quelques oxydes métalliques hydratés (thermogravimétrie et spectrographie infrarouge). Ann. Chem., 5: 905—960.

Chapman, D. and Nacey, J.F., 1958. A rapid spectroscopic method for the determination of water in glycerol. Analyst, 83: 377—379.

Cheng, F.F., 1959. Infrared analysis of inorganic anions. Diss. Abstr., 19: 2733.

Clark, S.P., 1957. Absorption spectra of some silicates in the visible and near infrared. Am. Mineralogist, 42: 732—742.

Dachille, F. and Roy, R., 1959. The use of infrared absorption and molar refractivities to check coordination. Z. Krist., 111: 462—470.

De, S.K., 1968. Infrared spectra of an alluvial soil profile. Indian J. Agric. Chem., 1: 35—42.

De, S.K., 1969. Infrared spectra of a red soil profile. Indian J. Agric. Chem., 2: 13—17.

De Keyser, W.L., 1965. Applications de la spectrométrie infrarouge à l'étude de matériaux céramiques. Bull. Soc. Fr. Céram., 68: 43—50.

De Mumbrum, L.E., 1960. Crystalline and amorphous soil minerals of the Mississippi coastal terrace. Proc. Soil Sci. Soc. Am., 24: 185—189.

Dospekhov, B.A. and Shaimukhametova, A.A., 1965. The study of soil particles less than 0.01 mm by thermographic and infrared spectroscopic methods. Izv. Timiryazevsk. S-kh. Akad.: 158—164.

Duval, C. and Lecomte, J., 1952. Quelques reactions à l'état solide, étudiées au moyen des spectres d'absorption infrarouges. Compt. Rend., 234: 2445—2447.

Duyckaerts, G., 1959. The infrared analysis of solid substances: A review. Analyst, 84 201—214.

Farmer, V.C., 1968. Infrared spectroscopy in clay mineral studies. Clay Miner., 7: 373—387.

Farmer, V.C., 1971. The characterization of absorption bands in clays by infrared spectroscopy. Soil Sci., 112: 62—68.

Farmer, V.C. and Russell, J.D., 1964. The infrared spectra of layer silicates. Spectrochim. Acta, 20: 1149—1173.

Farmer, V.C. and Russell, J.D., 1967. Infrared absorption spectrometry in clay studies. Proc. 15th Natl. Conf. Clays Clay Miner., Pittsburgh, 1966: 121—142.

Farmer, V.C., Russell, J.D., Ahlrichs, J.L. and Velde, B., 1967. Vibrations du groupe hydroxyle dans les silicates en couches. Bull. Groupe Fr. Argiles, 19: 5—10.

Farmer, V.C., Russell, J.D. and Ahlrichs, J.L., 1969. Characterization of clay mineral by infrared spectroscopy. Trans. 9th Int. Congr. Soil Sci., 3: 101—110.

Faust, G.T., Hathaway, J.C. and Millot, G., 1959. A restudy of stevensite and related minerals. Am. Mineralogist, 44: 342—370.

Ferraro, J.R., 1961. Inorganic infrared spectroscopy. J. Chem. Educ., 38: 201—208.

Fieldes, M., Walker, I.K. and Williams, P.P., 1956. Infrared absorption spectra of clay soils. N.Z. J. Sci. Technol., 38 B: 31—43.

Fieldes, M., Furkert, R.J. and Wells, N., 1972. Rapid determination of constituants of whole soils using infrared absorption. N.Z. J. Sci., 15: 615—627.

Flaig, W. and Beutelspacher, H., 1961. Infrarotspektren von anorganischen Bodenbestandteilen. Leitz. Mitt. Wiss. Tech., 1: 199—202.

Flaig, W., Beutelspacher. H. and Söchtig, H., 1962. Ein Beitrag aus dem Bereich der molekularen Dimension zur Morphologie des Bodens. Z. Pflanzenernähr. Düng. Bodenk., 98: 225—231.

Florinskaya, V.A. and Pechenkina, R.S., 1961. Application of infrared spectroscopy to the study of polycrystalline substances. Sov. Phys. Cryst., 6: 103—108.

Fripiat, J.J., 1960. Application de la spectroscopie infrarouge à l'étude des minéraux argileux. Bull. Gr. Fr. Argiles, 12: 25—41.

Garcia, S.G., Beutelspacher, H. and Flaig, W., 1956. Espectografia infrarroja de arcillas y minerales estructuralmente analogos. An. Real Soc. Esp. Fis. Quim., 52 B: 369—376.

Gillis, R.G., 1958. The intensity of infrared absorption bands - A bibliography. Aust. Commonw., Dept. Supply Def. Stand. Lab. Tech. Mem., 2: 48 pp.

Giménez, F.A. and Jaritz, G., 1966. Amorphe und kristalline Bestandteile einiger typischer Bodenbildungen Skandinaviens. Z. Pflanzenernähr. Düng. Bodenkd., 114: 27—46.

Gonzalez, G.S., Beutelspacher, H. and Flaig, W., 1956. Espectrografia infrarroja de arcillas y minerales estructuralmente analogos. An. Real Soc. Esp. Fis. Quim., 52 B: 396—376.

Gusev, V.V., Boldyrev, A.I. and Kalinina, Iu.M., 1969. The investigation of Kimmeridgian clays of the Kertch peninsula, investigated by the method of infrared spectroscopy. Dokl. Akad. Nauk S.S.S.R., 186: 181—184.

Hayashi, H. and Oinuma, K., 1964. Behaviors of clay minerals in treatment with hydrochloric acid, formamide and hydrogen peroxide. Clay Sci., 2: 75—91.

Hayashi, H. and Oinuma, K., 1967. Study on clay minerals by infrared analysis. Professor Hidekata Shibata Memorial Volume: 134—139.

Hayashi, H. and Oinuma, K., 1967. Recent progress in infrared study of clay minerals. J. Clay Sci., Japan, 6: 29—40.

Hidalgo, A. and Serratosa, J.M., 1955. Espectros de absorcion infrarroja de minerales de la arcille obtenidos mediante la technica de comprimidos de KBr. An. Edafol. Fisiol. Veg. (Madrid) 14: 269—293.

Hidalgo, A. and Serratosa, J.M., 1956. Espectros de absorcion infrarroja de minerales de la arcilla en la region de 600 a 350 K. An. Real Soc. Esp. Fis. Quim., 52 B: 101—104.

Hidalgo, A. and Serratosa, J.M., 1961. Espectros de reflexion infrarroja de minerales de la arcilla. An. Real Soc. Esp. Fis. Quim., 57 A: 225—230.

Hidalgo, A. and Serratosa, J.M., 1965. Application of infrared absorption spectra to the study of clay minerals. Ion, 16: 645—651.

Hidalgo, A., Serratosa, J.M. and Jubrias, M., 1956. Espectros de absorcion infrarroja de minerales de la arcilla sometidos a tratamiento termico. An. Edafol. Fisiol. Veg. (Madrid), 15: 607—627.

Hunt, J.M., 1950. Infrared spectra of reference clay minerals. III. Infrared spectra of clay minerals. Am. Pet. Inst. Proj. 49, Prelim. Rep., 8: 105—121.

Hunt, J.M. and Turner, D.S., 1953. Determination of mineral constituents of rocks by infrared spectroscopy. Anal. Chem., 25: 1169—1174.

Hunt, J.M., Wisherd, M.P. and Bonham, L.C., 1950. Infrared absorption spectra of minerals and other inorganic compounds. Anal. Chem., 22: 1478—1497.

Kalbus, G.E., 1957. Infrared examination of inorganic materials. Diss. Abstr., 17: 2413—2414.

Kanno, I., Tokudome, S., Arimura, S. and Onikura, Y., 1965. Genesis and characteristics of brown forest soils. Soil Sci. Plant Nutr., 11: 141—150.

Kato, Y., 1965. Mineralogical composition of silt and clay fractions. Soil Sci. Plant Nutr., 11: 62—73.

Keeling, P.S., 1963. Infrared absorption characteristic of clay minerals. Trans. Br. Ceram. Soc., 62: 549—563.

Keeling, P.S., 1965. The examination of clays and clay minerals by infrared spectrophotometry. Proc. Soc. Anal. Chem., 2: 125—126.

Keller, W.D. and Pickett, E.E., 1949. Absorption of infrared radiation by powdered silica minerals. Am. Mineralogist, 34: 855—868.

Keller, W.D. and Pickett, E.E., 1950. The absorption of infrared radiation by clay minerals. Am. J. Sci., 248: 264—273.

Keller, W.D., Spotts, J.H. and Biggs, D.L., 1952. Infrared spectra of some rock-forming minerals. Am. J. Sci., 250: 453—471.

Kendall, D.N., 1953. Identification of polymorphic forms of crystals by infrared spectroscopy. Anal. Chem., 25: 382—389.

Kiselev, A.V. and Lygin, V.I., 1959. Infrared absorption spectra and the structure of the hydroxyl coating of various hydrated silicas. Kolloidn. Zh. (U.S.S.R.), 21: 581—589.

Klages, M.G., 1969. Weathering of montmorillonite during formation of a solodic soil. II. Nature of the mixed layer products. Proc. Soil Sci. Soc. Am., 33: 543—546.

Kleber, W, and Moenke, H., 1964. Kristall-chemische Systematik der Mineralien auf Ultrarot spektroskopischer Grundlage. Monatsber. Dtsch. Akad. Wiss. Berlin, 6: 384—392.

Klimov, V.V., 1961. Infrared spectrometry of inorganic substances. Zav. Lab., 27: 292—294.

Klimov, V.V., Kagarlitskaya, N.V. and Shcherbov, D.B., 1960. Absorption spectra of some silicate minerals, 2-15 μ Tr. Kaz. Nauchn. Issled. Inst. Miner., Syr'ya, 1960: 312—317.

Kobayashi, K. and Oinuma, K., 1961. Clay mineral composition of the core sample obtained from the bottom of the Japan Trench by Takuyo in 1959. Contrib. Mar. Res. Lab. Hydrogr. Off., Japan, 2: 109—117.

Kolesova, V.A., 1954. Problem on the interpretation of vibration spectra of the silicates and the silicate glasses. Zh. Eksp. Teor. Fiz., 26: 124—127.

Kolesova, V.A., 1959. Infrared absorption spectra of the silicates. Opt. i Spektr., 6: 20—24.

Kolesova, V.A., 1959. Ultrarotabsorptionsspektren von aluminiumhaltigen Silikaten und einigen kristallinen Aluminaten. Opt. i Spektr., 6: 38—44.

Koltermann, M. and Müller, K.P., 1965. Kristallchemie und thermisches Verhalten von Sepiolith, Hectorit, Saponit, Bergleder und Anthophyllit. Tonind. Z. Keram. Rundsch., 89: 406—411.

Kumada, K. and Aizawa, K., 1959. The infrared absorption spectra of soil components. Soil Plant Food, 4: 181—188.

Launer, P.J., 1952. Regularities in the

infrared absorption spectra of silicate minerals. Am. Mineralogist, 37: 764—784.

Lawson, K.E., 1961. Infrared Absorption of Inorganic Substances. Bibliography, Reinhold, New York, 200 pp.

Lazarev, A.N., 1960. Infrared spectra of silicates with anions of the type $(Si_2O_7)^{6-}$ Opt. i Spektr., 9: 195—202.

Lazarev, A.N., 1961. Spectroscopic identification of Si_2O_7 groups in silicates. Sov. Phys. Cryst., 6: 101—103.

Lazarev, A.N., 1962. Interpretation of the spectra of silicates and germanates with ring anions. Opt. i Spektr., 12: 28—30.

Lazarev, A.N., 1971. Vibrational Spectra and Structure of Silicates. Plenum, New York, 250 pp.

Lazarev, A.N. and Tenisheva, T.F., 1961. Infrared absorption spectra of silicates and germanates with chain anions. Opt. i Spektr., 10: 79—85.

Lazarev, A.N. and Tenisheva, T.F., 1962. Vibrational spectra of silicates. V. Silicates with anions in ribbon form. Opt. i Spektr., 12: 215—219.

Lecomte, J., 1948. Applications analytiques des spectres infrarouges des poudres aux composés minéraux et à l'eau de cristallisation. Anal. Chim. Acta, 2: 727—733.

Lehmann, H. and Dutz, H., 1960. Ultrarotspektroskopische Untersuchungen an Gläsern und kristallinen Silikaten. Silic. Ind., 25: 559—566.

Lehmann, H. and Dutz, H., 1962. Infrared spectroscopy studies on the hydration of clinker minerals and cements. Natl. Bur. Std. (U.S.A.) Monogr., 43: 513—518.

Lindberg, J.D. and Snijder, D.G., 1972. Diffuse reflectance spectra of several clay minerals. Am. Mineralogist, 57: 485—493.

Lyon, R.J.P., 1962. Minerals in the Infrared - A Critical Bibliography. Stanford Res. Inst., Menlo Park, Calif., 76 pp.

Lyon, R.J.P., Tuddenham, W.M. and Thompson, C.S., 1959. Quantitative Mineralogy in 30 minutes. Econ. Geol., 54: 1047—1055.

Matossi, F., 1949. Vibration frequencies and binding forces in some silicate groups. J. Chem. Phys., 17: 679—685.

Milkey, R.G., 1960. Infrared spectra of some tectosilicates. Am. Mineralogist, 45: 990—1007.

Miller, F.A. and Wilkins, C.H., 1952. Infrared spectra and characteristic frequencies of inorganic ions. Their use in qualitative analysis. Anal. Chem., 24: 1253—1294.

Miller, F.A., Carlson, G.L., Bentley, F.F. and Jones, W.H., 1960. Infrared spectra of inorganic ions in the cesium-bromide region (700 - 300 cm^{-1}). Spectrochim. Acta, 16: 135—235.

Moenke, H., 1961. Ultrarotabsorption Spectralphotometrie und Silikatforschung. Silikattechnik, 12: 323—327.

Moenke, H., 1962. Entwicklung, Stand und Möglichkeiten der Ultrarotspektralphotometrie von Mineralien. Fortschr. Mineral., 40: 76—123.

Moenke, H., 1962. Die Ultrarotabsorption von Nesosilikaten im NaCl-Bereich von Prismenspektralphotometern. Silikattechnik, 13: 246—248.

Moenke, H., 1962. Mineralspektern, (aufgenommen mit dem Jenaer Spektralphotometer U.R.10). Akademie Verlag, Berlin, 42 pp., 355 charts.

Moenke, H., 1962. Spektralanalyse von Mineralien und Gesteinen. Akad. Verlagsges. Geest und Portig K.G., Leipzig, 222 pp.

Moenke, H., 1963. Mineralspektren, aufgenommen mit dem Jenaer Spektraphotometer. Am. Mineralogist, 48: 1425—1427.

Moenke, H., 1963. Ultrarotabsorption und das "Kapittel B der Kristallchemie der Silikate" nach N.W. Below. Silikattechn., 14: 65—68.

Moenke, H., 1963. Entwicklung, Stand und Möglichkeiten der Ultrarotspektralphotometrie von Mineralen. Fortschr. Mineral, 40: 70—123.

Moenke, H., 1966. Mineralspektren II (aufgenommen mit dem Jenaer Spektralphotometer U.R.10). Akademie Verlag, Berlin, 19 pp., 150 charts.

Monasterio, B.E., Serratosa, J.M. and Hidalgo, A., 1964. Espectres de absorcion infrarroja de la fraccion arcilla de suelos volcanicos de Chile. An. Edafol. Agrobiol (Madrid), 23: 293—303.

Müller-Hesse, H., Planz, E. and Schwiete, H., 1960. Die Infrarotspektroskopie in der Keramik unter besondere Berücksichtigung der Reaktionen im System BaO-Al_2O_3-SiO_2. Sprechsaal, Keram., Glas, Email, 93: 159—163.

Murthy, M.K. and Kirby, E.M., 1962. Infrared study of compounds and solid solutions in the system lithia-alumina-silica. J. Am. Ceram. Soc., 45: 324—329.

Nahin, P.G., 1955. Infrared analysis of clays and related minerals. Proc. 1st Natl. Conf. Clays Clay Technol., 1952. Dept. Natl. Res., Calif. State, Div. Mines, Bull., 169: 112—118.

Nahin, P.G., Merill, W.C., Grenall, A. and Crog, R.S., 1951. Mineralogical studies of California oil-bearing formations, 1. Identification of clays. J. Pet. Technol., 192: 151—158.

Nakamoto, K., 1970. Infrared Spectra of Inorganic and Coordination Compounds. Wiley, New York, 338 pp.

Naumann, A.W., Safford, G.J. and Mumpton, F.A., 1966. Low-frequency (OH) motions in layer silicate minerals. Proc. 14th Natl. Conf. Clays Clay Miner., Berkeley, 1965: 367—383.

Oinuma, K., 1966. Clay minerals composition of a long core from the Mid-Pacific Ocean. Bull. Natl. Sci. Mus. (Tokyo), 9: 53—59.

Oinuma, K. and Kodama, H., 1963. Infrared absorption spectra of some clay minerals. Nendo-Kagaku, Japan, 3: 211—221.

Oinuma, K. and Kodama, H., 1964. Infrared absorption spectra of clay minerals in the region from 2800 to 700 cm^{-1}. J. Tokyo Univ., Gen. Educ. (Natl. Sci.), 5: 1—22.

Oinuma, K. and Hayashi, H., 1966. Infrared study of clay minerals from Japan. J. Univ. Tokyo Gen. Educ. Natl. Sci, 6: 1—15.

Oinuma, K. and Kodama, H., 1967. Use of infrared absorption spectra for identification of clay minerals in sediments. J. Toyo Univ. Gen. Educ. (Natl. Sci.), 7: 1—23.

Oinuma, K. and Hayashi, H., 1968. Infrared spectra of clay minerals. Toyo Daigaku Kiyo KyoyoKatei Hen (Skizenkagaku) (J. Toyo. Univ. Japan), 9: 57—98.

Oinuma, K., Kobayashi, K. and Sudo, T., 1961. Procedure of clay mineral analysis. Clay Sci., 1: 23—28.

Omori, K., 1961. Infrared absorption spectra of some essential minerals. Sci. Rep. Tohoku Univ., Ser. III: 101—130.

Omori, K., 1964. Infrared studies of essential minerals from 11 to 25 microns. Sci. Rep., Tohoku Univ. Ser. III., 9: 65—97.

Orcharenko, F.D., 1966. The regulation of colloid-chemical properties of clay minerals. Proc. Int. Clay Conf., Jerusalem, 1: 299—310.

Palmieri, F., 1964. La spettrografia nell'Ultrarosso per la determinazione costitutiva della frazione più dispersa (minerali argillosi) del terreno. Ann. Fac. Agric. Portici, 29: 469—498.

Pawluk, S., 1963. Characteristics of 14 Å clay minerals in the B horizons of podzolized soils of Alberta. Proc. 11th Natl. Conf. Clays Clay Miner., Ottawa, 1962: 74—82.

Perez Rodriguez, J.L. and Martin Martinez, F., 1966. Infrared spectra of the clay fraction of some West Andalusian soils. 1. Soils over Tertiary material and limestones; 2. Terrace soils and Kemper marls. An. Edafol. Agrobiol., 26: 1069—1079; 1184—1192.

Plyusnina, I.I., 1967. Infrakrasnye Spektry Silikatov. (Infrared Spectra of Silicates.) Izd. Moskov. Univ., 189 pp.

Plyusnina, I.I. and Bokii, G.B., 1958. The infrared reflection spectra of the cyclosilicates in the wavelength interval from 7—15 μ. Sov. Phys. Cryst., 3: 761—764.

Prasad, J., 1965. Infrared absorption spectra of some clay minerals and mica-type structures. Trans. Indian Ceram. Soc., 24: 78—82.

Preu, E., 1965. Über die anorganischen Bestandteile der Kohlen. Freib. Forschungsh., A 373: 91—105.

Prima, A.M., 1960. Equations of the vibrations of planar silicon-oxygen rings in silicates. Opt. i Spektr., 9: 236—239.

Radwan, M.K., 1964. A suggested template to be used in characterization of some silicate minerals in the OH-region of the infrared spectrogram. J. Chem. U.A.R., 7: 55—67.

Ram, A., Bishui, B.M. and Prasad, J., 1960. Infrared spectra of clays from the regions Kusumpur, Kot Ransipur, Bhandak and Neyveli. Centr. Glass Ceram. Res. Inst. Bull., 7: 3—10.

Reh, H., 1965. Einsatz der Ultrarot-Apsorptionsspektroskopie für die Lösung mineralogisch-geologischer Probleme. Z. Angew. Geol., 11: 183—189.

Reh, H., 1966. Ein Beitrag zur klärung des mineralischen Aufbaus der Schiefer am SE-rand des Schwarzburger Sattels, vor allem durch Ultrarot Untersuchungen. Geologie, 15: 562—577.

Rhoads, D.C. and Stanley, D.J., 1966. Transmitted infrared radiation: a simple method for studying sedimentary structures. J. Sediment. Petrol., 36: 1144—1149.

Rudnitskaya, E.S., 1956. Use of infrared spectroscopy for the investigation of minerals. Zap. Vses. Mineral. Obshch., 85: 407—412.

Saksena, B.D., 1961. Infrared absorption studies of some silicate structures. Trans. Faraday Soc., 57: 242—258.

Sanchez Calvo, M.C., 1963. Kennzeichnung von Guinea-Tonen durch Röntgenstrahlen und ultrarote Absorption. Proc. Int. Clay Conf., Stockholm, 1: 155—166.

Scholze, H. and Dietzel, A., 1955. Infrarotuntersuchungen an Wasserhaltigen Silikaten. Naturwissenschaften., 42: 342—343.

Schwiete, H.E., Ludwig, U. and Niël, E.M., 1965. Studies on the beginning of hydration of clinker and cement. Proc. 7th Conf. Silic. Ind. Budapest, 1963.

Setkina, O.N., 1957. Infrared spectra of the clay minerals and their application. Tr. Leningr. Tekhnol. Inst. I.M. Lensoveta, 40: 155—162.

Setkina, O.N., 1959. Infrared spectra of minerals and their practical application. Zap. Vses. Mineral. Obshch., 88: 39—47.

Setkina, O.N. and Copstein, N.M., 1957. Infrared spectra of clay minerals. Tr. Leningr. Tekhnol. Inst. I.M. Lensoveta, 37: 79—90.

Sharp, D.W.A., 1962. The study of inorganic compounds by infrared spectroscopy. Spectrovision, 12: 1—12.

Stepanov, B.I. and Prima, A.M., 1958. The vibrational spectra of silicates. I: The computation of frequencies and intensities of the spectra bands of the silicates. Opt. i Spektr., 4: 734—749.

Stepanov, I.S., 1970. Infrared spectra of fine dispersed fractions of soils. Dokl. Akad. Nauk (U.S.S.R.), 193: 199—202.

Stubičan, V., 1960. Clay Mineral research at the Institute for Silicate Chemistry, Zagreb. Proc. 7th Natl. Conf. Clays Clay Miner., Washington, 1958: 295—302.

Stubičan, V. and Roy, R., 1959. Isomorphous substitution and infrared spectra of clays. Bull. Geol. Soc. Am., 70: 1682—1683.

Stubičan, V. and Roy, R., 1961. A new approach to assignment of infrared absorption bands in layer-lattice silicates. Z. Krist., 115: 200—214.

Stubičan, V. and Roy, R., 1961. Isomorphous substitution and infrared spectra of the layer lattice silicates. Am. Mineralogist, 46: 32—52.

Stubičan, V. and Roy, R., 1961. Infrared spectra of layer lattice silicates. J. Am. Ceram. Soc., 44: 625—627.

Szigetti, B., 1960. The infrared spectra of crystals. Proc. R. Soc., A 258: 377—401.

Tarasevich, Yu.I., 1968. Determination of the infrared spectra of clay minerals and substances absorbed on them. Ukr. Khim. Zh., 34: 439—446.

Tarte, P., 1960. Recherches sur le spectre infra-rouge des silicates; le comportement de la vibration antisymétrique de valence dans des solutions solides orthosilicates-orthogermanates. Bull. Acad. R. Belg. Cl. Sci., 46: 169—179.

Tarte, P., 1963. Applications nouvelles de la spectrometrie infrarouge à des problèmes de cristallochimie. Silic. Ind., 28: 345—354.

Tarte, P., 1963. Etude des silicates par spectrométrie infrarouge. Résultats actuels et perspectives d'avenir. Bull. Soc. Fr. Céram., 58: 13—34.

Tarte, P., 1965. Etude expèrimentale et interprétation du spectre infrarouge des silicates et des germanates. Application à des problèmes structuraux relatifs à l'état solide, I. Acad. R. Belg. Cl. Sci., Mem. Coll. in —8°, 35, 260 pp.

Touillaux, R., Fripiat, J.J. and Toussaint, F., 1960. Etude en spectroscopie infrarouge des mineraux argileux. Trans. 7th Int. Congr. Soil Sci., Madison, Wis., 7 (3): 460—467.

Tuddenham, W.M. and Lyon, R.J.P., 1960. Infrared techniques in the identification and measurement of minerals. Anal. Chem., 32: 1630—1634.

Tuddenham, W.M. and Zimmerley, S.R., 1960. Infrared analysis is quick and easy. Eng. Mining J., 161: 92—94.

White, J.L., 1971. Interpretation of infrared spectra of soil minerals. Soil Sci., 112: 22—31.

White, W.B. and Roy, R., 1964. Infrared spectra-crystal structure correlations, 2. Comparison of simple polymorphic minerals. Am. Mineralogist, 49: 1670—1687.

Willis, J.B., 1958. Vibrational spectroscopy and its application in structural inorganic chemistry. Rev. Pure Appl. Chem., 8: 101—128.

Wolff, R., 1965. Infrared absorption patterns (OH region) of several clay minerals. Am. Mineralogist, 50: 240—244.

Kaolin minerals

Angino, E.E., 1964. Far infrared spectra of montmorillonite, kaolinite and illite. Nature, 204: 569—571.

Beutelspacher, H. and Van der Marel, H.W., 1961. Kennzeichen zur Identifizierung von Kaolinit, "Fireclay"-Mineral und Halloysit, ihre Verbreitung und Bildung. Tonind. Z. Keram. Rundsch., 85: 570—582.

Beutelspacher, H. and Van der Marel, H.W., 1962. Elektronenoptische, röntgenographische, infrarotspektroskopische und thermoanalytische Untersuchungen von Kaolinit, "Fireclay"-Mineral und Halloysit. Klei Keram., 12: 237—250.

Bishui, B.M., Dhar, R.N. and Prasad, J., 1965. Estimation of kaolinite in clays by infrared method. Proc.

Natl. Inst. Sci. India, A 31: 459—464.

Biswas, M.A. and Basak, A.K., 1961. Infrared absorption analysis of some East Pakistani clays. Pakistan J. Sci. Ind. Res., 4: 118—120.

Brindley, G.W. and Zussman, J., 1959. Infrared absorption data for serpentine minerals. Am. Mineralogist, 44: 185—188.

Chukrov, F.V., Zvyagin, B.B., Rudnitskaya, E.S. and Ermilova, L.P., 1966. Nature and genesis of halloysite. Izv. Akad. Nauk S.S.S.R. Ser. Geol., 31: 3—20.

Chukhrov, F.V. and Zyvagin, B.B., 1966. Halloysite, a crystallochemically and mineralogically distinct species. Proc. Int. Clay Conf., Jerusalem, 1: 11—25.

De Keyser, W.L., Wollast, R. and De Laet, L., 1963. Contribution to the study of OH groups in kaolin materials. Proc. Int. Clay Conf., Stockholm, 2: 75—86.

De Kimpe, C., Gastuche, M.C. and Brindley, G.W., 1964. Low-temperature syntheses of kaolin minerals. Am. Mineralogist, 49: 1—16.

De Pablo, L., 1966. A disordered kaolinite from Conception de Buenos Aires Jalisco, Mexico. Proc. 13th Natl. Conf. Clays Clay Miner., 1964: 143—150.

Erdélyi, J. and Veniale, F., 1970. Idro-crisotile: un nuovo minerale del gruppo del serpentino. Rend. Soc. Ital. Mineral. Petrol., 26: 403—404.

Farmer, V.C., 1964. Infrared absorption of hydroxyl groups in kaolinite. Science, 145: 1189—1190.

Farmer, V.C. and Russell, J.D., 1964. The infrared spectra of layer silicates. Spectrochim. Acta, 20: 1149—1173.

Farmer, V.C. and Russell, J.D., 1966. Effects of particle size and structure on the vibrational frequencies of layer silicates. Spectrochim. Acta, 22: 389—398.

Farmer, V.C. and Russell, J.D., 1967. Infrared absorption spectrometry in clay studies. Proc. 15th Natl. Conf. Clays Clay Minerals, 27: 121—142.

Freund, F., 1967. Infrarotspektren von Kaolinit, Metakaolinit und Al-Si-Spinell. Ber. Dtsch. Keram. Ges., 44: 392—397.

Fripiat, J.J. and Toussaint, F., 1960. Predehydroxylation state of kaolinite. Nature, 186: 627—628.

Fripiat, J.J. and Toussaint, F., 1963. Dehydroxylation of kaolinite, II. Conductometric measurements and infrared spectroscopy. J. Phys. Chem., 67: 30—36.

Gribina, I.A. and Tarasevich, Yu.I., 1972. Spectral study of the interaction of heavy water with a kaolinite surface. Teor. Eksp. Khim., 8: 512—517.

Hayashi, H. and Oinuma, K., 1963. X-ray and infrared studies on the behaviours of clay minerals on heating. Clay Sci., 1: 134—154.

Jahanbagloo, I.C. and Zoltai, T., 1968. The crystal structure of hexagonal Al-serpentine. Am. Mineralogist, 53: 14—24.

Kakitani, S., 1956. Infrared absorption by some clay minerals. (OH stretching vibration of montmorillonite, kaolinite and halloysite.) Köbutsugaku Zasshi, 3: 49—52.

Kanno, I., Onikura, Y., Arimura, S. and Tokudome, S., 1965. Mineralogical and chemical characteristics of serpentine as a parent rock. Soil Sci. Plant Nutr., 11: 104—113.

Keeling, P.S., 1961. A new concept of clay minerals. Trans. Br. Ceram. Soc., 60: 449—474.

Keeling, P.S., 1961. Geochemistry of the common clay minerals. Trans. Br. Ceram Soc., 60: 678—689.

Keeling, P.S., 1965. Investigation of hydroxyl groups in kaolinitic clays by infrared spectrophotometer. Trans. Br. Ceram. Soc., 64: 137—151.

Kodama, H. and Oinuma, K., 1962. Identification of kaolin minerals in clays by X-ray and infrared absorption spectra. Clay Sci., 1: 113—119.

Kodama, H. and Oinuma, K., 1963. Identification of kaolin minerals in the presence of chlorite by X-ray diffraction and infrared absorption spectra. Proc. 11th Natl. Conf. Clays Clay Miner., Ottawa, 1962: 236—249.

Kunze, G.W. and Bradley, W.F., 1964. Occurrence of a tabular halloysite in a Texas soil. Proc. 12th Natl. Conf. Clays Clay Miner., Atlanta, 1963: 523—527.

Ledoux, R.L. and White, J.L., 1964. Infrared study of selective deuteration of kaolinite and halloysite at room temperature. Science, 143: 47—49.

Ledoux, R.L. and White, J.L., 1964. Infrared study of the OH groups in expanded kaolinite. Science, 143: 244—246.

Ledoux, R.L. and White, J.L., 1967. Etude infrarouge des complexes d'intercalation de la kaolinite. Silic. Ind., 32: 269—273.

Ledoux, R.L. and White, J.L., 1968. Infrared studies of the hydroxyl groups in intercalated kaolinite complexes. Proc. 13th Natl. Conf. Clays Clay Miner., Ottawa, 1962: 289—315.

Luce, R.W., 1971. Identification of serpentine varieties by infrared absorption. U.S. Geol. Surv. Pap., 750 B: 199—201.

Lyon, R.J.P. and Tuddenham, W.M., 1960. Infrared determination of the kaolin group minerals. Nature, 185: 835—836.

Maksimovic, Z., 1966. β-kerolite-pimelite series from Goleš mountain, Yugoslavia. Proc. Int. Clay Conf., Jerusalem, 1: 97—105.

Maksimovic, Z. and White, J.L., 1972. Infrared study of chromium-bearing halloysites. Int. Clay Conf., Madrid, 1: 87—100.

Miller, J.G., 1961. An infrared spectroscopic study of the iosthermal dehydroxylation of kaolinite at 470° C. J. Phys. Chem., 65: 800—804.

Montoga, J.W. and Baur, G.S., 1963. Nickeliferous serpentines, chlorites and related minerals found in two lateritic ores. Am. Mineralogist, 48: 1227—1238.

Newnham, R.E., 1961. A refinement of the dickite structure and some remarks on polymorphism in kaolin minerals. Mineral. Mag., 32: 683—704.

Oinuma, K. and Kodama, H., 1965. Identification of kaolin minerals in clays by infrared absorption spectra. Adv. Clay Sci., 5: 161—168.

Omori, K. and Yoshida, M., 1965. Infrared spectra for kaolinite and sericite in the wavelength region from 11 to 25 microns. Sci. Rep. Tohoku Univ., Ser. III, 9: 377—380.

Pampuch, R. and Wilkos, K., 1965. Assignment of infrared absorption bands in the hydroxyl region of kaolin group minerals of kaolinite clays. Proc. Conf. Silic. Ind., Budapest, 1963, 7: 179—182.

Parker, T.W., 1969. A classification of kaolinites by infrared spectroscopy. Clay Miner., 8: 135—141.

Parker, T.W., 1969. Infrared spectra: relationships and origins of kaolinites in the clays of South Devon. Proc. Br. Ceram. Soc., 13: 117—124.

Poncelet, G.M. and Brindley, G.W., 1967. Experimental formation of kaolinite from montmorillonite at low temperatures. Am. Mineralogist, 52: 1161—1173.

Radwan, M.K., 1964. A suggested template to be used in characterization of some silicate minerals in the OH region of the infrared spectrogram. J. Chem. U. A. R., 7: 55—67.

Romo, L.A., 1956. The exchange of hydrogen by deuterium in hydroxyls of kaolinite. J. Phys. Chem., 60: 987—989.

Roy, R., Roy, D.M. and Francis, E.E., 1955. New data on thermal decomposition of kaolinite and halloysite. J. Am. Ceram. Soc., 38: 198—205.

Scholze, H. and Dietzel, A., 1955. Infrarot Untersuchungen an wasserhaltigen Silikaten. Naturwissenschaften, 42: 342—343.

Scholze, H. and Dietzel, A., 1955. Infrarot-untersuchungen an Tonmineralen. Naturwissenschaften, 42: 575.

Serratosa, J.M., Hidalgo, A. and Vinas, J.M., 1962. Orientation of OH bands in kaolinite. Nature, 195: 486—487.

Serratosa, J.M., Hidalgo, A. and Vinas, J.M., 1963. Infrared study of the OH groups in kaolin minerals. Int. Clay Conf., Stockholm, 1: 17—26.

Speakman, K. and Majumdar, A.J., 1971. Synthetic deweylite. Mineral. Mag., 38: 225—234.

Stubičan, V., 1959. Residual hydroxyl groups in the metakaolin range. Mineral Mag., 32: 38—52.

Stubičan, V. and Roy, R., 1961. Proton retention in heated 1:1 clays studied by infrared spectroscopy, weight loss and deuterium uptake. J. Phys. Chem., 65: 1348—1351.

Tarasevich, Yu.I. and Ovcharenko, F.D., 1969. Infrared spectra of kaolinite. Kolloid. Zh., 31: 753—758.

Tuddenham, W.M., 1960. Infrared determination of the kaolin group minerals. Nature, 185: 835—836.

Van der Marel, H.W., 1966. Quantitative analysis of clay minerals and their admixtures. Contrib. Mineral Petrol., 12: 96—138.

Van der Marel, H.W. and Zwiers, J.H.L., 1959. OH stretching bands of the kaolin minerals. Silic. Ind., 24: 359—369.

Veniale, F. and Van der Marel, H.W., 1963. An interstratified saponite-swelling chlorite mineral as a weathering product of lizardite rock from St. Margherita (Pavia Province) Italy. Beitr. Mineral. Petrogr., 9: 198—245.

Wada, K., 1964. Ammonium chloride-kaolin complexes. I. Structural and bonding features. Clay Sci., 2: 43—56.

Wada, K., 1965. Intercalation of water in kaolin minerals. Am. Mineralogist, 50: 924—941.

Wada, K., 1967. A study of hydroxyl groups in kaolin minerals utilizing selective deuteration and infrared spectroscopy. Clay Miner., 7: 51—61.

White, J.L., Laycock, A. and Crusz, M., 1970. Infrared studies of proton delocalization in kaolinite. Bull. Gr. Fr. Argiles, 22: 157—165.

Wolff, R.G., 1963. Structural aspects of kaolinite using infrared absorption. Am. Mineralogist, 48: 390—399.

Wolff, R.G., 1965. Infrared absorption patterns (OH region) of several clay minerals. Am. Mineralogist, 50: 240—244.

Pyrophyllite

Farmer, V.C. and Russell, J.D., 1964. The infrared spectra of layer silicates. Spectrochim. Acta, 20: 1149—1173.

Farmer, V.C. and Russell, J.C., 1967. Infrared absorption spectrometry in clay studies. Proc. 15th Natl. Conf. Clays Clay Miner., 27: 121—142.

Faust, G.T., Hathaway, J.C. and Millot, G., 1959. A restudy of stevensite and allied minerals. Am. Mineralogist, 44: 342—370.

Hayashi, H., Koshi, K., Hamada, A. and Sakabe, H., 1962. Structural change of pyrophyllite by grinding and its effect on toxicity of the cell. Clay Sci., 1: 99—108.

Heller, L., Farmer, V.C., Mackenzie, R.C., Mitchell, B.D. and Taylor, H.F.W., 1962. The dehydroxylation and rehydroxylation of trimorphic dioctahedral clay minerals. Clay Miner. Bull., 5: 56—72.

Hennicke, H.W. and Niesel, K., 1965. Zur Kenntnis der Entwässerung des Pyrophyllits. Tonind. Z. Keram. Rundsch., 89: 496—503.

Kodama, H., 1958. Mineralogical study on some pyrophyllites in Japan. Mineral. J. (Japan), 2: 236—244.

Koltermann, M. and Müller, K.P., 1965. Kristallchemie und thermisches Verhalten von Sepiolith, Hectorit, Saponit, Bergleder und Anthophyllit. Tonind. Z. Keram. Rundsch., 89: 406—411.

Russell, J.D., Farmer, V.C. and Velde, B., 1970. Replacement of OH by OD in layer silicates and identification of the vibrations of these groups in infrared spectra. Mineral. Mag., 37: 869—879.

Talc

De Waal, S.A., 1970. Nickel minerals from Barberton, S. Africa. III. Willemsite a nickel-rich talc. Am. Mineralogist, 55: 31—42.

Farmer, V.C., 1958. The infrared spectra of talc, saponite and hectorite. Mineral. Mag., 31: 829—845.

Farmer, V.C. and Russell, J.D., 1964. The infrared spectra of layer silicates. Spectrochim. Acta, 20: 1149—1173.

Faust, G.T., Hathaway, J.C. and Millot, G., 1959. A restudy of stevensite and related minerals. Am. Mineralogist, 44: 342—370.

Imai, N., Otsuka, R., Nakamura, T., Tsunashima, A. and Sakamoto, T., 1973. "Hydrated talc" - an alteration product of wollastonite by reaction with magnesium-bearing hydrothermal solution. Clay Sci., 4: 175—191.

Koltermann, M. and Müller, K.P., 1965. Kristallchemie und thermisches Verhalten von Sepiolith, Hectorit, Saponit, Bergleder und Anthophyllit. Tonind. Z. Keram. Runsch., 89: 406—411.

Kurbatov, A.I. and Aleshin, S.N., 1964. Investigation of minerals of the hydroxide and chlorite groups by infrared spectroscopy. Dokl. Rossiisk. S-kh. Akad., 99: 5—10.

Wilkins, R.W.T. and Ito, J., 1967. Infrared spectra of some synthetic talcs. Am. Mineralogist, 52: 1649—1661.

Brucite

Benesi, H.A., 1959. Infrared spectrum of Mg $(OH)_2$. J. Chem. Phys., 30: 852.

Boutin, H. and Bassett, W., 1963. A comparison of OH motions in brucite and micas. Am. Mineralogist, 48: 659—663.

Freund, F., 1969. Dehydroxylation mechanism of clay minerals. The initial conversion of hydroxyl groups into water molecules by proton tunnelling. Int. Clay Conf., Tokyo, I: 121—128.

Glemser, O. and Hartert, E., 1956. Untersuchungen über die Wasserstoffbrückenbindung in kristallisierten Hydroxyden. Z. Anorg. Allgem. Chem., 283: 111—122.

Hartert, E. and Glemser, O., 1953. Zur Lage des Wasserstoffs im Gitter kristalliner Hydroxyde. Naturwissenschaften, 40: 199—200.

Hexter, R.M., 1958. On the infrared absorption spectra of crystalline brucite $Mg(OH)_2$ and portlandite $(Ca(OH)_2)$. J. Opt. Soc. Am., 48: 770—774.

Kurbatov, A.I. and Aleshin, S.N., 1964. Investigation of minerals of the hydroxide and chlorite groups by infrared spectroscopy. Dokl. Rossiisk. S-kh. Akad., 99: 5—10.

Mal'tseva, N.N. and Kharitonov, YU, Ya., 1962. Infrared absorption spectrum of magnesium hydride. Zh. Georg. Khim., 7: 947—948.

Mara, R.T. and Sutherland, G.B.B.M., 1953. The infrared spectrum of brucite $Mg(OH)_2$. J. Opt. Soc. Am., 43: 1100—1102.

Mara, R.T. and Sutherland, G.B.B.M., 1956. Crystal structure of brucite and portlandite in relation to infrared absorption. J. Opt. Soc. Am., 46: 464.

Petch, H.E. and Megaw, H.D., 1954. Crystal structure of brucite

$Mg(OH)_2$ and portlandite $Ca(OH)_2$ in relation to infrared absorption. J. Opt. Soc. Am., 44: 744—745.
Prasad, J., 1965. Infrared absorption spectra of some clay minerals and mica-type structures. Trans. Indian. Ceram. Soc., 24: 78—82.

Smectites

Angino, E.E., 1964. Far infrared spectra of montmorillonite, kaolinite and illite. Nature, 204: 569—571.
Chaussidon, J., 1970. Stretching frequencies of structural hydroxyls of hectorite and K-depleted phlogopite as influenced by interlayer cation and hydration. Clays Clay Miner., 18: 139—149.
Farmer, V.C., 1958. The infrared spectra of talc, saponite and hectorite Mineral. Mag., 31: 829—845.
Farmer, V.C. and Russell, J.D., 1964. The infrared spectra of layer silicates. Spectrochim. Acta, 20: 1149—1173.
Farmer, V.C. and Russell, J.D., 1967. Infrared absorption spectrometry in clay studies. Proc. 15th Natl. Conf. Clays Clay Miner., Pittsburgh, 1966: 121—142.
Farmer, V.C. and Russell, J.D., 1971. Interlayer complexes in layer silicates. The structure of water in lamellar ionic solutions. Trans. Faraday Soc., 67: 2737—2749.
Fripiat, J.J., Chaussidon, J. and Touillaux, R., 1960. Study of dehydration of montmorillonite and vermiculite by infrared spectroscopy. J. Phys. Chem., 64: 1234—1241.
Fripiat, J.J., 1964. Surface properties of alumino-silicates. Proc. 12th Conf. Clays Clay Miner., Atlanta, Georgia, 1963: 327—358.
Granquist, W.T. and Samner, G.G., 1959. Acid dissolution of a Texas bentonite. Proc. 6th Natl. Conf. Clays Clay Miner., Berkeley, 1957, 292—308.
Grover, G., Kühnel, R.A. and Roorda, H.J., 1974. High-nickel montmorillonite from Barro Alto (Brasil), in press.
Hamilton, J.D., 1971. Beidellitic montmorillonite from Swansea, N. S.Wales. Clay Miner., 9: 107—123.
Hayashi, H., 1963. Montmorillonite from some bentonite deposits in Yamagata Prefecture Japan. Clay Sci., 1: 176—182.
Hayashi, H. and Oinuma, K., 1963. X-ray and infrared studies on the behaviours of clay minerals on heating. Clay Sci., 1: 134—154.
Heller, L., Farmer, V.C., Mackenzie, R.C., Mitchell, B.D. and Taylor, H.F.W., 1962. The dehydroxylation and rehydroxylation of triphormic dioctahedral clay minerals. Clay Miner. Bull., 5: 56—72.
Kakitani, S., 1956. Infrared absorption by some clay minerals (OH-stretching vibration of montmorillonite, kaolinite and halloysite). Kobutsugaku-Zasshi, 3: 49—52.
Keeling, P.S., 1961. A new concept of clay minerals. Trans. Br. Ceram. Soc., 60: 449—474.
Keeling, P.S., 1961. Geochemistry of the common clay minerals. Trans. Br. Ceram. Soc., 60: 678—689.
Khamaraev, S.S., Yagudaev, M.R. and Aripov, E.A., 1965. Investigations about the building of structure in bentonite clays by means of infrared spectroscopy. Kolloid. Zh., 27: 121—124.
Matsui, T. and Yatsu, E., 1969. Iron-bearing montmorillonites from pedosphere in North Japan. Clay Sci., 3: 116—125.
Prasad, J., 1965. Infrared absorption spectra of some clay minerals and mica-type structures. Trans. Indian Ceram. Soc., 24: 78—82.
Prost, R. and Chaussidon, J., 1969. The infrared spectrum of water absorbed on hectorite. Clay Miner., 8: 143—149.
Radwan, M.K., 1964. A suggested template to be used in characterization of some silicate minerals in the OH region of the infrared spectrogram. J. Chem. U. A. R., 7: 55—67.
Roth, C.B. and Tullock, R.J., 1972. Deprotonation of nontronite resulting from chemical reduction of structural ferric iron. Int. Clay Conf., Madrid, 1972, I: 143—151.
Russell, J.D. and Farmer, V.C., 1964. Infrared spectroscopic study of the dehydration of montmorillonite and saponite. Clay Mineral. Bull., 5: 443—464.
Russell, J.D., Farmer, V.C. and Velde, B., 1970. Replacement of OH by OD in layer silicates and identification of the vibrations of these groups in infrared spectra. Mineral. Mag., 37: 869—879.
Russell, J.D. and Fraser, A.R., 1971. Infrared spectroscopic evidence for interaction between hydronium ions and lattice OH groups in montmorillonite. Clays Clay Miner., 19: 55—59.
Scholze, H. and Dietzel, A., 1955. Infrarot-untersuchungen an wasserhaltigen Silikaten. Naturwissenschaften, 42: 342—343.
Scholze, H. and Dietzel, A., 1955. Infrarot-Untersuchungen an Tonmineralen. Naturwissenschaften, 42: 575.
Schultz, L.G., 1966. Lithium and potassium absorption, differential thermal- and infrared properties of some montmorillonites. Proc. 13th Natl. Conf. Clays Clay Miner., Madison, 1964: 275—288.
Serratosa, J.M., 1960. Dehydration studies by infrared spectroscopy. Am. Mineralogist, 45: 1101—1104.
Serratosa, J.M., 1962. Dehydration and rehydration studies of clay minerals by infrared absorption spectra. Proc. 9th Natl. Conf. Clays Clay Miner., Purdue Univ., 1960: 412—418.
Serratosa, J.M. and Bradley, W.F., 1958. Determination of the orientation of OH bond axes in layer silicates by infrared absorption. J. Phys. Chem., 62: 1164—1167.
Sun, Ming-Shan, 1963. The nature of chrysocolla from Inspiration Mine, Arizona. Am. Mineralogist, 48: 649—658.
Tettenhorst, R., 1962. Cation migration in montmorillonites. Am. Mineralogist, 47: 769—773.
Weismiller, R.A., Ahlrichs, J.L. and White, J.L., 1967. Infrared studies of hydroxyl-aluminium interlayer material. Proc. Soil Sci. Soc. Am., 31: 459—463.
White, J.L., Bailey, G.W., Brown, C.B. and Ahlrichs, J.L., 1961. Infrared investigation of the migration of lithium ions into empty octahedral sites in muscovite and montmorillonite. Nature, 190: 342.
Wolff, R.G., 1965. Infrared absorption patterns (OH region) of several clay minerals. Am. Mineralogist, 50: 240—244.

Mica minerals

Ahlrichs, J.L., Fraser, A.R. and Russell, J.D., 1972. Interaction of ammonia with vermiculite. Clay Miner., 9: 263—273.
Angino, E.E., 1964. Far infrared spectra of montmorillonite, kaolinite and illite. Nature, 204: 569—571.
Arkhangelskii, I.V., Kolodiev, B.N., Kommissarova, L.N. and Fotchenkov, A.A., 1971. OH-Groups in monocrystals of synthetic fluorophlogopite. Dokl. Akad. Nauk S.S.S.R., 199: 1307—1309.
Arkhipenko, D.K., Grigor'eva, T.N. and Kovaleva, L.T., 1963. Comparison of oxonium content in various vermiculites by using X-ray diffraction and infrared spectrometry. Rentgenogr. Mineral. Syr'ya, Vses. Nauchn.-Issl. Inst. Mineral. Syr'ya, Akad. Nauk. S.S.S.R., 3: 79—84.
Bassett, W.A., 1958. Hydroxyl orientation in the micas. Geol. Soc. Am. Bull., 69: 1532—1533.
Bassett, W.A., 1960. Role of hydroxyl orientation in mica alteration. Bull. Geol. Soc. Am., 71: 449—456.

Boutin, H. and Bassett, W., 1963. A comparison of OH motions in brucite and micas. Am. Mineralogist, 48: 659—663.

Bradley, W.F. and Serratosa, J.M., 1960. A discussion of the water content of vermiculite. Proc. 7th Natl. Conf. Clays Clay Miner., Washington, 1958: 260—270.

Chaussidon, J., 1970. Stretching frequencies of structural hydroxyls of hectorite and K-depleted phlogopite as influenced by interlayer cation and hydration. Clays Clay Miner., 18: 139—149.

Chaussidon, J., 1972. The infrared spectrum of structural hydroxyls of K-depleted biotites. Clays Clay Miner., 20: 59—67.

Chaussidon, J., 1972. Le spectre infrarouge des biotites: vibrations d'élongation basse frequence des O-H du reseau. Int. Clay Conf., Madrid, I: 129—141.

Cloos, P., Fripiat, J.J. and Vielvoye, L., 1961. Mineralogical and chemical characteristics of a glauconitic soil of the Hageland region (Belgium). Soil Sci., 91:55—65.

Ehlmann, A.J., Hulings, N.C. and Glover, E.D., 1963. Stages of glauconite formation in modern foraminiferal sediments. J. Sediment. Petrol., 33: 87—96.

Farmer, V.C. and Russell, J.D., 1964. The infrared spectra of layer silicates. Spectrochim. Acta, 20: 1149—1173.

Farmer, V.C. and Russell, J.D., 1966. Effects of particle size and structure on the vibrational frequencies of layer silicates. Spectrochim. Acta, 22: 389—398.

Farmer, V.C. and Russell, J.D., 1967. Infrared absorption spectrometry in clay studies. Proc. 15th Natl. Conf. Clays Clay Miner., Pittsburgh, 1966: 121—142.

Farmer, V.C. and Russell, J.D., 1971. Interlayer complexes in layer silicates. The structure of water in lamellar ionic solutions. Trans. Faraday Soc., 67: 2737—2749.

Fripiat, J.J., Chaussidon, J. and Touillaux, R., 1960. Study of dehydration of montmorillonite and vermiculite by infrared spectroscopy. J. Phys. Chem., 64: 1234—1241.

Fripiat, J.J., Rouxhet, P. and Jacobs, H., 1965. Proton delocalization in micas. Am. Mineral., 50: 1937—2958.

Fripiat, J.J., Jacobs, H. and Rouxhet, P., 1966. Note sur la constance de l'orientation du vibrateur OH dans la muscovite chauffée. Silic. Ind., 31: 311—313.

Gastuche, M.C., 1963. Kinetics of acid dissolution of biotite. I. Interfacial rate process followed by optical measurement of the white silica rim. Int. Clay Conf., Stockholm, 1: 67—76.

Gilkes, R.J., Young, R.C. and Quirk, J.P., 1972. Oxidation of octahedral iron in biotite. Clays Clay Miner., 20: 303—315.

Hayashi, H. and Oinuma, K., 1963. X-ray and infrared studies on the behaviours of clay minerals on heating. Clay Sci., 1: 134—154.

Ishii, M. and Nakahira, M., 1969. Far infrared absorption spectra of micas. Int. Clay Conf., Tokyo, 1: 247—259.

Jörgensen, P., 1966. Infrared absorption of OH bonds in some micas and other phyllosilicates. Proc. 13th Natl. Conf. Clays Clay Miner., Madison, 1964: 263—273.

Jörgensen, P. and Rosenqvist, I.Th., 1963. Replacement and bonding conditions for alkali ions and hydrogen in dioctahedral and trioctahedral micas. Norsk Geol. Tidsskr., 43: 497—536.

Juo, A.S.R. and White, J.L., 1969. Orientation of the dipole moments of hydroxyl groups in oxidized and unoxidized biotite. Science, 165: 804—805.

Keeling, P.S., 1961. A new concept of clay minerals. Trans. Br. Ceram. Soc., 60: 449—474.

Keeling, P.S., 1961. Geochemistry of the common clay minerals. Trans. Br. Ceram. Soc., 60: 678—689.

Kimbara, K. and Shimoda, S., 1973. A ferric celadonite in amygdales of dolerite at Taiheizan, Akita Prefecture, Japan. Clay Sci., 4: 143—150.

Liese, H.C., 1963. Tetrahedrally coordinated aluminium in some natural biotites. An infrared absorption analysis. Am. Mineralogist, 48: 980—990.

Liese, H.C., 1967. Supplemental data on the correlation of infrared absorption spectra and composition of biotites. Am. Mineralogist, 52: 877—880.

Lyon, R.J.P. and Tuddenham, W.M., 1960. Determination of tetrahedral aluminium in mica by infrared absorption analysis. Nature, 185: 374—375.

Manghnani, M.H. and Hower, J., 1964. Glauconites: Cation exchange capacities and infrared spectra. I. The cation exchange capacity of glauconite. Am. Mineralogist, 49: 586—598.

Manghnani, M.H. and Hower, J., 1964. Infrared absorption characteristics of glauconites. Am. Mineralogist, 49: 1631—1642.

Newman, A.C.D., 1967. Changes in phlogopites during their artificial alteration. Clay Miner., 7: 215—227.

Noda, T. and Roy, R., 1956. OH-F exchange in fluorine phlogopite. Am. Mineralogist, 41: 929—932.

Nodo, T. and Ushio, M., 1965. Determination of OH content in fluorine-hydroxyl phlogopite by infrared spectrophotometry. Geochem. Int., 2: 599—606.

Oinuma, K. and Hayashi, H., 1965. Infrared study of mixed-layer clay minerals. Am. Mineralogist, 50: 1213—1227.

Omori, K. and Yoshida, M., 1965. Infrared spectra for kaolinite and sericite in the wavelength region from 11 to 25 microns. Sci. Rep. Tohoku Univ., Ser. III, 9: 377—380.

Prasad, J., 1965. Infrared absorption spectra of some clay minerals and micatype structures. Trans. Indian Ceram. Soc., 24: 78—82.

Radwan, M.K., 1964. A suggested template to be used in characterization of some silicate minerals in the OH region of the infrared spectrogram. J. Chem. U.A.R., 7: 55—67.

Rigault, G. and Aguilano, D., 1964. Infrared absorption of muscovite, examination in polarized transmitted and reflected light. Period. Mineral. (Rome), 33: 445—465.

Rimsaite, J., 1967. Biotites intermediate between dioctahedral and trioctahedral micas. Proc. 15th Natl. Conf. Clays Clay Miner., Pittsburgh, 1966: 375—393.

Rodriguez Pascual, M.C. and Garcia Vicente, J., 1967. A physicochemical study of some biotites. An. Edafol. Agrobiol. (Madrid), 26: 1129—1142.

Rosenqvist, I.Th., 1963. Studies in position and mobility of the H atoms in hydrous micas. Proc. 11th Conf. Clays Clay Miner. Ottawa, 1962: 117—135.

Rousseaux, J.M., Gomez Laverde, C., Nathan, Y. and Rouxhet, P.G., 1972. Correlation between the hydroxyl stretching bands and the chemical composition of trioctahedral micas. Int. Clay Conf., Madrid, I: 117—127.

Rouxhet, P.G., 1970. Hydroxyl stretching bands in mica: A quantitative interpretation. Clay Miner., 8: 375—388.

Rouxhet, P.G., 1970. Kinetics of dehydroxylation and of OH-OD exchange in macrocrystalline micas. Am. Mineralogist, 55: 841—853.

Saksena, B.D., 1964. Infrared hydroxyl frequencies of muscovite, phlogopite and biotite micas in relation to their structures. Trans. Faraday Soc., 60: 1715—1725.

Serratosa, J.M. and Bradley, W.F., 1958. Determination of the orientation of OH bond axes in layer silicates by infrared absorption. J. Phys. Chem., 62: 1164–1167.
Serratosa, J.M. and Bradley, W.F., 1958. Infrared absorption of OH bands in micas. Nature, 181: 111.
Serratosa, J.M. and Hidalgo, A., 1964. The infrared spectra of micas in the 1000 - 650 cm^{-1} region and its dependence on crystal orientation. Appl. Opt., 3: 315–316.
Smith Aitken, W.W., 1965. An occurrence of phlogopite and its transformation to vermiculite by weathering. Mineral. Mag., 35: 151–164.
Stubičan, V. and Roy, R., 1962. Boron substitution in synthetic micas and clays. Am. Mineralogist, 47: 1166–1173.
Tarasevich, Yu.I., 1969. Investigation of infrared spectra of heavy water sorbed by montmorillonite and vermiculite saturated with different cations. Int. Clay Conf., Tokyo, 1: 261–269.
Tarasevich, Yu.I. and Ovcharenko, F.D., 1969. Infrared-spectroscopic study of the thermal dehydration of cation-substituted vermiculite. Kolloid. Zh., 31: 451–458.
Tarasevich, Yu.I. and Ovcharenko, F.D., 1969. A spectral study of the absorption of heavy water by cation-substituted vermiculite. Dokl. Akad. Nauk S.S.S.R., 184: 142–143.
Titeica, R. and Palade, G., 1959. Spectre d'absorption infrarouge de certains micas. Rev. Phys., Acad. Rép. Pop. Roumaine, 4: 93–97.
Tomita, K. and Sudo, T., 1971. Transformation of sericite into an interstratified mineral. Clays Clay Miner., 19: 263–270.
Tsuboi, M., 1950. On the positions of the hydrogen atoms in the crystal structure of muscovite as revealed by the infrared absorption study. Bull. Chem. Soc. Japan., 23: 83–88.
Van der Marel, H.W., 1966. Quantitative analysis of clay minerals and their admixtures. Contrib. Mineral. Petrol., 12: 96–138.
Vedder, W., 1964. Correlations between infrared spectra and chemical composition of mica. Am. Mineralogist, 49: 736–768.
Vedder, W., 1965. Ammonium in muscovite. Geochim. Cosmochim. Acta, 29: 221–228.
Vedder, W. and McDonald, R.S., 1963. Vibrations of the OH-ions in muscovite. J. Chem. Phys., 38: 1583–1590.
Vedder, W. and Wilkins, R.W.T., 1969. Dehydroxylation and rehydroxylation, oxidation and reduction of micas. Am. Mineralogist, 54: 482–509.
Vergnoux, A.M., Théron, S. and Pouzol, M., 1954. Etude dans le proche infrarouge de la bande OH du mica. Compt. Rend., 238: 467–469.
White, J.L., Bailey, G.W., Brown, C.B. and Ahlrichs, J.L., 1961. Infrared investigation of the migration of lithium ions into empty octohedral sites in muscovite and montmorillonite. Nature, 190: 342.
White, J.L. and Burns, A.F., 1963. Infrared spectra of hydronium ion in micaceous minerals. Science, 141: 800–801.
Wilkins, R.W.T., 1967. The hydroxyl-stretching region of the biotite mica spectrum. Mineral. Mag., 36: 325–333.
Wilson, M.J., 1970. A study of weathering in a soil derived from a biotite-hornblende rock. 1. Weathering of biotite. Clay Miner., 8: 291–303.
Wise, W.S. and Eugster, H.P., 1964. Celadonite: syntheses, thermal stability and occurrence. Am. Mineralogist, 49: 1031–1083.
Wolff, R.G., 1965. Infrared absorption patterns (OH region) of several clay minerals. Am. Mineralogist, 50: 240–244.
Yamamoto, T. and Nakahira, M., 1966. Ammonium ions in sericites. Am. Mineralogist, 51: 1775–1778.

Chlorite minerals

Brydon, J.E. and Kodama, H., 1966. The nature of aluminium hydroxyde-montmorillonite complexes. Am. Mineralogist, 51: 875–889.
Chatterjee, N.D., 1966. On the widespread occurrence of oxidized chlorites in the Pennine zone of the Western Italian Alps. Contrib. Mineral. Petrol., 12: 325–339.
De Waal, S.A., 1970. Nickel minerals from Barberton, South Africa; II: Nimite, a nickel-rich chlorite. Am. Mineralogist, 55: 18–30.
Flehmig, W. and Menschel, G., 1972. Über die Lithiumgehalte und das Aufreten von Cookeit (Lithiumchlorit) in permischen Sandsteinen von Nordhessen, Contrib. Mineral. Petrol., 34: 211–223.
Gupta, G.C. and Malik, W.U., 1969. Fixation of hydroxyl-aluminium by montmorillonite. Am. Mineralogist, 54: 1625–1634.
Gupta, G.C. and Malik, W.U., 1969. Chloritization of montmorillonite by its coprecipitation with magnesiumhydroxyde. Clays Clay Miner., 17: 331–338.
Hayashi, H., 1961. Mineralogical study on alteration products from altered aureole of some "Kuroko" deposits. Kobutsugaku-Zasshi (J. Mineral. Soc. Japan), 6: 101–125.
Hayashi, H. and Oinuma, K., 1964. Aluminium chlorite from the Kamikita mine, Japan. Clay Sci., 2: 22–30.
Hayashi, H. and Oinuma, K., 1965. Relationship between infrared absorption spectra in the region of 450–900 cm^{-1} and chemical composition of chlorite. Am. Mineralogist, 50: 476–483.
Hayashi, H. and Oinuma, K., 1967. Si-O absorption band near 1000 cm^{-1} and OH absorption bands of chlorite. Am. Mineralogist, 52: 1206–1210.
Johnson, L.J., 1964. Occurrence of regularly interstratified chlorite-vermiculite as a weathering product of chlorite in a soil. Am. Mineralogist, 49: 556–572.
Kodama, H. and Oinuma, K., 1963. Identification of kaolin minerals in the presence of chlorite by X-ray diffraction and infrared absorption spectra. Proc. 11th Natl. Conf. Clays Clay Miner., Ottawa, 1962: 236–249.
Kurbatov, A.I. and Aleshin, S.N., 1964. Investigation of minerals of the hydroxyde and chlorite groups by infrared spectroscopy. Dokl. Rossiisk. S-kh. Akad., 99: 5–10.
Meyers, N.L. and Ahlrichs, J.L., 1972. Correlations of X-ray, Ir, Dta, Dtga and Cec observations on Al-hydroxyl interlayers. Int. Clay Conf., Madrid, 2: 243–254.
Montoga, J.W. and Baur, G.S., 1963. Nickeliferous serpentines, chlorites and related minerals found in two lateritic ores. Am. Mineralogist, 48: 1227–1238.
Post, J.L. and Plummer, C.C., 1972. Chlorite series of Flagstaff Hill area, California. A preliminary investigation Clays Clay Miner., 20: 271–283.
Serratosa, J.M. and Vinas, J.M., 1964. Infrared investigation of the OH bands in chlorites. Nature, 202: 999.
Tuddenham, W.M. and Lyon, R.J.P., 1959. Relation of infra-red spectra and chemical analysis for some chlorites and related minerals. Anal. Chem., 31: 377–380.
Van der Marel, H.W., 1966. Quantitative analysis of clay minerals and their admixtures. Contrib. Mineral. Petrol., 12: 96–138.
Weismiller, R.A., Ahlrichs, J.L. and White, J.L., 1967. Infrared studies of hydroxy-aluminium interlayer material. Proc. Soil Sci. Soc. Am., 31: 459–463.

Interstratified minerals

Behar, A., Hubenov, Gr. and Van der Marel, H.W., 1972. Highly interstratified clay minerals in salt affected soils. Contrib. Mineral. Petr., 34: 229—235.
Brown, G. and Weir, A.H., 1963. The identity of rectorite and allevardite. Proc. Int. Clay Conf., Stockholm, 1: 27—35.
Cranquist, W.T. and Pollack, S.S., 1967. Clay mineral synthesis. II. A randomly interstratified aluminium montmorillonite. Am. Mineralogist, 52: 212—226.
Derudder, R.D. and Beck, C.W., 1963. Stevensite and talc-hydrothermal alteration products of wollastonite. Proc. 11th Natl. Conf. Clays Clay Miner., 1962: 188—199.
Farmer, V.C. and Russell, J.D., 1964. The infrared spectra of layer silicates. Spectrochim. Acta, 20: 1149—1173.
Faust, G.T., Hathaway, J.C. and Millot, G., 1959. A restudy of stevensite and allied minerals. Am. Mineralogist, 44: 342—370.
Imai, N., Otsuka, R., Nakamura, T., Tsunashima, A. and Sakamoto, T., 1973. "Hydrated talc" — an alteration product of wollastonite by reaction with magnesium-bearing hydrothermal solution. Clay Sci., 4: 175—189.
Johnson, L.J., 1964. Occurrence of regularly interstratified chlorite-vermiculite as a weathering product of chlorite in a soil. Am. Mineralogist, 49: 556—572.
Kodama, H., 1966. The nature of the component layers of rectorite. Am. Mineralogist, 51: 1035—1055.
Kodama, H. and Brydon, J.E., 1966. Interstratified montmorillonite-mica clays from subsoils of the Prairie Provinces. Western Canada. Proc. 13th Natl. Conf. Clays Clay Miner., Madison, 1964: 151—173.
Nemecz, E., Varju, G.Y. and Barna, J., 1963. Allevardite from Királyhegy, Tokaj mountains, Hungary. Int. Clay Conf., Stockholm, 2: 51—67.
Oinuma, K. and Hayashi, H., 1965. Infrared study of mixed-layer clay minerals. Am. Mineralogist, 50: 1213—1227.
Russell, J.D. and White, J.L., 1966. Infrared study of the thermal decomposition of ammonium rectorite. Proc. 14th Natl. Conf. Clays Clay Miner., Berkeley, 1965: 181—191.
Shimoda, S., 1972. An interstratified mineral of mica and montmorillonite from the mineralized district at Niida near the Shakanai Mine, Akita Prefecture Japan. Clay Sci., 4: 115—125.
Shimoda, S. and Brydon, J.E., 1971. Infrared studies of some interstratified minerals of mica and montmorillonite. Clays Clay Miner., 19: 61—66.
Tomita, K. and Sudo, T., 1971. Transformation of sericites into an interstratified mineral. Clays Clay Miner., 19: 263—270.
Veniale, F. and Van der Marel, H.W., 1963. An interstratified saponite-swelling chlorite mineral as a weathering product of lizardite rock from St. Margherita (Pavia Province), Italy Beitr. Mineral. Petrol., 9: 198—245.
Veniale, F. and Van der Marel, H.W., 1968. A regular talc-saponite mixed-layer mineral from Ferriere, Nure Valley (Piacenza Province, Italy). Mineral. Petrol., 17: 237—254.
Veniale, F. and Van der Marel, H.W., 1969. Identification of some 1:1 regular interstratified trioctahedral clay minerals. Int. Clay Conf. Tokyo, 7: 233—244.

Pseudo layer silicates with chain structure (hormites)

Cannings, F.R., 1968. An infrared study of hydroxyl groups on sepiolite. J. Phys. Chem., 72: 1072—1074.
Hayashi, H., Otsuka, R. and Imai, N., 1969. Infrared study of sepiolite and palygorskite on heating. Am. Mineralogist, 54: 1613—1624.
Koltermann, M. and Müller, K.P., 1965. Kristallchemie und thermisches Verhalten von Sepiolith, Hectorit, Saponit, Bergleder und Anthophyllit. Tonind. Z. Keram. Rundsch., 89: 406—411.
Launer, P.J., 1952. Regularities in the infrared absorption spectra of silicate minerals. Am. Mineralogist, 37: 764—784.
Mendelovici, E., 1972. Infrared study of attapulgite and HCl treated attapulgite. Int. Clay Conf., Madrid, I: 193—201.
Mendelovici, E., 1973. Infrared study of attapulgite and HCl treated attapulgite. Clays Clay Miner., 21: 115—119.
Midgley, H.G., 1959. A sepiolite from Mullion, Cornwall. Clay Miner. Bull., 4: 88—93.
Otsuka, R., Hayashi, H. and Shimoda, S., 1968. Infrared absorption spectra of sepiolite and palygorskite. Mem. Sch. Sci. Eng., Waseda Univ., 32: 13—24.
Parry, W.T. and Reeves, C.C., 1968. Sepiolite from pluvial Mound Lake, Lynn and Terry Counties, Texas. Am. Mineralogist, 53: 984—993.
Pei-Lin-Tien, 1973. Palygorskite from Warren Quarry, Enderby, Leicestershire, England. Clay Miner., 10: 27—33.
Tarasevich, Yu.I. and Ovcharenko, F.D., 1965. Thermal dehydration of palygorskite as investigated with the aid of infrared spectroscopy. Dokl. Akad. Nauk S.S.S.R., 161: 1138—1141.
Tarasevich, Yu.I., 1970. Spectral study of the thermal dehydration of palygorskite. Dopov. Akad. Nauk Ukr. R.S.R., Ser. B., 32: 938—942.
Tarasevich, Yu.I. and Ovcharenko, F.D., 1971. Spectra investigation of the interaction between water and mountain leather surface. Dokl. Akad. Nauk S.S.S.R., 200: 897—900.
Van der Marel, H.W., 1966. Quantitative analysis of clay minerals and their admixtures. Contrib. Mineral. Petrol., 12: 96—138.

Iron minerals

Blyholder, G. and Richardson, E.A., 1962. Infrared and volumetric data on the absorption of ammonia, water and other gases on activated iron (III) oxide. J. Phys. Chem., 66: 2597—2602.
Blyholder, G. and Richardson, E.A., 1964. Infrared spectral observation of surface states. J. Phys. Chem., 68: 3882—3884.
Chukrov, F.V., Zvyagin, B.B., Ermilova, L.P. and Gorshkov, A.I., 1972. New data on iron oxides in the weathering zone. Int. Conf. Clays Clay Miner., Madrid, 1: 397—404.
Glemser, O., 1959. Binding of water in some hydroxides and hydrous oxides. Nature, 183: 943—944.
Glemser, O. and Hartert, E., 1953. Knickschwingungen der OH-Gruppe im Gitter von Hydroxyden. Naturwissenschaften, 40: 552—553.
Glemser, O. and Hartert, E., 1956. Untersuchingen über die Wasserstoffbrückenbindung in kristallisierten Hydroxyden. Z. Anorg. Allgem. Chem., 283: 111—122.
Glemser, O. and Rieck, G., 1958. Zur Bindung des Wassers in den Systemen Al_2O_3/H_2O, SiO_2/H_2O und Fe_2O_3/H_2O. Z. Anorg. Allg. Chem., 297: 175—188.
Hartert, E. and Glemser, O., 1953. Zur Lage des Wasserstoffs im Gitter kristalliner Hydroxyde. Naturwissenschaften, 40: 199—200.
Kojima, M., 1964. Infrared spectra of artificial iron minerals. Soil Sci. Plant Nutr., 10: 43.
Mackenzie, R.C. and Meldau, R., 1959. The ageing of sesquioxide gels: I. Iron oxide gels. Mineral. Mag., 32: 153—165.

Mitsuishi, A., Yoshinaga, H. and Fujita, S., 1958. The far infrared absorption of ferrites. J. Phys. Soc. Japan, 13: 1236—1237.
Moenke, H., 1962. Spektralanalyse von Mineralien und Gesteinen. Akad. Verlagsges. Geest und Portig, K.G., Leipzig, 222 pp.
Vratny, F., Dilling, M., Gugliotta, F. and Rao, C.N.R., 1961. Infrared spectra of metallic oxides, phosphates and chromates. J. Sci. Ind. Res., 20b: 590—593.
Waldron, R.D., 1955. Infrared spectra of ferrites. Phys. Rev., 99: 1727—1735.
White, W.B. and Roy, R., 1964. Infrared spectra-crystal structure correlations, 2 Comparison of simple polymorphic minerals. Am. Mineralogist, 49: 1670—1687.

Aluminium minerals

Brindley, G.W. and Choe, J.O., 1961. The reaction series, gibbsite → chi alumina → kappa alumina → corundum. Am. Mineralogist, 46: 771—785.
Cabannes-Ott, C., 1960. Etude de quelques oxydes hydratés, 7. Aluminium. Ann. Chim., Sér. 13, 5: 935—939.
Frederickson, L.D., 1954. Characterization of hydrated aluminas by infrared spectroscopy. Application to the study of bauxite ores. Anal. Chem., 26: 1883—1885.
Gastuche, M.C. and Herbillon, A., 1962. Etude des gels d'alumine: cristallisation en milieu désionisé. Bull. Soc. Chim. France, Sér. 5, 29: 1404—1412.
Glemser, O. and Hartert, E., 1953. Knickschwingungen der OH-Gruppe im Gitter von Hydroxyden. Naturwissenschaften, 40: 552—553.
Glemser, O. and Hartert, E., 1956. Untersuchungen über die Wasserstoffbrückenbindung in kristallisierten Hydroxyden. Z. Anorg. Allgem. Chem., 283: 111—122.
Glemser, O. and Rieck, G., 1956. Die Bindung des Wassers in den durch thermische Zersetzung von Aluminiumhydroxyden entstanden Phasen. Angew. Chem., 68: 182.
Glemser, O. and Rieck, G., 1958. Zur Bindung des Wassers in den Systemen Al_2O_3/H_2O, SiO_2/H_2O und Fe_2O_3/H_2O. Z. Anorg. Allg. Chem., 297: 175—188.
Harkins, T.R., Harris, J.T. and Schreven, D.D., 1959. Identification of pigments in paint products by infrared spectroscopy. Anal. Chem., 31: 541—545.
Hartert, E. and Glemser, O., 1953. Zur Lage des Wasserstoffs im Gitter kristalliner Hydroxyde. Naturwissenschaften, 40: 199—200.
Ignatieva, L.A., Chukin, G.D. and Bondarenko, G.V., 1968. Hydroxyl cover and reaction with water of gamma-Al_2O_3. Dokl. Akad. Nauk S.S.S.R., 181: 393—396.
Kolesov, V.A. and Ryskin, Ya.I., 1959. Infrarot absorptionsspektren von Hydrargillite $Al(OH_3)$. Opt. i Spektr., 7: 261—263.
Léonard, A.J., Van Cauwelaert, F. and Fripiat, J.J., 1967. Hydrated aluminas and transition aluminas. J. Phys. Chem., 71: 695—708.
Mackenzie, R.C., Meldau, R. and Gard, J.A., 1962. The ageing of sesquioxide gels, 2. Alumina gels. Mineral. Mag., 33: 145—157.
Omori, K., 1961. Infrared absorption spectra of sillimanite, andalusite, kyanite and diaspore. J. Japan Assoc. Mineral. Petrol. Econ. Geol., 46: 89—91.
Orsini, L. and Petitjean, M., 1953. Etude par spectrographie infrarouge d'une boehmite et de ses produits de deshydration. Compt. Rend., 237: 326—328.
Peri, J.B., 1965. Infrared and gravimetric study of the surface hydration of gamma alumina. J. Phys. Chem., 69: 211—219.
Peri, J.B. and Hannan, R.B., 1960. Surface hydroxyl groups on γ-alumina. J. Phys. Chem., 64: 1526—1530.
Sato, T., 1961. The transformation of bayerite into hydrargillite. J. Appl. Chem. (London), 11: 207—209.
Sato, T., 1962. Thermal transformation of alumina trihydrate, bayerite. J. Appl. Chem. (London), 12: 553—556.
Sato, T., 1963. Infrared study of hydrothermal conversion of alumina trihydrate into monohydrate. J. Appl. Chem. (London), 13: 316—319.
Schwarzmann, E. and Marsmann, H., 1966. Zur Deutung der IR-Spektren fester Hydroxyde mit Wasserstoffbrückenbinding. Naturwissenschaften, 53: 349—352.
Van der Marel, H.W., 1966. Quantitative analysis of clay minerals and their admixtures. Contrib. Mineral. Petrol., 12: 96—138.
Vratny, F., Dilling, M., Gugliotta, F. and Rao, C.N.R., 1961. Infrared spectra of metallic oxides, phosphates and chromates. J. Sci. Ind. Res., 20b: 590—593.
Wei, Y.K. and Bernstein, R.B., 1959. Deuterium exchange between water and boehmite (α-alumina monohydrate). Activation energy for proton diffusion in boehmite. J. Phys. Chem., 63: 738—741.
Weismiller, R.A., Ahlrichs, J.L. and White, J.L., 1967. Infrared studies of hydroxy-aluminium interlayer material. Proc. Soil Sci. Soc. Am., 31: 459—463.
White, W.B. and Roy, R., 1964. Infrared spectra-crystal structure correlations. 2. Comparison of simple polymorphic minerals. Am. Mineralogist, 49: 1670—1687.
Wickersheim, K.A. and Korpi, G.K., 1965. Interpretation of the infrared spectrum of boehmite. J. Chem. Phys., 42: 579—583.

Silica minerals

Bambauer, H.U., 1963. Merkmale des OH Spektrums alpiner Quarze (3 μ Gebiet). Schweiz. Mineral. Petrogr. Mitt., 43: 259—268.
Bappu, M.K.V., 1953. Spectroscopic study of amethyst quartz in the ultraviolet and infrared regions. Indian J. Phys., 27: 385—392.
Brügel, W., 1950. Das Reflexionsspektrum des Quarzglases bei 9 μ. Z. Physik., 128: 255—259.
Brunner, G., Wondratschek, H. and Laves, F., 1959. Über die Ultrarotabsorption des Quarzes im 3 μ Gebiet. Naturwissenschaften, 46: 664.
Brunner, G.O., Wondratschek, H. and Laves, F., 1960. Zur Ultrarotabsorption von Quarz und einigen Feldspaten im 3 μ Gebiet. Fortschr. Mineral., 38: 125—126.
Brunner, G.O., Wondratschek, H. and Laves, F., 1961. Ultrarotuntersuchung über den Einbau von H in natürlichen Quarz. Z. Electrochem., 56: 735—750.
Chester, R. and Elderfield, H., 1968. The infrared determination of opal in siliceous deep-sea sediments. Geochim. Cosmochim. Acta, 32: 1128—1140.
Dodd, D.M. and Fraser, D.B., 1965. The 3000 - 3900 cm^{-1} absorption bands and anelasticity in crystalline α-quartz. J. Phys. Chem. Solids, 26: 673—686.
Dodd, D.M. and Fraser, D.B., 1967. Infrared studies of the variation of H-bonded OH in synthetic α-quartz. Am. Mineralogist, 52: 149—160.
Florinskaya, V.A. and Pechenkina, R.S., 1952. Reflection and transmission spectra of different modifications of silica in the infrared region. Dokl. Akad. Nauk. S.S.S.R., 85: 1265—1268.
Gade, M. and Luft, K.F., 1959. Über ultrarotspektroskopische Quarzbestimmung in Grubenstäuben. Naturwissenschaften, 46: 315—316.
Giupseppetti, G. and Veniale, F., 1969. Relazioni tra natura dell'acqua morfologia e struttura degli

opali. Notta proliminare. Rend. Soc. Ital. Mineral. Petrol., 25: 407—437.

Haas, C., 1956. Vibration spectra of crystals. Spectrochim. Acta, 8: 19—26.

Haccuria, M., 1953. Spectre infrarouge de la silice amorphe, de la tridymite, de la cristobalite, du quartz et de la silice fondue. Bull. Soc. Chim. Belg., 62: 428—435.

Häfele, H.G., 1960. Über die Absorption des Quarzes im nahen Ultrarot. Z. Phys., 160: 420—430.

Hanna, R., 1965. Infrared absorption spectrum of silicon dioxide. J. Am. Ceram. Soc., 48: 595—599.

Haven, Y. and Kats, A., 1962. Hydrogen in α-quartz. Silic. Ind., 27: 137—140.

Hayashi, H., 1963. Procedure of mineral analysis of dusts in the lung by X-ray and infrared studies. Ind. Health, 1: 37—46.

Jones, J.B. and Segnit, E.R., 1969. Water in sphere-type opal. Mineral. Mag., 37: 357—361.

Kats, A., 1958. Absorption spectra of silica glass and quartz crystals containing contaminations by germaniun. Verres Refractaires, 12: 191—205.

Kats, A., 1962. Hydrogen in alpha quartz. Philips Res. Rep., 17: 135—195; 201—279.

Kats, A. and Haven, Y., 1960. Infrared absorption bands in α-quartz in the 3 μ-region. Phys. Chem. Glasses, 1: 99—102.

Kats, A., Haven, Y. and Stevels, J.M., 1962. Hydroxyl groups in α-quartz. Phys. Chem. Glasses, 3: 69—75.

Kleinman, D.A. and Spitzer, W.G., 1962. Theory of the optical properties of quartz in the infrared. Phys. Rev., 125: 16—30.

Lipinski, D. and Schwiete, H.E., 1963. Quantitative Bestimmung der SiO_2-Phasen mit Hilfe der Infrarotspektroanalyse. Glas Email Keramo Tech., 14: 325—329.

Lipinski, D. and Schwiete, H.E., 1964. Die Bildung der Cristobalits aus amorphen Siliziumdioxid unter verschiedenen Gasatmosphären. Tonind. Z. Keram. Rundsch., 88: 145—153, 217—225, 258—262.

Lippincott, E.R., Van Valkenberg, A., Weir, C.E. and Bunting, E.N., 1958. Infrared studies on polymorphs of silicon dioxide and germanium dioxide. J. Res. Natl. Bur. Std., 61: 61—70.

Lyon, R.J.P., Tuddenham, W.M. and Thompson, C.S., 1959. Quantitative mineralogy in 30 minutes. Econ. Geol., 54: 1047—1055.

McGinney, F.E., 1931. Infrared absorption of fused and crystalline quartz from 2 μ to 8 μ Proc. Univ. Durham Phil. Soc., 8: 337—350.

Milkey, R.G., 1960. Infrared spectra of some tectosilicates. Am. Mineralogist, 45: 990—1007.

Mitchell, E.W.J. and Rigden, J.P., 1957. The effects of radiation on the near infrared absorption spectrum of α-quartz. Phil. Mag., 2: 941—956.

Mitsudo, T., 1960. On acid clays and cristobalite from Itoigawa. J. Mineral. Soc. Japan, 4: 335—362.

Omori, K., 1967. Infrared study of mechanical mixtures of quartz, orthoclase and oligoclase from 11 to 25 microns. Miner. J., 5: 169—179.

Omori, K., 1968. Far infrared absorption spectrum of quartz and the analysis of absorption bands. Kobutsugaku Zasski, 9: 15—29.

Pelto, C.R., 1956. A study of chalcedony. Am. J. Sci., 254: 32—50.

Plendl, J.N., Mansur, L.C., Hadni, A., Brehat, F., Henry, P., Morlot, G., Naudin, F. and Strimer, P., 1967. Low temperature far infrared spectra of SiO_2 polymorphs. J. Phys. Chem. Solids, 28 (2): 1589—1597.

Ram, A., Bishui, B.M. and Dhar, R.N., 1963. Estimation of quartz in clays by infrared method. Centr. Glass Ceram. Res. Inst. Bull. (India), 10: 31—37.

Reitzel, J., 1955. Infrared spectra of SiO_2 from 400 to 600 cm^{-1}. J. Chem. Phys., 23: 2407—2409.

Rey, T., 1966. Ultrarot absorption von $AlPO_4$ und SiO_2 in Abhängigkeit von Fehlordnung und Temperatur. Z. Kristall., 123: 263—314.

Roberts, S. and Coon, D.D., 1962. Far infrared properties of quartz and sapphire. J. Opt. Soc. Am., 52: 1023—1029.

Saksena, B.D., 1958. The infrared absorption spectra of α-quartz between 4 and 15 microns. Proc. Phys. Soc. (London), 72: 9—16.

Saksena, B.D. and Narain, H., 1949. Raman and infrared spectra of β-quartz. Nature, 164: 583—584.

Scholze, H. and Franz, H., 1960. Eine Methode zur Aufnahme von Ultrarotspektren von körnigen Material und ihre Anwendung auf einige Feldspäte und SiO_2-Varietäten. Ber. Dtsch. Keram. Ges., 37: 420—423.

Sclar, C.B., Carrison, L.C. and Schwartz, C.M., 1962. Relation of infrared spectra to coordination in quartz and two high-pressure polymorphs of SiO_2. Science, 138: 525—526.

Sevchenko, N.A. and Florinskaya, V.A., 1956. The spectrum of reflection and transmission of different modifications of silica in the wavelength region 7 - 24 microns. Dokl. Akad. Nauk S.S.S.R., 109: 1115—1118.

Sevchenko, N.A. and Florinskaya, V.A., 1958. Reflection spectra of quartz crystal plates cut at different angles to the optical axis within the 7 - 24 μ wavelength range. Opt. i Spektr., 41: 261—264.

Shields, J.H. and Ellis, J.W., 1956. Dispersion of birefrigence of quartz in the near infrared. J. Opt. Soc. Am., 46: 263—265.

Sidorov, T.A., 1958. Infrared spectra at low temperatures and the structure of quartz and cristobalite. Opt. i Spektr., 4: 800—801.

Simon, J. and McMahon, H.O., 1953. Study of the structure of quartz, cristobalite and vitreous silica by reflection in infrared. J. Chem. Phys., 21: 23—30.

Soda, R., 1961. Infrared absorption spectra of quartz and some other silica modifications. Bull. Chem. Soc. Japan, 34: 1491—1495.

Spitzer, W.G. and Kleinmann, D.A., 1961. Infrared lattice bands of quartz. Phys. Rev., 121: 1324—1335.

Su, G.J., Borelli, N.F. and Miller, A.R., 1962. An interpretation of the infrared spectra of silicate glasses. Phys. Chem. Glasses, 3: 167—176.

Sun, Ming-Shan., 1962. Tridymite (low form) in some opal of New Mexico. Am. Mineralogist., 47: 1453—1455.

Van der Marel, H.W., 1966. Quantitative analysis of clay minerals and their admixtures. Contrib. Mineral. Petrol., 12: 96—138.

Vratny, F., Dilling, M., Gugliotta, F. and Rao, C.N.R., 1961. Infrared spectra of metallic oxides, phosphates and chromates. J. Sci. Ind. Res., 20b: 590—593.

Wentink, T. and Planet, W.G., 1961. Infrared emission spectra of quartz. J. Opt. Soc. Am., 51: 595—600.

Wilding, L.P., Brown, R.E. and Holowaychuk, N., 1967. Accessibility and properties of occluded carbon in biogenetic opal. Soil Sci., 103: 56—61.

Wood, D.L., 1957. Infrared absorption bands in α-quartz. J. Chem. Phys., 27: 1438—1439.

Wood, D.L., 1960. Infrared absorption of defects in quartz. J. Phys. Chem. Solids, 13: 326—336.

Zarzycki, J. and Naudin, F., 1962. Etude comparative des résaux vitreux des systèmes $2SiO_2,X_2O$ $2GeO_2.X_2O$ $2BeF_2$, XF. Proc. 4th Int. Meet. Adv. Molecul. Spectr., 3: 1071—1083.

7. Admixtures

Carbonates

Adler, H.H. and Kerr, P.F., 1962. Infrared study of aragonite and calcite. Am. Mineralogist, 47: 700—717.
Adler, H.H. and Kerr, P.F., 1963. Infrared absorption frequency trends for anhydrous normal carbonates. Am. Mineralogist, 48: 124—137.
Akhmanova, M.V. and Orlova, L.P., 1966. Investigation of rare-earth carbonates by infrared spectroscopy. Geochem. Int., 3: 444—451.
André-Louisfert, J., 1961. Le spectre d'absorption infrarouge des harmoniques et des combinaisons des vibrations internes de la calcite, de la dolomite et de la magnésite entre 1400 et 700 cm^{-1}. Compt. Rend., 252: 3565—3567.
Angino, E.E., 1967. Far infrared (500—30 cm^{-1}) spectra of some carbonate minerals Am. Mineralogist, 52: 137—148.
Baron, G., Caillère, S., Lagrange, R. and Pobequin. Th., 1959. Etude du Mondmilch de la grotte de la Clamouse et de quelques carbonates et hydrocarbonates alcalino-terreux. Bull. Soc. Fr. Minéral. Crist., 82: 150—158.
Caillère, S. and Pobeguin, Th., 1962. Comparaison des propriétés thermiques et du comportement dans l'infrarouge de quelques carbonates, nitrates, sulfates et séléniates simples. Bull. Soc. Fr. Minéral. Crist., 85: 48—57.
Carlson, E.T. and Berman, H.A., 1960. Some observations on the calcium aluminate carbonate hydrates. J. Res. Natl. Bur. Stand., 64 A: 333—341.
Chester, R. and Elderfield, H., 1967. The application of infrared absorption spectroscopy to carbonate mineralogy. Sedimentology, 9: 5—21.
Cifrulak, S.D., 1970. High pressure mid-infrared studies of calcium carbonate. Am. Mineralogist, 55: 815—824.
Decius, J.C., 1955. Coupling the out-of-plane bending mode in nitrates and carbonates of the aragonite structure. J. Chem. Phys., 23: 1290—1294.
De Keyser, W.L., 1965. Applications de la spectrométrie infrarouge à l'étude de matériaux céramiques. Bull. Soc. Fr., Céram., 68: 43—50.
Dutz, H. and Holland, H., 1961. Ultrarotspektroskopische Untersuchungen zur Hydratation von Brantkalk. Tonind. Z. Keram. Rundsch., 85: 58—60.
Elderfield, H., 1971. The effect of periodicity on the infrared absorption frequency ν_4 of anhydrous normal carbonate minerals. Am. Mineralogist, 56: 1600—1606.
Estep, P.A., Kovach, J.J., Hiser, A.L. and Karr, C., 1970. Characterisation of carbonate minerals in oil shales and coals by infrared spectroscopy. In: R.A. Friedel (Editor), Spectrometry of Fuels. Plenum Press, New York, N.Y., pp. 228—247.
Gatehouse, B.M., Livingstone, S.E. and Nyholm, R.S., 1958. The infrared spectra of some simple and complex carbonates. J. Chem. Soc.: 3137—3142.
Herman, H. and Dallemagne, M.J., 1964. Les carbonato-hydroxylapatites et le carbonate des os et dents étudiés par la spectrophotométrie dans l'infrarouge. Bull. Soc. Chim. Biol., 46: 373—383.
Hexter, R.M., 1958. High-resolution, temperature-dependent spectra of calcite. Spectrochim. Acta, 10: 281—290.
Huang, C.K. and Kerr, P.F., 1960. Infrared study of the carbonate minerals. Am. Mineralogist, 45: 311—324.
Meier, W. and Moenke, H., 1961. Über die Natur des Kalziumcarbonates in Gallensteinen. Naturwissenschaften, 48: 521.
Narayanan, P.S. and Lakshmanan, B.R., 1958. Infrared and Raman spectra of witherite ($BaCO_3$) and strontianite ($SrCO_3$). J. Indian Inst. Sci., 40 A: 1—11.
Pobeguin, T., 1959. Etude au moyen des rayons infrarouges de quelques concretions et specimens d'argiles recontrés dans les grottes. Compt. Rend., 248: 2220-2222.
Pobeguin, T., 1959. Détection au moyen des rayons infrarouges, des groupements OH et H_2O dans quelques hydrocarbonates et oxalates. Compt. Rend., 248: 3585—3587.
Pobeguin, Th. and Lecomte, J., 1953. Etude de quelques mélanges de phosphates et de carbonates de calcium, naturels ou artificiels au moyen de leurs spectres d'absorption infrarouges. Compt. Rend., 236: 1544—1547.
Posner, A.S. and Duyckaerts, G., 1954. Infrared study of the carbonate in bone, teeth and francolite. Experimentia, 10: 424—425.
Ramdas, A.K., 1953. The infrared absorption spectra of sodium nitrate and calcite. Proc. Indian Acad. Sci., 37a: 441—450.
Ruotsala, A.P., 1964. Determination of calcite-dolomite ratios by infrared spectroscopy. J. Sediment. Petrol., 34: 676—677.
Saumagne, P. and Josien, M.L., 1962. Etude par spectroscopie infrarouge des carbonates acides alcalins. Adv. Molecul. Spectr. Proc. 4th Int. Meet. Molecul. Spectr., 3: 1033—1038.
Schock, R.N. and Katz, S., 1968. Pressure dependence of the infrared absorption of calcite. Am. Mineralogist, 53: 1910—1917.
Schroeder, R.A., Weir, C.E. and Lippincott, E.R., 1962. Lattice frequencies and rotational barriers for inorganic carbonates and nitrates from low temperature infrared spectroscopy. J. Res. Natl. Bur. Stand., 66 A: 407—434.
Schutte, C.J.H. and Buys, K., 1961. A difference band of a fundamental and a lattice mode in the infrared spectra of lead and other carbonates. Nature, 192: 351—352.
Sterzel, W. and Chorinsky, E., 1968. Die Wirkung schwerer Kohlenstoffisotopen auf das Infrarotspektrum von Karbonaten. Spectrochim. Acta, 24 A: 353—360.
Szymanski, H.A. and Povinelli, R., 1961. Infrared spectra of water-soluble carbonates. Nature, 191: 64—65.
Underwood, A.L., Toribara, T.Y. and Neuman, W.F., 1955. An infrared study of the nature of bone carbonate. J. Am. Chem. Soc., 77: 317—319.
Van der Marel, H.W., 1966. Quantitative analysis of clay minerals and their admixtures. Contrib. Mineral. Petrol., 12: 96—138.
Weir, C.E., Lippincott, E.R., Van Valkenburg, A. and Bunting, E.N., 1959. Infrared studies in the 1—15 micron region to 30.000 Atmospheres. J. Res. Natl. Bur. Stand., 63 A: 55—62.
Weir, C.E. and Lippincott, E.R., 1961. Infrared studies of aragonite, calcite and vaterite type structures in the borates, carbonates and nitrates. J. Res. Natl. Bur. Stand., 65 A: 173—183.
White, W.B., 1971. Infrared characterization of water and hydroxyl ion in the basic magnesium carbonates. Am. Mineralogist, 56: 46—53.

Sulfur, sulfates, sulfides

Adler, H.H. and Kerr, P.F., 1965. Variations in infrared spectra, molecular symmetry and site symmetry of sulfate minerals. Am. Mineralogist, 50: 132—147.
Allen, H.C. and Cross, P.C., 1951. The hydrogen sulfide band at 1180 cm^{-1}. J. Chem. Phys., 19: 140.
Barrow, G.M., 1953. The infrared spectra of oriented rhombic sulphur

crystals with polarized radiation. J. Chem. Phys., 21: 219—222.
Bensted, J. and Prakash, S., 1968. Investigation of the calcium sulfate-water system by infrared spectroscopy. Nature, 219: 60—61.
Bernard, M.P., 1961. Etude par spectrométrie infrarouge du sulfate de glycocolle: fréquences des vibrations de valence des groupements OH...O. Compt. Rend., 252: 2093—2095.
Bernstein, H.J. and Powling, J., 1950. The vibrational spectra and structure of inorganic molecules. II. Sulfur S_8, sulfur chloride S_2Cl_2, phosphorous P_4. J. Chem. Phys., 18: 1018—1023.
Bernstein, H.J. and Powling, J., 1951. Erratum: The vibrational spectra and structure of inorganic molecules. II. Sulfur S_8, sulfur chloride S_2Cl_2, phosphorous P_4. J. Chem. Phys., 19: 139.
Caillère, S. and Pobeguin, Th., 1961. Sur les propriétés de quelques carbonates, sulfates et nitrates alkalino-terreux simples et anhydres. Compt. Rend. Congr. Soc. Savantes, Paris, 1960: 277—281.
Caillère, S. and Pobeguin, Th., 1962. Comparaison des propriétés thermiques et du comportement dans l'infrarouge de quelques carbonates, nitrates, sulfates et séléniates simples. Bull. Soc. Fr. Minéral. Crist., 85: 48—57.
Citron, I. and Underwood, A.L., 1960. Infrared determination of traces of sulfate in reagent chemicals. Anal. Chim. Acta, 22: 338—344.
Erd, R.C., Lyon, R.J.P. and Madsen, B.M., 1965. Infrared studies of saline sulfate minerals: Discussion. Bull. Geol. Soc. Am., 76: 271—282.
Gamo, I., 1961. Infrared spectra of water of crystallization in some inorganic chlorides and sulfates. Bull. Chem. Soc. Japan, 34: 760—764.
Gillespie, R.J. and Robinson, E.A., 1963. The sulphur-oxygen bond in sulphuryl and thionyl compounds: correlation of stretching frequencies and force constants with bond lengths, bond angles and bond orders. Can. J. Chem., 41: 2074—2085.
Hartert, E., 1956. Über den Einfluss der Bindung der Kristalwassers auf dessen Deformationsschwingung in Ultrarot. Naturwissenschaften, 43: 275—276.
Hass, M. and Sutherland, G.B.B.M., 1956. The infrared spectrum and crystal structure of gypsum. Proc. R. Soc. (London), A 236: 427—445.
Klimov, V.V., 1959. Absorption spectra of certain sulfates. Tr. Kazakh. Nauch. Issl. Inst. Mineral. Syr'ya, 1: 218—227.
Kronstein, M., Kraushaar, R.J. and Deacle, R.E., 1963. Sulfur as a standard of reflectance. J. Opt. Soc. Am., 53: 438—465.
Lohman, J.B., Reding, F.P. and Hornig, D.F., 1951. The fundamental vibrations and crystal structure of H_2S and D_2S. J. Chem. Phys., 19: 252—253.
Meyer, B., 1962. Vibration spectrum of trapped S_2. J. Chem. Phys., 37: 1577—1578.
Mischke, W., 1930. Die ultraroten Spektren von H_2O, H_2S, H_2Se. Z. Phys., 67: 106—126.
Mitsuishi, A., Yoshinaga, H. and Fujita, S., 1958. The far infrared absorption of sulfides, selenides and tellurides of zinc and cadmium. J. Phys. Soc. Japan, 13: 1235—1236.
Narasimham, N.A. and Apparao, K.V.S.R., 1966. Isotope shifts in the near infrared bands of diatomic sulphur. Nature, 210: 1034—1035.
Neff, V.D. and Walnut, T.H., 1961. Effect of temperature on the intensity and structure of bands in the infrared spectrum of rhombic sulphur. J. Chem. Phys., 35: 1723—1729.
Omori, K., 1968. Infrared diffraction and the far infrared spectra of anhydrous sulfates. Mineral. J., 5: 334—354.
Omori, K. and Kerr, P.F., 1962. Infrared study of sulfate minerals. Am. Mineralogist, 47: 198—199.
Omori, K. and Kerr, P.F., 1963. Infrared studies of saline sulfate minerals. Geol. Soc. Amer., 74: 709—734.
Omori, K. and Kerr, P.F., 1964. Infrared studies of sulfates from 11 to 25 microns. Sci. Rep. Tohoku Univ. 3th Ser., 9: 1—55.
Omori, K. and Kerr, P.F., 1964. Infrared studies of sulfates from 17 to 25 microns. Sci. Rep. Tohoku Univ., 3th Ser., 9: 65—97.
Omori, K. and Kerr, P.F., 1964. Infrared studies of sulfates from 11 to 25 microns. Sci. Rep. Tohoku Univ., 10: 1—55.
Omori, K., Hijikata, K., Igarashi, Y., Mizota, T., Tanabe, K. and Yagyu, T., 1964. Infrared studies of sulfates from 11 to 25 μ. Sci. Rep. Tohoku Univ., 3th Ser., 9: 57—64.
Ramdas, A.K., 1954. The infrared absorption spectrum of baryte. Proc. Indian Acad. Sci., 39 A: 81—89.
Razouk, R.I., Salem, A.Sh. and Mikhail, R.Sh., 1960. The sorption of water vapor on dehydrated gypsum. J. Phys. Chem., 64: 1350—1355.
Reding, F.P. and Hornig, D.F., 1957. Vibrational spectra of molecules and complex ions in crystals. X. H_2S and D_2S. J. Chem. Phys., 27: 1024—1030.
Schaack, G., 1963. Das Schwingungsspektrum des monoklinen $CaSO_4.2H_2O$. Phys. Kond. Materie, 1: 245—262.
Schäfer, K., 1948. Die Ermittlung der Verschiebung und Aufspaltung der Normalfrequenzen des SO_4-Ions in den verschiedenen wasserfreien Alkalisulfaten. Z. Elektrochem. Angew. Phys. Chem., 52: 98—103.
Schubert, K.D., 1965. Die qualitative und quantitative Analyse von natürlichen Mineralsalzen. Bergakademie, 17: 70—75.
Stekhanov, A.I., 1957. Structure of the OH band in crystals containing a hydrogen band. Izv. Akad. Nauk. S.S.S.R., Ser. Fiz., 21: 311—321.
Stekhanov, A.I., 1957. Interaction of intramolecular and intermolecular vibrations of the water molecule in the gypsum crystal. Vestn. Leningr. Univ., Ser. Fiz. Khim., 4: 62—69.
Stekhanov, A.I. and Popova, E.A., 1968. Evidence of structural imperfections of gypsum crystals and the vibrational spectra of the hydrogen bond. Opt. i. Spektr., 25: 378—381.
Takemoto, K. and Saiki, Y., 1962. Infrared absorption spectra of calcium sulfate dihydrate, hemihydrate and soluble and insoluble anhydrites. Sekko Sekkai, 61: 277—283.
Webber, D.S., 1954. Polarized infrared absorption spectrum of gypsum. Phys. Rev., 96: 846.
Wiegel, E. and Kirchner, H.H., 1966. Ultrarotspektroskopische Untersuchungen zur thermischen Dehydratation von Gips. Ber. Dtsch. Keram. Ges., 43: 718—723.
Zaitsev, G.A. and Neporent, B.S., 1956. Anisotropy of the absorption of gypsum crystals in the infrared. Zh. Eksp. Teor. Fiz., 29: 857—863.

Amphiboles and pyroxenes

Bancroft, G.M., Maddock, A.G., Burns, R.G. and Strens, R.G.J., 1966. Cation distribution in anthophyllite from Mössbauer and infrared spectroscopy. Nature, 212: 913—915.
Burns, R.G., 1966. Origin of optical pleochroism in orthopyroxenes. Mineral. Mag., 35: 715—719.
Burns, R.G. and Strens, R.G.J., 1966. Infrared study of the hydroxyl bands in clinoamphiboles. Science, 153: 890—892.
Burns, R.G. and Law, A.D., 1970. Hydroxyl stretching frequencies in the infrared spectra of anthophyllites and gedrites. Nature, 226: 73—75.

Burns, R.G. and Greaves, C., 1971. Correlations of infrared and Mössbauer site population measurements of actinolites. Am. Mineralogist, 56: 2010—2033.
Clark, S.P., 1957. Absorption spectra of some silicates in the visible and near infrared. Am. Mineralogist, 42: 732—742.
Ernst, W.G. and Wai, C.M., 1970. Mössbauer, infrared, X-ray and optical study of cation ordering and dehydrogenation in natural and heat-treated sodic amphiboles. Am. Mineralogist, 55: 1226—1258.
Koltermann, M. and Müller, K.P., 1965. Kristallchemie und thermisches Verhalten von Sepiolith, Hectorit, Saponit, Bergleder und Anthopyllit. Tonind. Z. Keram. Rundsch., 89: 406—411.
Launer, P.J., 1952. Regularities in the infrared absorption spectra of silicate minerals. Am. Mineralogist, 37: 764—784.
Lazarev, A.N. and Tenisheva, T.F., 1961. Vibrational spectra of silicates, 3. Infrared spectra of the pyroxenoids and other chain metasilicates. Opt. i Spektr., 11: 584—587.
Lazarev, A.N. and Tenisheva, T.F., 1962. Vibrational spectra of silicates. V. Silicates with anions in ribbon form. Opt. i Spektr., 12: 115—117.
Rowbotham, G. and Farmer, V.C., 1973. The effect of "A" site occupancy upon the hydroxyl stretching frequency in clinoamphiboles. Contrib. Mineral. Petrol., 38: 147—149.
Rutstein, M.S. and White, W.B., 1971. Vibrational spectra of high-calcium pyroxenes and pyroxenoids. Am. Mineralogist, 56: 877—887.
Strens, R.G.J., 1966. Infrared study of cation ordering and clustering in some (Fe-Mg) amphibole solid solutions. Chem. Communic.: 519—520.
White, W.B. and Keester, K.L., 1967. Selection rules and assignments for the spectra of ferrous iron in pyroxenes. Am. Mineralogist, 52: 1508—1514.
White, W.B., McCarthy, B.J. and Scheetz, B.E., 1971. Optical properties of chromium, nickel and cobalt containing pyroxenes. Am. Mineralogist, 56: 72—89.
Wilkins, R.W.T., 1970. Iron-magnesium distribution in the tremolite-actinolite series. Am. Mineralogist, 55: 1993—1998.
Wilkins, R.W.T. and Davidson, L.R., 1970. Occurrence and infrared spectra of holmquistite and hornblende from Mt.Marion near Kalgoorlie, Western Australia. Contrib. Mineral. Petrol., 28: 280—287.

Titano minerals

Dayal, B., 1952. The vibration spectrum of rutile. Proc. Indian Acad. Sci., 32A: 304—312.
Harkins, T.R., Harris, J.T. and Schreve, O.D., 1959. Identification of pigments in paint products by infrared spectroscopy. Anal. Chem., 31: 541—545.
Jones, P. and Hockey, J.A., 1971. Infrared studies of rutile surfaces, 1. Trans. Faraday Soc., 67: 2669—2678.
Jones, P. and Hockey, J.A., 1971. Infrared studies of rutile surfaces, 2. Hydroxylation, hydration and structure of rutile surface. Trans. Faraday Soc., 67: 2679—2685.
Kendall, D.N., 1953. Identification of polymorphic forms of crystals by infrared spectroscopy. Anal. Chem., 25: 382—389.
Launer, P.J., 1952. Regularities in the infrared absorption spectra of silicate minerals. Am. Mineralogist, 37: 764—784.
Matossi, F., 1951. The vibration spectrum of rutile. J. Chem. Phys., 19: 1543—1546.
Soffer, P.B., 1961. Studies of the optical and infrared absorption spectra of rutile single crystals. J. Chem. Phys., 35: 940—945.
Toubeau, G., 1961. Etude des minéraux opaques dans l'infrarouge proche. Bull. Soc. Belg. Géol., 70: 281—289.
Von Hippel, A., Kalnajs, J. and Westphal, W.B., 1962. Protons, dipols and charge carriers in rutile. J. Phys. Chem. Solids, 23: 779—799.
Vratny, F., Dilling, M., Gugliotta, F. and Rao, C.N.R., 1961. Infrared spectra of metallic oxides, phosphates and chromates. J. Sci. Ind. Res., 20b: 590—593.
White, W.B. and Roy, R., 1964. Infrared spectra-crystal structure correlations, 2. Comparison of simple polymorphic minerals. Am. Mineralogist, 49: 1670—1687.
Yates, D.J.C., 1961. Infrared studies of the surface hydroxyl groups on titanium dioxide and of the chemisorption of carbon monoxide and carbon dioxide. J. Phys. Chem., 65: 746—755.

Feldspars

Bradshaw, P.M.D., 1967. Measurement of the modal composition of granitic rock powder by point-counting, infrared spectroscopy and X-ray diffraction. Mineral. Mag., 36: 94—100.
Brunner, G.O., Wondratschek, H. and Laves, F., 1960. Zur Ultrarot—Absorption von Quarz und einigen Feldspaten im 3μ Gebiet. Fortschr. Mineralog., 38: 125—126.
Erd, R.C., White, D.E. Fahey, J.J. and Lee, D.E., 1964. Buddingtonite, an ammonium feldspar with zeolitic water. Am. Mineralogist, 49: 831—850.
Hafner, St. and Laves, F., 1957. Lage und Intensität der Ultrarotspektren von Alkalifeldspäten und sauren Plagioklasen in Abhängigkeit der chemischen Zusammensetzung und der Al/Si-Ordnung/Unordnung. Acta Cryst., 10: 820.
Hafner, St. and Laves, F., 1957. Variation der Lage und Intensität einiger Absorptionen von Feldspäten. Zur Struktur von Orthoklas und Adular. Z. Kristall., 109: 204—225.
Hunt, J.M. and Turner, D.S., 1953. Determination of mineral constituents of rocks by infrared spectroscopy. Anal. Chem., 25: 1169—1174.
Launer, P.J., 1952. Regularities in the infrared absorption spectra of silicate minerals. Am. Mineralogist, 37: 764—784.
Laves, F. and Hafner, St., 1956. Ordnung/Unordnung und Ultrarotabsorption, 1. (Al,Si)-Verteilung in Feldspäten. Z. Kristall., 108: 52—63.
Laves, F. and Hafner, St., 1962. Infrared absorption effects, nuclear magnetic resonance and structure of feldspars. Norsk. Geol. Tidsskr., 42: 57—71.
Lyon, R.J.P., Tuddenham. W.M. and Thompson, C.S., 1959. Quantitative mineralogy in 30 minutes. Econ. Geol., 54: 1047—1055.
Martin, R.F., 1970. Cell parameters and infrared absorption of synthetic high to low albites. Contrib. Mineral. Petrol., 26: 62—74.
Milkey, R.G., 1960. Infrared spectra of some tectosilicates. Am. Mineralogist, 45: 990—1007.
Moenke, H., 1962. Spektralanalyse von Mineralien und Gesteinen. Akad. Verlagsges. Geest und Portig K.G., Leipzig, 222 pp.
Müller-Hessem H., Planz, E. and Schwiete, H., 1960. Die Infrarotspektroskopie in der Keramik unter besondere Berücksichtigung der Reaktionen im System BaO-Al_2O_3-SiO_2. Sprechsaal Keram. Glas Email, 93: 159—163.
Omori, K., 1964. Infrared studies of essential minerals from 11 to 25 microns. Sci. Rep. Tohoku Univ., 9: 65—97.
Omori, K., 1967. Infrared study of mechanical mixtures of quartz, or-

thoclase and oligoclase from 11 to 25 microns. Mineral. J., 5: 169—179.
Scholze, H. and Franz, H., 1960. Eine methode zur Aufnahme von Ultrarot-Spektren von körnigem Material und ihre Anwendung auf einige Feldspäte und SiO_2 Varietäten. Ber. Dtsch. Keram. Ges., 37: 420—423.
Thompson, C.S. and Wadsworth, M.F., 1957. Determination of the composition of plagioclase feldspars by means of infrared spectroscopy. Am. Mineralogist, 42: 334—341.

Mangan minerals

Gattow, G. and Glemser, O., 1961. Darstellung und Eigenschaften von Braunsteinen. 2. Die γ und η Gruppen der Braunsteinen. Z. Anorg. Allg. Chem., 309: 20—36.
Gattow, G. and Glemser, O., 1961. Darstellung und Eigenschaften von Braunsteinen, 3. Die ϵ-, β- und α-Gruppe der Braunsteinen, über Ramsdellit und über die Umwandlungen der Braunsteinen. Z. Anorg. Allg. Chem., 309: 121—150.
Glemser, O., Gattow, G. and Meisiek., 1961. Darstellung und Eigenschaften von Braunsteinen. 1. Die δ-Gruppe der Braunsteinen. Z. Anorg. Allg. Chem., 309: 1—19.
Omori, K., 1964. Infrared studies of essential minerals from 11 to 25 microns. Sci. Rep. Tohoku Univ., 9: 65—97.
Preu, E., 1965. Über die anorganische Bestandteile der Kohlen. Freib. Forsch., A 373: 91—105.
Schwarzmann, E. and Marsmann, H., 1966. Zur Deutung der IR-Spektren fester Hydroxyde mit Wasserstoffbrückenbindung. Naturwissenschaften, 53: 349—352.
Vratny, F., Dilling, M., Gugliotta, F. and Rao, C.N.R., 1961. Infrared spectra of metallic oxides, phosphates and chromates. J. Sci. Ind. Res., 20b: 590—593.
White, W.B. and Roy, R., 1964. Infrared spectra crystal structure correlations, 2. Comparison of simple polymorphic minerals. Am. Mineralogist, 49: 1670—1687.

8. Amorphous materials

Silicic acid

Benesi, H.A. and Jones, A.C., 1959. An infrared study of the water-silica gel system. J. Phys. Chem., 63: 179—182.
Boehm, H.P. and Schneider, M., 1959. Über die Hydroxylgruppen an der Oberfläche des amorphen Siliziumdioxyds "Aerosil" und ihre Reaktionen. Z. Anorg. Allg. Chem., 301: 326—335.
Chevet, A., 1953. Recherches préliminaires sur les spectres d'absorption infrarouges de gels de silice. J. Phys. Radium, 14: 493—494.
Corradini, G. and Mariani, E., 1960. Spettrografia i.r. di silicati idrati di calcio sintetici. Ricerca Scientificia, 30: 1346—1358.
Elkington, P.A. and Curthoys, G., 1968. Hydrogen bonding and absorption on silica gel. J. Coll. Interface Sci., 28: 331—333.
Florinskaya, V.A. and Pechenkina, R.S., 1953. Spectres de réflexion et d'absorption des diverses variétés allotropiques de la silice dans l'infrarouge. Izvestija Akad. Nauk. S.S.S.R., Ser. Fiziskaja, 17: 649—653.
Fripiat, J.J., Gastuche, M.C. and Brichard, R., 1962. Surface heterogeneity in silica gel from kinetics of isotopic exchange OH—OD. J. Phys. Chem., 66: 805—812.
Fripiat, J.J., Leonard, A. and Barake, N., 1963. Relation entre la structure et texture des gels de silice Bull. Soc. Chim. France, 72: 122—140.
Garino-Canina, V., 1954. Bande d'absorption à 2.72μ de la silice vitreuse. Compt. Rend., 239: 705—706.
Glemser, O. and Rieck, G., 1958. Zur Bindung des Wassers in den Systemen Al_2O_3/H_2O, SiO_2/H_2O und Fe_2O_3/H_2O. Z. Anorg. Allg. Chem., 297: 175—188.
Haccuria, M., 1953. Spectre infrarouge de la silice amorphe, de la tridymite, de la cristobalite, du quartz et de la silice fondue. Bull. Soc. Chim. Belg., 62: 428—435.
Jaroslavskii, N.G., 1950. Spectres infrarouges d'absorption des absorbents du type gel de silice. Zh. Fiz. Him., 24: 68.
Lipinski, D. and Schwiete, H.E., 1964. Die Bildung des Cristobalits aus amorphen Siliziumdioxid unter verschiedenen Gasatmosphären. Tonind. Ztg. Keram. Rundsch., 88: 145—153, 217—225, 258—262.
McDonald, R.S., 1957. Study of the interaction between hydroxyl groups of aerosil silica and absorbed non-polar molecules by infrared spectrometry. J. Am. Chem. Soc., 79: 850—854.
McDonald, R.S., 1958. Surface functionality of amorphous silica by infrared spectroscopy. J. Phys. Chem., 62: 1168—1178.
Sevchenko, N.A. and Florinskaya, V.A., 1956. Reflexions und Transmissions Spektren verschiedener Kieselsaüre. Modifikationen im Wellenlängerbereich von 7 bis 24μ. Dokl. Akad. Nauk. S.S.S.R., 109: 1115—1118.
Simon, J. and McMahon, H.O., 1953. Study of the structure of quartz cristobalite and vitreous silica by reflection in infrared. J. Chem. Phys., 21: 23—30.
Terenin, A.N., Jaroslavsky, N.G., Karajakin, A.W. and Sidorova, A.I., 1955. Spectroscopie infrarouge des molécules absorbées sur verre poreux. Mikrochim. Acta, 467—470.
Wey, R. and Kalt, A., 1967. Synthèse d'une silice cristallisée. Compt. Rend., 265 (serie D): 1437—1440.
Wirzing, G., 1963. Untersuchungen am System SiO_2/H_2O im nahen Ultrarot. Naturwiss., 50: 466—469.
Yoshino, T., 1955. Infrared study of the absorption on silica gel. J. Chem. Phys., 23: 1564—1565.

Biogenous siliceous materials

Arimura, S. and Kanno. I., 1965. Some mineralogical and chemical characteristics of plant opals in soils and grasses of Japan. Bull. Kyushu Agric. Exp. Station., 11: 111—120.
Harkins, T.R., Harris, J.T. and Schreve, O.D., 1959. Identification of pigments in paint products by infrared spectroscopy. Anal. Chem., 31: 541—545.
Jones, R.L. and Beavers, A.H., 1963. Some mineralogical and chemical properties of plant opal. Soil Sci., 96: 375—379.

Obsidian, perlite, glass

Abou el Azm., A and Ashour, G., 1958. Infrared transmission measurements of a selected number of silicate, borate and cabalt glasses in relation to their structure. Egypt. J. Chem., 1: 313—325.
Adams, R.V., 1961. Infrared absorption due to water in glasses. Phys. Chem. Glasses, 2: 39—49.
Adams, R.V., 1961. Some experiments on the removal of water from glasses. Phys. Chem. Glasses, 2: 50—54.
Adams, R.V., 1961. Infrared absorption and the structure of glasses. Phys. Chem. Glasses, 2: 101—110.
Adams, R.V. and Douglas, R.W., 1959. Infrared studies on various samples of fused silica with special reference to the bands due to water. J. Soc. Glass Technol., 43: 147—158.
Anderson, S., 1950. Investigation of structure of glasses by their infrared reflection spectra. J. Am. Ceram. Soc., 33: 45—51.
Anderson, S., Bohon, R.L. and Kimpton, D.D., 1955. Infrared spectra and atomic arrangement in fused boron oxide and soda borate glasses. J. Am. Ceram. Soc., 38: 370—377.
Brügel, W., 1950. Das Reflexionsspektrum des Quarzglases bei 9μ. Z. Physik., 128: 255—259.
Bunch, T.E., Cohen, A.J. and Dence, M.R., 1967. Natural terrestrial maskelynite. Am. Mineralogist, 52: 244—253.
Cleek, G.W. and Scuderi, T.G., 1959. Effect of fluorides on infrared transmittance of certain silicate glasses. J. Am. Ceram. Soc., 42: 599—603.
Cleek, G.W., Villa, J.J. and Hahner, C.H., 1959. Refractive indices and transmittances of several optical glasses in the infrared. J. Opt. Soc. Am., 49: 1090—1095.
Čzerný, M., 1959. Über die Auswertung des Ultraroten Spektrums von Festkörpern. Glastech. Ber., 32: 265—266.
Day, D.E. and Rindone, G.E., 1962. Properties of soda aluminosilicate glasses. I. Refractive index, density, molar refractivity and infrared absorption spectra J. Am. Ceram. Soc., 45: 489—496.
Edwards, O.J., 1966. Optical transmittance of fused silica at elevated temperatures. J. Opt. Soc. Am., 56: 1314—1319.
Ellis, J.W., Lyon, W.K. and Drummond, D.G., 1936. The 2,73μ ab sorption band in fused quartz. Nature, 138: 248—249.
Florence, J.M., Allshouse, C.C., Glaze, F.W. and Hahner, C.H., 1950. Absorption of near infrared energy by certain glasses. J. Res. Natl. Bur. Stand., 45: 121—128.
Florence, J.M., Glaze, F.W. and Black, M.H., 1955. Infrared transmittance of some calcium aluminate and germanate glasses. J. Res. Natl. Bur. Stand., 55: 231—237.
Florinskaya, V.A., 1957. Transmission spectra of natural crystalline lead silicates and of crystallization produkts of two component lead silicate glasses in the region 1—13 μ. Opt. i Spektr., 2: 724—737.
Florinskaya, V.A. and Pechenkina, R.S., 1956. The infrared spectra of sodium silicate glasses in relation to their structure. Opt. i Spektr., 1: 690—709.

Florinskaya, V.A., 1960. Infrared reflection spectra of sodium silicate glasses and their relation to structure. Proc. All-Union Conf. Glassy State, Leningr.: 154—168.

Florinskaya, V.A. and Pechenkina, R.S., 1958. Spectra of simple glasses in the infrared range and their relations to the structure of glass. Glass Ind., 39:27—31; 93—96; 151—154; 168.

Florinskaya, V.A. and Pechenkina, R.S., 1958. The infrared spectra of simple glasses and their relation to glass structure. Proc. All-Union Conf. Classy State, Leningr.: 55—74.

Florinskaya, V.A. and Pechenkina, R.S., 1960. Investigation on the crystallization products of glasses in the system Na_2O-SiO_2 by infrared spectroscopy. Proc. All-Union Conf. Glassy State, Leningr.: 135—153.

Folman, M. and Yates, D.J.C., 1958. Infrared and length-change studies in absorption of H_2O and CH_3OH on porous silica glass. Trans. Faraday Soc., 54: 1684—1691.

Friedman, I. and Smith, R.L., 1958. The deuterium content of water in some volcanic glasses. Geochim. Cosmochim. Acta, 15: 218—228.

Fry, D.L., Mohan, P.V. and Lee, R.W., 1960. Hydrogen-Deuterium exchange in fused silica. J. Opt. Soc. Am., 50: 1321—1322.

Götz, J. and Vosáhlová, 1968. Beitrag zur quantitativen Bestimmung des Wassergehaltes in Glas mit Hilfe der Infraroten OH-Banden. Geotech. Ber., 41: 47—55.

Hanna, R. and Su, G.J., 1964. Infrared absorption spectra of sodium silicate glasses from 4 to 30μ. J. Am. Ceram. Soc., 47: 597—601.

Heaton, H.M. and Moore, H., 1957. Glasses consisting mainly of the oxides of elements of high atomic weight. J. Soc. Glass Technol., 41: 3—27; 28—71.

Hedden, W.A. and King, B.W., 1960. Antimonate glass for infrared-transmitting windows. J. Am. Ceram. Soc., 43: 387—388.

Hetherington, G. and Jack, K.H., 1962. Influence of "water" content on the properties of vitreous silica. Phys. Chem. Glasses, 3: 129—133.

Hockey, J.A. and Pethica, B.A., 1961. Surface hydration of silicas. Trans. Faraday Soc., 57: 2247—2262.

Houziaux, L., 1956. Spectres d'absorption infrarouge de quelques verres naturels entre 2 et 24 microns. Geochim. Cosmochim. Acta, 9: 298—300.

Houziaux, L., 1956. Spectres d'absorption de quelques verres de silice dans l'infrarouge entre 2 et 24 microns. Silic. Ind., 21: 491—500.

Janakirama-Rao, Bh.V., 1966. Infrared spectra of GeO_2-P_4O_{10}-V_2O_5 glasses and their relation to structure and electronic conduction, J. Am. Ceram. Soc., 49: 605—609.

Kats, A., 1958. Absorption spectra of silica glass and quartz crystal contaminations by germanium. Verres Refract., 12: 191—205.

Keller, W.D. and Pickett, E.E., 1954. Hydroxyl and water in perlite from Superior, Arizona. Am. J. Sci., 252: 87—98.

King, B.W. and Kelly, G.D., 1958. Infrared-transmitting glasses in the system K_2O—Sb_2O_3—Sb_2S_3. J. Am. Ceram. Soc., 41: 367—371.

Kolesova, V.A., 1954. Problem of the interpretation of vibration spectra of the silicates and the silicate glasses. Zh. Eksp. Teoret. Fiz., 26: 124—127.

Kolesova, V.A., 1957. Vibrational spectra for two-component phosphate glasses and for some crystalline phosphates. Fiz. Sb. L'Vov. Univ., 3: 461—465.

Kolesova, V.A., 1960. Structure of alkali aluminosilicata glasses based on their infrared absorption spectra. Proc. All-Union Conf. Glassy State, Leningr.: 177—179.

Lecomte, J., 1947. Infrared spectra. Proc. Int. Congr. Pure Appl. Chem., London, 11: 509—531.

Lehman, H. and Dutz, H., 1960. Ultrarotspektroskopische Untersuchungen an Gläsern und kristallinen Silikaten. Silic. Ind., 25: 559—566.

Lygin, V.I., 1958. Variation of the infrared absorption spectra of the hydroxyl groups of porous glass and silica gel during thermal dehydration. Vestn. Moskov Univ, Ser. Mat. Mekh. Astron. Fiz. Khim., 13: 223—226.

Mackenzie, J.D., McDonald, R.S. and Murphy, W.K., 1961. Infrared spectroscopy of melts and hygroscopic glasses. Rev. Sci. Instr., 32: 118—121.

Markin, E.P. and Sobolev, N.N., 1960. Infrared reflection spectrum of boric anhydride and fused quartz at high temperatures. Opt. i Spektr., 9: 309—312.

Moulson, A.J. and Roberts, J.P., 1961. Water in silica glass. Trans. Faraday Soc., 57: 1208—1216.

Pepperhoff, W., 1954. Infrared reflectance spectra of binary lead silicate glasses. Z. Elektrochem., 58: 520—522.

Scholze, H., 1959. Der Einbau des Wassers in Gläsern. Der Einfluss des im Glas gelösten Wasser auf das Ultrarotspektrum und die quantitative Ultrarotspektroskopische Bestimmung des Wassers in Gläsern. Glastech. Ber., 32: 81—88, 142—152, 278—281, 314—320, 381—386, 421—426.

Scholze, H., 1960. Zur Frage der Unterscheidung zwischen H_2O-Molekulen und OH-Gruppen in Gläsern und Mineralien. Naturwissenschaften, 47: 226—227.

Scholze, H. and Dietzel, A., 1955. Untersuchungen über den Wassergehalt von Gläsern durch Bestimmung der Ultrarotabsorption im Bereich von 1 bis 5μ. Glastech. Ber., 28: 375—380.

Scholze, H. and Franz, H., 1962. Zur Frage der Infrarot-Bande bei 4,25μ in Gläsern. Glastech. Ber., 35: 278—281.

Sevchenko, N.A. and Florinskaya, V.A., 1958. Infrared transmission spectra of porous and quartz-like glass. Opt. i Spektr., 4: 189—195.

Sevchenko, N.A. and Florinskaya, V.A., 1958. The transmission spectra of quartz glass in the region 2 to 24 mu. Opt. i Spektr., 5: 23—28.

Sidorov, A.N., 1960. Spectral investigation of the absorption of water on porous glass as a function of the degree of hydration of its surface. Opt. i Spektr., 8: 424—426.

Simon, I. and McMahon, H.O., 1953. Study of some binary silicate glasses by means of reflection in infrared. J. Am. Ceram. Soc., 36: 160—164.

Simon, J. and McMahon, H.O., 1953. A study of the structure of quartz, cristobalite and vitreous silica by reflection in infrared. J. Chem. Phys., 21: 23—30.

Stair, R. and Faick, C.A., 1947. Infrared absorption spectra of some experimental glasses containing rare earths and other oxides. J. Res. Natl. Bur. Stand., 38: 95—101.

Stepanov, B.I. and Prima, A.M., 1958. Bending vibrations of the silicates. Interpretation of the spectra of glass. Opt. i Spektr., 5: 15—22.

Su, G.J., Borelli, N.F. and McSwain, B.D., 1962. A comparison of the experimental techniques employed in the measurement of infrared spectra of borate glasses. Phys. Chem. Glasses, 3: 139—140.

Su, G.J., Borelli, N.F. and Miller, A.R., 1962. An interpretation of the infrared spectra of silicate glasses. Phys. Chem. Glasses, 3: 167—176.

Ulrich, D.R., 1964. Electrical and infrared properties of glasses in the system Bi_2O_3—TeO_2. J. Am. Ceram. Soc., 47: 595—596.

Weidel, R.A., 1959. The influence of some transition elements on visible and infrared transmission of cal-

cium aluminate glasses. J. Am. Ceram. Soc., 42: 408—412.
Zarzycki, J. and Naudin, F., 1962. Spectres infrarouges de la cristobalite et de quelques composés isostructuraux de SiO_2 ou BeF_2. Etude compérative des résaux vitreux des systèmes $2SiO_2$, X_2O, $2GeO_2$, X_2O, $2BeF_2$, XF. Adv. Molecul. Spectrosc. Proc. 4th Int. Meet Molecul. Spectrosc., 3: 1071—1083.
Zorina, M.L., Sarukhanishvili, A.V. and Setkina, O.N., 1966. Study of crystallization of multi-component ferruginous glasses by infrared spectroscopy. Izv. Akad. Nauk. S.S.S.R., Neorg. Mater., 2: 1846—1849.

Allophane (hisingerite), imogolite

Aomine, S. and Miyauchi, N., 1965. Imogolite of imogo-layers in Kyushu. Soil Sci. Plant Nutr. (Tokyo), 11: 212—219.
Aomine, S., Inoue, A. and Mizota, C., 1972. Imogolite of Chilean volcanic ash soils. Clay Sci., 4: 95—103.
Basila, M.R., 1962. An infrared study of a silica-alumina surface. J. Phys. Chem., 66: 2223—2228.
Besoain, E., 1969. Imogolite in volcanic soils of Chile. Geoderma, 2: 151—169.
Beutelspacher, H. and Van der Marel, H.W., 1961. Über die Amorphen Stoffe in den Tonen verschiedener Böden. Acta Univ. Carolinae, Geol. Suppl., 1: 97—114.
Briner, G.P. and Jackson, M.L., 1969. Allophanic material in Australian soils derived from pleistocene basalt. Austr. J. Soil Res., 7: 163—169.
Chukhrov, F.V., Berkhin, S.I., Ermilovo, L.P., Moleva, V.A.and Rudnitskaya, E.S., 1963. Allophanes from some deposits of the S.S.S.R. Int. Clay Conf., Stockholm 2: 19—28.
De Kimpe, C.R., 1967. Hydrothermal aging of synthetic alumino-silicate gels. Clay Miner., 7: 203—214.
De Mumbrum, L.E. and Chesters, G., 1964. Isolation and characterization of some soil allophanes. Proc. Soil Sci. Soc. Am., 28: 355—359.
Fieldes, M., 1955. Allophane and related mineral colloids. N. Z. J. Sci. Tech., 37B: 336—350.
Fieldes, M., Walker, I.K. and Williams, P.P., 1956. Infrared absorption spectra of clay soils. N. Z. J. Sci. Tech., 38B: 31—43.
Fieldes, M. and Furkert, R.J., 1966. Nature of allophane in soils, II. Difference in composition. N. Z. J. Sci., 9: 608—622.
Hayashi, H. and Oinuma, K., 1963. X-ray and infrared studies on the behaviours of clay minerals on heating. Clay Sci., 1: 134—154.
Jaritz, G., 1967. Ein Vorkommen von Imogolit in Bimsböden Westdeutschlands. Z. Pflanzenernähr. Düng. Bodenk., 117: 65—77.
Kobashi, T., 1967. Reaction of calcium hydroxyde with allophane-kaolinite clay minerals. Clay Sci., 3: 11—36.
Léonard, A., Suzuki, S., Fripiat, J.J. and De Krimpe, C., 1964. Structure and properties of amorphous silicoaluminas. I. Structure from X-ray fluorescence spectroscopy and infrared spectroscopy. J. Phys. Chem., 68: 2608—2617.
Mitchell, B.D. and Farmer, V.C., 1962. Amorphous clay minerals in some Scottish soil profiles. Clay Miner. Bull., 5: 128—144.
Miyauchi, N. and Aomine, S., 1964. Does "allophane B" exist in Japanese volcanic ash soils? Soil Sci. Plant Nutr., 10: 199—203.
Prasad, J., 1965. Infrared absorption spectra of some clay minerals and micatype structures. Trans. Indian. Ceram. Soc., 24: 78—82.
Russell, J.D., MacHardy, W.D. and Fraser, A.R., 1969. Imogolite: a unique aluminosilicate. Clay Miner., 8: 87—99.
Shoji, S. and Masui, J., 1969. Amorphous clay minerals of recent volcanic ash soils in Hokkaido, II. Soil Sci. Plant Nutr., 15: 191—201.
Snetsinger, K.G., 1967. High-alumina allophane as a weathering product of plagioclase. Am. Mineralogist, 52: 254—262.
Uytterhoeven, J. and Fripiat, J.J., 1962. Etude des hydroxyles d'une silicoalumine amorphe. Bull. Soc. Chim. Fr., 788—792.
Whelan, J.A. and Goldich, S.S., 1961. New data for hisingerite and neotocite. Am. Mineralogist, 46: 1412—1423.
Wolff, R.G., 1965. Infrared absorption patterns (OH region) of several clay minerals. Am. Mineralogist, 50: 240—244.
Wada, K., 1966. Deuterium exchange of hydroxyl groups in allophane. Soil Sci. Plant Nutr. (Tokyo), 12: 176—182.
Wada, K., 1967. A structural scheme of soil allophane. Am. Mineralogist, 52: 690—708.
Wada, K. and Greenland, D.J., 1970. Selectives dissolution and differential infrared spectroscopy for characterization of amorphous constituents in soil clays. Clay Miner., 8: 241—254.

9. Miscellany

Vivianite

Corbridge, D.E.C. and Lowe, E.J., 1954. Infrared spectra of some inorganic phosphorous compounds. J. Am. Chem. Soc., 76: 493—502.
Faye, G.H., Manning, P.G. and Nickel, E.H., 1968. The polarized optical absorption spectra of tourmaline, cordierite, chloritoid and vivianite: ferrous-ferric electronic interaction as a source of pleochroism. Am. Mineralogist, 53: 1174—1201.
Kravitz, L.C., Kingsley. J.D. and Elkin, E.L., 1968. Raman and infrared studies of coupled PO_4^{-3} vibrations. J. Chem. Phys., 49: 4600—4610.
Omori, K., 1964. Infrared studies of essential minerals from 11 to 25 microns. Sci. Rep. Tohoku Univ., 9: 65—97.
Omori, K. and Seki, T., 1960. Infrared studies of some phosphate minerals. Tohoku Univ. Sci. Rep., Ser. 3 (6): 397—403.
Pustinger, J.V., Cave, W.T. and Nielsen, M.L., 1959. Infrared spectra of inorganic phosphorus compounds. Spectrochim. Acta., 15: 909—925.
Tien, Pei-Lin and Waugh, T.C., 1969. Thermal and X-ray studies on earthy vivianite in Graneros shale (Upper Cretaceous), Kansas. Am. Mineralogist, 54: 1355—1362.

Olivine, forsterite

Clark S.P., 1957. Absorption spectra of some silicates in the visible and near infrared. Am. Mineralogist, 42: 732—742.
Duke, D.A. and Stephens, J.D., 1964. Infrared investigation of the olivine group minerals. Am. Mineralogist, 49: 1388—1406.
Hunt, J.M., Wisherd, M.P. and Bonham, L.C., 1950. Infrared absorption spectra of minerals and other inorganic compounds. Analyt. Chem., 22: 1478—1497.
Keller, W.D., Spotts, J.H. and Biggs, D.L., 1952. Infrared spectra of some rock forming minerals. Am. J. Sci., 250: 453—471.
Launer, P.J., 1952. Regularities in the infrared absorption spectra of silicate minerals. Am. Mineralogist, 37: 764—784.
Lehmann, H., Dutz, H. and Koltermann, M., 1961. Ultrarotspektroskopische Untersuchungen zur Mischkristallreihe Forsterite-Fayalit. Ber. Dtsch. Keram. Ges., 38: 512—514.
Moenke, H., 1962. Die Ultrarotabsorption von Neo-silikaten im NaCl-Bereich von Prismenspektralphotometern. Silikattechnik, 13: 246—248.
Saksena, B.D., 1961. Infrared absorption studies of some silicate structures. Trans. Faraday Soc., 57: 242—258.
Su, G.J., Borelli, N.F. and Miller, A.R., 1962. An interpretation of the infrared spectra of silicate glasses. Phys. Chem. Glasses, 3: 167—176.
Tarte, P., 1963. Structure du type olivine et monticellite. Spectrochim. Acta, 19: 25—47.

Mullite, sillimanite

Anderson, S., 1950. Investigation of structure of glasses by their infrared reflection spectra. J. Am. Ceram. Soc., 33: 45—51.
De Keyser, W.L., 1965. Applications de la spectrométrie infrarouge à l'étude de matériaux céramiques. Bull. Soc. Fr. Céram., 68: 43—50.
Muan, A., 1957. Phase equilibria at liquidus temperatures in the system iron oxide-Al_2O_3-SiO_2 in air atmosphere. J. Am. Ceram. Soc., 40: 121—133.
Müller-Hesse, H., 1960. Infrarotspektrographische Untersuchungen im System Al_2O_3-SiO_2. Fortschr. Mineral., 38: 173—176.
Murthy, M.K. and Kirby, E.M., 1962. Infrared study of compounds and solid solutions in the system lithia-alumina-silica. J. Am. Ceram. Soc., 45: 324—329.
Omori, K., 1961. Infrared absorption spectra of sillimanite, andalusite, kyanite and diaspore. J. Japan. Assoc. Mineral. Petrol. Econ. Geol., 46: 89—91.
Roy, R. and Francis, E.E., 1953. On the distinction of sillimanite from mullite by infrared techniques. Am. Mineralogist, 38: 725—728.
Tarte, P., 1959. La distinction mullite-sillimanite par spectrométrie infrarouge. Silic. Ind., 24: 7—13.

Chloritoid

Faye, G.H.,Manning, P.G. and Nickel, E.H., 1968. The polarized optical absorption spectra of tourmaline, cordierite, chloritoid and vivianite: ferrous-ferric electronic interaction as a source of pleochroism. Am. Mineralogist, 53: 1174—1201.
Harrison, F.W. and Brindley, G.W., 1957. The crystal structure of chloritoid. Acta Cryst., 10: 77—82.

Zeolites

Angell, C.L. and Schaffer, P.C., 1965. Infrared spectroscopic investigations of zeolites and absorbed molecules, 1. Structural OH groups. J. Phys. Chem., 69: 3436—3470.
Bertsch, L. and Habgood, H.W., 1963. An infrared spectroscopic study of the absorption of water and carbon-dioxide by Linde molecular sieve X. J. Phys. Chem., 67: 1621-1628.
Breger, I.A., Chandler, J.C. and Zubovic, P., 1970. An infrared study of water in heulandite and clinoptilolite. Am. Mineralogist, 55: 825—840.
Carter, J.L., Lucchesi, P.J. and Yates, D.J.C., 1964. The nature of residual OH groups on a series of near faujasite zeolites. J. Phys. Chem., 68: 1385—1391.
De Kimpe, C.R. and Fripiat, J.J., 1968. Kaolinite crystallization from H-exchanged zeolites. Am. Mineralogist, 53: 216—230.
Flanigen, E.M., Khatami, H. and Szymanski, H.A., 1971. Infrared structural studies of zeolite frameworks. Adv. Chem. Ser., 101. Molecul. Sieve Zeolites, 201—229: 526 pp.
Frohnsdorff, G.J.C. and Kington, G.L., 1958. A note on the thermodynamic and infrared spectra of sorbed water. Proc. R. Soc. (London), 247A: 469—472.
Graham, J., 1964. Absorbed water on clays. Rev. Pure Appl. Chem., 14: 81—90.
Habgood, H.W., 1965. Surface OH groups on zeolite X. J. Phys. Chem., 69: 1764—1768.
Harada, K., Iwamoto, S. and Kihara, K., 1967. Erionite, phyllipsite and gonnardite in the amygdales of altered basalt from Mazé, Niigata Prefecture, Japan. Am. Mineralogist, 52: 1785—1794.
Harada, K. and Tomita, K., 1967. A sodian stilbite from Onigajŏ, Mié Prefecture, Japan, with some experimental studies concerning the conversion of stilbite to wairakite at low water vapor pressures. Am. Mineralogist, 52: 1438—1450.
Kiselev, A.V. and Lygin, V.I., 1966. Infrared spectra and absorption by zeolites. In: L.H. Little (Editor), Infrared Spectra of Absorbed Species. Acad. Press. London-New York, pp 352—381.
Milkey, R.G., 1960. Infrared spectra of some tectosilicates. Am. Mineralogist, 45: 990—1007.
Oinuma, K. and Hayashi, H., 1967. Infrared absorption spectra of some zeolites from Japan. J. Tokyo Univ., Gen. Educ. (Natl. Sci.), 8: 1—12.
Omori, K., 1964. Infrared studies of essential minerals from 11 to 25 microns. Sci. Rep. Tohoku Univ., 9: 65—97.
Szymanski, H.A., Stamires, D.N. and Lynch, G.R., 1960. Infrared spectra

of water sorbed on synthetic zeolites. J. Opt. Soc. Am., 50: 1323–1328.

Ward, J.W., 1970. Infrared spectroscopic studies of zeolites. Proc. 2nd Int. Conf. Molecul. Sieve Zeolites, Worcester Polytech. Inst., Mass., 682–705.

Ward, J.W., 1971. Infrared spectroscopic studies of zeolites. Adv. Chem. Ser., 101, Molecul. Sieve Zeolites: 380–401.

White, J.L., Jelli, A.N., André, J.M. and Fripiat, J.J., 1967. Perturbation of OH groups in decationated Y-zeolites by physically absorbed gases. Trans. Faraday Soc., 63: 461–475.

Wolf, F. and Fürtig, H., 1966. Ultrarotspektroskopische Untersuchungen an Ionenausgetauschter Molekularsieben des Typs A. Tonind. Z., 90: 310–316.

Wolf, F. and Hädicke, U., 1967. Untersuchung von Strukturveränderungen des Zeoliths X unter Ionenaustausch - und Temperatureinflüssen. Tonind. Z. Keram. Rundsch., 91: 48–52.

Wright, A.C., Rupert, J.P. and Granquist, W.T., 1968. High- and Low-silica faujasites: A substitutional series. Am. Mineralogist, 53: 1293–1303.

Yukhnevich, G.V., Karyakin, A.V., Khitarov, N.I. and Senderov, E.E., 1961. Infrared spectroscopic study of some zeolites and the nature of the water bond in natrolite. Geochemistry: 937–944.

Zhdanov, S.P., Kiselev, A.V., Lygin, V.I. and Titova, T.I., 1963. Change in the infrared spectrum of X-zeolites owing to the thermal treatment in vacuo. Dokl. Akad. Nauk S.S.S.R., 150: 584–587.

Zhdanov, S.P., Kiselev, A.V., Lygin, V.I. and Titova, T.I., 1964. Infrared spectra of synthetic faujasites of different compositions and the water absorbed by them. Zh. Fiz. Khim., 38: 2408–2414.

Zhdanov, S.P., Kiselev, A.V., Lygin, V.I., Ovespyan, M.E. and Titova, T.I., 1965. Infrared spectra of synthetic zeolites of the NaA, NaX, NH_4X types and their decationated forms. Zh. Fiz. Khim., 39: 2454–2458.

Zhdanov, S.P., Kiselev, A.V., Lygin, V.I. and Titova, T.I., 1966. Characteristics of acidic centers of decationized zeolites examined by infrared spectroscopy. Zh. Fiz. Khim., 40: 1041–1047.

10. Organic matter

Coal

Adams, W.N. and Pitt, G.J., 1955. Examination of oxidized coal by infrared absorption methods. Fuel, 34: 383—384.

Alford, D.O., 1959. Infrared spectra of coal and some coal derivatives. Fuel, 38: 114—115.

Bent, R. and Ladner, W.R., 1965. A preliminary investigation into the use of alternated total reflectance for obtaining the infrared spectra of coals. Fuel, 44: 243—247.

Bergmann, G., Huch, G., Karweil, J. and Luther, H., 1954. Ultrarotspektren von Kohlen. Brennstoffchemie, 35: 175—176.

Bergmann, G., Huch, G., Karweil, J. and Luther, H., 1957. Infrarot spektren von Kohlen. Brenstoffchemie, 38: 193—199.

Brooks, J.D. and Sternhell, S., 1957. Chemistry of brown coals. I. Oxygen-containing functional groups in Victorian brown coals. Aust. J. Appl. Sci., 8: 206—211.

Brooks, J.D., Durie, R.A. and Sternhell, S., 1958. Chemistry of brown coals. II. Infrared spectroscopic studies. Aust. J. Appl. Sci., 9: 63—80.

Brooks, J.D., Durie, R.A. and Sternhell, S., 1958. Chemistry of brown coals. III. Pyrolytic reactions. Aust. J. Appl. Sci., 9: 303—320.

Brooks, J.D., Durie, R.A., Lynch, B.M. and Sternhell, S., 1960. Infrared spectral changes accompanying methylation of brown coals. Aust. J. Chem., 13: 179—183.

Brown, J.K., 1955a. Die Infrarotspektren von Kohlen. J. Chem. Soc. (London): 744—752.

Brown, J.K., 1955b. Infrarotuntersuchungen an verkokten Kohlen. J. Chem. Soc., London: 752—757.

Brown, J.K., 1959. Infrared spectra of solvent extracts of coals Fuel, 38: 55—63.

Brown, J.K. and Hirsch, P.B., 1955. Recent infrared and X-ray studies of coal. Nature, 175: 229—233.

Brown, J.K. and Wyss, W.F., 1955. Oxygen groups in bright coals. Chem. and Ind.: 1118.

Cannon, C.G., 1953. Infrared spectra of coals and coal products. Nature, 171: 308.

Czuchajowski, L., 1960. Infrared spectra of coals oxidized with H_2O_2 and HNO_3. Fuel, 39: 377—385.

Czuchajowski, L., 1961. Infrarotspektren von verkokten Kohlen und kohleähnlichen stoffen und einige Absorptionsveränderungen während einer nachfolgenden oxydation Fuel, 40: 361—374.

Durie, R.A. and Sternhell, S., 1958. Chemistry of brown coals. IV. Action of oxygen in presence of alkali. Aust. J. Appl. Sci., 9: 360—369.

Durie, R.A. and Sternhell, S., 1959. Some quantitative infrared absorption studies of coals, pyrolised coals and their acetyl derivatives. Aust. J. Chem., 12: 205—217.

Friedel, R.A., 1960. Spectra and the constitution of coal and derivatives. Proc. 4th Conf. Carbon, Buffalo, 1959: 321—336.

Friedel, R.A. and Pelipetz, M.G., 1953. Infrared spectra of coal and carbohydrate chars. Opt. Soc. Am., 43: 1051—1052.

Friedel, R.A. and Queiser, J.A., 1956. Die Untersuchung von Kohlen mit Hilfe der Ultrarotspektroskopie. Z. Anal. Chem., 152: 463—464.

Fujii, S., 1963. Infrared spectra of coal: the absorption band at 1600 cm^{-1}. Fuel, 42: 341—343.

Fujii, S. and Yokoyama, F., 1958. Infrared absorption spectra of coal. Nenryo Kyokaishi (Fuel Soc. Japan), 37: 643—647.

Gordon, R.R., Adams, W.N. and Jenkins, G.I., 1952. Infrared spectra of coals. Nature 170: 317.

Gordon, R.R., Adams, W.N., Pitt, G.J. and Watson G.H., 1954. Infrared spectra of coals. Nature, 174: 1098—1099.

Hadzi, D., 1951. Some infrared spectra of coal extracts and derivatives. Acad. Sci. Art, Sloven., Ljubljana, A3: 99—107.

Karr, Cl., Estep, P.A. and Kovach, J.J., 1967. Infrared analysis of minerals in coal using the 650 to 200 cm^{-1} region. Chem. Ind. (London), 9: 356—357.

Kinney, C.R. and Doucester, E.I., 1958. Infrared spectra of a coalification series from cellulose and lignin to anthracite. Nature, 182: 785—786.

Oelert, H.H., 1965. Quantitative infrared spectroscopy of coals. Erdöl Kohle, 18: 876—880.

Rao, H.S., Gupta, P.L., Kaiser, F. and Lahiri, A., 1962. The assignment of the 1600 cm^{-1} band in the infrared spectrum of coal. Fuel, 41: 417—423.

Urbanski, T., Hofman, W., Ostrowski, T. and Witanowski, M., 1959. Infrared absorption spectra of products of carbonization of cellulose. Bull. Acad. Polon. Sci., 7(12): 851—859.

Van Vucht, H.A., Rietveld, B.J. and Van Krevelen, D.W., 1955. Chemical structure and properties of coal. VIII. Infrared absorption spectra. Fuel, 34: 50—59.

Lignin

Björkman, A., 1954. Isolation of lignin from finely divided wood with neutral solvents. Nature, 174: 1057—1058.

Goulden, J.D.S. and Jenkinson, D.S., 1959. The infrared spectra of ligno-proteins isolated from compost. J. Soil Sci., 10: 264—270.

Hergert, H.L. and Kurth, E.F., 1953. The infrared spectra of lignin and related compounds. I. Characteristic carbonyl and hydroxyl frequencies of some flavanones, flavones, chalcones and acetophenones. J. Am. Chem. Soc., 75: 1622—1625.

Hergert, H.L., 1960. Conifer lignin and model compounds. J. Organ. Chem., 25: 405—413.

Kinney, C.R. and Doucette, E.I., 1958. Infrared spectra of a coalification series from cellulose and lignin to anthracite. Nature, 182: 785—786.

Kolboe, St. and Ellefsen, Ö., 1962. Infrared investigations of lignin. A discussion of some recent results. Tappi, 45: 163—166.

Smith, D.C.C., 1955. Ester groups in lignin. Nature, 176: 267—268.

Tollin, G. and Steelink, C., 1966. Electron paramagnetic resonance and infrared studies of melanin, tannin, lignin, humic acid and hydroxyquinones. Biochem. Biophys. Acta, 112: 377—379.

Ziechmann, W., 1964. Spectroscopic investigations of lignin, humic substances and peat. Geochim. Cosmochim. Acta, 28: 1555—1566.

Humus and humic acids

Aleshin, S.N. and Shafiryan, E.M., 1964. Infrared study of the qualitative changes of humus compounds of soddy-podzol soil under the influence of liming. Dokl. Ross. S-kh. Akad., 99: 33—37.

Beutelspacher, H. and Nigro, C., 1961. Propieta fisishe degli acidi umici contenutti nelle lignite italiane. Agrochimica, 6: 56—73.

Beutelspacher, H., Van der Marel, H.W. and Rietz, E., 1973. Bodenkunde. In: H. Volkmann (Editor), Handbuch der Infrarotspektroskopie. Verlag Chemie, pp. 297—327.

Boldyrev, A.I. and Aleshin, S.N., 1964. Use of infrared spectroscopy for study of soil humus. Dokl. Ross. S-kh., Akad., 99: 11—15.

Bolydrev. A.I. and Komarov, I.A., 1964. Recording infrared spectra of mineral and humic constituents of soil absorption complexes. Dokl. Ross. S-kh, Akad., 99: 25—31.

Boldyrev. A.I., Matyushenko, B.T. and Polyakov, A.A., 1969. Application of infrared spectra for the study of humus compounds of paddy soils. Pochvovedenie, 1969: 81—88.

Cêh, M. and Hadži, D., 1956. Infrared spectra of humic acids and their derivatives. Fuel, 35: 77—83.

Deuel, H. and Dubach, P., 1958. Extraktion und Fraktionierung decarboxylierbarer Humusstoffe. Helv. Chim. Acta, 41: 1310—1321.

Diakonova, K.V., 1964. The nature of humus substances of the soil solution, their dynamics and methods of studying. Pochvovedenie, 4: 57—66.

Dormaar, J.F., 1967. Infrared spectra of humic acids from soils formed under grass or trees. Geoderma, 1: 37—45.

Dormaar, J.F., 1967. Infrared absorption spectra of mineral matter derived from electrodialysed humic acids. Geoderma, 1: 131—138.

Dudas, M.J. and Pawluk, S., 1969/1970. Chernozem soils of the Alberta parklands. Geoderma, 3: 19—36.

Dupuis, Jh. and Jambu, P., 1969. Etude par spectrographie infrarouge des produits de l'humification en milieu hydromorphe calcique. Sci. Sol (H): 23—35.

D'Yakonova, K.V., 1964. Methods of investigating the nature and dynamics of humus substances in the soil solution. Sov. Soil Sci., 1964: 384—390.

Elofson, R.M., 1957. Infrared spectra of humic acids and related materials. Can. J. Chem., 35: 926—931.

Kumada, K., 1955. Absorption spectra of humic acids. Soil Plant Food (Tokyo), 1: 29—30.

Kumada, K. and Aizawa, K., 1958. The infrared spectra of humic acids Soil Plant Food, 3: 152—159.

Kumada, K. and Aizawa, K., 1959. The infrared absorption spectra of soil components. Soil Plant Food, 4: 181—188.

Kurbatov, A.I. and Aleshin, S.N., 1965. Use of visible and infrared spectroscopy in the study of humic compounds of some steppe soils Dokl. Mosk. S-kh. Akad., 103: 11—16.

Lowe, L.E., 1969. Distribution and properties of organic fractions in selected Alberta soils. Can. J. Soil Sci., 49: 129—141.

Lowe, L.E., 1969. Aspects of infrared spectra of humic acids in relation to methods of preparation. Can. J. Soil Sci., 49: 257—258.

Martin, F., Dubach, P., Mehta, N.C. and Deuel, H., 1963. Bestimmung der funktionellen Gruppen von Huminstoffen. Z. Pflanzenernähr. Düng. Bodenk., 103: 27—39.

McKeague, J.A., Schnitzer, M. and Heringa, P.K., 1967. Properties of an ironpan humic podzol from Newfoundland. Can. J. Soil Sci., 47: 23—32.

Mendez, J. and Stevenson, F.J., 1966. Reductive cleavage of humic acids with sodium amalgum. Soil Sci., 102: 85—93.

Mortensen, J.L. and Schwendinger, R.B., 1963. Electrophoretic and spectroscopic characterization of high molecular weight components of soil organic matter. Geochim. Cosmochim. Acta, 27: 201—208.

Moschopedis, S.E., 1962. The reaction of humic acids with diazonium salts. Fuel, 41: 425—435.

Orlov, D.S., Rozanova, O.N. and Matyukhina, S.G., 1962. Infrared absorption spectra of humic acid. Pochvovedenie, 1962 (1): 17—25.

Otsuki, A. and Hanya, T., 1966. Fractional precipitation of humic acids from recent sediment by use of N, N-dimethylformamide and its infrared absorption spectra. Nature, 212: 1462—1463.

Otsuki, A. and Hanya, T., 1967. Some precursors of humic acid in recent lake sediment suggested by infrared spectra. Geochim. Cosmochim. Acta, 31: 1505—1515.

Posner, A.M., Theng, B.K.G. and Wake, J.R.H., 1968. The extraction of soil organic matter in relation to humification. Trans. 9th Int. Congr. Soil Sci., Adelaide, Australia, 3: 153—162.

Rozanova, O.N. and Matyukhina, S.G., 1962. Infrared absorption spectra of humic acids. Sov. Soil Sci., 1: 15—21.

Falk, M. and Smith, D.G., 1963. Structure of carboxyl groups in humic acids. Nature, 200: 569.

Farmer, V.C. and Morrison, R.I., 1960. Chemical and infrared studies on Phragmites peat and its humic acids. Sci. Proc. R. Dublin Soc., Ser. A, 1 (4): 85—104.

Farmer, W.J. and Ahlrichs, J.L., 1969. Infrared studies of the mechanism of absorption of urea-d_4, methylurea-d_3 and 1,1-dimethylurea-d_2 by montmorillonite. Proc. Soil Sci. Soc. Am., 33: 254—258.

Ishiwatari, R., 1967. Infrared absorption band at 1540 cm^{-1} of humic acid from a recent lake sediment. Geochem. J., 1: 61—70.

Ishiwatari, R. and Hanyá, T., 1965. Infrared spectroscopic characteristics of humic substances extracted from some recent sediments. Nippon Kagaku Zasshi, 86: 1270—1274.

Ishiwatari, R., Kosaka, M. and Hanyá, T., 1966. Composition and optical characteristics of humic substances extracted from lake sediments. Nippon Kagaku Zasshi, 87: 557—566.

Johnston, H.H., 1961. Soil organic matter: II. Studies of the origin and chemical structure of soil humic acids. Proc. Soil Sci. Soc. Am., 25: 32—35.

Juste, C., 1966. Modifications in electrophoretic mobility and the infrared spectrum of iron or aluminium-enriched humic acids. Compt. Rend., Ser. D, 262: 2692—2695.

Juste C. and Dureau, P., 1964. Infrared spectroscopic study of the different fractions of humic acid extracts from a hummo-ferruginous podzol on a quartzose sand. Compt. Rend., 259: 612—614.

Kasatochkin, V.I. and Zilberbrand, O.I., 1956. X-ray and infrared spectroscope method applied to the study of structure of humus substances. Pochvovedenie, 1956 (5): 80-85.

Kasatochkin, V.I., Kononova, M.M. and Zilberbrand, O.I., 1958. Infrared absorption spectra of humus substances of soil. Dokl. Akad. Nauk S.S.S.R., 119: 785—788.

Khan, S.U., 1969/1970. Humic acid fraction of a gray wooded soil as influenced by cropping systems and fertilizers. Geoderma, 3: 247—254.

Kleinhempel, D. and Hieke, W., 1965. Zur Trennung von Huminsäuren durch Geldiffusion. Albrecht Thaer Arch., 9: 165—172.

Kononova, M.M. and Belchikova, N.P., 1950. Experimental characterization of the nature of humic acids of soil by means of spectrophotometry. Dokl. Akad. Nauk S.S.S.R., 72: 125—128.

Kumada, K., 1955. Absorption spectra of humic acids. J. Sci. Soil Manure, 25: 217—221.

Sawyer, C.D. and Pawluk, S., 1963. Characteristics of organic matter in degrading chernozemic surface soils. Can. J. Soil. Sci., 43: 275—286.

Scharpenseel, H.W., 1966. Hydrothermalsynthese von Ton-Huminsäure sowie anderen organomineralischen Komplexen; Untersuchungen am Röntgengerät, Infrarot spektrometer und Elektronmikroskop. Z. Pflanzenernähr. Düng. Bodenk., 114: 188—203.

Scharpenseel, H.W. and Albersmeyer, W., 1960. Infrarotspektroskopische Untersuchungen an Huminsäuren, Huminsäureaufschlüssen und phenolisch-chinoiden Vergleichssubstanzen. Z. Pflanzenernähr. Düng. Bodenk., 88: 203—211.

Scharpenseel, H.W., König, E. and Menthe, E., 1964. Infrarot- und Differential – Thermo – Analyse an Huminsäureproben aus verschiedenen Bodentypen aus Wurmkot und Streptomyceten. Z. Pflanzenernähr. Düng., Bodenk., 106: 134–150.
Scheffer, F. and Welte, E., 1950. Die Anwendung der Absorptionsspektrographie in der Humusforschung. Z. Pflanzenernähr. Düng. Bodenk., 48: 250–263.
Scheffer, F. and Schlüter, H., 1959. Über Aufbau und Eigenschafter der Braun- und Grau-huminsäuren. Z. Pflanzenernähr. Düng. Bodenk., 84: 184–193.
Schnitzer, M., 1965. The application of infrared spectroscopy to the analysis of soil humic compounds. Can. Spectrosc., 10: 121–127.
Schnitzler, M., Shearer, D.A. and Wright, J.R., 1959. A study in the infrared of high-molecular weight organic matter extracted by various reagents from a podzolic B-horizon. Soil Sci., 87: 252–257.
Schnitzer, M. and Skinner, S.I.M., 1963. Reactions between a number of metalions and the organic matter of a podzol B_h horizon. Soil Sci., 96: 86–93.
Schnitzer, M. and Gupta, U.C., 1964. Some chemical characteristics of the organic matter extracted from the O and B_2 horizons of a gray wooded soil. Proc. Soil Sci. Soc. Am., 28: 374–377.
Schnitzer, M. and Skinner, S.I.M., 1964. Properties of iron- and aluminium organic-matter complexes, prepared in the laboratory and extracted from a soil. Soil Sci., 98: 197–203.
Schnitzer, M. and Desjardins, J.G., 1965. Carboxyl and phenolic hydroxyl groups in some organic soils and their relation to the degree of humification. Can. J. Soil Sci., 45: 257–264.
Schnitzer, M. and Skinner, S.I.M., 1965. Carboxyl and hydroxyl groups in organic matter and metal retention. Soil Sci., 99: 278–284.
Scholz, H., 1959. Die absorptionschromatographische Auftrennung von Huminsäuren. Z. Pflanzenernähr. Düng. Bodenk., 84: 159–169.
Shivrina, A.N., 1962. Chemical and spectrophotometric characteristics of watersoluble humin-like compounds formed by the fungus Inonotus obliquus (pers.) pil. Sov. Soil Sci.,: 1260–1266.
Smith, D.G. and Lorimer, J.W., 1964. An examination of the humic acids of Sphagnum peat. Can. J. Soil Sci., 44: 76–87.
Sokolov, D.F. and Sudnitsyna, T.N., 1961. Composition and optic properties of humic acids of certain forest soils. Dokl. Akad. Nauk S.S.S.R., 138: 931–934.
Tan, K.H. and McCreery, R.A., 1970. The infrared identification of a humo-polysaccharide ester in soil humic acid. Soil Sci. Plant Anal., 1: 75–84.
Theng, B.K.G., Wake, J.R.H. and Posner, A.M., 1966. The infrared spectrum of humic acids. Soil Sci., 102: 70–72.
Theng, B.K.G., Wake, J.R.H. and Posner, A.M., 1967. The humic acids extracted by various reagents from a soil, 2. Infrared, visible and ultraviolet absorption spectra. J. Soil Sci., 18: 349–363.
Theng, B.K.G. and Posner, A.M., 1967. Nature of the carbonyl groups in soil humic acids. Soil Sci., 104: 191–201.
Thompson, S.O. and Chesters, G., 1970. Infrared spectra and differential thermograms of lignins and soil humic materials saturated with different cations. J. Soil Sci., 21: 265–272.
Tokudome, S. and Kanno, I., 1964. Characterization of humus of humic allophane soils in Japan. Humic acids (ch)-to- fulvic acids (cf) ratios. Kyushu Nogyo Shik. Iho Sta., 10: 185–193.
Tokudome, S. and Kanno, I., 1964. Some physicochemical properties of humic and fulvic acids. Bull. Kyushu Agric. Exp. Sta., 10: 195–204.
Tokudome, S. and Kanno, I., 1965. Some physico-chemical properties of humic and fulvic acids. Soil Sci. Plant Nutr., Tokyo, 11: 193–199.
Tokudome, S. and Kanno, I., 1968. Nature of the humus of some Japanese soils. Trans. 9th Int. Congr. Soil Sci., Adelaide, Australia, 3: 163–173.
Tollin, G. and Steelink, C., 1966. Electron paramagnetic resonance and infrared studies of melanin, tannin, lignin, humic acid and hydroxyquinones. Biochem. Biophys. Acta, 112: 377–379.
Visser, S.A., 1964. A physico-chemical study of the properties of humic acids and their changes during humification. J. Soil Sci., 15: 202–219.
Wada, K. and Inoue, T., 1967. Retention of humic substances derived from rotted clover leaves in soils containing montmorillonite and allophane. Soil Sci. Plant Nutr., Tokyo, 13: 9–16.
Wagner, G.H. and Stevenson, F.J., 1965. Structural arrangement of functional groups in soil humic acid as revealed by infrared analysis. Proc. Soil Sci. Soc. Am., 29: 43–48.
Wright, J.R. and Schnitzer, M., 1963. Metallo-organic interactions associated with podzolization. Proc. Soil Sci. Soc. Am., 27: 171–176.
Ziechmann, W., 1958. Untersuchungen über das Infrarotspektrum von Huminsäuren. Brenstoff Chem., 39: 353–359.
Ziechmann, W., 1959. Die Darstellung von Huminsäuren im heterogen System mit neutraler Reaktion. Z. Pflanzenernähr. Düng. Bodenk., 84: 155–159.
Ziechmann, W., 1964. Spectroscopic investigations of lignin, humic substances and peat. Geochim. Cosmochim. Acta, 28: 1555–1566.
Ziechmann, W. and Pawelke, G., 1959. Zum Vergleich natürlicher und synthetischer Huminsäuren und ihre Vorstufen. Z. Pflanzenernähr. Düng. Bodenk., 84: 174–184.
Ziechmann, W. and Scholz, H., 1960. Spektroskopische Untersuchungen an Huminsäuren. Naturwissenschaften, 47: 193–196.
Ziechmann, W. and Rochus, W., 1971. Über Bildung und Eigenschaften von Huminstoffen. Tonind. Z., 95: 69–74.

Fulvic acids

Kobo, K. and Tatsukawa, R., 1959. On the colored material of fulvic acid. Z. Pflanzenernähr. Düng. Bodenk., 84: 137–147.
Orlov, D.S. and Zub, V.T., 1963. Optical properties and elemental composition of fulvic acids of different origin. Nauchn. Dokl. Vysshei. Shkoly. Biol. Nauki, 3: 182–188.
Schnitzer, M., 1969. Reactions between fulvic acid, a soil humic compound and inorganic soil constituents. Proc. Soil Sci. Soc. Am., 33: 75–81.
Schnitzer, M. and Kodama, H., 1972. Reactions between fulvic acids and Cu^{2+} montmorillonites. Clays Clay Miner., 20: 359–369.
Tokudome, S. and Kanno, I., 1964. Humic acids (ch) – to fulvic acids (cf) ratios. Kyushu Nogyo Shik. Iho Sta., 10: 185–193.
Tokudome, S. and Kanno, I., 1964. Some physical properties of humic and fulvic acids. Kyushu Nogyo Shik. Iho Sta., 10: 195–204.
Tokudome, S. and Kanno, I., 1965. Some physico-chemical properties of humic- and fulvic acids. Soil Sci. Plant Nutr., Tokyo, 11: 193–199.

Wright, J. and Schnitzer, M., 1963. Metallo-organic interactions associated with podzolization. Proc. Soil Sci. Soc. Am., 27: 171—176.

Graphite, carbon black

Hallum, J.V. and Drushel, H.V., 1958. The organic nature of carbon black surfaces. J. Phys. Chem., 62: 110—117.

Bitumen

Friedel, R.A. and Queiser, J.A., 1956. Infrared analysis of bituminous coals and other carbonaceous materials. Anal. Chem., 28: 22—30.
Friedel, R.A. and Nalwalk, A.J., 1968. Characterization of carbonaceous material from the Puerto Rico trench of the Atlantic Ocean. Nature, 217: 345—347.
Glebovskaia, E.A. and Maltsanskaia, T.N., 1970. Classification of dispersed organic matter in terms of the infrared spectra of bitumenoids. Dokl. Acad. Sci. U.S.S.R., 190: 212—215.
Glebovskaia, E.A. and Maltsanskaia, T.N., 1970. Principles underlying the dispersion of organic matter according to infrared spectra of bitumenoids. Dokl. Akad. Nauk S.S.S.R., 190: 1444—1447.
Knotnerus, J., 1970. Infrarotanalyse bituminöser Materialen. Erdöl Kohle, 23: 341—347.
Murchison, D.G., 1963. Infrared spectrum of resinite in bituminous coal. Nature, 198: 254—255.
Qudrat-i-Khudo, M., Samad, S.A. and Rashid, A., 1965. Faridpur peat. II. Fraction and infrared studies of peat bitumen. Pakistan J. Sci. Ind. Res., 8: 240—250.
Thomas, J.D.R., 1962. Chemistry of peat bitumen: fractionation and infrared studies. J. Appl. Chem. (London), 12: 289—294.
Tokudome, S. and Kanno, I., 1965. Some physico-chemical properties of humic and fulvic acids. Soil Sci. Plant Nutr., 11: 193—199.

11. Inorganic (clay) — organic matter

Abramov, V.N., Kiselev, A.V. and Lygin, V.I., 1963. Nature of the absorption by zeolites. Infrared spectrum for benzene absorbed by zeolites of the B X and 10 X type. Zh. Fiz. Khim., 37: 1156—1160.

Abramov, V.N., Kiselev, A.V. and Lygin, V.I., 1964. Infrared spectroscopic study of the absorption of phenol, aniline and nitrobenzene by aerosil and zeolite. Zh. Fiz. Khim., 38: 1044—1047.

Alekseev, A.V., Filimonov, V.N. and Terenin, A.N., 1962. Infrared spectra of N O absorbed on synthetic zeolites. Dokl. Akad. Nauk. S.S.S.R., 147: 1392—1395.

Angell, C.L. and Schaffer, P.C., 1966. Infrared spectroscopy investigation of zeolites and absorbed molecules. II Absorbed carbon monoxide. J. Phys. Chem., 70: 1413—1418.

Arshad, M.A. and Lowe, L.E., 1966. Fractionation and characterisation of naturally occurring organo-clay complexes. Proc. Soil Sci. Soc. Am., 30: 731—735.

Barraclough, C.G., Bradley, D.C., Lewis, J. and Thomas, I.M., 1961. The infrared spectra of some metal alkoxides, trialkylsilyloxides and related silanols. J. Chem. Soc., 1961: 2601—2605.

Basila, M., 1961. Hydrogen bonding interaction between absorbate molecules and surface hydroxyl groups on silica. J. Chem. Phys., 35: 1151—1158.

Beck, C.W. and Brunton, G., 1960. X-ray and infrared data on hectorite-guanidines and montmorillonite-guanidines. Proc. 8th Natl. Conf. Clays Clay Miner., Norman, Okla., 1959: 22—38.

Bergmann, K. and O'Konski, C.T., 1963. A spectroscopic study of methylene blue monomer, dimer and complexes with montmorillonite. J. Phys. Chem., 67: 2169—2177.

Blackmann, L.C.F. and Harrop, R., 1965. Infrared spectra of quaternary ammonium compounds absorbed on silica-gel "Aerosil". Nature, 208: 777—778.

Bodenheimer, W., Heller, L., Kirson, B. and Yariv, Sh., 1963. Copper-polyamine-clay complexes. Proc. Int. Clay Conf., Stockholm, 2: 351—363.

Bodenheimer, W., Heller, L. and Yariv, Sh., 1966. Infrared study of copper-montmorillonite-alkylamines. Proc. Int. Clay Conf., Jerusalem, 2: 171—177.

Boldyrev, A.I. and Komarov, I.A., 1964. Recording infrared spectra of mineral and humic constituents of soil absorption complexes. Dokl. Ross. S-kh. Akad., 99: 25—31.

Brindley, G.W. and Ray, S., 1964. Complexes of Ca-montmorillonite with primary monohydric alcohol. Am. Mineralogist, 49: 106—115.

Calvet, R., 1963. Organic complexes of clay; a bibliographical review. Ann. Agron., 14: 31—117.

Chassin, P., 1969. Absorption du glycocolle par la montmorillonite. Bull. Gr. Fr. Argiles, 21: 71—88.

Chaussidon, J., Calvet, R., Helsen, J. and Fripiat, J.J., 1962. Catalytic decomposition of cobalt hexamine cations on the surface of montmorillonite. Nature, 196: 161—162.

Chaussidon, J. and Calvet, R., 1965. Evolution of amine cations absorbed on montmorillonite with dehydration of the mineral. J. Phys. Chem., 69: 2265—2268.

Chumaevski, N.A., 1961. Characteristic absorption bands in the infrared spectra of silicon—organic compounds. Opt. i Spektr., 10: 33—37.

Courtois, M. and Teichner, S.J., 1962. Etude par spectroscopie infrarouge des gaz CO, O_2, CO_2 et de leurs produits d'interaction absorbés chimiquement à l'oxyde de nickel. J. Chim. Phys., 59: 272—288.

Damm, K. and Noll, W., 1958. Über die Bestimmung von Silanolgruppen in Organopolysiloxan und ihr Verhalten bei thermischer Kondensation. Kolloid-Z., 158: 97—108.

Davydov, V. Ya., Kiselev, A.V. and Lygin, V.I., 1963. Investigation of the absorption of ethyl ether on silica by the method of infrared spectroscopy. Kolloidn. Zh., 25: 152—158.

Davydov, V. Ya. And Kiselev, A.V., 1968. Reactions of hydroxyl groups of a silica surface with water and alcohol molecules. Kolloid Zh., 30: 353—358.

De Mumbrum, L.E. and Jackson, M.L., 1956. Infrared absorption evidence on exchange reaction mechanism of copper and zinc with layer silicate clays and peat. Proc. Sci. Soc. Am., 20: 334—337.

Deuel, H., 1957. Organische Derivate von Silikaten. Angew. Chem., 69: 270.

Deuel, H., Huber, G. and Günthard, Hs.H., 1952. Untersuchungen an Phenyl montmorilloniten. Helv. Chim. Acta, 35: 1799—1802.

Dowdy, R.H. and Mortland, M.M., 1967. Alcohol water interactions on montmorillonite surfaces. 1. Ethanol. Proc. 5th Natl. Conf. Clays Clay Miner., Madison, Wisc., 1966: 259—271.

Dudas, M.J. and Pawluk, S., 1969/1970. Naturally occurring organo-clay complexes of orthic black chernozems. Geoderma, 3: 5—17.

Farmer, V.C. and Mortland, M.M., 1965. An infrared study of complexes of ethylamine with ethylammonium and copper ions in montmorillonite. J. Phys. Chem., 69: 683—686.

Farmer, V.C. and Mortland, M.M., 1966. An infrared study of the coordination of pyridine and water to exchangeable cations in montmorillonite. J. Chem. Soc., (A): 344—351.

Farmer, W.J. and Ahlrichs, J.L., 1969. Infrared studies of the mechanism of absorption of urea-d4, methylurea-d3 and 1,1-dimethylurea-d2 montmorillonite. Proc. Soil. Sci. Soc. Am., 33: 254—258.

Fenn, D.B. and Mortland, M.M., 1972. Phenol complexes in smectites. Proc. Int. Clay Conf., Madrid, 2: 311—325.

Folman, M. and Yates, D.J.C., 1958. Infrared and length-change studies in absorption of H_2O and CH_3OH on porous silica glass. Trans. Faraday Soc., 54: 1684—1691.

Folman, M. and Yates, D.J.C., 1959. Infrared studies of physically absorbed polar molecules and of the surface of a silica absorbent containing hydroxyl groups. J. Phys. Chem., 63: 183—187.

Friedlander, H.Z., 1963. Organized polymerization. I. Olefins on a clay surface. J. Polymer. Sci., 4: 1291—1301.

Fripiat, J.J., Uytterhoeven, J., Schobinger, U. and Deuel, H., 1960. Etude physicochimique de quelques dérivés organiques d'un silicagel. Helv. Chim. Acta, 43: 176—181.

Fripiat, J., Servais, A. and Léonard, A., 1962. La nature de la liaison amine-montmorillonite. Bull. Soc. Chim. France, 29: 635—644.

Fripiat, J.J., Cloos, P., Calicis, B. and McKay, K., 1966. Identification of absorbed species and decay products by infrared spectroscopy. Proc. Int. Clay Conf., Jerusalem, 1: 233—245.

Fripiat, J.J. and Mendelovici, E., 1968. Le derivé methyle du chrysotile. Bull. Soc. Chem. Fr., 2: 483—492.

Fripiat, J.J., Pennequin, M., Poncelet, G. and Cloos, P., 1969. Influence of the Van der Waals force on the infrared spectra of short aliphatic alkylammonium cations held on montmorillonite. Clay Miner., 8: 119—134.

Gentili, R. and Deuel, H., 1957. Or-

ganic derivatives of clay minerals. 5. Degradation of phenyl montmorillonite. Helv. Chim. Acta, 40: 106–113.
Greenland, D.J., 1965. Absorption of soil organic compounds and its effect on soil properties. Soil Fertil., 28: 521–532.
Hair, M.L., 1967. Infrared Spectroscopy in Surface Chemistry. M. Dekker, New York, 315 pp.
Harvey, M.C. and Nebergall, W.H., 1962. The silicon-phenyl asymmetrical stretching vibration. Appl. Spectrosc., 16: 12–14.
Heller-Kallai, L., Yariv, S. and Riemer, M., 1972. Effect of acidity on the sorption of histidine by montmorillonite. Proc. Int. Clay Conf., Madrid, 2: 381–393.
Hoffman, R.W. and Brindley, G.W., 1961. Infrared extinction coefficients of ketones absorbed on Ca-montmorillonite in relation to surface coverage. J. Phys. Chem., 65: 443–448.
Holmes, R.M. and Toth, S.J., 1957. Physico-chemical behavior of clay-conditioner complexes. Soil Sci., 84: 479–487.
Horak, M., Bazant, V. and Chvalovsky, V., 1960. Bestimmung der Basizität von Siliciumgebundenem Sauerstoff mittels der Infrarot Spektroskopie. Coll. Czechoslov. Chem. Communic., 25: 2822–2830.
Jang Sung Do and Condrate, R.A., 1972. Infrared spectra of lysine absorbed on several cation-substituted montmorillonites. Clays Clay Miner., 20: 79–82.
Kakudo, M., Kasai, P.N. and Watase, T., 1953. Crystal structure of silanols and their infrared spectra. J. Chem. Phys., 21: 1894–1895.
Kiselev, A.V. and Lygin, V.I., 1961. Investigation of benzene and hexane absorption on silica by the method of infrared spectroscopy. Kolloid Zh., 23: 574–581.
Kiselev, A.V., 1964. Molecular interactions at short distances. Zh. Fiz. Khim., 38: 1501–1513.
Kiselev, A.V., Kubelkova, L. and Lygin, V.I., 1964. Study by infrared spectroscopy of absorption of methanol on synthetic faujasites. Zh. Fiz. Khim., 38: 2719–2725.
Kiselev, A.V., Lygin, V.I. and Titova, T.I., 1964. Study of specific absorption of ammonia on silica and zeolite by the infrared spectra. Zh. Fiz. Khim., 38: 2730–2733.
Kohl, R.A. and Taylor, S.A., 1961. Hydrogen bonding between the carbonyl groups and Wyoming bentonite. Soil Sci.; 91: 223–227.
Kreshkov, A.P., Mikhailenko, Y.Y. and Yakimovich, G.F., 1954. Qualitative analysis of silicon organic compounds by means of infrared absorption spectroscopy. Zh. Anal. Khim., 9: 208–216.
Kriegsmann, H., 1957. Spektroskopische Untersuchungen zur Si-X-Si-Brückenbindung im Hexamethyldisiloxan und seinem NH-, CH_2- und S-Analogen. Z. Spektrochem., 61: 1088–1094.
Kriegsmann, H., 1958. Die Schwingungsspektren der Derivate $(CH_3)_2SiX_2$. Z. Elektrochem., 62: 1033–1037.
Kriegsmann, H. 1958. Die Schwingungsspektren des $(CH_3)_3SiOH$, $(CH_3)_3SiF$, und $(CH_3)_3SiSH$. Z. Anorg. Allg. Chem., 294: 113–119.
Kriegsmann, H., 1959. Die Schwingungsspektren einiger Tetramethyldisiloxane. Z. Anorg. Allg. Chem., 299: 78–86.
Kriegsmann, H., 1959. Über die Bindungsverhältnisse in der Siliciumchemie. Z. Anorg. Allg. Chem., 299: 138–150.
Kriegsmann, H., 1960. Die Raman- und IR/Spektren linearer Methylpolysiloxane. Z. Elektrochem., 64: 541-545.
Kriegsmann, H., 1960. Die Schwingungsspektren Kettenförmiger Methylhydrogen polysiloxane. Z. Elektrochem., 64: 848–852.
Kriegsmann, H. and Licht, K., 1958. Spektroskopische Untersuchungen an Titan- und Kieselsäure-Estern. Z. Elektrochem., 62: 1163–1174.
Kriegsmann, H. and Schowtka, K.H., 1958. Vibrational spectra of phenylsilanes, phenylhalosilanes and phenylhalosiloxanes. Z. Phys. Chem., 209: 261–297.
Kriegsmann, H. and Clauss, H., 1959. Schwingungsspektren und Struktur des Tetramethyl-cyclo-disilthians und des hexamethyl-cyclo-trisilthians. Z. Anorg. Allg. Chem., 300: 210–220.
Kriegsmann, H. and Engelhardt, G., 1961. Schwingungsspektren und Bindungsverhältnisse einiger Disilazane und verwandter Verbindungen Z. Anorg. Allg. Chem., 310: 320–327.
Kriegsmann, H. and Beyer, H., 1961. IR- und Ramanspektren einiger substituierter Disilylacetylene. Z. Anorg. Allg. Chem., 311: 180–185.
Kukolev, G.V., Nemetz, I.I. and Semchenko, G.D., 1970. Peculiar features of the interaction between some surface-active substances and kaolinite clay. Dokl. Akad. Nauk S.S.S.R., 194: 1141–1144.
Larson, G.O. and Sherman, L.R., 1964. Infrared spectrophotometric analysis of some carbonyl compounds absorbed on bentonite clay. Soil Sci., 98: 328–331.
Lazarev, A.N., 1960. Vibrations of the SiO_4 groups in the spectra of tetra alkoxysilanes. Opt. i Spektr., 8: 270–272.
Lazarev, A.N. and Voronkov, M.G., 1958. Infrared spectra of esthers of orthosilicic acid. Opt. i Spektr., 4: 180–188.
Lazarev, A.N. and Voronkov, M.G., 1960. Vibrations of the silicon-oxygen chains in the spectra of polyalkoxysiloxanes. Opt. i Spektr., 8: 325–329.
Ledoux, R.L. and White, J.L., 1966. Infrared studies of hydrogen bonding of organic compounds on oxygen and hydroxyl surfaces of layer lattice silicates. Int. Clay Conf., Jerusalem, 1: 361–374.
Ledoux, R.L. and White, J.L., 1966. Infrared studies of hydrogen bonding of organic compounds on oxygen and hydroxyl surfaces of layer lattice silicates. Int. Clay Conf., Jerusalem, 2: 233–234.
Ledoux, R.L. and White, J.L., 1966. Infrared studies of hydrogen-bonding interaction between kaolinite surfaces and intercalated potassium acetate, hydrazine, formamide and urea. J. Colloid. Interface Sci., 21: 127–152.
Ledoux, R.L. and White, J.L., 1967. Etude infrarouge des complexes d'intercalation de la kaolinite. Silic. Ind., 32: 269–273.
Leonard, A., Servais, A. and Fripiat, J.J., 1962. Absorption of amines by montmorillonites. I.-The chemical processes. II.-Structure of the complexes. Bull. Soc. Chim. Fr.,: 617–625.
Little, L.H. and Amberg, C.H., 1962. Infrared spectra of carbon monoxide and carbon dioxide absorbed on chromia-alumina and on alumina. Can. J. Chem., 40: 1997–2006.
Little, L.H., 1966. Infrared Spectra of Absorbed Species. Acad. Press, London, 428 pp.
Lucchesi, P.J., Carter, J.L. and Yates, D.J.C., 1962. An infrared study of the chemisorption of ethylene on aluminium oxide. J. Phys. Chem., 66: 1451–1456.
Lynch, D.L., Wright, L.M., Hearns, E.E. and Cotnoir, L.J., 1957. Some factors affecting the absorption of cellulose compounds, pectins and hemicellulose compounds on clay minerals. Soil Sci., 84: 113–126.
McDonald, R.S., 1957. Study of the interaction between hydroxyl groups of aerosil silica and absorbed non-polar molecules by infrared spectrometry. J. Am. Chem. Soc., 79: 850–854.

Mortensen, J.L., 1962. Absorption of hydrolized poly-acrylonitrile on kaolinite Proc. 9th Natl. Conf. Clays Clay Miner., Purdue Univ., 1960: 530—545.

Mortland, M.M., 1965. Nitric oxide absorption by clay minerals. Proc. Soil Sci. Soc. Am., 29: 514—519.

Mortland, M.M., 1966. Urea complexes with montmorillonite: an infrared absorption study. Clay Miner., 6: 143—156.

Mortland, M.M., 1970. Clay-organic complexes and interactions. Adv. Agronom., 22: 75—117.

Mortland, M.M., Fripiat, J.J., Chaussidon, J. and Uytterhoeven, J., 1963. Interaction between ammonia and expanding lattices of montmorillonite and vermiculite. J. Phys. Chem., 67: 248—258.

Mortland, M.M. and Meggit, W.F., 1966. Interaction of Ethyl N.N.-dipropyl Thiolcarbonate with montmorillonite. J. Agric. Food Chem., 14: 126—129.

Mortland, M.M., Farmer, V.C. and Russell, J.D., 1969. Reactivity of montmorillonite surface with weak organic bases. Proc. Soil Sci. Soc. Am., 33: 818.

Nicholson, D.E., 1960. Infrared of ammonia absorbed on silica-alumina catalysts. Nature, 186: 630—631.

Olejnik, S., Posner, A.M. and Quirk, J.P., 1971. The complexes of formamide. n-methylformamide and dimethylformamide. Clays Clay Miner., 19: 83—94.

Olejnik, S., Posner, A.M. and Quirk, J.P., 1971. Infrared spectrum of kaolinite-pyridine N-oxide complex. Spectrochim. Acta, A, 27: 2005—2009.

Ovcharenko, F.D., Tarasevich, Yu.I. and Radul, M.M., 1967. Study of allyl alcohol absorption on montmorillonites by infrared spectroscopy. Dopor Akad. Nauk. Ukr. R.S.R., Ser. B, 29: 631—634.

Parfitt, R.L. and Mortland, M.M., 1968. Ketone absorption on montmorillonite. Proc. Soil Sci. Soc. Am., 32: 355—363.

Parfitt, R.L. and Greenland, D.J., 1970. The absorption of polyethylene glycols on clay minerals. Clay Miner., 8: 305—315.

Peri, J.B., 1965. Infrared study of absorption of ammonia on dry γ alumina. J. Phys. Chem., 69: 231—239.

Peri, J.B., 1966. Infrared study of OH and NH_2 groups on the surface of a dry silica aerogel. J. Phys. Chem., 70: 2937—2945.

Pliskin, W.A. and Eischens, R.P., 1956. Infrared spectra of chemisorbed olefins and acetylene. J. Chem. Phys., 24: 482—483.

Raman, K.V. and Mortland, M.M., 1969. Proton transfer reactions at clay mineral surfaces. Proc. Soil Sci. Soc. Am., 33: 313—317.

Richards, R.E. and Thompson, H.W., 1949. Infrared spectra of compounds of high molecular weight. Part IV, Silicones and related compounds. J. Chem. Soc., 1: 124—132.

Russell, J.D., 1965. Infrared study of the reactions of ammonia with montmorillonite and saponite. Trans. Faraday Soc., 61: 2284—2294.

Sánches, A., Hidalgo, A. and Serratosa, J.M., 1972. Absorption des Nitriles dans la Montmorillonite. Proc. Int. Clay Conf., Madrid, 2: 339—349.

Scharpenseel, H.W., 1966. Hydrothermalsynthese von Ton-Huminsäure sowie anderen Organomineralischen Komplexen; Untersuchungen am Röntgengerät, IR-Spektrometer und Elektronenmikroskop. Z. Pflanzenernähr. Düng. Bodenk., 114: 188—203.

Schnitzer, M., 1969. Reactions between fulvic acid, a soil humic compound and inorganic soil constituents. Proc. Soil Sci. Soc. Am., 33: 75—81.

Schnitzer, M., 1970. Reactions of fulvic acid, a water-soluble humic fraction, with metals and organic molecules. Am. Chem. Soc. Div. Water Waste Chem. Gen. Pap., 10: 165—169.

Schnitzer, M. and Skinner, S.I.M., 1964. Organo-metallic-interaction in soils: Properties of iron- and aluminium- organic matter complexes, prepared in the laboratory and extracted from a soil. Soil Sci., 98: 197—203.

Schnitzer, M. and Kodama, H., 1967. Reactions between a podzol fulvic acid and Na-montmorillonite. Proc. Soil Sci. Soc. Am., 31: 632—636.

Schnitzer, M. and Kodama, H., 1972. Reactions between fulvic acid and Cu^{2+} montmorillonite. Clays Clay Miner., 20: 359—367.

Serratosa, J.M., 1965. Use of infrared spectroscopy to determine the orientation of pyridine sorbed on montmorillonite. Nature, 208: 679—681.

Serratosa, J.M., 1966. Infrared analysis of the orientation of pyridine molecules in clay complexes. Proc. 14th Natl. Conf. Clays Clay Miner., Berkeley, 1965: 385—391.

Serratosa, J.M., 1968. Infrared study of benzonitrile (C_6H_5CN)-montmorillonite complexes. Am. Mineralogist, 53: 1244—1251.

Serratosa, J.M., 1968. Infrared study of the orientation of chlorobenzene sorbed on pyridinium-montmorillonite. Clays Clay Miner., 16: 93—97.

Serratosa, J.M., Johns, W.D. and Shimoyama, A., 1970. Infrared study of alkylammonium vermiculite complexes. Clays Clay Miner., 18: 107—113.

Servais, A., Fripiat, J.J. and Leonard, A., 1962. Etude de l'absorption des amines par les montmorillonites, 1: Les processus chimiques. Bull. Soc. Chim. Fr., 617—625.

Sheppard, N., 1959. Infrared spectra of absorbed molecules. Spectrochim. Acta, 14: 249—260.

Simon, I. and McMahon, H.O., 1952. Infrared spectra of some alkyl silanes and siloxanes in gaseous, liquid and solid phases. J. Chem. Phys., 20: 905—907.

Smith, I.T., 1964. Infrared spectra of polar molecules absorbed on titanium dioxide pigments. Nature, 201: 67—68.

Stepanov, I.S., 1971. Investigation of the system bentonit + humic acid by the method of infrared spectroscopy. Dokl. Akad. Nauk S.S.S.R., 198: 1192—1194.

Swoboda, A.R. and Kunze, G.W., 1966. Infrared techniques for studying the absorption of volatile vapors by clays. Soil Sci., 101: 373—377.

Swoboda, A.R. and Kunze, G.W., 1968. Reactivity of montmorillonite surface with weak organic bases. Proc. Soil Sci. Soc. Am., 32: 806—811.

Swoboda, A.R. and Kunze, G.W., 1968. Infrared study of pyridine absorbed on montmorillonite surface. Proc. 13th Natl. Conf. Clays Clay Miner., Madison Wis., 1966: 277—288.

Swoboda, A.R., Kunze, G.W., 1969. Acidity of clay surfaces. Proc. Soil Sci. Soc. Am., 33: 819.

Tahoun, S.A. and Mortland, M.M., 1966. Protonation of amides on the surface of montmorillonite. Soil Sci., 102: 248—254.

Tahoun, S.A. and Mortland, M.M., 1966. Coordination of amides on the surface of montmorillonite. Soil Sci., 102: 314—321.

Tarasevich, Yu.I., 1968. Determination of the infrared spectra of clay minerals and substances absorbed on them. Ukr. Khim. Zh., 34: 439—446.

Tarasevich, Yu,I., Radul, N.M. and Ovcharenko, F.D., 1967. Absorption of trimethyl-carbinole upon montmorillonite investigated by means of infrared spectroscopy. Dokl. Akad. Nauk S.S.S.R., 173: 615—617.

Tarasevich, Yu.I., Telichkum, V.P. and Ovcharenko, F.D., 1968. Infrared spectra of acetonitrile absorbed on montmorillonite. Dokl. Akad. Nauk S.S.S.R., 182: 141–143.
Tarasevich, Yu.I., Radul, N.M. and Ovcharenko, F.D., 1968. n-Propyl alkohol absorption on montmorillonite as studied by infrared spectroscopy. Kolloid. Zh., 30: 137–143.
Tarasevich, Yu.I. and Ovcharenko, F.D., 1969. Infrared spectra of deuterized methanol absorbed on montmorillonite. Dokl. Akad. Nauk. S.S.S.R., 187: 372–375.
Tarasevich, Yu. I., Telichkun, V.P. and Ovcharenko, F.D., 1970. Spectral study of the interaction of pyridine with a montmorillonite surface. Teor. Eksp. Khim., 6: 804–809.
Taymurazova, L.Kh. and Ignat'yva, L.A., 1965. Study of the interaction between minerals and polymers by infrared spectroscopy. Vestn. Mosk. Univ., Ser. VI., Biol. Pochvoved., 20: 81–86.
Tensmeyer, L.G., Hoffmann, R.W. and Brindley, G.W., 1960. Infrared studies of some complexes between ketones and calcium montmorillonite. J. Phys. Chem., 64: 1655–1662.
Terenin, A.N., Jaroslavsky, N.G., Karajakin, A.W. and Sidorova, A.I., 1955. Spectroscopie infrarouge des molécules absorbées sur verre poreux. Mikrochim. Acta, 1955: 467–470.
Terenin, A.N., 1957. Infrared spectra of surface compounds on silicate absorbents. Poverkhnost. Khim. Soedinen. Rol Yavl. Absorb., Sb. Tr. Konf. Absorb., pp. 206–222.
Uytterhoeven, J., Fripiat, J.J. and Dockx, L., 1959. L'Acetylation des aluminosilicates. Bull. Sci. Acad. R. Belg., 45: 611–624.
Uytterhoeven, J. and Fripiat, J.J., 1961. Application de la spectroscopy I-R à l'étude de quelques dérives organique des alumino-silicates et du silicagel. Rep. 21th Int. Geol. Congr. Copenhagen, 1960, 24: 80–87.
Wartmann, J. and Deuel, H., 1958. Organische Derivate des Silikagels. Chimia (Switz.) 12: 82–83.
Weiss, A., 1963. Mica-type layer with alkylammonium ions. Proc. 10th Natl. Conf. Clays Clay Miner., Austin, Texas, 1961: 191–224.
Wright, N. and Hunter, M.J., 1947. Infrared spectra of the methylpolysiloxanes. J. Am. Chem. Soc., 69: 803–809.
Yariv, S., Russell, J.D. and Farmer, V.C., 1966. Infrared study of the absorption of benzoic acid and nitrobenzene in montmorillonite. Israel J. Chem., 4: 201–213.
Yariv, S., Heller, L. and Kaufherr, N., 1969. Effect of acidity in montmorillonite interlayers on the sorption of aniline derivatives. Clays Clay Miner., 17: 301–308.
Yates, D.J.C., 1954. Infrared studies of the surface hydroxyl groups on titanium dioxide and of the chemisorption of carbon monoxide and carbon dioxide J. Phys. Chem., 58: 746–753.
Yates, D.J.C., 1961. Infrared studies of the surface hydroxyl groups on titanium dioxide and of the chemisorption of carbon monoxide and carbon dioxide. J. Phys. Chem., 65: 746–753.

SUBJECT INDEX

1. Minerals (text)

[1] dt. = deuterated; ht. = heated.

2. Minerals (Figures)

3. Organic